Complex Molecular Surfaces and Interfaces

Sheffield, UK

24–26 July 2017

FARADAY DISCUSSIONS

Volume 204, 2017

ROYAL SOCIETY OF CHEMISTRY

The Faraday Division of the Royal Society of Chemistry, previously the Faraday Society, was founded in 1903 to promote the study of sciences lying between chemistry, physics and biology.

Faraday Discussions (Print ISSN 1359-6640, Electronic ISSN 1364-5498) is published 10 times a year by the Royal Society of Chemistry, Thomas Graham House, Science Park, Milton Road, Cambridge, UK CB4 0WF.

Volume 204 ISBN 13: 978-1-78262-944-3

2017 annual subscription price: print+electronic £1010 US $1782; electronic only £962, US $1693. Customers in Canada will be subject to a surcharge to cover GST. Customers in the EU subscribing to the electronic version only will be charged VAT.

All orders, with cheques made payable to the Royal Society of Chemistry, should be sent to the Royal Society of Chemistry Order Department, Royal Society of Chemistry, Thomas Graham House, Science Park, Milton Road, Cambridge, CB4 0WF, UK
Tel +44 (0)1223 432398; E-mail **orders@rsc.org**

If you take an institutional subscription to any Royal Society of Chemistry journal you are entitled to free, site-wide web access to that journal. You can arrange access via Internet Protocol (IP) address at **www.rsc.org/ip**

Customers should make payments by cheque in sterling payable on a UK clearing bank or in US dollars payable on a US clearing bank.

Printed in the UK

MIX
Paper from responsible sources
FSC www.fsc.org FSC® C013604

Faraday Discussions are unique international discussion meetings that focus on rapidly developing areas of chemistry and its interfaces with other scientific disciplines.

Complex Molecular Surfaces and Interfaces

Faraday Discussions

www.rsc.org/faraday_d

A General Discussion on Complex Molecular Surfaces and Interfaces was held in Sheffield, UK on the 24th, 25th and 26th of July 2017.

CONTENTS

ISSN 1359-6640; ISBN 978-1-78262-944-3

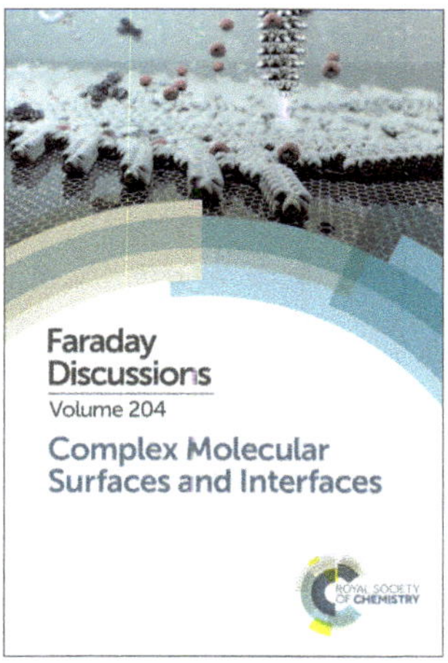

Cover
See B. E. Hirsch, K. P. McDonald, S. L. Tait and A. H. Flood, *Faraday Discuss.*, 2017, **204**, 159–172.

Anions (colored red) are held at surfaces by molecular receptors until the electric fields used for scanning tunneling microscopy also act to eject them.

Front cover image courtesy of Amar Flood, Indiana University. Image credit: Danny McMurray.

INTRODUCTORY LECTURE

9 **Molecular assembly at surfaces: progress and challenges**
R. Raval

PAPERS AND DISCUSSIONS

35 **A surprising way to control the charge transport in molecular electronics: the subtle impact of the coverage of self-assembled monolayers of floppy molecules adsorbed on metallic electrodes**
Ioan Bâldea

53 **Kinetic control in the temperature-dependent sequential growth of surface-confined supramolecular copolymers**
Sabrina Engel, Daniel Spitzer, Leona Lucas Rodrigues, Eva-Corinna Fritz, David Straßburger, Monika Schönhoff, Bart Jan Ravoo and Pol Besenius

69 **Supramolecular chiral self-assemblies of Gly–Pro dipeptides on metallic fcc(110) surfaces**
C. Méthivier, H. Cruguel, C.-M. Pradier and V. Humblot

83 **Pt–dipyridyl tetrazine metal–organic network on the Au(100) surface: insights from first principles calculations**
Duy Le and Talat S. Rahman

97 **Probing the interplay between geometric and electronic structure in a two-dimensional K–TCNQ charge transfer network**
P. J. Blowey, L. A. Rochford, D. A. Duncan, D. A. Warr, T.-L. Lee, D. P. Woodruff and G. Costantini

111 **Self-assembly of a binodal metal–organic framework exhibiting a demi-regular lattice**
Linghao Yan, Guowen Kuang, Qiushi Zhang, Xuesong Shang, Pei Nian Liu and Nian Lin

123 **Supramolecular effects in self-assembled monolayers: general discussion**

159 **Physical and chemical model of ion stability and movement within the dynamic and voltage-gated STM tip–surface tunneling junction**
Brandon E. Hirsch, Kevin P. McDonald, Steven L. Tait and Amar H. Flood

173 **Sequential nested assembly at the liquid/solid interface**
Baharan Karamzadeh, Thomas Eaton, David Muñoz Torres, Izabela Cebula, Marcel Mayor and Manfred Buck

191 **Molecular self-assembly of substituted terephthalic acids at the liquid/solid interface: investigating the effect of solvent**
A. Della Pia, D. Luo, R. Blackwell, G. Costantini and N. Martsinovich

215 **Transfer of chiral information from a chiral solvent to a two-dimensional network**
Iris Destoop, Andrea Minoia, Oleksandr Ivasenko, Aya Noguchi, Kazukuni Tahara, Yoshito Tobe, Roberto Lazzaroni and Steven De Feyter

233 **Peptide binding to metal oxide nanoparticles**
S. P. Schwaminger, S. A. Blank-Shim, I. Scheifele, P. Fraga-García and S. Berensmeier

251 **Elucidating "screw dislocation"-driven film formation of sodium thiosulphate with complex hierarchical molecular assembly**
Deepak Dwivedi, Katerina Lepkova, Thomas Becker and Matthew R. Rowles

271 **Supramolecular systems at liquid–solid interfaces: general discussion**

297 **A new on-surface synthetic pathway to 5-armchair graphene nanoribbons on Cu(111) surfaces**
Kewei Sun, Penghui Ji, Haiming Zhang, Kaifeng Niu, Ling Li, Aixi Chen, Qing Li, Klaus Müllen and Lifeng Chi

307 **Surface induced crystallization of polymeric nano-particles: effect of surface roughness**
Ahmad Jabbarzadeh and Xin Chen

331 **What can be inferred from moiré patterns? A case study of trimesic acid monolayers on graphite**
Saskia Spitzer, Oliver Helmle, Oliver Ochs, Joshua Horsley, Natalia Martsinovich, Wolfgang M. Heckl and Markus Lackinger

349 **The influence of nearest-neighbour interactions and assembly dynamics on the transport properties of porphyrin supramolecular assemblies on Au(111)**
Alison A. Pawlicki, Ayelet Vilan, Matthew Jurow, Charles Michael Drain and James D. Batteas

367 **The culture of HaCaT cells on liquid substrates is mediated by a mechanically strong liquid–liquid interface**
D. Kong, K. D. Q. Nguyen, W. Megone, L. Peng and J. E. Gautrot

383 **Surface-bound quadruple H-bonded dimers: formation and exchange kinetics**
Digvijay Gahtory, Rickdeb Sen, Maarten M. J. Smulders and Han Zuilhof

395 **Preparing macromolecular systems on surfaces: general discussion**

419 **Generic nature of long-range repulsion mechanism on a bulk insulator?**
J. L. Neff, A. Richter, H. Söngen, C. Venturini, A. Gourdon, R. Bechstein and A. Kühnle

429 **Erecting buckybowls onto their edge: 2D self-assembly of terphenylcorannulene on the Cu(111) surface**
Quirin S. Stöckl, Tsun-Cheng Wu, Anaïs Mairena, Yao-Ting Wu and Karl-Heinz Ernst

439 **Exploring catalyst passivation with NMR relaxation**
Neil Robinson, Lynn F. Gladden and Carmine D'Agostino

453 **The role of halogens in on-surface Ullmann polymerization**
Gianluca Galeotti, Marco Di Giovannantonio, Josh Lipton-Duffin, Maryam Ebrahimi, Stefano Tebi, Alberto Verdini, Luca Floreano, Yannick Fagot-Revurat, Dmitrii F. Perepichka, Federico Rosei and Giorgio Contini

471 **The dynamics of benzene on Cu(111): a combined helium spin echo and dispersion-corrected DFT study into the diffusion of physisorbed aromatics on metal surfaces**
M. Sacchi, P. Singh, D. M. Chisnall, D. J. Ward, A. P. Jardine, W. Allison, J. Ellis and H. Hedgeland

CONCLUDING REMARKS

487 **Complex molecular surfaces and interfaces: concluding remarks**
David B. Amabilino and Steven L. Tait

DISCUSSION

503 **Probing properties of molecule-based interface systems: general discussion and Discussion of the Concluding Remarks**

ADDITIONAL INFORMATION

531 **Poster titles**

533 **List of participants**

Faraday Discussions

PAPER

Molecular assembly at surfaces: progress and challenges

R. Raval

Received 14th July 2017, Accepted 14th July 2017
DOI: 10.1039/c7fd90072d

Molecules provide versatile building blocks, with a vast palette of functionalities and an ability to assemble *via* supramolecular and covalent bonding to generate remarkably diverse macromolecular systems. This is abundantly displayed by natural systems that have evolved on Earth, which exploit both supramolecular and covalent protocols to create the machinery of life. Importantly, these molecular assemblies deliver functions that are reproducible, adaptable, finessed and responsive. There is now a real need to translate complex molecular systems to surfaces and interfaces in order to engineer 21st century nanotechnology. 'Top-down' and 'bottom-up' approaches, and utilisation of supramolecular and covalent assembly, are currently being used to create a range of molecular architectures and functionalities at surfaces. In parallel, advanced tools developed for interrogating surfaces and interfaces have been deployed to capture the complexities of molecular behaviour at interfaces from the nanoscale to the macroscale, while advances in theoretical modelling are delivering insights into the balance of interactions that determine system behaviour. A few examples are provided here that outline molecular behaviour at surfaces, and the level of complexity that is inherent in such systems.

Introduction

The past two decades have witnessed a step-change in the creation of organised molecular assemblies at surfaces. The scope and versatility of these 2D organic architectures have been captured *via* scanning probe microscopies, advanced surface spectroscopies and diffraction techniques, as is amply demonstrated by all the contributions to this Faraday Discussion. The variety and abundance of 2D molecular assembly observed at surfaces has illustrated the potential power of 'bottom-up' construction strategies, which combine nanometer-precision and highly parallel fabrication, as a synthesis protocol in new nanotechnologies for sensors, molecular electronics, smart coatings, organic solar cells, catalysts, medical devices, optoelectronics, *etc.* Such molecule–surface systems are already demonstrating an array of sophisticated functions such as molecular recognition,

Surface Science Research Centre, Department of Chemistry, University of Liverpool, Liverpool, L69 3BX, UK. E-mail: Raval@liverpool.ac.uk

organisation, responsivity, reactivity, chiral selectivity, energy harvesting, light manipulation, information storage, and nanolocomotion. The variety of systems, assembly protocols, structures and functions are clearly showcased by researchers contributing to these Proceedings.

However, despite the scientific success that has undoubtedly been achieved, it has to be recognised that the roadmap to creating robust, molecule–surface-based devices that can be routinely used for applications is still in its infancy and a number of challenges lie ahead. Chief amongst these is creating a fundamental understanding and control of molecular behaviour and assembly at surfaces and interfaces. Here, the flexibility and adaptability of the molecular state needs to be recognised and tracked. In particular, the manifold of molecule–surface and molecule–molecule interactions need to be captured in order to understand the system from the single molecule perspective to the macroscale collective, from the detailed structure of a single phase to the broader polymorphic phase diagram, from behaviour in controlled conditions to complex environments. This will require the deployment of numerous tools with which to fabricate and interrogate the interface across multiple length scales, and close collaborations between experimentalists and theorists. Only then will the generic factors emerge with which to design complex interfaces with intelligent and sophisticated capabilities. Finally, the field will need engineers who can translate this knowledge base into working devices.

In this Introductory Lecture, I will draw on work that has emerged from my group and collaborations to discuss aspects of molecular behaviour at surfaces, from supramolecular assembly to covalent synthesis. The importance of the nanoscale details in determining global assembly are evident from this work, as is the salutary lesson that, even in controlled environments, the molecule–surface system invariably exhibits complex behaviour.

Highly organised molecular assembly at surfaces

The behaviour of molecules at defined single crystal metal surfaces, HOPG, oxide layers, *etc.* has been instrumental in demonstrating the organisational ability of molecules at interfaces, displaying highly ordered assemblies with precise molecular arrangements that can propagate from the nanoscale to the macroscale.[1–6] Generally, these organised assemblies fall into two main groups: (i) those which involve weak interactions with the surface, with supramolecular interactions dominating system behaviour; and, (ii) those that involve strong bonding to the surface, with the resulting behaviour determined by a balance of molecule–surface, molecule–molecule and intra-surface interactions. I will concentrate on the latter systems, but point here to the excellent examples of both types of systems by other contributions in this Faraday Discussion. Both sets of systems show aspects that are challenging to the experimentalist and the theorist. Foremost, a significant proportion of the systems display polymorphism where the nature of the molecular entity and its organisation vary significantly as surface coverage, temperature and deposition flux are varied.[2] These aspects often remain unexplored, with researchers focussing on structure determination of specific organised phases only. Even when studies are targeted at a single phase populated by just one type of molecular species, many structural models can often be suggested, with only slight variations in energy. Such aspects constitute a central debate in 3D molecular crystals[7] and these considerations are now beginning to emerge in the study of 2D systems. Two examples described below on the assembly of tartaric acid and the amino acid, proline, on the Cu(110) surface illustrate the complex nature of supramolecular assembly at surfaces, especially when the molecules have a strong chemisorption interaction with the surface. In addition, both molecules are chiral, adding further complexity arising from the nature of enantiomeric interactions and their self-assembly behaviour.

A second advance in the field is the creation of covalent macromolecular structures at surfaces, bringing the domain of organic covalent synthesis to the surface and the interface. This is a very important development in the field, with rapid progress since the first reports emerged.[8–11] I will again draw on our work on dehydrogenative coupling *via* C–H activation and scission to illustrate some of the state-of-the-art in the field. Finally, the importance of kinetics and transport in dictating the assembly that emerges under particular conditions surely remains as an outstanding contemporary challenge, and will require development of both experimental and theoretical approaches.

Chiral supramolecular assembly of tartaric acid on Cu(110): some recent developments

The assembly of chiral molecules at surfaces has become an important area in its own right, with surface spectroscopies and scanning probe microscopy delivering important insights on chiral ordering, recognition, heterogeneous enantioselectivity and mirror-symmetry breaking to produce organised homochiral assemblies.[2,3,12–33]

An exemplar in the field of surface chirality is provided by the tartaric acid/Cu(110) system.[12–18] This system displays a high degree of polymorphism with the nature of the molecular building block dynamically changing between the intact bi-acid, the singly dehydrogenated monotartrate and the doubly dehydrogenated bitartrate, as revealed by a combination of Reflection Absorption Infrared Spectroscopy (RAIRS), Temperature Programmed Desorption (TPD), Low Energy Electron Diffraction (LEED) and Scanning Tunnelling Microscopy (STM) experiments.[2,13,22] One phase, in particular, created by the bitartrate species has attracted attention because it displays global space group homochirality for the enantiopure system,[12] with (*R*,*R*)-bitartrate assembling into the (1 2, −9 0) structure and (*S*,*S*)-bitartrate into the mirror (9 0, −1 2) assembly (Fig. 1a and b), with hierarchical transfer of homochirality from the single-molecule to the macroscale assembly. The racemic system displays Pasteurian chiral resolution at the surface (Fig. 1c), with enantiomers segregating into mirror domains of (1 2, −9 0) and (9 0, −1 2) structure. Recently, we re-investigated this system[18] using STM and dispersion corrected periodic DFT calculations that are now able to handle the large unit cell of the system. This work showed that the original calculated structure[14,15] for this system was incorrect, unable to either account for the intensity variation observed in the STM images, or the chiral segregation observed for the racemic system. This work also revealed the level of detailed analysis required to understand complex chiral molecule assembly at surfaces.

The starting point for analysing molecular assembly at a surface is to understand the nature of the adsorbed molecular species. Calculations[18] for the enantiopure system show that (*R*,*R*)-bitartrate chemisorbs strongly to the Cu(110)

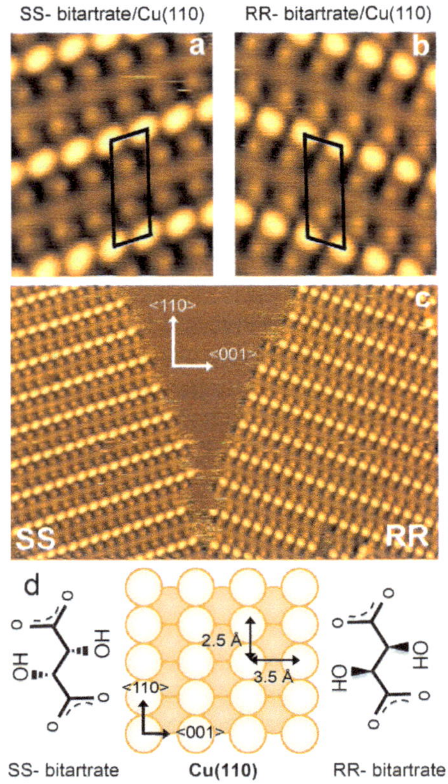

Fig. 1 STM images of bitartrate on Cu(110). Top panel: STM image of the global homochiral assembly of (a) (*S,S*)-bitartrate on Cu(110), (45 × 52 Å^2, $I_t = 0.23$ nA, $V_t = 0.541$ V) and (b) (*R,R*)-bitartrate on Cu(110) (45 × 52 Å^2, $I_t = 0.23$ nA, $V_t = 0.541$ V.) (c) STM image of the chiral segregation observed after adsorption of racemic tartaric acid on Cu(110), showing domains of (9 0, −1 2) (*S,S*) and (1 2, −9 0) (*R,R*) bitartrate (277 × 193 Å^2, $I_t = 0.21$ nA, $V_t =$ 0.557 V). (d) Structures of the two enantiomers of bitartrate and the Cu(110) surface.

surface by bridging across two close packed Cu rows to create a 4-point bonded adsorption motif, with both carboxylate groups bonded to the surface by four Cu–O bonds (Fig. 2). Interestingly, it is the detail of the adsorption footprint that holds the key to system behaviour, specifically the chiral segregation that is observed for the racemic mixture. Three distinct adsorption motifs are possible, one transcribing a rectangular footprint (R_{ec}) and two producing oblique footprints (O_a or O_b), which are mirror images. Each change in footprint leads to significant local changes in the adsorbed molecules. Adsorption in the rectangular footprint leads to strong hydrogen bonds between the α-hydroxyls and the adjacent carboxylate oxygen, causing a chiral distortion in the molecular backbone, with a C–C torsion angle of −28°. This chiral distortion is increased further with a C–C torsion angle of −50° if the O_a oblique footprint is adopted. Concomitantly, there is significant restructuring of the internal hydrogen bonds. This adsorption motif is just 0.083 eV less stable than the rectangular footprint model. If (*R,R*)-bitartrate adopts the mirror O_b oblique footprint, the carbon skeleton distorts in the opposite direction with a C–C torsion angle of +41°, the

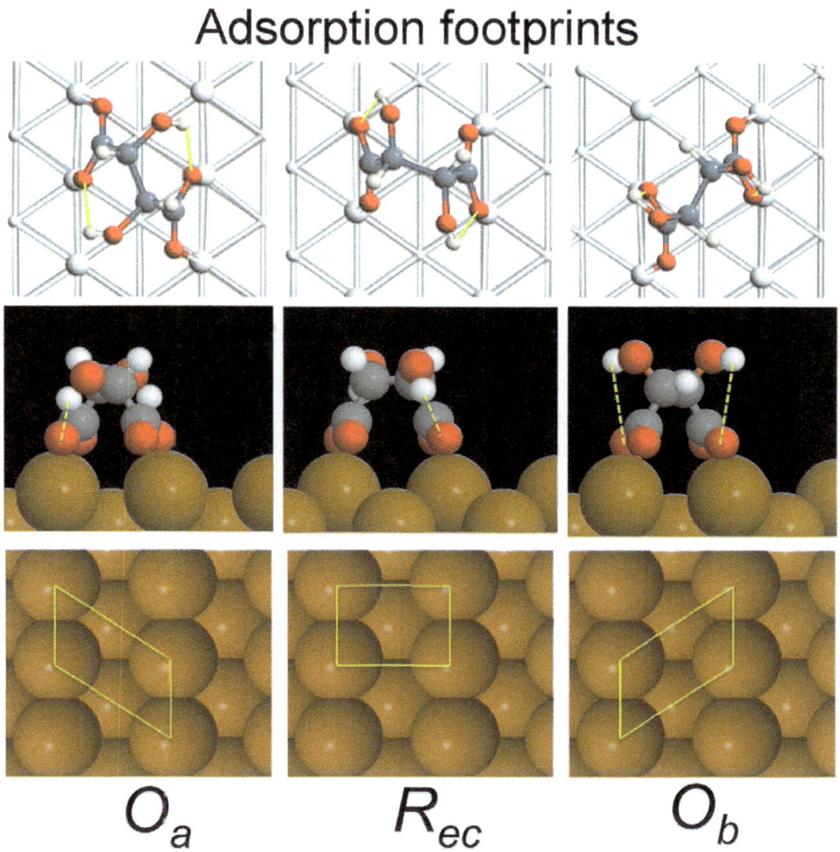

Fig. 2 Calculated adsorption structures for (*R*,*R*)-bitartrate on Cu(110), adopting three different adsorption sites, projecting O_a, R_{ec} and O_b footprints.[18] The top two panels show the adsorption geometries with the different atoms identified as follows: carbon: grey; oxygen: red; hydrogen: white. Intra-molecular H-bonds are depicted with yellow lines. The bottom panel shows the adsorption footprints created by the four O–Cu chemisorption bonds.

internal hydrogen bonds are weakened, and the O_b structure is 0.42 eV less stable than the rectangular R_{ec} footprint. This is a high energy penalty, so this (*R*,*R*)-bitartrate conformation can be excluded from consideration for self-assembly. However, the R_{ec} and O_a footprints are both candidates for 2D assembly of (*R*,*R*)-bitartrate into the experimentally observed (1 2, −9 0) structure. Therefore, all allowed permutations of the R_{ec} and O_a conformations need to be considered and, in order to discriminate between the various structural models, three major questions need to be addressed: (i) Is the assembly structure energetically preferred? (ii) Does the structure predict the chiral segregation observed for the racemic mixture? (iii) Does the structure produce the observed STM image?

Experimental observations show that the enantiopure (1 2, −9 0) (*R*,*R*)-bitartrate structure possesses rows of bitartrate trimers that assemble with high fidelity across the entire surface. Each trimer is aligned along a chiral direction, and each row of trimers also assembles in a chiral direction with respect to the surface, as shown in Fig. 1 and 3 (top row). Four distinct assemblies using the R_{ec} and the O_a footprints need to be considered, shown as structures 1–4 in Fig. 3. The inserted table in Fig. 3 lists the computed energies of the modelled structures.[18] Structure 4 with a $O_aO_aO_a$ footprint combination can be excluded since it has a significantly higher energy compared to the other three. Structure 2 has the lowest energy, with 1 and 3 only slightly higher. Therefore, all three are viable candidates on energetic grounds.

To distinguish between these possibilities, we turned to the additional two questions outlined above. Specifically, we considered whether the experimentally observed chiral segregation of the racemic mixture is predicted.[18] We, therefore, investigated the effect of substituting (*S*,*S*) and mixed (*R*,*R*)/(*S*,*S*) enantiomer combinations within the three minimum energy structures 1, 2 and 3 calculated for (*R*,*R*)-bitartrate, as shown in Fig. 3. The results for the originally proposed $R_{ec}R_{ec}R_{ec}$ structure 1 are particularly interesting since this assembly is, in fact, slightly more stable when (*S*,*S*)-bitartrate is substituted into the assembly

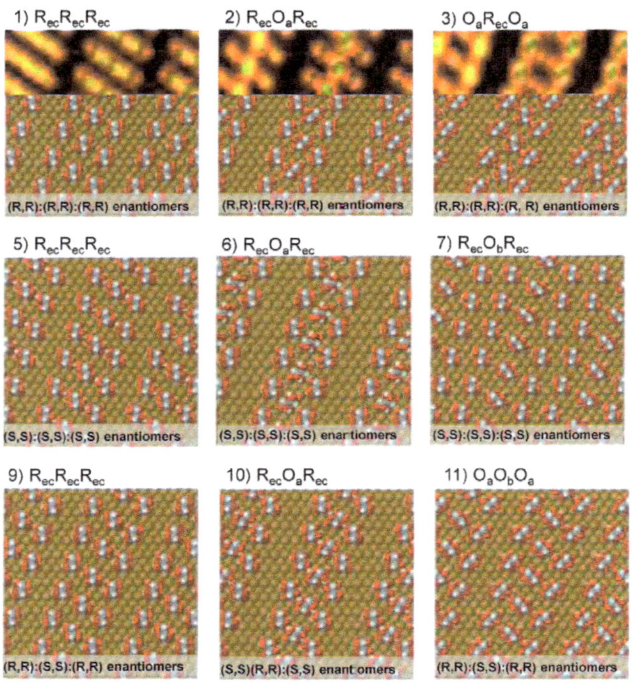

Structure	E_{ad} (eV)	ΔE_{ad} (eV)
1	-2.152	+0.123
2	-2.275	0.0
3	-2.244	+0.031
4	-2.079	+0.196
5	-2.163	+0.112
6	-1.816	+0.460
7	-2.115	+0.160
8	-2.081	+0.194
9	-2.161	+0.114
10	-2.157	+0.118
11	-2.165	+0.110

Fig. 3 Minimum energy structures calculated using DFT for various bitartrate arrangements on Cu(110).[18] The top row displays the enantiopure (*R*,*R*)-bitartrate overlayers created using the combination of footprints indicated at the top of each panel. Structures 1–3 have their simulated STM images (at bias of −0.5 V) superimposed on the panels. The second row shows occupation of the same footprints by the enantiopure (*S*,*S*)-bitartrate system. The bottom row shows occupation of the footprint template by mixed (*R*,*R*) and (*S*,*S*) enantiomers, with the most favourable mixed enantiomer combinations shown. The inset table shows E_{ad} of each structure and also ΔE_{ad}, the adsorption energy relative to structure 2.

(structure 5). This prediction is the reverse of the experimentally observed structures for enantiopure adsorption. As a further consideration, a mixture of (*R*,*R*) and (*S*,*S*) enantiomers were incorporated within the $R_{ec}R_{ec}R_{ec}$ arrangement (structure 9). Again, the adsorption energy for mixed enantiomer occupation is slightly lower than that of the enantiopure (*R*,*R*) structure 1, and similar to that of the (*S*,*S*) structure 5. These calculations suggest that if the $R_{ec}R_{ec}R_{ec}$ footprint arrangement is adopted, the enantiomorph preference for the enantiopure systems would be flipped and no chiral segregation should be observed for the racemic system. This prediction clearly contradicts the experimental result (Fig. 1), and reveals the limitations of the previously calculated structural model[14,15] that had been accepted for bitartrate assembly on Cu(110) for over 15 years.

A similar enantiomer substitution exercise can be conducted for the other two viable footprint arrangements that can create the (1 2, −9 0) structure (structures 2 and 3). When (*S*,*S*)-bitartrate occupation is considered for the $R_{ec}O_aR_{ec}$ footprint combination (structure 6), a high penalty energy of +0.46 eV arises, largely because the (*S*,*S*)-enantiomer is forced to adopt the O_a footprint, rather than its preferred O_b footprint. This result illustrates the central role played by the adsorption footprint in generating an enantiospecific response during assembly, *i.e.*, the R_{ec} footprint is insensitive to the handedness of the enantiomer, while the O_a/O_b footprints are highly enantiospecific. This strong enantiospecificity also rules out (*S*,*S*)-bitartrate occupation of the $O_aR_{ec}O_a$ footprint arrangement, which would be energetically unviable. We, therefore, also examined whether it was possible for enantiopure (*S*,*S*)-bitartrate to generate a (1 2, −9 0) assembly by adopting a $R_{ec}O_bR_{ec}$ footprint arrangement, which allows the central (*S*,*S*)-bitartrate to occupy its favoured oblique footprint, but with the trimer rows still directed along the same chiral direction exhibited by the (*R*,*R*)-bitartrate assembly. Such an arrangement requires the trimer unit to expand in order to accommodate the (*S*,*S*) enantiomer in the O_b footprint, but it is possible to create an ordered (1 2, −9 0) assembly (structure 7). This reduces the energy penalty, but structure 7 is still less stable than the enantiopure (*R*,*R*)-bitartrate structure 2 by ~0.1 eV, which is sufficient to lead to the enantiomorph selection displayed by the enantiopure assembly and the chiral segregation of the racemic mixture at 300 K. A similar exercise shows that enantiopure (*S*,*S*)-bitartrate occupation of the $O_bR_{ec}O_b$ trimer (structure 8) or a (*R*,*R*):(*S*,*S*):(*R*,*R*) occupation of the $O_aO_bO_a$ combination (structure 11) are still less stable than the enantiopure ((*R*,*R*)-bitartrate structure 2).

To further establish whether structure 2 is the observed experimental phase, we simulated the STM images[18] that would be generated from (*R*,*R*)-bitartrate assembly with structures 1, 2 and 3 (Fig. 3 inset). It can be seen that only structure 2 reproduces the experimental data, with the middle molecule in the trimer imaging brighter than the two outer ones. Therefore, overall, structure 2, in which two types of footprints are occupied, provides a better model for the chiral assembly of bitartrate on Cu(110). Crucially, the adoption of the oblique footprint within the trimer is key to creating an energetic preference for the (*R*,*R*)-bitartrate to assemble into the (1 2, −9 0) structure, and drive chiral segregation into distinct enantiomorphs for the racemic system.

What this discussion outlines is that the nature and diversity of the individual nanoscale adsorption motifs largely determines the plethora of assemblies that could arise and, therefore, needs careful consideration. When the assembly of chiral molecules is considered, then the effect of enantiomer substitution into this large manifold of structures also needs to be considered in order to understand the chiral behaviour of the system. Finally, it should be recognised that although the diversity of viable adsorption forms adds to the complexity of analysis, the flexibility to adopt a range of energetically near-equivalent conformations enables the system to respond dynamically to its environment and make local adjustments that could drive important functions, such as adaptation, host–guest accommodation, chiral recognition[20] and heterogeneous enantioselective catalysis.[21,32]

Creating spaces within molecular assemblies

As important as where the molecules are in a particular assembly, it is equally important to know where the gaps and spaces are generated within an assembly, since pores and channels provide environments for important functions such as sensing, reactivity, storage, separation, selective transport, *etc.* Therefore, it is important to understand the factors that create spaces within molecular assemblies at surfaces. The bitartrate/Cu(110) system provides an example of surface stress-induced gap creation within an assembly.[14,18] Modelling such effects has only become possible with recent advances in computational power, allowing large unit cells to be addressed. The structural model 2 of (*R*,*R*)-bitartrate on Cu(110) shows trimer rows separated by gaps where 3 Cu atoms in the $[1\bar{1}0]$ direction remain exposed (Fig. 3). The origin of this gap arises directly from the adsorption of the molecule *via* the 4 oxygens of the carboxylate groups. As each carboxylate group bonds onto the close-packed row, there is a 2.5% increase in the Cu–Cu distance of the surface atoms binding to the molecule. This creates a stress within the surface, which is relieved by breaks in the assembly. Calculations[18] tracking the associated energy change as the gap between the trimer rows is increased progressively from 1 atom to 4 atoms show that the energy of the system falls steadily until a gap of 3 atoms is reached, after which no energy advantage accrues if the gap is increased. Elsewhere in these Proceedings, papers by Nian Lin and Manfred Buck illustrate beautiful molecular assemblies leading to the creation of a wide variety of pores, while the group of de Feyter have created impressive multicomponent assemblies *via* a core–shell approach in which multiple components are incorporated *via* host–guest interactions.[34]

The importance of disorder in determining ordering

The literature on molecular assembly at surfaces, by and large, focusses on the ordered arrangements that are observed. However, our work on the adsorption of scalemic mixtures of (*R*,*R*)- and (*S*,*S*)-bitartrate on Cu(110) provides an important lesson in illustrating the importance of the disordered phase.[17] Adsorption of the racemic 50 : 50 mixture produces chiral segregation as shown in Fig. 4a, with enantiomorphous domains of (*R*,*R*)-bitartrate and (*S*,*S*)-bitartrate formed in equal measure at the surface. This is shown in LEED data for the system, where diffraction spots arising from each structure are present with equal intensity (Fig. 4a). When enantiomerically unbalanced adsorption is carried out, the system behaviour changes radically. For example, the 60 : 40 mixture shows LEED spots arising from the majority species only (Fig. 4b). In fact, the organisational behaviour of the two

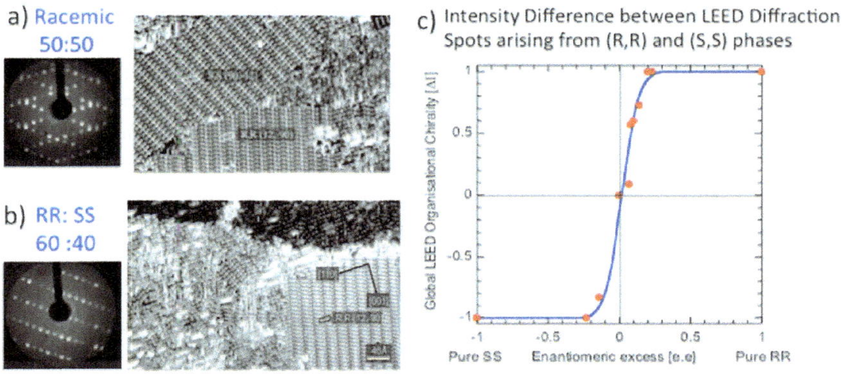

Fig. 4 Non-linear behaviour in the organisation of the (*R*,*R*)-bitartrate and (*S*,*S*)-bitartrate phases on Cu(110) as the enantiomeric ratios are changed. (a) LEED pattern and experimental STM image from racemic adsorption showing that the organised phases from each enantiomer are formed in equal abundance; (b) LEED pattern and experimental STM image from a 60 : 40 *RR* : *SS* mixture showing that only the majority (*R*,*R*) enantiomer creates an organised phase, giving rise to the LEED pattern of one enantiomorphous domain only, with STM showing that the minority enantiomer is restricted to forming very small organised clusters; (c) highly non-linear behaviour in the formation of the two enantiomorphous domains (measured by LEED intensity) as the enantiomeric excess (e.e) is varied.

enantiomers becomes highly non-linear as shown in Fig. 4c, with the organised phase for the minority species rapidly disappearing in the LEED data even for slight deviations from the racemic state. STM data for a 60 : 40 *RR* : *SS* system echo the LEED experiment, with large domains of the majority (*R*,*R*)-bitartrate visible, whereas the organised (*S*,*S*)-bitartrate phase is almost annihilated, appearing only in very small domains containing tens of molecules (Fig. 4b). This was a very surprising result; the expectation would be that domains of each enantiomer would be observed with a 60 : 40 ratio. Kinetic Monte Carlo calculations undertaken by A. P. J. Jansen[17] revealed that the system behaviour is actually driven by the configurational entropy of the disordered 'sea' surrounding the organised domains. The greatest configurational entropy arises when the disordered phase is racemic with an equal population of the two enantiomers. This aspect has a profound influence on how the organised phases evolve. For a 60 : 40 adsorption mixture, there is initial preferential 2D crystallisation of the majority enantiomer to create a racemic disordered sea. Once the excess enantiomer has organised, both enantiomers start to crystallise. However, in order to maintain a racemic disordered sea *via* a dynamic equilibrium with the organised phase, the domain boundaries of the segregated enantiomers must be equal. This is only possible if the majority enantiomer organises in larger domains, and the minority in smaller domains. This leads to the highly non-linear behaviour observed, with drastic deviation in the crystallisation propensity and domain size exhibited by the two enantiomers.

Amino acid assembly from the single molecule and single footprint perspective: proline on Cu(110)

The bitartrate/Cu(110) system discussed above showed the importance of considering adsorption footprints created when molecules chemisorb at a surface. Monte Carlo simulations on other systems[23] demonstrate that the evolution of molecular surface organisation and local patterns is strongly influenced by the adsorption footprints the system adopts. There is, therefore, a real need to experimentally identify the adsorption footprints that are adopted within an assembly. This is a challenging task, especially for the small molecules considered here, which image as single features in STM. The importance of footprints reaches another level when considering the adsorption of amino acids on the Cu(110) surface, as exemplified by our work on the amino acid proline and its derivatives on Cu(110).[24–28] The amino group of proline is incorporated within a pyrrolidine ring (Fig. 5iii), leading to greatly increased structural rigidity. This rigidity is instrumental in enabling identification of different adsorption footprints, since each adsorption conformation triggers a specific reorientation of the attached pyrrolidine ring, which leads to image contrasts in STM, as described below.

Adsorption of enantiopure (*S*)-proline on Cu(110) leads to a (4 × 2) assembly containing the dehydrogenated prolate species.[25,26] STM images, in Fig. 5ib, show the (4 × 2) unit cell contains two distinct prolate conformers, which image as

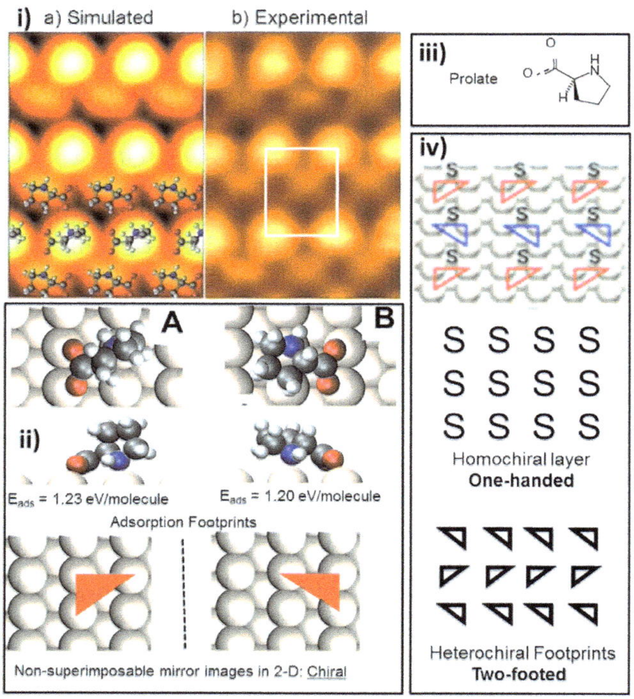

Fig. 5 (i) DFT simulated and experimental STM image of the (4 × 2) overlayer of (*S*)-prolate on Cu(110); (ii) DFT calculated adsorption geometries for conformers A (bright spot in simulated image) and B (faint spot in STM images) of adsorbed prolate with views from above, the side, and in terms of their triangular adsorption footprint (atoms are depicted by: red: O, blue: N, black: C, and white: H); (iii) structure of the prolate molecule; (iv) schematic diagram showing the (4 × 2) adlayer identifying the molecular and footprint chirality at each position. This information is further deconvoluted to depict the enantiomer and footprint arrays that are formed.

bright and faint features. DFT calculations reveal that the two observed conformers arise as a direct consequence of the different adsorption footprints that are adopted by the molecule. For both conformers, the two carboxylate oxygen atoms bind to adjacent copper atoms in the close-packed row. In addition, a third bonding contact occurs with the nitrogen binding to a copper atom in the neighbouring row. It is the position of this bond that distinguishes the two conformers. When the nitrogen atom bonds to the copper atom to the left, a left-handed triangular adsorption footprint is described and the pyrrolidine ring tilts significantly away from the surface (conformer A, Fig. 5ii), imaging as a bright protrusion in STM. However, if the nitrogen bonds to the right, a right-handed triangular adsorption footprint is described, the ring lies flat (conformer B, Fig. 5ii) and images as a faint protrusion. Importantly, the left and right-handed footprints are non-superimposable mirror images in two-dimensions and are, therefore, chiral. This means that, effectively, a second level of chirality is imprinted at the surface in the form of the adsorption footprints.

The two conformers have similar adsorption energies and are both involved in the molecular assembly, and eight distinct (4×2) arrangements can be constructed for (*S*)-prolate on Cu(110).[25] By calculating adsorption energies for the different assemblies and comparing their simulated STM images with the experimental data (Fig. 5i), it can be determined that the enantiopure (*S*)-prolate assembly possesses an ordered heterochiral footprint arrangement (Fig. 5iv). Experimentally, the STM contrast enables chirality to be tracked at the single footprint level and confirms the theoretical prediction.

The molecule-by-molecule analysis of the (*S*)-prolate/Cu(110) assembly demonstrates that the molecule–surface interface now possesses two manifestations of chirality, one arising from the molecular handedness and the other from the footprint chirality, which we have chosen to label as 'footedness'.[24,28,29] In cases where single-handed molecules project footprints of one chirality only, *i.e.* the molecules are single-footed, one can confine the description of the interface chirality with respect to the molecular handedness only. However, if a single-handed molecule can project footprints of either chirality, *i.e.* it is two-footed (Fig. 6a), then the chiral description of the interface needs to capture this dual aspect. We have, therefore, introduced an overall chirality descriptor,[29] $\mathbf{C}_{\mathrm{F}}^{\mathrm{H}}$, which describes the chiral ordering at the molecule–surface interface from both the molecular handedness (H, superscript) and the adsorption footedness (F, subscript) viewpoints (Fig. 6b). Here, the classical descriptions of global chiral ordering such as enantiopure assembly, racemic conglomerate, racemic compound and random solid solution are given a label (En, Cn, Rc and Rn, respectively), as shown in Fig. 6c. Thus, the *S*-prolate/Cu(110) assembly has an overall chirality of $\mathbf{C}_{\mathrm{Rc}}^{\mathrm{En}}$.

Determining the overall chirality of such an interface is particularly difficult to determine experimentally, since techniques such as reflection absorption infrared spectroscopy (RAIRS), near edge X-ray absorption fine structure (NEXAFS) and photoelectron diffraction (PhD) are averaged across the whole surface and do not provide information at the single-molecule level. Here, the rigidity of the prolate and the re-orientation of the ring as the adsorption footprint is altered provides direct mapping of this aspect from the STM images.

Such analysis can be taken further by considering adsorption of the racemic mixture, where four diastereomers of handedness and footedness can arise

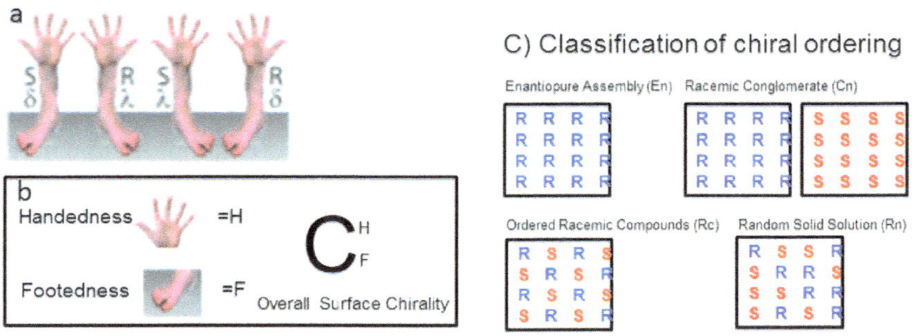

Fig. 6 (a) Depiction of the dual manifestation of chirality at surfaces created by both handedness (*R* and *S*) and footedness (λ and δ); (b) description of overall surface chirality as C_F^H where the organisation of handedness is depicted by (H, superscript) and of footedness by (F, subscript); (c) the ordering description is based on the different classes of chiral ordering and is indicated as follows: enantiopure (En); racemic conglomerate (Cn), racemic compound (Rc) and random solid solutions (Rn).

(Fig. 6a). The complexity of the overlayer that could be generated by such a system is evident from the STM image of racemic prolate on Cu(110),[26] Fig. 7i. Although the arrangement of the molecules follows the (4 × 2) pattern observed for the enantiopure system, there appears a random distribution of bright and faint protrusions. This complex image can be deconvoluted by considering high contrast images obtained from the enantiopure systems, Fig. 7ii, which show 4 distinct features, each associated with a particular handedness–footedness diastereomer. This now provides a key to read the entire STM image, as shown in Fig. 7ib–d, where a portion of the STM image is analysed in terms of chirality at the single-molecule and single-footprint level.

The ensuing arrangement can be deconvoluted into separate maps of molecular chirality and maps of adsorption footprints (Fig. 7id). From this, it can be seen that the enantiomer arrangement within the adlayer is entirely random, *i.e.* the molecular assembly is a two-dimensional random solid solution. However, the adsorption footprints adopt the strict heterochiral template that is exhibited by the enantiopure (*S*)-prolate system. Therefore, racemic prolate/Cu(110) produces an overall chirality of C_{Rc}^{Rn}. Clearly in this system, the chirality of the adsorption footprint, and not that of the molecule, drives the arrangements of both enantiopure and racemic proline at the surface.

The complexity of assembly exhibited in these and other amino acids on Cu(110) has recently[29] been rationalised using 3 simple rules, described below:

(i) The carboxylate rule. '*The carboxylate groups of neighbouring molecules avoid repeated binding in adjacent sites*'. This rule arises because electrostatic repulsion and surface strain create a large energy penalty if carboxylate groups are placed at nearest neighbour positions along an infinitely repeating 1D chain. As a result, for periodic (3 × 2) and (4 × 2) assemblies observed for amino acids on Cu(110), the carboxylate groups are staggered in both high symmetry directions of the surface.

(ii) The footedness rule. '*The allowed ordering at both the handedness and footedness levels is determined by the chirality of the adsorption footprints that can be created at the surface and the manner in which individual enantiomers may occupy these footprints*'. The footedness rule encapsulates two aspects: (a) the generation

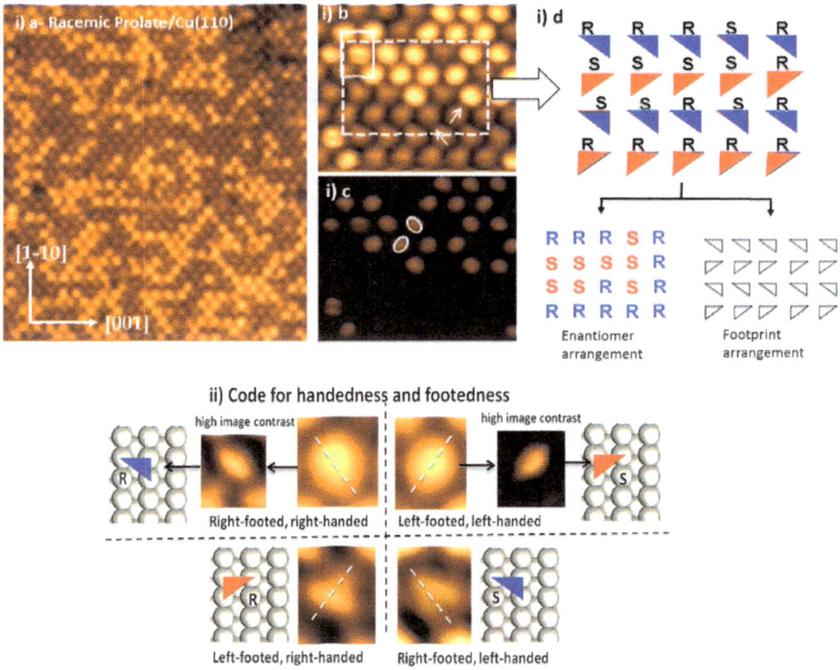

Fig. 7 (i) (a) STM image of the racemic (*R,S*)-prolate overlayer on Cu(110) (158 × 125 Å^2, $I(t) = -0.62$ nA, $V(t) = -1238$ mV); (i) (b) and (i) (c) high-resolution images showing the bright and faint features in the STM overlayer 44 × 50 Å^2, at differing image contrast; (i) (d) schematic diagram identifying the molecular and footprint chirality at each position within the (4 × 2) adlayer contained within the outlined square. This information is further deconvoluted to give the enantiomer and footprint arrangements at the surface. (ii) Code for interpreting each of the four distinct STM features in terms of molecular and footprint chirality.

of allowed footprint templates; and, (b) occupation of these templates by enantiomers.

(iii) The interactions rule. '*Of the outputs allowed by the carboxylate and footedness rules, the overall chirality that a molecular overlayer ultimately adopts will be determined by a balance between the optimization of molecule–substrate interactions and intermolecular interactions*'. These interactions will clearly be system-specific.

For the prolate/Cu(110) system, eight organised (4 × 2) templates can be suggested by the carboxylate rule, of which 2 are homochiral and 6 are heterochiral.[29] DFT calculations show that the specific heterochiral (4 × 2) footprint arrangement adopted is one that minimises repulsive carboxylate–carboxylate interactions and maximises intermolecular hydrogen bonding.[25,26] The reason that the C_{Rc}^{En} and the C_{Rc}^{Rn} assemblies emerge for enantiopure and racemic prolate adsorption, respectively, arises from the footedness rule. Fig. 8a compares how the footprints in a template would be occupied by enantiomers for a single-footed and a two-footed system. Such considerations allow a 'surface chirality generator' to be run for each template. Fig. 8b shows this for the preferred heterochiral template adopted by the prolate/Cu(110) system. It can be seen that the creation

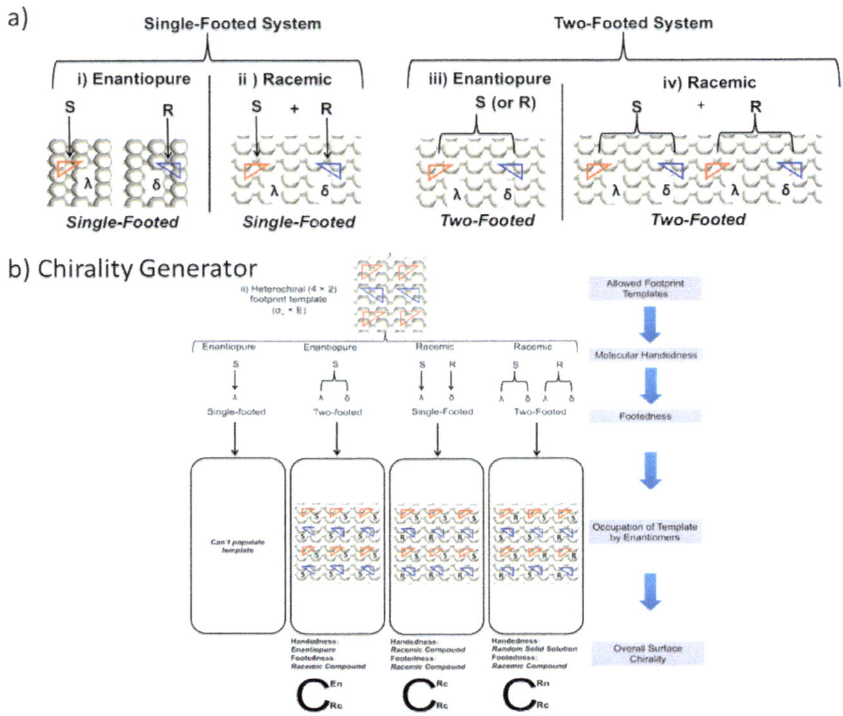

Fig. 8 (a) The occupation of footprints by chiral molecules [ref. 29] in: (i) and (ii) a single footed system with the (*S*)-enantiomer placing the λ-footprint and the (*R*)-enantiomer placing the δ-footprint; (iii) and (iv) a two-footed system where each enantiomer may place either the λ or δ-footprint (red triangles = left λ-footprint, blue triangles = right δ-footprint). The behaviour for both enantiopure and racemic systems is shown. (b) Depiction of the surface chirality generator [ref. 29], which predicts the overall chiral ordering that would arise for the heterochiral (4 × 2) footprint template, as the molecular handedness and footedness of the molecules are altered.

of the C^{En}_{Rc} overlayer for enantiopure adsorption and the C^{Rn}_{Rc} overlayer for racemic adsorption are now self-evident. The two-footed aspect for the racemic assembly is further confirmed by calculations that show that within the adopted heterochiral footprint arrangement, the upright conformer of (*R*)-proline is equivalent to the flat-lying conformer of (*S*)-proline, with respect to the formation of hydrogen-bonded chains. Thus, each adsorption position in the footprint template can be occupied by either enantiomer resulting in a 2-D random solid solution at the surface.[26]

The simple rules derived for the amino acid assemblies on Cu(110)[29] provide a framework for capturing the allowed manifold of assemblies that could potentially arise, and are a step forward in creating design rules that target the creation of specific arrangements and outcomes. For example, one of the conclusions of our analysis is that chiral segregation is a very particular output of systems that favour homochiral templates and are single-footed.[29] Note that the bitartrate/Cu(110) system falls in this category because there is a strong energy penalty for an enantiomer to occupy a mirror footprint. Currently, the literature

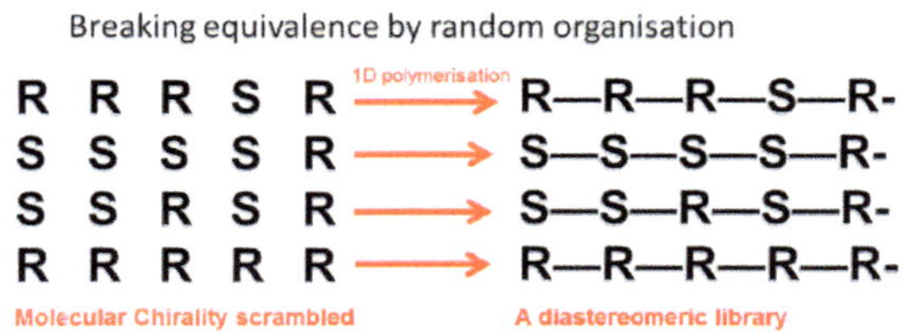

Fig. 9 Schematic showing how 1D polymerisation of a random solid solution of a racemic mixture would lead to a diastereomeric library of polymers.

on amino acids shows that many prefer heterochiral footprint templates and are also two-footed. This leads to an interesting input into debates on the origin of homochirality of life,[35,36] where most mirror-symmetry breaking scenarios involve chiral resolution or amplification mechanisms to create homochiral domains of monomers as a starting point for creating homochiral polymers of life. However, for almost all of these scenarios there is an equal probability of creating homochiral pools of either enantiomer, opening up equivalent pathways for creating right- and left-handed polymers. In an alternative scenario, if adsorption of a racemic mixture leads to a random solid solution of enantiomers at a surface, an inherent mechanism for mirror-symmetry breaking emerges. If the randomly organised array of enantiomers were subject to polymerisation, a large diastereomeric library of polymers would emerge (Fig. 9), each different with distinct properties. If homochiral polymers provided a particularly favoured attribute, these would be selected by evolutionary pressure. However, the chance of a randomly organised assembly creating a homochiral polymer becomes progressively smaller as the length of polymer is increased. For such a scenario, the probability of creating two mirror image homochiral polymers becomes vanishingly small as the polymer length increases beyond even a small number of units, thus providing an inherent mirror-symmetry breaking process. This idea now leads us to consider whether the surface provides an appropriate environment to covalently couple molecules into macromolecules.

On-surface covalent assembly of macromolecules

The field of on-surface synthesis has moved at great pace since the first reports of macromolecular creation *via* direct coupling of individual molecular components at a surface.[8–11] Today, a range of approaches have been translated successfully from organic synthesis methodology to on-surface synthesis, including condensation reactions,[11,37–39] Ullmann coupling,[8,10,40–43] Glaser coupling[44–46] and dehydrogenative coupling,[9,47–53] producing a variety of covalent structures at the surface, ranging from on-surface synthesis of complex molecules,[41,54,55] to linear polymers,[8,47,49,56] extended networks[10,45,55] and graphene nanoribbons.[10,57,58] Examples of on-surface synthesis are also illustrated in this Faraday Discussion by Lifeng Chi, Nian Lin and Federico Rosei.

In collaboration with David Amabilino, we have explored using the C–H group as the functional synthon and exploiting the surface to activate, dissociate and couple the components. Of course, C–H bond activation is well known for being challenging in organic synthesis and heterogeneous catalysis, and our choice of the Cu(110) surface provided a level of reactivity which is sufficient to enable reactions to occur, but not so great as to decompose the molecular building blocks. In addition, the anisotropy of the surface allowed regioselective behaviour to be identified. Finally, this approach opens up access to the huge range of starting materials that possess C–H groups and, also, delivers clean coupling reactions in which C–H bond scission leads to H atoms that recombine and desorb cleanly from the surface. Below, the success of this dehydrogenative coupling approach to covalently link a range of molecules at surfaces is outlined, with examples ranging from homocoupling to heterocoupling reactions involving components of different shape, size, connectivity and topology.

Covalent coupling to create linear poly-porphyrin organometallic oligomers

The reactions of the unfunctionalised free base H_2-porphyrin and H_2-diphenylporphyrin on Cu(110)[47,48] provide a good example of dehydrogenative coupling, where the reaction is triggered simply by heating. This generates linear oligomers of porphyrins, aligned along the [100] direction (Fig. 10a), accompanied by the evolution of hydrogen from the surface. Complex functions invariably require complex molecular assembly, so an important step in on-surface synthesis is to advance from homocoupling reactions to heterocoupling scenarios, where different building blocks can be combined. A successful example[49] of this is provided by the heterocoupling of H_2-porphyrin and Zn-diphenylporphyrin on Cu(110), to give linear co-oligomers (Fig. 10b). The STM images provide a direct map of oligomer composition and sequence, with H_2-porphyrin imaging as squares and Zn-diphenylporphyrin showing serrated features arising from the two phenyl groups. It can, therefore, be seen that a random co-polymer is created when the system is heated rapidly to 650 K. By altering synthesis protocols, block co-polymers in which sections of one monomer are followed by sections of the other are produced (Fig. 10b).

STM data show that all the homocoupled and heterocoupled oligomers possess a 10.8 Å monomer-to-monomer distance. DFT calculations[47,48] confirm this to be consistent with the creation of organometallic C–Cu–C bonds arising from cleavage of the sp^2 C–H bond at the 3, 5 and 7 positions. The calculated model and the STM simulation for a 2 Cu-atom and a 3 Cu-atom bonded polymer are shown in Fig. 10c. High-resolution STM data, *e.g.* Fig. 10a and d, are in good agreement with the theoretical images, showing the expected core-to-core distance of 10.8 Å, with the organometallic coupling Cu atoms imaging bright. It is noted that organometallic polymers are an important class of hybrid materials, combining the properties of organic systems with those of metals, with applications in molecular wires, photoelectronics and sensors.

This work reveals a number of important factors: (i) the C–H group can certainly be activated at a Cu surface, leading to C–H bond scission; (ii) the reaction involves recombinative H_2 desorption, indicating clean synthesis that does not leave side-products at the surface; (iii) organometallic coupling occurs at the surface, leading to [porphyrin-Cu]$_n$ organocopper oligomers, a reaction which has no counterpart in general organometallic synthesis, *i.e.*, the surface creates a unique reaction environment to create new types of functional matter;

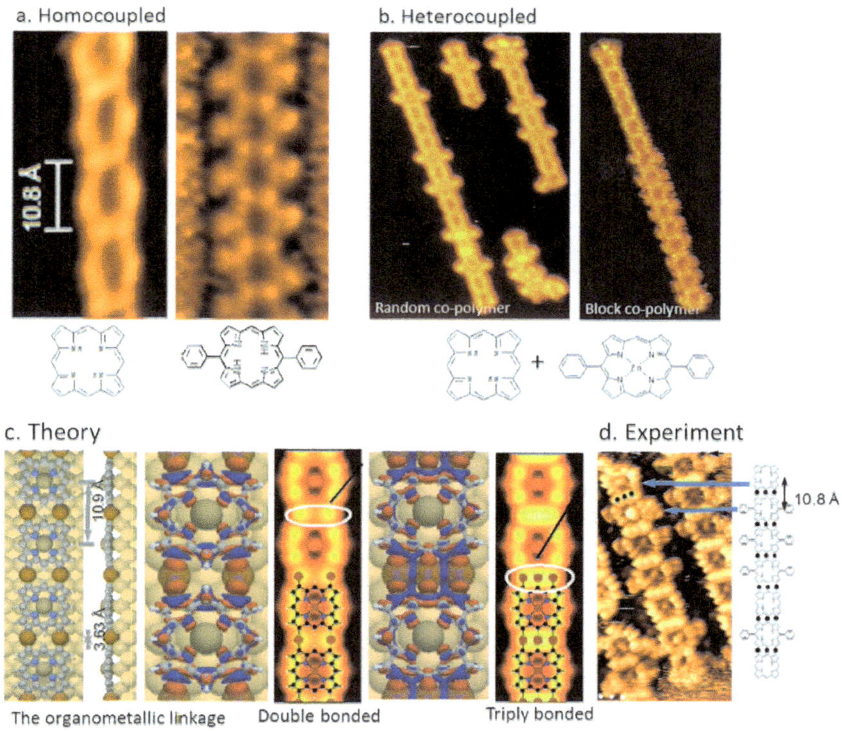

Fig. 10 Covalently bonded linear oligomers created from the reaction of H_2-porphyrin, H_2-diphenylporphyrin and Zn(II) diphenylporphyrin. (a) Homocoupled structures left, $I_t = 0.31$ nA, $V_t = 0.30$ V and right, $I_t = 0.41$ nA, $V_t = -1.68$ V; (b) heterocoupled structures, left: 100×190 Å, $I_t = 0.13$ nA, $V_t = -1.68$ V, and right: 65×170 Å, $I_t = 0.14$ nA, $V_t = -0.57$ V; (c) theoretically calculated polymer structures on Cu(110) and the corresponding simulated STM images showing the doubly and triply bonded Cu-porphyrins chains and the charge accumulation (red) and depletion (blue) that occurs when the polymer is formed. (d) High resolution experimental STM image (35×70 Å, $I_t = 0.17$ nA, $V_t = -0.83$ V) showing a heterocoupled oligomer formed by reaction of H_2-porphyrin and Zn(II)-diphenylporphyrin on Cu(110) where the chemical nature of each individual monomer and the number of interconnecting C–Cu–C bonds can be identified, as shown in the structural model.

(iv) directional macromolecules aligned along a specific surface direction are synthesised, despite the absence of functional groups to stereodirect the reaction.

Control of product directionality at a surface will be important in future molecular devices, and DFT modelling by Hanke *et al.*[48] of the H_2-porphyrin/Cu(110) system shows that each additional organometallic bond formed along the [001] direction leads to an energetically more stable structure, with the most stable structures corresponding to the triply-coupled unit with Cu insertion at the 1,3,5 positions of the porphyrin, or the doubly-coupled unit with C–Cu–C coupling at the 1 and 5 position (Fig. 11). An analysis of the product configurations reveals that the commensurability of the oligomer structure with the underlying surface structure along the [001] direction is also particularly

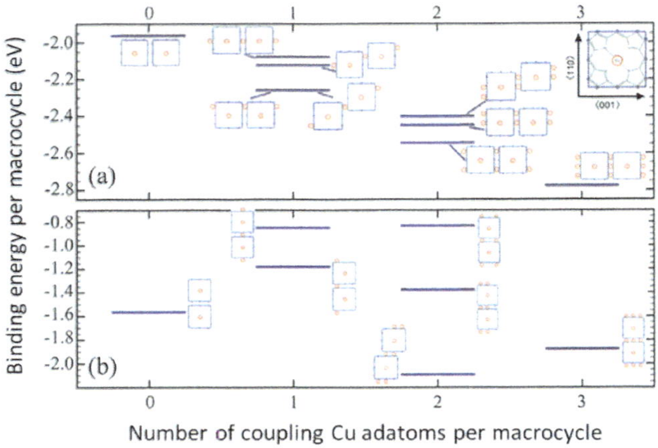

Fig. 11 Binding energies per porphyrin core for covalently bonded structures aligned along the [001] direction (top panel) and along the [1$\bar{1}$0] direction (bottom panel) of the Cu(110) surface. Reproduced with permission from *ACS Nano*, 2011, **5**(11), 9093 [ref. 48]. Copyright (2011) American Chemical Society.

favourable, accommodating the coupling Cu atoms on preferred 4-fold sites of the underlying surface as 'added rows' and causing very little distortion of the porphyrin core. However, coupling in the orthogonal [1$\bar{1}$0] direction is not favourable because the product dimensions are less compatible with surface geometry, leading to distortion and buckling of the porphyrin ring which, in turn, decreases the bonding interaction. An energy advantage only appears when the triply-bonded or the doubly-bonded offset units are created (Fig. 11), however, both still compare unfavourably with products aligned in the [001] direction.

Heterocoupling of unlike molecules

Coupling of two very different molecules was demonstrated when H_2-porphyrin and pentacene were co-adsorbed on Cu(110) and reacted together by heating to 650K.[49] STM data showed that a diverse set of products is synthesised at the surface *via* modular construction where dimer and monomer pentacene bind to the 1D porphyrin organometallic wires (Fig. 12a). Inter-module connections occur either at the side of the oligoporphyrin long edge to produce simple and complex key topologies or at the porphyrin chain ends to give capped ladder topologies. When pentacene caps the porphyrin oligomers, it acts as a terminating group as shown in Fig. 12b and c. Bright regions imaged at the inter-module boundaries are indicative of copper atoms linking the two covalent modules together. The STM image in Fig. 12a also shows discrete homocoupled pentacene dimers, where dehydrogenation along one length of each pentacene enables 5 organometallic linkages to be formed, which are accommodated as a local 'added row' structure. This aligns the pentacene long-axis parallel to the Cu(110) rows, which agrees with DFT calculations and STM observation (Fig. 12d).

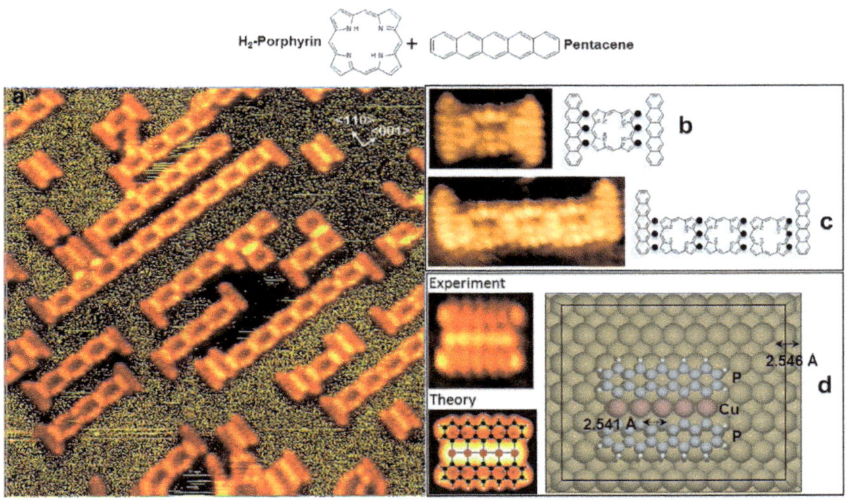

Fig. 12 (a) STM image (150 × 150 Å^2, I_t = 1.35 nA, V_t = 0.47 V) showing the diversity of macromolecules formed by coupling reactions of H_2-porphyrin and pentacene on Cu(110);[49] (b) and (c) high resolution STM showing two types of pentacene–porphyrin products, alongside chemical structural models where the black atoms represent the coupling Cu atoms; (d) experimentally measured and theoretically calculated STM image of a covalently bonded pentacene dimer (I_t = 0.19 nA, V_t = −0.034 V, theory V_t = −1.34 V).

Covalent coupling to create C–C coupled porphyrins

Dehydrogenative coupling can also create C–C covalent linkages, first reported for tetra(mesityl)porphyrin on Cu(110).[9] This reaction is particularly interesting because the 4-methyl groups act as unique connection points, displaying highly regioselective C–C bond creation. This system has recently been modelled in great detail by state-of-the-art density functional theory (DFT) and Nudged Elastic Band (NEB) calculations, to identify the thermodynamic and kinetic factors that determine the unique selectivity in covalent bonding.[53] This study deconvoluted the hierarchy of C–H bond strengths within the molecule, and demonstrated how the adsorption site and adsorbate conformation influence this hierarchy, leading to the selectivity in C–H activation and de-hydrogenation propensity at different sites on the molecule. In addition, thermodynamic and kinetic effects, including diffusion anisotropies and product accommodation at the surface determine which intermolecular C–C coupling reactions and coupled products are favoured at the surface. These insights are invaluable in unravelling the detail of the covalent assembly at surfaces and a step towards knowledge-based synthetic rules. Finally, the diversity of molecules that can be homocoupled and hetero-coupled at the Cu(110) surface is shown schematically in Fig. 13. This demonstrates that the C–H synthon provides a highly generic route for clean, on-surface synthesis. This approach is particularly useful for parallel synthesis of multivariate libraries of covalent matter at a surface for systems where the design parameters to deliver specific functions are unknown (which, at present, is generally the case).

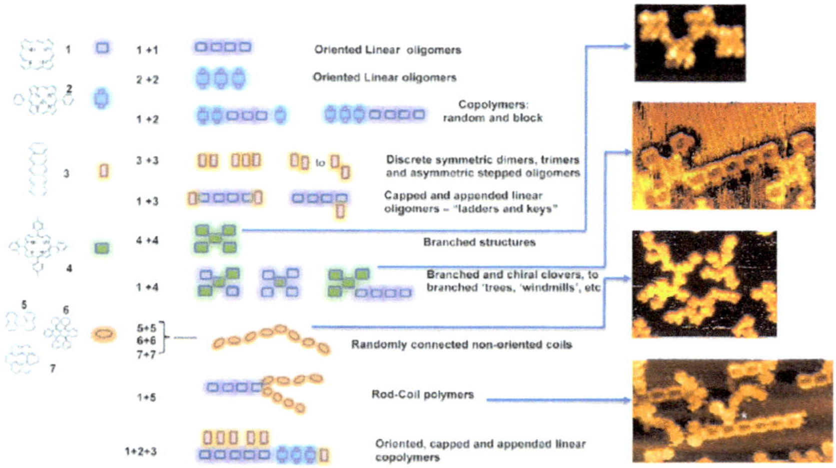

Fig. 13 Summary and overview (with selected STM images) of successfully covalently coupled molecules on Cu(110).[49] 2*H*-Porphyrin (1), Zn(II)-diphenylporphyrin (2), pentacene (3), tetramesitylporphyrin (4), perylene (5), rubrene (6) and coronene (7). Reproduced with permission from *ACS Nano*, 2014, **8**(9), 8856 [ref. 49]. Copyright (2014) American Chemical Society.

Creating functional molecular systems

We are still at the earliest stages of creating 'bottom-up' functional molecular devices at surfaces. Nanopore sequencing technology provides a glimpse of the future potential,[59,60] however this successful approach requires the use of a biological entity, *e.g.* the α-hemolysin pore, as the chief operational component. Elsewhere, molecular electronics has utilised single molecule components and self-assembled molecular layers to deliver function.[61,62] In terms of *de novo* design of molecular devices, the field of molecular machines certainly stands out, with the 2016 award of the Nobel Prize in Chemistry to Sauvage, Stoddart and Feringa. An important achievement is to translate such entities at a surface and this has been successfully demonstrated recently for selected systems.[63–65]

A future direction is 'bottom-up' assembly of simple components at a surface to create a device. The example below[66] shows an approach we created in collaboration with David Amabilino and Lluisa Pérez-Garcia, with insights into system behaviour modelled in collaboration with Lev Kantorovitch. The aim of the study was to create a surface system in which a simple molecular walker is confined to walk along a specified path between two locations. Such a system would be a mimic of the rotaxane system,[67] where a macrocycle shuttles along an axis confined between two stations, or the biological linear protein motor kinesin on microtubule filaments.[68,69] An important aim was to achieve room temperature operation. Therefore, a divalent bis(imidazolyl) molecule, Fig. 14a, was synthesised, with the imidazolyl functionality providing attachment 'feet' that would chemically bond and detach from the close-packed Cu(110) rows, which would act as a linear track for motion. To create barriers on the track, porphyrin oligomers were synthesised directly on the surface in an orthogonal orientation to the track,

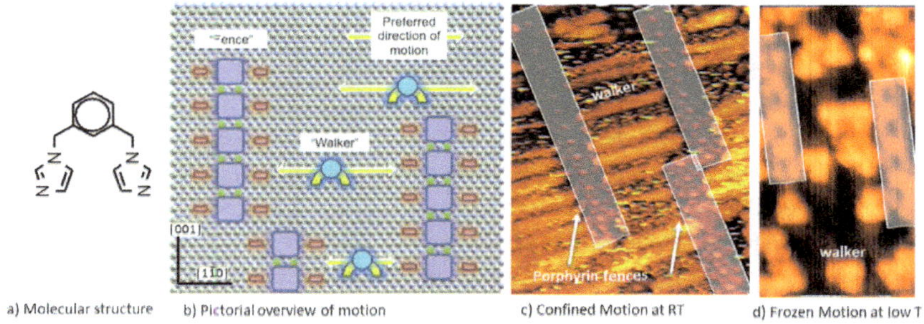

Fig. 14 (a) The molecule structure of the 'walker' (1,3-bis(imidazol-1-ylmethyl)benzene). (b) Pictorial overview of the two-component molecular system designed to deliver confined, directional motion of a walker molecule between covalent porphyrin oligomer fences. (c) STM image of the system at 300 K showing diffusion streaks associated with the motion of the walker molecule along the [$\bar{1}$10] direction, confined between the porphyrin fences, which are highlighted by rectangles in the images. (d) STM image of the system at 100 K when the diffusion of the walker is frozen and each molecule is observed as a triangular feature.

thus creating a 'fence' to confine the motion. Fig. 14b depicts a schematic of the proposed system while Fig. 14c shows the STM image of the system at room temperature. The motion of the molecular walker is imaged as intensity streaks, which are clearly aligned along the [1$\bar{1}$0] surface axis showing directional motion. It is also clear that the motion is confined between the immobile porphyrin fences. At low temperature, the motion of the walker molecule can be frozen and each individual molecule can be imaged (Fig. 14d). DFT and NEB calculations show that the walker moves by attaching and detaching its imidazolyl feet in an inchworm fashion along the [1$\bar{1}$0] direction, with directionality of motion arising from specific chemisorption interaction with the surface and from highly anisotropic diffusion barriers, which make motion in the orthogonal [001] direction energetically expensive. Given the ability of porphyrins to heterocouple with a number of different molecules, Fig. 10 and 12, the opportunity now opens up to create non-equivalent "stations" which could incorporate recognition or stimuli-triggered functions that could initiate responsive motions.

Conclusions

Molecular assembly at surfaces and interfaces has made rapid advances, delivering ever more complex systems and utilising an increasing toolbox of molecules and coupling approaches. Concurrently, the deployment of high resolution imaging, spectroscopic and diffraction techniques and advanced theoretical modelling is yielding increasingly detailed insights. What is clear from this burgeoning body of work is that the molecule–surface system possesses great diversity and polymorphism, arising from the manifold of energetically viable structures the system can adopt. This aspect presents a real challenge to both the experimentalist and theoretician. Generally, the global assembly that is adopted can be traced back to effects that are already encoded into the system at the single molecule level. Here the local flexibility and response of the molecule with respect to the balance of molecule–surface and molecule–molecule interactions plays a critical role. A corollary of this behaviour is that such systems possess responsivity, adaptability and tunability. However, knowledge-based engineering of such finely balanced systems will require multiscale analysis of the system, from the nanoscale to the macroscale. For example, little is understood of the influence and balance of the many interactions that drive molecular assembly at surfaces, such as molecule–molecule interactions encompassing non-directional van der Waals and directional H-bonding interactions, alongside strong electrostatic and covalent interactions. Equally, the role of surface–molecule interactions, ranging from van der Waals to strong electrostatic charge transfer, direct metal–molecule bonding and surface reconstructions need to be understood. There is also ample evidence that the kinetics of self-assembly is a major factor in determining the final structure, often driving polymorphic behaviour and leading to widely varied outcomes, depending on the conditions of formation. Thus experimental and theoretical approaches need to be progressed in order to capture system kinetics, including diffusion barriers and transition states. This will require time-resolved techniques and computational methods that can encompass reasonable time domains. Finally, it is pertinent to note that it took many decades of effort to create the synthetic principles of organic chemistry. It will also take our field an equivalent time to lay down the design principles of molecular assembly at surfaces.

Acknowledgements

This article draws on the work of many fine researchers and collaborators it has been my pleasure to work with. Their work has been referenced within the manuscript and particular acknowledgement goes to members of my group: Matthew Forster, Sam Haq, Mendel In't Veld, Ning Liu, Andrew Mark and Bart Wit; and to my collaborators and colleagues: David Amabilino, George Darling, Matthew Dyer, Felix Hanke, Tonek Jansen, Andreas Floris, Andrew Hodgson, Lev Kantorovitch, Mats Persson and Lluisa Pérez-Garcia. I would like to thank them all for their contributions as we made this scientific journey. I also thank EPSRC (EP/J019364/1, EP/K039687/1, EP/F00981X/1) and EU FP7 projects RESOLVE (NMP4-SL-2008-214340) and SMALL (PITN-GA-2009-238804) programmes for supporting this research.

Notes and references

1 *Supramolecular Chemistry at Surfaces*, ed. D. B. Amabilino, Royal Society of Chemistry, 2016, ISBN: 978-1-84973-952-8.

2 S. M. Barlow and R. Raval, *Surf. Sci. Rep.*, 2003, **50**, 201.

3 J. A. A. W. Elemans, I. De Cat, H. Xu and S. De Feyter, *Chem. Soc. Rev.*, 2009, **38**, 722.

4 T. Balandina, K. Tahara, N. Sändig, M. O. Blunt, J. Adisoejoso, S. Lei, F. Zerbetto, Y. Tobe and S. De Feyter, *ACS Nano*, 2012, **6**(9), 8381.

5 L. Bartels, *Nat. Chem.*, 2010, **2**, 87.

6 F. Kling, R. Bechstein, P. Rahe and A. Kühnle, *Self-Assembly of Organic Molecules on Insulating Surfaces*, Springer Series, Heidelberg, 2015.

7 G. R. Desiraju, *J. Am. Chem. Soc.*, 2013, **135**(27), 9952.

8 L. Grill, M. Dyer, L. Lafferentz, M. Persson, M. V. Peters and S. Hecht, *Nat. Nanotechnol.*, 2007, **2**, 687.
9 M. In't Veld, P. Iavicoli, S. Haq, D. B. Amabilino and R. Raval, *Chem. Commun.*, 2008, 1536.
10 J. Cai, P. Ruffieux, R. Jaafar, M. Bieri, T. Braun, S. Blankenburg, M. Muoth, A. P. Seitsonen, M. Saleh, X. Feng, K. Müllen and R. Fasel, *Nature*, 2010, **466**, 470.
11 M. Matena, T. Riehm, M. Stöhr, T. A. Jung, A. Thomas and L. H. Gade, *Angew. Chem., Int. Ed.*, 2008, **47**, 2414.
12 M. O. Lorenzo, C. J. Baddeley, C. Muryn and R. Raval, *Nature*, 2000, **404**(6776), 376.
13 M. O. Lorenzo, S. Haq, T. Bertrams, P. Murray, R. Raval and C. J. Baddeley, *J. Phys. Chem. B*, 1999, **103**(48), 10661.
14 L. Barbosa and P. Sautet, *J. Am. Chem. Soc.*, 2001, **123**(27), 6639.
15 C. G. M. Hermse, A. P. van Bavel, A. P. J. Jansen, L. Barbosa, P. Sautet and R. van Santen, *J. Phys. Chem. B*, 2004, **108**(30), 11035.
16 R. Fasel, J. Wider, C. Quitmann, K.-H. Ernst and T. Greber, *Angew. Chem., Int. Ed.*, 2004, **116**, 2913.
17 S. Haq, N. Liu, V. Humblot, A. P. J. Jansen and R. Raval, *Nat. Chem.*, 2009, **1**, 409.
18 G. Darling, M. Forster, C. Lin, N. Liu, R. Raval and A. Hodgson, *Phys. Chem. Chem. Phys.*, 2017, **19**, 7617.
19 The overlayer unit mesh is given in real space matrix notation as follows and quoted in the text as (m_{11} m_{12}, m_{21} m_{22}): $\begin{pmatrix} a_0 \\ b_0 \end{pmatrix} = \begin{pmatrix} m_{11} & m_{12} \\ m_{21} & m_{22} \end{pmatrix} \begin{pmatrix} a_s \\ b_s \end{pmatrix}$ where a_0, b_0 are the overlayer net vectors and a_s, b_s are the underlying metal surface mesh vectors.
20 N. Liu, S. Haq, G. R. Darling and R. Raval, *Angew. Chem., Int. Ed.*, 2007, **46**(40), 7613.
21 A. J. Gellman, W. T. Tysoe and F. Zaera, *Catal. Lett.*, 2015, **145**(1), 220.
22 M. Ortega-Lorenzo, V. Humblot, P. Murray, C. J. Baddeley, S. Haq and R. Raval, *J. Catal.*, 2002, **205**, 123.
23 P. Szabelski, *Chem.–Eur. J.*, 2008, **14**, 8312.
24 M. Forster, M. S. Dyer, M. Persson and R. Raval, *J. Am. Chem. Soc.*, 2011, **133**(40), 15992.
25 M. Forster, M. S. Dyer, M. Persson and R. Raval, *J. Am. Chem. Soc.*, 2009, **131**(29), 10173.
26 M. Forster, M. S. Dyer, M. Persson and R. Raval, *Angew. Chem., Int. Ed.*, 2010, **49**(13), 2344.
27 M. Forster, M. Dyer, S. D. Barrett, M. Persson and R. Raval, *J. Phys. Chem. C*, 2011, **115**, 1180.
28 A. G. Mark, M. Forster and R. Raval, *ChemPhysChem*, 2011, **12**, 1474.
29 M. Forster and R. Raval, *Chem. Commun.*, 2016, **52**, 14075.
30 R. Fasel, M. Parschau and K.-H. Ernst, *Nature*, 2006, **439**, 449.
31 K. Tahara, H. Yamaga, E. Ghijsens, K. Inukai, J. Adisoejoso, M. O. Blunt, S. De Feyter and Y. Tobe, *Nat. Chem.*, 2011, **3**, 714.
32 J.-C. Lemay, Y. Dong, M. N. Groves, V. Demers-Carpentier, G. Goubert, R. Lafleur-Lambert, J. Boukouvalas, B. Hammer and P. H. McBreen, *Science*, 2016, **646**, 13.

33 M. Lingenfelder, G. Tomba, G. Costantini, L. C. Ciacchi, A. De Vita and K. Kern, *Angew. Chem., Int. Ed.*, 2007, **46**, 4492.
34 R. G. Velpula, T. Takeda, J. Adisoejoso, K. Inukai, K. Tahara, K. S. Mali, Y. Tobe and S. De Feyter, *Chem. Commun.*, 2017, **53**, 1108.
35 M. Avalos, R. Babiano, P. Cintas, J. L. Jimenez and J. C. Palacios, *Origins Life Evol. Biospheres*, 2004, **34**, 391.
36 I. Weissbuch, G. Bolbach, L. Leiserowitz and M. Lahav, *Origins Life Evol. Biospheres*, 2004, **34**, 79.
37 M. Treier, N. V. Richardson and R. Fasel, *J. Am. Chem. Soc.*, 2008, **130**(43), 14054.
38 A. C. Marele, R. Mas-Ballesté, L. Terracciano, J. Rodríguez-Fernández, I. Berlanga, S. S. Alexandre, R. Otero, J. M. Gallego, F. Zamora and J. M. Gómez-Rodríguez, *Chem. Commun.*, 2012, **48**(54), 6779.
39 J. F. Dienstmaier, D. D. Medina, M. Dogru, P. Knochel, T. Bein, W. M. Heckl and M. Lackinger, *ACS Nano*, 2012, **6**(8), 7234.
40 L. Lafferentz, V. Eberhardt, C. Dri, C. Africh, G. Comelli, F. Esch, S. Hecht and L. Grill, *Nat. Chem.*, 2012, **4**(3), 215.
41 Q. Fan, J. Dai, T. Wang, J. Kuttner, G. Hilt, J. M. Gottfried and J. Zhu, *ACS Nano*, 2016, **10**, 3747.
42 M. Kittelmann, P. Rahe, M. Nimmrich, C. M. Hauke, A. Gourdon and A. Kühnle, *ACS Nano*, 2011, **5**(10), 8420.
43 R. Gutzler, L. Cardenas, J. Lipton-Duffin, M. el Garah, L. E. Dinca, C. E. Szakacs, C. Fu, M. Gallagher, M. Vondrácek, M. Rybachuk, D. F. Perepichka and F. Rosei, *Nanoscale*, 2014, **6**(5), 2660.
44 Y.-Q. Zhang, N. Kepcija, M. Kleinschrodt, K. Diller, S. Fischer, A. C. Papageorgiou, F. Allegretti, J. Björk, S. Klyatskaya, F. Klappenberger, M. Ruben and J. V. Barth, *Nat. Commun.*, 2012, **3**, 1286.
45 H.-Y. Gao, H. Wagner, D. Zhong, J.-H. Franke, A. Studer and H. Fuchs, *Angew. Chem., Int. Ed.*, 2013, **52**(14), 4024.
46 H. Zhou, J. Liu, S. Du, L. Zhang, G. Li, Y. Zhang, B. Z. Tang and H. J. Gao, *J. Am. Chem. Soc.*, 2014, **136**(15), 5567.
47 S. Haq, F. Hanke, M. S. Dyer, M. Persson, P. Iavicoli, D. B. Amabilino and R. Raval, *J. Am. Chem. Soc.*, 2011, **133**(31), 12031.
48 F. Hanke, S. Haq, R. Raval and M. Persson, *ACS Nano*, 2011, **5**(11), 9093.
49 S. Haq, F. Hanke, J. Sharp, M. Persson, D. B. Amabilino and R. Raval, *ACS Nano*, 2014, **8**(9), 8856.
50 Q. Sun, C. Zhang, H. Kong, Q. Tan and W. Xu, *Chem. Commun.*, 2014, **50**, 11825.
51 A. Wiengarten, K. Seufert, W. Auwärter, D. Ecija, K. Diller, F. Allegretti, F. Bischoff, S. Fischer, D. A. Duncan, A. C. Papageorgiou, F. Klappenberger, R. G. Acres, T. H. Ngo and J. V. Barth, *J. Am. Chem. Soc.*, 2014, **136**(26), 9346.
52 D. Zhong, J.-H. Franke, S. K. Podiyanachari, T. Blomker, H. Zhang, G. Kehr, G. Erker, H. Fuchs and L. Chi, *Science*, 2011, **334**(6053), 213.
53 A. Floris, S. Haq, M. In't Veld, D. B. Amabilino, R. Raval and L. Kantorovich, *J. Am. Chem. Soc.*, 2016, **138**(18), 5837.
54 C. J. Villagómez, T. Sasaki, J. M. Tour and L. Grill, *J. Am. Chem. Soc.*, 2010, **132**(47), 16848.
55 M. Abel, S. Clair, O. Ouurdjini, M. Mossayon and L. Porte, *J. Am. Chem. Soc.*, 2011, **133**, 11203.

56 J. A. Lipton-Duffin, J. A. Miwa, M. Kondratenko, F. Cicoira, B. G. Sumpter, V. Meunier, D. F. Perepichka and F. Rosei, *Proc. Natl. Acad. Sci. U. S. A.*, 2010, **107**, 11200.
57 J. Björk, S. Stafström and F. Hanke, *J. Am. Chem. Soc.*, 2011, **133**(38), 14884.
58 Y.-C. Chen, D. G. de Oteyza, Z. Pedramrazi, C. Chen, F. R. Fischer and M. F. Crommie, *ACS Nano*, 2013, **7**, 6123.
59 D. Branton, *et al.*, *Nat. Biotechnol.*, 2008, **26**, 1146.
60 Y. Feng, Y. Zhang, C. Ying, D. Wang and C. Du, *Genomics, Proteomics Bioinf.*, 2015, **13**, 4.
61 J. R. Heath, *Annu. Rev. Mater. Res.*, 2009, **39**, 1.
62 X. Chen, M. Roemer, L. Yuan, W. Du, D. Thompson, E. del Barco and C. A. Nijhuis, *Nat. Nanotechnol.*, 2017, **12**, 797–803.
63 Y. Shirai, A. J. Osgood, Y. Zhao, K. F. Kelly and J. M. Tour, *Nano Lett.*, 2005, **5**, 2330.
64 T. Kudernac, N. Ruangsupapichat, M. Parschau, B. Macia, N. Katsonis, S. R. Harutyunyan, K. H. Ernst and B. L. Feringa, *Nature*, 2011, **479**, 208.
65 J. Bern, D. A. Leigh, M. Lubomska, S. M. Mendoza, E. M. Perez, P. Rudolf, G. Teobaldi and F. Zerbetto, *Nat. Mater.*, 2005, **4**, 704.
66 S. Haq, B. Wit, H. Sang, A. Floris, Y. Wang, J. Wang, L. Pérez-García, L. Kantorovitch, D. B. Amabilino and R. Raval, *Angew. Chem., Int. Ed.*, 2015, **54**, 7101.
67 J. A. Bravo, F. M. Raymo, J. F. Stoddart, A. J. P. White and D. J. Williams, *Eur. J. Org. Chem.*, 1998, **11**, 2565.
68 A. Yildiz, M. Tomishige, R. D. Vale and P. R. Selvin, *Science*, 2004, **303**(5658), 676.
69 V. Balzani, A. Credi and M. Venturi, *ChemPhysChem*, 2008, **9**, 202.

Faraday Discussions

PAPER

A surprising way to control the charge transport in molecular electronics: the subtle impact of the coverage of self-assembled monolayers of floppy molecules adsorbed on metallic electrodes†

Ioan Bâldea ‡*

Received 10th March 2017, Accepted 19th May 2017
DOI: 10.1039/c7fd00101k

Inspired by earlier attempts in organic electronics aiming at controlling charge injection from metals into organic materials by manipulating the Schottky energy barrier using self-assembled monolayers (SAMs), recent experimental and theoretical work in molecular electronics showed that metal–organic interfaces can be controlled *via* changes in the metal work function that are induced by SAMs. In this paper we indicate a different route to achieve interface-driven control over the charge transfer/transport at the molecular scale. It is based on the fact that, in floppy molecule based SAMs, the molecular conformation can be tuned by varying the coverage of the adsorbate. We demonstrate this effect with the aid of benchmark molecules that are often used to fabricate nanojunctions and consist of two rings that can easily rotate relative to each other. We show that, by varying the coverage of the SAM, the twisting angle φ of the considered molecular species can be modified by a factor of two. Given the fact that the low bias conductance G scales as $\cos^2 \varphi$, this results in a change in G of over one order of magnitude for the considered molecular species. Tuning the twisting angle by controlling the SAM coverage may be significant, *e.g.*, for current efforts to fabricate molecular switches. Conversely, the lack of control over the local SAM coverage may be problematic for the reproducibility and interpretation of the STM (scanning tunneling microscope) measurements on repeatedly forming single molecule break junctions.

1 Introduction

Molecules adsorbed at interfaces may exhibit unique behaviors that can be very different from those in bulk situations. In the context of molecular electronics,

Theoretische Chemie, Universität Heidelberg, Im Neuenheimer Feld 229, D-69120 Heidelberg, Germany. E-mail: ioan.baldea@pci.uni-heidelberg.de

† Electronic supplementary information (ESI) available. See DOI: 10.1039/c7fd00101k

‡ Also at National Institute for Lasers, Plasmas, and Radiation Physics, Institute of Space Sciences, Bucharest, Romania.

one important reason for this uniqueness is that, in contrast to the (practically) fixed density in bulk cases, molecules in a self-assembled monolayer (SAM) on a substrate electrode have a density ("coverage") that can be widely varied/controlled. Ubiquitous molecular junction architectures like STM (scanning tunneling microscope) and CP-AFM (conducting probe atomic force microscope) are based on SAMs adsorbed on substrates. SAMs are formed and packed onto the surface of the substrate in a specific way. Nevertheless, common theoretical approaches to electron nanotransport completely ignore the impact (to be considered in the present paper) of SAM coverage and SAM arrangement.

For transport calculations, current implementations like TRANSIESTA and ATK (Atomistic Tool Kit) apply periodic boundary conditions in the (x, y)-directions which are perpendicular to the transport z-direction and pay attention in choosing sufficiently large supercells; otherwise, periodic replicas transverse to the transport direction would not have negligible interaction and the computed currents would not reflect the transport through single molecule junctions. At the same time, however, in doing that, such approaches implicitly restrict themselves to cases of extremely low SAM coverage.

Interactions between the SAMs and electrodes are known to play an important part in molecular electronic devices, as expressed by, *e.g.*, changes in the metal work function and the Fermi level pinning effects, as revealed in experiments[1–5] and justified theoretically.[6–11] Other important effects at the SAM-electrode interface are brought about by applied voltages, modeled in some cases by fractional charges acquired by the adsorbed molecules, and they turn out to cause significant changes in the molecular electronic and vibrational properties.[12–15]

What is particularly interesting is the effect of adding fractional charge to floppy molecules; the molecular conformation can be changed to some extent in this way, as we showed recently.[16] Still, however interesting this impact on the molecular conformation is, it is neither as easily feasible experimentally nor as spectacular as that of varying the coverage of the SAM.

This effect will be demonstrated in the present paper with the aid of benchmark floppy molecular species often used to fabricate molecular junctions[17–29] that consist of two rings that can easily rotate relative to each other. Specifically, we will consider two molecular species: 4-methyl-4′-mercaptobiphenyl CH_3–$(C_6H_4)_2$–SH (BPMT) and 4,4′-dithiol-1,1′-biphenyl HS–$(C_6H_4)_2$–SH (BPDT). A schematic picture is presented in Fig. 1, wherein the atom labeling used below is also indicated.

In the charge transfer across such molecules the twisting angle φ is important. The hopping integral t between the two π-orbitals characterizing the delocalized electrons of the two rings is proportional to $\cos\varphi$.[30] Because within the tight-binding approximation the electron transmission is proportional to the square of t, the ohmic conductance G

$$G \propto \cos^2 \varphi \tag{1}$$

is highly sensitive to the twisting angle φ and can be efficiently controlled by this quantity.

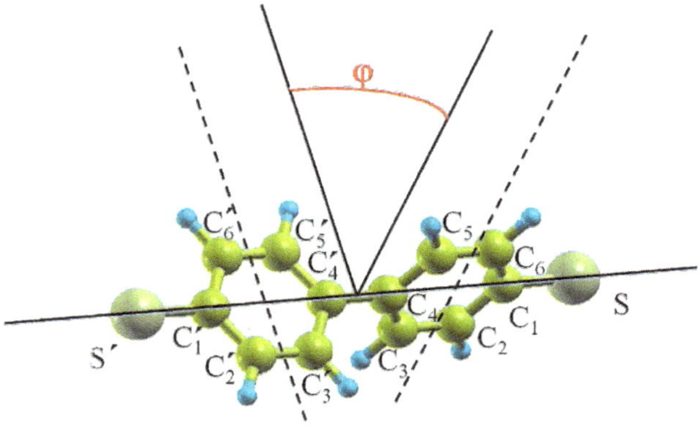

Fig. 1 Schematic representation of a floppy molecule consisting of two rings that can easily be rotated relative to each other. The torsional (twisting) angle φ represents the most salient structural feature of such a molecule. As indicated in the legend, φ affects the charge transfer across the molecule. The atom labeling indicated here is used in the main text with reference to the molecular species considered in this study (BPMT and BPDT). The figure was generated using XCrySDen.[31]

2 Computational details

Calculations done in conjunction with the present study on SAMs consisting of BPMT and BPDT molecules adsorbed on a face centered cubic (fcc) Au (111) surface are based on density functional theory (DFT) and have been performed by running the SIESTA[32] 4.0 package[33] in parallel.[34]

In contrast to "all electron" programs, SIESTA uses norm-conserving pseudopotentials to represent core electrons. For non-metal atoms we used the pseudopotentials obtained by translating the ABINIT pseudopotential database which is available to download.[33] For reasons indicated in Section 3.2, for gold atoms we preferred to use the pseudopotentials in ref. 35. Valence electrons were treated within the LCAO (linear combination of atomic orbitals) approximation using numerical localized atom centered basis sets of various qualities (SZP, DZP and TZP; see the list below and in the ESI†).

We report the results of the computations for slabs consisting of a SAM with a (6×6) structure as well as with a herringbone arrangement adsorbed on several layers of the fcc Au (111) surface. Because SIESTA is a code based on periodic boundary conditions in all three directions, in the direction transverse to the slabs we introduced a large vacuum region of at least 40 Å in order to make the electrostatic interactions between repeated slabs negligible. We checked that the slab dipole corrections – that can be accounted for by introducing a dipole layer in the middle of the vacuum layer and are needed, *e.g.*, for computing work functions[8] – are not necessary for the present geometry optimizations. We used a fine mesh cut-off of 300 Ry (~4082 eV) and *k*-sampling adapted to the SAM supercell to consider $5 \times 5 \times 1$ and $6 \times 9 \times 1$ Monkhorst–Pack grids of *k*-points for the (6×6)

and herringbone superstructures, respectively. The choice of these Monkhorst–Pack grids is not critical, it is merely for consistency with ongoing work on other related projects. For isolated molecules treated with SIESTA (Section 3.1 and 3.2), Γ-point calculations were performed using large (~40 Å) cubic unit cells.

As demonstrated by a comparison between the various methods utilized, which is presented in detail in Section 3.3, we found that a satisfactory description of the relaxed SAMs considered in this study can be obtained:

• By using split valence polarized double zeta (DZP) basis sets for the non-metal atoms and polarized single zeta (SZP) basis sets for metal atoms;

• By including only three gold layers in the calculations, even without the need to relax the gold atoms in the top layer;

• By performing relaxation until the forces on all atoms are smaller than 0.02 eV $Å^{-1}$;

• By performing DFT calculations without accounting for van der Waals dispersion corrections. To achieve this, we have compared the results obtained by using the PBE (Perdew–Burke–Ernzerhof) exchange correlation functional[36] – a parameter free functional that belongs to the broader class of GGA (conjugated gradient approximation) functionals (sometimes also called non-local functionals) – to those based on the van der Waals (VDW) functional (DRSLL) proposed by Dion *et al.*[37] which is based on the specific implementation of ref. 38.

As a test of the correctness of the relaxed geometries obtained by using SIESTA, for the envisaged isolated molecules we also compared the results of the SIESTA calculations to quantum chemistry approaches based on standard DFT/B3LYP, MP2 (second order Møller–Plesset perturbation expansion) and CCSD (coupled cluster singles and doubles) methods as implemented in GAUSSIAN 09.[39]

3 Results and discussion

Because the tables presented in the main body of this paper and in the ESI,† which contain very detailed results and information on the methods employed, are both self-explanatory and facilitate an interested reader to reproduce them, in the sections that follow we will only briefly emphasize the main ideas.

To make the paper more easily accessible to the reader who is not an expert in computational chemistry or not interested in computational details, only results obtained by the "reference" method (specified by the acronym "0.02", see the ESI†) will be presented in the main text. Very detailed numerical results obtained at various levels of theory are included in the ESI.†

3.1 Isolated molecular species

For molecules consisting of rings that can easily rotate relative to each other, the torsional degree of freedom represents the most salient structural feature. As is well known, it is not only the theoretical approach but also the interpretation of the experimental measurements[40–42] related to this floppy degree of freedom and the associated low torsional energy barrier that may raise delicate issues, even for isolated molecules in the gas phase. Therefore, the reliability of the adopted theoretical SIESTA-based approach represents an important aspect to be considered. Insight can be gained by inspecting the bond metric data obtained by several approaches.

Tables 1, S1 and S2† show the bond metric data of the isolated BPMT and BPDT molecules obtained by SIESTA calculations at the DFT/GGA-PBE level as well as those obtained by GAUSSIAN 09 calculations at the DFT/PBE and DFT/B3LYP levels and by *ab initio* MP2 and CCSD approaches.

All of the approaches listed in Tables S1 and S2† predict that the inter-ring bond is the longest C–C bond. Its length, $\lesssim$ 1.5 Å, which is compatible with a single bond, is almost 0.1 Å longer than the aromatic inter-ring bonds. Out of the latter, the bonds parallel to the molecular axis are the shortest C–C bonds. In isolated BPMT, out of the C–C bonds parallel to the molecular axis, those closer to the thiol group (denoted as C_2C_3 in Table S1†) are ~ 0.002–0.003 Å shorter than those opposite the thiol group, denoted as $C'2$ $C'3$ in Table S1.†This slight shortening of the C_2C_3 bonds parallel to the molecular axis can be attributed to the effect of the thiol end group, because their length is very close to that of the C_2C_3 bonds in the BPDT molecule (*cf.* Tables 1 and S2†). Inspection of Tables 1, S1 and S2† reveals that, in general, the bond metric data of the thiol-terminated ring in the BPMT molecule practically coincide with the (identical) rings in the BPDT molecule.

By and large, the results obtained by the SIESTA calculations for the considered isolated molecules agree with those obtained by the calculations with the GAUSSIAN 09 suite of programs[39] using the same (PBE) functional (g09 keyword PBEPBE). Moreover, there is also good agreement with the g09 results based on the hybrid functional B3LYP. It is worth noting that true hybrid exchange correlation functionals, which are known to provide better quantitative description, are not available in SIESTA.

3.2 Impact of a single gold atom

Tables 2, S3 and S4† reveal the impact of attaching a gold atom to the thiol end group after removing its H atom from the two molecular species considered in this study. Results were obtained in several ways and are presented in these tables.

The results show that the effect of the basis sets is not considerable. This is apparent by comparing the results obtained using the basis sets DZP and SZP for the light and gold atoms, respectively, with those obtained using the basis sets TZP and DZP for the light and gold atoms, respectively.

What is more significant is the impact of the pseudopotentials. This can be seen by inspecting the lengths of the S–Au bonds that are computed using the

Table 1 Optimized geometries of isolated BPMT and BPDT molecules. The results were obtained by various methods listed in the beginning of Section 3. The twisting angle φ and several bond lengths XY between various atoms X and Y were labeled according to Fig. 1. For the nearly symmetric bonds C_1–C_2 and C_1–C_6, C_2–C_3 and C_5–C_4 and C_3–C_4 and C_5–C_4, we only show their average lengths, which are denoted in the table by C_1C_2, C_2C_3 and C_3C_4, respectively, and this is similar to the other ring whose atoms are labeled with the prime symbol. Throughout, the lengths are given in angstroms and the angles in degrees

Molecule	C_1C_2	C_2C_3	C_3C_4	$C_4C'_4$	$C'_1C'_2$	$C'_2C'_3$	$C'_3C'_4$	SC_1	SC'_1	φ
BPMT	1.4173	1.4052	1.4189	1.4915	1.4105	1.4079	1.4199	1.7833	8.9871	38.855
BPDT	1.4170	1.4050	1.4194	1.4883	1.4170	1.4050	1.4149	1.7841	8.9995	38.823

Table 2 Optimized geometries of isolated BPMT and BPDT molecules where the H atom of the thiol group was replaced by a gold atom. In addition to the data shown in Tables 1, S1 and S2,† the Au–S bond lengths are indicated here. See the caption of Tables 1, S1 and S2† and Fig. 3 for the notations. Lengths are shown in angstroms and angles in degrees

	C_1C_2	C_2C_3	C_3C_4	SC_1	SC'_1	AuS
BPMT–Au	1.4213	1.4039	1.4211	1.7784	8.9750	2.3449
BPDT–Au	1.4209	1.4036	1.4223	1.7801	8.9926	2.3472
	$C'_1C'_2$	$C'_2C'_3$	$C'_3C'_4$	$C_4C'_4$	φ	
BPMT–Au	1.4108	1.4076	1.4200	1.4897	36.976	
BPDT–Au	1.4176	1.4044	1.4205	1.4875	34.714	

pseudopotentials obtained by translating the ABINIT pseudopotential database, which is available to download,[33] with those computed by using the pseudopotentials in ref. 35 for the metal atoms. For the former, Au–S bond lengths which are longer by ~ 0.2 Å than those in Tables 2, S3 and S4† are obtained. As can be seen, using the pseudopotential in ref. 35 for the gold atoms yields results in reasonable agreement with the other methods listed in Tables 2, S3 and S4.† So, this was our choice for the gold pseudopotential in all of the calculations reported in this paper; the Au–S bond lengths estimated for the SAMs considered below (*cf.* Tables 3, S5, 4, S6 and S7†) are comparable with the values of 2.677 Å (ref. 43) and 2.7 Å (ref. 28) reported earlier for related situations.

The most encouraging message of the results presented in Section 3.1 and 3.2 is that, contrary to what one might have suspected, the value of the twisting angle φ is not substantially dependent on the method utilized. This makes us confident that the SAM geometries optimized by SIESTA, in particular the values of the torsional angles φ, can be trusted even when (layers of) gold atoms are considered.

3.3 Self-assembled monolayers

In this section we will consider the SAMs of BPMT and BPDT adsorbed on the (111) surface of a face-centered cubic (fcc) structured gold crystal.

Table 3 Results obtained when the SAMs consisting of BPMT and BPDT molecules arranged in a (6 × 6) adsorbate unit cell on the fcc Au (111) surface (*cf.* Fig. 2a). $d^{(1,2,3)}$ represent the distances of the sulfur atom to the closest gold atoms in the top layer. L is the length of the molecular backbone measured between the S and C′ (or S′) atoms at the opposite ends of the BPMT (or BPDT) molecule. φ is the twisting angle and z is the distance from the S atom to the top layer of the gold slab. The SAM height Z is defined as the distance between the top layer and the most distant H atom of the molecule. The tilt angle θ is the angle between the molecular axis (the line through the S and C'_1 atoms for BPMT and through the S and S′ atoms for BPDT) and the gold surface. Lengths are shown in angstroms and angles in degrees

Method	φ	$d^{(1)}$	$d^{(2)}$	$d^{(3)}$	$L = SC$	z	Z	θ
BPMT	36.615	2.6695	2.6768	2.9802	8.9634	2.1981	10.2616	32.727
BPDT	33.866	2.6459	2.7968	2.8011	10.7372	2.1810	12.6341	13.867

Table 4 Results obtained when the SAMs of BPMT formed a regular herringbone superstructure on an fcc Au (111) surface (*cf.* Fig. 2b). Here, $d_1^{(1,2)}$, $d_2^{(1,2)}$ and $d_3^{(1,2)}$ represent the shortest three S–Au bond lengths (subscripts 1, 2, 3) corresponding to the sulfur atoms of the two crystallographically nonequivalent BPMT molecules (superscripts 1, 2) that have twisting angles denoted by $\varphi_{1,2}$. $\theta_{1,2}$ are the angles between the metal surface and the lines passing through the S and C′ atoms at opposite molecular ends, $z_{1,2}$ are the distances of the two S atoms to the electrode surface, $Z_{1,2}$ are molecular heights, and $L_{1,2}$ are the lengths of the molecular backbones measured between the S and C′ atoms at opposite molecular ends. HH, CC and SS denote the distances between the atoms belonging to the two molecules of the supercell, as shown in Fig. 3, where the angles $\beta_{u,l}$ between oppositely placed rings in the two nonequivalent molecules are also depicted

Molecule	φ_1	θ_1	$d_1^{(1)}$	$d_2^{(1)}$	$d_3^{(1)}$	L_1	z_1	Z_1
BPMT	75.976	5.697	2.6564	2.6852	2.7880	8.9714	2.1430	12.1772
BPDT	75.036	10.006	2.6287	2.7851	2.8833	10.7074	2.1973	12.8746
Molecule	φ_2	θ_2	$d_1^{(2)}$	$d_2^{(2)}$	$d_3^{(2)}$	L_2	z_2	Z_2
BPMT	75.935	5.670	2.6698	2.6860	2.7928	8.9721	2.1348	12.1700
BPDT	75.561	10.244	2.6185	2.7865	2.8827	10.7056	2.1925	12.8523
Molecule	β_u	β_l	XX_u	SS_l	HH_{uu}	HH_{ul}	HH_{lu}	HH_{ll}
BPMT (X = C)	74.994	76.422	4.9670	4.9574	2.9802	2.9891	2.9943	2.9951
BPDT (X = S)	74.070	72.510	4.9710	5.7271	2.8833	2.9055	2.922	3.3043

For thiol terminated systems like the presently considered molecules, it is widely accepted that, after the hydrogen atom of the thiol terminal group desorbs, the sulfur atom of the resulting radical strongly binds to the gold surface.[43] It is this scenario that we have adopted with all of the calculations reported here.

Attempts to experimentally resolve the preferential adsorption position of the thiol group on Au (111) yielded controversial results.[44–50] Although some controversy also exists from the theoretical side,[51–56] the prevailing view is that adsorption occurs at hollow sites.[7–9,43]

This turned out to be the case for the two structures of the adsorbed SAMs we are going to discuss below, which are depicted in Fig. 2. To target the global minimum, for the coverages examined, we conducted geometry optimization starting from the optimized geometry of the isolated molecule placed in or slightly perturbed from atop, bridge and hollow locations.

The first example to be examined is a SAM characterized by a (6 × 6) unit cell of the adsorbate (*cf.* Fig. 2a). This corresponds to a very low coverage situation, which would fulfil the requirement of the weakly interacting periodic replicas perpendicular to the transport direction, so that the computed currents reflect the transport properties of the single molecule junctions.

For these (6 × 6) SAMs we found that the hollow site is preferential. More specifically, this preferential hollow site turned out to be somewhat displaced towards the bridge site in the case of BPMT and towards the top site in the case of BPDT; these situations correspond to $d_1 \approx d_2 > d_3$ and to $d_1 < d_2 \approx d_3$, respectively, as can be seen in Tables 3 and S5.† Inspection of Tables 3 and S5† shows robust numerical values with little sensitivity to the theoretical method utilized. It

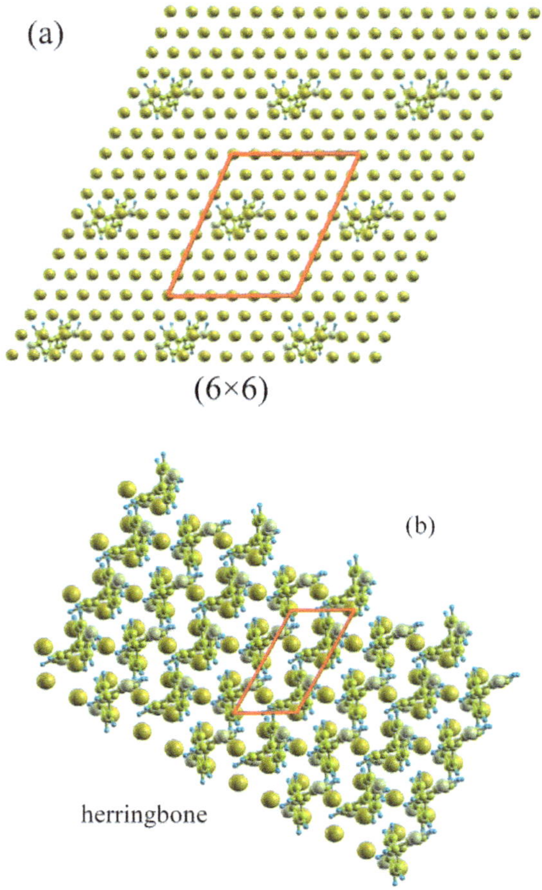

Fig. 2 Top view of the SAMs with (a) (6 × 6) structure and (b) herringbone arrangement examined in the present paper. The supercells used in calculations are depicted in red. The figure was generated using XCrySDen.[31]

is worth noting that, for both BPMT- and BPDT-based (6 × 6) SAMs the values of the twisting angle φ do not appreciably differ from the values of the isolated molecules, compared to the values in Tables 1, S1 and S2.†

Let us next consider the case of the SAMs with herringbone arrangements (*cf.* Fig. 2b). Herringbone patterns represent a common assumption for adsorbed SAMs based on the presently considered BPMT and BPDT molecules as well as on a variety of other related molecular species.[7–9,57–64] In this densely packed phase, molecules are allowed to form a regular $p(\sqrt{3} \times 2\sqrt{3})R30°$ structure (highlighted in red in Fig. 2b) on the Au (111) surface, wherein two translationally nonequivalent molecules per unit cell (Fig. 3) form a characteristic herringbone pattern.

The results collected in Tables 4, S6 and S7† offer a detailed characterization of the optimized structures of the SAMs absorbed on an fcc Au (111) surface consisting of BPMT and BPDT molecules forming an ordered overlayer with herringbone structure.

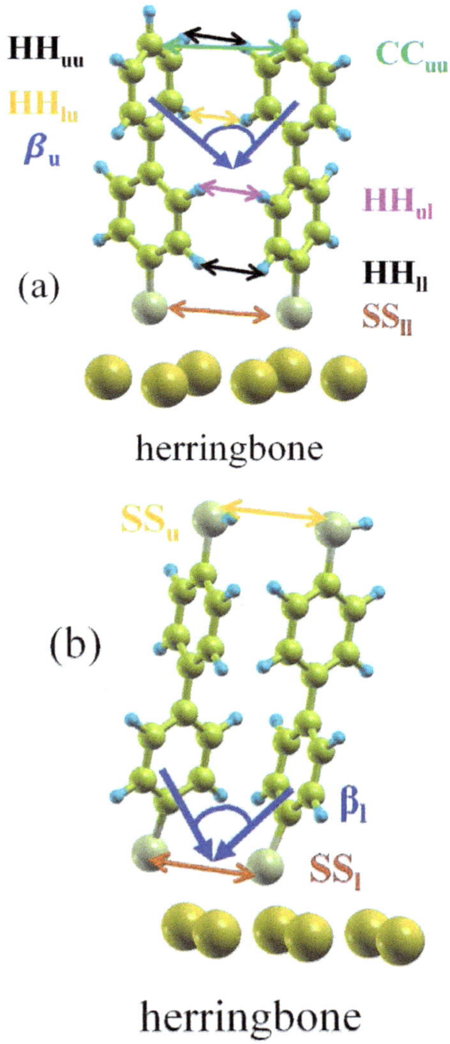

Fig. 3 The two crystallographically nonequivalent molecules (a) BPMT and (b) BPDT in the herringbone supercells utilized in the SIESTA calculations. The meaning of the various lengths and angles indicated in Tables 4, S6 and S7† is visible here. The figure was generated using XCrySDen.[31]

These data show that for both BPMT- and BPDT-based SAMs the "lower" (subscript l in Tables 4, S6 and S7†) rings of the two crystallographically nonequivalent molecules that are closer to the electrodes are not strictly perpendicular; the angle β_l between their planes (*cf.* Fig. 3) $\approx 74°$. This also applies for the angle β_u (*cf.* Fig. 3) between the "upper" (subscript u in Tables 4, S6 and S7†) rings at the other end of the molecules. They are arranged such that the upper ring of one molecule is nearly parallel to the lower ring of the other molecule (*cf.* Fig. 3).

From the perspective of the present study, the most important information included in Tables 4, S6 and S7† is the fact that, roughly, the two twisting angles characterizing the two crystallographically nonequivalent molecules have φ values ($\varphi_1 \approx \varphi_2 \approx 74°$) two times larger than the values ($\varphi \approx 36°$) that are characteristic for isolated molecules (*cf.* Tables 1, S1, S2, 2, S3 and S4†) or for the low coverage (6 × 6) SAMs.

We believe that the finding that the twisting angle of the SAMs with a herringbone arrangement is substantially different to those of isolated molecules and of low coverage SAMs is an important result. In view of its importance for the present study, we have tested this result in a number of ways. Let us give a few examples for illustration:

(i) We have checked that using DZP and TZP basis sets for gold and light atoms instead of SZP and DZP, respectively, only yields changes in angles of a few degrees and changes in bond lengths in the order of 10^{-2} Å; compare the differences between the second and the third lines of Table S6† for the BPMT molecule and Table S7† for the BPDT molecule. These differences are comparable to or even smaller than the differences between various levels of theory for isolated molecules, as illustrated in Tables 1, S1, S2, 2, S3 and S4.† A similar basis set effect occurs in the case of the (6 × 6) SAMs; compare the second and fourth lines of the upper and lower panels of Table S5†.

(ii) We have considered the effect of dispersion corrections, which we have included by using the van der Waals (VDW) functional (DRSLL) reported by Dion *et al.*[37] which is based on the specific implementation of the work in ref. 38. The negligible impact of dispersion corrections can be seen by comparing the second and the fourth lines of the three panels in Tables S6 and S7.†

(iii) We have verified that the imposition of tighter conditions for geometry relaxation does not significantly affect the results. Compare the second line of the panels in Table S7,† obtained by imposing a maximum force of 0.02 eV $Å^{-1}$ on all of the atoms in the BPDT molecule, with the fifth (sixth) line obtained by 0.01 eV $Å^{-1}$ (0.005 eV $Å^{-1}$). This effect is similar to that encountered in the case of the (6 × 6) SAMs; compare the second and the third lines of the upper and lower panels in Table S5†.

(iv) We have also checked that increasing the number of metallic layers does not significantly affect the results. Compare the second and fifth lines of the panels in Table S6† as well as the second and the seventh lines of the panels in Table S7,†which refer to slabs consisting of three and six gold layers, respectively.

(v) Finally, we have considered the effect of metal geometry relaxation, by also optimizing the positions of the gold atoms in the top layer. By comparing the second and sixth lines of Table S6,† we conclude that this relaxation effect is negligible for a slab consisting of a BPMT SAM with a herringbone structure adsorbed on three gold layers. The same conclusion holds for BPMT-based SAMs adsorbed on six gold layers; compare the fifth and seventh lines of Table S6.† The relaxation of the metallic surface is also negligible for BPDT-based SAMs with herringbone structures. By comparing the second and eighth lines of Table S7,† we conclude that this relaxation effect is negligible for a slab consisting of a BPDT SAM with a herringbone structure adsorbed on three gold layers. For slabs with six gold layers, we arrive at the same conclusion after comparing the seventh and ninth lines of Table S7.†

A comment is still in order concerning metal relaxation. In the present calculations, we did not optimize the gold lattice constant a. This would have required the use of different a-values for different exchange-correlation functionals and basis sets, which in turn, would have raised questions on the relevance of comparing bond metric data for BPMT- and BPDT-based SAMs computed at different levels of theory. To facilitate this comparison, we have set the gold lattice constant to the experimental value, $a = 4.078$ Å. In this way, however, the relaxed positions of the gold atoms at the interface are not correctly estimated; they are affected by the interaction of these gold atoms with the rest of the gold atoms in the slab which were frozen at the periodic sites of a lattice whose constant was not optimized. If, without molecules adsorbed on it, we optimized the top layer of gold by freezing the other "bulk" gold atoms in the slab at the periodicity defined by the experimental (not optimized) gold lattice constant, we would have found, nonphysically, a nonvanishing relaxation of the top layer. This is the reason why, for cases where the top layer of the gold slab was allowed to relax, some entries (*e.g.* the distances $d^{(1,2,3)}$ of the sulfur atom to the closest gold atoms) are not given in the last two lines of Tables S6 and S7.† The values of the tilt angles, $\theta_{1,2}$, which are slightly larger for cases where the top layers were allowed to relax (*cf.* Tables S6 and S7†), reflect the fact that the position of the S atom close to the slab is somewhat affected by this relaxation effect.

A question that naturally arises in the context of the foregoing results is whether variations from $\varphi \sim 36°$ to $\varphi \sim 74°$ in the torsional angle smoothly occur as the intermolecular distance(s) D decrease or, in the cases where this change occurs suddenly, how close can the molecules be brought before the (large) value of φ sets in. Restricting oneself to SAM superstructures commensurate with the underlying crystal lattice means that it is impossible to answer whether or not φ varies smoothly with the variation in the (discrete) variable D. What we can state here is that, for (3×3)- and (3×2)-SAMs (the results are not presented here because of their similarity to the (6×6) superstructure), φ only differs by a few degrees from the values characterizing the isolated species of the (6×6)-SAMs discussed above. Investigating the dependence on the (quasi-)continuum variable D requires the consideration of incommensurate SAM superstructures, but such studies are computationally prohibitive at present.

3.4 Additional remarks

The main purpose of this paper was to demonstrate that the twisting angle φ of two benchmark molecules – BPMT and BPDT – utilized to fabricate molecular junctions can be substantially modified when the coverage of a SAM adsorbed on an electrode changes. To achieve this, as seen in Section 3.3, it was sufficient to consider and compare two typical situations: one – a (6×6) superstructure of the adsorbate – which was an "ideal" situation for "good" transport calculations, the other – a herringbone arrangement – which was often assumed/utilized to analyze experimental measurements.

In both cases, the molecular backbones of both of the considered species were found to stand nearly upright on a metallic surface. Values of the tilt angle θ in Tables 3, S5, S6 and S7† are small and are consistent with experimental values reported in many studies on BPMT and BPDT SAMs in the range of $0° \lesssim \theta \lesssim 23°$.[3,59,65,66] Contrary to SAMs based on alkanethiolates, wherein the substantial tilting is driven by the re-establishment of interchain van der Waals interactions, the almost perpendicular molecular orientation of the SAMs based on molecular species, including the BPMT and BPDT molecules considered here, was often observed experimentally and assigned as a probable effect of the maximization of π–π interactions.[66,67] Concerning the experimental data processing, we should note that (i) to correctly estimate the value of θ, information on the twisting angle φ is needed[65] and (ii) assuming that the value of φ for isolated molecules can also be used for the SAM in question, which is a common practice in the field, is an assumption that the present study does not support.

Given the main purpose of this paper as formulated here, we did not attempt to exhaustively investigate the various adsorption sites (atop, bridge, hollow) nor the impact of the SAM coverage.

Concerning the first aspect, we note that we did not attempt to systematically constrain the calculations to converge to the lowest energy geometries corresponding to the atop and bridge configurations. Still, in all cases examined, we also started the calculations from situations close to the atop or bridge sites. In most cases, however, those relaxation calculations converged to hollow configurations; therefore, we accepted the latter as corresponding to the optimized structure. To refer to a few exceptions, wherein the calculations did not converge to the hollow position, we can mention that for BPMT-SAMs with (6×6) superstructures the bridge configuration turned out to be 0.26 eV higher in energy than the hollow configuration. For the (3×3) and (3×2) superstructures (not discussed here because they are similar to that of (6×6)), we found energy differences of 0.55 eV and 0.58 eV, respectively. More importantly from the present standpoint is that even the less stable configurations had torsional angles that only differed by a few degrees to those of the hollow configurations.

Regarding the second aspect, let us mention in passing that our calculations (to be reported elsewhere) indicate that coverages exist, wherein the molecular axes are substantially tilted away from the normal to the electrode surface, as is also occasionally observed experimentally.[59,61] Also in passing we can note that transport calculations that implicitly assume low SAM coverages are also problematic because the metallic work functions – which influence the alignment of the relevant molecular orbitals relative to the metallic Fermi level – depend on coverage.

4 Relevance of the experiments

4.1 General considerations

According to the SIESTA-based results reported in this paper, the two rings of the BPMT and BPDT molecules that form SAMs with a characteristic herringbone superstructure are almost orthogonal to each other. Therefore, from the perspective of the present results, the herringbone arrangement which is often assumed in order to analyze experiments may not be favorable for molecular electronic devices; it can be associated with a very low conductance.

A comprehensive experimental investigation on CP-AFM molecular junctions based on BPDT carried out very recently,[68,69] which includes measurements of the coverage of SAMs using nuclear RBS (Rutherford back-scattering) and NRA (nuclear reaction analysis) methods,[70] provides strong support to the present theoretical finding. These experimental data unambiguously demonstrate that the coverage of the BPDT-based SAMs used to fabricate CP-AFM junctions, which possess high conductance values, is clearly different from the value characterizing the ubiquitously assumed herringbone arrangement. We will return to the case of BPDT-based junctions in Section 4.2, where specific aspects will be analyzed.

As is obvious and has been long recognized in molecular electronics, the dependence of the twisting angle φ on the coverage of the SAM is important from the perspective of molecular switches. Because in the case of such molecular species the transmission (and hence the conductance) scales as $\cos^2 \varphi$, the nearly orthogonal molecular conformation corresponds to an almost perfect OFF state in a single molecule switch. By contrast, a nearly planar conformation acts as an ON state.

Less obvious and, to the best of our knowledge, not pointed out so far in the literature is the fact that the dependence of the twisting angle on the (local) SAM coverage may raise problems and can be a source of controversy in interpreting or comparing transport data obtained by different experimental groups and/or measured with different instruments.[71] In this context, let us refer to the case of repeatedly formed STM single molecule break junctions.[19] In a typical current measurement,[19] the STM tip (one electrode) crashes into the substrate (second electrode) on which a SAM is adsorbed. After mechanical contact, where the molecules of the SAM bridge both the tip and substrate, the STM tip is immediately retracted. During this process the molecules bridging the two electrodes start to break until, ultimately, a single ("active") molecule bridges the two electrodes and forms a single molecule junction, through which the current to be studied is measured. It should be clear how invasive this process – which is repeated thousands of times[19,72] – is for the local homogeneity of the SAM coverage, and how different the twisting angle of the "last" molecule can be in an event of junction formation from another event, out of thousands of events of junction formation. It should also be clear that, besides the fully relaxed SAM arrangement investigated in the present study, the consideration of other (presumably) metastable SAM configurations can be relevant.

From reasons delineated above it is understandable that controversies[71] can arise in interpreting transport data through STM break junctions based on floppy molecules consisting of two rings that can easily rotate relative to each other and how important detailed characterization of the SAMs is. For the same reasons, it should be clear that junctions fabricated using a CP-AFM architecture are not (or at least much less) affected by the aforementioned process; the transport measurements in this case require a smooth approach of the AFM tip to the SAM which is a process that is far less perturbing than the repeated formation of the STM break junctions. This is why exceptionally small statistical variations in the transport properties can be achieved in the experiments on CP-AFM junctions.[68] Provided good quality SAMs are fabricated, large-area junctions are even less affected than the CP-AFM junctions, due to the significant "signal averaging" that occurs over millions or billions of molecules "connected" in parallel.

4.2 A specific analysis: conductance data for molecular junctions based on BPDT

In Section 4.1 we anticipated that in the experimental evidence that exists for the recently fabricated molecular junctions based on BPDT,[68] which are characterized by exceptionally small statistical variances in the transport properties, a herringbone structure can be safely ruled out.

Indeed, for the SAMs of BPDT on gold, RBS and NRA measurements[68] yielded a coverage of $\Sigma \approx 3.3$ molecule per nm^2. This corresponds to an area of $\mathscr{A}_{\text{exp}} = 1/\Sigma \approx 30$ Å^2 per molecule. By taking the value of $d = a/\sqrt{2} = 2.884$ Å for the gold–gold distance, the area per molecule in the herringbone structure is $\mathscr{A}_{\text{hb}} = (3 \times \sqrt{3})d^2/2 = 21.6$ Å^2. The difference between these values of $\mathscr{A}_{\text{hb}}$ and $\mathscr{A}_{\text{exp}}$ is too large to be assigned to the experimental error of the latter, which does not exceed $\sim$3%.[68]

It may be interesting to compare the conductance data reported in ref. 68 with earlier experimental data for molecular junctions based on BPDT.[72,73] For the BPDT-based junctions fabricated under a conducting probe atomic force microscopy (CP-AFM) platform, a conductance of $G_{\text{CP-AFM}}^{\text{junction}}/G_0 \simeq 0.48$ was reported.[68] Here $G_0 = 2e^2/h \simeq (12.9\ \text{k}\Omega)^{-1}$ is the conductance quantum. Because BPDT-bundles that form the CP-AFM junctions in ref. 68 consist of $N = 82$ molecules (*cf.* Table 1 in ref. 68), this translates into an effective conductance per molecule of $G_{\text{CP-AFM}} = G_{\text{CP-AFM}}^{\text{junction}}/N \simeq 5.9 \times 10^{-3}\ G_0$. Using STM-techniques, earlier studies by two different groups reported the values of $G/G_0 \approx 1.7 \times 10^{-4}$ (ref. 73) and $G/G_0 \approx 1.9 \times 10^{-4}$ (ref. 72). They are in reasonable agreement with each other but are substantially smaller than the value of $G_{\text{CP-AFM}}/G_0 \simeq 5.9 \times 10^{-3}$ by a factor of $\sim$30. Although a conductance (per molecule) in the CP-AFM junctions that is somewhat larger than that of STM junctions can be expected in view of the more stable molecule-tip contact in the former, it is unlikely that this could result in such a large discrepancy. Given the fact that SAMs with herringbone structures can safely be excluded in the case of ref. 68 (see above), one may speculate that SAMs with herringbone arrangements that characterize the junctions in ref. 72 and 73 could be responsible for the substantially lower conductance measured there.

We do not necessarily endorse such an explanation for the specific case analyzed above. Still, the foregoing analysis suggests that a more complete characterization of the molecular junctions (including the SAM) than what is commonly done in most experimental studies in molecular electronics is highly desirable.

5 Conclusion

In the present paper, based on SIESTA-DFT calculations, we have found that coverages exist for which the twisting angle φ, which characterizes self-assembled monolayers (SAMs) based on 4-methyl-4′-mercaptobiphenyl CH_3–$(C_6H_4)_2$–SH (BPMT) and 4,4′-dithiol-1,1′-biphenyl HS–$(C_6H_4)_2$–SH (BPDT) floppy molecules adsorbed on fcc Au (111) gold surfaces, can vary substantially. In SAMs with a (6×6) structure, wherein the coverage is very low, the twisting angle does not notably differ from the value of $\varphi \approx 36°$ which is characteristic for isolated molecules. On the contrary, in densely packed SAMs with herringbone arrangements, SIESTA calculations yielded considerably larger values of $\varphi \approx 74°$. Because the ohmic conductance G of a molecule consisting of two rings scales as $\cos^2 \varphi$, this change in φ (from $\sim$36° to $\sim$74°), which amounts to a factor of $\sim$2, translates according to eqn (1) into a variation in G of over one order of magnitude ($\cos^2 36° = 0.65$, $\cos^2 74° = 0.076$).

Phenomena related to torsional angles φ in molecules consisting of rings that can be easily twisted are a challenging problem even for isolated species. Therefore, in spite of taking various precautions (*cf.* Section 3.3), it is hard to exclude the fact that calculations at higher levels of theory – *e.g.* those based on hybrid exchange-correlation functionals or more elaborate *ab initio* levels that are, at the moment, too expensive computationally – can yield certain quantitatively significant corrections of the φ-values reported here. Nevertheless, we believe that the basic idea emerging from the present investigation, that of controlling the charge transfer at the metal–organic interface by tuning the conformation of floppy molecules *via* changes in SAM coverage, is worth pursuing in companion with experimental efforts that the present theoretical investigation aims to stimulate.

Acknowledgements

The author thanks Dan Frisbie for permission to refer to recent unpublished experimental data obtained in his group. Financial support for this research provided by the Deutsche Forschungsgemeinschaft (grant BA 1799/3-1) and computational support by the State of Baden-Württemberg through bwHPC and the German Research Foundation (DFG) through grant no. INST 40/467-1 FUGG are gratefully acknowledged.

Notes and references

1 J. M. Beebe, B. Kim, C. D. Frisbie and J. G. Kushmerick, *ACS Nano*, 2008, **2**, 827–832.

2 B. Kim, S. H. Choi, X.-Y. Zhu and C. D. Frisbie, *J. Am. Chem. Soc.*, 2011, **133**, 19864–19877.

3 Z. Xie, I. Bâldea, C. Smith, Y. Wu and C. D. Frisbie, *ACS Nano*, 2015, **9**, 8022–8036.

4 (*a*) I. Bâldea, Z. Xie and C. D. Frisbie, *Nanoscale*, 2015, **7**, 10465–10471; (*b*) I. Bâldea, *Org. Electron.*, 2017, **49**, 19–23.

5 Z. Xie, I. Bâldea, S. Oram, C. E. Smith and C. D. Frisbie, *ACS Nano*, 2017, **11**, 569–578.

6 V. De Renzi, R. Rousseau, D. Marchetto, R. Biagi, S. Scandolo and U. del Pennino, *Phys. Rev. Lett.*, 2005, **95**, 046804.

7 G. Heimel, L. Romaner, J.-L. Brédas and E. Zojer, *Phys. Rev. Lett.*, 2006, **96**, 196806.

8 G. Heimel, L. Romaner, E. Zojer and J.-L. Brédas, *Nano Lett.*, 2007, **7**, 932–940.

9 L. Romaner, G. Heimel and E. Zojer, *Phys. Rev. B: Condens. Matter Mater. Phys.*, 2008, **77**, 045113.

10 C. Van Dyck, V. Geskin and J. Cornil, *Adv. Funct. Mater.*, 2014, **24**, 6154–6165.

11 S. Osella, D. Cornil and J. Cornil, *Phys. Chem. Chem. Phys.*, 2014, **16**, 2866–2873.

12 F. Avila, D. J. Fernandez, J. F. Arenas, J. C. Otero and J. Soto, *Chem. Commun.*, 2011, **47**, 4210–4212.

13 I. Bâldea, *RSC Adv.*, 2016, **6**, 93715–93721.

14 I. Bâldea, *Chem. Phys.*, 2017, **482**, 311–318.

15 I. Bâldea, *J. Phys. Chem. A*, 2017, **121**, 2282–2287.

16 I. Bâldea, *RSC Adv.*, 2016, **6**, 111903–111907.

17 S. Woitellier, J. Launay and C. Joachim, *Chem. Phys.*, 1989, **131**, 481–488.
18 T. Wandlowski, K. Ataka and D. Mayer, *Langmuir*, 2002, **18**, 4331–4341.
19 B. Xu and N. J. Tao, *Science*, 2003, **301**, 1221–1223.
20 H. Kondo, J. Nara, H. Kino and T. Ohno, *J. Chem. Phys.*, 2008, **128**, 064701.
21 H. Kondo, J. Nara, H. Kino and T. Ohno, *J. Phys.: Condens. Matter*, 2009, **21**, 064220.
22 F. Pauly, J. K. Viljas, J. C. Cuevas and G. Schön, *Phys. Rev. B: Condens. Matter Mater. Phys.*, 2008, **77**, 155312.
23 L. Venkataraman, J. E. Klare, C. Nuckolls, M. S. Hybertsen and M. L. Steigerwald, *Nature*, 2006, **442**, 904–907.
24 L. Cui, B. Liu, D. Vonlanthen, M. Mayor, Y. Fu, J.-F. Li and T. Wandlowski, *J. Am. Chem. Soc.*, 2011, **133**, 7332–7335.
25 M. Bürkle, J. K. Viljas, D. Vonlanthen, A. Mishchenko, G. Schön, M. Mayor, T. Wandlowski and F. Pauly, *Phys. Rev. B: Condens. Matter Mater. Phys.*, 2012, **85**, 075417.
26 M. Bürkle, L. A. Zotti, J. K. Viljas, D. Vonlanthen, A. Mishchenko, T. Wandlowski, M. Mayor, G. Schön and F. Pauly, *Phys. Rev. B: Condens. Matter Mater. Phys.*, 2012, **86**, 115304.
27 J. Balachandran, P. Reddy, B. D. Dunietz and V. Gavini, *J. Phys. Chem. Lett.*, 2012, **3**, 1962–1967.
28 A. Tan, J. Balachandran, B. D. Dunietz, S.-Y. Jang, V. Gavini and P. Reddy, *Appl. Phys. Lett.*, 2012, **101**, 243107.
29 I. Bâldea, *Faraday Discuss.*, 2014, **174**, 37–56.
30 J. C. Cuevas and E. Scheer, *Molecular Electronics: An Introduction to Theory and Experiment*, World Scientific Publishers, 2010.
31 A. Kokalj, *Comput. Mater. Sci.*, 2003, **28**, 155–168.
32 J. M. Soler, E. Artacho, J. D. Gale, A. Garcia, J. Junquera, P. Ordejon and D. Sanchez-Portal, *J. Phys.: Condens. Matter*, 2002, **14**, 2745–2779.
33 https://www.departments.icmab.es/leem/siesta/.
34 bwHPC program supported by the State of Baden-Württemberg and the German Research Foundation (DFG) through grant no. INST 40/467-1 FUGG.
35 P. Rivero, V. M. Garca-Surez, D. Pereiguez, K. Utt, Y. Yang, L. Bellaiche, K. Park, J. Ferrer and S. Barraza-Lopez, *Comput. Mater. Sci.*, 2015, **98**, 372–389.
36 J. P. Perdew, K. Burke and M. Ernzerhof, *Phys. Rev. Lett.*, 1996, **77**, 3865–3868.
37 M. Dion, H. Rydberg, E. Schröder, D. C. Langreth and B. I. Lundqvist, *Phys. Rev. Lett.*, 2004, **92**, 246401.
38 G. Román-Pérez and J. M. Soler, *Phys. Rev. Lett.*, 2009, **103**, 096102.
39 M. J. Frisch, G. W. Trucks, H. B. Schlegel, G. E. Scuseria, M. A. Robb, J. R. Cheeseman, G. Scalmani, V. Barone, B. Mennucci, G. A. Petersson, H. Nakatsuji, M. Caricato, X. Li, H. P. Hratchian, A. F. Izmaylov, J. Bloino, G. Zheng, J. L. Sonnenberg, M. Hada, M. Ehara, K. Toyota, R. Fukuda, J. Hasegawa, M. Ishida, T. Nakajima, Y. Honda, O. Kitao, H. Nakai, T. Vreven, J. A. Montgomery Jr, J. E. Peralta, F. Ogliaro, M. Bearpark, J. J. Heyd, E. Brothers, K. N. Kudin, V. N. Staroverov, T. Keith, R. Kobayashi, J. Normand, K. Raghavachari, A. Rendell, J. C. Burant, S. S. Iyengar, J. Tomasi, M. Cossi, N. Rega, J. M. Millam, M. Klene, J. E. Knox, J. B. Cross, V. Bakken, C. Adamo, J. Jaramillo, R. Gomperts, R. E. Stratmann, O. Yazyev, A. J. Austin, R. Cammi, C. Pomelli, J. W. Ochterski, R. L. Martin, K. Morokuma, V. G. Zakrzewski, G. A. Voth, P. Salvador, J. J. Dannenberg,

S. Dapprich, A. D. Daniels, O. Farkas, J. B. Foresman, J. V. Ortiz, J. Cioslowski and D. J. Fox, *Gaussian 09, Revision B.01*, Gaussian, Inc., Wallingford CT, 2010, http://www.gaussian.com.

40 A. Almenningen, O. Bastiansen, L. Fernholt, B. N. Cyvin, S. J. Cyvin and S. Samdal, *J. Mol. Struct.*, 1985, **128**, 59–76.

41 O. Bastiansen and S. Samdal, *J. Mol. Struct.*, 1985, **128**, 115–125.

42 Á. J. Pérez-Jiménez, J. C. Sancho-García and J. M. Pérez-Jordá, *J. Chem. Phys.*, 2005, **123**, 134309.

43 S. Piccinin, A. Selloni, S. Scandolo, R. Car and G. Scoles, *J. Chem. Phys.*, 2003, **119**, 6729–6735.

44 F. T. Arce, M. E. Vela, R. C. Salvarezza and A. J. Arvia, *J. Chem. Phys.*, 1998, **109**, 5703–5706.

45 F. Terán Arce, M. E. Vela, R. C. Salvarezza and A. J. Arvia, *Langmuir*, 1998, **14**, 7203–7212.

46 P. Fenter, A. Eberhardt and P. Eisenberger, *Science*, 1994, **266**, 1216–1218.

47 P. Fenter, F. Schreiber, L. Berman, G. Scoles, P. Eisenberger and M. Bedzyk, *Surf. Sci.*, 1998, **412–413**, 213–235.

48 G. J. Kluth, C. Carraro and R. Maboudian, *Phys. Rev. B: Condens. Matter Mater. Phys.*, 1999, **59**, R10449–R10452.

49 H. Kondoh, M. Iwasaki, T. Shimada, K. Amemiya, T. Yokoyama, T. Ohta, M. Shimomura and S. Kono, *Phys. Rev. Lett.*, 2003, **90**, 066102.

50 M. S. Yeganeh, S. M. Dougal, R. S. Polizzotti and P. Rabinowitz, *Phys. Rev. Lett.*, 1995, **74**, 1811–1814.

51 T. Hayashi, Y. Morikawa and H. Nozoye, *J. Chem. Phys.*, 2001, **114**, 7615–7621.

52 J. Gottschalck and B. Hammer, *J. Chem. Phys.*, 2002, **116**, 784–790.

53 Y. Morikawa, T. Hayashi, C. Liew and H. Nozoye, *Surf. Sci.*, 2002, **507–510**, 46–50.

54 J. Nara, S. Higai, Y. Morikawa and T. Ohno, *J. Chem. Phys.*, 2004, **120**, 6705–6711.

55 A. Bilić, J. R. Reimers and N. S. Hush, *J. Chem. Phys.*, 2005, **122**, 094708.

56 S. Kim, K. Ihm, T.-H. Kang, S. Hwang and S.-W. Joo, *Surf. Interface Anal.*, 2005, **37**, 294–299.

57 E. Sabatani, J. Cohen-Boulakia, M. Bruening and I. Rubinstein, *Langmuir*, 1993, **9**, 2974–2981.

58 T. Ishida, W. Mizutani, U. Akiba, K. Umemura, A. Inoue, N. Choi, M. Fujihira and H. Tokumoto, *J. Phys. Chem. B*, 1999, **103**, 1686–1690.

59 T. Leung, P. Schwartz, G. Scoles, F. Schreiber and A. Ulman, *Surf. Sci.*, 2000, **458**, 34–52.

60 W. Azzam, P. Cyganik, G. Witte, M. Buck and C. Wöll, *Langmuir*, 2003, **19**, 8262–8270.

61 W. Azzam, C. Fuxen, A. Birkner, H.-T. Rong, M. Buck and C. Wll, *Langmuir*, 2003, **19**, 4958–4968.

62 J. J. Stapleton, P. Harder, T. A. Daniel, M. D. Reinard, Y. Yao, D. W. Price, J. M. Tour and D. L. Allara, *Langmuir*, 2003, **19**, 8245–8255.

63 W. Azzam, Self-assembled monolayers on gold made from organothiols containing anoligophenyl-backbone, PhD thesis, Ruhr-Universität Bochum, 2003.

64 P. Cyganik, M. Buck, W. Azzam and C. Wöll, *J. Phys. Chem. B*, 2004, **108**, 4989–4996.

65 S. Frey, V. Stadler, K. Heister, W. Eck, M. Zharnikov, M. Grunze, B. Zeysing and A. Terfort, *Langmuir*, 2001, **17**, 2408–2415.
66 J. F. Kang, A. Ulman, S. Liao, R. Jordan, G. Yang and G.-Y. Liu, *Langmuir*, 2001, **17**, 95–106.
67 J. F. Kang, S. Liao, R. Jordan and A. Ulman, *J. Am. Chem. Soc.*, 1998, **120**, 9662–9667.
68 Z. Xie, I. Bâldea, A. T. Demissie, C. E. Smith, Y. Wu, G. Haugstad and C. D. Frisbie, *J. Am. Chem. Soc.*, 2017, **139**, 5696–5699.
69 Z. Xie and C. D. Frisbie, private commuication.
70 A. T. Demissie, G. Haugstad and C. D. Frisbie, *J. Am. Chem. Soc.*, 2015, **137**, 8819–8828.
71 I. Bâldea, *Nanoscale*, 2013, **5**, 9222–9230.
72 S. Guo, J. Hihath, I. Diez-Pérez and N. Tao, *J. Am. Chem. Soc.*, 2011, **133**, 19189–19197.
73 A. Mishchenko, D. Vonlanthen, V. Meded, M. Bürkle, C. Li, I. V. Pobelov, A. Bagrets, J. K. Viljas, F. Pauly, F. Evers, M. Mayor and T. Wandlowski, *Nano Lett.*, 2010, **10**, 156–163.

Faraday Discussions

PAPER

Kinetic control in the temperature-dependent sequential growth of surface-confined supramolecular copolymers†

Sabrina Engel,[a] Daniel Spitzer,[b] Leona Lucas Rodrigues,[b] Eva-Corinna Fritz,[a] David Straßburger,[b] Monika Schönhoff,[c] Bart Jan Ravoo *[a] and Pol Besenius *[b]

Received 9th March 2017, Accepted 12th April 2017
DOI: 10.1039/c7fd00100b

We report the sequential growth of supramolecular copolymers on gold surfaces, using oppositely charged dendritic peptide amphiphiles. By including water-solubilising thermoresponsive chains in the monomer design, we observed non-linear effects in the temperature-dependent sequential growth. The step-wise copolymerisation process is characterised using temperature dependent SPR and QCM-D measurements. At higher temperatures, dehydration of peripheral oligoethylene glycol chains supports copolymer growth due to more favourable comonomer interactions. Both monomers incorporate methionine amino acids but remarkably, desorption of the copolymers *via* competing sulphur gold interactions with the initial monomer layer is not observed. The surface-confined supramolecular copolymers remain kinetically trapped on the metal surface at near neutral pH and form viscoelastic films with a tuneable thickness.

Introduction

The development of surface-confined stimuli-responsive supramolecular and macromolecular architectures is of growing interest.[1] Control over self-assembly processes at interfaces whilst integrating functional groups is an important strategy in order to tailor surface properties, particularly for interfacial oligopeptides, proteins and synthetic macromolecules, ranging from wettability, antifouling properties, biocompatibility to cell adhesion. Pioneering work in the surface-assisted or surface-confined assembly of oligopeptide-based supramolecular soft matter

[a]*Organic Chemistry Institute, Westfälische Wilhelms-University Münster, Corrensstraße 40, 48149 Münster, Germany. E-mail: b.j.ravoo@uni-muenster.de*

[b]*Institute of Organic Chemistry, Johannes Gutenberg-University Mainz, Duesbergweg 10-14, 55128 Mainz, Germany. E-mail: besenius@uni-mainz.de*

[c]*Institute of Physical Chemistry, Westfälische Wilhelms-University Münster, Corrensstraße 28-30, 48149 Münster, Germany*

† Electronic supplementary information (ESI) available: Full experimental details, synthetic procedures, molecular characterisation and instrumentation details. See DOI: 10.1039/c7fd00100b

relies on the preparation of peptide supramolecular nanotubes, by crystallisation or vapour deposition techniques using hydrophobic diphenylalanine, as shown by the Gazit lab,[2] or the step-wise addition of oppositely charged cyclic peptides into vertically aligned peptide nanotubes by N. Ashkenasy.[3] In efforts to prepare interfacial synthetic photosystems, Matile and co-workers have developed a dynamic non-covalent and covalent hybrid strategy, using the ring-opening of cyclic disulphide monomers, referred to as self-organizing surface-initiated polymerisation (SOSIP).[4] Our contribution to the field has been a recently disclosed concept that combines charge-regulated self-assembly and supramolecular copolymerisation techniques.[5] Our approach can be likened to solid phase synthesis,[6] whereby the sequential addition of oppositely charged peptide-based monomers produces chiral supramolecular alternating copolymers that are grafted off Au surfaces.[5] To this end, the gold surface was first functionalised with a cationic N-terminal thioether derivative and then exposed sequentially to glutamic acid and lysine-based peptides in a bottom-up approach. Besides attractive electrostatic interactions, hydrophobic shielding of phenylalanine groups and hydrogen bonding along the β-sheet encoded sequences enabled the stepwise growth of ordered and height adjustable copolymer brushes perpendicular to the Au surface. Generally, the kinetically controlled self-assembly of supramolecular polymers remains a formidable challenge, particularly in the surface confinement of dynamic supramolecular materials.[7] In order to expand our recently reported strategy for the kinetic control in multicomponent surface-confined supramolecular polymers, we designed cationic and anionic dendritic peptide amphiphiles, using an alternating lysine–methionine or glutamic acid–methionine sequence. Similar to our previously reported phenylalanine-rich oligopeptide sequences,[5] the herein presented oppositely charged methionine comonomers are able to lead to one-dimensional supramolecular copolymers *via* a combination of attractive Coulomb interactions, hydrogen bonding and hydrophobic shielding. In principle both comonomers are able to bind to metal surfaces and can then act as initiated self-assembled monolayers. Analogous to our previous studies related to the solution self-assembly of supramolecular homo- and copolymers, which were equipped with hydrophilic tetraethylene glycol chains as water-solubilising groups,[8] we further integrated N-terminal dendritic triethylene glycol chains in order to obtain surface-confined copolymers that are much more soluble, and can be desorbed more conveniently post-assembly. Supported by the large body of work on LCST type polymeric materials,[9] we were further interested in investigating the thermoresponsive behaviour of the surface confined organic materials, since short triethylene glycol chains often give rise to thermal desolvation between 30 °C and 40 °C.[9] By carrying out temperature dependent surface plasmon resonance (SPR) and dissipative quartz microcrystal balance (QCM-D) experiments, we aimed to elucidate the kinetic parameters that dictate the surface confined self-assembly, as well as the viscoelastic properties of the produced supramolecular copolymer brushes.

Results and discussion

Comonomer design and synthesis

Our synthetic strategy for comonomers **1** and **2** relied on our previously established convergent procedures for C_3-symmetric peptide amphiphiles obtained *via* amidation of 1,3,5-benzenetricarbonyl trichloride with an amine functional peptide building block, followed by the final deprotection step of the peptide side chain functional groups.[5,8b,8e,10]

In this work, we shortened the synthetic routes significantly using Cu(I)-catalysed azide–alkyne cycloaddition (CuAAC) chemistry.[11] We first employed standard Fmoc solid-phase peptide synthesis using an azide functionalised glycine unit as the final N-terminal amino acid to prepare two peptides N_3-Gly-Phe-Glu(O^{t}Bu)-Met-Glu(O^{t}Bu)-Met-OH (**12**) and N_3-Gly-Phe-Lys(Boc)-Met-Lys(Boc)-Met-OH (**13**) (Scheme 1). After amidation of the C-terminus with a triethylene glycol functionalised Newkome-type dendron[12] extended with a hexyl spacer, the obtained amphiphilic structures **14** and **15** were ready to use for CuAAC with a 1,3,5-triethynylbenzene core. Using the N-terminal azide enabled us to circumvent additional orthogonal protecting groups (*e.g.* Fmoc or benzyloxylcarbonyl), which simplified the synthetic routes for both comonomers **1** and **2**. We performed the CuAAC in DMSO, using concentrated solutions of the peptidic building blocks. According to ^{1}H-NMR reaction control in d_6-DMSO, full conversion was reached within 90 minutes yielding 80% and 94% of pure protected C_3-symmetric monomers **1a** and **2a**, respectively, after precipitation as reaction work-up. The final acidic deprotection, in the presence of triethylsilane as a quencher to prevent alkylation of the thioether side chains of methionine, yielded the anionic and cationic comonomers **1** and **2**, respectively (Fig. 1A, Scheme 1). Including this final deprotection, the C_3-symmetric amphiphilic

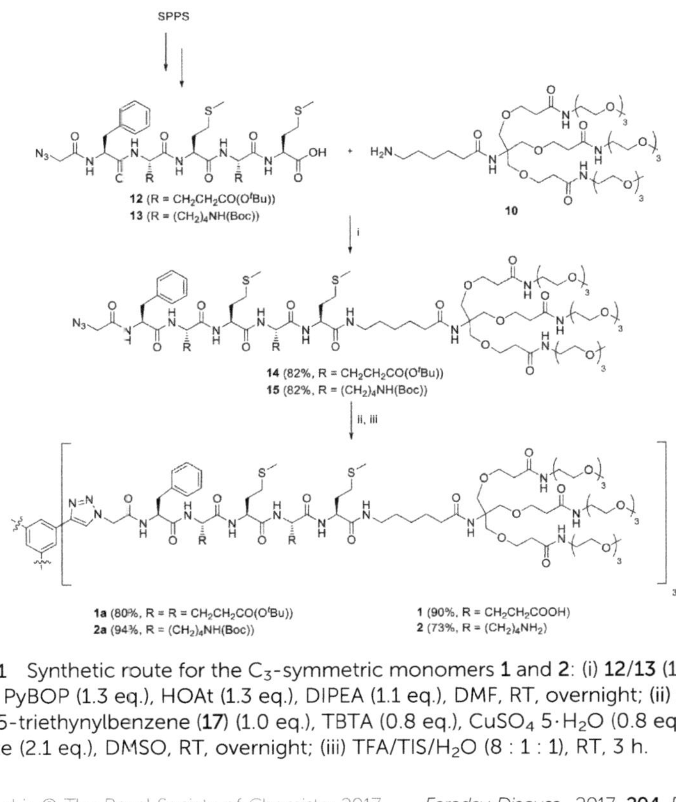

Scheme 1 Synthetic route for the C_3-symmetric monomers **1** and **2**: (i) **12**/**13** (1.0 eq.), **10** (1.2 eq.), PyBOP (1.3 eq.), HOAt (1.3 eq.), DIPEA (1.1 eq.), DMF, RT, overnight; (ii) **14**/**15** (3.3 eq.), 1,3,5-triethynylbenzene (**17**) (1.0 eq.), TBTA (0.8 eq.), $CuSO_4$ 5·H_2O (0.8 eq.), sodium ascorbate (2.1 eq.), DMSO, RT, overnight; (iii) TFA/TIS/H_2O (8 : 1 : 1), RT, 3 h.

monomers can be synthesised in four steps starting from conventional materials. Detailed synthetic procedures and characterisation is provided in the (ESI†).

Temperature dependent SPR spectroscopy of sequentially grown supramolecular copolymers

Surface plasmon resonance (SPR) spectroscopy is a powerful method to detect binding processes on surfaces by measuring the changes in the refractive index over a gold surface evoked by interactions of analytes with the surface, surface-immobilised molecules or the formation of thin organic films ranging from self-assembled monolayers, to multilayer assemblies prepared by the Langmuir/Blodgett/Kuhn technique or by alternate polyelectrolyte deposition.[13] Thus, SPR spectroscopy is a reliable method to monitor the sequential growth of peptidic supramolecular alternating copolymerisation in real-time and to investigate the temperature-dependent kinetics of the copolymerisation. All SPR experiments were performed at a flow rate of 50 μL min^{-1} in degassed phosphate buffer (10 mM, pH 6.0) as eluent and monomer concentrations of 20 μM. The copolymerisation on the surface begins with the immobilisation of **2** as the initiating comonomer. **2** was repeatedly added to a blank gold surface until saturation of the SPR signal was obtained corresponding to a saturated self-assembled monolayer (data is not shown and excluded from the SPR sensorgrams in Fig. 2–4). Typically, an overall injection time of about 45 min was applied. In contrast to our previous protocol,[5] we opted for a direct attachment of the initiator layer and avoided the competitive displacement of a weakly bound ethylene glycol thioether monolayer as a first step in the copolymerisation. The aim was to produce a high surface

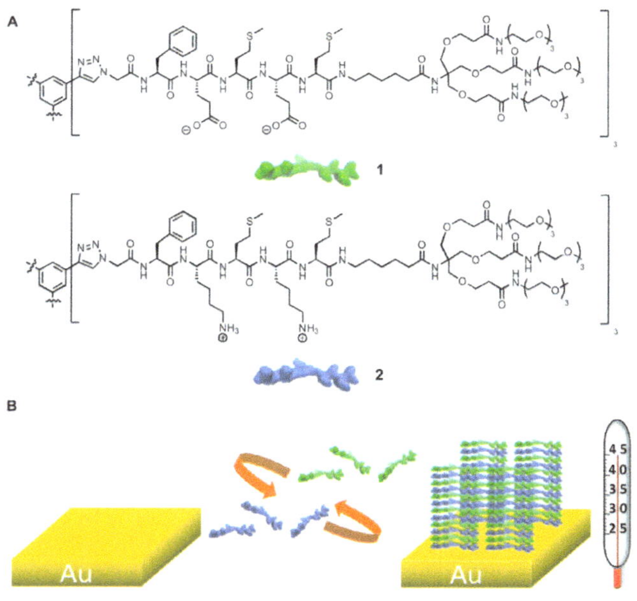

Fig. 1 (A) Molecular structure of the oppositely charged water-soluble dendritic monomers **1** and **2**. (B) Step by step sequential approach for the preparation of supramolecular copolymer brushes on bare Au surfaces at different temperatures.

coverage of **2** and maximise interaction opportunities for complementary comonomer **1** on the gold surface. Sequential growth of the supramolecular copolymer brushes occurred *via* successive additions of the comonomer solutions using identical time intervals of 4 min (200 μL) each followed by 5 min buffer rinsing (see Fig. 2, green and blue arrows).

First, copolymerisation kinetics were investigated at 25 °C. Comparable to our previous results, each comonomer binding resulted in an instantaneous and step-wise signal increase of 0.8 ± 0.3 mRIU, which indicates saturation of the growing copolymer chain ends and thus a selective interaction between the surface-bound cationic and anionic comonomer. After 18 sequential additions of comonomers **1** and **2** (degree of polymerisation DP = 18), an overall plateau is reached at 13 mRIU, which remains constant even after extensive washing with buffer for 1 h. In agreement with our previous report,[5] this observation suggests that stable and kinetically trapped surface-bound copolymers are formed. Only upon successive washing steps of the functionalised surface with aqueous solutions of pH 2 and 12, disassembly of the copolymers and a drop in the SPR signal is observed. At acidic and basic pH, the glutamate side chains are protonated and the ammonium pendant groups of lysine deprotonated, which switches off the attractive Coulomb interactions and leads to a controlled depolymerisation, in full agreement with our previous studies on surface[5] and solution self-assembled multi-component supramolecular copolymers.[5,7a,8b,8e,10]

These observations demonstrate that the ethylene glycol terminated comonomers presented in this work underlie a similar growth mechanism to form pH responsive copolymer brushes on gold surfaces, compared to the previously published system lacking any water solubilising side groups.[5] We have achieved kinetically trapped and surface-confined copolymers, since the repeated addition of the complementary comonomers, which could act as chain stopper, does not lead to desorption of surface bound supramolecular structures. The monomer

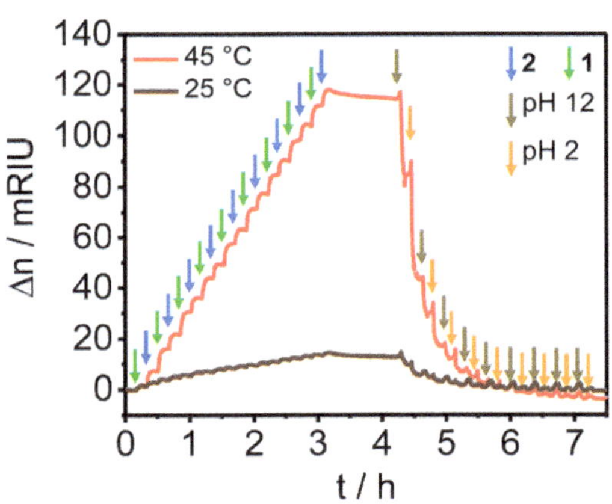

Fig. 2 SPR sensorgram of the copolymerisation based on successive addition of the complementary comonomers **1** and **2** (green and blue arrows, respectively; after surface saturation with comonomer **2**), with subsequent polymer removal by the addition of buffer at pH 12 (brown arrows) and pH 2 (orange arrows) at 25 °C (brown SPR sensorgram) and 45 °C (red SPR sensorgram).

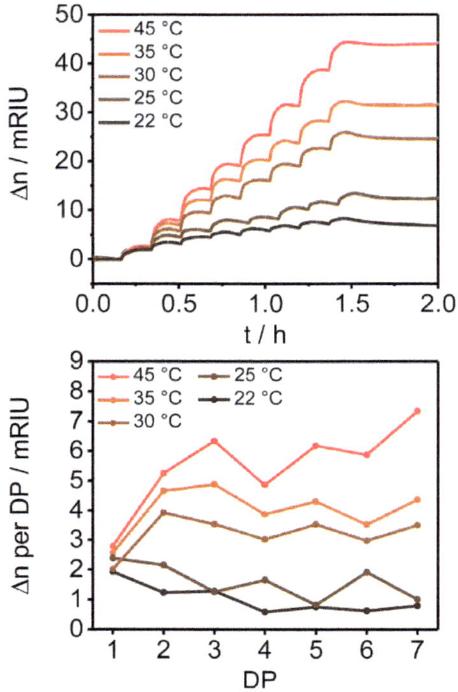

Fig. 3 SPR sensorgrams of the temperature-dependent copolymerisation of comonomers **1** and **2** (after surface saturation with comonomer **2**) between 22 °C and 45 °C (top) and the SPR signal change per monomer addition step (bottom).

complexation kinetics on the active chain end of the growing copolymer are much faster, compared to competitive binding within the copolymer domains located closer to the solid substrate. Kinetics of the disassembly process are therefore

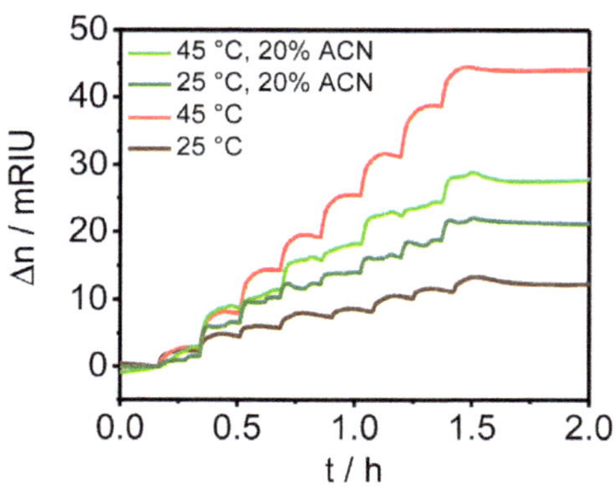

Fig. 4 SPR sensorgram of the copolymerisation based on successive addition of the complementary comonomers **1** and **2** (after surface saturation with comonomer **2**), in phosphate buffer (10 mM, pH 6.0) at 25 °C (brown curve) and 45 °C (red curve), and in phosphate buffer (10 mM, pH 6.0, 20% v/v CH_3CN) at 25 °C (dark green curve) and 45 °C (light green curve).

slowed down considerably. Note that in contrast to our previous work, the incorporation of methionine amino acids into both comonomers could potentially lead to descrption of the surface bound polymers, not just *via* Coulomb interactions and hydrogen bonding, but also *via* competitive displacement of the thioether–gold interactions in the initial monomer layer. Previously we have used C-terminal thioether functions only in the very first self-assembled monomer layer. In consequence, we suggest that kinetic trapping of the copolymers is also aided by the high surface coverage of the supramolecular structures which protects the initial monomer layer and peptide gold interactions.

Secondly, the sequential growth of surface-confined copolymers with DP = 18 was performed at 45 °C, applying identical polymerisation parameters as described above. Remarkably, the successive addition of the comonomers at elevated temperatures also resulted in a step-wise signal increase, but with tenfold higher values in the refractive index change of 6.6 ± 2.0 mRIU, yielding a final plateau value after 18 comonomer additions of 115 mRIU. Again, the immobilised copolymer brushes were stable when washing with buffer for more than 1 h. A complete depolymerisation was only induced by applying successive basic and acidic washes (pH 12 and pH 2). These results are indicative of a temperature dependent growth, which most likely originates from two effects. First, higher temperatures possibly lead to desolvation of the triethylene glycol chains, which favours monomer interactions due to a decrease in the steric demand and due to an increase of the hydrophobic shielding after desolvation of the hydration layer. Secondly, one would expect the entropy-driven multivalent Coulomb interaction between oppositely charged monomers and the associated release of counterions to be more dominant at higher temperatures, which would increase the thermodynamic driving force for comonomer binding.

Further detailed temperature-dependent SPR studies of the sequential supramolecular copolymerisation were performed in order to give insights into the temperature dependent growth and thermoresponsive behaviour of the copolymer brushes. The copolymer growth with DP = 8 using comonomers **1** and **2** was monitored at five different temperatures ranging from 22 °C to 45 °C and the stepwise increasing SPR signal was evaluated (Fig. 3). Regarding the first binding event of the comonomer **1** to the cationic monolayer of **2**, a temperature-independent increase of the SPR signal of 2.4 ± 0.4 mRIU (averaged over all temperatures) was observed (see value for DP = 1 in Fig. 3, bottom). This suggests that in our protocol a densely-packed monolayer of cationic comonomer **2** is formed on the gold surfaces, irrespective of the temperature, and that the self-assembly of the second layer of anionic comonomer **1** is either temperature independent, or the resolution of the SPR set-up is not high enough in order to detect significant differences. However, the further addition of a cationic comonomer **2** reveals a binding event that very clearly shows an unequal increase of the SPR ranging from 1.2 mRIU at 22 °C up to 5.3 mRIU at 45 °C (Fig. 3, top, and bottom for DP = 2), indicative of temperature dependent behaviour. The unequal increase between DP = 1 and DP = 2 is particularly pronounced between 25 °C and 30 °C (Fig. 3, bottom). In further binding events DP > 2, the temperature dependent SPR signal increase per monomer addition remains very pronounced (0.9 ± 0.3 mRIU for 22 °C, and 6.0 ± 0.9 mRIU for 45 °C), with the largest difference between 25 °C and 30 °C (Fig. 3). This non-linear temperature-dependent behaviour of the change in the refractive index is indicative of thermoresponsive behaviour of surface confined supramolecular copolymers, and a desolvation effect of the triethylene glycol shell of the copolymers which favour stronger supramolecular interactions and monomer binding events.

To investigate the impact of desolvation effects in the temperature dependent sequential copolymer growth further, we performed SPR measurements using a small amount of added CH_3CN (20% v/v) in the phosphate buffer as eluent, in order to suppress the observed thermoresponsive behaviour. From previous studies in aqueous solutions of peptidic supramolecular polymers equipped with dendritic oligoethylene glycol chains, we knew that a polar organic solvent and hydrogen-bond acceptor like CH_3CN would be a thermodynamically good solvent for the monomers.[8a,8c,8d,14] Supported by the large body of work on LCST type polymeric materials,[9] we suspected that this would suppress desolvation in the relevant temperature range of 22 °C to 45 °C. Note, that 20% v/v CH_3CN in the aqueous buffer does not lead to a significant destabilisation of hydrophobic supramolecular polymers.[8d] Fig. 4 summarizes the sequential copolymer growth with DP = 8 using comonomers **1** and **2**, monitored at two different temperatures of 25 °C and 45 °C, both in pure phosphate buffer (10 mM, pH 6.0), and phosphate buffer (10 mM, pH 6.0, 20% v/v CH_3CN). The sequential increase of the SPR signal is still comparably high, meaning that the step-wise growth of the copolymer brushes is still possible in the presence of the organic solvent. Given that CH_3CN has a higher refractive index than H_2O, it is not surprising that both, the 25 °C and 45 °C sensorgrams with added CH_3CN in the buffered medium give rise to more pronounced step-wise increase of the SPR signals, compared to the 25 °C curve in pure buffer. However, it is apparent that the addition of CH_3CN in the 45 °C sequential growth profile leads to a decrease in the signal intensity. This is therefore good evidence that upon the addition of the polar organic solvent, the oligoethylene glycol based solvation shell of the supramolecular structures remains swollen and intact. Thus, these experiments support the notion that the thermoresponsive behaviour of the copolymers in pure buffer facilitates their sequential growth as discussed above.

Temperature dependent QCM-D studies of the sequentially grown supramolecular copolymers

The SPR results provided strong evidence for the temperature-controlled growth of supramolecular copolymers on a gold surface, which operates under kinetic control. To prove that the SPR observations are not only due to a temperature dependent change of the refractive index and to further collect more experimental details about the physical properties of the resulting surface-confined copolymers, quartz crystal microbalance with dissipative monitoring (QCM-D) was performed. The surface-confined sequential copolymerisation process can be monitored *in situ* by the QCM-D technique, which determines the changes in the resonance frequency (Δf) of a quartz crystal that is coated by the polymeric film.[15] Additionally, the instrument can measure changes in dissipation (ΔD), a value which is influenced by the viscoelastic properties of the surface coverage and inversely proportional to the decay time of the oscillation.

Fig. 5 shows the frequency change and the respective change in dissipation as a function of time for the overtones 5–11. At a temperature of 25 °C a frequency decrease combined with a dissipation increase were observed, indicating the immobilisation of soft matter on the surface. The mass increment per layer is very regular, indicative of a well-controlled linear increase of surface coverage with DP. The deviation in Δf for the different overtones is attributed to the viscoelastic behaviour of the film and indicates a soft, hydrated film. This is consistent with the dissipation, as both the deviation of Δf between overtones as well as dissipation values ΔD are increasing with DP. Again, the linear increase of ΔD proves that the polymerization steps are regular and reproducible.

At 45 °C the absolute values of Δf are considerably higher, pointing out significantly larger layer thickness. In addition, the dissipation values are more than twice as high for copolymers grown at 45 °C compared to 25 °C, suggesting that very soft copolymer films are formed on the surface. The graphs describing the sequential growth for copolymers at 45 °C further show that the different harmonics deviate from each other, even more so with increasing copolymer length. With each monomer addition step, pronounced changes in the dissipation ΔD are obtained, reaching values of $\sim 10^{-5}$ after DP = 10. Based on previously reported detailed viscoelastic modelling of the layer-by-layer self-assembly of biopolymer (peptide and saccharide) based polyelectrolyte systems,[16] our results suggest the formation of a viscoelastic copolymer film on the gold surface. Thus,

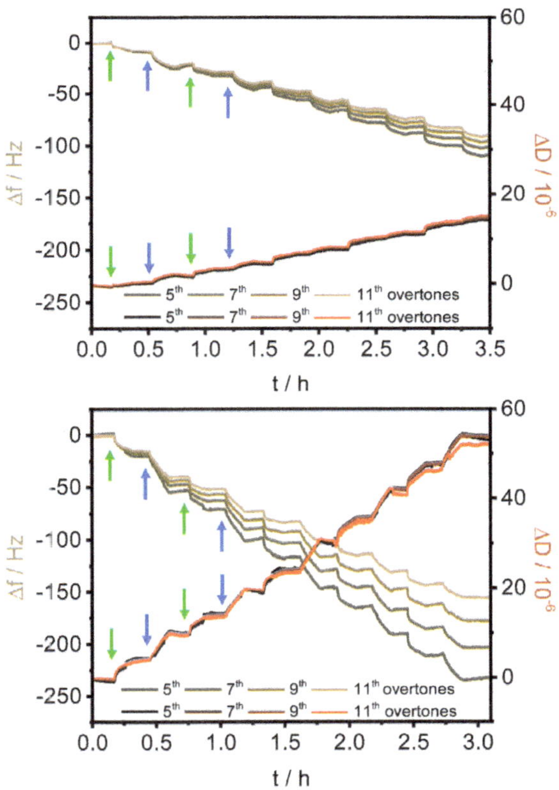

Fig. 5 QCM-D measurements of the sequential copolymerisation of **1** and **2** (green and blue arrows; after surface saturation with comonomer **2**) with DP = 10 at 25 °C (top) and 45 °C (bottom).

the data obtained from QCM-D studies corroborate our SPR experiments and demonstrate that at higher temperature more monomer molecules are adsorbed on the surface, yielding a thicker as well as softer polymer brush as compared to the polymer films that are generated at lower temperatures.

Conclusions

In a nutshell, we have extended the charge-regulated supramolecular copolymerisation of oppositely charged dendritic peptide amphiphiles on surfaces, by including water-solubilising thermoresponsive chains in the monomer design that are based on short biocompatible triethylene glycol units. Both monomers were synthesised using a facile and high yielding synthetic approach. By performing SPR measurements in the range of 22 °C to 45 °C, we observed sequential growth of the cationic and anionic comonomers to yield surface-confined copolymers. We observed non-linear temperature-dependent behaviour of the refractive index. This behaviour is indicative of the desolvation of the triethylene glycol shell surrounding the peptidic core of the copolymers. At higher temperatures the desolvation of the oligoethylene glycol chains, the entropy-driven multivalent Coulomb interaction and the release of counterions most likely promote stronger comonomer binding. In the SPR experiment, 18 sequential additions of comonomers are performed fairly quickly, and at 45 °C lead to an increase of the refractive index by 115 mRIU, which is more than an order of magnitude higher compared to the same protocol performed at 25 °C.

Critically, the surface bound copolymers remain kinetically trapped on the surface, even after flushing a buffer of intermediate ionic strength and near neutral pH for more than 1 hour. The copolymers can only be desorbed after the addition of acidic or basic aqueous solutions, which quenches the intermolecular Coulomb interactions *via* protonation or deprotonation of the carboxylate or ammonium side chains. Additional temperature dependent QCM-D measurements corroborate the conclusions drawn from SPR experimental data. Furthermore, they provide evidence for the self-assembly of surface bound viscoelastic materials.

In contrast to our previous work, the incorporation of methionine into both oligopeptide based comonomers means that every addition of monomer could potentially lead to desorption of the surface bound polymers *via* competitive displacement of the thioether–gold interactions in the initial monomer layer. Desorption is not observed, which further supports our hypothesis that kinetically trapped copolymers are produced at near neutral pH. We suggest that kinetic trapping is achieved *via* a combination of charge regulation on the active end of the growing copolymer which slows down the kinetics of the self-assembly process and the high surface coverage which protects the initial monomer layer and peptide gold interactions. Our approach is attractive for the design of new stimuli-responsive biomimetic materials on solid substrates. Thus, further investigations will focus on investigating the viscoelastic properties of the surface-grafted copolymer brushes, and exploit potential applications in optoelectronics and sensor applications.

Experimental

For full experimental details, synthetic procedures, molecular characterisation and instrumentation details we refer to the ESI†.

SPR experiments were performed on the spectrometer SR7000DC (Reichert). SPR sensorchips (1.44 $cm^2 \times 0.3$ mm) consist of glass with a 50 nm thick gold layer and have a refractive index of 1.5168 (Xantec SC Au, Xantec bioanalytics). All SPR measurements were performed at a flow rate of 50 $\mu L\ min^{-1}$.

QCM experiments were performed using a QCM-D (E4, QSense, Göteborg, Sweden) with an ISMATEC IPC model ISM 935 pump system. Gold coated quartz crystals with a base frequency of 5 MHz (QSense) were used as substrates. All measurements were performed simultaneously in two flow cells with a flow rate of 50 $\mu L\ min^{-1}$.

Synthesis of monomers 1 and 2: CuAAC reactions and deprotection

Compound 1a. N_3-L-Gly-L-Phe-L-Glu(O^{t}Bu)-L-Met-L-Glu(O^{t}Bu)-L-Met-Ahx-D_3-TEG (**14**) (259 mg, 148 μmol, 3.3 eq.) was dissolved in DMSO-d_6 (2 mL) under argon. TBTA (19 mg, 36 μmol, 0.8 eq.) and $CuSO_4\ 5 \cdot H_2O$ (9.0 mg, 36 μmol, 0.8 eq.) were dissolved in DMSO-d_6 (0.6 mL) and added to the reaction mixture. Subsequently, sodium ascorbate (18.7 mg, 94 μmol, 2.1 eq.) and lastly 1,3,5-triethynylbenzene (6.74 mg, 44.9 μmol, 1.0 eq.) were added. The solution was stirred at rt over night, precipitated in H_2O and washed twice with H_2O (40 mL). The crude product was purified *via* FC (MeOH : $CHCl_3$ → 1 : 9 to 2 : 8, $R_f = 0.23$ in $CHCl_3$: MeOH → 9 : 1.2) followed by SEC ($CHCl_3$: MeOH → 2 : 1). The solvent was removed *in vacuo* to yield a colourless solid.

Yield: 194 mg, 36 μmol, 80%.

Chemical formula: $C_{248}H_{415}N_{39}O_{78}S_6$.

^{1}H-NMR (400 MHz, 296 K, DMSO-d$_6$): δ [ppm] = 8.69 (d, J = 8.1 Hz, 3H, α-NH^{Phe}), 8.49 (s, 3H, CH^{triazole}), 8.42 (d, J = 7.6 Hz, 3H, α-NH), 8.28 (s, 3H, CH^{BTA}), 8.12 (d, J = 7.4 Hz, 3H, α-NH), 8.06 (d, J = 7.5 Hz, 3H, α-NH), 7.94 (d, J = 8.0 Hz, 3H, α-NH), 7.90 (t, J = 5.7 Hz, 9H, NH^{TEG}), 7.88–7.84 (m, 3H, NH^{hexyl}), 7.28–7.18 (m, 15H, CH^{Phe}), 6.98 (s, 3H, NHC$_q$), 5.21 (d, J = 16.5 Hz, 1H, CH_2^{Gly}), 5.11 (d, J = 16.5 Hz, 1H, CH_2^{Gly}), 4.71–4.60 (m, 3H, α-CH^{Phe}), 4.40–4.20 (m, 12H, α-CH), 3.60–3.36 (m, 126H, CH_2O/CH_2^{TEG}), 3.24 (s, 27H, OCH_3^{TEG}), 3.23–3.17 (m, 18H, CH_2^{TEG}), 3.10–2.92 (m, 9H, $NHCH_2^{hexyl}$/CH_2^{Phe}), 2.82–2.74 (m, 3H, CH_2^{Phe}), 2.47–2.36 (m, 12H, CH_2S), 2.30 (t, J = 6.4 Hz, 18H, $OCH_2CH_2C{=}O$), 2.27–2.18 (m, 12H, $CH_2C{=}O^{Glu}$), 2.10–2.00 (m, 6H, $CH_2C{=}O^{hexyl}$), 2.03 (s, 9H, SCH_3), 2.02 (s, 9H, SCH_3), 1.95–1.82 (m, 12H, CH_2CH_2S), 1.82–1.70 (m, 12H, $CH_2CH_2C{=}O^{Glu}$), 1.48–1.41 (m, 12H, CH_2^{hexyl}), 1.38 (s, 27H, $C_qCH_3^{Glu}$), 1.37 (s, 27H, $C_qCH_3^{Glu}$), 1.26–1.18 (m, 6H, CH_2^{hexyl}).

MALDI-MS (MeOH) (m/z): calculated for $[C_{248}H_{415}N_{39}O_{78}S_6K]^+$: 5420.77, found 5418.56.

Compound 1. **1a** (159 mg, 29.5 μmol, 1.0 eq.) was dissolved in a mixture of TFA : TIS : H_2O (8 : 1 : 1, 15 mL) and stirred over 3 h at rt. The solvent was removed *in vacuo* and redissolved in a small amount of $CHCl_3$. After precipitation in pentane (45 mL) and washing with pentane (2 × 45 mL) the precipitate was dissolved in H_2O. After lyophilisation, the resulting solid was purified by SEC ($CHCl_3$: MeOH → 2 : 1, $R_f = 0.05$ in $CHCl_3$: MeOH → 9 : 1.2) to give a colourless fluffy solid after freeze-drying out of a small amount of water.

Yield: 134 mg, 26.6 μmol, 90%.

Chemical formula: $C_{224}H_{367}N_{39}O_{78}S_6$.

^{1}H-NMR (400 MHz, 296 K, DMSO-d$_6$): δ [ppm] = 12.09 (s, 6H, COOH), 8.68 (d, J = 8.1 Hz, 3H, α-NH^{Phe}), 8.49 (s, 3H, CH^{triazole}), 8.43 (d, J = 7.4 Hz, 3H, α-NH), 8.28

(s, 3H, CH^{BTA}), 8.13 (d, $J = 7.2$ Hz, 3H, α-NH), 8.06 (d, $J = 7.3$ Hz, 3H, α-NH), 7.97 (d, $J = 8.1$ Hz, 3H, α-NH), 7.90 (t, $J = 5.6$ Hz, 9H, NH^{TEG}), 7.88–7.83 (m, 3H, NH^{hexyl}), 7.31–7.14 (m, 15H, CH^{Phe}), 6.98 (s, 3H, NHC_q), 5.20 (d, $J = 16.6$ Hz, 1H, CH_2^{Gly}), 5.11 (d, $J = 16.7$ Hz, 1H, CH_2^{Gly}), 4.70–4.61 (m, 3H, α-CH^{Phe}), 4.39–4.16 (m, 12H, α-CH), 3.57–3.33 (m, 126H, CH_2O/CH_2^{TEG}), 3.23 (s, 27H, $OCH_3{}^{TEG}$), 3.22–3.16 (m, 18H, CH_2^{TEG}), 3.10–2.90 (m, 9H, $NHCH_2^{hexyl}/CH_2^{Phe}$), 2.83–2.71 (m, 3H, CH_2^{Phe}), 2.30 (t, $J = 6.5$ Hz, 18H, $OCH_2CH_2C{=}O$), 2.34–2.21 (m, 12H, CH_2S), 2.09–2.00 (m, 24H, $CH_2C{=}O^{hexyl}/SCH_3$), 1.96–1.70 (m, 24H, $CH_2CH_2S/CH_2CH_2C{=}O$), 1.53–1.12 (m, 18H, CH_2^{hexyl}).

MALDI-MS (MeOH) (*m*/*z*): calculated for $[C_{224}H_{367}N_{39}O_{78}S_6K]^+$: 5084.40, found 5082.22.

Compound 2a. N_3-L-Gly-L-Phe-L-Lys(Boc)-L-Met-L-Lys(Boc)-L-Met-Ahx-D_3-TEG (**15**) (232 mg, 127 μmol, 3.3 eq.) was dissolved in DMSO-d_6 (2 mL) under argon. TBTA (16 mg, 30 μmol, 0.8 eq.) and $CuSO_4\ 5\cdot H_2O$ (7.7 mg, 31 μmol, 0.8 eq.) were dissolved in DMSO-d_6 (1 mL) and dropped into the reaction mixture. Afterwards sodium ascorbate (15.9 mg, 80 μmol, 2.1 eq.) and lastly 1,3,5-triethynylbenzene (5.75 mg, 38 μmol, 1.0 eq.) were added. The solution was stirred at rt over night. The crude product was purified *via* SEC ($CHCl_3$: MeOH → 2 : 1, $R_f = 0.57$ in $CHCl_3$: MeOH → 9 : 1). The solvent was removed *in vacuo* and the product was obtained as a colourless solid.

Yield: 203 mg, 36 μmol, 94%.

Chemical formula: $C_{260}H_{445}N_{45}O_{78}S_6$.

^{1}H-NMR (400 MHz, 296 K, DMSO-d_6): δ [ppm] = 8.67 (d, $J = 8.2$ Hz, 3H, α-NH^{Phe}), 8.49 (s, 3H, $CH^{triazole}$), 8.38–8.33 (m, 3H, α-NH), 8.28 (s, 3H, CH^{BTA}), 8.07–7.92 (m, 18H, α-NH), 7.90 (t, $J = 5.7$ Hz, 9H, NH^{TEG}), 7.85–7.79 (m, 3H, NH^{hexyl}), 7.39–7.14 (m, 15H, CH^{Phe}), 6.99 (s, 3H, NHC_q), 6.78–6.68 (m, 6H, NH^{Lys}), 5.21 (d, $J = 16.7$ Hz, 3H, CH_2^{Gly}), 5.11 (d, $J = 16.6$ Hz, 3H, CH_2^{Gly}), 4.75–4.60 (m, 3H, α-CH^{Phe}), 4.45–4.07 (m, 12H, α-CH), 3.58–3.36 (m, 126H, CH_2O/CH_2^{TEG}), 3.24 (s, 27H, OCH_3^{TEG}), 3.23–3.16 (m, 18H, CH_2^{TEG}), 3.10–2.90 (m, 9H, $NHCH_2^{hexyl}/CH_2^{Phe}$), 2.90–2.84 (m, 12H, CH_2NH^{Lys}), 2.82–2.64 (m, 3H, CH_2^{Phe}), 2.48–2.35 (m, 12H, CH_2S), 2.30 (t, $J = 6.4$ Hz, 27H, $OCH_2CH_2C{=}O$), 2.10–2.01 (m, 6H, $CH_2C{=}O^{hexyl}$), 2.04 (s, 9H, SCH_3), 2.02 (s, 9H, SCH_3), 1.95–1.71 (m, 12H, CH_2CH_2S), 1.68–1.42 (m, 12H, α-$CHCH_2{}^{Lys}$), 1.36 (s, 27H, $C_qCH_3{}^{Lys}$), 1.33 (s, 27H, $C_qCH_3{}^{Lys}$), 1.30–1.18 (m, 24H, $CH_2CH_2CH_2NH^{Lys}$).

MALDI-MS (MeOH) (*m*/*z*): calculated for $[C_{260}H_{445}N_{45}O_{78}S_6K]^+$: 5679.03, found 5677.46.

Compound 2. 2a (229 mg, 40.5 μmol, 1.0 eq.) was dissolved in a mixture of TFA : TIS : H_2O (8 : 1 : 1, 15 mL) and stirred over 3 h at rt. The solvent was removed *in vacuo* and redissolved in a small amount of $CHCl_3$. After precipitation in pentane (45 mL) and washing with pentane (2 × 45 mL) the precipitate was dissolved in H_2O. After lyophilisation, the resulting solid was purified by SEC ($CHCl_3$: MeOH → 2 : 1, $R_f = 0.05$ in $CHCl_3$: MeOH → 9 : 1.2) to give a colourless fluffy solid after freeze-drying out of a small amount of water.

Yield: 170 mg, 29.6 μmol, > 73%.

Chemical formula: $C_{230}H_{379}N_{45}O_{66}S_6$.

^{1}H-NMR (400 MHz, 296 K, DMSO-d_6): δ [ppm] = 8.71 (d, $J = 8.1$ Hz, 3H, α-NH^{Phe}), 8.49 (s, 3H, $CH^{triazole}$), 8.41 (d, $J = 7.7$ Hz, 3H, α-NH), 8.27 (s, 3H, CH^{BTA}), 8.10 (d, $J = 7.6$ Hz, 3H, α-NH), 8.03 (d, $J = 7.6$ Hz, 3H, α-NH), 7.95–7.90 (m, 3H, α-NH), 7.91 (t, $J = 5.6$ Hz, 9H, NH^{TEG}), 7.71 (s, 18H, NH_3^{Lys}), 7.31–7.19 (m, 15H,

CH^{Phe}), 7.00 (s, 3H, NHC_q), 5.21 (d, $J = 16.6$ Hz, 3H, CH_2^{Gly}), 5.11 (d, $J = 16.5$ Hz, 3H, CH_2^{Gly}), 4.70–4.60 (m, 3H, α-CH^{Phe}), 4.40–4.16 (m, 12H, α-CH), 3.56–3.34 (m, 126H, CH_2O/CH_2^{TEG}), 3.23 (s, 27H, OCH_3^{TEG}), 3.22–3.15 (m, 18H, CH_2^{TEG}), 3.13–2.88 (m, 9H, $NHCH_2^{hexyl}/CH_2^{Phe}$), 2.86–2.69 (m, 12H, $CH_2NH_3^{Lys}$), 2.68–2.64 (m, 3H, CH_2^{Phe}), 2.48–2.34 (m, 12H, CH_2S), 2.30 (t, $J = 6.4$ Hz, 10H, $OCH_2CH_2C{=}O$), 2.04 (s, 9H, SCH_3), 2.02 (s, 9H, SCH_3), 2.10–1.99 (m, 6H, $CH_2C{=}O^{hexyl}$), 1.98–1.61 (m, 4H, CH_2CH_2S), 1.59–1.46 (m, 4H, α-$CHCH_2^{Lys}$), 1.30–1.17 (m, 8H, $CH_2CH_2CH_2NH^{Lys}$).

MALDI-MS (MeOH) (m/z): calculated for $[C_{230}H_{379}N_{45}O_{66}S_6Na]^+$: 5062.74, found 5055.39.

Acknowledgements

We thank the Fonds der Chemischen Industrie for a doctoral fellowship [S. E.]. This work was supported by the Deutsche Forschungsgemeinschaft (SFB 858) [E. C. F.]. The University of Mainz is acknowledged for financial support.

References

1 (*a*) S. T. Milner, *Science*, 1991, **251**, 905–914; (*b*) M. Husseman, E. E. Malmström, M. McNamara, M. Mate, D. Mecerreyes, D. G. Benoit, J. L. Hedrick, P. Mansky, E. Huang, T. P. Russell and C. J. Hawker, *Macromolecules*, 1999, **32**, 1424–1431; (*c*) S. Edmondson, V. L. Osborne and W. T. S. Huck, *Chem. Soc. Rev.*, 2004, **33**, 14–22; (*d*) W. Senaratne, L. Andruzzi and C. K. Ober, *Biomacromolecules*, 2005, **6**, 2427–2448; (*e*) M. A. C. Stuart, W. T. S. Huck, J. Genzer, M. Müller, C. Ober, M. Stamm, G. B. Sukhorukov, I. Szleifer, V. V. Tsukruk, M. Urban, F. Winnik, S. Zauscher, I. Luzinov and S. Minko, *Nat. Mater.*, 2010, **9**, 101–113; (*f*) X. Liu, Y. Liang, F. Zhou and W. Liu, *Soft Matter*, 2012, **8**, 2070–2086; (*g*) Q. Wei, T. Becherer, S. Angioletti-Uberti, J. Dzubiella, C. Wischke, A. T. Neffe, A. Lendlein, M. Ballauff and R. Haag, *Angew. Chem., Int. Ed.*, 2014, **53**, 8004–8031; (*h*) B. B. Hsu, K. S. Jamieson, S. R. Hagerman, E. Holler, J. Y. Ljubimova and P. T. Hammond, *Angew. Chem., Int. Ed.*, 2014, **53**, 8093–8098; (*i*) C. Hu, F. Tian, Y. Zheng, C. S. Y. Tan, K. R. West and O. A. Scherman, *Chem. Sci.*, 2015, **6**, 5303–5310.

2 (*a*) M. Reches and E. Gazit, *Nat. Nanotechnol.*, 2006, **1**, 195–200; (*b*) L. Adler-Abramovich, D. Aronov, P. Beker, M. Yevnin, S. Stempler, L. Buzhansky, G. Rosenman and E. Gazit, *Nat. Nanotechnol.*, 2009, **4**, 849–854.

3 M. Mizrahi, A. Zakrassov, J. Lerner-Yardeni and N. Ashkenasy, *Nanoscale*, 2012, **4**, 518–524.

4 N. Sakai, M. Lista, O. Kel, S.-i. Sakurai, D. Emery, J. Mareda, E. Vauthey and S. Matile, *J. Am. Chem. Soc.*, 2011, **133**, 15224–15227.

5 H. Frisch, E.-C. Fritz, F. Stricker, L. Schmüser, D. Spitzer, T. Weidner, B. J. Ravoo and P. Besenius, *Angew. Chem., Int. Ed.*, 2016, **55**, 7242–7246.

6 (*a*) R. B. Merrifield, *J. Am. Chem. Soc.*, 1963, **85**, 2149–2154; (*b*) A. A. Hariri, G. D. Hamblin, Y. Gidi, H. F. Sleiman and G. Cosa, *Nat. Chem.*, 2015, **7**, 295–300.

7 (*a*) J. M. Lehn, *Proc. Natl. Acad. Sci. U. S. A.*, 2002, **99**, 4763–4768; (*b*) X. Wang, G. Guerin, H. Wang, Y. Wang, I. Manners and M. A. Winnik, *Science*, 2007, **317**,

644–647; (*c*) N. Sakai, R. Bhosale, D. Emery, J. Mareda and S. Matile, *J. Am. Chem. Soc.*, 2010, **132**, 6923–6925; (*d*) Y. Tidhar, H. Weissman, S. G. Wolf, A. Gulino and B. Rybtchinski, *Chem.–Eur. J.*, 2011, **17**, 6068–6075; (*e*) P. A. Korevaar, S. J. George, A. J. Markvoort, M. M. J. Smulders, P. A. J. Hilbers, A. P. H. J. Schenning, T. F. A. De Greef and E. W. Meijer, *Nature*, 2012, **481**, 492–496; (*f*) P. A. Korevaar, C. Grenier, A. J. Markvoort, A. P. H. J. Schenning, T. F. A. de Greef and E. W. Meijer, *Proc. Natl. Acad. Sci. U. S. A.*, 2013, **110**, 17205–17210; (*g*) P. A. Korevaar, C. J. Newcomb, E. W. Meijer and S. I. Stupp, *J. Am. Chem. Soc.*, 2014, **136**, 8540–8543; (*h*) S. Ogi, K. Sugiyasu, S. Manna, S. Samitsu and M. Takeuchi, *Nat. Chem.*, 2014, **6**, 188–195; (*i*) A. G. L. Olive, N. H. Abdullah, I. Ziemecka, E. Mendes, R. Eelkema and J. H. van Esch, *Angew. Chem., Int. Ed.*, 2014, **53**, 4132–4136; (*j*) S. Ogi, V. Stepanenko, K. Sugiyasu, M. Takeuchi and F. Würthner, *J. Am. Chem. Soc.*, 2015, **137**, 3300–3307; (*k*) J. Kang, D. Miyajima, T. Mori, Y. Inoue, Y. Itoh and T. Aida, *Science*, 2015, **347**, 646–651; (*l*) A. Pal, M. Malakoutikhah, G. Leonetti, M. Tezcan, M. Colomb-Delsuc, V. D. Nguyen, J. van der Gucht and S. Otto, *Angew. Chem., Int. Ed.*, 2015, **54**, 7852–7856; (*m*) E. Krieg, M. M. C. Bastings, P. Besenius and B. Rybtchinski, *Chem. Rev.*, 2016, **116**, 2414–2477; (*n*) A. Aliprandi, M. Mauro and L. De Cola, *Nat. Chem.*, 2016, **8**, 10–15; (*o*) P. Besenius, *J. Polym. Sci., Part A: Polym. Chem.*, 2017, **55**, 34–78.

8 (*a*) M. von Gröning, I. de Feijter, M. C. A. Stuart, I. K. Voets and P. Besenius, *J. Mater. Chem. B*, 2013, **1**, 2008–2012; (*b*) H. Frisch, Y. Nie, S. Raunser and P. Besenius, *Chem.–Eur. J.*, 2015, **21**, 3304–3309; (*c*) R. Appel, S. Tacke, J. Klingauf and P. Besenius, *Org. Biomol. Chem.*, 2015, **13**, 1030–1039; (*d*) R. Appel, J. Fuchs, S. M. Tyrrell, P. A. Korevaar, M. C. A. Stuart, I. K. Voets, M. Schönhoff and P. Besenius, *Chem.–Eur. J.*, 2015, **21**, 19257–19264; (*e*) P. Ahlers, H. Frisch and P. Besenius, *Polym. Chem.*, 2015, **6**, 7245–7250; (*f*) B. Kemper, M. von Gröning, V. Lewe, D. Spitzer, T. Otremba, N. Stergiou, D. Schollmeyer, E. Schmitt, B. J. Ravoo and P. Besenius, *Chem.–Eur. J.*, 2017, **23**, 6048–6055.

9 (*a*) J.-F. Lutz, Ö. Akdemir and A. Hoth, *J. Am. Chem. Soc.*, 2006, **128**, 13046–13047; (*b*) J.-F. Lutz, *J. Polym. Sci., Part A: Polym. Chem.*, 2008, **46**, 3459–3470; (*c*) P. H. J. Kouwer, M. Koepf, V. A. A. Le Sage, M. Jaspers, A. M. van Buul, Z. H. Eksteen-Akeroyd, T. Woltinge, E. Schwartz, H. J. Kitto, R. Hoogenboom, S. J. Picken, R. J. M. Nolte, E. Mendes and A. E. Rowan, *Nature*, 2013, 651–655; (*d*) G. Vancoillie, D. Frank and R. Hoogenboom, *Prog. Polym. Sci.*, 2014, **39**, 1074–1095.

10 H. Frisch, J. P. Unsleber, D. Lüdeker, M. Peterlechner, G. Brunklaus, M. Waller and P. Besenius, *Angew. Chem., Int. Ed.*, 2013, **52**, 10097–10101.

11 (*a*) H. C. Kolb, M. G. Finn and K. B. Sharpless, *Angew. Chem., Int. Ed.*, 2001, **40**, 2004; (*b*) W. G. Lewis, L. G. Green, F. Grynszpan, Z. Radic, P. R. Carlier, P. Taylor, M. G. Finn and K. B. Sharpless, *Angew. Chem., Int. Ed.*, 2002, **41**, 1053–1057; (*c*) C. W. Tornøe, C. Christensen and M. Meldal, *J. Org. Chem.*, 2002, **67**, 3057–3064.

12 G. R. Newkome and C. Shreiner, *Chem. Rev.*, 2010, **110**, 6338–6442.

13 (*a*) P. Schuck, *Annu. Rev. Biophys. Biomol. Struct.*, 1997, **26**, 541–566; (*b*) W. Knoll, *Annu. Rev. Phys. Chem.*, 1998, **49**, 569–638.

14 P. Ahlers, H. Frisch, D. Spitzer, Z. Vobecka, F. Vilela and P. Besenius, *Chem.–Asian J.*, 2014, **9**, 2052–2057.

15 (*a*) G. Sauerbrey, *Z. Phys.*, 1959, **155**, 206–222; (*b*) M. Rodahl, F. Höök and B. Kasemo, *Anal. Chem.*, 1996, **68**, 2219–2227.

16 (*a*) T. Crouzier and C. Picart, *Biomacromolecules*, 2009, **10**, 433–442; (*b*) D. Bütergerds, C. Cramer and M. Schönhoff, *Adv. Mater. Interfaces*, 2017, **4**, 1600592; (*c*) D. Bütergerds, C. Kateloe, C. Cramer and M. Schönhoff, *J. Polym. Sci., Part B: Polym. Phys.*, 2017, **55**, 425–434.

Faraday Discussions

PAPER

Supramolecular chiral self-assemblies of Gly–Pro dipeptides on metallic fcc(110) surfaces†

C. Méthivier,[a] H. Cruguel,[b] C.-M. Pradier[a] and V. Humblot *[a]

Received 24th March 2017, Accepted 10th May 2017
DOI: 10.1039/c7fd00116a

Adsorption of the Glycine–Proline (Gly–Pro) dipeptide has been investigated using surface science complementary techniques on Au(110) and Ag(110), showing some interesting differences both in the chemical form and surface organization of the adsorbed peptide. On Au(110), Gly–Pro mainly adsorbs in neutral form ($COOH/NH_2$), at low coverage or for a short interaction time; the surface species become zwitterionic at a higher coverage or longer interaction time. These changes are accompanied by a complete reorganization of the molecules at the surface. On Ag(110), only anionic molecules (COO^-/NH_2) were detected on the surface and only one type of arrangement was observed. These results will be compared to some previously obtained on Cu(110), thus providing a unique comparison of the adsorption of the same di-peptide on three different metal surfaces; the great influence of the substrate on both the chemical form and the arrangement of adsorbed di-peptides was made clear.

Introduction

Surface science techniques enable detailed characterization of model bio-interfaces, for instance the adsorption of amino acids or small peptides on model metal surfaces.[1–3] Adsorption of glycine, methionine,[17] alanine,[3] glutamic acid,[37,38] and lysine[12,13] *etc.* has been well described in recent literature thanks to studies in UHV conditions, mainly on copper and less frequently on gold; very few studies on silver surfaces could be found.[4–19]

From these well documented studies, it appears that amino acids mostly undergo a deprotonation of their acid groups when interacting with copper;[4,12,13,20,21] they bind to the surface *via* the oxygen atoms of the carboxylate groups. Conversely, such a deprotonation process is not systematic on gold; the latter metal seems to also favour an interaction of the NH_2 group with the metal,

[a]*Sorbonne Universités, UPMC Université Pierre et Marie Curie, Univ Paris 6, Laboratoire de Réactivité de Surface, UMR CNRS 7197, 4 place Jussieu, F-75005 Paris, France. E-mail: vincent.humblot@upmc.fr*

[b]*Sorbonne Universités, UPMC Université Pierre et Marie Curie, Univ Paris 6, Institut des NanoScience de Paris, 4 place Jussieu, F-75005 Paris, France*

† Electronic supplementary information (ESI) available. See DOI: 10.1039/c7fd00116a

inducing an easily detectable electronic transfer from the N atom to the metal, the latter is evidenced by a shift of the XPS N 1s peak at a Binding Energy (BE) below 399.5 eV.[17,22,23]

As for small peptides, the number of investigations addressing the issue of peptide interactions with metal surfaces is of course smaller;[24] the same tendencies can be deduced, an easy deprotonation of the acid groups on copper and a frequent interaction of nitrogen with gold. The image is, however, more complex having in mind that, for instance the GSH tri-peptide (Gly–Pro–Glu) preferentially interacts *via* its S atoms at low coverage on gold[25,26] and copper, the Gly–Pro dipeptide was observed to bind *via* oxygen and nitrogen atoms on copper and overall, the geometry, point of interaction, and sometimes, the chemical form of the peptide itself varies with coverage!

These findings originate from an examination of numerous studies addressing the interaction of one amino acid or one peptide with a metal, not from a rigorous comparison of the adsorption of one species with a surface of identical geometry but of various metals. This is precisely the objective of the present investigation where the di-peptide, Glycine–Proline (Gly–Pro), *cf.* Fig. 1, was adsorbed, under similar conditions on the fcc(110) surfaces of gold and silver.

A variety of experimental surface science techniques of characterization have been applied, namely Polarisation Modulation Reflection Absorption Infrared Spectroscopy (PM-RAIRS), X-ray Photoelectron Spectroscopy (XPS) and Scanning Tunneling Microscopy (STM). Let's underline that PM-RAIRS analyses were taken during exposure to Gly–Pro ($P = 10^{-9}$ Torr), while XPS and STM analyses were performed in another chamber after exposure.

The comparison will be enriched by the results of recent studies on the adsorption of the same dipeptide on Cu(110).[23,27]

The Gly–Pro dipeptide bears a nitrogen atom on its cycle, C=O amide groups and only one COOH–NH_2 group; this makes the IR and XPS characteristic signals easily identifiable. And, as a matter of fact, on Cu(110), two binding points, the COO^- groups and the N atom of the NH_2 groups, have been identified by

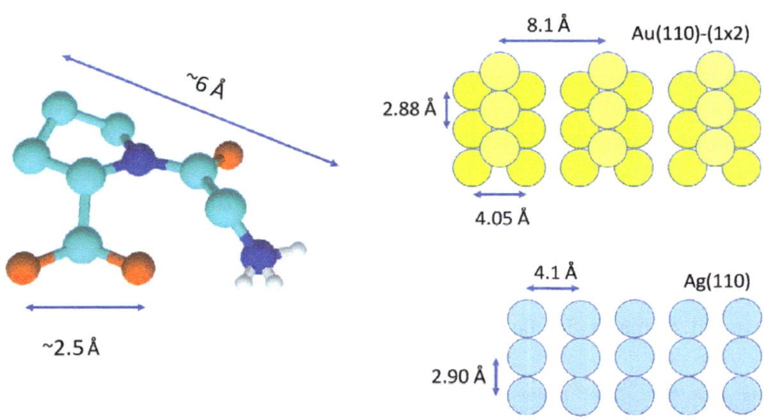

Fig. 1 (Left) Molecular model for Gly–Pro in its zwitterionic chemical form, together with calculated molecular dimensions in the gas phase; (right) atomic dimensions of the Au and Ag surfaces.

combining PM-RAIRS and XPS. Importantly, long exposure (more than 60 min under $P = 10^{-9}$ Torr) did not lead to the formation of more than one saturated monolayer.[23]

One will see in the present work that, on Au(110), Gly–Pro is mainly in neutral or zwitterionic form; its orientation and surface organization are coverage and time-dependent forming a line of dimers or trimmers first, followed by dense heptamer arrays. Conversely, on Ag(110), one observes a significant deprotonation of the acid group thus leading to a different mode of interaction with the surface. From low values of coverage, Gly–Pro molecules form chiral domains on Ag(110) with the creation of "zig-zag" lines of molecules.

Results and discussion

Gly–Pro on Au(110)-(1 × 2)

PM-RAIRS spectra of the Au(110) surface recorded with continuous exposure of Gly–Pro are shown in Fig. SI1.† After 10 min exposure (bottom spectrum) and 35 min of exposure (top spectrum), the peaks are not very intense, but are well defined. All bands increase with increasing exposure to Gly–Pro molecules, leading to the following main IR bands: the most intense, observed at 1668 cm^{-1}, is attributed to the elongation vibration νC=O of the amide I bands, the features at 1723 and 1054 cm^{-1} are respectively assigned to the vibrations νC=O and νC–OH of the carboxylic acid groups, while the bands at 1596 and 1415 cm^{-1} are attributed to the asymmetric and symmetric stretching of the carboxylate (COO^-) functions, respectively.[26] Moreover, the presence of lower absorption bands at 1645 and 1519 cm^{-1} can be attributed to asymmetric and symmetric deformations of the NH_3^+ of the protonated amine groups, and finally a band observed at 1562 cm^{-1} is associated with the δNH_2 deformation band.[12] This first result suggests the presence of several chemical species on the surface, most probably the coexistence of Gly–Pro molecules in the neutral $COOH/NH_2$ or anionic NH_2/COO^-, and zwitterionic NH_3^+/COO^- forms.

The high resolution XPS spectra of the N 1s and O 1s core levels are shown in Fig. 2 for two different exposure times: 10 and 35 minutes. Two contributions are identified in the N 1s region of Gly–Pro at 400.2 (400.4) ± 0.1 eV and 401.6 (402.1) ± 0.1 eV, which correspond, respectively to the nitrogen atoms in the NH_2 amine and NH amide groups, and N in the NH_3^+ protonated amine groups. The O 1s region shows three contributions, at 531.5 (531.7) ± 0.1 eV, 532.4 (532.6) ± 0.1 eV and 533.5 (533.7) ± 0.1 eV, attributed to the oxygen atoms of deprotonated acid and amide groups (**COO**$^-$/HNC=**O**), of the oxygen atoms of the protonated acid groups (HO–C=**O**) and of the oxygen atoms of the OH acid groups (**OH**–C=O), respectively.[17,28,29] The presence of protonated acid groups is confirmed by XPS data of the C 1s region, Fig. SI5,† showing a contribution at high Binding Energy (BE), *ca.* 290 eV.

Note that, the ratio of the NH_3^+/NH_2 XPS contributions passes from 41% after 10 minutes of exposure to nearly 65% after 35 minutes, *cf.* Table SI1,† thus indicating a net increase of the zwitterionic species. Considering that, simultaneously, the fraction of the O 1s contribution at higher BE (533.2 eV) is quasi-constant, one must admit that, the fraction of zwitterionic molecules mainly increases at the expense of anionic molecules. One may even estimate that the fraction of the latter decreases from 39 to 35%. This is in agreement with the

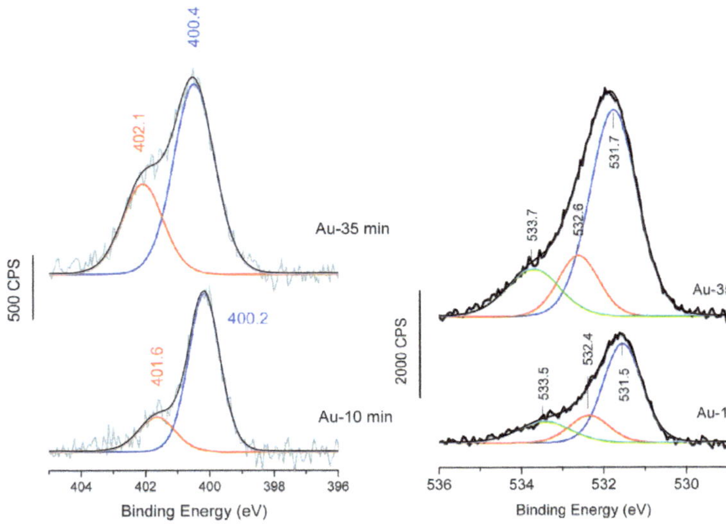

Fig. 2 XPS high resolution spectra for the N 1s and O 1s regions for a 10 and 35 minute dose of Gly–Pro adsorbed on Au(110).

growth of the peak at *ca.* 1723–1756 cm^{-1} which originates from the C=O stretch of the acid groups. As a matter of fact, looking now at the IR band intensities and their variations with time of exposure, it helps to understand the adsorption process of Gly–Pro on Au(110), even if this task is rather complex due to the coexistence of several chemical species on the surface. Finally, XPS data also allow us to estimate the local coverage for both dosing times: after 10 min, the local coverage is calculated at 1.6 molecules per nm^2, and 2.4 molecules per nm^2 after 35 minutes; knowing that a saturated monolayer is reached for a local coverage of 4 molecules per nm^2, one can estimate the coverage to be around 40% and 60% of a saturated monolayer for 10 and 35 minutes, respectively.

We then investigated the possible 2D arrangements that the adsorption of Gly–Pro on Au(110)-(1 × 2) implies. Fig. 3 shows STM data obtained after 10 minutes of dosing, which distinctly show an arrangement of lines along the [110] axis, and a second oblique arrangement located in between these lines of molecules. At first, one is tempted to believe that these two different 2D structures result from the adsorption of Gly–Pro in different chemical forms. Thus, by scanning the same given area, Fig. SI2,† the evolution of the 2D arrangements is clear, in particular, the mobility of the molecules constituting the linear arrangements, passing from simple lines to double or triple lines. Moreover, one can distinguish that the island of molecules having an oblique arrangement sees its area, and thus its number of molecules, increase as a function of the relaxation time (or time of observation); namely, the area is multiplied by 4 after 150 minutes without any further dosing. This allows us to conclude that the linear arrangements, forming first, are most likely made up of Gly–Pro molecules in the neutral or anionic chemical form; as a consequence, the oblique arrangement whose fraction increases with time is likely constituted of molecules of Gly–Pro in their zwitterionic form.

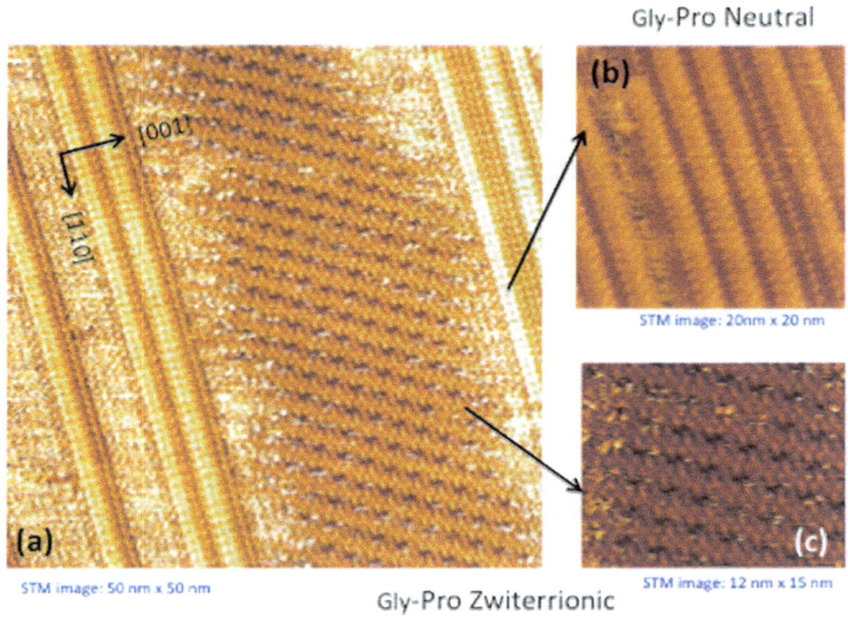

Fig. 3 STM images showing the 2 domains of Gly–Pro coexisting in surfaces (a) and each of the domains separated, (b) Gly–Pro in neutral and anionic forms and (c) Gly–Pro in zwitterionic form.

We then analyzed each of these 2D arrangements in greater detail. Fig. 3(a) shows STM images where the lines of molecules that appear to be organized in groups of at least two lines are seen in more detail, while for the oblique 2D arrangement, small groups of molecules self-organize at an angle of 30° with the main [001] crystallographic axis. In the case of the Gly–Pro molecule in its chemical neutral/anionic forms, Fig. 3(b), the in-line arrangement is reminiscent of that already observed in the case of Gly–Pro on Cu(110).[27] By measuring the distances between the lines and between each of the neutral Gly–Pro molecules, the latter are organized in lines of 2 or 3 units, with a non-chiral 2D repeating pattern, similar to a rectangular unit cell (3×6), Fig. 4(a).

From XPS data, we were able to estimate the average thickness of the adlayer after 10 minutes of dosing; the thickness was estimated at 0.3 nm, which corresponds to about 1.6 Gly–Pro molecules per nm^2, this slightly overestimates what can be seen on the model in Fig. 4(a) with a bit more than one molecule within a 1 nm^2 dashed line square. Let's recall that the model presented is only for double lines, knowing from the STM images that they could be triple lines or more.

Turning now to the second 2D structure created by Gly–Pro in its zwitterionic chemical form, Fig. 3(c), we realize the complexity of the 2D arrangement composed of an hexagonal structure with repetitive patterns along two non-orthogonal axes and disoriented with respect to the crystallographic axes; this last point suggests the creation of a chiral over-structure created by the zwitterionic Gly–Pro molecules. Measurements taken using the STM images allow us to have more information on the intermolecular spacing and on the position of the

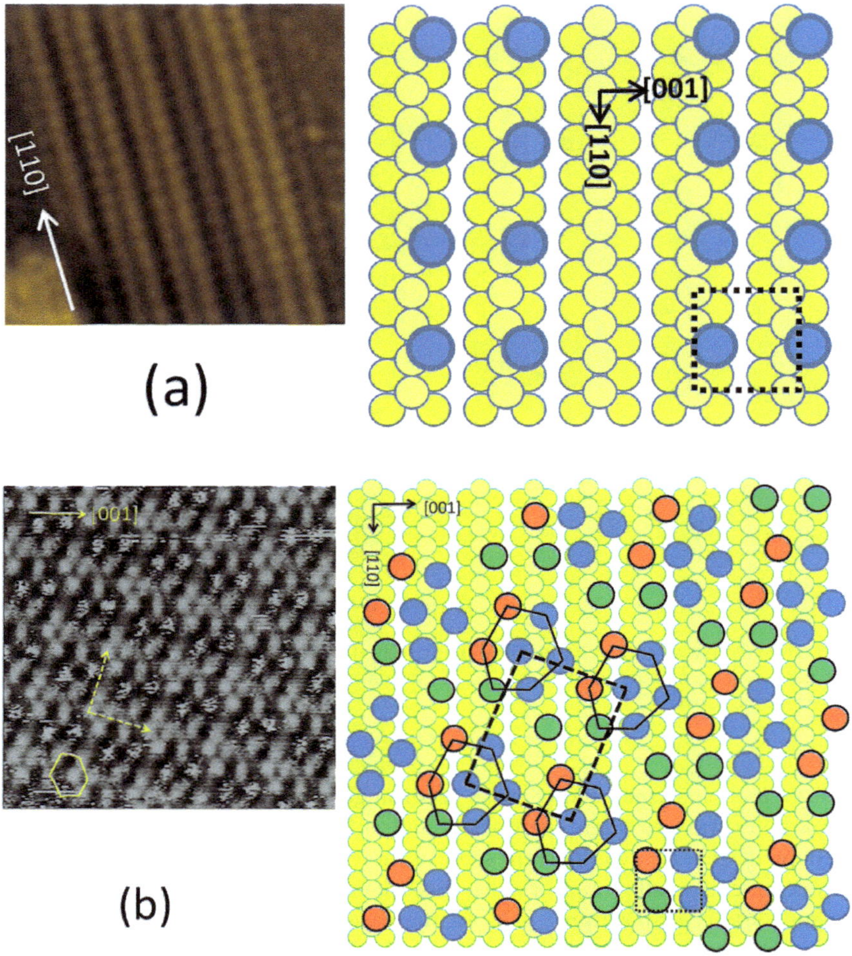

Fig. 4 (a) Adsorption model for Gly–Pro adsorbed as neutral molecules on Au(110) showing a (3 × 6) unit cell. The insert shows the corresponding STM image. (b) Adsorption model for Gly–Pro adsorbed as zwitterionic molecules on Au(110) showing a complex chiral (4 5, −6 1) unit cell composed of heptamers of Gly–Pro molecules arranged as a distorted hexagonal motif. The insert shows the corresponding STM image. On both schematic adsorption models, the dashed line square represents an area of 1 nm^2.

atoms with respect to the gold surface. Molecules of Gly–Pro are organized in hexagonal arrangements, around a central molecule, thus creating groups of heptamers, Fig. 4(b). A particular colour code has been used in this adsorption model to help the reader visualize the unique formation of the heptamer. One should note that all GP molecules are equivalents regardless of the colour that represents them. Nevertheless, a group of 4 molecules painted in blue seems to compose the centre of the heptamer motif, while the red and green molecules compose the surrounding GP forming the hexagonal patches. Finally, it can be noted that in the model depicted in Fig. 4(b), one green GP molecule is not part of any heptamer, but is centred in the middle of 4 hexagonal groups, possibly suggesting a more complex organisation within the GP film. In addition, these heptamers are themselves organized in a chiral unit cell identified as a (4 5, −6 1) unit cell. One may probably say that surface chirality is expressed in two ways in this 2D arrangement, both within a given heptamer group and at the extended level of the heptamer arrays. Again, from XPS data, we were able to estimate the average thickness after 35 minutes of dosing as 0.46 nm, which corresponds to about 2.4 Gly–Pro molecules per nm^2, or around 3 molecules within the square depicted on the model of Fig. 4(b).

Gly–Pro on Ag(110)

PM-RAIRS spectra of the Ag(110) surface recorded with continuous exposure of Gly–Pro are shown in Fig. SI3.† After 10 min of exposure (bottom spectrum), the peaks are not very intense, but well defined. One can isolate two main features at 1383 and 1619 cm^{-1} assigned respectively to the symmetric and asymmetric stretching vibration of the deprotonated carboxylate group νCOO^-; the δNH_2 signature is visible through a weak feature shouldered at 1619 cm^{-1}. Further, other features associated with δCH_2 and $\nu C{=}O$ of the amide bond are observed at 1457 cm^{-1} and 1651 cm^{-1}, respectively. Taking into account the features and the absence of a spectral signature for the free acidic groups (COOH) and protonated amine groups (NH_3^+), we can concluded that after 10 minutes of exposure of Gly–Pro on Ag(110), the molecules are mainly present as anionic species, COO^-/NH_2. Following coverage increase, the PM-RAIRS spectra profiles change as intensity increases together with the appearance of new features at 1693 cm^{-1} and 1550 cm^{-1}. The first is assigned to $\nu C{=}O$ of an acidic protonated group, the second feature could be the amide II band signature ($\nu CN + \delta NH$) which was not previously observed. We can therefore conclude, that within a monolayer of anionic Gly–Pro molecules (COO^-/NH_2), a small fraction of neutral ($COOH/NH_2$) molecules are present on the Ag surface.

High resolution XPS data have also been recorded at various coverages of Gly–Pro adsorbed on Ag(110). In Fig. 5, the O 1s region exhibits a single peak at 531.9 ± 0.1 eV after 10 minutes, attributed to the oxygen atoms of the amide and carboxylate groups (HN-C=**O**/CO**O**$^-$) showing only one chemical form for the acidic groups of Gly–Pro adsorbed at the surface; at higher coverage, after 35 minutes, this same peak is slightly downshifted to 531.7 eV ± 0.1 eV.

Concerning the N 1s region, the analysis of the data is a bit more complicated, since the plasmon loss peak of Ag 3d can be found around 400 eV.[30] As a consequence, the N 1s region already exhibits features on the clean Ag surface at 399 and 401 eV, which are usually assigned to different protonation states of amine groups and amide groups, Fig. SI4.† Therefore, assignments of the different contributions observed for 10 and 35 minutes of exposure are not straightforward. In fact, one cannot simply subtract the contribution of the Ag plasmon loss to obtain the contributions of the nitrogen atoms from Gly–Pro. However, by applying certain constraints we were able to determine that most of the contribution due to adsorbed Gly–Pro appears as a single peak at 400.6 ± 0.1 eV (**N**H_2/OC–**N**H) after 10 minutes of exposure, and a small additional contribution at 399.7 eV for an exposure of 35 minutes. XPS data thus confirm the chemical form of adsorbed Gly–Pro species at the Ag(110) surface: mainly as anionic molecules (NH_2/COO^-) for the monolayer regime. When looking at the C 1s XPS data

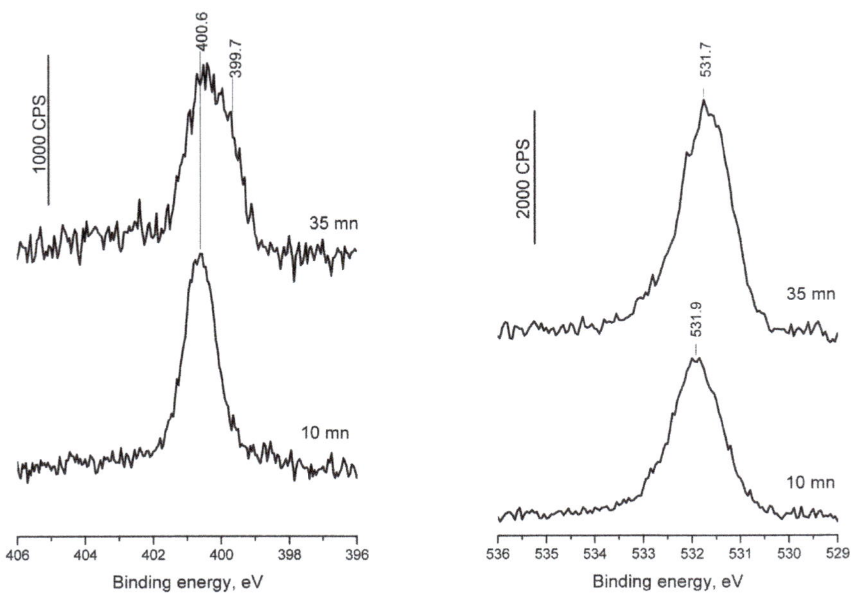

Fig. 5 XPS high resolution spectra for the N 1 s and O 1s regions for a 10 and 35 minute dose of Gly–Pro adsorbed on Ag(110).

following adsorption of Gly–Pro on Ag(110), Fig. SI5,† one should note the absence of a strong contribution at high BE around 290 eV, showing the absence of protonated acidic species. These results confirmed that the Gly–Pro molecules are present mainly as anionic species when adsorbed on Ag(110).

STM experiments were also performed following adsorption of anionic Gly–Pro molecules on Ag(110). Fig. 6 shows STM data obtained after 10 minutes of dose, which distinctly shows an arrangement in line along the axis [110]; these lines are arranged on a zig-zag fashion. When measuring the intermolecular distances on STM profiles, one can draw an adsorption model with zig-zag lines separated by

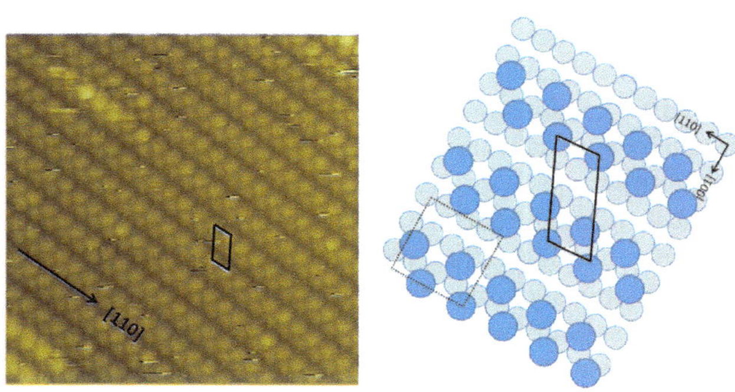

Fig. 6 STM image of anionic Gly–Pro molecules adsorbed on Ag(110) showing a (3 2, 0 2) chiral unit cell. Corresponding adsorption model for Gly–Pro adsorbed as anionic molecules on Ag(110). On the schematic adsorption model, the dashed line square represents an area of 1 nm^2.

an empty atomic row in the [001] crystallographic direction, forming a (3 2, 0 2) chiral unit cell. Finally, from XPS data, we were able to calculate a local molecular density between 1.4 and 2.4 Gly–Pro molecules per nm^2, which correspond to local coverage of 35% and 60% of a saturated monolayer of adsorbed GP molecules.

Discussion

Adsorption of Gly–Pro on Au(110) and Ag(110), characterized by XPS, PM-RAIRS and STM shows clear remarkable differences in the behaviours of these two metals.

On Au(110), after a short time of exposure (10 min), the Gly–Pro molecules are mainly in the NH_2/COOH or NH_2/COO^- chemical form, with a minor amount of NH_3^+/COO^-; RAIRS signals are weak, suggesting that the molecule lies rather flat on the surface, a reasonable configuration for adsorbed molecules is that they bind to the gold surface *via* the peptidic C=O group and one of the oxygen atoms of the carboxylate, in agreement with the low intensity of the IR symmetric COO^-. We correlate this chemical information to the complex bi-phasic domains imaged by STM: rows of dimers, coexisting with heptamers of Gly–Pro molecules. After a longer time of exposure (35 min), the average coverage attains a value of 2.4 Gly–Pro molecules per nm^2, still below one monolayer (which can be estimated at 4 molecules per nm^2 for maximum local coverage); a higher coverage goes with a net increase of the N 1s contribution at high energy, which indicates a higher fraction of molecules in zwitterionic form. The hexagonal structure, now dominating on the surface, is likely formed by zwitterionic molecules, this can be explained by a higher proximity of the molecules to each other, arranged into heptamers, which facilitates proton exchange between COOH and NH_2 groups, thus leading to zwitterionic molecules. As a matter of fact, and this is a key message of this contribution, an increase in the coverage of Gly–Pro on Au(110) changes not only the chemistry but also the organisation of the molecules, with aligned rows of molecules rearranging into a denser patches of chiral hexagonal units. This example tends to show the tendency of gold to favour proton mobility from one molecule to another, provided that the coverage is high enough.

On Ag(110), adsorption of Gly–Pro is very different from what is observed on Au(110); after 10 min of exposure, both XPS and RAIRS data suggest the presence of mainly anionic molecules with a small fraction of neutral molecules; they organize in zig-zag lines, forming chiral cells. Longer exposure (35 min), induces a significant increase in the coverage (1.4 to 2.4 Gly–Pro molecules per nm^2), with neither chemical change, nor modification of the arrangement of the molecule.

At this point, it is of course tempting to recall the data on Cu(110): under similar conditions, Gly–Pro also adsorbs in anionic form, and forms rows of aligned molecules;[27] the small difference lies in the interaction of the surface with the nitrogen atom of the terminal NH_2 group, occurring from the beginning of the adsorption on copper, but only after a longer time of exposure on silver. These findings show that Ag and Cu, both reactive metals, facilitate acid deprotonation followed by the direct interaction of the oxygen atoms of the carboxylate groups; they also favour the interaction of nitrogen with the surface.

The interaction of amino acids in their ionic form with copper is, according to us, a rather general rule.[4,10] It has indeed been shown for glycine, alanine, leucine, lysine, histidine, to cite a few, on Cu(110).[6,7,20,22,31–33] Note the particular example of proline, which has an N atom in the pyrrolidine rings and binds to the Cu(110) surface *via* the oxygen atoms of the deprotonated carboxylate group and also *via* the N atom of the ring, thus confirming the affinity of both oxygen and nitrogen with copper.[34]

Interestingly, methionine, an amino acid bearing a sulfur atom along its side chain, was also adsorbed on both Au(111) and Cu(110), and Cu(111) and Ag(111),[17–19,35,36] thus enabling us to tackle the influence of the nature of the substrate on the adsorption mode (not forgetting the fact that differences in the surface structure may also account for differences in the adsorption modes). As expected, on copper, methionine undergoes a deprotonation of its COOH groups at any coverage, thus adopting an anionic form. However, on Au(111), methionine adopts a zwitterionic form and interacts preferentially *via* one oxygen of the carboxylate but not *via* the nitrogen atom.[17,35] The 2D structures of adsorbed methionine on these surfaces is interesting; for all (111) surfaces, the molecules aligned as lines of dimers along the crystallographic axes, regardless of their chemistry: zwitterionic for Au and Ag, against anionic for Cu. This clearly shows that the structure of the surface, along with the anchoring geometry leads to the same 2D arrays. While the difference in structure between Cu(111) and Cu(110), at constant chemistry (*i.e.* anionic) goes from lines of dimers to random adsorption, respectively.[17,18] Finally, glutamic acid (Glu) is another amino acid whose adsorption process was carefully characterized on well defined Ag surfaces. On Ag(110), Glu adsorbs in its anionic form, with only one out of the two carbonyl groups being deprotonated; XPS data also show the interaction of the N atom of the NH_2 group with the surface.[37,38] Note that, on Ag(111), glutamic acid appears to be preferentially in a zwitterionic form. Once again, one sees the possible role of the surface structure on the chemistry of adsorption.

Conclusions

We would not end this paper without recommending being very cautious in the prediction or comparison of the behaviours of amino acids or small peptides with metals. Chemistry and structural arrangement are likely to very much depend on experimental conditions, pressure and time of exposure, surface defects, and time of relaxation of the system before analysis. Let's cite V. Feyer and his group who performed a large study on the adsorption of amino acids and of small peptides on metals; in particular, they showed that Glycil–Glycil and L-histidine interact with Cu(110) *via* their carboxylate groups (and nitrogen in some cases), they also proposed that on Au(111), though less reactive than copper, histidine molecules also interact with their carboxylate groups.[39] We have no explanation for the behaviour of histidine on Au(111), in terms of acid deprotonation, different from that of Gly–Pro on Au (110) (the surface geometry is different, but could it be the only reason?).

Another example, in disagreement with our finding is that of glycine which, according to Zhao *et al.*, does not chemisorb on clean Ag at 300 K. The authors ascribe this to the inability of this substrate to deprotonate the acid group.[40] These systems are indeed complex!

Though a number of questions are still pending, it is the first time, to our knowledge, that one small peptide was adsorbed on the same planar surface of three different metals. Analyses by XPS and RAIRS data provided us with an undoubtful characterization of the adsorbed species, neutral, anionic or zwitterionic, which depends on the nature of the metal, and on the coverage as far as gold is concerned.

Moreover, STM images enabled us to propose reasonable pictures of Gly–Pro layer structures. On Au(110) and Ag(110), Gly–Pro molecules form lines and create chiral cells; on Au(110), and only on that surface, a longer time of exposure or a longer time on the surface induces a rearrangement, which is likely more stable, into heptamers where molecules are in close proximity to each other. Such a reorganization might be correlated to the reconstruction of the Au(110) surface.

These examples also demonstrate the need to combine complementary surface science techniques to get a reasonable picture of the behaviour of peptides at a metal surface.

Experimental

Materials

Gly–Pro from Bachem, was used as received. For sublimation evaporation, the peptide was deposited in a small glass tube and resistively heated with a W/Th wire wrapped around the crucible. The evaporator was initially separated from the main chamber by a gate valve and differentially pumped using a turbomolecular pump. Before sublimation, the Gly–Pro powder was outgassed at 390 K. It was then heated to 400 K and introduced in the chamber, where the glass tube was placed in front of the gold, silver and copper single crystals. The dosing pressure was maintained around 2×10^{-9} Torr during the deposition of the adlayer.

The gold Au(110) crystal (10 mm diameter, 2 mm thick) and the silver Ag(110) crystal (12 mm diameter, 2 mm thick) were provided by Surface Preparation Laboratory (The Netherlands) with a purity of 99.99% (4 N), and an alignment accuracies of 0.1°.

STM experiments

Scanning Tunneling Microscopy experiments were conducted in an Omicron Vakuumphysik chamber XA VT-STM with facilities for STM, LEED, AES and sample cleaning, and a base pressure of 5×10^{-11} Torr. The single crystals were cleaned by cycles of Ar^+ ion sputtering ($P_{Ar} = 7 \times 10^{-5}$ Torr, 600 V during 15 minutes), flashing and annealing to 850 K for 30 minutes. Before STM experiments, the deposition of Gly–Pro was carried out at room temperature. STM experiments were carried out with the sample at room temperature, images were acquired in constant current mode, and no correction was applied except a plane fit correction.

PM-RAIRS and XPS experiments

The Au(110) and Ag(110) single crystals were mounted in a multi-technique Ultra High Vacuum UHV chamber, base pressure 1×10^{-10} Torr, with PM-RAIRS, LEED, and XPS facilities. The crystals were cleaned by cycles of Ar^+ ion sputtering ($P_{Ar} = 5 \times 10^{-5}$ Torr, 3 kV during 5 minutes), flashing, and annealing to 850 K for 10 minutes. The surface structure and cleanliness were monitored by LEED and XPS before and after adsorption experiments.

Adsorption of Gly–Pro was monitored *in situ* using PM-RAIRS; a Nicolet 5700 spectrometer was interfaced with the UHV chamber, together with a liquid nitrogen-cooled MCT detector. All spectra were recorded *in situ*, after stopping the dosing of Gly–Pro by closing a gate valve between the UHV chamber and Knüdsen cell. All spectra were recorded at a resolution of 8 cm^{-1} by co-addition of 1024 scans (time of acquisition: 10 min). The modulating frequency was set for a maximal sensitivity at 1500 cm^{-1}.

The sample was analyzed using X-ray photoemission spectroscopy with a SPECS GmbH (Berlin, Germany) Phoibos 100-1D delay line detector hemispherical analyzer and a monochromatized AlKα X-ray source (1486.6 eV). After recording a broad range spectrum (pass energy 100 eV), high resolution spectra were recorded for the N 1s, C 1s and O 1s core levels (pass energy 20 eV). High-resolution XPS conditions were fixed: "Fixed Analyser Transmission" analysis mode, and a 7 × 20 mm entrance slit, leading to a resolution of 0.1 eV for the spectrometer, and an electron beam power of 150 W (15 kV and 10 mA). The spectra were fitted using Casa XPS v.2.3.17 Software [Casa Software Ltd., UK] and applying a Gaussian/Lorentzian ratio G/L equal to 70/30.

Notes and references

1 K. H. Ernst, *Phys. Status Solidi B*, 2012, **249**, 2057–2088.
2 J. J. Gray, *Curr. Opin. Struct. Biol.*, 2004, **14**, 110–115.
3 B. Kasemo, *Surf. Sci.*, 2002, **500**, 656–677.
4 S. M. Barlow and R. Raval, *Surf. Sci. Rep.*, 2003, **50**, 201–341.
5 Y. J. Chabal and S. B. Christman, *Phys. Rev. B: Condens. Matter Mater. Phys.*, 1984, **29**, 6974–6976.
6 Q. Chen, D. J. Frankel and N. V. Richardson, *Langmuir*, 2002, **18**, 3219–3225.
7 Q. Chen, D. J. Frankel and N. V. Richardson, *Surf. Sci.*, 2002, **497**, 37–46.
8 Q. Chen and N. V. Richardson, *Nat. Mater.*, 2003, **2**, 324–328.
9 S. F. Chen, L. Y. Li, C. L. Boozer and S. Y. Jiang, *Langmuir*, 2000, **16**, 9287–9293.
10 D. Costa, C.-M. Pradier, F. Tielens and L. Savio, *Surf. Sci. Rep.*, 2015, **70**, 449–553.
11 J. Hasselström, O. Karis, M. Weinelt, N. Wassdahl, A. Nilsson, M. Nyberg, L. G. M. Pettersson, M. G. Samant and J. Stöhr, *Surf. Sci.*, 1998, **407**, 221–236.
12 V. Humblot, C. Méthivier and C.-M. Pradier, *Langmuir*, 2006, **22**, 3089–3096.
13 V. Humblot, C. Méthivier, R. Raval and C.-M. Pradier, *Surf. Sci.*, 2007, **601**, 4189–4194.
14 M. O. Lorenzo, S. Haq, T. Bertrams, P. Murray, R. Raval and C. J. Baddeley, *J. Phys. Chem. B*, 1999, **103**, 10661–10669.
15 E. M. Marti, C. Methivier, P. Dubot and C. M. Pradier, *J. Phys. Chem. B*, 2003, **107**, 10785–10792.
16 E. M. Marti, C. Methivier and C. M. Pradier, *Langmuir*, 2004, **20**, 10223–10230.
17 C. Méthivier, V. Humblot and C.-M. Pradier, *Surf. Sci.*, 2015, **632**, 88–92.
18 A. Schiffrin, J. Reichert, Y. Pennec, W. Auwarter, A. Weber-Bargioni, M. Marschall, M. Dell'Angela, D. Cvetko, G. Bavdek, A. Cossaro, A. Morgante and J. V. Barth, *J. Phys. Chem. C*, 2009, **113**, 12101–12108.

19 A. Schiffrin, A. Riemann, W. Auwarter, Y. Pennec, A. Weber-Bargioni, D. Cvetko, A. Cossaro, M. Alberto and J. V. Barth, *Proc. Natl. Acad. Sci. U. S. A.*, 2007, **104**, 5279–5284.
20 S. M. Barlow, K. J. Kitching, S. Haq and N. V. Richardson, *Surf. Sci.*, 1998, **401**, 322–335.
21 M. Nyberg, M. Odelius, A. Nilsson and L. G. M. Pettersson, *J. Chem. Phys.*, 2003, **119**, 12577–12585.
22 V. Feyer, O. Plekan, T. Skala, V. Chab, V. Matolin and K. C. Prince, *J. Phys. Chem. B*, 2008, **112**, 13655–13660.
23 C. Méthivier, V. Humblot and C.-M. Pradier, *J. Phys. Chem. C*, 2016, **120**(48), 27364–27368.
24 D. Costa, L. Savio and C.-M. Pradier, *J. Phys. Chem. B*, 2016, **120**, 7039–7052.
25 A. Vallée, V. Humblot, C. Méthivier and C.-M. Pradier, *Surf. Sci.*, 2008, **602**, 2256–2263.
26 A. Vallée, V. Humblot, C. Méthivier and C.-M. Pradier, *J. Phys. Chem. C*, 2009, **113**, 9336–9344.
27 C. Méthivier, H. Cruguel, D. Costa, C.-M. Pradier and V. Humblot, *Langmuir*, 2016, **32**, 13759–13763.
28 C. Méthivier, V. Lebec, J. Landoulsi and C.-M. Pradier, *J. Phys. Chem. C*, 2011, **115**, 4041–4046.
29 C. Méthivier, V. Humblot and C.-M. Pradier, *J. Phys. Chem. C*, 2016, **120**, 27364–27368.
30 J. F. Moulder, W. F. Stickle, P. E. Sobol and K. D. Bomben, *Handbook of X-ray Photoelectron Spectroscopy*, Eden Prairie, MN, 1992.
31 S. M. Barlow, S. Haq and R. Raval, *Langmuir*, 2001, **17**, 3292–3300.
32 M. Canepa, L. Lavagnino, L. Pasquali, R. Moroni, F. Bisio, V. De Renzi, S. Terreni and L. Mattera, *J. Phys.: Condens. Matter*, 2009, **21**, 264005.
33 V. Feyer, O. Plekan, N. Tsud, V. Lyamayev, V. Chab, V. Matolin, K. C. Prince and V. Carravetta, *J. Phys. Chem. C*, 2010, **114**, 10922–10931.
34 E. Mateo Marti, S. M. Barlow, S. Haq and R. Raval, *Surf. Sci.*, 2002, **501**, 191–202.
35 V. Humblot, F. Tielens, N. B. Luque, H. Hampartsoumian, C. Méthivier and C.-M. Pradier, *Langmuir*, 2014, **30**, 203–212.
36 A. Naitabdi and V. Humblot, *Appl. Phys. Lett.*, 2010, **97**, 223112.
37 M. Smerieri, L. Vattuone, T. Kravchuk, D. Costa and L. Savio, *Langmuir*, 2011, **27**, 2393–2404.
38 M. Smerieri, L. Vattuone, M. Rocca and L. Savio, *Langmuir*, 2013, **29**, 6867–6875.
39 V. Feyer, O. Plekan, N. Tsud, V. Chab, V. Matolin and K. C. Prince, *Langmuir*, 2010, **26**, 8606–8613.
40 X. Y. Zhao, H. Yan, X. W. Tu, R. G. Zhao and W. S. Yang, *Langmuir*, 2003, **19**, 5542–5545.

Faraday Discussions

PAPER

Pt–dipyridyl tetrazine metal–organic network on the Au(100) surface: insights from first principles calculations†

Duy Le[a] and Talat S. Rahman *[ab]

Received 9th March 2017, Accepted 25th April 2017

DOI: 10.1039/c7fd00097a

Metal–organic coordination networks with active metal centers are a promising class of materials for next-generation catalysts. Motivated by experimental observations of the formation of a Pt–Dipyridyl Tetrazine (DT) metal–organic network on the Au(100) surface [D. Skomski *et al.*, *J. Am. Chem. Soc.*, 2014, **136**, 9862], we carried out density functional theory based calculations on the same system. In this discussion, we demonstrate that the strong interaction between DT ligands and Pt metal centers makes the network stable and that the Pt centers become positively charged by donating their electrons to the DT ligands, resulting in +2 oxidation states for the Pt centers. We further show that the Au substrate withdraws electrons from and hybridizes with the d_{z^2} orbital of the Pt centers, altering their electronic structure and related properties. Furthermore, we find that the Pt centers can absorb SO_2 *via* donor–acceptor interactions, leading to the formation of σ-bonds in which Pt d_{z^2} orbitals act as electron donors, and that the strength of the resultant σ-bond depends on the registry of the Pt centers with the Au(100) surface. Finally, we identify factors, such as the specificity of the ligands and the substrate, and the fullness of the outer shell of the metal centers, that may affect the chemical properties of the metal centers. We suggest modifications (and replacement) of these factors as one of the ways to tune and design metal–organic coordination networks for next-generation catalysts.

Introduction

Traditional heterogeneous catalysts, despite being the favorite in industry, do not possess ideal properties for efficient performance. Their bulky and complex structure often leads to various side reactions, reducing the selectivity of the catalyst. Moreover, only a small fraction of their volume, mostly on the surface, actually participates in reactions as a catalyst. A more efficient design for next-

[a]*Department of Physics, University of Central Florida, Orlando, FL 32816, USA. E-mail: Talat.Rahman@ucf.edu*

[b]*Donostia International Physics Center (DIPC), 20018 San Sebastián, Spain*

† Electronic supplementary information (ESI) available: Survey of structure of Pt–DT/Au(100); adsorption of DT on Au(100); Pt monolayer on Au(100); oxidation number of Pt atoms. See DOI: 10.1039/c7fd00097a

generation heterogeneous catalysts not only requires fewer atoms (especially those of precious metals) but also a high product selectivity. One way to achieve such a catalyst design is *via* well-dispersed atoms on a based structure, *e.g.* a metal or metal oxide surface, as has been proposed: single atom alloys[1] and singly dispersed atom catalysts.[2,3]

Another candidate for such catalyst design is the metal–organic coordination network (MOCN) in which metal coordination atoms are connected and stabilized by organic ligands adsorbed on a relatively flat surface. Compared to single atom alloys and singly-dispersed atom catalysts, MOCNs have a number of advantages. Firstly, similar to self-assembled monolayers (SAMs), the self-correction ability of MOCNs encourages the formation of highly uniform networks, which often results in arrays containing well-defined metal sites. Secondly, the organic ligand may itself be designed to obtain the desired MOCN structure, providing a further method for fine-tuning the catalytic performance of the metal centers. On the other hand, MOCNs often encounter challenges related to their stability at (high) temperatures at which they perform most effectively as realistic catalysts. Finally, the attainment of metal centers that are indeed simultaneously reaction active and product selective, while simple in concept, remains illusive in reality.

With the above points in mind, a good number of MOCNs have already been successfully created.[4–20] Among them, the Pt–Dipyridyl Tetrazine (DT) MOCN synthesized by Skomski *et al.*[20] has attracted our attention since, despite being a simple network, it provides hints for the successful design of a MOCN with realistic catalytic applications, particularly because of the availability of Pt sites for the adsorption of reactants. In this MOCN, Pt atoms and DT ligands form linear chains, in which single-site Pt(II) atoms are stabilized and separated by the ligands on a reconstructed Au(100) surface. In their experiments, they find the chains to be stable at relatively high temperatures (~150 °C).

In this discussion, we report the insights that we have obtained for the electronic structure and stability of the Pt–DT chains on the reconstructed Au(100) surface through the application of density functional theory (DFT) based calculations. We present a detailed analysis of the atomic structure and energetics of the Pt–DT chain which point to their stability at the temperatures used in the experiments. Furthermore, our examination of the electronic and charge states of the system helps confirm that the metal center indeed has the characteristics of Pt^{2+}. In addition, we investigate some possible applications of the Au(100)-supported Pt–DT chain in molecular adsorption and suggest strategies to tune MOCNs for good catalytic performance.

Details of the simulations

We used the Vienna Ab initio Simulation Package (VASP)[21,22] to perform DFT-based calculations employing the projector-augmented wave (PAW)[23,24] and plane wave basis set methods. We utilized the optB88-vdW[25,26] version of the van der Waals density functional (vdW-DF)[27,28] because of its accuracy in the treatment of dispersion bonded systems[26] and reproduction of results in good agreement with the random phase approximation (RPA).[29] We used the PAW pseudopotential, generated using the Perdew–Burke–Ernzerhof functional (PBE),[30] which was provided in the code VASP. This process of using PAW-PBE pseudopotential for vdW-DF-based calculations has already been shown to produce reliable results.[25]

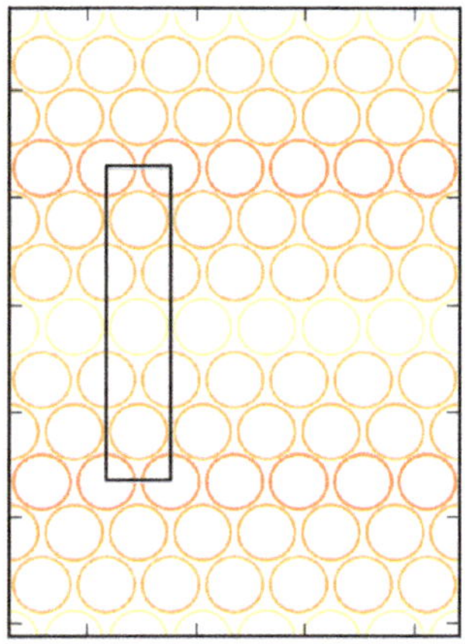

Fig. 1 The topmost layer of the (5 × 1) reconstructed Au(100) surface used in this work. The height of each surface Au atom is color coded from the lowest (lightest) to the highest (darkest), with the maximum height variation being about 0.69 Å. The rectangular inset indicates the surface unit cell, underneath which is an unreconstructed (1 × 5 × 5) Au(100) slab.

The Román-Pérez and Soler[31] algorithm was used to speed up the calculations.[25] We set the energy cutoff for plane-wave expansion at 500 eV. We modelled the reconstructed Au(100) surface with a 5-layer Au(100) slab, with the top layer configured to resemble that of Au(111) to account for its (5 × 1) hexagonal reconstruction. Specifically, to depict the reconstructed Au(100) slab, we first optimized the unreconstructed (1 × 5 × 5) Au(100) slab on top of which we placed a (1 × 6 × 1) Au(100) layer of ~15 Å in a vacuum. The topmost layer of the relaxed structure of the resultant slab, referred herein as reconstructed (1 × 5 × 5) Au(100), is shown in Fig. 1. In the system consisting of the DT ligands and Pt adatoms on the Au(100) surface, our supercell consists of 3 Pt atoms, 3 DT ligands, and the reconstructed (9 × 5 × 5) Au(100) slab. All together, we thus have 360 atoms in the supercell. Given the large number of atoms in the supercell, we sampled the Brillouin Zone only at the Γ point, for computational feasibility. Atomic charges were evaluated using the Bader analysis.[32,33]

Results and discussion

In this section, we first consider the formation of the Pt–DT complex in the absence of the Au(100) support. This is followed by an analysis of the registry of the Pt–DT complex on the Au(100) surface and a subsequent detailed investigation of the electronic and geometric structure of the most preferred Pt–DT/Au(100) configuration, including an evaluation of charge transfer amongst the constituent parts and the effect of the Au substrate on the electronic density of states of the metal centers. We conclude the section with considerations of the binding of a prototype molecule, SO_2, on the metal center to ascertain its propensity for chemical activity.

Binding site on the DT molecules

When considering the binding sites on molecules such as DT, our attention focuses on the N lone pairs, which act as Lewis bases and should be able to transfer their electrons to a Lewis acid, such as a transition metal (TM) ion. Fig. 2a shows the lone pairs of N in DT which are the possible binding sites for the TM metal atoms in our system of interest. Based on the positions of the lone pairs, there are 3 possible motifs for Pt atoms to form a chain with the ligands, as shown in Fig. 2a. The formation energies (the difference in the total energy of the system and its individual components when they are far apart) for the structures in MOTIF #1, MOTIF #2, and MOTIF #3 are −7.38, −6.25, and −4.05 eV, respectively, indicating that MOTIF #1 is strongly favored over the other two.

The Pt–DT chain

We next build the linear Pt–DT chain in which the Pt atoms are coordinated by DT ligands, as suggested by the preferred binding sites (MOTIF #1 in Fig. 2a). The results of structural relaxation, including the in-plane components of the supercell vectors, interestingly show that the Pt–Pt distance is 6.77 Å, which is very close to the experimental measurement of 6.5 ± 0.6 Å when the chain is on the Au(100) surface. The angle between the DT molecular axis and the chain direction is found to be ~58°, once again in excellent agreement with the angle of 55 ± 5° found experimentally when the chain is on the Au(100) surface. Note that our results so far are for the Pt–DT chain in the absence of the Au(100) support. We find the formation energy of the chain to be −7.34 eV, indicating a strong interaction between Pt and DT. In this chain, the Pt atoms form a planar-coordinated structure with the four N atoms *via* Pt–N bonds whose bond lengths are either 2.01 or 2.05 Å. The fact that these bond lengths are quantitatively typical of those found for Pt–N in other compounds/structures[34] is a further indication of strong Pt–N bonds and Pt–DT interaction in the chains considered here.

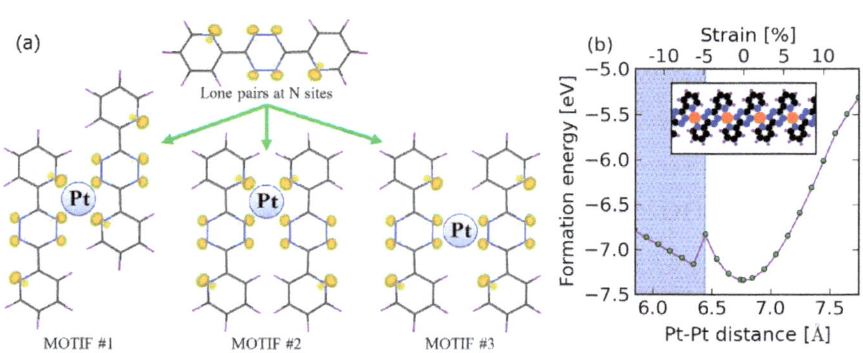

Fig. 2 (a) Wire-frame model of the dipyridyl tetrazine molecule together with the lone pairs at the nitrogen sites (top) and three motifs of the Pt–DT bonding structures. (b) Variation of formation energy of the Pt–DT chain as a function of the Pt–Pt distance in the chain. The region with a light blue background indicates the non-flat chain (see text). The inset is the model of the Pt–DT chain for the lowest formation energy structure. Black, blue, magenta, and red circles represent C, N, H, and Pt atoms, respectively.

As seen in Fig. 2b, the formation energy of the chain is a function of the Pt–Pt distance, d. Another important conclusion from Fig. 2b is that within a strain of 5%, the energy of the Pt–DT chain varies by about 7%, indicating that the chain is flexible within this range. The strength of the Pt–N interaction and the flexibility of the chain suggest that it could remain stable at high temperatures, as shown in experiments by Skomski *et al.*[20] It is worth noting that with a compressive strain of more than 5%, illustrated by the light blue background in Fig. 2b, the chain begins to corrugate and transform into a non-flat structure. Note that determination of the exact transition point at which the PT–DT structure becomes corrugated is not of interest here, thus we do not provide a precise value of the strain that deforms the Pt–DT chain.

Registry of the Pt atoms with the Au(100) surface atoms

With hints for the determination of alignment angle from the experimental scanning-tunneling microscopy (STM) and atomic-force microscopy (AFM) images,[20] we placed the Pt–DT chains (inset in Fig. 2b) on the (5 × 1) hex-reconstructed Au(100) surface so that the chains align at 45° with respect to the reconstruction rows. Note that we slightly adjusted the Pt–Pt distance to 6.95 Å (by ~2.7%) to make the chain commensurate with the surface, with the least strain on the chain. This slight expansion increases the formation energy of the chain by ~100 meV. As mentioned earlier, in our computational model, there are 3 Pt atoms and 3 DT molecules in a supercell. One of the Pt atoms is initially placed either in registry with a surface Au atom, in a 3-fold hollow site on the reconstructed Au(100) surface, or in the Au–Au bridge site at the highest or lowest Au rows, giving rise to 9 such configurations (see Fig. S1 in the ESI†). After ionic relaxation of these structures, we find that the configuration in which a Pt atom is in registry with an Au atom in the highest Au row (Au atoms are color-coded in Fig. 1 and 3, according to their height in the surface layer) is slightly favored, while that in which the Pt atom is in registry with the Au atom in the lowest Au row

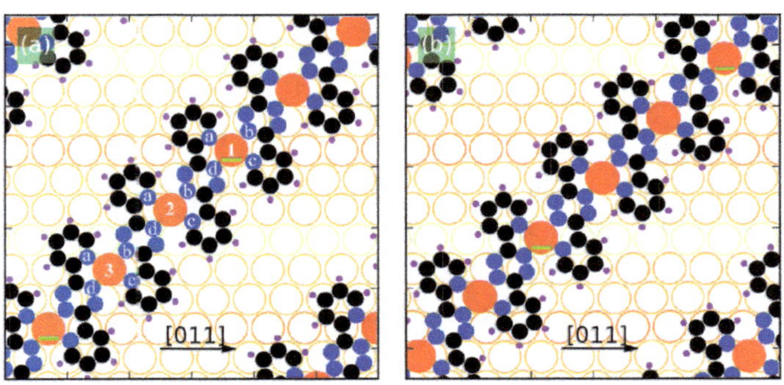

Fig. 3 Configuration of the Pt–DT chain with the lowest (a) and the highest (b) formation energy. Open circles represent the surface Au atoms which are color coded according to their height (the lowest are yellow while the highest are dark orange). Black, blue, magenta, and red solid circles represent C, N, H, and Pt atoms, respectively. The Pt atoms in registry with the highest Au row (a) or lowest Au row (b) are marked with a yellow underline. In (a), we label the three Pt atoms as 1, 2, and 3 and their neighboring N atoms with the letters a, b, c, and d, for discussions later in the text.

is the least favored. Fig. 3 shows the two configurations of the chain at the two different registries with the surface Au atoms. The formation energies per Pt–DT unit for these two configurations in Fig. 3 are −9.76 eV and −9.67 eV, respectively. Because of the small energy difference between the two configurations, it is possible that both are present in the experiments. In fact, the other seven structures presented in the ESI† also have similar formation energies and may all be present in the experiments. However, to simplify our discussion, in the following text we focus on the lowest energy configuration (Fig. 3a).

Note from Fig. 2b that the energy of the chain increases as the Pt–Pt distance changes slightly. However, once the chain is on the surface, this increase is compensated by the formation of Pt–Au bonds which eventually lead to a stable structure. Indeed, the formation energy per Pt–DT unit (−9.76 eV) of the Pt–DT chain on Au(100) is ~0.5 eV less than the sum of the binding energy of an individual DT molecule (−1.90 eV) on Au(100) and the formation energy per Pt–DT unit of the unsupported Pt–DT chain (−7.33 eV). We can thus attribute about −0.5 eV to the Pt–Au interaction, suggesting that the Pt–Au interaction is strong enough to counteract the strain applied on the chain, which would have increased the formation energy of the chain by ~100 meV, as discussed earlier. It should also be noted that the formation energy of the Pt–DT chain on the Au(100) surface is 2.41 eV (per Pt–DT unit) less than that of the segregated DT ligands and Pt monolayer (average binding energy of −5.45 eV per Pt atom) on Au(100) (details in the ESI†), implying that the segregation of Pt atoms and DT molecules is not energetically favoured on Au(100).

Evidence for the strong Pt–Au interaction also comes from Fig. 4a in which we plot the valence electron density of the system. There are signs of formation of covalent bonds between all three sets of Pt and Au atoms, especially between the Pt atom in registry with an Au surface atom (Pt #1 and Au #1, respectively). The charge density difference plot near these atoms, Fig. 4b, further emphasizes the

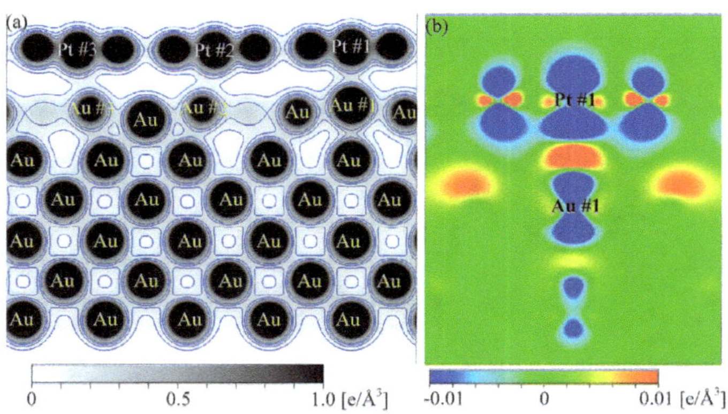

Fig. 4 (a) Valence electron density plotted along the vertical plane passing through Pt atoms. Au and Pt labels indicate the position of the respective ion cores. The labels 1, 2, and 3 are now extended to the relevant Au atoms. The four contours are drawn for 0.1, 0.2, 0.3, and 0.4 e Å^{-3}. (b) Charge density difference plot of the region near the Pt_1 and Au_1 atoms shown in (a). Negative and positive values indicate the depletion and accumulation of electrons.

electron accumulation in the region between Pt and Au atoms, confirming the formation of the Pt–Au σ bond, albeit weakly.

Charge transfer

Bader analysis indicates that on average, a Pt atom loses 0.74 e, a DT molecule gains 0.19 e, and the Au(100) surface receives 0.56 e from one Pt atom and one DT molecule. This charge of the Pt atoms in the chain is substantially larger than the close-to-zero value of −0.06 e of Pt in a Pt monolayer on the reconstructed Au(100) surface. The former value is consistent with the results of the same analysis for Pt^{2+} in PtO (see ESI†), suggesting that the oxidation state of Pt in the Pt–DT chain is +2, while the latter value (−0.06 e) indicates that without DT molecules, Pt has an oxidation state of 0, in excellent agreement with the X-ray photoelectron spectroscopy (XPS) data.[20] Our calculations also indicate a 2.1 eV shift of the 4f core level of Pt in the Pt–DT chain on Au(100) with respect to that in the Pt bulk. This shift once again affirms the +2 oxidation state of the Pt centers.

Fig. 5a shows the differences between the valence electrons of the individual atoms of the DT ligands in the Pt–DT chain structure on the Au(100) surface from that in the isolated DT ligands. It indicates that the tetrazine rings (dominantly red) gain electrons while the pyridine rings (slightly more blue) lose electrons. On average, each pyridine ring loses ∼1/10 e while the tetrazine ring gains ∼4/10 e. Fig. 5b shows the difference between valence electrons of the individual atoms of the Pt–DT chain when supported on Au(100) and when the Pt–DT chain is unsupported. Most atoms of the Pt–DT chain now appear blue, indicating that the Au(100) substrate withdraws electrons from the chain. In particular, each pyridine, tetrazine ring, and Pt atom loses approximately ∼1/5, ∼1/10, and ∼1/20 e, respectively, to the Au surface.

Effect of the Au substrate on the electronic density of states of Pt centers

Since the Pt–N interactions are strong compared to the Pt–Au interactions, one would expect the Pt–DT interactions to dictate the properties of the Pt centers. Detailed analysis of the partial density of states (PDOS) of the N atoms shows that

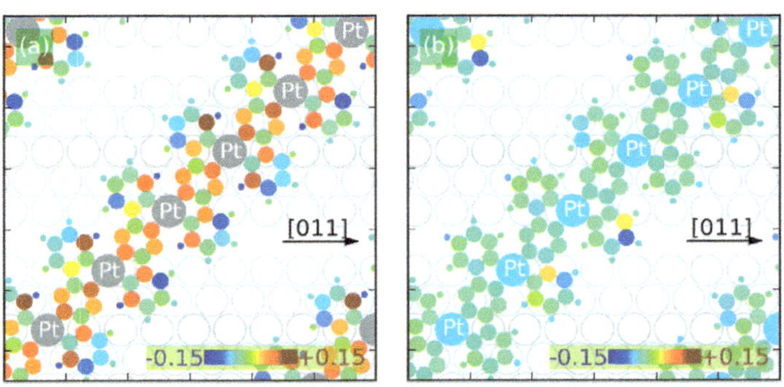

Fig. 5 Electrons gained (red) and lost (blue) by individual DT ligand atoms in the Pt–DT chain on Au(100) as compared to those (a) in the isolated ligand and (b) unsupported by Au(100). Electrons gained or lost by Pt atoms (filled gray circles) in (a) and by Au atoms (unfilled gray circles) in (a) and (b) are not displayed.

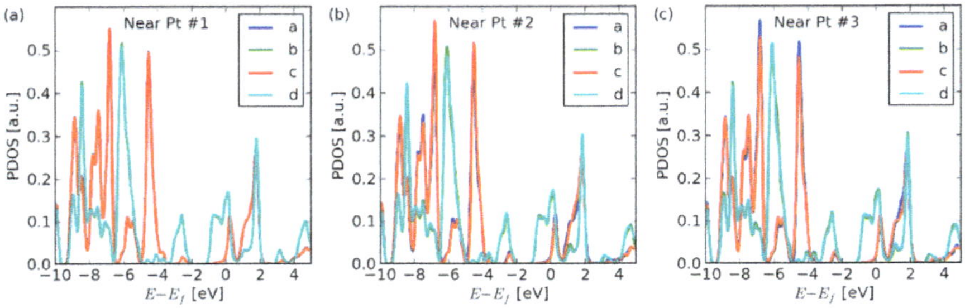

Fig. 6 PDOS (in arbitrary units) of the N atoms surrounding the Pt atoms. N atoms are denoted by the letters a, b, c, and d, and Pt atoms are denoted by the numbers 1, 2, and 3, as shown in Fig. 3a.

their electronic properties are almost identical (Fig. 6) despite being linked to Pt atoms that are in different registry with the Au(100) surface, suggesting that Pt–DT interactions are not strongly dependent on the registry of Pt atoms with the Au(100) surface. The above result is understandable since the Pt–DT interactions are mainly controlled by in-plane components of the Pt and N orbitals, which should not depend strongly on the registry of the Pt atoms with the Au(100) surface (Fig. 7) and should not have an impact on the reactivity of these Pt atoms. But, as we shall see, the Pt–Au interactions play a more important role in determining the catalytic performance of the Pt centers.

As shown in Fig. 4b, the charge depletion region around Pt and Au atoms adapts the shape of the d_{z^2} orbital, suggesting that a σ-type bond is formed between the Pt and Au atoms. Fig. 8a and b display the hybridization, as indicated

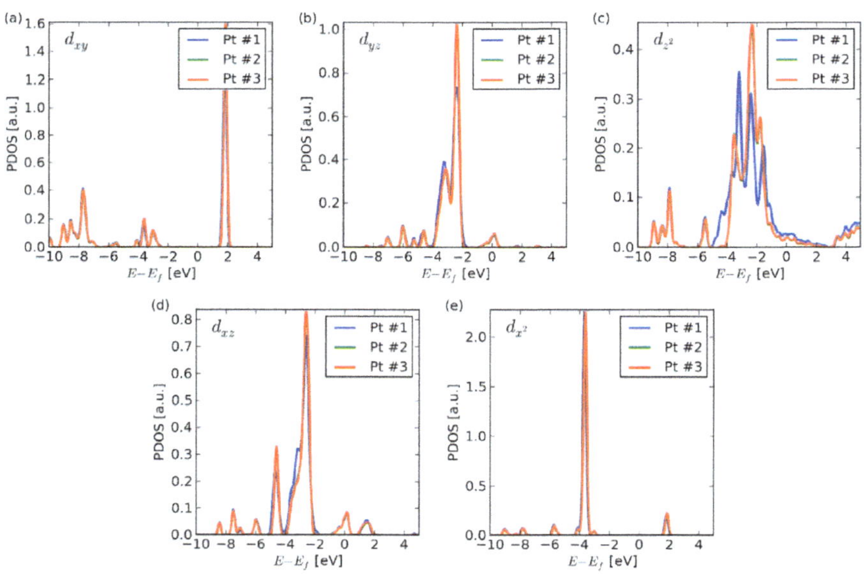

Fig. 7 Five components of the d states (in arbitrary units) of the Pt atoms. Pt atoms are denoted with the numbers 1, 2, and 3, as shown in Fig. 3a.

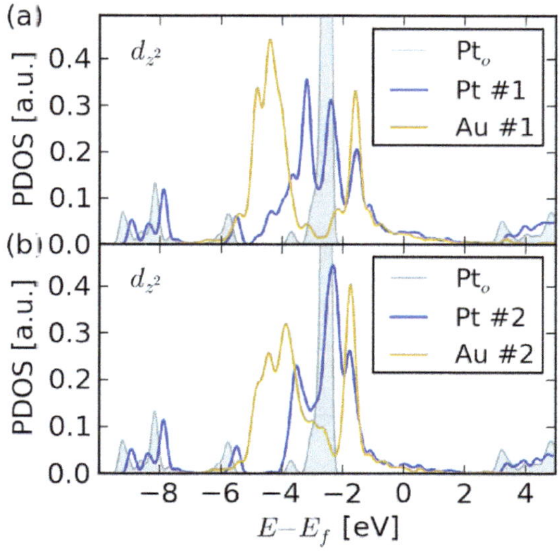

Fig. 8 d_{z^2} component of the density of states of the Pt and Au atoms. Pt and Au are labelled as in Fig. 3a and 4a. Pt_o indicates Pt atoms in the Pt–DT chain without Au(100) support.

by the overlap of states at the same energy, between the d_{z^2} states of the Pt and Au atoms, which now depends on the registry of the Pt and Au atoms. For instance, more hybridization is seen when Pt is in registry with the Au atom (*i.e.* Pt #1 and Au #1 in Fig. 4a) in Fig. 8a than is the case for Pt #2 and Au #2 in Fig. 8b. Note also that the hybridization between d_{z^2} of the Pt and Au atoms and the electron depletion around the d_{z^2} orbital of the Pt atoms shifts the Pt d_{z^2} states closer to the Fermi level, when supported on Au(100), as compared to those of unsupported Pt (Pt_o in Fig. 8a and b). Such a shift towards the Fermi level is expected to have an impact on the reactivity of the Pt centers.

Adsorption of SO_2

The occupied Pt d_{z^2} orbital discussed above is an ideal candidate for the donor–acceptor interaction in which Pt d_{z^2} acts as a donor. To test the ability of the Pt atoms in the chain structures considered above to facilitate such an interaction, we choose SO_2 as a probe molecule, as there is prior evidence that SO_2 has an affinity for Pt in planar coordination structures.[18,35] Our DFT calculations show that SO_2 indeed binds to Pt #1, #2, and #3 centers with binding energies of −0.63, −0.59, and −0.59 eV, respectively. Here, binding energy is calculated as the difference between the total energy of the system in which SO_2 adsorbs on Pt–DT/Au(100) and the sum of the total energy of SO_2 and Pt–DT/Au(100) when they are far apart. The small difference in the binding energies of SO_2 at the three different Pt centers on Au(100) shows the effect of the substrate, albeit rather small in this case. The charge density difference plot shown in Fig. 9 clearly indicates the donation of electrons from the d_{z^2} orbital of the Pt center to the $1\pi^*$ orbital[35] of SO_2 to form the Pt–S σ-bond, making the Pt–SO_2 unit non-planar, rather pyramidal, similar to the case of Ir–SO_2.[36,37] The Pt–S bond length is found to be 2.79 Å,

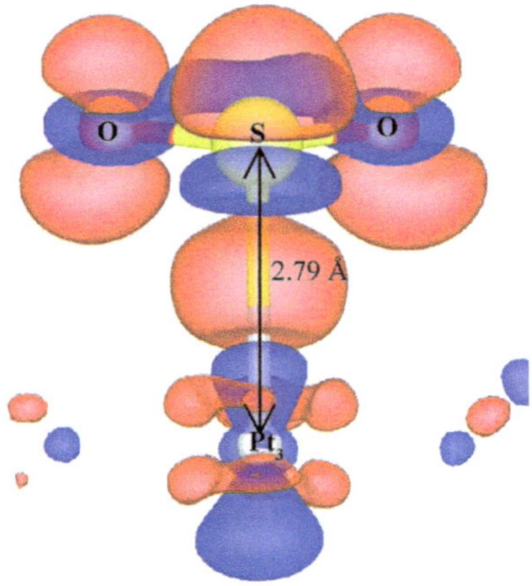

Fig. 9 Charge density difference plot showing σ-bond formation between Pt and S atoms. Red and blue iso-surfaces indicate electron depletion and accumulation regions, respectively. The iso-surface value is 0.01 e Å^{-3}.

which is much larger than that (2.48–2.61 Å) found in L(CH_3)Pt·(SO_2),[35] indicating that a relatively weak Pt–S bond forms between SO_2 and the Pt–DT chain. Recall that in the Pt–DT chain, Pt has an oxidation state of +2, which makes it positively charged and less prone to donating electrons. We also calculated the binding energy of SO_2 on the Pt–DT chain without the presence of the Au(100) substrate and found that it is −0.72 eV, indicating that the Au(100) substrate weakens the strength of the Pt–S bond. This is easy to understand as the Au substrate withdraws electrons from the Pt d_{z^2} orbital (Fig. 4b), reducing their availability for donation to SO_2.

Discussion

Despite being shown to be stable at high temperatures,[20] owing to the formation of strong Pt–N bonds, the Pt centers in Pt–DT chains do not display the characteristics of active metal centers, as reflected by their relatively weak affinity towards molecules such as SO_2. One way to understand this is that the only available orbital for binding with the adsorbates, d_{z^2}, is already occupied, making it only available for the formation of a donor–acceptor σ-bond in which Pt acts as an electron donor. It is thus not surprising that in our testing of the strength of the interaction between the Pt–DT chain and a set of other molecules including CO, CO_2, H_2, H_2O, and NH_3, we found that none of them bind chemically to the Pt center. On the other hand, this result may imply that the Pt–DT chain and MOCNs, in general, may be good candidates for highly selective catalysts because in the MOCN environment, only a very limited number of orbitals of metal centers are available for interaction with adsorbates, reducing the possibility of side reactions.

The ability to tune the electronic properties of metal centers of MOCNs on a substrate makes MOCNs a promising material for next-generation catalysts. Because the ligands control the charge states of the metal centers, the tuning of properties may be carried out *via* ligand modification or replacement, as long as they interact strongly with the metal centers to make the resultant MOCN stable at a high temperature. It may also be possible to tune the activity of the metal center by varying the substrates, as the coupling with the substrate atom will play a role. We find a small indication of the effect of the substrate in the system considered here, but not a very strong one. One aspect that we have not considered in this work is the choice of a metal center other than Pt. It is very likely that the level of completeness of the outer shells of the metal atom would impact the stabilization of the metal coordination center with the desired occupancy of their active orbitals, *e.g.* d_{z^2}, which is considered to be the most important one[38] in catalytic reactions. In fact, in our ongoing work in which we replace the Pt metal center in the MOCN considered here with a number of other transition metals, we are finding that V, Fe, Cr, and Co metal centers provide better opportunities for catalytic reactions.[39] These results are, however, preliminary and hence not included here.

Conclusions

From detailed analysis based on DFT simulations of the observed Pt–DT MOCN on the Au(100) surface,[20] we were able to demonstrate that the strong interaction between the DT ligands and the Pt metal centers makes the network stable at temperatures at which the experiments were carried out ($\sim$150 °C). Furthermore, we found that the Pt centers donate their electrons to the DT ligands making them positively charged, resulting in a +2 oxidation state for the Pt centers. Interestingly, the Au substrate withdraws electrons from and hybridizes with the d_{z^2} orbital of the Pt centers, altering the electronic structure of the latter. In fact, this Au–Pt hybridization makes the Pt centers not very reactive towards a large number of small molecules that we considered. As a case in point we showed that Pt centers can absorb SO_2, but weakly, *via* donor–acceptor interactions to form σ-bonds in which Pt d_{z^2} orbitals act as electron donors, and that the strength of the resultant σ-bond depends on the registry of Pt centers with the Au(100) surface. We also identify possible factors that affect the chemistry of the metal centers in MOCNs. These include the nature of the ligands, the substrate, and the completeness of the outer shell of the metal centers. Modification and replacement of these factors are ways to tune and design MOCNs for next generation catalysts. We look forward to further experimental and theoretical investigations of systems, such as the ones considered here, to enable the derivation of design rules of MOCNs for catalytic applications.

Acknowledgements

The authors thank Steven Tait for fruitful discussions. This work is supported in part by the National Science Foundation Grant CHE-1310327. Calculations were performed on the Extreme Science and Engineering Discovery Environment (XSEDE), which is supported by the National Science Foundation Grant ACI-1053575.

Notes and references

1 G. Kyriakou, M. B. Boucher, A. D. Jewell, E. A. Lewis, T. J. Lawton, A. E. Baber, H. L. Tierney, M. Flytzani-Stephanopoulos and E. C. H. Sykes, *Science*, 2012, **335**, 1209–1212.

2 S. R. Zhang, L. Nguyen, J. X. Liang, J. J. Shan, J. Y. Liu, A. I. Frenkel, A. Patlolla, W. X. Huang, J. Li and F. Tao, *Nat. Commun.*, 2015, **6**, 7938.

3 E. J. Peterson, A. T. Delariva, S. Lin, R. S. Johnson, H. Guo, J. T. Miller, J. H. Kwak, C. H. F. Peden, B. Kiefer, L. F. Allard, F. H. Ribeiro and A. K. Datye, *Nat. Commun.*, 2014, **5**, 4885.

4 X. Q. Shi, C. S. Lin, C. Minot, T. C. Tseng, S. L. Tait, N. Lin, R. Q. Zhang, K. Kern, J. I. Cerda and M. A. Van Hove, *J. Phys. Chem. C*, 2010, **114**, 17197–17204.

5 S. L. Tait, Y. Wang, G. Costantini, N. Lin, A. Baraldi, F. Esch, L. Petaccia, S. Lizzit and K. Kern, *J. Am. Chem. Soc.*, 2008, **130**, 2108–2113.

6 H. H. Yang, Y. H. Chu, C. I. Lu, T. H. Yang, K. J. Yang, C. C. Kaun, G. Hoffmann and M. T. Lin, *ACS Nano*, 2013, **7**, 2814–2819.

7 S. Fabris, S. Stepanow, N. Lin, P. Gambardella, A. Dmitriev, J. Honolka, S. Baroni and K. Kern, *Nano Lett.*, 2011, **11**, 5414–5420.

8 A. Shchyrba, M. T. Nguyen, C. Wackerlin, S. Martens, S. Nowakowska, T. Ivas, J. Roose, T. Nijs, S. Boz, M. Schar, M. Stohr, C. A. Pignedoli, C. Thilgen, F. Diederich, D. Passerone and T. A. Jung, *J. Am. Chem. Soc.*, 2013, **135**, 15270–15273.

9 T. Lin, X. S. Shang, P. N. Liu and N. A. Lin, *J. Phys. Chem. C*, 2013, **117**, 23027–23033.

10 T. Classen, G. Fratesi, G. Costantini, S. Fabris, F. L. Stadler, C. Kim, S. de Gironcoli, S. Baroni and K. Kern, *Angew. Chem., Int. Ed.*, 2005, **44**, 6142–6145.

11 S. L. Tait, A. Langner, N. Lin, S. Stepanow, C. Rajadurai, M. Ruben and K. Kern, *J. Phys. Chem. C*, 2007, **111**, 10982–10987.

12 A. Langner, S. L. Tait, N. Lin, C. Rajadurai, M. Ruben and K. Kern, *Proc. Natl. Acad. Sci. U. S. A.*, 2007, **104**, 17927–17930.

13 Q. H. Yuan, L. J. Wan, H. Jude and P. J. Stang, *J. Am. Chem. Soc.*, 2005, **127**, 16279–16286.

14 S. Stepanow, N. Lin and J. V. Barth, *J. Phys.: Condens. Matter*, 2008, **20**, 184002.

15 J. V. Barth, *Surf. Sci.*, 2009, **603**, 1533–1541.

16 C. S. Kley, J. Cechal, T. Kumagai, F. Schramm, M. Ruben, S. Stepanow and K. Kern, *J. Am. Chem. Soc.*, 2012, **134**, 6072–6075.

17 S. Vijayaraghavan, D. Ecija, W. Auwarter, S. Joshi, K. Seufert, M. Drach, D. Nieckarz, P. Szabelski, C. Aurisicchio, D. Bonifazi and J. V. Barth, *Chem.–Eur. J.*, 2013, **19**, 14143–14150.

18 M. Albrecht, R. A. Gossage, A. L. Spek and G. van Koten, *Chem. Commun.*, 1998, 1003–1004, DOI: 10.1039/a708630j.

19 D. Skomski, C. D. Tempas, B. J. Cook, A. V. Polezhaev, K. A. Smith, K. G. Caulton and S. L. Tait, *J. Am. Chem. Soc.*, 2015, **137**, 7898–7902.

20 D. Skomski, C. D. Tempas, K. A. Smith and S. L. Tait, *J. Am. Chem. Soc.*, 2014, **136**, 9862–9865.

21 G. Kresse and J. Furthmuller, *Comput. Mater. Sci.*, 1996, **6**, 15–50.

22 G. Kresse and J. Hafner, *Phys. Rev. B: Condens. Matter Mater. Phys.*, 1993, **47**, 558–561.

23 G. Kresse and D. Joubert, *Phys. Rev. B: Condens. Matter Mater. Phys.*, 1999, **59**, 1758–1775.
24 P. E. Blöchl, *Phys. Rev. B: Condens. Matter Mater. Phys.*, 1994, **50**, 17953–17979.
25 J. Klimeš, D. R. Bowler and A. Michaelides, *Phys. Rev. B: Condens. Matter Mater. Phys.*, 2011, **83**, 195131.
26 J. Klimeš, D. R. Bowler and A. Michaelides, *J. Phys.: Condens. Matter*, 2010, **22**, 022201.
27 T. Thonhauser, V. R. Cooper, S. Li, A. Puzder, P. Hyldgaard and D. C. Langreth, *Phys. Rev. B: Condens. Matter Mater. Phys.*, 2007, **76**, 125112.
28 M. Dion, H. Rydberg, E. Schröder, D. C. Langreth and B. I. Lundqvist, *Phys. Rev. Lett.*, 2004, **92**, 246401.
29 F. Mittendorfer, A. Garhofer, J. Redinger, J. Klimes, J. Harl and G. Kresse, *Phys. Rev. B: Condens. Matter Mater. Phys.*, 2011, **84**, 201401.
30 J. P. Perdew, K. Burke and M. Ernzerhof, *Phys. Rev. Lett.*, 1996, **77**, 3865–3868.
31 G. Román-Pérez and J. M. Soler, *Phys. Rev. Lett.*, 2009, **103**, 096102.
32 R. F. W. Bader, *Atoms in Molecules: A Quantum Theory*, Oxford University Press, New York, 1994.
33 W. Tang, E. Sanville and G. Henkelman, *J. Phys.: Condens. Matter*, 2009, **21**, 084204.
34 F. Basolo and R. G. Pearson, in *Prog. Inorg. Chem.*, John Wiley & Sons, Inc., 1962, pp. 381–453, DOI: 10.1002/9780470166055.ch6.
35 J. B. Li and A. Y. Rogachev, *Phys. Chem. Chem. Phys.*, 2015, **17**, 1987–2000.
36 G. J. Kubas, *Inorg. Chem.*, 1979, **18**, 182–188.
37 S. J. Laplaca and J. A. Ibers, *Inorg. Chem.*, 1966, **5**, 405.
38 S. Stolbov and S. Zuluaga, *J. Phys. Chem. Lett.*, 2013, **4**, 1537–1540.
39 N. Ud din, D. Le and T. S. Rahman, to be published.

Faraday Discussions

PAPER

Probing the interplay between geometric and electronic structure in a two-dimensional K–TCNQ charge transfer network†

P. J. Blowey,[ab] L. A. Rochford,‡[c] D. A. Duncan,[b] D. A. Warr,[c] T.-L. Lee,[b] D. P. Woodruff*[a] and G. Costantini*[c]

Received 6th March 2017, Accepted 31st March 2017
DOI: 10.1039/c7fd00093f

Scanning tunnelling microscopy (STM), low energy electron diffraction (LEED), ultraviolet and soft X-ray photoelectron spectroscopy (UPS and SXPS) have been used to characterise the formation of a coadsorption phase of TCNQ and K on Ag(111), while the normal incident X-ray standing waves (NIXSW) technique has been used to obtain quantitative structural information. STM and LEED show that an ordered incommensurate phase is formed in which the K atoms are surrounded by four TCNQ molecules in a 'windmill' motif, characteristic of other metal/TCNQ phases, in which the nominal TCNQ : K stoichiometry is 1 : 1. UPS and SXPS data indicate the TCNQ is in a negatively-charged state. NIXSW results show that the carbon core of the TCNQ is essentially planar at a height above the Ag(111) surface closely similar to that found without coadsorbed K. In the presence of TCNQ the height of the K ions above the surface is significantly larger than on clean Ag(111), and the ions occupy sites above 'holes' in the TCNQ network. NIXSW data also show that the N atoms in the molecules must occupy sites with at least two different heights above the surface, which can be reconciled by a tilt or twist of the TCNQ molecules, broadly similar to the geometry that occurs in bulk TCNQ/K crystals.

Introduction

The use of thin molecular films in organic electronic devices (OEDs) has attracted a significant amount of interest in recent years, offering a low-cost, transparent, flexible and light-weight alternative to typical inorganic semiconductor devices.[1,2]

[a]Department of Physics, University of Warwick, Coventry, CV4 7AL, UK. E-mail: G.Costantini@warwick.ac.uk
[b]Diamond Light Source, Harwell Science and Innovation Campus, Didcot, OX11 0QX, UK
[c]Department of Chemistry, University of Warwick, Coventry, CV4 7AL, UK. E-mail: D.P.Woodruff@warwick.ac.uk

† Electronic supplementary information (ESI) available. See DOI: 10.1039/c7fd00093f

‡ Present address: School of Chemistry, University of Birmingham, Birmingham, B15 2TT, UK.

Although these devices have already seen some commercial success, further optimisation is required to meet the demands of current electronic applications. The performance of OEDs is highly dependent on the interfacial energy-level alignment, particularly between the metallic electrodes and the active organic material, as this defines the height of the barrier for the transport of charge carriers.[3,4] Consequently, the overall efficiency of a device can be improved by optimising the energy-level alignment across this metal–organic interface. One common approach to achieve this is to add interfacial layers designed to tune the local electrostatic dipole and thus the relative energy of the frontier molecular orbitals with respect to the electrode Fermi level. In the case of organic photovoltaic devices, ultra-thin films of materials with high electron affinity can be used as a hole-extraction layer to raise the work function of the metal electrode and reduce the barrier for charge transport across the metal–organic interface.[5–8] This hole-extraction layer should be as thin as possible, as a thicker layer may itself hinder charge transfer across the interface, and ideally would also allow the work function shift to be tuned to allow the hole-extraction layer to be optimised for the particular electrode and active material in the device.[5] Recently, charge-transfer networks, consisting of small molecular electron acceptors and donors, have also been identified as promising candidates for use as hole-extraction layers.[9–12] These charge-transfer networks offer the advantage of tuneable electronic properties through synthetic chemical modification of the electron donors and acceptors.[11,12] Furthermore, such charge-transfer networks form ordered layers through supramolecular self-assembly, which is an effective method for fabricating nanoscale structures.[13] It has also been suggested that through this approach, tuneable hole-extraction layers that have a thickness of only a single molecular layer (thus minimising the hindrance on charge carrier transport across the interface) could be a realistic possibility.[12] One example of this is the two-dimensional (2D) networks that are formed between the strong electron acceptor molecule tetracyanoquinodimethane (TCNQ) and alkali metals.[11,12]

Scanning tunnelling microscopy (STM) studies show that TCNQ molecules co-adsorbed with alkali metals assemble into stable and well-defined metal–organic structures. Structures observed for Cs–TCNQ networks on Ag(100)[11] and for Li–TCNQ and Na–TCNQ networks on Au(111)[14,15] are all characterised by a common 'windmill' motif in which a central alkali metal ion is bound to four TCNQ molecules. Density functional theory (DFT) calculations performed for the Cs–TCNQ/Ag(100) system indicate that the alkali metal ions play a dual role, both directing the self-assembly of the molecules through ionic bonding and having a significant impact on the electronic properties of the surface.[11,12] The high electron affinity of TCNQ molecules means that they become charged on metallic substrates, and the combination of the charged molecules and their image charges in the metallic substrate creates relatively strong dipole moments perpendicular to the surface. The mutual repulsion of these dipoles might be expected to suppress self-assembly, but the addition of metallic adatoms[16–18] – in particular alkali adatoms[11,14,15] – produces strong dipoles of opposite polarity, stabilising the metal–organic networks, and also reducing the work function increase caused by the negatively charged TCNQ molecules in isolation. One surprising feature of the network structure predicted by DFT calculations is that despite the alkali ions occupying 'pores' or 'holes' in the molecular assembly that are large enough to comfortably accommodate them in a coplanar bonding arrangement, they occupy sites above the molecular plane.[12] This effect has been rationalised as being due to the larger positive dipole moment that arises from the increased height of the alkali metal ion above the surface, which stabilises the negative molecular dipoles more effectively.[12] The DFT calculations therefore predict a rather interesting result indicating a close relationship between the height of the alkali metal ion above the substrate, and the electronic properties of the surface.

So far there have been no experimental quantitative structural studies of this phenomenon, although our recent structural investigation of TCNQ on Ag(111) has shown clear evidence for the stabilising effect of Ag adatoms in the molecular self-assembly.[19] Here, we report on a significant extension of this work to investigate the structure formed by coadsorption of TCNQ and potassium (K) on Ag(111), using a combination of techniques to probe the interplay between the adsorption structure and the electronic properties of the surface. STM and low energy electron diffraction (LEED) were used to determine the molecular ordering, to obtain precise surface unit mesh dimensions and to ascertain whether the molecular overlayer is commensurate with respect to the underlying substrate. Soft X-ray and ultraviolet photoelectron spectroscopy (SXPS, UPS) provided complementary information on the composition and chemical and electronic properties of the surface. Most significantly, we have also used the normal incident X-ray standing wave technique (NIXSW), to obtain quantitative structural information, and specifically to determine the height of the constituent atoms within the alkali-molecule network above the underlying surface.

Methods

All characterisation experiments in this work were conducted at room-temperature under ultra-high vacuum (UHV) conditions. STM measurements were performed in an UHV system at the University of Warwick and NIXSW, SXPS and UPS experiments were performed in the UHV end-station installed on beamline I09 of the Diamond Light Source storage ring. This beamline is equipped with both a crystal monochromator to provide the 'hard' X-rays needed to perform the NIXSW experiments and a grazing incidence plane grating monochromator providing soft X-rays to perform high-resolution SXPS. LEED patterns were recorded both at Warwick and at Diamond, providing a cross-reference for the preparation of comparable samples in both UHV chambers. Low current LEED optics were used in both UHV systems (Warwick: Omicron SPECTALEED, I09: OCI BDL800IR-MCP LEED) to avoid beam damage to the molecular overlayer. LEED pattern simulations were obtained using LEEDpat software.[20] A single crystal Ag(111) substrate (cutting precision of 0.1°) was prepared *in situ* using cycles of sputtering with 1 keV Ar^+ ions for 30 minutes followed by annealing to ~500 °C for another 30 minutes. The sample temperature was measured indirectly by a thermocouple in contact with the sample plate onto which the crystal was mounted. STM images were recorded at ambient temperature in constant current mode using electrochemically etched polycrystalline tungsten tips. All STM images were plane corrected and flattened using the open source image-processing software Gwyddion.[21]

The NIXSW technique[22,23] exploits the X-ray standing wave formed by the interference of incident and scattered X-rays at a Bragg condition. The strong backscattering out of the crystal leads to a finite extinction length of the X-rays and consequently to the strong reflectivity existing in a finite range of incident X-ray energies. Within this range the standing wave, which has a periodicity in intensity equal to that of the scattering planes, shifts in phase relative to the scattering planes in a systematic fashion. Monitoring the X-ray absorption at atoms immersed in the standing wave (within or above the crystal) through this photon energy range thus provides a profile characteristic of the height of the absorbing atoms above the scattering planes. In the present case, X-ray absorption at the C and N atoms of TCNQ, as well as the co-adsorbed K atoms, was monitored, whilst sweeping the photon energy through the (111) Bragg condition at near-normal incidence to the surface, by measuring the intensity of the C 1s, N 1s and K 2p photoelectron spectra collected using a VG Scienta EW4000 HAXPES hemispherical electron analyser (angular acceptance range $\pm$ 30°) mounted at 90° to the incident photon beam. The NIXSW absorption profiles were corrected for non-dipolar effects in the angular distribution of the photoemission, values for the backward-forward asymmetry parameter Q (ref. 24) were obtained from theory-based computations of the angular distribution parameters.[25] UP spectra were recorded using a monochromated He-I$_\alpha$ source ($h\nu = 21.23$ eV), work functions were determined using a sample bias of -8.00 V in order to measure the secondary electron cut-off.

TCNQ powder (98% pure; Sigma Aldrich) was triply purified by thermal gradient sublimation and thoroughly degassed before being deposited onto the Ag(111) substrate from an organic molecular beam deposition source at an operating temperature of 400 K. K was deposited *via* resistive heating of an SAES alkali metal dispenser using a current of 6.3 A. The K–TCNQ networks were prepared *in situ* by first depositing a nominally saturated single layer (defined here as ~1 monolayer (ML)) of TCNQ onto Ag(111) and post annealing the substrate to 550 K to yield a sub-monolayer adsorption phase.[19] K was then deposited onto this layer before annealing the sample to 470 K.

Results

STM, LEED and photoemission characterisation

STM of TCNQ co-deposited with K on Ag(111) shows the formation of large ordered domains, with bright, oblong protrusions identified as TCNQ molecules (Fig. 1a). The molecules assemble into windmill-like structures, consisting of four TCNQ molecules spiralling around a central point. This 'windmill' motif is common for TCNQ on surfaces and is usually brought about by the presence of metal adatoms, either from the underlying substrate[19] or co-deposited along with the molecules.[11,14,15,17,18] This packing also bears a strong resemblance to the bulk crystal structure of K–TCNQ in which layers of TCNQ molecules arrange in similar four-fold coordinated windmill structures in layers on either side of a plane of K^+ ions (see ESI, Fig. S1†).[26] This suggests that on Ag(111) the TCNQ molecules form two-dimensional networks with the codeposited K atoms *via* ionic interactions with their cyano groups. This possibility is supported by STM images taken at larger negative voltage biases which show bright circular protrusions at the centre of the windmills; these we interpret as the locations of the K atoms (Fig. 1b). This interpretation is consistent with that of previous STM studies of Na–TCNQ networks on Au(111)[15] and Cs–TCNQ networks on Ag(100),[11] which show similar

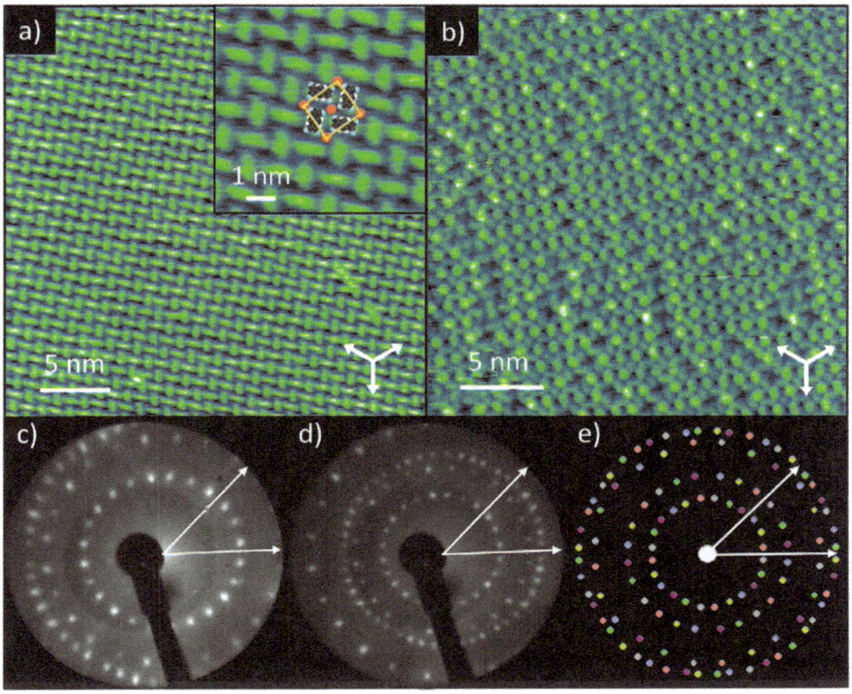

Fig. 1 TCNQ co-deposited with K on Ag(111). (a) Representative STM image showing an ordered domain of TCNQ molecules arranged into windmill-like structures ($V_{samp} = 0.1$ V, $I = 175$ pA, white arrows indicate the substrate ⟨110⟩ directions). Inset: expanded region of the image with superimposed scaled molecular models and K atoms and the overlayer unit mesh indicated. (b) STM image showing alternative contrast with bright protrusions, interpreted to be K atoms, visible between the TCNQ molecules ($V_{samp} = -1.5$ V, $I = 300$ pA, white arrows indicate the substrate ⟨110⟩ directions). (c) LEED pattern recorded at a kinetic energy of 14.5 eV with two substrate ⟨211⟩ directions shown as white arrows. (d) LEED pattern recorded at a kinetic energy of 23.5 eV with two substrate ⟨211⟩ directions shown as white arrows. (e) Simulated LEED pattern derived from the STM measured unit mesh with beams originating from the six distinct domains coloured separately and two substrate ⟨211⟩ directions shown as white arrows.

bright protrusions. The K–TCNQ coadsorption phase is described by a unit mesh containing two TCNQ molecules and two alkali metal atoms with vectors of length $\boldsymbol{a} = (13.7 \pm 0.4)$ Å and $\boldsymbol{b} = (13.0 \pm 0.4)$ Å, having an included angle of $(88 \pm 2)°$; the vector $\boldsymbol{a}$ is offset by $(9 \pm 2)°$ from the substrate ⟨110⟩ directions. Fig. 1e shows the simulated LEED pattern obtained using this unit mesh (fine-tuned within the error ranges of the STM measurements). As the overlayer mesh does not share the three-fold rotational and ⟨211⟩ mirror symmetries of the substrate there are 6 symmetrically equivalent domains of the overlayer, and the simulated pattern shows the beams from each domain in a different colour. The experimental LEED patterns (Fig. 1c and d) are in excellent agreement with the simulated pattern, confirming, by a technique not susceptible to instrumental drift or calibration errors, the accuracy of the STM-derived unit mesh. The LEED patterns also indicate that this overlayer phase is incommensurate with respect to the substrate

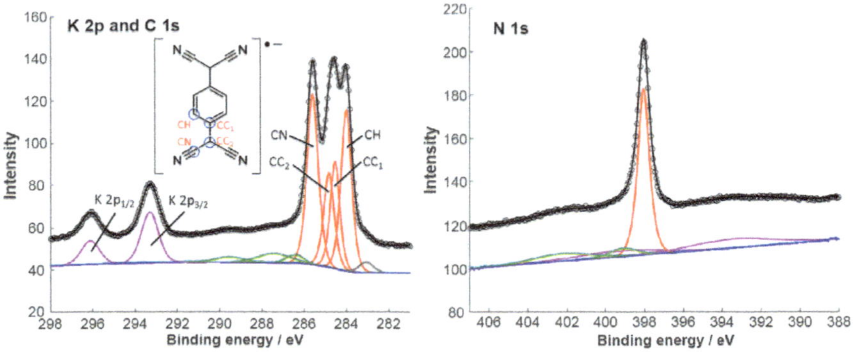

Fig. 2 Soft X-ray photoelectron spectra of the K–TCNQ windmill phase on Ag(111). The K 2p/C 1s spectrum was recorded with a photon energy of 435 eV, with measured intensity (black circles) and the sum of fitted components (black line) comprising a background (blue), four main C 1s components (red), shakeup features (green), a beam damage feature (grey) and a single split K 2p component (magenta). The N 1s spectrum was recorded with a photon energy of 550 eV, with measured intensity (black circles), and the sum of fitted components (black line) comprising of a linear background (blue), one main component (red), two N shakeup features (green) and two Ag 3d plasmon features (magenta).

as simulations of the closest commensurate structure yield a distinctly different simulated LEED pattern (see ESI, Fig. S2†).

The STM images indicate that the ordered phase can accommodate one K atom per TCNQ molecule, but the images recorded at larger bias voltages appear to show that there is a significant concentration of K atom vacancies. Stoichiometry estimates based on the measured K 2p and C 1s photoemission intensities (corrected for the difference in photoemission cross-sections) also indicate that there were fewer K atoms than TCNQ molecules on the samples measured with NIXSW. STM images showed the same ordering when the surface was prepared using a lower K exposure, albeit with an increased concentration of vacancies (see ESI, Fig. S3†), which indicates that this ordered phase of TCNQ, which can accommodate up to 1 K per TCNQ molecule, is favourable for a range of K : TCNQ stoichiometric ratios. Similar structures have been reported for Na–TCNQ and Li–TCNQ networks on Au(111)[14,15] although on the Ag(100) surface, Cs–TCNQ networks with a smaller stoichiometric ratio of 1 : 4, are found to form.[11]

Fig. 2 shows high-resolution SXP spectra recorded from the ordered surface phase. The C 1s spectrum clearly shows at least three distinct peaks, which have been fitted with four components (Table 1), corresponding to the four chemically inequivalent C species in TCNQ (see the inset of the K 2p and C 1s spectrum). In previous reports, C 1s XPS has been used to deduce the charge state of TCNQ, with the relative binding energies and overall lineshape of the spectrum being characteristically different when TCNQ is negatively charged compared to when it is

Table 1 XPS measured binding energies of the main C 1s, N 1s and K 2p peaks

Component	CH	CC_1	CC_2	CN	K $2p_{3/2}$	K $2p_{1/2}$	N
Binding energy/eV	284.0	284.5	284.8	285.6	293.3	296.1	398.1

neutral.[14,27] Based on this interpretation it has previously been shown that TCNQ accepts electrons from the Ag(111) substrate, becoming negatively charged.[14,19] This is reinforced by UPS work function measurements, measured both here and in a previous study,[14] that show an increase of 0.4 eV when TCNQ is deposited on clean Ag(111) (data not shown). This is consistent with the molecules being negatively charged; a neutral molecule would be expected to decrease the work function *via* the 'push-back' effect.[3] Here, upon the addition of K, the overall line shape of the C 1s spectrum remains the same, albeit with a small shift of ∼0.2 eV in the binding energies compared to TCNQ alone,[19] indicating that the TCNQ molecules retain their negative charge. The value of the N 1s binding energy of 398.1 eV is also in good agreement with other systems in which TCNQ is believed to be negatively charged.[14,27,28] The K 2p SXPS shows binding energies of 293.3 eV and 296.1 eV for the K $2p_{3/2}$ and K $2p_{1/2}$ components respectively, which are consistent with typical binding energies for K^+ in ionic systems.[29] This is a negative shift of 1.2 eV from K deposited alone on Ag(111) which gives binding energies of 294.5 eV and 297.3 eV for these two components and is more consistent with the binding energies of elemental K.[30] These values can be reconciled with the charge state of K, DFT calculations predicting a relatively small charge of 0.16–0.24*e* for K on Ag(111)[31] whereas for K–TCNQ networks on Ag(100), similar to those presented here, the alkali metal is predicted to completely donate its valence electron.[12] The implication is that K interacts ionically in the networks with TCNQ on the Ag(111) surface. The measured work function of the surface shows a 0.5 eV decrease from that of TCNQ/Ag(111) with

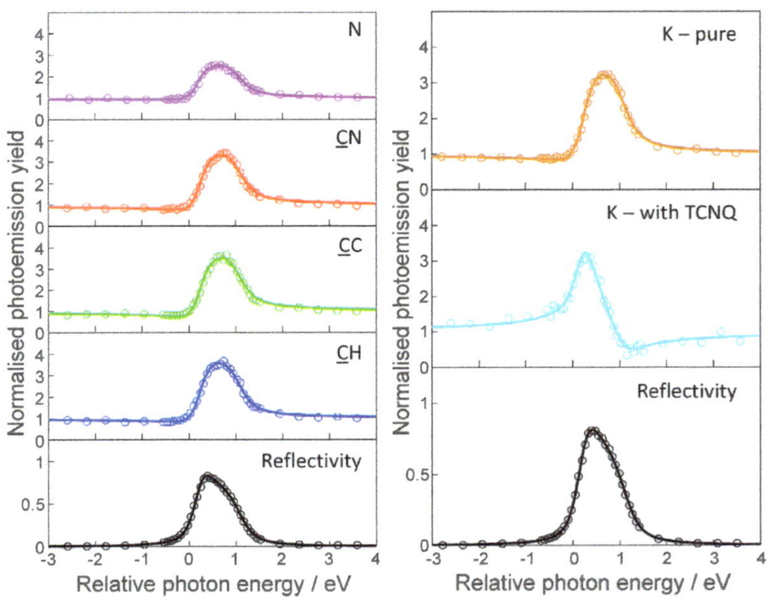

Fig. 3 Sample set of NIXSW photoemission yield curves obtained using the [111] reflection for C, N, and K in the K–TCNQ coadsorption phase, in addition to pure K on Ag(111). Photon energies are quoted relative to the Bragg energy of 2.630 keV. Least square fits (solid lines) to the photoemission yields (circles) were obtained to extract the coherent fractions and coherent positions.

a net 0.1 eV decrease relative to clean Ag(111) (data not shown). This is consistent with K becoming positively charged on the surface and also illustrates how the addition of an alkali metal can be used to tune the surface work function.

NIXSW and structural models

As described in the Methods section, the NIXSW technique offers the ability to determine the height of X-ray absorbing atoms above the scattering planes of the chosen Bragg reflection. So, by using normal incidence to the surface, the (111) scattering planes are parallel to the surface, and the measured height is relative to the outermost atomic layer of the surface providing that there is no significant relaxation of the outermost substrate layers. For the specific case of fcc(111) surfaces these surface relaxations are known to be extremely small ($<\sim 0.02$ Å) at the clean surface, while such relaxations of all surfaces are generally reduced in the presence of adsorbates. By using core level photoemission to monitor the absorption we can not only identify the atomic species of the absorber through its photoelectron binding energy, but also distinguish absorption at atoms of the same elemental species in chemically distinct sites through the chemical shifts in these binding energies.

Measured NIXSW absorption profiles can be uniquely fitted by two structural parameters known as the coherent position, P, and the coherent fraction, f (Fig. 3). If the absorbing atoms occupy a single well-defined site, the value of P is a direct measure of the atom's height relative to the nearest bulk-extended scattering plane (in units of the interplanar spacing). In this simple case, f is essentially an order parameter which is equal to unity in a system with perfect static and dynamic order, although in practice values of ≥ 0.80 are typical for absorbers occupying a single site. However, the interpretation of NIXSW results becomes significantly more complex when the value of f is much lower, as this can generally not be attributed to disorder and almost certainly indicates that the atomic species occupies two or more distinctly different heights. For example, in the extreme case of two sites that are equally occupied and separated by exactly half of the scattering plane spacing, the value of f is zero, and the measured P is meaningless, despite the system being highly ordered. This effect proves to be important in interpreting aspects of the data from the present system.

Table 2 shows the results of NIXSW measurements recorded for the K–TCNQ coadsorption phase, for pure K on Ag(111), and for the pure TCNQ phase measured by us on Ag(111).[19] The values of P have been converted to layer spacings in Å units, shown as $D_{(111)}$, adding one bulk layer spacing where appropriate to ensure one has physically plausible interatomic distances. The reported values of the coherent fractions have been adjusted to account for non-linearity in the electron analyser (as described in detail in the ESI†). For the K–TCNQ network, the NIXSW at all three chemically-distinct C atoms yields high coherent fractions (>0.8), indicating that the molecules occupy a single well-defined height above the surface. The corresponding coherent positions show the molecules adopt a relatively flat conformation $\sim$2.8 Å above the surface, with the peripheral nitrile C atoms only slightly lower (<0.1 Å) than the central C ring. In contrast to this, the N NIXSW yields a low coherent fraction of 0.37, clearly indicating that the N atoms must occupy two or more different heights above the surface. In this case the N coherent position of 2.79 Å, which is a weighted average of the different

Table 2 NIXSW fitting parameters for the K–TCNQ phase, TCNQ on Ag(111)[19] and K on Ag(111). Separate fitting parameters were obtained for the different C 1s photoemission components as defined in Fig. 2 (due to lower resolution at the photon energies required for NIXSW, components CC_1 and CC_2 were fitted together in a single component here). The coherent position $D_{(111)}$ is given as $D_{(111)} = (P + 1)d_{111}$ where P is the fractional coherent position and d_{111} is the spacing of the [111] lattice planes (2.359 Å)

	K–TCNQ		TCNQ only[19]		K only	
	f	$D_{(111)}$/Å	f	$D_{(111)}$/Å	f	$D_{(111)}$/Å
K	0.71(8)	3.56(3)	—	—	0.74(8)	2.81(1)
CH	0.90(10)	2.85(1)	0.86(9)	2.86(2)	—	—
CC	0.90(10)	2.79(1)	0.89(10)	2.78(2)	—	—
CN	0.82(9)	2.79(1)	0.80(9)	2.76(2)	—	—
N	0.37(4)	2.79(5)	0.35(4)	2.75(3)	—	—

contributing heights, is almost identical to the height of the C atoms. This indicates that the two or more different N atom heights must be distributed relatively evenly about the height of the carbon backbone of the molecule. The C and N NIXSW measurements for the K–TCNQ networks therefore show very similar behaviour to that of TCNQ deposited alone on the Ag(111) surface, with the coherent positions and fractions of the two systems being closely related. For TCNQ on Ag(111) without coadsorbed K, the results of our combined DFT and NIXSW study show that the low N coherent fraction can be rationalised by the inclusion of Ag adatoms within the molecular adsorption structure; this results in some N atoms bonding to nearby adatoms, leading to their associated CN bonds pointing up out of the surface, whereas the other N atoms – located further from the Ag adatoms – bond to the underlying surface Ag atoms, causing their associated CN bonds to point down towards the substrate.[19] This creates two different N heights and results in a reduction in the N coherent fraction.

In the case of the K–TCNQ network, the closely similar N NIXSW parameters are more difficult to rationalise. The windmill structure of four TCNQ molecules coordinated to a single K ion is somewhat similar to the structure formed by TCNQ molecules and Ag adatoms in the absence of adsorbed K, but the 1 : 1 K to TCNQ stoichiometry of the K–TCNQ network means that every CN group should be in close proximity to a K ion. One might therefore expect that all of the N atoms would bond to K ions in a similar way and thus that all the N atoms would have the same height above the surface. One possible solution to this dilemma may lie in the fact that the STM images (Fig. 1b) indicate that there is a significant concentration of K vacancies, at which N atoms would only be able to bond to the lower Ag atoms of the substrate. However, an estimate of the concentration of these vacancies in the NIXSW experiments can be obtained from the relative intensities of the XPS peaks from the K atoms and the TCNQ atomic components, which indicate a K vacancy concentration of ~20%. In this case, even if the height difference of the N atoms at the vacancies and the K bonding sites was exactly half the spacing of the [111] planes (*i.e.* a height difference of 1.18 Å), the coherent fraction would only be reduced to 0.6. A smaller or larger height difference would lead to an even higher coherent fraction. The presence of these vacancies is thus insufficient to account for the much lower measured N coherent fraction. We return to this problem below.

The coherent fractions for the K atoms, both alone and coadsorbed with TCNQ, are high, indicating that they occupy a single well-defined height above the surface in each phase. In the absence of TCNQ, the coherent position of the K atoms corresponds to a height of 2.81 Å above the surface, in excellent agreement with the results of a previous quantitative LEED structure determination of K on Ag(111).[32] In the K–TCNQ networks, the K ions occupy sites significantly higher (3.56 Å) above the surface. Presumably this greater height is because the K ions are lying on top of the molecular layer rather than interacting directly with the underlying Ag substrate, and indeed they lie 0.77 Å above the C backbone of the molecules. This qualitative conclusion is consistent with the findings of an earlier DFT study of TCNQ/alkali coadsorption phases on the Ag(100) surface.[12] As remarked in the introduction, on the (100) surface the alkali ions are expected to occupy sites significantly above the 'holes' in the TCNQ network,[12] despite the fact that simple consideration of atoms sizes would indicate that they could 'fall through' these holes to bond directly to the substrate. For the present coadsorption system on Ag(111), in which the pores in the molecular layer are smaller than those calculated on the Ag(100) surface, it is interesting to establish if the height of the K ions may be reconcilable with simple considerations of the expected N–K interaction distance.

Fig. 4a shows a model of the K–TCNQ phase on Ag(111) that is consistent with the STM and LEED results. The molecules in this model are assumed to lie flat and coplanar, consistent with the NIXSW results for the C atoms. Here the K ions lie above a pore in the molecular layer that is 5.23 Å in diameter (red circle in Fig. 4a). The bulk crystal structure of K–TCNQ has K–N interatomic distances of 2.93 Å,[26] which corresponds almost exactly to the sum of the K^+ ionic radius (1.37 Å)[33] and the N van der Waals radius (1.55 Å).[34] If the same interatomic distance occurs in an adsorbed K–TCNQ monolayer, a pore with a diameter of 5.84 Å or larger would be required to accommodate the K ion in a site coplanar with the molecules. As the pore size in the K–TCNQ coadsorption phase (5.23 Å) is smaller than this, the K ions occupying this pore site should lie 1.30 Å above the molecules, a value significantly greater than the measured height difference of 0.77 Å. However, this simple model assumes that all of the N atoms are at the same height above the surface, which is inconsistent with the measured low N coherent fraction of 0.37. If this low coherent fraction is attributed to the influence of two equally occupied N sites at different heights – the simplest multi-site model – a height difference of 0.87 Å is required to reduce the coherent fraction from 0.89, the largest possible value expected for an absorbing atom when accounting for vibrations, to the observed value of 0.37. In this model the measured N coherent position (2.79 Å) corresponds to the average height of the two sites, so half of the N atoms must have a height of 2.36 Å whereas the others are 3.23 Å above the surface. The high coherent fraction of the K ions rules out the possibility of multiple K adsorption heights, excluding for example the possibility of two different fourfold coordination sites surrounded by 'all-up' or 'all-down' N atoms. Instead, both N heights must be present surrounding each K site. In this situation, there are two possibilities: either pairs of N atoms opposite each other around the K coordination site have the same height, or adjacent pairs of N atoms around this site have the same height. In the former case the height of the upper

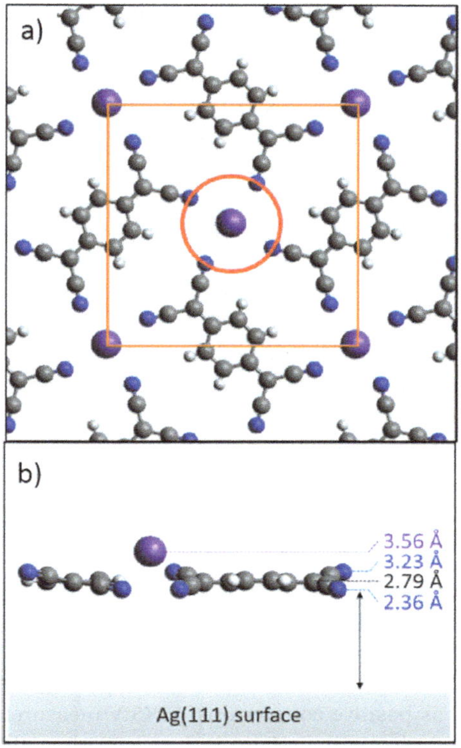

Fig. 4 (a) Top view of the K–TCNQ adsorbed phase. The orange square depicts the unit mesh with sides 13.4 Å in length. The red circle cuts through the centre of the four N atoms surrounding a K ion and has a diameter of 5.23 Å. (b) Side-on view of the proposed 'twist' model with the heights of the K ion, N atoms and the centre of the molecule highlighted.

pairs of N atoms determines the height of the K ions, which would be forced up to 4.53 Å, much higher than the experimentally measured value. By contrast, if adjacent pairs of N atoms surrounding the K atoms have different heights, the K ions could lie closer to the surface. While not strictly necessary, this type of N height pairing would be satisfied if the TCNQ molecules were tilted with respect to the surface by rotation about the long molecular axis; this phenomenon has been reported for other small, nominally planar, molecules on metal surfaces.[35–37] Interestingly, in the bulk crystal structure of K–TCNQ (in which TCNQ arranges into similar windmill structures around out of plane K^+ ions – see Fig. S1†), the molecules are tilted about the long molecular axis in just this fashion, with a tilt angle of between 8.9–15.8° depending on whether the crystal is in its high or low temperature structural phase.[26] A height difference for the two inequivalent N sites of 0.87 Å on the surface (the maximum value estimated above to be required to match the measured coherent fraction on Ag(111)) would require a tilt of 11.5°. This tilt would also lower the predicted coherent fractions of the C atoms in the molecule, but to a lesser extent, with expected values of 0.71, 0.89 and 0.71 for the CH, CC and CN coherent fractions, respectively. The experimental values for CC and CN are consistent with this model, but the measured coherent fraction for the

CH component is significantly higher than predicted for this model, indicating that the central carbon ring does lie more closely parallel to the surface. A solution to this inconsistency is provided if the peripheral CN groups are twisted with respect to the central ring, rather than the full planar molecule being tilted. This twist model is depicted in Fig. 4b. Other studies have found that TCNQ can lose its planarity upon becoming negatively charged, leading to the peripheral C atoms becoming sp^3 hybridised and the central ring aromatising.[18,38–40] This gives the molecule improved flexibility and allows TCNQ to bend and distort from a planar geometry, so such a twist in the molecule is certainly plausible. It should be noted, however, that whilst the twist model can satisfy all of the experimental results, other factors, such as the influence of K vacancies described above, may also contribute to the low coherent fraction of N; the actual difference in height of the inequivalent N atoms surrounding occupied K sites that is required to account for the low coherent fraction may therefore be lower than the suggested 0.87 Å, thereby reducing the required twist or tilt angle.

Conclusions

In summary, we have shown using STM and LEED that the codeposition of TCNQ and K on the Ag(111) surface leads to the formation of an incommensurate 'windmill' phase with a nominal K : TCNQ stoichiometry of 1 : 1. SXPS and UPS data indicate that the TCNQ molecules are negatively charged on the surface, with the K atoms acting as positive counterions. NIXSW measurements indicate that the TCNQ molecules adsorb in a relatively flat geometry with K ions lying above the molecular plane. However, a low measured coherent fraction for the N atoms clearly shows that the N atoms occupy two or more different heights above the surface. The simplest model that appears to be compatible with all of these experimental results has the central TCNQ ring almost exactly parallel to the surface, with the peripheral moieties twisted such that out of the four CN groups coordinating to each K ion, one adjacent pair points up and the other pair points down, in a similar manner to the geometry found in the bulk crystal structure of K–TCNQ. As yet, this is the only quantitative structural study of any TCNQ/alkali coadsorption phase and it would certainly be of considerable interest to establish the generality of the structural implications we have observed.

Acknowledgements

G. C. and D. A. W. acknowledge financial support from the EU through the ERC Grant "VISUAL-MS".

References

1 S. R. Forrest, *Nature*, 2004, **428**, 911–918.
2 P. Peumans, A. Yakimov and S. R. Forrest, *J. Appl. Phys.*, 2003, **93**, 3693–3723.
3 S. Braun, W. R. Salaneck and M. Fahlman, *Adv. Mater.*, 2009, **21**, 1450–1472.
4 H. Ishii, K. Sugiyama, E. Ito and K. Seki, *Adv. Mater.*, 1999, **11**, 605–625.
5 S. Lattante, *Electronics*, 2014, **3**, 132.

6 I. Hancox, L. A. Rochford, D. Clare, M. Walker, J. J. Mudd, P. Sullivan, S. Schumann, C. F. McConville and T. S. Jones, *J. Phys. Chem. C*, 2013, **117**, 49–57.

7 N. Koch, S. Duhm, J. P. Rabe, A. Vollmer and R. L. Johnson, *Phys. Rev. Lett.*, 2005, **95**, 237601.

8 M. Kröger, S. Hamwi, J. Meyer, T. Riedl, W. Kowalsky and A. Kahn, *Appl. Phys. Lett.*, 2009, **95**, 123301.

9 G. Bae, H. Jung, N. Park, J. Park, S. Hong and W. Park, *Appl. Phys. Lett.*, 2012, **100**, 183102.

10 E. L. Hanson, J. Guo, N. Koch, J. Schwartz and S. L. Bernasek, *J. Am. Chem. Soc.*, 2005, **127**, 10058–10062.

11 N. Abdurakhmanova, A. Floris, T.-C. Tseng, A. Comisso, S. Stepanow, A. De Vita and K. Kern, *Nat. Commun.*, 2012, **3**, 940.

12 A. Floris, A. Comisso and A. De Vita, *ACS Nano*, 2013, **7**, 8059–8065.

13 J. V. Barth, G. Costantini and K. Kern, *Nature*, 2005, **437**, 671–679.

14 C. Wackerlin, C. Iacovita, D. Chylarecka, P. Fesser, T. A. Jung and N. Ballav, *Chem. Commun.*, 2011, **47**, 9146–9148.

15 T. R. Umbach, I. Fernández-Torrente, M. Ruby, F. Schulz, C. Lotze, R. Rurali, M. Persson, J. I. Pascual and K. J. Franke, *New J. Phys.*, 2013, **15**, 083048.

16 M. N. Faraggi, N. Jiang, N. Gonzalez-Lakunza, A. Langner, S. Stepanow, K. Kern and A. Arnau, *J. Phys. Chem. C*, 2012, **116**, 24558–24565.

17 T.-C. Tseng, N. Abdurakhmanova, S. Stepanow and K. Kern, *J. Phys. Chem. C*, 2011, **115**, 10211–10217.

18 S. Barja, D. Stradi, B. Borca, M. Garnica, C. Díaz, J. Rodriguez-García, M. Alcamí, A. Vázquez de Parga, F. Martín and R. Miranda, *J. Phys.: Condens. Matter*, 2013, **25**, 484007.

19 P. J. Blowey, S. Velari, L. A. Rochford, D. A. Duncan, D. A. Warr, T.-L. Lee, A. De Vita, G. Costantini and D. P. Woodruff, to be published.

20 K. E. Hermann and M. A. Van Hove, *LEEDpat*, 2014, http://www.fhi-berlin.mpg.de/KHsoftware/LEEDpat/index.html.

21 D. Nečas and P. Klapetek, *Cent. Eur. J. Phys.*, 2012, **10**, 181–188.

22 D. P. Woodruff, *Prog. Surf. Sci.*, 1998, **57**, 1–60.

23 D. P. Woodruff, *Rep. Prog. Phys.*, 2005, **68**, 743.

24 C. J. Fisher, R. Ithin, R. G. Jones, G. J. Jackson, D. P. Woodruff and B. C. C. Cowie, *J. Phys.: Condens. Matter*, 1998, **10**, L623.

25 M. B. Trzhaskovskaya, V. I. Nefedov and V. G. Yarzhemsky, *At. Data Nucl. Data Tables*, 2001, **77**, 97–159.

26 M. Konno, T. Ishii and Y. Saito, *Acta Crystallogr., Sect. B: Struct. Sci.*, 1977, **33**, 763–770.

27 R. Precht, R. Hausbrand and W. Jaegermann, *Phys. Chem. Chem. Phys.*, 2015, **17**, 6588–6596.

28 R. Precht, S. Stolz, E. Mankel, T. Mayer, W. Jaegermann and R. Hausbrand, *Phys. Chem. Chem. Phys.*, 2016, **18**, 3056–3064.

29 W. E. Morgan, J. R. Van Wazer and W. J. Stec, *J. Am. Chem. Soc.*, 1973, **95**, 751–755.

30 L. G. Petersson and S. E. Karlsson, *Phys. Scr.*, 1977, **16**, 425.

31 K. Doll, *Phys. Rev. B: Condens. Matter*, 2002, **66**, 155421.

32 G. S. Leatherman, R. D. Diehl, P. Kaukasoina and M. Lindroos, *Phys. Rev. B: Condens. Matter*, 1996, **53**, 10254–10260.

33 R. Shannon, *Acta Crystallogr., Sect. A: Found. Crystallogr.*, 1976, **32**, 751–767.
34 A. Bondi, *J. Phys. Chem.*, 1964, **68**, 441–451.
35 L. Duy, A. Maral, K. Adam and S. R. Talat, *J. Phys.: Condens. Matter*, 2012, **24**, 222001.
36 W. Liu, A. Tkatchenko and M. Scheffler, *Acc. Chem. Res.*, 2014, **47**, 3369–3377.
37 X. Stammer, K. Tonigold, A. Bashir, D. Kafer, O. Shekhah, C. Hulsbusch, M. Kind, A. Groß and C. Woll, *Phys. Chem. Chem. Phys.*, 2010, **12**, 6445–6454.
38 B. Milián, R. Pou-Amérigo, R. Viruela and E. Ortí, *J. Mol. Struct.: THEOCHEM*, 2004, **709**, 97–102.
39 D. Stradi, B. Borca, S. Barja, M. Garnica, C. Diaz, J. M. Rodriguez-Garcia, M. Alcami, A. L. Vazquez de Parga, R. Miranda and F. Martin, *RSC Adv.*, 2016, **6**, 15071–15079.
40 L. Romaner, G. Heimel, J.-L. Brédas, A. Gerlach, F. Schreiber, R. L. Johnson, J. Zegenhagen, S. Duhm, N. Koch and E. Zojer, *Phys. Rev. Lett.*, 2007, **99**, 256801.

Faraday Discussions

PAPER

Self-assembly of a binodal metal–organic framework exhibiting a demi-regular lattice

Linghao Yan,[a] Guowen Kuang,[a] Qiushi Zhang,[a] Xuesong Shang,[b] Pei Nian Liu [b] and Nian Lin *[a]

Received 2nd March 2017, Accepted 19th April 2017
DOI: 10.1039/c7fd00088j

Designing metal–organic frameworks with new topologies is a long-standing quest because new topologies often accompany new properties and functions. Here we report that 1,3,5-tris[4-(pyridin-4-yl)phenyl]benzene molecules coordinate with Cu atoms to form a two-dimensional framework in which Cu adatoms form a nanometer-scale demi-regular lattice. The lattice is articulated by perfectly arranged twofold and threefold pyridyl–Cu coordination motifs in a ratio of 1 : 6 and features local dodecagonal symmetry. This structure is thermodynamically robust and emerges solely when the molecular density is at a critical value. In comparison, we present three framework structures that consist of semi-regular and regular lattices of Cu atoms self-assembled out of 1,3,5-tris[4-(pyridin-4-yl)phenyl]benzene and trispyridylbenzene molecules. Thus a family of regular, semi-regular and demi-regular lattices can be achieved by Cu–pyridyl coordination.

Introduction

In his book "Harmonices Mundi",[1] Johannes Kepler illustrated three types of Euclidean tessellations made out of regular polygons: regular, semi-regular, and demi-regular, indicated with red, green and blue frames, respectively in Scheme 1. Regular (colored in red) and semi-regular (colored in green) tessellations have one type of distinct transitive vertex, while demi-regular (colored in blue) tessellations feature more than one type of vertex ($n > 1$).[2,3] Tiling is not only a mathematic theme but is also expressed in crystalline materials. Regular tessellations are pervasively found in natural and synthetic crystals including metal–organic frameworks (MOFs).[4] Semi-regular tessellations are rare in nature but have recently been reported in various man-made materials, such as ABC star polymer assemblies,[5,6] liquid crystalline networks,[7] self-assembled DNA structures,[8]

[a]Department of Physics, The Hong Kong University of Science and Technology, Hong Kong, China. E-mail: phnlin@ust.hk

[b]Shanghai Key Laboratory of Functional Materials Chemistry and Institute of Fine Chemicals, East China University of Science and Technology, Meilong Road 130, Shanghai, China

nanoparticle and colloidal supperlattices,[9,10] MOFs,[11–13] and in interfacial self-assembly.[14–17] Demi-regular tessellations are commonly seen in artistic works. Some of these structures feature 12-fold local rotational symmetry. Crystalline materials expressing such rotational symmetry are expected to possess novel properties, for example, wide-angle optical isotropy.[18] However, none of the 20 2-uniform tessellations have been observed in crystalline materials. The major challenge is that the demi-regular lattices must comprise at least two types of distinctive bonding modes that are arranged in a specific ordered pattern.

Here, we demonstrate the use of interfacial coordination self-assembly to synthesize a two-dimensional demi-regular metal–organic coordination framework. The structure features a nanometer scale atomic lattice that resembles one of the twenty non-Archimedean lattices of (3^6; 4, 3^2, 4, 3), the pattern "o" in Kepler's drawing (highlighted by the thick blue frame in Scheme 1). The key ingredient is to make use of the co-existing twofold and threefold Cu–pyridyl coordination modes that are present at a critical molecular density. The lattice consists of perfectly arranged twofold and threefold pyridyl–Cu coordination motifs in a 6 : 1 ratio. This structure is self-organized *via* interfacial supramolecular assembly at a critical in-plane compression. The lattice bears a local dodecagonal symmetry, which may result in interesting physical and chemical properties. Together with our former work, in which Cu–pyridyl coordination structures can also exhibit semi-regular and demi-regular tiling, a complete

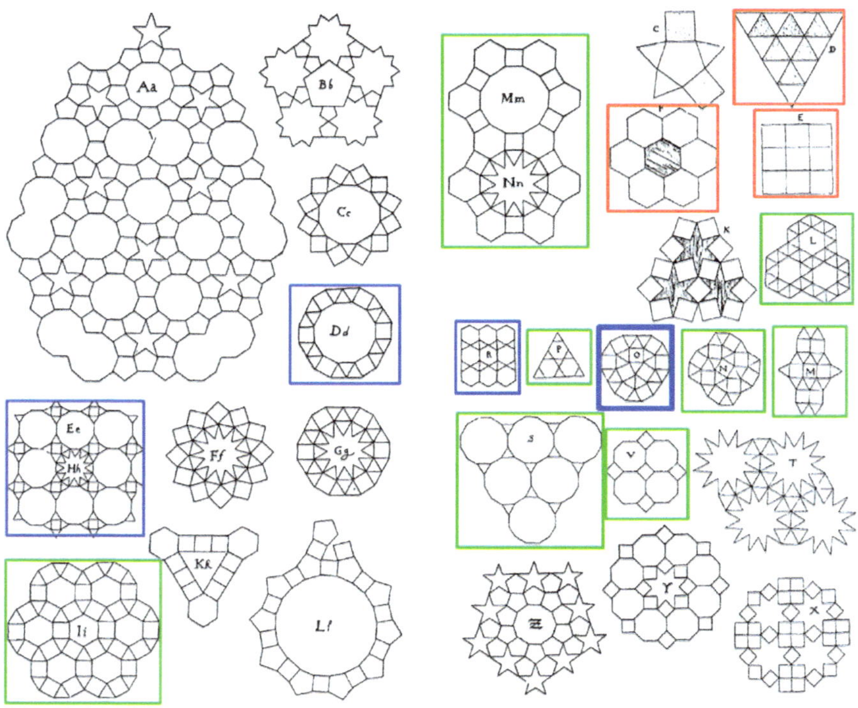

Various more or less regular tilings of the plane by regular polygons, reproduced from J. Kepler's book "Harmonices Mundi", published in 1619.

Scheme 1

family of tiling is achieved. We believe this work represents not only a realization of an unprecedented MOF structure, but also a breakthrough in material design which opens a door for synthesizing two-dimensional metal–organic materials with high-level complexity in the future.

Method

Single-crystal Cu(111) and Au(111) substrates were cleaned through repeated cycles of Ar^+ sputtering and subsequently annealed at 800 K. The samples were characterized using a commercial ultra-high vacuum (10^{-10} mbar) STM system (Omicron GmbH) at cryogenic temperature (5 K or 78 K) or room temperature. 1,3,5-tris[4-[pyridin-4-yl]phenyl]benzene (ext-TPyB) molecules were deposited at 310 °C while 1,3,5-trispyridylbenzene (TPyB) was deposited at 200 °C on the Cu(111) surface or Au(111) surface held at room temperature. The samples were then annealed at 120–150 °C to let ext-TPyB or TPyB and Cu atoms self-assemble into a two-dimensional metal–organic coordination network.

Results and discussion

The molecule used in this study is 1,3,5-tris[4-(pyridin-4-yl)phenyl]benzene (ext-TPyB, Fig. 1a). The ext-TPyB molecule or its smaller analogue, 1,3,5-tris[pyridyl]

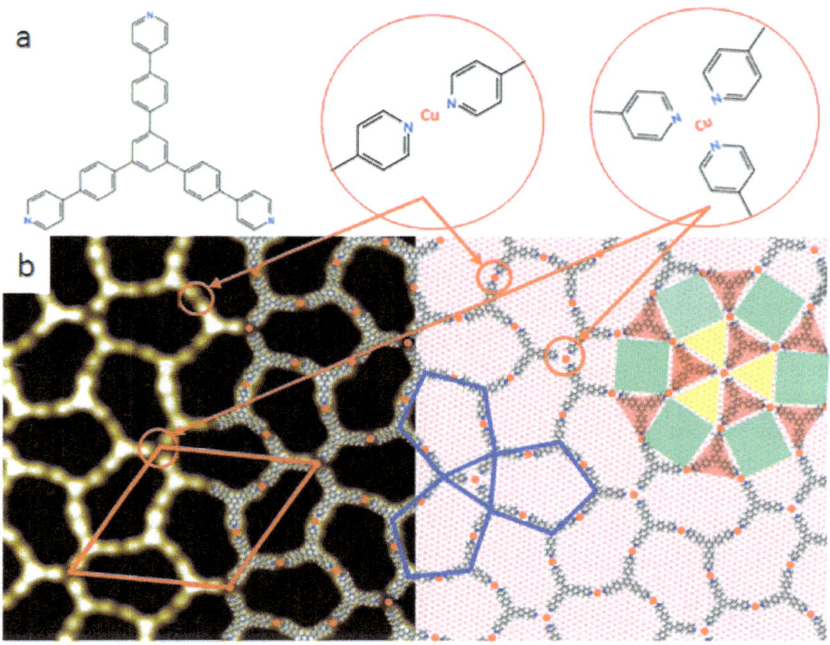

Fig. 1 (a) Chemical structure of the ext-TPyB molecule and twofold and threefold Cu–pyridyl coordination motifs. (b) High-resolution STM image (15 nm × 15 nm, imaged at 5k) of the Cu–ext-TPyB framework self-assembled on Cu(111) and the structural model. The red diamond-shaped frame marks a unit cell. The edges of the colored squares and triangles link the neighboring Cu adatoms to highlight the (3^6; 4, 3^2, 4, 3) non-Archimedean demi-regular tessellation.

benzene (TPyB), forms Cu-coordinated networks on a Cu(111) or Au(111) surface exhibiting a honeycomb or a triangular structure.[19–22] Twofold and threefold Cu–pyridyl coordination motifs, as illustrated in Fig. 1a, are present in these networks. In this work, we increased the molecular concentration on the surface and found that new phases emerged. Fig. 1b shows the self-assembled structure comprising three-arm star-like entities that match ext-TPyB molecules in their size and shape.[20] The molecules are connected with one another at their three tips, forming a complex yet periodic two-dimensional (2D) framework articulated with twofold and threefold intermolecular connections (Fig. 1b, circles). Apparently, these connections delineate twofold and threefold Cu–pyridyl coordination motifs. The Cu atoms are not resolved by STM, which is often the case in surface-supported metal–organic structures due to the weak electronic state of the central metal atoms.[23,24]

A molecular model lies on top of the STM topograph in Fig. 1b. The projected Cu–N distance in the model is 2.1 ± 0.1 Å for the twofold coordination and 2.3 ± 0.1 Å for the threefold coordination. Both values agree well with the literature values.[19–22] The diamond-shaped frame in Fig. 1b depicts a unit cell, which has an inner angle of 60° and a side length of 5.69 ± 0.07 nm. It is commensurate with the Cu(111) lattice and renders a supercell of $(13\sqrt{3} \times 13\sqrt{3})R30^\circ$. The unit cell contains five ext-TPyB molecules and seven Cu adatoms, six of which are twofold coordinated and one is threefold coordinated. The network consists of pentagonal 2D pores, as indicated by the blue lines linking the adjacent ext-TPyB molecules. To highlight the 2D Cu adatom lattice, the neighboring Cu adatoms are linked by the edges of the colored squares and triangles in the right part of Fig. 1b. The lattice consists of equilateral triangles and squares fused together in an edge-to-edge manner. The nearest-neighbor Cu–Cu distance is 2.0 ± 0.05 nm. In the lattice, the central Cu atom is a vertex of six triangles, and the surrounding six Cu atoms are vertices of two squares and three triangles each. The two types of vertices can be named as 3^6 and 4, 3^2, 4, 3, respectively, in terms of the number and type of the polygons surrounding them. Accordingly, the lattice can be defined as (3^6; 4, 3^2, 4, 3), corresponding to one of the 20 2-uniform demi-regular tessellations.[3] Hereafter we refer to this structure as the D-phase.

The D-phase homogeneously covers the entire surface. The single domains of the D-phase extend over full atomic terraces, such that the domain boundaries exist only at atomic steps. Within the single domains, the D-phase is nearly defect-free (Fig. 2a). A typical defect is shaded in green, which comprises three molecules that are rotated by 60° around the central Cu atom from their normal position. Such defects amount to less than 4% of the total area. It is worth pointing out that such defects do not disrupt the Cu lattice structure. Globally, the Cu lattice bears *p6m* symmetry; locally, one can recognize 12-fold rotational symmetry (*cf.* the blue frame in Fig. 2a). The inset in Fig. 2a is a 2D Fourier transformation (FT) of the Cu adatom lattice. The 12 spots are circled in the FT pattern, manifesting dodecagonal symmetry. Similar FT characteristics have been found in colloidal and polymer systems exhibiting local 12-fold symmetry.[25–28] Fig. 2c is a 2D FT of a perfect lattice of the (3^6; 4, 3^2, 4, 3) tessellation shown in Fig. 2b. The two FT patterns are nearly identical. The FT power spectra of the experimental structure and the ideal (3^6; 4, 3^2, 4, 3) lattice are plotted in Fig. 2d for comparison. We focus on three peaks at 0.55, 1.15 and 1.49 nm^{-1}, which correspond to real-space

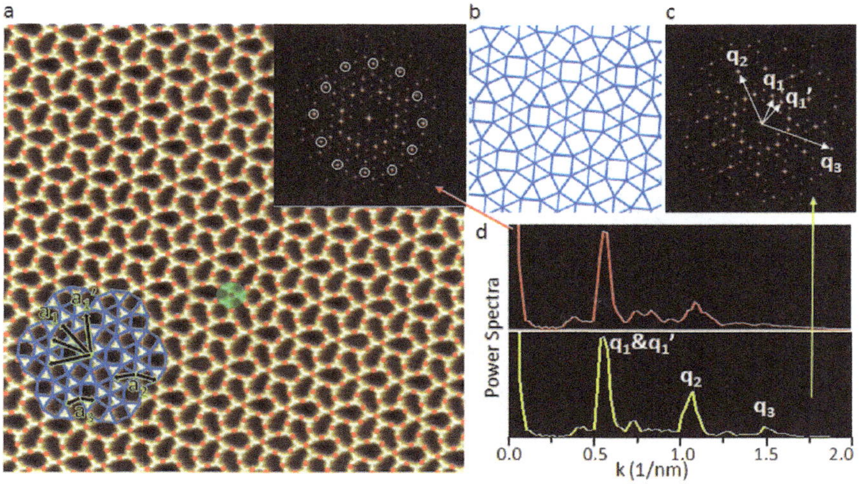

Fig. 2 Comparison of the real-space arrangement and reciprocal pattern of the Cu adatom lattice and a perfect (3^6; 4, 3^2, 4, 3) tessellation. (a) STM topograph of the D-phase (60 nm × 60 nm, imaged at 5k). Cu adatoms are marked by red dots. A defect is shaded in green. Inset: 2D FT of the Cu adatom lattice. (b) A perfect (3^6; 4, 3^2, 4, 3) tessellation. (c) 2D FT of the point lattice in (b). (d) FT power spectra of the inset in (a) and (c).

distances of 5.7, 2.8 and 2.1 nm, respectively. The strongest peak at 0.55 nm^{-1} is attributed to two sets of six-fold diffraction spots q_1 and q_1'. In real space, they are associated with two sets of 5.7 nm vectors a_1 and a_1' in Fig. 2a. The peak at 1.15 nm^{-1} arises from the 12-fold spots q_2 which are associated with the diagonals of the squares orientated in 12 directions, as represented by vectors a_2 in Fig. 2a. The peak at 1.49 nm^{-1} arises from another set of 12-fold spots q_3. The real-space vectors a_3 correspond to the nearest-neighbor Cu–Cu distance in 12 directions. In the experimental FT power spectrum, the first two peaks are well documented, while the q_3 peak is not as strong as the perfect lattice. Overall, the fair agreements confirm that the Cu adatoms in the D-phase represent a factual manifestation of the (3^6; 4, 3^2, 4, 3) lattice.

Next we discuss the underlying driving force that forms the D-phase. When molecular density is low (molecular density [M] < 0.145 molecules per nm^2), ext-TPyB molecules self-assemble as sub-monolayer hexagonal frameworks (defined as the H-phase) that are stabilized by twofold Cu–pyridyl coordination (Fig. 3a).[20] At a density [M] of 0.145 mol nm^{-2}, the surface is fully covered by the H-phase. Above this density, threefold coordination motifs begin to emerge, introducing defects in the H-phase. At a density [M] of 0.15 mol nm^{-2}, small patches of the D-phase appear. As the density increases further, the area of the D-phase expands while the area of the H-phase shrinks. At a density [M] of 0.16 mol nm^{-2}, the D-phase covers 44% of the whole surface (Fig. 3b). At a density [M] of 0.18 mol nm^{-2}, the entire surface is covered by the D-phase. Fig. 3c shows a 0.5 μm × 0.5 μm area that is fully covered by the D-phase. In the absence of annealing, small domains of the H-phase, D-phase and other disordered phases are present on the surface of this sample. The homogeneous D-phase forms after 120 °C annealing. This structure persists after annealing at 200 °C, indicating that it is a thermodynamically favored structure. At higher densities above [M] = 0.18 mol nm^{-2}, a new

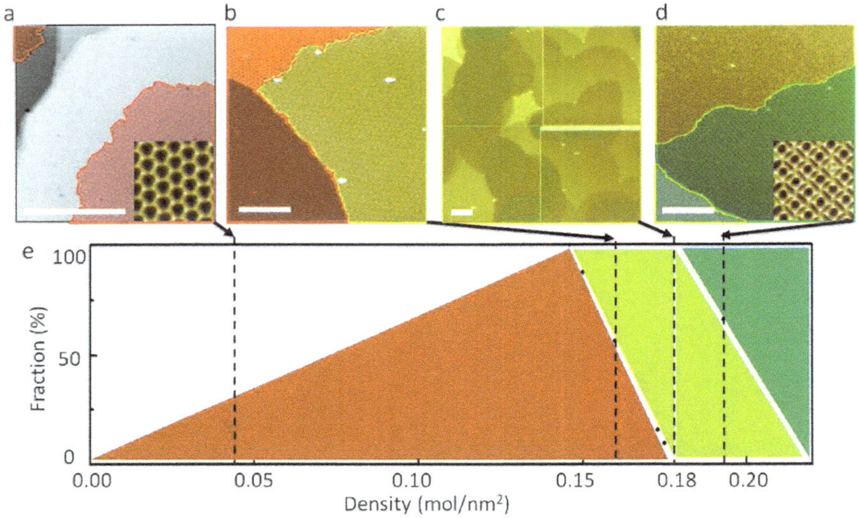

Fig. 3 Evolution of the area fractions of various phases against molecular density. (a) Sub-monolayer H-phase. (b) Mixed H-phase and D-phase. (c) Full-monolayer D-phase. (d) Mixed D-phase and S-phase (scale bars: 50 nm). Insets: high-resolution STM topographs (15 nm × 15 nm) of the H-phase and S-phase (all imaged at 78k). (e) Area percentage of various phases as a function of molecular coverage.

structure emerges which contains periodically-arranged squares made of four ext-TPyB molecules (inset in Fig. 3d, defined as the S-phase). At a density $[M]$ of 0.195 mol nm^{-2}, the D-phase and the S-phase coexist and occupy respectively 55% and 45% of the surface (Fig. 3d).

We systematically varied the molecular density and analyzed the total area occupied by each of the three phases. The area percentage is plotted against molecular density in Fig. 3e, which shows that the D-phase appears within a narrow window of 0.15 mol nm^{-2} < $[M]$ < 0.22 mol nm^{-2} and occupies the entire surface at $[M]$ = 0.18 mol nm^{-2}. Accompanied by increasing molecular density, two-phase transformations occur: first from the H-phase to the D-phase, then from the D-phase to the S-phase. Given that increasing molecular density accounts for increasing in-plane compression, the phase transformations can be thought of as driven by in-plane compression.[22,29] At low densities when $[M]$ < 0.15 mol nm^{-2}, only twofold coordination forms, indicating that twofold coordination is energetically more favorable than threefold coordination under low in-plane compression. However, as in-plane compression increases, the twofold coordination motifs transform into threefold ones to release compression stress. When the ratio of threefold to twofold coordination reaches a critical value of 0.133, the molecules reorganize at the annealing temperature to form the D-phase. Presumably, coordination bond energy and stress are balanced in the D-phase to minimize the total free energy.

It is worth noting that another binodal structure was observed in our previous study of the coordination assembly of TPyB with Cu on a Au(111) surface.[22] We discovered that, when the molecular density is below a critical value, Cu and TPyB formed a honeycomb structure, and increasing the molecular density above the critical value results in multistep nonreversible structural transformations involving pentagonal, rhombic, zigzag, and eventually triangular structures. Likewise, the structural transformations are attributed to the in-plane compression pressure. Along with the structural changes, the coordination configuration also undergoes a change from the twofold Cu–pyridyl binding in the honeycomb, pentagonal, rhombic and zigzag structures to the threefold Cu–pyridyl coordination in the triangular structure. Fig. 4a shows a binodal structure which features hexagons formed by twofold Cu–pyridyl coordination and the neighboring hexagons are joined by threefold Cu–pyridyl coordination modes in a side-by-side manner. As shown in Fig. 4b, the Cu adatoms in this structure exhibit the semi-regular "L" pattern in Kepler's drawing. Unlike the D-phase, this structure does not cover the entire surface, but co-exists with other structures that consist of a single coordination motif of twofold- or threefold Cu–pyridyl coordination.[22]

Besides the demi-regular lattice expressed in the D-phase, the Cu–pyridyl coordination also generates structures exhibiting semi-regular and regular lattices. A semi-regular lattice is made up of two or more regular polygons that are arranged with the same vertex. There are eight semi-regular lattices. Trihexagonal lattices (also known as the Kagomé lattice), defined as (3, 6, 3, 6) (the "p" pattern in Kepler's drawing), can be made by Cu–TPyB coordination and Cu–ext-TPyB coordination on Au(111) and Cu(111) surfaces.[19–22,30] On the Cu(111) surface, by depositing low-dosage TPyB/ext-TPyB molecules so the adsorbed molecules partially cover the surface, a honeycomb network structure is formed.[20,21] The Cu adatoms in this structure form a trihexagonal lattice. The Cu–ext-TPyB coordination structures on the Cu(111) surface were discussed as the H-phase formed at a molecular density $[M] \leq 0.145$ mole nm^{-2}. Fig. 5a shows a large area H-phase with the Cu adatoms highlighted by red dots and the trihexagonal lattice (Kagomé lattice) highlighted by the blue frame. Fig. 5b is a high resolution STM image overlaid with ext-TPyB molecules and Cu adatoms. Similar honeycomb TPyB/ext-TPyB network structures can be formed by depositing low-dosage TPyB/ext-TPyB and copper atoms on the Au(111) surface.[19,22]

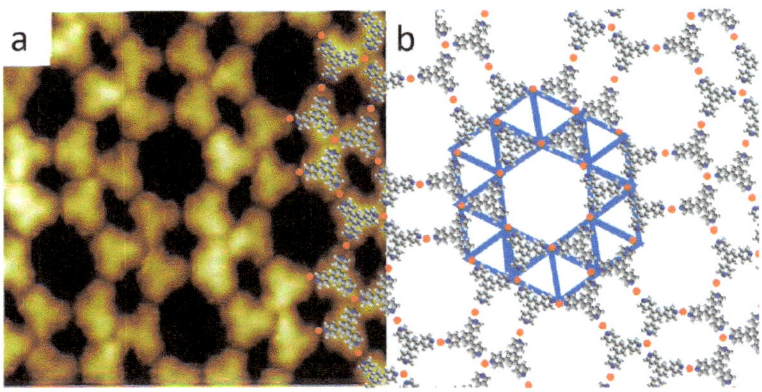

Fig. 4 (a) STM image of the Cu–TPyB coordination assembly on an Au(111) surface (10 nm × 10 nm) reproduced from ref. 22, with permission from the American Chemical Society. (b) A structural model in which Cu adatoms are marked by red dots. The blue lines highlight the semi-regular lattice.

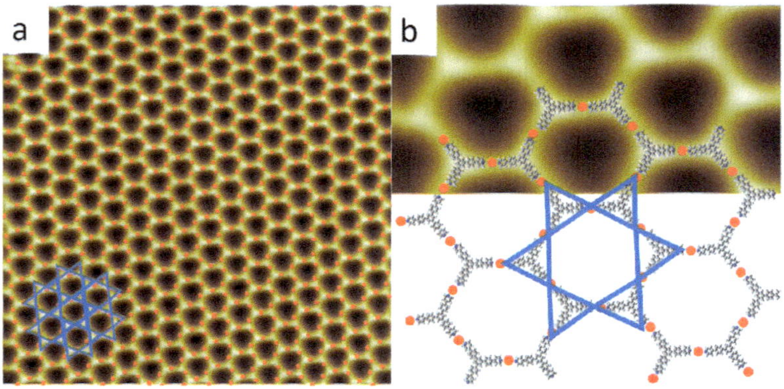

Fig. 5 (a) STM topograph of the H-phase self-assembled on Cu(111) (60 nm × 60 nm). Cu adatoms are marked by red dots. (b) High-resolution STM image (15 nm × 7.5 nm) of the Cu–ext-TPyB framework and the structural model. The blue frames highlight the (3, 6, 3, 6) trihexagonal lattice.

A regular lattice consists of one type of regular polygons connected by the same vertex. There are three regular lattices: equilateral triangles, squares and regular hexagons, which are the "d", "e" and "f" patterns in Kepler's drawing. Fig. 6 shows that the Cu–TPyB coordination framework assembled on the Au(111) surface consists of a triangular lattice of Cu adatoms formed by the threefold Cu–pyridyl coordination. The density of TPyB molecules in the triangular structure is 0.59 TPyB per nm^2, which is the highest in all phases formed. This phase grew larger with an increasing molecular density. It is noteworthy that even at the highest molecular density, the triangular structure did not grow into extended large domains but was partitioned by domain boundaries as shown in Fig. 6. The domain boundaries exist as hexagonal corner holes and linear defects as shown in Fig. 6a, which consist of twofold Cu–pyridyl coordination.

2D coordination frameworks consisting of other lattices were reported in the literature. Fig. 7 exemplifies two structures formed by ditopic linear molecules

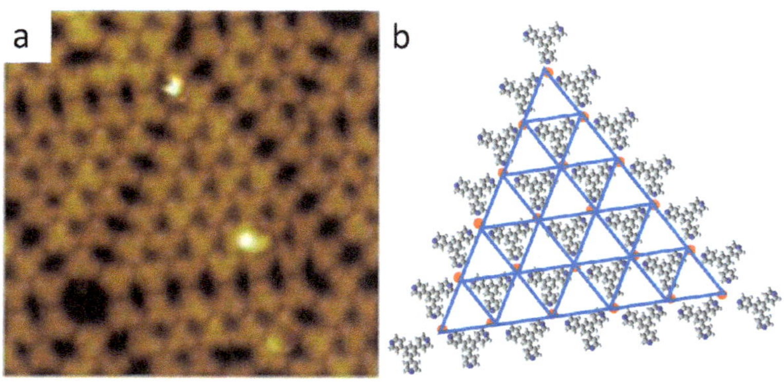

Fig. 6 (a) STM topograph of the Cu–TPyB coordination assembly on a Au(111) surface (15 nm × 15 nm) reproduced from ref. 22, with permission from the American Chemical Society. (b) The structural model. Cu adatoms are marked by red dots. The blue frame highlights the triangular lattice.

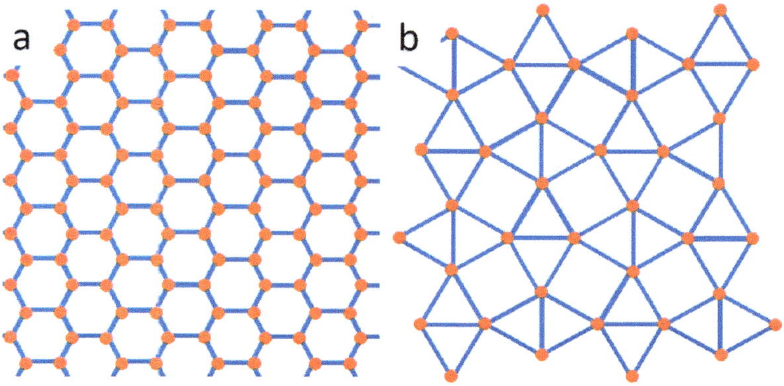

Fig. 7 Schematic models of (a) hexagonal[31] and (b) snub square lattices.[17] Metal atoms and molecular linkers are highlighted by red dots and blue frames, respectively.

(dicarbonitrile-polyphenyl) coordinated with metal atoms. Fig. 7a and b feature a regular hexagonal lattice of Cu adatoms[31] and a semi-regular snub square (3, 3, 4, 3, 4) lattice of Ce adatoms,[17] which correspond to the "f" and "n" patterns in Kepler's drawing, respectively.

Conclusions

In summary, we have demonstrated the assembly of a non-Archimedean demi-regular lattice of Cu atoms in an interfacial metal–organic framework. The co-existing and perfectly arranged twofold and threefold Cu–pyridyl coordination modes emerging from critical in-plane compression are the key ingredients for forming this structure. We anticipate that this strategy may be used to design other non-Archimedean lattices. The unique lattice symmetry and high stability are expected to lead to novel physical properties that deserve further study. Some of these structures feature 12-fold local rotational symmetry. Crystalline materials expressing such rotational symmetry are expected to possess novel properties, for example, wide-angle optical isotropy.[18] Together with our former work reporting that Cu–pyridyl coordination structures exhibit semi-regular and regular lattices, we conclude that regular, semi-regular and demi-regular lattices can be achieved by Cu–pyridyl coordination on surfaces.

Acknowledgements

This work is supported by Hong Kong RGC 16303514, the National Natural Science Foundation of China (Project No. 21172069, 21372072 and 21190033) and the Science Fund for Creative Research Groups (21421004).

References

1 J. Kepler, *Harmonices Mundi*, Johannes Planck, Linz, 1619.

2 O. Krotenheerdt, *Die homogenen Mosaike n-ter Ordnung in der euklidischen Ebene*, Wiss. Z. Martin-Luther-Univ. Math.-natur. Reihe, Halle-Wittenberg, 1970, vol. 19, pp. 97–122.

3 B. Grünbaum and G. Shephard, *Tilings and patterns*, WH Freeman, New York, 1987.

4 M. O'Keeffe and B. G. Hyde, *Philos. Trans. R. Soc., A*, 1980, **295**, 553–618.

5 S. Sioula, N. Hadjichristidis and E. L. Thomas, *Macromolecules*, 1998, **31**, 8429–8432.

6 A. Takano, W. Kawashima, A. Noro, Y. Isono, N. Tanaka, T. Dotera and Y. Matsushita, *J. Polym. Sci., Part B: Polym. Phys.*, 2005, **43**, 2427–2432.

7 B. Chen, X. B. Zeng, U. Baumeister, G. Ungar and C. Tschierske, *Science*, 2005, **307**, 96–99.

8 F. Zhang, Y. Liu and H. Yan, *J. Am. Chem. Soc.*, 2013, **135**, 7458–7461.

9 Q. Chen, S. C. Bae and S. Granick, *Nature*, 2011, **469**, 381–384.

10 J. A. Millan, D. Ortiz, G. van Anders and S. C. Glotzer, *ACS Nano*, 2014, **8**, 2918–2928.

11 T. Paik, B. T. Diroll, C. R. Kagan and C. B. Murray, *J. Am. Chem. Soc.*, 2015, **137**, 6662–6669.

12 C. Volkringer, M. Meddouri, T. Loiseau, N. Guillou, J. Marrot, G. Férey, M. Haouas, F. Taulelle, N. Audebrand and M. Latroche, *Inorg. Chem.*, 2008, **47**, 11892–11901.

13 B.-Q. Song, X.-L. Wang, Y.-T. Zhang, X.-S. Wu, H.-S. Liu, K.-Z. Shao and Z.-M. Su, *Chem. Commun.*, 2015, **51**, 9515–9518.

14 K. Tahara, S. Furukawa, H. Uji-i, T. Uchino, T. Ichikawa, J. Zhang, W. Mamdouh, M. Sonoda, F. C. De Schryver, S. De Feyter and Y. Tobe, *J. Am. Chem. Soc.*, 2006, **128**, 16613–16625.

15 Z. Shi and N. Lin, *J. Am. Chem. Soc.*, 2009, **131**, 5376–5377.

16 V. Stepanenko, R. Kandanelli, S. Uemura, F. Wuerthner and G. Fernandez, *Chem. Sci.*, 2015, **6**, 5853–5858.

17 D. Ecija, J. I. Urgel, A. C. Papageorgiou, S. Joshi, W. Auwaerter, A. P. Seitsonen, S. Klyatskaya, M. Ruben, S. Fischer, S. Vijayaraghavan, J. Reichert and J. V. Barth, *Proc. Natl. Acad. Sci. U. S. A.*, 2013, **110**, 6678–6681.

18 M. E. Zoorob, M. D. B. Charlton, G. J. Parker, J. J. Baumberg and M. C. Netti, *Nature*, 2000, **404**, 740–743.

19 Z. Shi, J. Liu, T. Lin, F. Xia, P. N. Liu and N. Lin, *J. Am. Chem. Soc.*, 2011, **133**, 6150–6153.

20 S. Wang, W. Wang, L. Z. Tan, X. G. Li, Z. Shi, G. Kuang, P. N. Liu, S. G. Louie and N. Lin, *Phys. Rev. B: Condens. MatterMater. Phys.*, 2013, **88**, 245430.

21 W. Wang, X. Shi, S. Wang, J. Liu, M. A. Van Hove, P. N. Liu, R.-Q. Zhang and N. Lin, *Phys. Rev. Lett.*, 2013, **110**, 046802.

22 J. Liu, T. Lin, Z. Shi, F. Xia, L. Dong, P. N. Liu and N. Lin, *J. Am. Chem. Soc.*, 2011, **133**, 18760–18766.

23 S. Stepanow, N. Lin, D. Payer, U. Schlickum, F. Klappenberger, G. Zoppellaro, M. Ruben, H. Brune, J. V. Barth and K. Kern, *Angew. Chem., Int. Ed.*, 2007, **46**, 710–713.

24 T. Classen, G. Fratesi, G. Costantini, S. Fabris, F. L. Stadler, C. Kim, S. de Gironcoli, S. Baroni and K. Kern, *Angew. Chem., Int. Ed.*, 2005, **44**, 6142–6145.

25 W. Steurer, *Z. Kristallogr.*, 2004, **219**, 391–446.

26 X. B. Zeng, G. Ungar, Y. S. Liu, V. Percec, S. E. Dulcey and J. K. Hobbs, *Nature*, 2004, **428**, 157–160.
27 K. Hayashida, T. Dotera, A. Takano and Y. Matsushita, *Phys. Rev. Lett.*, 2007, **98**, 195502.
28 S. Fischer, A. Exner, K. Zielske, J. Perlich, S. Deloudi, W. Steurer, P. Lindner and S. Foerster, *Proc. Natl. Acad. Sci. U. S. A.*, 2011, **108**, 1810–1814.
29 F. Vidal, E. Delvigne, S. Stepanow, N. Lin, J. V. Barth and K. Kern, *J. Am. Chem. Soc.*, 2005, **127**, 10101–10106.
30 R. Zhang, G. Lyu, C. Chen, T. Lin, J. Liu, P. N. Liu and N. Lin, *ACS Nano*, 2015, **9**, 8547–8553.
31 G. E. Pacchioni, M. Pivetta and H. Brune, *J. Phys. Chem. C*, 2015, **119**, 25442–25448.

Faraday Discussions

DISCUSSIONS

Supramolecular effects in self-assembled monolayers: general discussion

David Amabilino, Ioan Bâldea, James Batteas, Pol Besenius, Peter Beton, Manfred Buck, Lifeng Chi, Giovanni Costantini, Philip Davies, Steven De Feyter, Yuri Diaz Fernandez, Deepak Dwivedi, Karl-Heinz Ernst, Amar Flood, Brandon Hirsch, Vincent Humblot, Robert Jones, Angelika Kühnle, Markus Lackinger, Nian Lin, Trolle R. Linderoth, Claire-Marie Pradier, Talat Rahman, Rasmita Raval, Neil Robinson, Marco Sacchi, Sebastian Schwaminger, Steven L. Tait, Phil Woodruff and Han Zuilhof

DOI: 10.1039/C7FD90073B

Steven L. Tait opened the discussion of the introductory lecture by Rasmita Raval: You made an analogy in your presentation between polymorphism and the many chemical functions that happen in parallel in a cell. In studies of molecular self-assemblies, there is often a tendency to look for a single low energy structure that could potentially have a distinct, single chemical function. Do we need to reconsider this mindset in order to take advantage of potential benefits of polymorphic systems that could potentially be developed as multi-functional systems? What suggestions do you have for designing and characterizing polymorphic molecular systems that have useful multi-functionality?

Rasmita Raval replied: Polymorphism is often ignored entirely, as most researchers continue to concentrate on solving a defined phase. I feel strongly that polymorphism is a fantastic attribute that complex molecules display at surfaces. Each phase will have its unique chemical and geometric structure, with a distinct functionality. Therefore, one now has the possibility to create many functions from a single molecule-single surface system, simply by changing conditions such as coverage, temperature, pH, *etc.* which lead to different thermodynamically preferred assemblies, or *via* kinetic capture of metastable states. Natural systems use such attributes to create highly responsive systems, or to switch certain functions on and off. We should be prepared to 'roam' across the phase-space, map the phases that are created and, most importantly, explore the functions that each phase could deliver. Such information is the basis for exploiting polymorphism to create smart surface systems that harness the flexibility of the system and display different functions by design. One can imagine that such approaches can then be extended by creating local environments which favour specific phases, thus providing a route towards a system that concurrently displays multiple functions in local domains across a surface. One could take this further whereby the function of one domain acts as impulse or feedback for a neighbouring domain, creating a communicating multifunctional surface, which would be a 2D mimic of the multifunctional cell.

Talat Rahman commented: You've raised a number of interesting issues; one which somehow we've not yet taken into account is entropic effects. When you were showing the slide with the arrangements of your enantiomers, differences in their energies were of the order of tens or hundreds of millivolts, which is about where entropic contributions can be expected to play a role. The other thing, which we're finding now, and your disordered systems allude to that, is the issue with adsorption sites. We typically consider those that are the most obvious. Yet recent findings using machine learning, for example, reveal possible adsorption sites that one may not assume naturally. On second thought they make sense, because most of the time adsorption is determined by the nature of the orbitals in the bonding layer. Would you agree then that the disorder seen in your experiments may be related to the system having to choose amongst a number of adsorption sites with similar binding energies?

Rasmita Raval answered: Entropic effects are largely ignored at present, since they are difficult to measure experimentally in surface systems, and difficult to calculate. You are absolutely right to point out that the enthalpic energies calculated for different phases often vary by very small amounts. Therefore, entropic factors will play a key role in determining which phase is exhibited at a particular temperature. This also increases the polymorphism of a system and, with it, the opportunities to design system function as I have outlined in my reply to the previous question from Steven L. Tait. Our work has also shown that entropy plays a critical role in determining both the disordered and ordered phases at a surface.[1] In the case we studied, entropic effects determined the ordering propensity of each enantiomer. Your second point on adsorption sites is also important. Given the complexity of the molecule–surface system, there is a tendency to iterate calculations around obvious adsorption configurations. However, you are right that different adsorption sites often differ by small energy differences and your suggestion of using machine learning to fully sample the parameter space to capture the variety of configurations available to a single adsorbed species is a good way forward. In addition, we need to recognise that the relative energies for the different adsorption configurations can change significantly when translating from the single adsorbed molecule to the organised phase. Therefore, there is also a need to explore all the permutations that could occur in the ordered phase. We certainly saw the importance of this when we revisited the tartaric acid/Cu(110) system (as described in my lecture and in ref. 2). Finally, does the disordered phase include a number of different adsorption configurations that are close in energy? This is difficult to prove experimentally except in large molecular systems where each adsorbate could potentially be mapped. Clearly, there will be a Boltzmann distribution of different species within the disordered phase and entropic factors will become important for such systems. Undoubtedly, a fascinating area of discovery for the field of complex molecules at surfaces!

1 S. Haq, N. Liu, V. Humblot, A. P. J. Jansen and R. Raval, *Nat. Chem.*, 2009, **1**, 409.
2 G. R. Darling, M. Forster, C. Lin, N. Liu, R. Raval and A. Hodgson, *Phys. Chem. Chem. Phys.*, 2017, **19**, 7617.

Karl-Heinz Ernst asked: Rasmita, you mentioned that one has to take all options into account when resolving the structure of an adsorbate or a self-assembled domain, *e.g.* the Cu(110)/TA-(9 0, 12) phase. How about surface reconstruction or Cu adatoms? Shouldn't these options be taken into account as well?

Rasmita Raval replied: You are absolutely right. The response of the surface to the molecule is a critical factor. We had demonstrated molecule-induced surface reconstruction back in 1990 when we showed that the CO molecules could reconstruct the Pd(110) surface once a critical coverage had been reached.[1] Even more surprising was the fact that as the coverage is increased, the reconstruction is lifted and a new surface assembly is created. This system certainly showed not only that surface reconstruction should be considered, but also that the reconstruction can be surface phase-dependent and can be induced and lifted during the course of adsorption and assembly. We certainly considered the possibility of surface reconstruction for the Cu(110)/TA-(9 0, 12) phase, especially given the strength of chemisorption of the bitartrate to the surface. However, so far, our experimental or theoretical data do not indicate any drastic reconstruction.

1 R. Raval, S. Haq, M. A. Harrison, G. Blyholder and D. A. King, *Chem. Phys. Lett.*, 1990, **167**, 391–398.

Claire-Marie Pradier said: With your molecular assemblies and supramolecular systems, you're very close to the molecular machines (Nobel prize this year). To go further, could molecules which have the same functions together display coherent motions on the surface? Please could you comment on that?

Rasmita Raval replied: Yes, indeed, the 2016 Nobel Prize in Chemistry has been really inspirational! For many future technologies, we will require molecular machines at surfaces. Here, we are going to have to combine a number of aspects *e.g.* the motion of single molecular entities, the creation of supramolecular or covalent architectures that confine or direct the motion, the ability to switch systems, creating recognition sites for docking, *etc.* Furthermore, all of this will need to be achievable at operable temperatures, *e.g.* room temperature. Our paper[1] describes a simple surface system that we created in collaboration with David Amabilino and Lluisa Pérez-García, where a simple walker molecule walks along a particular surface track, with the motion confined by covalent porphyrin fences created by on-surface synthesis. One can say that this system has analogies to the rotaxane molecule where travel of the macrocycle ring occurs along a defined track and is confined by two end stations. A particular aspect of our work is that the walker molecule chemisorbs to the surface *via* two 'feet', which then detach and attach in an inchworm fashion as shown by the calculations undertaken by Lev Kantorovitch's group. This strong interaction means the system can operate at room temperature. This is just the first step – we now need to design systems that will actually perform work!

1 S. Haq *et al.*, *Angew. Chem. Int. Ed.*, 2015, **54**, 7101.

Manfred Buck enquired: You presented an example of one-dimensional motion of molecules between barriers and, in this context, molecular machines were mentioned. Since, for the system presented, the motion is a random walk in 1D, are there any ideas/concepts apart from scanning probe-based ones which might lead to a directed motion?

Rasmita Raval responded: You are right, the simple surface molecular machine system I discussed shows 1D directional motion, but with random stochastic steps in either direction. Tip-induced directed motion has been demonstrated for surface species. Away from scanning probe-based solutions, one will have to use approaches such as using functional groups that can be activated *e.g.* by light, pH or redox cycles, incorporating catalytic centres that provide fuel, designing in asymmetry or Brownian motion ratchet approaches used in natural systems, utilising chemical gradients, anisotropic physical fields and non-equilibrium systems that direct motion. What is clear is that there is plenty of science that remains to be tackled!

Steven De Feyter asked: Probably a more philosophical question: to what extent are we biased by the things that we can "see" using our techniques? To what extent do we really grasp reality if we are biased to look at "static features" rather than "dynamic systems" because of the limitations of the instrumentation or methodology that we use?

Rasmita Raval responded: A truly philosophical question for all science! We construct our scientific knowledge on measurable things and are largely blind to that which is unmeasurable. As our techniques advance, our knowledge base comes under scrutiny (the evolution of quantum mechanics is a fine example). However, it is true that order, symmetry and static structures are often given the greatest attention in many great fields of science. I am convinced that we need to re-balance our perspectives and to actively seek disorder, non-equilibrium and transient aspects in our field. This will need both a change in scientific attitude (by scientists, editors and the funding authorities!) and the development of better spatial- and time-resolved techniques and theory.

David Amabilino commented: A really interesting point was raised about disorder in supramolecular structure in layers being ignored sometimes. If you look at "disorder" from a covalent chemistry perspective being the formation of non-desired molecules, when synthesis is performed in solution we might get a 60% yield, so 60% is what you wanted, what was the rest? That 40% goes mainly unidentified. This is a great opportunity for surface science to try and understand and direct chemistry, where techniques could be eventually used to identify each product on a surface and control reaction pathways through concentrations and temperature. How far away in time do you see this dream coming true?

Rasmita Raval answered: You raise a really important and interesting point here. In this context, one can deploy the local analytical techniques that are now at the disposal of a surface scientist and observe model covalent reactions at a surface and identify the variety of products created, even quantify the ratios. Now that on-surface covalent organic synthesis has become part of our repertoire, it would be interesting to undertake test reactions and address this question directly. Specifically, it would be instructive to expand our observations from products we want to see, to the others that demonstrate the diversity of outputs, including messy, disordered products. As you state, the opportunity exists to alter reaction conditions and observe their effect on product distribution, with the prize of delivering the parameters that guide reactions. How far away are we from this dream? Well, I think we are actually ready to start straightaway on surfaces! We will need to create close conversations between organic chemistry, surface science and theory to design systems that will allow us to observe, learn and then engineer outputs. Will this throw light on organic synthesis in solution? Maybe not directly, but I have no doubt that generic principles will arise that can be translated. Finally, I also point out that the recent advances in local probing at surfaces means that the time is ripe to re-visit heterogeneous catalysis and address the central questions of activity and selectivity in surface reactions.

Talat Rahman noted: Just a quick comment about the interpretation of experimental data; looking at scan images, the tip is playing a major role there, and it has to be included in nucleation. We are seeing what is happening, but our interpretation is not there yet.

Rasmita Raval answered: Tip effects are, of course, known to be important in imaging and progress is being made on modeling different types of tips. We also know that in some systems, the tip appears to be actively involved in nucleating an island, or dispersing an existing island. There is little understanding or interpretation of exactly what is happening here. It would be interesting to understand this phenomenon. Perhaps we could then exploit it to nucleate and write organised molecular structures where we wanted them.

Talat Rahman opened a general discussion of the paper by Ioan Bâldea: Your results are sensitive to the coverage of the molecule. Did you perform the calculations for several coverages? Also, van der Waals interactions are expected to be important, both for molecule–surface and inter-molecular interactions. Did you try out different ways of incorporating these interactions?

Ioan Bâldea replied: Thanks for your two-fold question. To your first point: yes, I performed calculations for more coverages than I've shown in the paper. The (6×6) arrangement is shown as a typical case of very low coverage. Increasing the coverage by a factor of 6 in a (3×2) arrangement – which corresponds to the unit cell of the herringbone ordering wherein one out of the two molecules is removed – yields no notable change in the twisting angle, the value of which is basically that of isolated molecules. The problem with variable coverages is that the coverage cannot be varied continuously; so, one cannot investigate whether the increase in twisting angles from $\sim 36°$ to $\sim 76°$ is abrupt or gradual. Varying the coverage (quasi) continuously amounts to investigating incommensurate SAM superstructures, situations that can be mimicked only by choosing very large SAM unit cells, which are computationally prohibitive. To the second point: I show in the paper results for SAMs adsorbed on Au(111) demonstrating that van der Waals corrections are not quantitatively significant: please compare the results of the "standard" calculations without dispersion corrections (acronym "0.02") with those obtained by including dispersion corrections (labeled "VDW-DRSLL") in Tables S6 and S7 of the ESI. To this aspect I must still note that several earlier works on some SAMs adsorbed *via* thiol groups already found that dispersion corrections are negligible.

James Batteas remarked: Could you comment on the energy barrier for the rotation in this system? You've isolated these two main states, but there will always be some fluctuations due to thermal effects. Could you comment on any relative impacts from thermal motions here on the transport?

Ioan Bâldea answered: Thank you, this is certainly a relevant aspect. Unfortunately, the state-of-the-art of the theory does not allow accurate values of the torsional energy barrier to be achieved. Even the torsional energy barrier of the isolated biphenyl molecule (amounting to ~0.1 eV according to experiments[1]) could hardly be reproduced theoretically;[2] quantum chemical calculations at such elaborate levels[2] for molecules forming SAMs adsorbed on metals are hopeless at present. Nevertheless, values of ~0.1 eV are still four times larger than the thermal energy at room temperature, and for this reason it is hard to imagine that, albeit significant, thermal fluctuations would completely wash out the impact of torsional angle on conductance. In fact, due to steric hindrance I would expect higher torsional energy barriers for molecules densely packed in SAMs than for isolated molecules. In addition to this, for studies on the related bipyridine molecule[3,4] it is known that the torsional motion is highly anharmonic. To compute currents, ensemble averaging over the various molecular conformations has to be performed,[5,6] and this anharmonicity acts to suppress fluctuations away from the averaged value of the torsional angle.

1 O. Bastiansen and S. Samdal, *J. Mol. Struct.*, 1985, **128**, 115.
2 M. P. Johansson and J. Olsen, *J. Chem. Theor. Comp.*, 2008, **4**, 1460.
3 I. Bâldea, *EPL*, 2012, **99**, 47002.
4 I. Bâldea, *J. Phys. Chem. C*, 2014, **118**, 8676.
5 I. G. Medvedev, *Electrochim. Acta*, 2008, **53**, 6545.
6 I. Bâldea, *J. Phys. Chem. C*, 2013, **117**, 25798.

Trolle R. Linderoth enquired: What is the energy difference between the two conformational states? For instance, in your condensed, herringbone layer of molecules where you have a preferred intramolecular angle of 76° between the two benzene rings, what would happen if as a defect you introduced a molecule with the 36° angle preferred for the isolated molecules? Information on the energetics would allow to estimate to what extent these states are populated in the experimental situation at room temperature.

Ioan Bâldea answered: Thanks for your interesting question. The unit cell used in calculations for SAMs of BPMT and BPDT with herringbone arrangement has 62 atoms and 64 atoms, respectively. To simulate a concentration of defects amounting to 1/9 (so, a pretty high value), given the fact that SIESTA employs periodic boundary conditions, one would need unit cells containing at least $3 \times 3 \times 62 = 558$ and $3 \times 3 \times 64 = 576$ atoms for BPMT- and BPDT-based SAMs, respectively. With computational resources available to date, such calculations are prohibitive for any theoretical approach that could be trusted. One could expect that the energy to create such a defect is something of the order of the torsional energy barrier of isolated biphenyl (~0.1 eV, *cf.* ref. 1). For elaborate *ab initio* theoretical approaches it was already a big challenge to reproduce the experimental value for this case (one isolated biphenyl molecule!). Johansson and Olsen[2] drew attention to the fact that using basis sets of at least augmented quadruple zeta quality is a crucial prerequisite. To reach the accuracy of *ab initio* methods, hybrid exchange–correlation functionals (B3LYP) are needed in DFT calculations, which – parenthetically – are not implemented in SIESTA. Concerning the temperature effect: albeit not extraordinary large, energy barriers of ~0.1 eV (in fact, for molecules forming SAMs adsorbed on electrodes I would expect more than for isolated molecules) are still significantly larger than the thermal energy of ~26 meV at room temperature. Therefore, although thermal fluctuations can be significant, it is hard to believe that the effect will be wiped out.

1 O. Bastiansen and S. Samdal, *J. Mol. Struct.*, 1985, **128**, 115.
2 M. P. Johansson and J. Olsen, *J. Chem. Theor. Comp.*, 2008, **4**, 1460.

Robert Jones asked: In the paper you refer to a herringbone surface structure, *e.g.* in Fig. 2 and associated text. Clean Au(111) undergoes a herringbone reconstruction. It is unclear to me whether the substrate gold (111) surface has the herringbone reconstruction, or whether it is the SAM that has the herringbone structure on top of a bulk terminated Au(111) surface. Could you clarify this please?

Ioan Bâldea responded: Thanks for your interesting question. In the "standard" calculations (acronym "0.02") I took gold layers of fcc Au(111) with gold–gold separation frozen at the experimental value $d = 2.88367$ Å. The twisting angles obtained in this way do not notably differ from the values obtained by also optimizing the positions of the gold atoms in the top layers (acronyms "CAB-R" and "CABCAB-R"); compare the $\varphi_{1,2}$ values of the second line with those of the last two lines in Tables S6 and S7 of the ESI (DOI: 10.1039/C7FD00101K).

Manfred Buck commented: For thiol layers, the binding of the head group to the substrate exerts a decisive influence on the structure of the SAM. In addition, the Au surface significantly restructures upon SAM formation which also involves Au adatoms. Has the influence of the interface structure on the SAM structure been looked at, *i.e.*, have unreconstructed and adatom-containing models been compared?

Ioan Bâldea replied: The results of my calculations – whether without or with reconstruction, in cases where I included it (*cf.* the last two lines in Tables S6 and S7 of the ESI, DOI: 10.1039/C7FD00101K) – support the prevailing view that adsorption occurs at hollow positions. Still, to reiterate what I said in the paper, results on the preferential adsorption position of both experimental and theoretical studies published so far are controversial.

Trolle R. Linderoth remarked: Your conclusion seems to be that the key effect of two conformational states is rather independent of the details of the interactions to the gold substrate; could the important effects be modelled in a 2D system without the substrate, and perhaps thereby allow the intermolecular distance to be varied continuously?

Ioan Bâldea replied: This is a very interesting question. Work done in this direction since I submitted the manuscript is encouraging. A two-dimensional model seems to be able to capture the essential aspects of the interactions, allowing at the same time a continuous variation of the intermolecular spacing.

Manfred Buck said: In the SAM calculations, does the torsion angle between the aromatic rings of biphenyl depend on the starting conditions? Is there a difference in the final SAM structure if the calculations start from a significantly twisted conformation of the biphenyl as found in an isolated molecule or a coplanar conformation?

Ioan Bâldea replied: Thanks for raising an interesting point. Occasionally, starting with twisting angles very different from the most stable configuration (*e.g.*, near coplanarity), I found situations where optimization ended with configurations wherein one molecule had a torsional angle of $\sim 40^\circ$ and was substantially tilted (tilt angle $\theta \sim 40–50^\circ$ with the normal to the electrode surface), while the second molecule (I am referring to the herringbone unit cell) had a twisting angle $\varphi \sim 76^\circ$, and stood nearly upright on the metal surface. However, such configurations (metastable or transition states?) were by a few tenth of eV higher in energy than the lowest energy extremum (corresponding to $\varphi_{1,2} \sim 76^\circ$ and nearly "vertical" molecules), which was recovered by slightly perturbing that starting conformation.

Rasmita Raval opened the discussion of the paper by Pol Besenius: You have shown beautiful supramolecular assemblies for your system. Often, such assemblies are probed by techniques such as CD that give the average response of the system. However, it would seem that most of the structural models are derived from molecular dynamic calculations. What experimental methods are currently being applied to such systems to get independent verification of structure, or is this currently too challenging?

Pol Besenius replied: We used sum frequency generation and IR vibrational spectroscopy experiments to show that the structures are ordered in our first report on these sequentially-grown copolymers.[1] SFG experiments suggested that the degree of order increased with the number of added monomers. In IR spectra of the surface-grafted copolymers the amide I and II bands were characteristic for hydrogen-bonded secondary structures, which were similar to a material that was self-assembled in solution and then deposited on a Au surface. Typically, in solution studies of the copolymers, we perform CD, WAXS, FRET, as well as TEM, cryoTEM and PFG-NMR.[2–4]

1 H. Frisch *et al.*, *Angew. Chem. Int. Ed.*, 2016, **55**, 7242–7246.
2 R. Appel *et al.*, *Chem. Eur. J.*, 2015, **21**, 19257–19264.

3 H. Frisch *et al.*, *Chem. Eur. J.*, 2015, **21**, 3304–3309.
4 H. Frisch *et al.*, *Angew. Chem. Int. Ed.*, 2013, **52**, 10097–10101.

Trolle R. Linderoth asked: The state of your monomer, is that known? Should we think of that as 2D flat or globular, and how does this affect the self-assembly on the surface?

Pol Besenius replied: We have previously shown with TEM that the monomers flatten out on a surface.[1] A manuscript has just been submitted where we looked at the copolymer formation in solution with multi-angle dynamic light scattering. In isolated solutions of physiological ionic strength, both monomers have a hydrodynamic radius of 1.9 nm and are globular. However, from unpublished work, we know that the dendritic core is important. If it is too floppy, the growth into anisotropic nanorod-like objects is not favourable.

1 H. Frisch *et al.*, *Chem. Eur. J.*, 2015, **21**, 3304–3309.

Claire-Marie Pradier commented: You mentioned in the title surface-confined supramolecular copolymers. In fact, you build confined surface polymers because you bind them to initial layers with a density that is controlled by yourself; why is such a control of the multilayer density (confinement at the surface) so important?

Pol Besenius responded: The sequential growth of copolymers occurs on the surface-bound layers. In that sense we manage to confine the assembly on the surface, and solution self-assembly is avoided. I agree, we do not describe a self-assembly process between two states that are both bound or confined on the surface, but we use the surface as a means to confine the copolymer growth. The density of the initial layer is critical. If it is not high enough, the rods that grow off the surface start to collapse or flatten out on the surface.

Trolle R. Linderoth said: To help clarify the question I think all the surface scientists in the audience have in mind: your cartoons show distinct columns of alternating molecules. Is there such a degree of 2D order, or should we think of your structures as alternating disordered layers?

Pol Besenius replied: In our first report on these sequentially-grown copolymers, we used sum frequency generation and IR vibrational spectroscopy experiments to show that the structures are ordered.[1] SFG data suggested that the degree of order increased with the number of added monomers or molecular layers. In IR spectra of the surface-grafted copolymers the amide I and II bands were characteristic for hydrogen-bonded secondary structures. These were similar to a material that was self-assembled in solution and then deposited on a Au surface.

1 H. Frisch *et al.*, *Angew. Chem. Int. Ed.*, 2016, **55**, 7242–7246.

Sebastian Schwaminger commented: I have a question concerning the QCM measurement. You assigned the viscoelastic effects to a soft polymer. Is it possible that the viscoelastic effect at 45 °C, or at least part of this behaviour, is caused by temperature effects? Or is this effect dependent on the copolymer structure exclusively?

Pol Besenius replied: The experimental data obtained from temperature- and solvent-dependent SPR experiments combined with the QCM-D studies suggest that at higher temperatures more monomer molecules are adsorbed on the surface, which leads to a thicker and softer polymer matrix, compared to the ones at low temperatures. You are right, though, in suggesting that temperature effects in the viscoelastic properties of the polymer film cannot be excluded.

Amar Flood said: A very nice talk. I need some clarification on the data recorded at room temperature relative to higher temperatures, as shown in Fig. 2 of your paper (DOI: 10.1039/C7FD00100B). It seems that you get a greater SPR response at higher temperatures. Is this because more material is going down in a layer-by-layer manner at higher temperature? I would assume yes. However, as I understand layer-by-layer, you always need equal charge balance to achieve regular steps with each layer, just as you observed. So, do you know why more material needs to go down at higher temperature to balance the charge than at lower temperatures?

Pol Besenius replied: This is a good point. I agree with your interpretation of the temperature-dependent SPR data. The differences in the incremental increase in the SPR signal (Fig. 3 of the paper, DOI: 10.1039/C7FD00100B) suggest non-linear temperature-dependent behaviour, which is indicative of desolvation of the triethylene glycol shell of the copolymers. As you suggest, charges must be neutralised after each sequential monomer addition to avoid accumulation which would inhibit structure formation. However, in our sample preparation we make sure that for all temperatures we start the kinetic measurements with the same first dense layer of cationic monomers, thus a surface that is saturated as pointed out in the experimental part. We therefore assume that the structure formation is more efficient at higher temperatures, due to a combination of lower steric demand of the desolvated dendritic triethylene glycol chains and the more pronounced entropic contribution originating from the counterion release upon electrostatic binding. The data in Fig. 4 of the paper support this notion, since the addition of acetonitrile suppresses thermoresponsive behaviour of the side chains. Indeed, the difference in the SPR data at 25 °C and 45 °C becomes very small after the addition of 20 vol% acetonitrile to the aqueous buffer.

Yuri Diaz Fernandez commented: Regarding the remarkable temperature dependence observed in your system for both SPR and QCM data, how will the exchange of dendritic molecules adsorbed in the first layer affect the subsequent growing of the multilayers? Expecting an increase in the exchange rate with temperature, could the formation of mixed layers explain the increase in mass per single functionalisation step observed at higher temperatures?

Pol Besenius responded: Since in the current system every monomer has six thioether groups, and could displace the surface-bound oligomers or polymers, I would expect that higher temperatures and an increase in the exchange dynamics would decrease the driving force for the formation of supramolecular multilayers.

This is not supported by the experimental data. We have therefore assigned the temperature-dependency to the LCST behaviour and desolvation effects of the dendritic ethylene glycol chains. We have shown that the addition of 20% acetonitrile to the aqueous comonomer solutions supports this notion, since the LCST effect vanishes and the temperature-dependency is much less pronounced.

Han Zuilhof said: Your idea of a one-on-one ordered structure based on the SPR data hinges on the refractive indices. Given the sizes of these molecules (identical, but for one small functional group), it is likely that the refractive indices are nearly equal. However, the additions of compounds **1** and **2** are really not equal (especially in Fig. 4 of your paper, DOI: 10.1039/C7FD00100B, is this clearly observable); what are your opinions about the refractive indices of these materials, and do you have evidence to suggest that it must indeed be the difference in refractive indices, rather than a loss of order, that must cause this effect?

Pol Besenius responded: I agree, the differences in the incremental refractive index changes for the anionic compared to the cationic monomer are surprising. This observation cannot be explained just by the small difference in molecular weight of both species. Your point about the difference in supramolecular order caused by the addition of the monomer is valid. In our initial report,[1] the SFG data suggested that the order increases with the layer thickness. Intriguingly, when focussing on the C=O signal of the glutamic acid side chain modes, we observed an increase even after the addition of the lysine-containing comonomer. This suggests that the order of the surface-bound stacks is dictated by interactions between both monomers and that the order within the stack, or at least in the active chain end, changes after the addition of each species.

1 H. Frisch *et al.*, *Angew. Chem. Int. Ed.*, 2016, **55**, 7242–7246.

Han Zuilhof asked: Relating to that one-on-one order you refer to, could you see in IR measurements hydrogen bonding on your surface as you have in solution? And do you have AFM data that would indicate the step-by-step growth of the thickness, rather than only for an extended series of alternations?

Pol Besenius responded: As I mentioned, we used IR spectroscopy to show that the structures are ordered.[1] In the IR spectra of the surface-bound copolymers, the amide I and II bands were characteristic for hydrogen-bonded secondary structures. These were similar compared to a material that was self-assembled in solution and then deposited on a Au surface. With respect to the AFM data, in that same paper we performed experiments for a sequential number (n) of monomer addition steps ($n = 10, 20, 30$) and looked at the change in the height profiles. From a linear regression we estimated an increase of 0.68 ± 0.05 nm per monomer. So far, we have not performed AFM characterisation in smaller incremental steps.

1 H. Frisch *et al.*, *Angew. Chem. Int. Ed.*, 2016, **55**, 7242–7246.

Steven De Feyter enquired: Could you explain to what extent your approach is different from what is known as electrostatic self-assembly, or layer-by-layer assembly of polyelectrolytes? What is different, if so, in your approach and outcome?

Pol Besenius replied: I agree, our strategy resembles the well-known layer-by-layer self-assembly of oppositely charged polyelectrolytes into multilayered films. However, using β-sheet encoded amphoteric peptides, a combination of hydrogen bonding and hydrophobic shielding, in addition to the Coulomb attractive interactions, allow us to tune and control the produced architectures at multiple levels. This is not easily achieved for the non-directional layer-by-layer deposition of charged polymers. I particularly emphasise that the chiral β-sheets direct the self-assembly perpendicular to the Au surface, and the number of monomer addition steps defines the height of the copolymer and resolution is achieved on a sub-nanometer level. Both of these factors will be important in the optoelectronic applications we are currently pursuing.

Karl-Heinz Ernst opened a general discussion of the paper by Vincent Humblot: In your talk and in many other papers, zwitterionic states of amino acids are reported. I followed this for many years, and from what I've seen, especially on gold surfaces, this could be questionable. Many people work with the N 1s XPS peak at roughly 402 eV and assign them to NH_3^+ (or equivalent if secondary or tertiary amino groups). But I've seen papers where people adsorb nitrogen-containing compounds on gold, where we see this peak and there's no way that we have any acidity such that protonation of the nitrogen can occur. Gold is the metal with the highest electronegativity and may be the acceptor for the lone pair from the nitrogen. Shouldn't we discuss, at least for nitrogen-containing compounds on gold, other possibilities than zwitterions? I think we have to be careful about assigning a number of 402 eV to zwitterionic states.

Vincent Humblot responded: This is quite a good and a tricky question all together. We agree that a peak around 402 eV has been observed for molecules that cannot have a positively-charged nitrogen such as quinolone and cinchonidine adsorbed on Au(111) surfaces.[1] However, these molecules are quite different from amino acids and small peptides due to their peculiar structure and the presence of unsaturated benzene rings. In the cases of amino acids, and more precisely in our Gly–Pro case, there is no presence of such conjugated systems due to aromatic compounds. In addition, our assignment of the high-energy peak at 402 eV is not made blindly and is supported by literature results (see for instance reviews ref. 2 or 3), and by comparing the several peaks that we can obtain for our molecules. We, for instance, take great care at looking at the energy differences, especially the splitting between the NH amide group and the second peak at higher energy with a difference of 1.4 to 1.7 eV following adsorption of GP on Au(110), with a peak at 400.2 eV for NH, NH_2 and another one at 401.6 eV for NH_3^+ species. Another example lies with the adsorption of Gly–Pro on a Cu(110) surface under two very interesting chemical forms: anionic (top panel of Fig. 1) and zwitterionic (bottom panel of Fig. 1).[4] In that example, one can clearly see the splitting of the nitrogen N 1s peak with a second component at higher energy arising for the zwitterionic adsorbed molecule at an energy higher by 0.8 eV, centred at 401.1 eV. Finally, I would like to cite three more examples where the same molecule (methionine) was adsorbed as zwitterionic or anionic species on

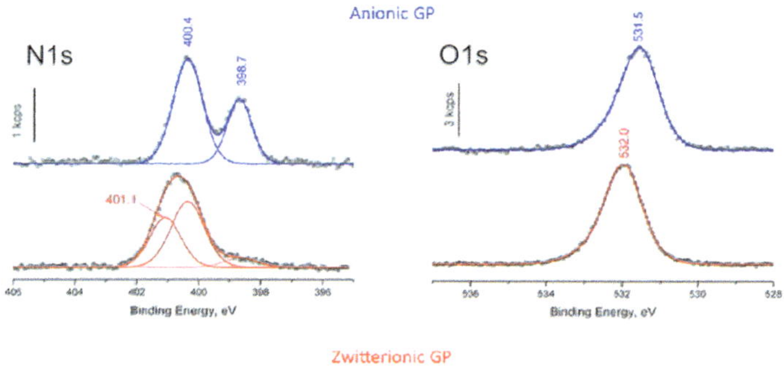

Fig. 1 N 1s and O 1s XPS data of anionic and zwitterionic Gly–Pro molecules adsorbed on Cu(110). Reproduced with permission from the American Chemical Society. Copyright American Chemical Society 2016.

three very different surfaces: Cu, Ag and Au. For Cu(111), anionic species gave a single peak at 399.5 eV with a second one at 401.3 eV for the zwitterionic molecules,[5] while on Cu(110) anionic molecules exhibited a single peak at 400.1 eV.[6] On Ag(111), methionine adsorbed as zwitterions and presented a single N 1s peak at 401.15 eV,[7] and its equivalent study on zwitterions on methionine/Au(111) ended up also with a single peak at 401.6 eV.[8] To conclude, taking into account all these examples we can fairly assume, even confirm, that in the present case, the peak observed at high binding energy (401.6 eV) following adsorption of Gly–Pro/Au(111) is unambiguously assigned to NH_3^+ species.

1 B. Behzadi *et al.*, *Appl. Surf. Sci.*, 2007, **253**, 3480–3484.
2 K.-H. Ernst, *Phys. Status Solidi B*, 2012, **249**, 2057–2088.
3 S. M. Barlow and R. Raval, *Surf. Sci. Rep.*, 2003, **50**, 201–341.
4 C. Méthivier *et al.*, *J. Phys. Chem. C*, 2016, **120**, 27364–27368.
5 A. Schiffrin *et al.*, *J. Phys. Chem. C*, 2009, **113**, 12101–12108.
6 C. Méthivier *et al.*, *Surf. Sci.*, 2015, **632**, 88–92.
7 A. Schiffrin *et al.*, *PNAS*, 2007, **104**, 5279–5284.
8 V. Humblot *et al.*, *Langmuir*, 2014, **30**, 203–212.

Karl-Heinz Ernst asked: In your RAIR spectra, how unambiguous is your assignment of vibrations of a NH_3^+ group?

Vincent Humblot responded: We agree that our PM-RAIR spectra could be ambiguous to analyse for two main reasons: the low intensity of the signal (small surface and very thin film) and the multiplicity of vibrations in the 1500–1700 cm^{-1} region. However, by doing cross-checking in between the different IR data we have obtained for Gly–Pro adsorbed on Cu(110) as anionic and zwitterionic molecules[1] or our very recent results on the adsorption of aspartic acid on Cu(110) [2] and the literature available on several amino acids adsorbed on coinage metals, we can be fairly confident with our assignments of the IR data. Nevertheless, one can never be 100% sure of these assignments, especially when in a 100 cm^{-1} region, one can find so many vibrational bands: NH_3^+ and NH_2 deformation mode, COO^- stretching mode, amide I and II bands (in the case of peptide).

1 C. Méthivier *et al.*, *J. Phys. Chem. C*, 2016, **120**, 27364–27368.
2 R. Totani *et al.*, *J. Phys. Chem. C*, 2017, **121**, 15842–15850.

Philip Davies said: Please could you tell us what you calibrated your XPS to for both surfaces?

Vincent Humblot answered: For our XPS experiments, the calibration was carried out on the Au 4f7/2 peak at 84.0 eV and on the Ag 3d5/2 at 368.0 eV. Using such calibrations on our very thin films (thickness below 1 nm) ends up with the C 1s C–C, C–H position at 284.8 eV for all studied surfaces.

Philip Davies asked: I acknowledge the problems caused by the silver plasmon peak, but why does the signal-to-noise ratio in the N 1s region of Fig. 5 of your paper (DOI: 10.1039/C7FD00116A) get worse as the exposure time to the adsorbate increases? One would expect the opposite. Also in Fig. 5 of your paper, can you assign the low binding energy peak that develops in the N 1s region as the exposure increases?

Vincent Humblot replied: The N 1s data presented in Fig. 5 of our paper are the results of the substraction of the XPS signal before (Ag plasmon peak contribution only) and after adsorption of the dipeptide (contribution of Ag plasmon and nitrogen of the peptide). This substraction was weighted by the attenuation of the silver 3d signal, thus this induces some noise on the resulting peaks, especially for the high coverage one. In the figure shown here (Fig. 2), I present both raw and corrected XPS data. About the low energy peak appearing at higher coverage at 399.7 eV, the assignment is rather difficult taking into account both sets of data; it could well be an artefact due to the substraction of the Ag plasmon, or it could be due to the particular amide bond NH of the proline ring.

Han Zuilhof enquired: About four years ago we had a paper[1] predicting carbon 1s spectra from simple DFT calculations and Koopmans' theorem applied to core

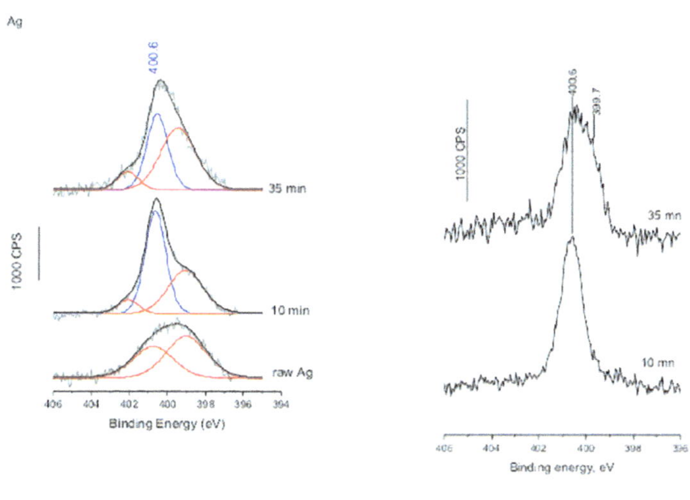

Fig. 2 Raw and corrected N 1s XPS data of Gly–Pro molecules adsorbed on Ag(110).

electrons. This yielded an average deviation of about 0.3 eV. We also extended this to N 1s calculations, yielding similarly small errors.[2] Have you ever attempted something like this to help your assignment, either direct N 1s energy calculations, or to use the (more reliable) C 1s calculations to state something on the charge of adjacent N atoms?

1 M. Giesbers *et al.*, *Langmuir*, 2013, **29**, 4782.
2 J. Zhao, F. Gao, S. P. Pujari, H. Zuilhof and A. V. Teplyakov, *Langmuir*, 2017, DOI: 10.1021/acs.langmuir.7b02301.

Vincent Humblot responded: We haven't yet used DFT calculations to evaluate the splitting of XPS peaks within a given region. This is something we will certainly investigate in the near future. With the average deviation calculated of 0.3 eV, in the case of N 1s, with bands splitting between 1.3 and 2.0 eV, DFT calculations would unambiguously answer these questions.

Han Zuilhof commented: You are referring to energy differences between two N atoms of roughly 2 eV. This is clearly outside the error of the relative energies DFT calculations we've used, pointing to the potential of DFT calculations to sort out this unclarity.

Karl-Heinz Ernst said: I agree that in bulk samples or in thicker films of amino acids there are zwitterions, no doubt, but, again, experiments and calculations of aromatic N-containing molecules (*e.g.* quinoline) on gold show this N 1s peak at 402 eV, but no acidic group around. Lone pair electrons binding to the gold may show this peak as well. One should be more careful with just saying 402 eV is zwitterionic N 1s.

Vincent Humblot answered: Again, we agree with that experimental result on quinolone adsorbed on gold. However, as I stated in a previous answer, we are not assigning the 402 eV peak blindly to a NH_3^+ moiety, we support that assignment with the splitting observed between both contributions of the N 1s ranging between 1.2 and 2.0 eV. For more details, please see my answer to your earlier question.

Giovanni Costantini remarked: I imagine that if the dipeptide really becomes a zwitterion on the surface, then the picture is that it first adsorbs as a neutral molecule and stays neutral until it encounters another (neutral) molecule, and then the proton moves from one to the other. If this is what happens for Gly–Pro on Au(110), then you should see a clear signature for only neutral molecules at low coverage. Evidently, this is not the situation for your 10 min deposition. Have you tried to deposit much less? I think this would be a very important test to corroborate (or alternatively to disprove) the hypothesis that the molecules get in a zwitterionic state.

Vincent Humblot answered: We do believe that this is actually the phenomenon that is occurring at the surface, as our XPS data as a function of increasing coverage show an increase of the zwitterionic molecules at the surface between a 10 min dose and a 35 min dose, with the % of zwitterions going from 40 to 60%, approximately. However, we should indeed investigate lower dosing time to confirm our hypothesis.

Giovanni Costantini commented: Assuming that Gly–Pro is zwitterionic on Au(110) at higher coverage, the relative intensities of the components in the C 1s and N 1s XPS spectra should be in a well-defined ratio. In the paper you give the "fraction of the O 1s contribution at higher BE (533.2 eV)", which, however, cannot be directly related to the NH_3^+/NH_2 ratio because the O signal also contains the HNC=O contribution. You should, however, be able to calculate the ratios between neutral, zwitterionic and anionic photoemission signals correctly (and independently) evaluated from the C 1s and N 1s spectra, and check that these are compatible. I think this would be a further essential test to corroborate (or alternatively to disprove) the hypothesis that the molecules get in a zwitterionic state on Au(110).

Vincent Humblot replied: We agree with you, and these calculations have been performed, even though they are only briefly mentioned in the present paper. The calculation method was applied to several of the systems we have investigated, and we can assume that our method is quite reliable. To answer more precisely to the question, we have found by cross-checking our data for N 1s, O 1s and C 1s that after a 10 min dose, the surface is covered with 40% of zwitterionic GP molecules and 60% of neutral GP molecules. The presence of anionic molecules was ruled out by the O 1s data. After 35 min of exposure, the calculations are more complex, and we end up with 65% of adsorbed molecules being zwitterionic and 35% of the GP molecules being either neutral or anionic. However, in this last case, our experimental data do not allow the differentiation between neutral and anionic molecules, and it is more than likely that DFT calculations suggested by Prof. Zuilhof could help us in deciphering this particular question.

Rasmita Raval said: Some very important issues have been raised regarding the structural determination of complex molecules at surfaces. My comment is that, first, such systems invariably need a multi-technique approach that pools data from spectroscopy, imaging, diffraction techniques and theoretical modelling. However, what is becoming increasingly clear is that the databases used to interpret the experimental techniques, *e.g.* vibrational frequencies and XPS binding energies, also need to advance. A good example is the number of years surface scientists took to understand how the vibrational frequency of CO at surfaces is affected by adsorption, the adsorption site and due to assembly as dipole and chemical shifts come into play. As we make rapid progress in mapping complex molecule assemblies at surfaces, there is a need to understand how spectroscopic information from such systems can be interpreted correctly. This is a huge gap and represents a field of analytical science in its own right, and is crucial if we are going to advance towards knowledge-based design of functional surfaces.

Phil Woodruff communicated in reply: I completely agree. As you mention, there was a history of over-interpreting vibrational spectroscopy data in terms of structural assignments for the simple adsorbate CO and NO in the past, but there is clear evidence of similar things happening with scanning probe techniques and XPS for other molecules. Early in the history of surface science we learnt the need to apply several complementary methods to achieve a reliable understanding of adsorption systems, but in recent years this lesson seems to have been unlearnt. Both scanning probe microscopies and advances in DFT methods have played a huge positive role in understanding these problems, but are not sufficient in isolation. We need independent determination of surface composition, spectral fingerprinting, and the application of quantitative structural methods to complement this information.

Vincent Humblot answered: We do agree with Prof. Raval; even using a huge number of techniques (experimental and computational) to investigate our systems (RAIRS, XPS, LEED, MS, NEXAFS, STM and MD, DFT) there are still gaps that need to be understood. We think that calculations on STM, RAIRS and XPS data (more and more present in the literature for the last 10 years) should help us increase our knowledge. Nevertheless, systematic experimental studies should still be carried out for fundamental knowledge to be increased.

Phil Woodruff addressed Vincent Humblot: The discussion of XPS core level shifts in terms of only the initial state charge completely fails to take account of the final state relaxation effects. The fact that Prof. Ernst mentions DFT calculations showing a N 1s shift inconsistent with expected charge transfer is not surprising; on the contrary, it is entirely consistent with this view. Proper DFT calculations of core level shifts account for both initial and final state effects. These core level shifts are never a reliable guide to initial state charge transfer.

Vincent Humblot responded: We do agree with you that DFT calculations can be a real support when considering initial and final states relaxation effects; however, experimental results could never be replaced by theoretical calculations, and theoretical results should be compared carefully with experimental data. Nevertheless, in the present paper we only take into account the final states and all our analyses are performed and compared to final states data, from us or from others.

Talat Rahman commented: I was also wondering, perhaps in your STM data, do you also find adatoms? There's a big difference between Au and Cu surfaces in their response to the peptides; could that be because in some cases you have more adatoms?

Vincent Humblot answered: We can never rule out the possibility of adatom superstructures with these kinds of molecules. We have indeed evidenced the presence of adatoms with a slightly bigger molecule, a tripeptide Gly–Pro–Glu, which is binding on a chelating mode on Au surfaces and induces mass transportation.[1,2] In the present case of GP, we have not seen any evidence of adatoms on Cu(110) or Ag(110). However, on Au(110), we do not distinguish the (1×2) reconstruction upon adsorption of GP, due to adlayer or to surface reconstruction? We cannot tell only with STM data; some complementary techniques would be needed (XRD, NEXAFS, *etc.*)

1 V. Humblot *et al.*, *J. Am. Chem. Soc.*, 2012, **134**, 6579–6583.
2 V. Humblot *et al.*, *Surface Science*, 2014, **628**, 21–29.

Angelika Kühnle asked: Talking about zwitterionic species, I have three questions regarding the deprotonation of the carboxylic acid group. Firstly, you mention that the tendency to deprotonate is a function of the metal substrate, but it should also be a function of the specific molecule and its acidity. Is there a study addressing this point? Secondly, what happens in the gas phase? Are the molecules more prone to becoming zwitterionic in the gas phase than others? Finally, on the surface, where does the proton go?

Vincent Humblot responded: To our knowledge, there is yet no specific study addressing the peculiar point of the relation between the nature of the molecule *vs.* the nature of the surface adsorbing the given molecule when deprotonation/dehydrogenation occurs. Some molecules do indeed have a tendency to be present as zwitterions either under a crystalline form or in the gas phase. This would probably influence the following adsorption and hence anchoring mode when this molecule arrives in the vicinity of a metal surface. In the case of Gly–Pro, as powder the molecule is neutral (COOH/NH_2) and so we believe it is also in the gas phase upon sublimation. We are currently trying to analyse the chemical nature of the molecular beam before the molecules land at the surface, but as you can imagine, it is tricky to implement. Finally, to where the proton goes after "deprotonation" of the carboxylic group, that remains a tricky point indeed. On the copper surface, protons are moving around until they find another for coupling and thus desorb as molecular H_2. That answers the case for anionic species, if one considers that Au and Ag will react similarly to Cu surfaces. In the case of zwitterionic molecules, the answer could be simpler: the proton leaving the carboxylic acid ends up on the amine group that thus become protonated, and the charge equilibrium present in the molecular gas phase is maintained for adsorbed molecules.

Trolle R. Linderoth enquired: To follow up on the earlier comment from Giovanni Costantini which suggested that by going to low coverage one could isolate the neutral state of the molecule: Is it correct that the proton exchanges only occur on the surface? It could also occur, say, in the bulk state in the molecular evaporator?

Giovanni Costantini answered: I definitively think it is possible, actually I think that the starting material, the dipeptide powder, will probably be in the zwitterionic state (we have measured by XPS different dipeptide powders and found that this is generally true). However, I do not think that zwitterions would survive the sublimation. In fact, the very formation of a zwitterion in the solid (powder) state and on a surface is the result of the transfer of a proton from one molecule to a neighbouring one (possibly *via* the substrate for the on-surface case). So, when the molecules are separated (I guess it is safe to assume that what sublimes is individual molecules), I do not see how the zwitterionic state can be sustained.

David Amabilino communicated in response: It should be remembered that proton transfer could in principle be intramolecular, and does not necessarily take place between neighbouring molecules.

Angelika Kühnle said: I guess it is quite possible that the molecules form zwitterions in the sublimator. Whether or not they survive sublimation might again be a function of the specific molecule? When adsorbed on the surface, I could envision that also a single molecule can transform into a zwitterion, again depending on molecule and surface. My point is that the proton, being a positively-charged species, cannot leave the surface as neutral hydrogen. If it recombines to leave as hydrogen from the surface, it needs to get back the electron from somewhere. Can you comment on this process?

Vincent Humblot replied: I agree with you that the proton is becoming *de facto* a charged species and even when creating molecular H_2, it needs to recover an electron. The question can be reversed to the carboxylate group; what happens to the delocalised electron once the oxygen atoms are "chemisorbed" at the surface? Is it possible that this same proton finds its way to the H^+ radical lying nearby at the surface? Or is it more realistic/plausible that the metal surfaces act as a "sea of electrons", thus providing the negative charge when needed?

Giovanni Costantini communicated: Molecular coverage is evidently an important parameter in your study. How did you compare the coverage across the three different techniques you used? Surely just using the exposure time cannot be a good way, because it is well known that the reproducibility of typical molecular sublimation sources used in surface science is rather poor, not to speak about the comparison between different sources in different UHV systems.

Vincent Humblot communicated in reply: This comment is correct, it is always very difficult to compare surface coverage by taking into account only the evaporation times. However, two of our techniques (XPS and RAIRS) are performed on the same UHV chamber with the same evaporation geometry, which limits the differences that can be observed. In addition, several identical experiments have been performed allowing us to estimate only a very small dispersion of coverage. As for STM experiments, carried out on a second and hence different experimental UHV, it is clear that identical evaporation time cannot be used as such. However, a calibration of the evaporation time was done by dosing small amounts of molecules until reaching full monolayer saturation. Again, several experiments were performed, allowing us to obtain quite a good correlation between dosing time used on both experimental setups.

Giovanni Costantini communicated: Because of the typical poor reproducibility of molecular sublimation sources, it would be better that for the XPS studies the actual coverage (evaluated by analysing the intensity of the core level signals) was quoted instead of the exposure time. Even if the absolute coverage values might still be questionable, at least relative values should be rather robust.

Vincent Humblot communicated in response: Again, we do agree with you and that is the reason why, in the present paper, for each coverage and for every surface (Ag and Au), as well as on a Cu surface,[1] the relative coverages in terms of monolayer (ML) fraction were calculated. For instance, on Au(110), a 10 min dose corresponds to 40% of a saturated monolayer, while after 35 min, the relative coverage is close to 60% of a ML (page 4 of the manuscript, DOI: 10.1039/C7FD00116A). As for Ag, for dosing times of 10 and 35 min, we estimated the relative local coverages to be 35% and 60% of a monolayer, respectively (page 9 of the manuscript).

1 C. Méthivier *et al.*, *J. Phys. Chem. C*, 2016, **120**, 27364–27368.

Giovanni Costantini communicated: How did you check that the Gly-Pro dipeptide is adsorbed intact and is not modified/fragmented by thermal sublimation?

Vincent Humblot communicated in reply: During sublimation, the flux of molecules was followed by mass spectrometry, showing no specific molecular fragments that could suggest a breaking of the molecule upon sublimation. Our sublimation power was very low (around 0.3 and 0.5 W) and should avoid any fragmentation of the molecules. In addition, XPS data were carefully analyzed showing a perfect stoichiometry for each C, N and O region, as well as all together.

Talat Rahman addressed Pol Besenius: In your paper, there was a detailed look at the system's electronic structure, nature of bonding, *etc.* Are there simple models that you can point to, based on electronic structure calculations, that help one understand what's going on in your systems? It would be nice to have simple models motivated by detailed calculations.

Pol Besenius responded: We are currently collaborating with a number of soft matter physicists and theoreticians in the Netherlands and Germany to pursue both avenues, to help us in supporting structure formation in solution and on surfaces. One manuscript is currently in preparation and will be submitted in due course.

Phil Woodruff opened the discussion of the paper by Talat Rahman: I just want to return to a topic discussed in the previous session, namely core level shifts in XPS. You do a full calculation of this adsorbate system and determine the charge transfer using a Bader analysis; whatever its imperfections might be, this is far more meaningful than any experiment can achieve. You then do a proper calculation of the XPS core level shift taking account of both initial and final state effect. You then spoil this nice story by arguing that the core level shift that you determine is consistent with experimental values found in oxides with the metal atom in a known charge state and argue that this confirms your calculated charge transfer. This is a *non sequitur*. You are nicely calculating the final state effect and then promptly ignoring it in making this comparison.

Talat Rahman responded: It was probably fortuitous that final state effects did not play a major role for the system on hand.

Steven L. Tait said: I would like to respond to the comment by Phil Woodruff about not being able to deduce charge state *a priori* from XPS. As Prof. Woodruff pointed out, we agree that the Pt 4f binding energy shift in the XPS data alone is not sufficient to prove a change in charge state of the metal centers, as there are other potential differences that could lead to a binding energy change, particularly a difference in final state effects. For this reason, multiple control studies were conducted to try to examine the binding energy changes.[1] We conducted XPS experiments on the same surface with Pt nanoparticles to examine the Pt(0) condition. We also tested that charge state by co-depositing Pt with a ligand that does not induce a charge transfer. We compare the Pt(II) state to a Pt-porphyrin on the same surface as a second reference point. In later experiments,[2] we also verified the complexation using high-resolution electron energy loss spectroscopy (HREELS).

1 D. Skomski *et al.*, *J. Am. Chem. Soc.*, 2014, **136**, 9862.
2 C. G. Williams *et al.*, *J. Phys. Chem. C*, 2017, **121**, 13183–13190.

Phil Woodruff commented: If one sees the same chemical shifts for a number of different systems, the implication is presumably that if the charge state is the same then the screening in the final state must also be closely similar. Strictly, you can't say anything about the separate initial and final state effects.

Steven L. Tait responded: Another important point regarding the on-surface redox activity of this system is that the system was designed to have a high likelihood of this type of reaction. The ligand used here, dipyridyl-tetrazine, is an excellent electron acceptor, and ligands similar to this are known to have a strong oxidizing ability as they coordinate metal atoms in non-surface environments. Thus, it is not surprising that the charge transfer and redox reaction would be active, although this was the first report demonstrating that this reaction could be achieved in an on-surface environment with this type of ligand.[1]

1 D. Skomski *et al.*, *J. Am. Chem. Soc.*, 2014, **136**, 9862.

Ioan Bâldea addressed Talat Rahman: You used Bader charge analysis to see charge transfer between adatoms. Have you also looked more microscopically at this charge transfer to see which orbitals are involved and changes in orbital occupations?

Talat Rahman replied: As you know, Bader charge analysis does not resolve the orbitals involved in charge transfer. It is an intuitive way to divide molecule or solid structures into atoms by defining zero-flux surfaces that separate the atoms. We have not analyzed in detail the orbitals involved in the charge transfer. However, from preliminary charge difference plots (not included in the paper) we can see that in-plane components of Pt orbitals ($d_{x^2-y^2}$, d_{xy}) transfer an electron to the pi orbitals of the tetrazine rings. Moreover, from Fig. 4b of the paper (DOI: 10.1039/C7FD00097A), based on the shapes of the depletion region (blue), we can conclude that the Au substrate has withdrawn charge from the pi orbital of the ligands and the d_{z^2} orbital of the Pt atoms. The withdrawal from the d_{z^2} orbital of the Pt atom is also evident by its shift closer to the Fermi level.

Markus Lackinger commented: The exact adsorption site of the Pt atom seems to be a crucial point. I was wondering if there is any experimental basis for this?

Talat Rahman answered: As we summarize in the ESI (DOI: 10.1039/C7FD00097A), we did try a good number of possible adsorption sites for Pt centers on Au(100) and found nine which were within 100 meV in total energy. We chose from this set the one with the lowest total energy. However, I should point out that we have not included entropic contributions which might also make a difference. At the moment we do not have experimental confirmation of our calculated adsorption sites.

Steven L. Tait said: To respond to the previous question about the adsorption site: We do not have a precise experimental measurement of the adsorption site. The STM experiments were conducted at room temperature or at higher temperatures. We have a precise measurement of the spacing along the chains, but not of the registry with the substrate. The spacing along the chains and the chain angle relative to the reconstruction rows of Au(100) indicate that the Pt atom sites would not all sit in the same adsorption site, but approximately every third Pt could be in the same adsorption site.[1] A virtually identical chain structure was observed using vanadium metal centers.[2] It is also interesting to note that although Pt nanoislands are able to lift the surface reconstruction,[3] the Pt–DPTZ chains do not.

1 D. Skomski *et al.*, *J. Am. Chem. Soc.*, 2014, **136**, 9862.
2 D. Skomski *et al.*, *J. Am. Chem. Soc.*, 2015, **137**, 7898–7902.
3 C. D. Tempas, D. Skomski and S. L. Tait, *Sur. Sci.*, 2016, **654**, 33–38.

Trolle R. Linderoth addressed Talat Rahman: The structures were realised on hex-reconstructed Au(100). Would the situation have been different for the simpler situation of Au(111)?

Talat Rahman responded: Yes. On Au(111) we would not have found the corrugation that is present on Au(100) which accounted for homogeneity in the Au–Pt interactions in this system. Au(111) has the herringbone reconstruction, which might have added another type of complexity, but my hunch is that the Pt–Au interaction would have been weaker than in the present case.

Neil Robinson asked: From a heterogeneous catalysis point of view, you stress in the introduction of your paper the need to incorporate fewer precious metal atoms in the catalysts we design. However, you have then grown your metal–organic network on a gold surface. Not only do charge transfer effects between the active platinum sites and this gold surface render the structure relatively inactive (as alluded to towards the end of your paper), but such a system obviously involves a huge number of inactive precious metal atoms. Is it possible to form such a network on a less expensive surface, such as an industrially relevant oxide material, and do these structures provide evidence for enhanced catalytic activity?

Talat Rahman answered: Our work was motivated by the experiments of Tait *et al.* and hence the focus on Au(100) as the support for metal–organic structures.

In a related theoretical study, we have compared the viability of a number of other metal atoms as the coordination center for the dipyridyl tetrazine network and found that several others (Mo, Cr, V, Fe and Co) may serve as a more active site for adsorption of molecules such as CO than what we have found for Pt. Of course, much computational work remains to be done before we can be confident in predicting that any one of these other metal atoms would serve as a more catalytically active site than what we have reported for Pt.

Steven L. Tait added: In response to the question about whether this has been tried on metal oxides: yes, we have attempted to grow these metal–organic chain structures on metal oxides. In fact, we have recently been conducting experiments to assemble metal–organic structures on high surface area oxide powders to explore their chemical and catalytic activity. We are preparing a manuscript now to report the first successful results from those experiments.

Deepak Dwivedi said: Thank you for the nice presentation. We know from the earlier studies that crystallographic morphology and surface energy of crystal faces have a great impact on the adsorption. I wanted to ask, could we characterize these aspects for a catalyst which has large surface heterogeneities? Is it possible to investigate these aspects under different operating conditions such as under different pressure and temperature? Can we calculate the effect of surface energy on the catalytic performance of a complex catalyst surface such as CNT or graphene *etc.*?

Talat Rahman answered: Your question has two important parts. The first is whether it is possible to characterize molecular adsorption characteristics for a surface with a large number of inhomogeneities such as steps, kinks, and terraces of finite width. The answer is yes, if one knows *a priori* the nature of the inhomogeneity under experimental conditions. Alternatively, one could calculate adsorption characteristics for vicinal surfaces which have regularly patterned steps and kinks (represented by high Miller indices). Of course, these calculations require more computational resources than low Miller index surfaces. The other part of your question concerns calculations of adsorption characteristics under varying temperature and pressure. The answer to this is again in the affirmative, as one then calculates Gibbs free energy as a function of the chemical potential and temperature. You can find several examples of such calculations. One that comes quickly to my mind is in the calculations that we carried out for hydrogen adsorption on Co nanoparticles as a function of pressure, temperature and coverage.[1] Here again, dealing with adsorption of a molecule such as dipyridyl-tetrazine (DPTZ) is going to be much more computationally intensive than hydrogen.

1 E. A. Lewis *et al.*, *J. Phys. Chem. C*, 2012, **116**, 25868.

Deepak Dwivedi asked: Just out of curiosity and for making my understanding clear, could you let me know whether you have observed the effect of crystallographic re-orientation on the catalyst performance (for example, gold or aluminum which has an FCC structure)? How easy or difficult is it to answer this question through first principle-based studies?

Talat Rahman answered: We have not carried out similar calculations for any other Au surface, neither have we looked at the adsorption of DPTZ on any other metal surface. We have confined ourselves to the reconstructed Au(100) since we were most interested in helping understand the experimental data obtained in Professor Tait's laboratory. We could extend these first principles calculations to the surfaces you mention. As we have shown in the present work, the charge transfer between the surface atoms and the Pt and ligand atoms plays an important role in determining the characteristics of the metal–organic network. I thus expect the results to depend on the nature (elemental, crystallographic) of the surface that we choose to study. It would be interesting to see the trends as a function of the nature of the substrate in future studies.

Karl-Heinz Ernst enquired: I'm interested in the activation of CO_2 activation and hydrogenation and in making it more efficient than today's energy-consumptive catalysts. We follow the same approach by testing metal-coordination compounds as potential catalysts, but as in your example, no activation is observed. What kind of metal would you put in a coordination compound in order to eject an electron into a CO_2 species for activation?

Talat Rahman responded: We have carried out calculations of CO_2 sequestration and also hydrogenation for a few novel catalysts, and the results have been gratifying as there is experimental evidence for the same. Interestingly, it is a metal-free catalyst, namely defect-laden hexagonal boron nitride (h-BN) that appears to be an appropriate one. The results of our initial joint experimental and theoretical work on hydrogenation of alkenes to alkanes are already in print.[1] We are in the process of submitting the results for CO_2 hydrogenation on h-BN.

1 D. J. Nash *et al.*, *ACS Omega*, 2016, **1**, 1343–1354.

Steven L. Tait opened a general discussion of the paper by Giovanni Costantini: There have been examples in the literature of ionic assemblies on copper surfaces that involve deprotonated carboxylic acid with cesium[1] or with sodium.[2] In those cases, the terephthalic acid deprotonates upon adsorption on the Cu surface and is, presumably, in an anionic state even before the alkali metal is deposited to the surface. In your case, do you think the charge transfer is occurring in a single step or does the TCNQ have a significant amount of charge transfer before the complexation occurs? Does the TCNQ state change again when the potassium is deposited on the surface? What measurements do you think would be the most useful in examining this process in the future?

1 S. Stepanow *et al.*, *ACS Nano*, 2010, **4**, 1813–1820.
2 D. Skomski *et al.*, *J. Am. Chem. Soc.*, 2012, **134**, 14165–14171.

Giovanni Costantini answered: TCNQ deposited on its own (*i.e.* without K) on Ag(111) forms metal–organic structures with Ag adatoms and is negatively charged. In this case, the charge is transferred from silver (from both adatoms and substrate). This system is described in ref. 19 (in preparation) of our *Faraday Discussions* paper, DOI: 10.1039/C7FD00093F. In order to form the K-TCNQ metal–organic structures that are the subject of the present paper, K is deposited on top of the Ag-TCNQ phase and the sample is annealed to 470 K. This results in only slight changes in the SXPS spectra of TCNQ (C 1s and N 1s), while the K 2p SXP spectra are negatively shifted by 1.2 eV from K deposited alone on Ag(111), indicating that K gets positively charged.

Robert Jones said: The normal incidence X-ray standing wave data (NIXSW) was taken using the (111) Bragg reflection and provides positions for the various molecular components along the surface normal, relative to the (111) Bragg planes. Reciprocal space data indicated that the ad-layer was incommensurate with the gold substrate. As stated in the paper (DOI: 10.1039/C7FD00093F), this means that (for a rigid ad-layer) all positions determined by NIXSW relative to planes at an angle to the surface normal, such as (−111) reflections, should give a coherent fraction of zero; *i.e.*, all adatoms are at all possible positions relative to the angled Bragg planes. Were any NIXSW data obtained using such angled planes, and did they show the expected zero coherent fraction? Would the authors care to speculate whether it is possible that an overlayer comprising a complex molecule with some flexibility, such as here, could be incommensurate overall, but with some of the more strongly bonding entities (*e.g.* N in CN) pulled into higher symmetry sites such that they are partially ordered relative to the substrate? If this were the case, then (−111) NIXSW may show a non-zero coherent fraction.

Phil Woodruff responded: Because the overlayer was known to be incommensurate, no triangulation measurements of the (−111) NIXSW were made. It is perfectly true, of course, that even in an incommensurate molecular overlayer there could be local distortions of the molecule such that, in the case of TCNQ for example, the N atoms that bond to the Ag surface may try to adopt slightly more favourable local sites. However, with four N atoms per molecule in symmetrically inequivalent sites relative to the substrate, none of which are likely to lie very close to high-symmetry local adsorption sites, the (−111) coherent fraction (averaged over all symmetrically-equivalent azimuthal rotations and the many inequivalently located molecules in the surface unit mesh) can be expected to be very low even if local distortions do occur. In this situation it is difficult to see that the information gained would really be interpretable in terms of a specific structural model.

Phil Woodruff commented: It's perhaps worth stressing that a LEED pattern provides valuable complementary information to STM in understanding the long-range order. Of course, in STM it's easy to measure the approximate size of the unit mesh of an ordered surface phase, but unless it's also possible to image clean areas of the surface at the same time it's very difficult to make a reliable assignment as to whether this phase is commensurate or incommensurate because of problems of drift and/or piezo calibration. By contrast, LEED always shows the substrate diffraction and the overlayer diffraction and the exact relationship of the two can be established.

David Amabilino asked: Regarding Fig. 1b in your article (DOI: 10.1039/C7FD00093F), the nice pattern you see and assign to the K ions on top: when you look at the TCNQ image, that looks extremely well ordered, but when you look at the K image, the domains look a lot smaller (less well ordered?) and there are differences in contrast. Could that be reflecting the different directions of the cyano groups or what is that reflecting?

Giovanni Costantini responded: At the specific scanning conditions (tip state and negative bias voltage) shown in Fig. 1b of the article, K atoms are appearing as bright protrusions. The apparent lower order in Fig. 1b is thus most probably the result of the absence of K atoms in some of the windmill structures within the network. We checked this by intentionally reducing the K exposure, which results in a higher number of K vacancies (see ESI Fig. S3).

David Amabilino said: So what you're saying is the irregularity is because there is no K ion there? It seems that even in the ordered domains of the potassium cation there is some irregularity; did you look at the contour line along rows of atoms to see if there was variance, possibly because of the cyano group orientation?

Giovanni Costantini responded: Yes, the irregularity is not in the TCNQ molecules, these are still ordered in a regular interdigitated windmill network (please note that the TCNQ molecules can still be seen in Fig. 1b of the paper but are less apparent because of the higher apparent height of the K ions). The irregularity is given by the fact that some of the K ions are missing, and in the imaging mode shown in Fig. 1b the K atoms are more evident. I guess that the "irregularity" you refer to that should exist "even in the ordered domains of the potassium cation" is related to the STM apparent height and not to the lateral position (which seems to be quite regular to me). It is true that not all the K ions appear to have the same height, but this is actually also true for the TCNQ molecules as seen under the imaging conditions displayed in Fig. 1a. We have no idea about the origin of this slight contrast difference and we have not examined this in detail. Interpreting the measured height of STM images is however quite tricky, as one needs to have a full knowledge of the metal–organic interface and of the tip. We can, for example, not exclude that the incommensurability between the molecular adlayer and the Ag(111) substrate could play a role. However, we would exclude that the different apparent height of some of the K ions is caused by the fact that these are corresponding to (a majority of) those cyano groups that have a specific orientation. In fact, if this was true, one would expect that there existed (at least) two different K adsorption sites, while the coherent fractions determined from our NIXSW measurements indicate a single adsorption site.

Rasmita Raval enquired: Given that the system is incommensurate and does not adopt specific adsorption sites at the surface, would you expect to see a range of N environments and heights in your system? A number of N environments and heights may also arise from molecules that are not imaged, *e.g.* within the disordered 'sea' surrounding the ordered domains. In addition, these molecules are likely to have significant mobility, regardless of charge. After all, the charged molecules in your system self-assemble, which implies mobility across the surface. Would these lead to the low coherent fractions seen in the N 1s NIXSW?

Giovanni Costantini replied: To the first question: "The fact that the metal–organic superstructure is incommensurate with the substrate implies a range of N environments and heights". The fact that TCNQ molecules have several non-equivalent adsorption sites will surely add to the reduction of the coherent fraction *f*. However, we think that this will be a secondary effect. The presence of K vacancies (for which we have experimental evidence) is expected to contribute more to reduction of *f*. However, as discussed in the paper, even the presence of K vacancies is not enough to account for the measured low value of *f*. The model we propose in the paper is the simplest we can think of that explains the experimental data. To your second question: "Mobile molecules that are not imaged in the STM data could lead to the low coherent fractions seen in the N 1s NIXSW". The NIXSW measurements were always performed on systems prepared in full monolayer conditions for which the number of mobile molecules can be expected to be extremely small.

Phil Woodruff commented: As I understand it, Prof. Raval is suggesting that molecules in disordered regions or where the molecules are mobile could account for the low coherent fractions seen in the N 1s NIXSW. However, it's notable that the C 1s NIXSW shows a high coherent fraction, so the implication is that there is no significant variation in the heights of the molecules above the surface. Notice that the coherent fractions for these (111) reflections are a measure of the 'order' in the heights of the constituent atoms, and it's hard to see why lateral disorder of these large molecules should have much effect on this. As the standing wave technique is not widely known, it's perhaps worth commenting on the proper interpretation of the coherent fraction. It is generally thought of as an order parameter, and it is certainly true that disorder (*e.g.* thermal vibrations) does lower the coherent fraction. However, it's also true that if a system is perfectly ordered but has two equally-occupied atomic sites that differ in height by exactly half the substrate interlayer spacing, then the coherent fraction is exactly zero.

Steven L. Tait asked: The high diffusivity of this system is interesting. There are examples of organic adsorbates with high diffusivity, including some where increasing the size of the adsorbate leads to a significant increase in diffusivity due to a mismatch of binding units in the molecule with the spacing of the surface atoms.[1] In this case, it seems that the diffusivity is high, even after the deposition of potassium on the surface. Would it be reasonable to consider units of $K^+(TCNQ^-)$ diffusing across the surface as a zwitterionic complex?

1 D. Skomski and S. L. Tait, *J. Phys. Chem. C*, 2013, **117**, 2959–2965.

Giovanni Costantini answered: It might be the case, but unfortunately we do not have any evidence or measurement regarding the diffusion of TCNQ molecules or TCNQ-K or, another possibility, TCNQ-Ag complexes.

David Amabilino enquired: As a follow-up on my previous question and relating to the point about the nitrogen position, if your coverage of K is not 100%, can you see the difference between a TCNQ in proximity to the silver substrate and a charged TCNQ modified by the presence of the potassium cation by that technique? How sure can you be that your different contributions from the standing wave technique are not coming from different charged species, one closer to the surface, the other slightly pulled up to the potassium cation?

Giovanni Costantini responded: I agree that part of the N low coherent fraction measured by NIXSW might (actually will) be due to the presence of K vacancies. But, as we demonstrate in the paper (DOI: 10.1039/C7FD00093F), the measured density of defects (around 20%) is not enough to explain the measured low coherent fraction.

Rasmita Raval asked: Were image charges accounted for in the description of charged molecular systems at surfaces?

Giovanni Costantini replied: We clearly expect that the highly polarizable metallic substrate will tend to screen localized charges on both the TCNQ molecules and the K ions. In fact, in the paper (DOI: 10.1039/C7FD00093F) we write: "The high electron affinity of TCNQ molecules means that they become charged on metallic substrates, and the combination of the charged molecules and their image charges in the metallic substrate creates relatively strong dipole moments perpendicular to the surface. The mutual repulsion of these dipoles might be expected to suppress self-assembly, but the addition of metallic adatoms – in particular alkali adatoms – produces strong dipoles of opposite polarity, stabilising the metal–organic networks, and also reducing the work function increase caused by the negatively charged TCNQ molecules in isolation."

Lifeng Chi opened the discussion of the paper by Nian Lin: We have a quite similar system but with hydrogen bonding. The structures are formed by 2-fold and 3-fold hydrogen bonding, strongly coverage dependent. But in our case, they did not mix together.

Nian Lin responded: Interesting. This phenomenon shall be very general, not limited to metal–organic systems.

Lifeng Chi asked: How many coordination bonds can the Cu atom form?

Nian Lin responded: Two or three.

Lifeng Chi enquired: The spectroscopy you did, was that on molecules or on coordinated Cu atoms? If it was on molecules, can you see the difference in HOMO–LUMO with the 2-fold and 3-fold coordination?

Nian Lin responded: The STS data were acquired on both the molecules and Cu atoms. We cannot conclusively determine HOMO or LUMO in STS. DFT-calculated PDOS might be useful for us to assign the STS features to HOMO/LUMO.

Trolle R. Linderoth said: In a recent paper[1] we showed that surface MOCNs can involve Cu-trimers rather than single adatoms at the interaction nodes. Could Cu-trimers be responsible for the three-fold coordination nodes you observe, and help to explain why both two-fold and three-fold coordination nodes are observed in the network ?

1 F. Bebensee *et al.*, *Angew. Chem. Int. Ed.*, 2014, **53**, 12955–12959.

Nian Lin replied: The distance between neighboring pyridyl ligands, as determined by the STM image and molecular modelling, is too short to accommodate three Cu adatoms. A single Cu adatom coordinated by three pyridyl units in a three-fold motif renders a Cu–N distance of about 2.1 Å.

Giovanni Costantini remarked: The ext-TPyB molecule has 3-fold symmetry and shows both 2-fold and 3-fold coordination. I do not think that the extremely high degree of regularity, the high porosity, and the extension of the observed metal–organic networks it forms in the D-phase are compatible with a random "choice" of the Cu ions in the network to coordinate either two or three pyridyl groups. I rather think there must be some type of energetic gain, such that if one of the pyridyl groups of an ext-TPyB molecule is already involved in a 3-fold coordination node, the chances of the two remaining pyridyl groups of the same molecule to be also involved in a 3-fold coordination node are essentially zero. Is there any evidence for this? Do you have plans to look deeper into this possibility?

Nian Lin replied: This is an interesting thought. I do not know any physical/chemical mechanism that can cause the coordination of one terminal group to affect those of others that are located quite some distance apart. It's worthwhile to search the literature or ask chemist colleagues to see if such phenomena are known in chemistry. Experimentally, we can examine the behavior of normal TPyB. If your proposed mechanism is at work, it shall be expressed also in TPyB.

Talat Rahman asked: Could your structures also be driven by the strain in the system? The underlying substrate could be playing a role.

Nian Lin replied: Yes, we believe the structure is driven by an in-plane compression due to high molecular coverage. Yes, all Cu atoms in the structure are sitting at the Cu(111) lattice epitaxially.

Trolle R. Linderoth remarked: In relation to the question of the mechanism behind formation of your metal–organic coordination networks: Your STM images show the perfect extended structures. Can information on the growth mechanism and underlying interactions be inferred from images during early stages of growth or from defects/edge terminations?

Nian Lin answered: To follow the growth process will provide insights into the formation mechanism. Experimentally, however, this is not so easy using STM. LEEM might be an ideal technique.

Talat Rahman said: I am very glad to see you attain networks with several different symmetries at the metal coordination center. This is something that my colleague Ludwig Bartels and I were trying to look for, motivated by our findings of pattern formation of arenethiol films on Cu(111),[1] but have not succeeded so far.

1 K.-Y. Kwon *et al.*, *J. Am. Chem. Soc.*, 2009, **131**, 5540–5545.

Nian Lin responded: I am very much looking forward to the deep insights of this structure being revealed by theory. I can send you our data if you are interested.

Talat Rahman enquired: You did have some coverage dependence?

Nian Lin responded: Yes, as shown in Fig. 3 of our paper (DOI: 10.1039/C7FD00088J).

Talat Rahman asked: I think the question is whether the coordination is taking place first locally and then progressing gradually, or that the ordering takes place all of a sudden for a particular coverage?

Nian Lin replied: Very important question. Since experimentally STS does not see the "real" growth, but the static structures formed after annealing, we cannot answer this question. LEEM or kMC might be able to shed some light.

Steven L. Tait said: In the structure that you found to be thermodynamically stable (D-phase), there are a significant number of three-fold coordination sites of pyridyl around Cu atoms. In prior studies of pyridyl–Cu coordination on surfaces, the two-fold motif is usually seen,[1,2] but not the three-fold. Can you comment on why the three-fold coordination structure is not typically seen in surface studies of Cu coordination to pyridyl groups, but was observed here?

1 Y. Li *et al.*, *J. Am. Chem. Soc.*, 2012, **134**, 6401–6408.
2 S. L. Tait *et al.*, *J. Phys. Chem. C*, 2007, **111**, 10982–10987.

Nian Lin replied: You are right that two-fold coordination is energetically more favored than three-fold. At the coverage of forming D-phase, however, a structure consisting exclusively of two-fold coordination would be under huge in-plane compression. I speculate that this compression shifts the free energy landscape since the structure with three-fold coordination occupies less space.

Steven L. Tait asked: Would you please comment on how the surface structure differs post-annealing between low coverage and high coverage?

Nian Lin answered: The structures shown in the paper are all after annealing. So they are thermodynamically equilibrium structures at the specific coverage.

Talat Rahman commented: I worked with arenethiol molecules on Cu(111) some time ago. We found these 3-fold structures in STM images taken by Ludwig Bartels' group. We found in our calculations that it was arising from Cu adatoms.[1]

1 K.-Y. Kwon *et al.*, *J. Am. Chem. Soc.*, 2009, **131**, 5540–5545.

Nian Lin responded: Very interesting.

Marco Sacchi asked: Do you think you could observe similar self-assembly structures on other metal surfaces?

Nian Lin replied: In principle, yes, as far as a metal can coordinate with a ligand in two- and three-fold coordination simultaneously.

Marco Sacchi said: There are other metals that could coordinate either two or three ligands, for instance iron.

Nian Lin answered: To my knowledge, iron can be of three- or two-fold coordination with different ligands. Recently, we indeed found that Eu can coordinate with CN in 2-, 3-, 4-, 5- and 6-fold coordination, which leads to a quasicrystalline network.[1]

1 J. Urgel, D. Ecija, G. Lyu, R. Zhang, C. A. Palma, W. Auwärter, N. Lin and J. V. Barth, *Nat. Chem.*, 2016, **8**, 657.

Steven L. Tait enquired: How do you account for the striking difference in the scanning tunnelling spectroscopy measurements on the two-fold *vs.* three-fold Cu centers? Does that agree with expectations of the electronic states based on what is known from other metal–organic systems (including those not at surfaces)?

Nian Lin responded: We need theory (DFT) to answer this question. PDOS shall be able to tell the difference.

Peter Beton said: You have one defect in one of your images, is that a missing molecule? I was interested in what sort of defects you do see, and also, what you see at the domain boundaries. You've got this different coordination, when you see these ordered structures you'd expect to see some faceted edges. What kind of facet would you expect to see at the edges of this kind of island?

Nian Lin replied: The defect is not a missing molecule, but three molecules that coordinate to a central Cu atom undergo a 60° rotation around the central Cu. We saw domain boundaries when the molecular coverage is below or above the perfect coverage of forming the D-phase. In the D-phase, each 3-fold coordination is surrounded by six 2-fold coordinations and each 2-fold coordination is surrounded by three 2-fold and one 3-fold coordinations. What happens at the domain boundaries is the 2-fold and 3-fold coordination motifs are not perfectly arranged in such a manner. The edges of the D-phase always are boundaries since this phase is formed when the entire surface is covered by the molecular networks. The domain boundaries are not straight and often irregular, so it's not clear what facets they are.

Peter Beton asked: Can the phases grow continuously into one another? Can the hexagonal network grow continuously on an edge of this demi-regular network without any obvious boundaries?

Nian Lin answered: Experimentally we cannot see the real "growth". So, even though experimentally we did adjust the coverage gradually, what we "see" is the structures formed after annealing the sample at each coverage, which does not represent "continuous growth". This is the limit of STM. LEEM could be an ideal tool to answer your question.

Brandon Hirsch enquired: Prof. Lin, my question pertains to probing the kinetic pathways involved in the phase transition from the demi-regular lattice with a combination of 2- and 3-fold Cu centres to the S phase with exclusively 3-fold Cu centres. What would you expect if you formed a full monolayer of the demi-regular lattice, then added more of the ligand without annealing and analysed it structurally with STM? In principle, the molecules would coordinate at the 2-fold sites to form 3-fold Cu centres, but this would simultaneously close off the neighbouring pores.

Nian Lin replied: Very good suggestion. We have not done that, but it is definitely worthwhile to try it.

Rasmita Raval commented: Are the pore size and patterns observed in your surface tilings similar to those seen in zeolites and MOFs? Can you comment on the possibility of creating very different patterns and pores in a surface 2D system, which would lead to different functions compared to 3D porous materials?

Nian Lin responded: Yes and no. The underlying chemistry and assembly mechanism of the 2D porous structures are very similar to the MOFs. However, in zeolites and MOFs, there are pores for channels and compartments that can transport or store molecules. The 2D porous structure does not have such functions. In 2D systems, guest molecules are confined in 2D compartments. The reduced freedom might be exploited for stereo-selective reactions. In addition, one unique feature of the 2D systems is that the metal centers in these structures are unsaturated and easily accessible, which may be useful in heterogeneous catalysis.

Talat Rahman said: It would be a good idea to do it on another surface that does not interact strongly with the molecules. Just to see if it's a strain and geometric effect, or if it's really electronically driven that would be nice to know.

Nian Lin answered: To distinguish the two effects, we need an inert surface. BN might be a choice. We will look into it.

Trolle R. Linderoth opened a general discussion of the topics raised at the session: Turning to the more general discussion of characterisation of metal–organic coordination networks, the NIXSW technique used to determine the adsorption height of K-atoms in the K-TCNQ network is very elegant. Could it be employed for the other networks we have discussed in this session? In particular, would it be applicable to the hex-reconstructed Au(100) surface where the topmost hexagonal Au layer breaks the periodicity of the underlying bulk structure which sets up the X-ray standing wave pattern?

Nian Lin answered: Definitely. In particular, to determine the height of metal adatoms is extremely important.

Giovanni Costantini replied: The answer to the first, more general, question is yes, the NIXSW technique can (and indeed I think it should) be used for other molecular structures at surfaces, including some of those presented in this session. The NIXSW technique is not "disrupted" by a reconstructed surface, in the sense that the wavelength of the standing wave would still be that of the bulk interlayer spacing, irrespectively of a reconstructed surface. However, the NIXSW technique only tells about the height of an atom above the scattering planes of the chosen Bragg reflection which, in the case of a reconstructed surface, do not coincide with the outermost atomic layer of the surface. As such, it would be more difficult to determine surface–molecule bond lengths for a molecule adsorbed on a reconstructed surface.

Phil Woodruff commented: The standing wave is established in many layers of the substrate over a thickness of ~microns, so a reconstruction of the outermost layer or layers has no effect on the viability of the technique. However, what one measures is the height of an absorbing atom relative to the nearest extended bulk layer, not relative to the nearest actual layer, so surface reconstruction or relaxation means it is not trivial to relate the NIXSW height to adsorbate–substrate bond lengths. If the atoms in a reconstructed layer had sufficiently large surface core level shifts relative to the bulk then NIXSW could be used to measure this effect, but in practice this never seems to be the case.

Phil Woodruff said: The issue of how to deal with systems for which DFT calculations indicate several structures have very similar total energies is one we have recently had to address in a study of TCNQ (without coadsorbed alkali atoms) on Ag(111) and is presented in the poster by Phil Blowey. The ordered phase studied is commensurate with three molecules per unit mesh, and the combination of our NIXWS results and those of DFT calculations is that Ag adatoms are also involved in this phase, but the question is whether there are one, two or three adatoms per unit mesh. DFT calculations show two possible models, in particular, to have very similar energies that are significantly lower than most of the other structures. We have therefore compared our experimental structure parameter values with those based on a model in which the different possible structures are assumed to be co-occupied with probabilities given by the appropriate Boltzmann factors. This seems to lead to good agreement. Notice, though, that this treatment does not include entropic effects.

Giovanni Costantini said: I have a general comment on the need for experimental quantitative structural characterisation of surface-based molecular structures. The great majority of the work in this area available in the literature is based on a combination of STM and (typically) DFT-based calculations. While STM is an invaluable tool for determining the lateral arrangement and periodicity of a supramolecular structure and sometimes can also give information on the orientation of individual molecules, it cannot determine the actual structure of the molecule–substrate interface (*i.e.* the height and lateral position of the atoms constituting the molecule with respect to the substrate atoms), or (very often) determine whether the molecular superstructure is commensurate with the substrate or not. In most of the literature available in the field, the only quantitative structural information on an adsorbed molecular structure is that which is obtained directly from theory. This is particularly dangerous (unreliable) if the simulation needs to be done within a cell with periodic boundary conditions and if the simulation cell is chosen based only on the STM results. To properly resolve and understand a molecule–substrate structure, independent, experimental, quantitative structural determination is essential. Unfortunately this is not what is typically done, but the good news is that a number of techniques are available, such as NIXSW, LEED *I*–*V*, PhD. Performing and interpreting these experiments might not always be easy and straightforward, but it is definitively needed if a proper understanding of these systems is what we are looking for. Moreover, high quality experimental quantitative structural determination is the best way to benchmark DFT calculations, which is essential when dispersive interactions are included.

Phil Woodruff replied: I certainly echo these comments. The combination of STM and DFT alone is not a reliable way of determining surface structures. One problem is certainly the issue of the reliability of the atomic coordinates determined by DFT, particularly in systems where van der Waals interactions are important and the choice of the appropriate dispersion-corrected version of DFT is not clear. Quantitative structural information can provide valuable benchmark data to help to address this problem. In addition, however, it's important to remember that DFT, like quantitative experimental surface structural methods, can only find the correct structure if the initial structural model is correct. STM images can certainly be helpful in trying to identify possible structural models, but are not a substitute for more quantitative experimental structural data.

Talat Rahman responded: One should also note that most DFT calculations of STM images are also based on approximations and do not take into account the nature of the tip, which may have an impact on what is being imaged. The good news is that computational codes that employ more realistic DFT calculations are now emerging.

Talat Rahman commented: There was a time when we were getting a lot of vibrational information about systems, STS measurements, vibrational spectroscopy, providing a lot more information. We can't rely solely on DFT or any one structural method.

Marco Sacchi said: I strongly agree with what you are saying. We have been measuring benzene dynamics and comparing HeSE results with DFT calculations. It is remarkable that, as far as I know, there is still not a very accurate height measurement for such an important molecule, but just an indirectly deduced value. Indeed, as a theoretician, I would like to have this kind of quantitative measurements of structural parameters, since they are essential in order to benchmark dispersion-corrected DFT schemes, for instance.

Phil Woodruff replied: There are quite a number of fully quantitative experimental structure determinations for benzene on various surfaces. These include our own determinations of the local structure of benzene at low coverage (no long-range order) on Ni(111) and Ni(110) and an ordered phase on Ni(111).[1,2] There are also a number of structure determinations using quantitative LEED, initially from the Somorjai group starting in the 1980s, on Co, Pd, Pt and Ru surfaces.

1 O. Schaff *et al.*, *Surf. Sci.*, 1996, **348**, 89.
2 J.-H. Kang *et al.*, *Surf. Sci.*, 2000, **448**, 23.

Talat Rahman responded: There are some relatively old experimental data (LEED) for benzene on Pt. See, for example, ref. 1. But you are right that the experimental value of the height of benzene over other metals is not readily available.

1 A. Wander *et al.*, *Surf. Sci.*, 1991, **249**, 21.

Trolle R. Linderoth commented: Our interesting discussion of metal–organic coordination networks has shown that further understanding of the metal–ligand interaction nodes (symmetries, binding energies) as well as the nucleation mechanism of the networks are very interesting questions for future studies. Are there other questions that the community sees as particularly relevant to address?

Talat Rahman said: I think we are fortunate that with the foundation laid over four decades by surface science and with a plethora of recent advances, we are getting a lot of information, so much data is available from experiments and computational modelling. This is great as these provide good tests for theoretical models. We also recognize that future advances in DFT need to include considerations of dynamics and kinetics. It is fine to use structures obtained from DFT as the starting point, but we have to understand and accept the limits of DFT. It is, after all, a technique for determining the system's ground state at zero temperature. We should encourage those looking at many other factors affecting ordering at surfaces, including coverage dependence, kinetics, vibrational entropy, *etc.*

Nian Lin remarked: Self-assembly, by its nature, is a dynamic process. However, so far we primarily use STM and DFT to study the static states. We need some new tools to resolve the dynamics, for example, molecular dynamics and fast XPS.

Talat Rahman commented: One more thing I wanted to add, which I alluded to earlier, is that while there is the need to continue to develop more accurate and robust methods beyond DFT, which can describe reliably the properties of nanostructures in complex environments, out-of-equilibrium and nonadiabatic, there is also the need to build some simple models that explain the behaviour of complex molecules as shown in so many nice presentations here. With the information from DFT, one could use, for example, coarse graining and related methods from statistical mechanics to understand how peptides order on surfaces, whether they form dimers or other structures. It is important that our theory and modelling in concert with experimental data tells us something about the nature of the interactions responsible for complex molecular ordering on surfaces on a global, general level, as not everyone can or should perform these expensive DFT calculations for all systems. Nature is inherently simple, and physical reasoning allows us to break complex structures into simple parts for better understanding. I believe more efforts along such directions would be good.

Faraday Discussions

PAPER

Physical and chemical model of ion stability and movement within the dynamic and voltage-gated STM tip–surface tunneling junction†

Brandon E. Hirsch, Kevin P. McDonald, Steven L. Tait* and Amar H. Flood*

Received 12th March 2017, Accepted 21st March 2017
DOI: 10.1039/c7fd00104e

The interaction and mobility of ions in complex systems are fundamental to processes throughout chemistry, biology, and physics. However, nanoscale characterization of ion stability and migration remains poorly understood. Here, we examine ion movements to and from physisorbed molecular receptors at solution–graphite interfaces by developing a theoretical model alongside experimental scanning tunneling microscopy (STM) results. The model includes van der Waals forces and electrostatic interactions originating from the surface, tip, and physisorbed receptors, as well as a tip–surface electric field arising from the STM bias voltage (V_b). Our model reveals how both the electric field and tip–surface distance, d_{tip}, can influence anion stability at the receptor binding sites on the surface or at the STM tip, as well as the size of the barrier for anion transitions between those locations. These predictions agree well with prior and new STM results from the interactions of anions with aryl-triazole receptors that order into functional monolayers on graphite. Scanning produces clear resolution at large magnitude negative surface biases (−0.8 V) while resolution degrades at small negative surface biases (−0.4 V). The loss in resolution arises from frequent tip retractions assigned to anion migration within the tip–surface tunneling region. This experimental evidence in combination with support from the model demonstrates a local voltage gating of anions with the STM tip inside physisorbed receptors. This generalized model and experimental evidence may help to provide a basis to understand the nanoscale details of related chemical transformations and their underlying thermodynamic and kinetic preferences.

Department of Chemistry, Indiana University, 800 E. Kirkwood Avenue, Bloomington, IN 47405, USA. E-mail: tait@indiana.edu; aflood@indiana.edu

† Electronic supplementary information (ESI) available. See DOI: 10.1039/c7fd00104e

Introduction

The movement of ions in solution under the influence of electric fields is fundamental to a significant number of processes across science and technology. These include the movement of ions across membranes[1,2] to maintain processes critical to life, such as the biosynthesis of adenosine triphosphate[3] and nerve propagation[4] occurring in aqueous solutions. In organic media, ion movement plays a critical role in Li-ion batteries[5] that use lithium salts of anions like hexafluorophosphate in solvents of high and low-dielectric character. In all these examples, the electric fields either emanate from interfaces (electrodes) or are maintained across interfaces (membranes). These interfaces can be examined by local probe microscopies for which scanning tunneling microscopy (STM) provides both a source of the electric field as well as a tool for feedback-controlled imaging on molecular length scales. The imaging relies on quantum mechanical electron tunneling across the tip–surface junction. STM images are the foundation of many recent reports of STM-induced phase changes,[6–10] chemical transformations[11] and switches[12–24] attributed to effects between the STM tip and the surface, whose character is not always clear. An obvious effect is the electric field,

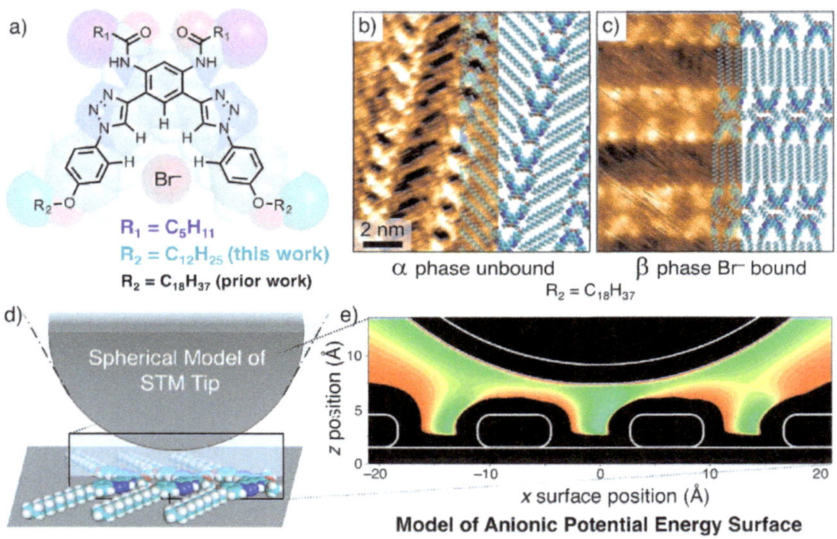

Fig. 1 (a) Molecular model and chemical structure of the aryl-triazole oligomer (ATO) molecules that behave as supramolecular receptors for anionic guests such as Br^-. (b) STM topography image and packing model of the α phase self-assembly of ATO–C_{18} ($I_t = 100$ pA, $V_b = -0.7$ V). (c) The anion-bound β phase assembly of ATO–C_{18} ($I_t = 100$ pA, $V_b = -0.7$ V). (d) Schematic representation of the electric field model for three physisorbed α phase receptors on the graphite surface underneath a spherical STM tip of radius 2.5 nm. The black box indicates the area of the calculation of the energy of a bromide anion. (e) 2D-contour plot of the anion potential energy representing a single slice in the *xz* plane through the central binding pocket of the ATO receptors. The top semicircle feature is the tip, the central phenylene rings of the ATO appear on the surface as lobed features. The ATOs are centered 3.4 A above the surface, which is also the position of the anion binding position in the receptor (at $x = 0$ A). Parts (b) and (c) were adapted with permission from *ACS Nano* 2014, **8**, 10858. Copyright 2014 American Chemical Society.

which acts over longer length scales than tunneling. Continuum models of these electric fields between the tip and surface do not generally include molecular-level details, such as the inclusion of physisorbed molecules, and detailed experimental support. To address this situation, we performed a parallel experimental–theoretical investigation on the movement of anions inside electric fields in tip–surface junctions.

Previous experimental work using STM at the solution–solid interface,[23,25,26] has demonstrated the ability to bind and eject anions using a physisorbed array of receptors. Observation of anion binding was achieved using aryl-triazole oligomer (ATO) anion receptors[27–30] (Fig. 1a) with long alkyl chains ($R_2 = C_{18}H_{37}$) assembled at the octanoic acid–graphite interface. The free receptors assembled into a 2D crystalline array of anti-parallel lamellae, called the α phase (Fig. 1b). Subsequent addition of bromide or iodide ions as tetrabutyl ammonium salts triggered a switch from the α phase to a different packing arrangement called the β phase (Fig. 1c) with anions clearly bound to the receptors. Critically, the voltage between the STM tip and the surface could be used to predictably induce mobility in the ions. We found that tetrabutyl ammonium iodide ion pairs could be ejected from the surface-arrayed receptors upon repeated scanning at −0.7 V. At larger negative surface biases ($V_b = -1.5$ V), ejection of ions from the β phase allowed formation of the α phase at the site of scanning. Alternatively, positive surface biases (+1.1 V) regenerated the β phase. This work represents a starting set of observations involving sub-molecular resolution of ion coordination and movement under electrically perturbed conditions.

The voltage-controlled phase switching coincident with anion movement was considered an ideal starting point to generate a microscopic model of the electric field between the tip and the surface. The model described herein combines well established physical and chemical principles to provide a picture of the tip–surface junction containing surface-arrayed ATO receptors. The model yields new insights into experimental phenomena of bias-dependent imaging that could not previously be explained. Experimentally, we employed a new member of the aryl-triazole receptor class, ATO–C_{12}, with $R_2 = C_{12}H_{25}$ (Fig. 1a). With these surface-arrayed receptors in the α-phase, addition of anions does not drive a switch to the β phase. Instead, we observed dramatic changes in the clarity of the STM imaging with bias that only emerge in the presence of added anions. The change in image quality is not simply a resolution loss. It occurs on account of frequent rapid tip retractions controlled by the feedback loop. These sharp tip retractions are correlated to anion movements occurring within the tip–surface tunneling region.

Together, this combined theoretical and experimental approach enables rational explanations and predictions for the movements of ions inside electric fields. Naturally, large negative surface biases destabilize the anion at the surface whereas moderate biases generate bistability in the position of the anion (bound at the surface or at the tip). The dependence of the anion's stability as a function of tip–surface distance was also investigated. While impacting bistability, it has a more dramatic effect on the barrier to anion movement between surface and tip. The dependence of the STM imaging parameters on anion stability outlined in this work may aid in the rationalization of a number of other switching and tip-dependent observations.[10] This work also contributes fundamental understanding of the microscopic details underlying thermodynamic and kinetic preferences related to ion-dependent chemical transformations.[31]

Results and discussion

Model of ions in the STM tip–surface tunneling junction

We constructed a model (Fig. 1d and e) of the potential energy surface that an anion experiences in the STM tip–surface tunneling region. It is based on standard coulombic and Lennard-Jones treatment of electrostatic (ES) and van der Waals (vdW) interactions, respectively. Existing descriptions of electric fields between a sphere and a plane were used to describe the electric field between the tip and surface.[32] The two generalized categories of interactions (vdW and ES) are used to calculate the potential energy of an anion in the proximity of the tip, surface, and molecular ATO receptors.

The STM tip is modeled as a Pt sphere of radius 25 Å. Experimentally, the tip composition is an 80/20% Pt/Ir wire that is mechanically cut, thus the exact apex dimensions and composition in the experiment are unknown. While overly simplified, the spherical and compositional assumptions of the STM tip provide an opportunity to calculate the general anion stability/lability effects in a straightforward manner. The STM instrument is operated in a feedback-controlled constant-current mode. Thus, the distance between the tip and surface is constantly changing throughout the STM imaging experiments. As such, the model includes four tip–surface distances ($z = 8.5, 10, 12, 100$ Å). While the exact distance between the tip and surface is unknown, previous models have reported between 3.0 Å and 10.0 Å,[33] depending on the substrate and tip compositions. Thus, the distances selected agree well with these expectations. The largest distance $z = 100$ Å was used to simulate tip retracted conditions.

The surface was treated as a semi-infinite continuum plane at a position of $z = 0$ Å (center of topmost atomic layer). The treatment yields straightforward calculations of the ES and vdW contributions to anion stability near the surface. The Lennard-Jones 6–12 potential[34,35] includes components for the weak stabilizing vdW forces and the short-range Pauli repulsion interactions that help define the spatial boundaries accessible to the anion (Fig. 2a and S1†). The attractive ES interactions the anion feels as it approaches a conductive surface were described using the image charge model (Fig. 2d and S2†). Detailed descriptions of these interactions can be found in the ESI.†

Both vdW and ES contributions of the anion with the tip were modelled. Again, the vdW stabilizing interactions with the tip are weak, but the core nature of the Pauli repulsion is effectively captured in the Lennard-Jones treatment (Fig. 2b and S3†). In a similar fashion to the surface, the ES interaction of the anion was captured using an image charge treatment (Fig. 2e and S4†).

Interactions of the anions with each ATO receptor were calculated using atomic coordinates and atomic partial charges that were determined from a geometry-optimized density functional theory (DFT) calculation (Fig. S5†). The atomic coordinates from the DFT calculation were used to calculate the vdW interactions of the anion with the molecule (Lennard-Jones treatment) (Fig. S7†). The ES interactions of the molecule with the anion were determined by summing the coulombic interactions of the anion (treated as a point charge) with all of the partial charges on the molecular receptor (Fig. S7†). Combining the ES and vdW interactions produces the binding of the anion to the ATO molecule (Fig. 2g). A 1D row of 25 molecules was used in the calculation to minimize any local potential

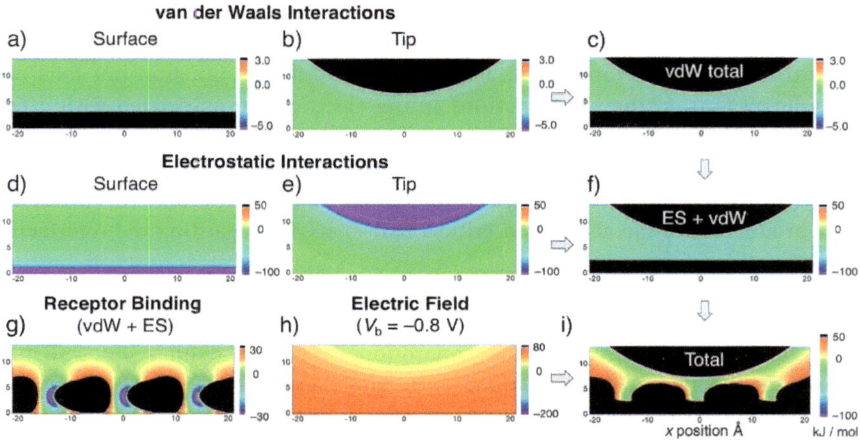

Fig. 2 2D-contour plots of anion location (Å) and corresponding energy (kJ mol^{-1}) based on the theoretical model incorporating physisorbed ATO inside the STM tip–surface tunneling gap with tip position at $z = 10$ Å and $V_b = -0.8$ V. 2D-contour plots shown are as follows (a) surface vdW interactions, (b) tip vdW interactions, (c) total sum vdW interactions, (d & e) surface and tip ES interactions respectively, (f) sum of the vdW and ES interactions, (g) the physisorbed receptor ES binding potential (including vdW forces), (h) the repulsive interaction present between the anion interacting with the electric field from $V_b = -0.8$ V, and (i) the total potential energy surface simulated from our model created by summing all of the interactions discussed.

variations from longer range forces such as the anion interaction with the molecular dipole moment (Fig. S6†). Further descriptions of the receptor anion interactions can be found in the ESI.†

In our model, the combined use of ES and vdW to model the receptor afforded a bromide binding energy of ~36 kJ mol^{-1} in solution. Previously reported values of related receptors in 50/50 v/v% dichloromethane/octanoic acid solutions display binding energies of −21 kJ mol^{-1}.[25] However, the interfacial and periodic arrangement of the ATO receptors observed in the liquid–solid studies is expected to deviate from solution phase values. These binding energies are also expected to differ for each particular crystalline lattice depending on the organization, orientation, and local packing environment of the receptors. If the contribution from the lattice increases or decreases the anion affinity inside the receptor pocket, as previously predicted,[25] it will affect the overall energy wells and barriers. We believe the theoretical treatment employed here provides a sufficiently plausible binding affinity for the ATO receptors to allow for a discussion of general trends.

In total, the three components (surface, molecular receptors, and tip) were each represented by vdW and ES interactions (six in total). The final component included in the model involves the interaction of the anion with the electric field present between the STM tip and the surface (Fig. 2h). In the experiment and in our model, the STM tip is grounded and the surface is biased. The positive and negative bias applied to the surface can support attractive or repulsive interactions, respectively, with the anion in the physisorbed receptor. Critically, the assumptions of a spherical STM tip and an infinite plane surface allow a facile calculation of the electric field between surface and tip using a well-established sphere-plane model of image charges.[32] The electric field between the STM tip and the surface is strong in the local area of the tunnel junction but falls off rapidly away from the tip. Variations in the electric field associated with a tip positioned at 12 Å and a range of surface biases (+0.8, +0.4, −0.4, −0.8, −1.1, and −1.5 V) can be found in the ESI (Fig. S9†). These variations in the electric field strength as a function of tip–surface separation distance and bias contribute to the production of bistability and critical barrier trends regarding the interfacial coordination of anions.

Taken together, the anion's vdW interactions with the tip and the surface provide only weak stabilization (note the modest energy scale; −5 to 3 kJ mol^{-1}, Fig. 2c). A repulsive term in the Lennard-Jones potential defines the available space for the anion (Fig. 2c). Stabilizing ES interactions between the anion and the conductive surface and tip are much greater (−100 to 50 kJ mol^{-1}) and thus create the underlying potential energy surface for the anion within the tip–surface junction (Fig. 2f). Addition of the receptor's anion recognition properties (±30 kJ mol^{-1}) and the tip–surface electric field (+0.8, −0.8 V ≡ −80, +80 kJ mol^{-1} at the surface, $z = 0$ Å) act to modify this base level potential energy surface (for $V_b = -0.8$ V, −40 to −20 kJ mol^{-1} at surface and tip respectively, Fig. 2i).

The model composed of the surface-arrayed receptors underneath a spherical STM tip contains many assumptions. First, the model is static. As such, it does not account for charge relaxation upon anion movement or any other dynamic relaxation of charge. The solvent, which screens the ES interactions, was accounted for by using a continuum model employing the dielectric constant for octanoic acid ($\varepsilon = 2.85$). Local variations in solvent, which may be important in the narrow tip–surface region, were not accounted for. To simplify the model, the counter cations were ignored. Nevertheless, dynamic ion pairs have been previously assigned in STM images containing the larger iodide anion paired with the counter cation tetrabutyl ammonium in these surface-arrayed receptors.[25] The presence of such ion pairs would be expected to shield the existing electrostatic interactions. Such shielding was not considered here.

Model of the tip–surface region: physical and chemical effects

The model allows us to make predictions on the expected behaviors when changing the magnitude and direction (polarity) of the tip–surface electric field. From these variations we are able to examine trends associated with the stability of the anions at real space locations in the tip–surface junction by inspection of the potential energy contour plots (Fig. 3). By comparing different surface biases (left-to-right in Fig. 3), one can ascertain stability thresholds and barrier variations that indicate three different anion potential energy landscapes: stable receptor binding, unstable binding (transition to near-tip region), and an intermediate bistable landscape. Decreasing the tip–surface distance (top-to-bottom in Fig. 3) affects the strength of the binding region and also shows changes in stability, particularly in binding sites away from the tip center that will impact "sweeping" of the anion by the tip (see below). Many voltage and tip–surface distance trends emerge visually in these representations.

The first primary observation from the model is the change in stability introduced with a shift in the field's polarity (stable at +0.8 V, bistable at −0.8 V and

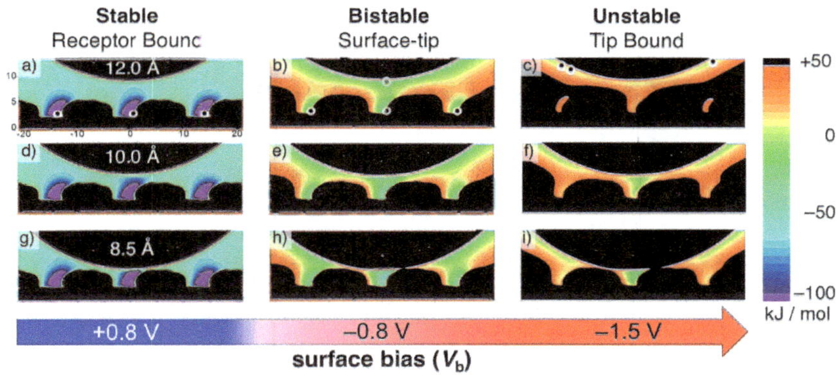

Fig. 3 (a–i) 2D-contour plots of the energies that emerge from the theoretical model where the bias applied the to the surface (V_b) is varied from +0.8, (a, d, g); −0.8, (b, e, h); and −1.5 V, (c, f, i) along the horizontal axis. Along the vertical axis the tip–surface distance is altered between z = 8.5, (g–i); 10.0, (d–f); and 12.0 Å, (a–c). Color scale ranges from −100 (purple) to +50 (black) kJ mol^{-1}.

unstable at −1.5 V). Positive surface biases produce a large global minimum (observed as deep purple color) involving anion binding within the surface adsorbed ATO receptors (Fig. 3a, d and g). While difficult to see with the common energy scale, the stability of the anion inside the surface-arrayed receptor is enhanced (~60 kJ mol^{-1}) when a surface bias of +0.8 V is applied compared to when no surface bias is applied to the surface, *i.e.*, 0.0 V, (Fig. S10†). Alternatively, when negative surface biases are modelled, this single global minimum is replaced with two local minima creating conditions of bistability. The new secondary minimum, which corresponds to the anion positioned at the tip, emerges from the electric field negative surface bias and the vdW and ES attraction to the tip. This repulsion destabilizes the anion in the surface-adsorbed receptors and creates a new minimum at the tip (Fig. 3b, e, and h). Thus, large negative biases shift stability to the tip (Fig. 3c, f, and i).

Modelling: voltage dependence

Anion stability fluctuations can be examined in great detail for all the tip simulated heights by extracting slices through the 2D contour plots at the point of maximum binding potential, $x = 0$ (Fig. 4). Each plot represents an individual applied surface bias value, but contains all four simulated tip height positions. The plots are arranged together and plotted on a single axis for comparison across the bias values. For example, with $z = 10$ Å, the stability of the central binding pocket at the best anion binding position of $(x, z) = (0, 3.4$ Å$)$ changes from approximately −60 to −40 kJ mol^{-1} when changing the bias from −0.4 to −0.8 V (compare red traces in Fig. 4d and e).

Under the assumptions of our model, large negative biases ($V_b = -1.5$ V, Fig. 4f) shift the global stable state for the position of the anion from the surface-bound receptors to the tip for all tip–surface separation distances. This result agrees remarkably well with our previous phase switching observations[25] made with ATO–C_{18} and iodide anions. Unfortunately, a strict prediction of a threshold

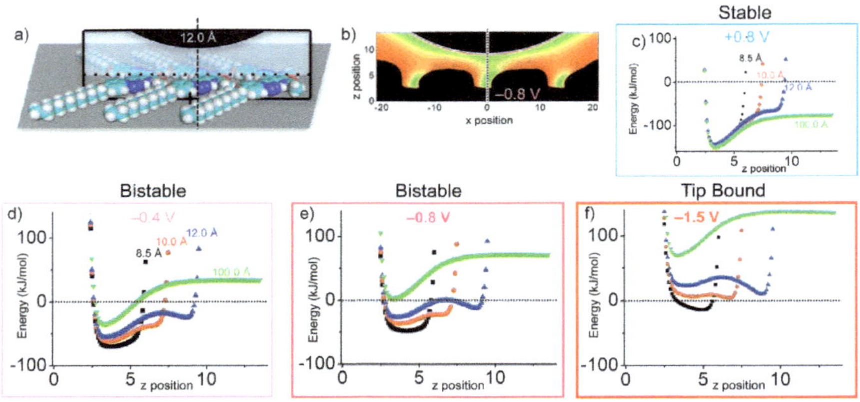

Fig. 4 (a) Schematic representation of three physisorbed α phase receptors on the graphite surface underneath a spherical STM tip positioned above the surface. The black box with bold outline indicates the modeled area shown in the other figures. (b) 2D-contour plot of the anion potential energy in the area highlighted in (a) for a tip position of $z = 12.0$ Å and $V_b = -0.8$ V. (c–f) The energy (kJ mol^{-1}) is plotted against the z-position of the anion at $x = 0.0$ Å (binding picket in receptor) for four different tip heights. The four z-positions of the tip are color coded in each individual plot $z = 8.5$ (black squares), 10.0 (red circles), 12.0 (blue triangles), and 100 Å (green downward triangles). The four different biases are $V_b =$ (c) +0.4, (d) −0.4, (e) −0.8, and (f) −1.5 V.

voltage at which this stability tips over from surface to tip cannot be confidently assigned on account of uncertainties in the tip state and dynamic tip–surface region in the experiments. However, the distance dependence of the electric field, its impact on the anion stability, and trends in behavior under different conditions can be drawn from the model to allow insight into the experimental environments and behavior.

Modeling: tip–surface distance-dependence

At any bias value, a larger tip–surface distance decreases the impact of the electric field on the relative anion binding stability in the receptor compared to at the tip. With negative surface biases, larger tip–surface distances cause the anions to be more stable within the surface-adsorbed receptors relative to the tip.

The barrier between the two minima corresponding to the anion at the surface (stable) or at the tip (metastable) displays pronounced changes with distance. When the tip is near (8.5 Å) the negatively biased surface, the barrier between surface and tip disappears (black traces, Fig. 4d–f) and the anion would be able to easily transit between the receptor and the tip. This broad energy well slopes toward the surface at small negative surface biases (−0.4 V). The shape of this potential well arises from the electric field with large contributions from the binding affinity in the receptor molecule. At −0.8 V, that shape is relatively flat, while at −1.5 V the slope tilts toward the tip as the strength of the field overwhelms the binding advantage of the receptor. This broad energy well is a convolution of the electric field with the attractive interactions (mostly ES) involving the surface, receptor, and tip. As the tip–surface distance decreases, the ES interactions of the tip and the receptor/surface begin to merge together. This overlap of attractive interactions reduces the barrier as the tip nears the surface. In addition, the slope of the potential energy resulting from the electric field gets steeper as the tip gets closer to the surface. At negative biases, this effect acts simultaneously alongside the overlapping tip and surface ES interactions to reduce the barrier to bistability at small tip–surface separation distances. Conversely, when those distances increase as the tip moves away from the surface, the barrier between the two locations grows.

Experimentally, the tip–surface distance is constantly being altered in STM imaging experiments as a consequence of the feedback loop. This situation means that for any specific bias, the barrier is constantly rising and falling as the STM seeks to maintain the constant current setpoint.

Bias dependent imaging of surface-arrayed receptors with anions

The theoretical model helped establish general stability and dynamics trends expected for STM imaging of surface-arrayed ATO receptors in the presence of size-complementary anions at liquid–solid interfaces. In the experimental investigation of these trends, such as bias-dependent binding with negative surface biases, we discovered dynamic effects that are consistent with the expectations developed from modeling the system with changing barriers.

Previous reports[25] with ATO–C_{18} displayed anion-dependent voltage-driven switching between different surface phases (Fig. 1b and c). Here, we investigated how the STM bias voltage controls the stability of the anions without causing switching between two phases. Based on studies exploring the role of the alkyl chain length on the relative stability of the two phases, we predicted that shortened chain lengths would inhibit phase switching.[36] We found the $C_{12}H_{25}$ chains in ATO–C_{12} (Fig. 1a) fulfilled this goal. Surface assembles of ATO–C_{12} generate the expected α phase at the 1-octanoic acid/graphite interface using solution concentrations of 100 μM (Fig. S11†), as described elsewhere.[36] The unit cell measured along the inter-row spacing (5.35 ± 0.21 nm) for ATO–C_{12} is smaller than that for ATO–C_{18} (6.49 ± 0.13 nm).

Addition of one equivalent of *n*-tetrabutylammonium bromide (TBABr) to the ATO–C_{12} receptor solution prior to deposition yielded no change in the packing structure for the α phase. As anticipated, the structure observed matched the one seen for the assembly from solutions of the receptor alone. Attempts to resolve anions bound within pockets of the receptors arrayed in the α phase structure by altering the scan parameters were unsuccessful (Fig. S12†). Under all the conditions examined that produced clear STM images, the surface-arrayed receptors displayed an empty binding pocket towards solution. Insufficient image contrast was observed with positive biases in these experiments, which was consistent with all of our prior work with ATO molecules,[23,36] thus all of the imaging was conducted with negative biases. Despite the results from our model, no anion STM observations left reasonable doubt that the anions could actually bind to the physisorbed receptors. Specifically, the model shows that the binding pockets of the surface-adsorbed receptors serve as global minima for all but the largest negative biases. The expectation, therefore, is that less negative biases would allow for the anions to remain in the receptor pockets for imaging.

At large negative surface biases, the α phase structure of ATO–C_{12} could be clearly imaged (Fig. 5a and d). However, as the magnitude of the negative surface

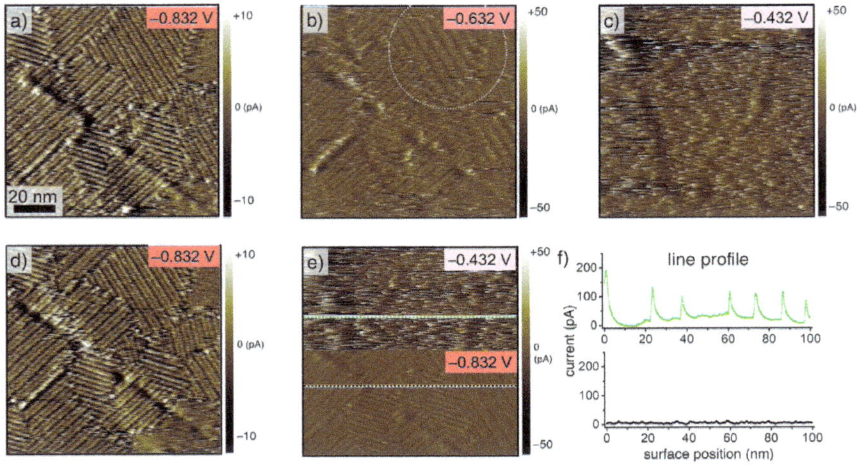

Fig. 5 (a–e) STM current image sequence (obtained with different surface biases) of the α phase self-assembly of the ATO–C_{12} with the *ex situ* addition of one equivalent of TBABr. Images at −0.832 V (panels a and d) show clear resolution of the α phase physisorbed receptors before and after imaging with moderate (b) −0.632 V and low (c) −0.432 V bias. These surface biases foster large tip retractions, as indicated by the line profile in (f), which corresponds to the green line in (e). Relatively little tip retraction is found in line profiles at −0.832 V (f), black line. Note these height profiles measure the deviation from the current setpoint. Reversibility in the voltage resolution control is demonstrated in (e) where the bias was switched from −0.832 V to −0.432 V in the middle of the image. Tunneling parameters: $I_t = 100$ pA for each image with biases as marked. Circled area in (b) highlights the tip retractions occurring along the α phase crystal lattice.

bias voltage is reduced, *i.e.*, less negative surface biases, we observed a loss in clarity of the α phase in the STM image (Fig. 5b and c). Loss in clarity is accompanied by frequent and large tip retractions (Fig. 5f) corresponding to large ~100–300 pA jumps in current. These current spikes result in withdrawing tip motion of ~2 nm (spike in current and topography channels, respectively, Fig. S13†). Such tip retractions are often assigned to dynamics, such as, the adsorption–desorption or surface rearrangement of molecules. However, upon returning to the same well-resolved high-bias scan conditions, we see that the physisorbed receptors, their domain boundaries and all the defects are in exactly the same positions (compare Fig. 5a and e). The tip retractions could be switched on and off reversibly by the magnitude of the STM bias (Fig. 5e). Importantly, this trend of unstable imaging and smaller negative biases was not observed when no anions were present in the solution (Fig. S14†). Thus, the presence of the anion (and corresponding counter cations) in solution leads to dynamic motion to cause frequent tip retractions and a consequent loss in STM image clarity. These results represent an experimental observation involving the movement of individual anions in the electric field present in the tip–surface tunneling junction.

Further observations that support our assignment of dynamics related to the anion can be found in the STM images obtained with $V_b = -0.632$ V (Fig. 5b). Here, we see how the clarity loss tip retractions follows the lamellae arrangement of ATO receptors (circled domain). The coincidence of the tip retractions with the positions of the physisorbed receptors also supports the conclusion that the tip retractions are caused by the movement of the anions within the tip–surface STM tunneling junction.

The theoretical model helps provide some rationalization of the bias-dependent STM image quality. In the model, when sufficiently large negative biases are used, *e.g.*, −1.5 V, the anions are fully ejected from the adsorbed ATO receptors and the α phase can be resolved clearly. However, when weaker electric fields are present under moderate negative surface bias, *e.g.*, −0.4 V, the anions are believed to be bound to the surface array of ATO receptors. However, when the intense electric field present at the apex of the tip (tunneling region) is close to the anion, dynamics are observed. Any decomplexation events are likely accompanied by anion migration towards the tip. Anion migration occurring in the tunneling gap is believed to produce enhanced current levels, arising from a transient modification of the tunneling gap. This subsequently triggers the constant-current feedback loop to retract the tip.

When the magnitude of the negative surface biases becomes smaller (with the same current setpoint), the tip will, on average, move closer to the surface. If interfacial anion binding has occurred, a closer tip–surface distance will produce more frequent anion decomplexation. The energy difference between anion binding at the surface and at the tip equalize as the barrier between them decreases. For example, the black trace in Fig. 4d represents the anion potential energy at a bias of −0.4 V and a tip–surface distance of 8.5 Å. Decomplexation corresponds to a virtually barrierless movement (~10 kJ mol^{-1} uphill). At 298 K, this is in the diffusion-controlled limit. Individual anion decomplexation events could have a measurable impact on the density of states, enhancing the tunneling current to cause tip retraction. Given results from the model, anion dynamics in the tip–surface tunneling junction can be reasonable well correlated to the experimental observations involving sharp tip retractions.

With large negative surface bias (−1.5 V), the global minimum is at the tip and the anions are cleanly removed from the receptors with the STM images of the ATO–C_{12} α phase appearing clear (Fig. 5a and e). At these larger biases, the stronger electric fields can eject anions from receptors over further distances along the *z* direction; note that there is a pathway for anion ejection from the side receptor sites, *i.e.*, before the anion in question is in the tunneling region (Fig. 3c). Thus, the location of anion ejection and tunneling are not coincident and with the anions swept out of the receptors prior to imaging, the images do not appear distorted. Interestingly, trace and retrace differences indicate tip asymmetry effects likely to affect the anion ejection (Fig. S13†).

We have previously reported[25] success (Fig. 1b and c) with imaging ions in the binding pockets of related receptors (ATO–C_{18}) when self-organized in a head-to-head (β phase) arrangement rather than a head-to-tail arrangement (α phase). In these experiments, bromide anions were clearly imaged in the β phase (Fig. 1c). Separately, iodide ions paired with their tetrabutylammonium counter cations were also observed.[25] Our calculations show[25] the ATO receptor likely has stronger anion binding for the β phase arrangement of receptors relative to the head-to-tail α phase. The α phase, co-localizes destabilizing carbonyl oxygen atoms proximal to the binding pocket. The enhanced stability of the bromide anion seen in the head-to-head β phase may facilitate more consistent STM imaging of the bound anion. For the tetrabutylammonium iodide, we almost exclusively imaged ion pairs. This ion pairing certainly has a stabilizing effect on the anion in the electric field. Nevertheless, ion pairs are expected to have a tendency to dynamically reorient (requiring motion out of the binding pocket) in the electric field at negative surface bias. Some of these dynamics have been captured previously,[25] related work involving macrocylic versions of the aryl-triazole receptors,[26] which naturally foster enhanced binding, were also found to facilitate high resolution STM imaging. These anion-recognition studies allow design to be translated from receptors that bind by themselves as mono-molecular systems in solution to crystalline arrays, such as the films studied here, and porous networks like metal–organic frameworks (MOFs).[37–40] Such higher-order design elements are vital in the application of ion receptors to materials.[41]

Conclusions

We have presented a straightforward and semi-quantitative model of ions in the tip–surface junction involving physisorbed receptors at solution–surface interfaces that agrees well with trends from experiment. The model of an anion in the STM tip–surface junction reveals how the interplay of various interactions determines the specific behaviors of such ionic systems at electrified surfaces. Experimentally, we have observed that higher bias voltages can "sweep" anions from their molecular receptors before they are able to interfere with the tunneling junction, allowing high image quality. Modelling low biases predicts anion stability at the surface sites, yet the smaller tunneling distances involved with these conditions equalize the stabilities of the tip and surface, and this lowers the energetic barriers to anion movement from the surface receptors to the tip. Experimentally, the movement of anions in the tunneling region is believed to transiently enhance the tunneling current and result in tip retraction. This is observed as a loss in imaging quality. While this is the first step to better understand the role of bias, tip–surface distance, and electric field, a higher level theoretical treatment of this problem may reveal more details on these ionic systems. For this purpose, it will be important to develop a better understanding of solvation effects and the character of the solvent in the near surface region, particularly in the tunneling gap. Finally, given that the movement of ions depends on bias and distance, these effects might promote or inhibit chemical transformations in which the reactant or product is an ion, such as for deprotonation or diazonium chemistry.

Acknowledgements

We thank the National Science Foundation for support (DMR 1533988).

References

1 O. Scheel, A. A. Zdebik, S. Lourdel and T. J. Jentsch, *Nature*, 2005, **436**, 424–427.
2 M. O. Jensen, V. Jogini, D. W. Borhani, A. E. Leffler, R. O. Dror and D. E. Shaw, *Science*, 2012, **336**, 229–233.
3 P. D. Boyer, *Annu. Rev. Biochem.*, 1997, **66**, 717–749.
4 K. J. Swartz, *Nature*, 2008, **456**, 891–897.
5 K. Xu, *Chem. Rev.*, 2014, **114**, 11503–11618.

6 K. S. Mali, D. Wu, X. Feng, K. Müllen, M. Van der Auweraer and S. De Feyter, *J. Am. Chem. Soc.*, 2011, **133**, 5686–5688.
7 S.-B. Lei, K. Deng, Y.-L. Yang, Q.-D. Zeng, C. Wang and J.-Z. Jiang, *Nano Lett.*, 2008, **8**, 1836–1843.
8 A. Semenov, J. P. Spatz, M. Moller, J. M. Lehn, B. Sell, D. Schubert, C. H. Weidl and U. S. Schubert, *Angew. Chem., Int. Ed.*, 1999, **38**, 2547–2550.
9 S. L. Lee, Y. Fang, G. Velpula, F. R. Cometto, M. Lingenfelder, K. Mullen, K. S. Mali and S. De Feyter, *ACS Nano*, 2015, **9**, 11608–11617.
10 J. A. A. W. Elemans, *Adv. Funct. Mater.*, 2016, **26**, 8932–8951.
11 A. Ciesielski, S. Lena, S. Masiero, G. P. Spada and P. Samorì, *Angew. Chem., Int. Ed.*, 2010, **49**, 1963–1966.
12 G. Schull, L. Douillard, C. Fiorini-Debuisschert, F. Charra, F. Mathevet, D. Kreher and A. J. Attias, *Adv. Mater.*, 2006, **18**, 2954–2957.
13 S. B. Lei, K. Tahara, X. L. Feng, S. H. Furukawa, F. C. De Schryver, K. Mullen, Y. Tobe and S. De Feyter, *J. Am. Chem. Soc.*, 2008, **130**, 7119–7129.
14 M. M. S. Abdel-Mottaleb, N. Schuurmans, S. De Feyter, J. Van Esch, B. L. Feringa and F. C. De Schryver, *Chem. Commun.*, 2002, 1894–1895, DOI: 10.1039/b205240g.
15 M. Surin, P. Samori, A. Jouaiti, N. Kyritsakas and M. W. Hosseini, *Angew. Chem., Int. Ed.*, 2007, **46**, 245–249.
16 L. Piot, R. M. Meudtner, T. El Malah, S. Hecht and P. Samorì, *Chem.–Eur. J.*, 2009, **15**, 4788–4792.
17 P. Zell, F. Mogele, U. Ziener and B. Rieger, *Chem.–Eur. J.*, 2006, **12**, 3847–3857.
18 R. Gutzler, T. Sirtl, J. F. Dienstmaier, K. Mahata, W. M. Heckl, M. Schmittel and M. Lackinger, *J. Am. Chem. Soc.*, 2010, **132**, 5084–5090.
19 S.-L. Lee, Z. Yuan, L. Chen, K. S. Mali, K. Mullen and S. De Feyter, *J. Am. Chem. Soc.*, 2014, **136**, 4117–4120.
20 K. Tahara, K. Inukai, J. Adisoejoso, H. Yamaga, T. Balandina, M. O. Blunt, S. De Feyter and Y. Tobe, *Angew. Chem., Int. Ed.*, 2013, **52**, 8373–8376.
21 D. Bleger, A. Ciesielski, P. Samori and S. Hecht, *Chem.–Eur. J.*, 2010, **16**, 14256–14260.
22 H. Maeda, T. Shirai and S. Uemura, *Chem. Commun.*, 2013, **49**, 5310–5312.
23 B. E. Hirsch, S. Lee, B. Qiao, C.-H. Chen, K. P. McDonald, S. L. Tait and A. H. Flood, *Chem. Commun.*, 2014, **50**, 9827–9830.
24 S. Yokoyama, T. Hirose and K. Matsuda, *Langmuir*, 2015, **31**, 6404–6414.
25 B. E. Hirsch, K. P. McDonald, B. Qiao, A. H. Flood and S. L. Tait, *ACS Nano*, 2014, **8**, 10858–10869.
26 S. Lee, B. E. Hirsch, Y. Liu, J. R. Dobscha, D. W. Burke, S. L. Tait and A. H. Flood, *Chem.–Eur. J.*, 2016, **22**, 560–569.
27 Y. Li and A. H. Flood, *Angew. Chem., Int. Ed.*, 2008, **47**, 2649–2652.
28 Y. Li and A. H. Flood, *J. Am. Chem. Soc.*, 2008, **130**, 12111–12122.
29 Y. Hua and A. H. Flood, *Chem. Soc. Rev.*, 2010, **39**, 1262–1271.
30 S. Lee, Y. Hua, H. Park and A. H. Flood, *Org. Lett.*, 2010, **12**, 2100–2102.
31 E. Bekyarova, M. E. Itkis, P. Ramesh, C. Berger, M. Sprinkle, W. A. de Heer and R. C. Haddon, *J. Am. Chem. Soc.*, 2009, **131**, 1336.
32 F. F. Dall'Agnol and V. P. Mammana, *Rev. Bras. Ensino Fis.*, 2009, **31**, 3503.
33 M. Yankowitz, K. Watanabe, T. Taniguchi, P. San-Jose and B. J. LeRoy, *Nat. Commun.*, 2016, **7**, 13168.
34 J. E. Jones, *Proc. R. Soc. London, Ser. A*, 1924, **106**, 463–477.

35 J. N. Israelachvili, *Intermolecular and Surface Forces*, Academic Press, Elsevier, Amsterdam, 3rd edn, 2011.
36 B. E. Hirsch, K. P. McDonald, A. H. Flood and S. L. Tait, *J. Chem. Phys.*, 2015, **142**, 101914.
37 R. Custelcean and B. A. Moyer, *Eur. J. Inorg. Chem.*, 2007, 1321–1340.
38 Q. W. Li, W. Y. Zhang, O. S. Miljanic, C. H. Sue, Y. L. Zhao, L. H. Liu, C. B. Knobler, J. F. Stoddart and O. M. Yaghi, *Science*, 2009, **325**, 855–859.
39 B. L. Chen, S. C. Xiang and G. D. Qian, *Acc. Chem. Res.*, 2010, **43**, 1115–1124.
40 M. M. Wanderley, C. Wang, C. D. Wu and W. B. Lin, *J. Am. Chem. Soc.*, 2012, **134**, 9050–9053.
41 K. P. McDonald, B. Qiao, E. B. Twum, S. M. Lee, P. J. Gamache, C. H. Chen, Y. Yi and A. H. Flood, *Chem. Commun.*, 2014, **50**, 13285–13288.

Faraday Discussions

PAPER

Sequential nested assembly at the liquid/solid interface†

Baharan Karamzadeh,[a] Thomas Eaton,[b] David Muñoz Torres,[b] Izabela Cebula,‡[a] Marcel Mayor *[bcd] and Manfred Buck *[a]

Received 23rd March 2017, Accepted 26th April 2017

DOI: 10.1039/c7fd00115k

Studying the stepwise assembly of a four component hybrid structure on Au(111)/mica, the pores of a hydrogen bonded bimolecular network of 3,4,9,10-perylenetetracarboxylic diimide (PTCDI) and 1,3,5-triazine-2,4,6-triamine (melamine) were partitioned by three and four-armed molecules based on oligo([biphenyl]-4-ylethynyl)benzene, followed by the templated adsorption of either C_{60} fullerene or adamantane thiol molecules. The characterisation by ambient scanning tunneling microscopy (STM) reveals that the pore modifiers exhibit dynamics which pronouncedly depend on the molecular structure. The three-armed molecule 1,3,5-tris([1,1′-biphenyl]-4-ylethynyl)benzene (3BPEB) switches between two symmetry equivalent configurations on a time scale fast compared to the temporal resolution of the STM. Derivatisation of 3BPEB by hydroxyl groups substantially reduces the switching rate. For the four-armed molecule configurational changes are observed only occasionally. The observation of isolated fullerenes and small clusters of adamantane thiol molecules, which are arranged in a characteristic fashion, reveals the templating effect of the trimolecular supramolecular network. However, the fraction of compartments filled by guest molecules is significantly below one for both the thermodynamically controlled adsorption of C_{60} and the kinetically controlled adsorption of the thiol with the latter causing partial removal of the pore modifier. The experiments, on the one hand, demonstrate the feasibility of templating by nested assembly but, on the other hand, also pinpoint the requirement for the energy landscape to be tolerant to variations in the assembly process.

[a]*EastCHEM School of Chemistry, University of St Andrews, North Haugh, St Andrews, UK. E-mail: mb45@st-andrews.ac.uk*

[b]*Department of Chemistry, University of Basel, St. Johannsring 19, CH 4056 Basel, Switzerland. E-mail: marcel.mayor@unibas.ch*

[c]*Institute of Nanotechnology, Karlsruhe Institute of Technology (KIT), P.O.Box 3640, D-76021 Karlsruhe, Germany*

[d]*Lehn Institute of Functional Materials (LIFM), Sun Yat-Sen University (SYSU), Guangzhou, P. R. China*

† Electronic supplementary information (ESI) available: Details of the chemical synthesis of **4BPEB-2OH** and additional STM images. See DOI: 10.1039/c7fd00115k

‡ Present address: Department of Chemical and Process Engineering, University of Strathclyde, James Weir Building, 75 Montrose Street, Glasgow G1 1XJ, UK.

Introduction

Supramolecular assembly at interfaces has been widely studied both in ultrahigh vacuum (UHV) and at the solid–liquid interface. Providing convenient access to the sub-5 nm length scale with atomic precision,[1–25] 2D networks are of particular interest as they enable ultraprecise templating by exploiting the interactions of guest species with the network structure and/or defining adsorption sites through the confinement of guest entities to network pores.[5,18,26–40]

Strategies for the generation of structures comprise either the simultaneous assembly of multicomponent systems to complex patterns[41] or a sequential step by step procedure.[28,34] The former is accomplished under thermodynamically controlled conditions involving well designed interactions and geometries of the individual components. The latter can be performed either under thermodynamic or kinetic control, thus offering more degrees of freedom. Notably, guest species which are thermodynamically more stable than the network, and thus ultimately displace it, can be templated.[42] Furthermore, very different types of steps can be combined[43] and the network can serve as a temporary sacrificial mask.[44] However, apart from the underpotential deposition of Cu following pore filling of a network by a thiol[43] and the generation of a binary SAM by replacement of the network,[44] sequential assembly has not been going beyond the two steps of network formation and adsorption of a guest species. Since the supramolecular template defines the arrangement of the guest in this case, any changes will require the adaptation of the network by varying the pore size and/or geometry[28] or functionalising the pores by derivatisation of the network components.[45,46]

A complementary, yet unexplored strategy for the stepwise generation of surface based nanostructures is nested self-assembly. As illustrated in Fig. 1, the structure of a porous molecular network which defines the base frame, is modified prior to the adsorption of guest species. While this strategy offers an additional level of structural control, it will require precise matching of components as regards both geometries and interaction energies. Important aspects in this context are the affordable tolerances in the dimensions of the pore modifier and the influence of a highly confined space on the assembly process with the latter being particularly important for the solid/liquid interface where the adsorption of species involves an exchange of solvent molecules. The studies reported here build on our previous work where, in experiments examining a series of threefold symmetric star-shaped molecules differing in size, tris([biphenyl]-4-ylethynyl) benzene (**3BPEB-H**, see Fig. 1) was identified as the one fitting the pores of a hydrogen bonded honeycomb network.[47] Having observed that **3BPEB-H** exhibits a dynamic behaviour, one focus of the present study is on the underlying mechanism and the factors affecting it. This includes modification of the molecules by both chemical functionalisation and the addition of a fourth arm (see Fig. 1). It is noted that there are two equivalent threefold degenerate configurations for the threefold symmetric star molecules (ignoring the asymmetry introduced by the OH groups) whereas the degeneracy is lifted for **4BPEB-2OH**, thus resulting in six equivalent positions. The other focus is on the templating of the star/network structure using C_{60} fullerene and adamantane thiol. C_{60}, a commonly used guest molecule in surface based self-assembly,[6,27,28,30–35,39,46,48–51]

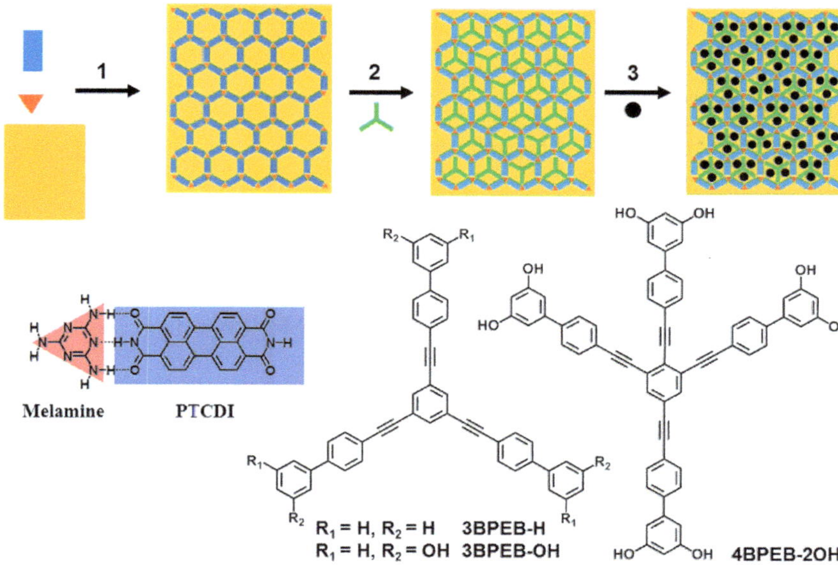

Fig. 1 Top: scheme of sequential molecular assembly comprising the formation of a bimolecular honeycomb network of PTCDI and melamine (1), compartmentation of the network pores (2), and the adsorption of guest molecules into the subpores (3). Bottom: structures of the molecules used.

and the much less studied thiols[42,44,45,52] have been chosen as representatives for assembly under thermodynamic and kinetic control.

Results and discussion

Synthesis

The syntheses of the star-shaped guest molecules are based on Pd catalysed coupling chemistry.[53] While the syntheses of the two three-armed stars **3BPEB-H** and **3BPEB-OH** were already reported elsewhere,[47] the assembly of the four-armed star **4BPEB-2OH** is sketched in Scheme 1. For the central unit, a fourfold 1,2,3,5-ethynyl-substituted benzene was required, which was obtained from commercially available 2,4,6-tribromoaniline **6**, which was converted to 1,3,5-tribromo-2-iodobenzene **7** using a variation of the Sandmeyer reaction.[54] Subsequent treatment with a twelvefold excess of 2-methylbut-3-yn-2-ol under Sonogashira coupling conditions gave the fourfold ethynyl substituted benzene **8**, which provided the central building block, **9**, after deprotection. Treatment of 1,2,3,5-tetraethynylbenzene, **9**, with four equivalents of iodo-biphenyl derivative **5** again under Sonogashira conditions provided the four-armed star **4PBEP-2OH** as a brown solid in 45% isolated yield after column chromatography. The assembly of 4′-iodo-[1,1′-biphenyl]-3,5-diol **5** is displayed in Scheme 1 and is based on Suzuki coupling combined with suitable masking strategies.

The four-armed star **4PBEP-2OH** was characterized using ^{1}H- and ^{13}C-NMR spectroscopy and mass spectrometry. The identity of the structure was further corroborated by its appearance in the STM experiments reported in this paper.

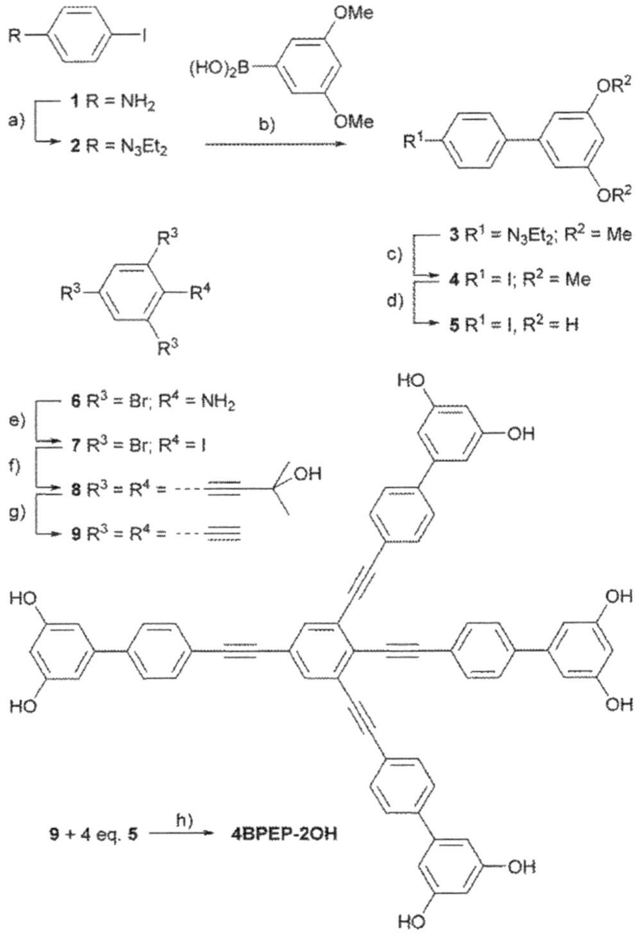

Scheme 1 Synthesis of the four-armed star structure **4BPEB-2OH**. Reagents and conditions: (a) conc. HCl, aq. $NaNO_2$, Et_2NH, K_2CO_3, 0 °C to RT, 53%; (b) $Pd(PPh_3)_2Cl_2$, K_2CO_3, $C_6H_5CH_3$, EtOH, MicroWave, 120 °C, 45 min, 56%; (c) MeI, 130 °C, 90%; (d) BBr_3, CH_2Cl_2, −78 °C, 79%; (e) (1) pTsOH·H_2O, $NaNO_2$, KI, CH_3CN, H_2O, 15 °C, 10 min; (2) 20 °C, 2 h, 68%; (f) 2-methyl-3-butyn-2-ol, $Pd(PPh_3)_2Cl_2$, CuI, DMF, Et_2NH, 80 °C, 24 h, 49%; (g) NaH, $C_6H_5CH_3$, 120 °C, 3 h, 49%; (h) $Pd(PPh_3)_4$, CuI, THF, iPr_2NH, 45 °C, 16 h, 45%.

The synthetic protocols and analytical data of all new chemical structures are provided in ESI.†

Surface studies

I. Compartmentation of network pores. Exposure of the PTCDI–melamine network to **3BPEB-H** yields images like those depicted in Fig. 2. While the network is well resolved with both the melamine and PTCDI components clearly seen, the characteristic star shape of the pore modifier appears only occasionally like in the marked pore in the upper half of Fig. 2a. In such cases where even the individual benzene rings of the **3BPEB-H** molecule are resolved, the occurrence of the star

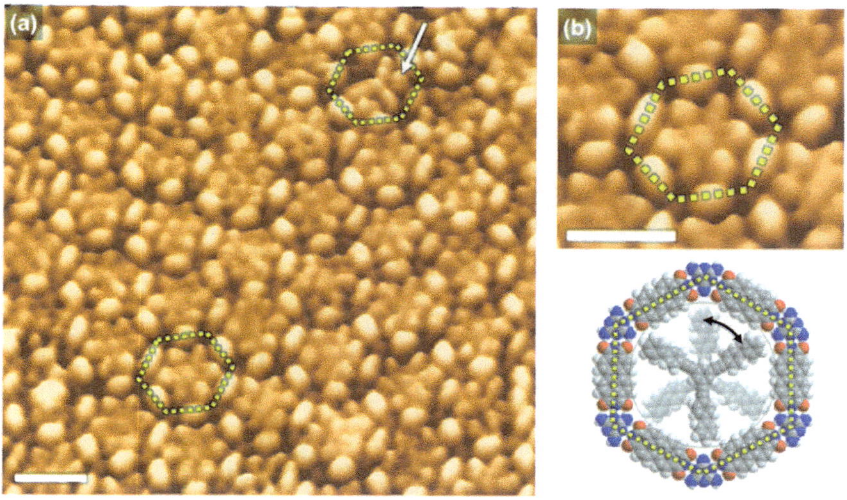

Fig. 2 High resolution STM images in 3D representation of **3BPEB-H** adsorbed in the pores of a supramolecular network composed of PTCDI and melamine. (a) Larger scale image with two pores highlighted, one showing the star molecule (top), the other one showing a sixfold pattern. The arrow marks an additional species trapped in the star modified network pore. (b) Magnified image of the pore exhibiting a sixfold pattern with the structural model indicating a switch between the two equivalent positions of the star molecule. Scale bars: 3 nm.

shape is typically accompanied by at least one additional protrusion like the one marked by the arrow in Fig. 2a which indicates the presence of additional trapped species.

However, in more than 90% of the pores other patterns are seen with the highly symmetric sixfold pattern shown in Fig. 2b being the dominating one. An obvious interpretation of the sixfold pattern is, as illustrated by the molecular model in Fig. 2b, fast switching of the molecule between the two equivalent positions, so a superposition of both configurations is seen. To verify this, experiments were performed with **3BPEB-OH**. The three additional OH groups compared to **3BPEB-H** should increase the interaction of the star molecule with the network and/or the substrate and thus, affect the switching rate of the molecule, similar to a monocomponent system,[55,56] which demonstrated a packing dependent influence of steric factors and intermolecular interaction on molecular rotation. This is fully confirmed as evidenced by the STM image shown in Fig. 3a-i. The star shape of the molecule is now clearly visible in most of the pores whereas the sixfold pattern, highlighted by the circle in Fig. 3a-i, is observed at a level of less than 10%. In some of the pores which display threefold geometry, additional species are trapped. As illustrated in Fig. 3a-ii, these can be sufficiently big, like a residual PTCDI, to cause a deformation of the star.[47] However, there are also features visible such as the cross-like geometry shown in Fig. 3a-iii, which do not correlate with the structure of the star molecule in an obvious way. While the clear appearance of the shape of the molecule in Fig. 3a-i suggests that the molecules are now locked into place, a closer look at the patterns reveals that **3BPEB-OH** also exhibits dynamic behaviour. As documented in Fig. 3b, which

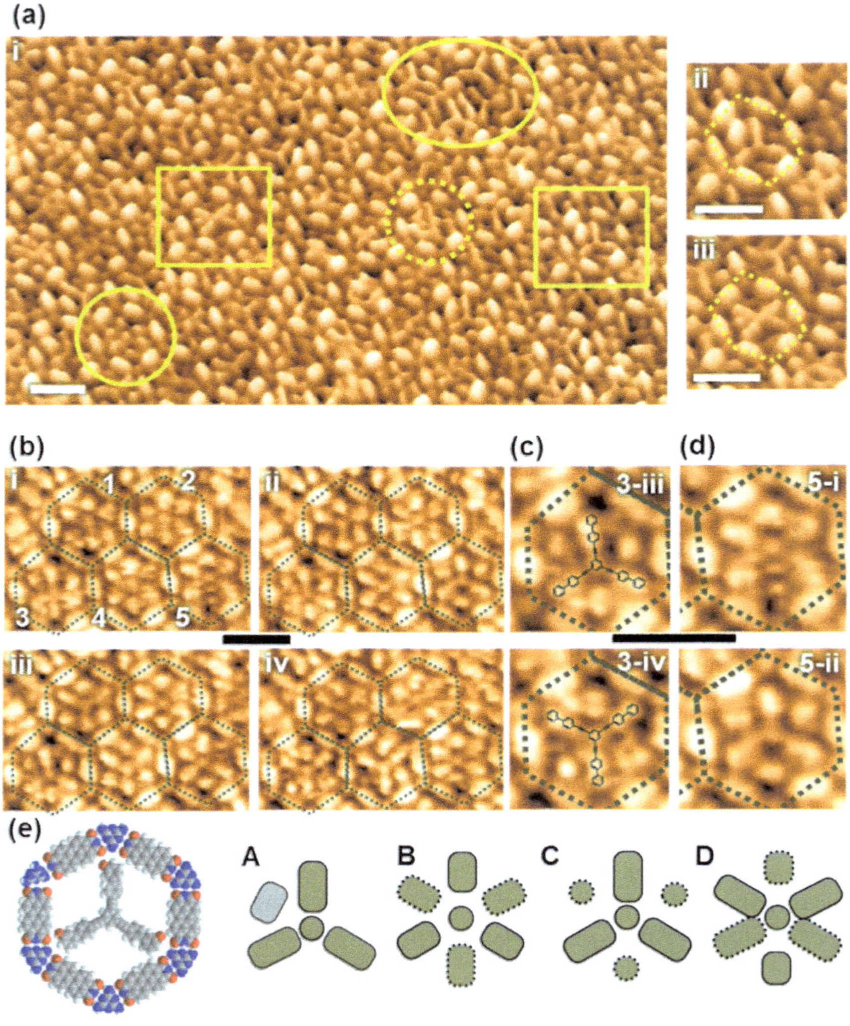

Fig. 3 STM images of **3BPEB-OH** adsorbed in the pores of the PTCDI–melamine network. (a) 3D representation of the structure with rectangles framing the pores shown on the right on an enlarged scale. Solid and dotted circles highlight a sixfold pattern and an irregular pattern, respectively, and the ellipse marks the area of a network defect. (b) Sequence of four STM images of the same area of the **3BPEB-OH**/network structure recorded over a period of time of about 6 min. The network is indicated by the dotted lines. (c and d) Close-ups of pores 3 and 5 for the images as indicated. For pore 3 the images are overlaid with the structure of **3BPEB-OH**. (e) Structural model of **3BPEB-OH** in the network pore and schematic representations of pore patterns (A–D) observed for **3PBEB-H** and **3PBEB-OH**. Scale bars: 3 nm.

shows a series of images of the same pores recorded over a period of time of about 6 min, significant changes occur. Except for pore 2 where the sixfold pattern is essentially preserved, the appearance of the other pores change markedly, yet in pronouncedly different ways. In pores 1 and 5 the star shape appears intermittently (1-iii, 5-ii) among a sixfold pattern. A different case is seen in pore 3 where the star shape is seen throughout the sequence but a switch in orientation occurs between images (iii and iv) as highlighted by the magnified images of pore 3 seen in Fig. 3c. Another noteworthy feature of this pore is that in some images (iii and iv) the biphenyl moiety appears to be resolved whereas in most other cases only more or less elongated protrusions are seen. Pore 4 also exhibits an interesting time behaviour as it evolves from a sixfold symmetry in image (i) to a threefold symmetry in image (iv) with a mirror symmetric appearance in image (ii), caused by the slightly elongated shape of four of the protrusions which resembles the cross-like geometry depicted in Fig. 3a-iii. Such a temporary cross-like feature is also seen for pore 5 as seen from the comparison of the enlarged images of the pore in Fig. 3d. From this it can be concluded that the cross-type shape arises from the dynamics and imaging of the molecule and not from a different molecular species. Noting that some of the pores exhibit irregular features (for an example see the dotted circle in Fig. 3a-i), Fig. 3e summarizes the different regular patterns observed for **3BPEB-H** and **3BPEB-OH** together with a structural model.

Of the various patterns, A and B are straightforward to explain with A, as discussed previously,[47] showing a star molecule locked into place by an additional species and B, as mentioned above, resulting from the superposition of two equivalent configurations. To produce pattern B, which is prevalent in **3PBEB-H**, the switching rate has to be fast compared to the temporal resolution of the STM. While our ambient experiments do not allow us to determine the switching rate, an idea of its value may be obtained by comparing our images with those from UHV studies where switching rates have been quantified and Arrhenius parameters be determined for a porphyrin derivative on top of a supramolecular network pore,[57] dibutyl sulfide on Au,[58] and a sexiphenyl-trimer inside a hexagonal network pore.[59] The comparison is based on the observation that at sufficiently high temperatures the images appear as a smooth superposition of the different configurations, whereas at lower temperatures the images become frizzy or even the individual configurations are seen. The smooth sixfold pattern seen for **3BPEB-H** would be compatible with a switching rate in the range of 10^4–$10\ s^{-1}$ by taking Fig. 2 of ref. 58 and Fig. 3 of ref. 59 as reference.

The discrete sixfold pattern suggests that the two equivalent configurations, where the arms of the molecule point to the vertices of the pore, are long lived compared to any intermediate configurations. Otherwise one would expect a different pattern such as a toroidal structure as seen for hexa-*tert*-butyl decacyclene[60] or coronene[61] as resulting from a continuous and random rotational motion. This raises the question of how the transition between the two states occurs.

Looking at the models shown in Fig. 4, a purely rotational step is not possible for steric reasons, as indicated by the dotted circle in structure (i), if one assumes that the molecule is rigid and the molecule is confined to two dimensions, *i.e.*, all parts of the molecule are always in contact with the surface. Even if one takes into account that the ethynyl bonds exhibit considerable flexibility,[63–66] they would have to be deformed to an extent (more than 60°) which seems unreasonable (structure ii). Including the third dimension by allowing the molecule to partially detach, the deformation of the molecule during a rotational step can be limited to its normal flexibility by one arm passing above a PTCDI molecule (structure iii). The latter scenario is corroborated by studies on the dynamical behaviour of sexiphenyldicarbonitrile molecules in a honeycomb network where hopping

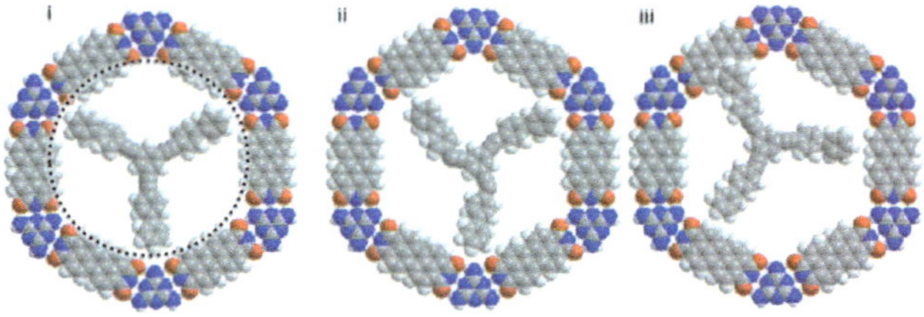

Fig. 4 Models of **3BPEB-H** in one of the two stable configurations (i), a transition state with the molecule confined to two dimensions (ii), and a transition state allowing part of the molecule to detach from the surface (iii). The size of the undistorted **3BPEB-H** star in (i) is based on crystallographic data[62] and drawn at the scale of the network pore. The dotted circle in (i) frames the area assuming a rigid geometry of the star molecule.

between pores was reported to already occur well below room temperature.[67] This system is also interesting as these molecules form a trimer structure with a threefold symmetry, thus yielding a caged structure very similar in geometry to ours. In a detailed experimental study[59] rotations of the trimeric entity already at temperatures well below 100 K were observed. Molecular dynamics simulations reveal the details of the rotational mechanism.[67] In this case it can be restricted to two dimensions since, in contrast to our covalent star structure, the trimer structure allows for a dynamical adaptation to the geometry of the network during the transition between two states. This is afforded by weak supramolecular interactions which provide more degrees of freedom such as frustrated rotations of the individual components and changes in the phenyl hydrogen/CN bonds.

As far as the patterns **C** and **D** (Fig. 3e) are concerned they are less straightforward to explain and at this point we can only speculate on their origin. Pattern **C** still exhibits a threefold symmetry which suggests that it derives from a rotational motion of the star molecule. It can be viewed as a weighted six-fold pattern where the degeneracy of the two configurations is lifted, so the average dwell time of the star molecule is now different for the two configurations. Possible reasons for lifting the degeneracy could be a small deformation of the pore or a contamination such as a solvent molecule trapped at the vertex of a pore. An example of such a pattern is pore 1 in Fig. 3b-iii. It is noted that small changes in geometry could also affect the tunnelling contrast and, thus, the appearance of the molecule. This could explain why in some cases the threefold symmetry seems broken like in image (iv) of pore 3 (Fig. 3b and c). The explanation of pattern **D** which exhibits mirror symmetry is even more speculative. Deviating from the threefold symmetry of the system an imaging effect is the most obvious explanation if opposing vertices are associated with a loss of contrast. This can arise from a dynamic effect, *i.e.*, the arm pointing towards this direction retains some mobility due to a weaker interaction with the substrate.

Based on the proposed mechanism of rotation, the decrease in the rate of rotation for **3BPEB-OH** compared to **3BPEB-H** is associated with the increased interaction of the OH group with the network and the substrate, thus resulting in a higher activation barrier for detachment of one arm. It is noted that such a detachment should also involve a rotation around the long axis of the arm which would result in a dynamic statistical distribution of all OH groups across the 12 equivalent positions.

Extending the experiments to **4BPEB-2OH**, the STM image presented in Fig. 5a shows that this less symmetric molecule also adsorbs in the network pores. In the large majority of pores the shape of the molecule with its four arms is clearly resolved. In contrast to the three-armed molecules presented above, no sixfold symmetric states are seen which indicates a further slow-down of the dynamics owing to the additional interactions of the extra arm and the second OH groups of **4BPEB-2OH**. However, the molecules still undergo rotation as illustrated in the sequence of images shown in Fig. 5b. Comparing images (i and iii), changes in the orientation of the molecules occur in pores 1, 3, and 6. The lower symmetry of the molecules reveals (modulo 6) the number of rotational steps and the direction of rotation. The former is one and the latter has to be random as the system is in thermodynamic equilibrium. It is noted that the feature seen in image (ii) of pore 1 does not reflect the molecular shape but shows a cross-like geometry similar to the one observed for **3PBEB-OH**. Again, the fact that one sees the molecular structure in image (i) and (iii) points to a dynamical and/or imaging effect.

II. Adsorption of guest species. The hybrid structure composed of the PTCDI–melamine network and the **3BPEB-OH** or **4BPEB-2OH** molecules were exposed to either C_{60} fullerene or adamantane thiol. Looking first at the results for C_{60} which are compiled in Fig. 6 and 7 for **3BPEB-OH** and **4BPEB-2OH**,

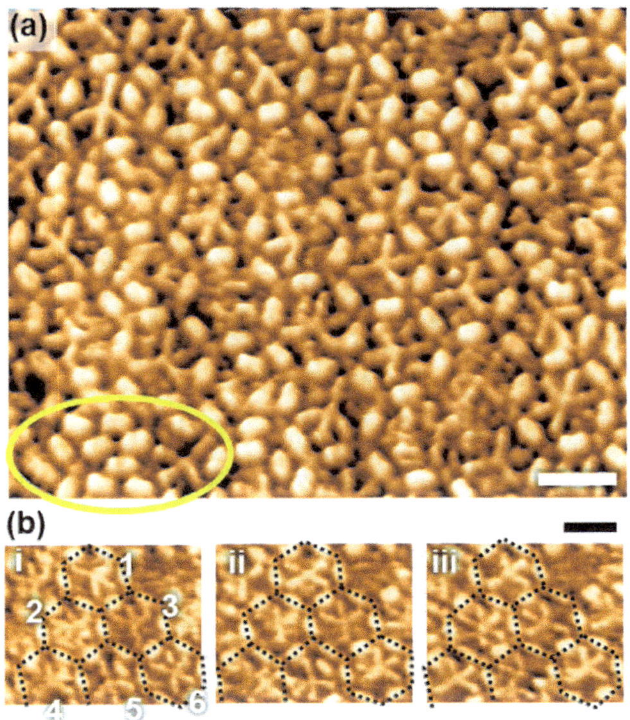

Fig. 5 STM images of 4BPEB-2OH adsorbed in the pores of a PTCDI–melamine network. (a) Overview showing the four-armed molecule locked in the pores. The yellow ellipse marks a defect in the network. (b) Illustration of the dynamics of the molecules by a sequence of images of the same pores recorded over a period of time of 2.5 min. Scale bars: 3 nm.

respectively, both systems do not produce any obvious regular arrangement of the fullerene molecules. However, the Fourier transforms show peaks which indicates that the distribution of the C_{60} molecules is not random.

Since the network itself is not resolved when the C_{60} is present, pores without a star molecule serve as reference points as they are revealed by characteristic clusters of C_{60} involving 6–7 molecules like those shown in Fig. 6b. As seen from the overlay images (Fig. 6c and 7) the positions of the fullerenes fit nicely to the network/star structure, thus, revealing the templating effect of the star/network hybrid structure. The reason why the images look irregular is the lack of perfection as regards the filling of the compartments. Instead of the maximum number of fullerenes per network pore, which is three for **3BPEB-OH** and two for **4BPEB-2OH**, the majority of pores show only 1 or 2 fullerenes in the case of **3BPEB-OH** and only 1 for **4BPEB-2OH.** While the low number of fullerenes in the pore already indicates the presence of the pore modifier, further evidence comes from the distance between the fullerenes. They are well separated by about 18 Å which is distinctly more than the 10 Å of paired fullerenes as seen in the small pores of

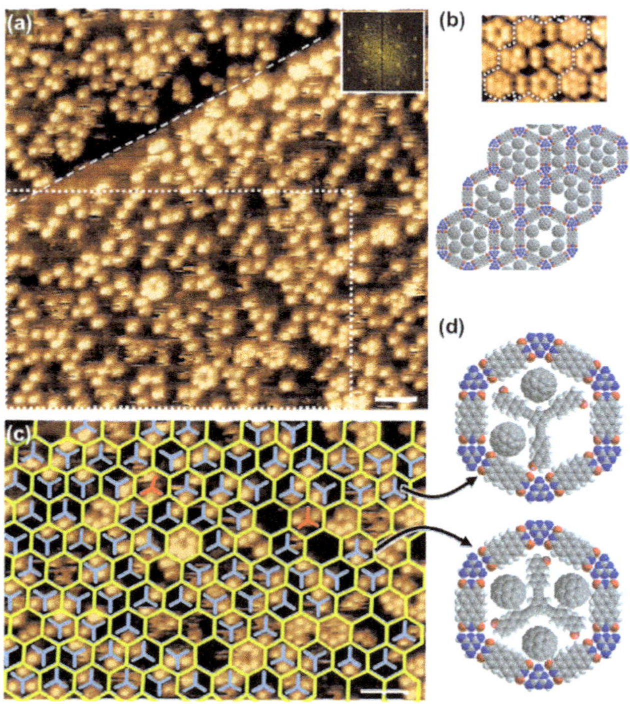

Fig. 6 STM images of C_{60} adsorbed onto a Au/mica substrate patterned by a **3BPEB-OH**/ network structure. (a) Overview image with inset showing the Fourier transform. The dashed line marks a substrate step. (b) STM image (same scale as image (a)) and model of the area framed by the dotted line showing various arrangements of fullerenes in a network without star molecules. Small pores at packing faults of the network can accommodate two close-packed fullerenes. (c) Enlarged image of the section marked by the dotted rectangle in (a) with network structure and star molecules overlaid. Red stars indicate compartmentation where the **3BPEB-OH** molecules have to be deformed to fit. (d) Models of star modified pores accommodating two or three fullerenes. Scale bars: 5 nm.

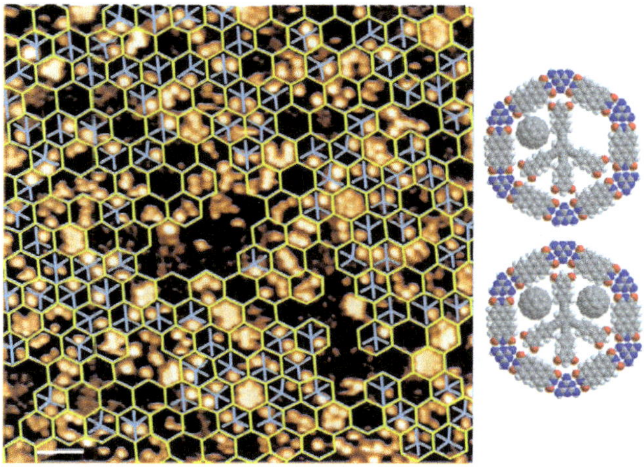

Fig. 7 STM image of C_{60} adsorbed onto a Au/mica substrate patterned by a **4BPEB-2OH**/ network structure with network and pore modifier overlaid in areas where the fullerenes map the network pattern. For the original image and its Fourier transform see Fig. ESI1.† On the right, models of network pores filled with **4BPEB-2OH** and one or two isolated fullerenes are shown.

Fig. 6b. It is noted that, due to the lack of intermolecular stabilisation, imaging of the isolated C_{60} is more delicate compared to pairs or clusters of C_{60} as depicted in Fig. 6b. This is reflected in the occasional occurrence of short streaks or the imaging of a fraction of a C_{60} molecule before it is moved by the tip.

The filling of only a fraction of the compartments can be explained by sites being blocked by another species such as a trapped solvent molecule, an unknown contamination due to the handling in an ambient environment, or residual components of the PTCDI–melamine network. Importantly, this does not only mean a complete steric blocking of an adsorption site but includes any factors which influence the adsorption energy of the C_{60}. Without the stabilising interactions between C_{60} molecules adsorbed in larger pores,[34,45,46] minor changes in the environment can affect the energy landscape to the extent that adsorption of an isolated C_{60} molecule becomes unfavourable.[6] While it is also dependent on the preparation conditions, it is noted in this context that templated adsorption of isolated fullerenes with high yield is rather the exception[21,30,35] than the rule.[27,28,31–33,46,51]

The adsorption of C_{60} being sensitively dependent on the energy landscape, it is of interest to study guest species which have a higher driving force for adsorption. Choosing adamantane thiol which forms a rather strong S–Au bond and is sterically less demanding than C_{60}, the question arises how this fits into the energy landscape of the pore modified hybrid structure. In contrast to C_{60}, which interacts with the surface and the network/star structure relatively weakly and, thus, can be adsorbed under thermodynamically controlled conditions, adamantane thiol, which has already been successfully adsorbed into the pores of an unmodified PTCDI–melamine network,[42,44] requires kinetic control. While adamantane thiol does not displace the network,[42] it can displace the star molecules as seen in tests on uniform layers of the star molecules. Indeed, adsorption at room temperature does not produce a clear signature of the compartmentation by the **3BPEB-OH** star but clusters of adamantane thiol molecules are seen which are similar to those observed for the network without the star molecules. In line with the above discussed dynamics involving a thermally induced temporal detachment of the arms, we conclude that adamantane thiol quite easily displaces the pore modifier. In an attempt to reduce the dynamics, thiol adsorption was performed at a lower temperature, *i.e.*, the melting temperature of the solvent. As seen from Fig. 8, adamantane thiol which is molecularly resolved as bright protrusions, easily adsorbs also at cryogenic temperatures. Overlaying the hexagonal network based on the original image and the Fourier filtered image which are provided in Fig. ESI2,† reveals that a significant fraction of the network pores display clusters of 10–15 molecules, thus indicating removal of the star molecules even at this low temperature. However, in a number of pores the adamantane thiol is present as either an individual molecule or small aggregates of 2 to 4 molecules. A closer look reveals that specific arrangements are realised which indicate compartmentation of the hexagonal pore by the star molecules. Different cases can be identified which are marked by the green, red, and white stars. In pores with green stars the arrangement of the protrusions is in agreement with **3BPEB-OH** being present in an undistorted geometry while pores labelled by the red stars can only be described by a slightly distorted geometry. Some configurations, labelled by the white stars, require an even larger displacement with one arm detached. However, this interpretation has to be taken with a grain of salt as such cases are not very frequent, thus, raising the question of other possibilities such as the combination of some contamination and thiol molecules coincidentally resulting in a pattern suggesting the presence

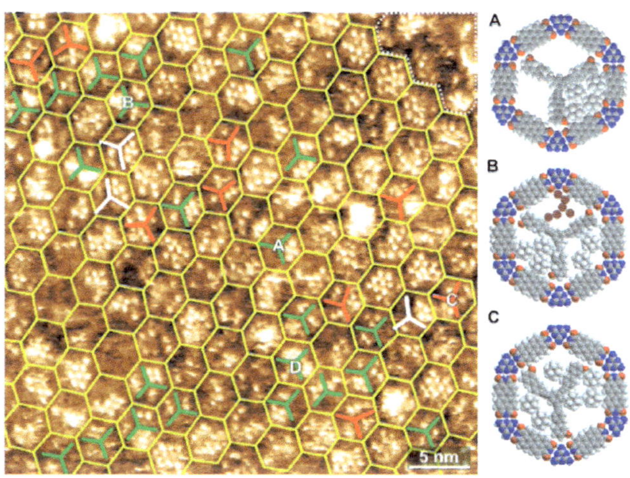

Fig. 8 Left: STM image of a **3BPEB-OH** modified PTDCI–melamine network on Au/mica after immersion in a 10 μM solution of adamantane thiol in ethanol at a temperature of −111 °C and an immersion time of 45 s. The PTCDI–melamine network is indicated by yellow hexagons. The dotted white line in the upper right corner marks an area where no network is discernible. Compartmentation by undistorted and distorted **3BPEB-OH** molecules is indicated by green and red stars, respectively. White stars indicate cells where star molecules, if present, must be dislocated. Right: models of specific configurations with the letters A–C labelling the respective pores in the STM image. For details see text.

of the star molecule. In pores with the undistorted star geometry, one to four thiol molecules are observed per compartment. Models labelled A and B in Fig. 8 provide specific examples of how the thiols can be arranged. While the maximum of four molecules per compartment packs the thiols densely and thus enforces a diamond shaped geometry, variations in the positions are seen for 1–3 molecules. It is noted that the maximum number of 12 molecules per pore, *i.e.*, four in each compartment, have not been observed. Rather a mix involving one to four molecules is seen or even only one compartment being filled like in the case of pore A. However, the latter does not necessarily mean that the other two compartments are empty. It is conceivable that for less than 4 molecules any features are smeared out due to a fast diffusion and rearrangement of the thiols. A dynamic behaviour can also explain brighter features which either lack the distinct pattern of densely packed thiol molecules (*e.g.* pore D) or exhibit protrusions like the ones indicated by the circles in model B, which are too close to each other to be explained by discrete molecules. Analogous to the above discussion of the dynamics of the star molecules, the different features are interpreted to arise from different time scales involved in the dynamical behaviour of the molecules.

The experiments presented above demonstrate the principle of refinement of a structure by nesting and templating under either thermodynamic (C_{60}) or kinetic control (adamantane thiol). At this point we would like to briefly discuss some aspects which are crucial for the further development of this approach towards the design of functional structures. Relying on a sequence of spontaneous assembly processes one necessarily goes down the free energy ladder and, therefore, the whole process is critically dependent on the energy landscape of the systems, *i.e.*, the enthalpy difference between steps and the activation barriers associated with each processing step. For rather weakly interacting systems such as C_{60}, the available enthalpy range is small, thus, limiting the opportunities for a yield-optimised capturing of guest molecules. Minimisation of pore residues by tweaking the preparation conditions of the supramolecular network and the purity of chemicals are the options in this case but it is unclear whether, for the simple solution based processing applied here, the variations can be kept within the affordable tolerances. For this reason more strongly adsorbing systems like thiols appear better suited. However, exploiting kinetic control in such cases requires careful design of the energy landscape and, clearly, in the present case the facile displacement of the star molecule is the bottleneck.

While increasing the stability of the pore modifier by a stronger bonding to the substrate seems an obvious solution to the problem, this raises more fundamental questions as to what extent an increased stability affects the dynamics and how the change in dynamics influences the assembly process. Since adsorption at the liquid solid interface is an exchange process between solvent molecules and the adsorbate, adsorption sites become available either by desorption of a solvent molecule of by surface diffusion. Therefore, the rate of exchange is expected to strongly depend on dimensions. In a highly confined space such as a single compartment of the star-network hybrid structure, lateral diffusion is impeded. If the lateral pathway becomes unavailable, exchange can only occur *via* the third dimension, a process which is also strongly dependent on the steric configuration. Thus it might well be that a too rigid compartment geometry defined by a strongly adsorbed pore modifier is not the optimal solution but a compromise between flexibility and stability is more advisable.

Experimental

Synthesis

The syntheses of **3BPEB-H** and **3BPEB-OH** have been described before.[47] Details for **4BPEB-2OH** are described in the ESI.†

Network preparation and adsorption of star molecules

Preparation of the PTCDI/melamine network was performed as described previously.[47,68] In short, Au on mica substrates were flame-annealed in an oxygen rich natural gas flame before use and immersed in a PTCDI/melamine solution at 100 °C for 1 minute after which they were quickly rinsed with dimethyl formamide (DMF) at room temperature and immediately dried in a nitrogen stream.

The star molecules were adsorbed either from 1,2,4-trichlorobenzene (TCB) or DMF by immersion of the network modified substrate at 60 °C. During the adsorption time of about two minutes the solution was cooled down. After removal from the solution, substrates were quickly blown dry in a stream of nitrogen. Due to the small amounts of substances available, the concentrations of the solutions of the star molecules were not exactly known and, therefore, immersion times were determined empirically by verifying that the adsorption times are sufficiently long for the formation of a full monolayer of star molecules on a clean Au substrate.

For the adsorption of C_{60} the sample was immersed in a TCB solution of the fullerene at 60 °C with the solution cooling down during the exposure time of about 2 minutes. The adsorption of adamantane thiol was performed at the melting point of ethanol. For this, a 10 μM solution of the thiol in ethanol was put in a vial and cooled down with liquid nitrogen. The sample was placed on the solidified mixture and when exposing the vial to room temperature the sample was immersed upon melting of the thiol/ethanol mixture. To keep the temperature at its minimum, immersion was done while the thiol/ethanol mixture was in the two phase solid/liquid state.

Characterisation

Samples were imaged in an ambient environment using a PicoPlus STM (Molecular Imaging). Tips were cut from Pt/Ir wire (80 : 20, 0.25 mm diameter, Advent Research Materials Ltd.). Bias and current were typically in the range of ±100–500 mV and 10–100 pA.

Conclusions

The stepwise spontaneous assembly of a four component molecular system involving the compartmentation of pores of a hydrogen bonded network as the key step followed by the templated adsorption of guest molecules into the sub-pores was investigated. The structural refinement of a porous network structure by nesting with another molecule introduces additional degrees of freedom in the ultraprecise design of surface based molecular structures. Furthermore, it eliminates the uncertainties associated with the usual approach of network modification where the use of derivatised network components might result in a change of the network structure and/or polymorphism due to changes in interactions. The pore modifier adsorbs into the pores in high yield, so the approach taken here should be rather straightforwardly extendable to different geometries and functionalisation of the pore modifier. The latter will also be important for controlling the dynamic behaviour of the oligo-(biphenyl-ethynyl) benzene molecules.

Concluding from the experiments that the pore modifier exhibits a dynamics which involves an intermittent partial detachment of the molecule, it is considered crucial for the stability of the system and, thus, the control of adsorption of guest molecules, that the mobility of the pore modifier is restricted to the surface plane. Showing that the templating of guest molecules by the pore modified network structure is feasible, a move towards functional structures will require an improvement in the yield of the templated adsorption. Our experiments suggest that the level of perfection, which can be achieved by adsorbing weakly interacting guest species under thermodynamically controlled conditions, will be limited due to imperfections in the synthesis of the network and the solution based processing. While kinetic control offers a solution to the problem, the adsorption strength of the pore modifier has to be improved in order to avoid its displacement by guest molecules which currently limits the adsorption conditions and the extent of guest templating.

Acknowledgements

This work was supported in part by The Leverhulme Trust (grant no. RPG-2013-177). The synthetic chemistry team thanks the Swiss National Science Foundation (SNF, grant no. 200020-159730) for continuous and generous financial support. We are grateful to A. Khlobystov for the provision of C_{60}. BK acknowledges EaStCHEM and the Funds for Women Graduates (FfWG) for postgraduate studentships.

Notes and references

1 C.-A. Palma, M. Cecchini and P. Samori, *Chem. Soc. Rev.*, 2012, **41**, 3713–3730.
2 N. Lin, S. Stepanow, F. Vidal, K. Kern, M. S. Alam, S. Strömsdörfer, V. Dremov, P. Müller, A. Landa and M. Ruben, *Dalton Trans.*, 2006, 2794–2800.
3 J. V. Barth, *Annu. Rev. Phys. Chem.*, 2007, **58**, 375–407.
4 A. Ciesielski, C.-A. Palma, M. Bonini and P. Samorì, *Adv. Mater.*, 2010, **22**, 3506–3520.
5 J. Teyssandier, S. D. Feyter and K. S. Mali, *Chem. Commun.*, 2016, **52**, 11465–11487.
6 S. Yoshimoto, Y. Honda, O. Ito and K. Itaya, *J. Am. Chem. Soc.*, 2008, **130**, 1085–1092.
7 H. Liang, Y. He, Y. C. Ye, X. G. Xu, F. Cheng, W. Sun, X. Shao, Y. F. Wang, J. L. Li and K. Wu, *Coord. Chem. Rev.*, 2009, **253**, 2959–2979.
8 N. A. Wasio, R. C. Quardokus, R. P. Forrest, C. S. Lent, S. A. Corcelli, J. A. Christie, K. W. Henderson and S. A. Kandel, *Nature*, 2014, **507**, 86–89.
9 L.-J. Wan, *Acc. Chem. Res.*, 2006, **39**, 334–342.

10 M. El Garah, A. Dianat, A. Cadeddu, R. Gutierrez, M. Cecchini, T. R. Cook, A. Ciesielski, P. J. Stang, G. Cuniberti and P. Samori, *Small*, 2016, **12**, 343–350.

11 S. Uemura, R. Tanoue, N. Yilmaz, A. Ohira and M. Kunitake, *Materials*, 2010, **3**, 4252–4276.

12 S.-L. Lee, Y. Fang, G. Velpula, F. P. Cometto, M. Lingenfelder, K. Müllen, K. S. Mali and S. De Feyter, *ACS Nano*, 2015, **9**, 11608–11617.

13 A. G. Slater, P. H. Beton and N. R. Champness, *Chem. Sci.*, 2011, **2**, 1440–1448.

14 J. Elemans, S. B. Lei and S. De Feyter, *Angew. Chem., Int. Ed.*, 2009, **48**, 7298–7332.

15 J. Lu, D.-L. Bao, H. Dong, K. Qian, S. Zhang, J. Liu, Y. Zhang, X. Lin, S.-X. Du, W. Hu and H.-J. Gao, *J. Phys. Chem. Lett.*, 2017, **8**, 326–331.

16 D. Nieckarz and P. Szabelski, *Chem. Commun.*, 2014, **50**, 6843–6845.

17 P. Szabelski, W. Rzysko, T. Panczyk, E. Ghijsens, K. Tahara, Y. Tobe and S. De Feyter, *RSC Adv.*, 2013, **3**, 25159–25165.

18 T. Kudernac, S. B. Lei, J. Elemans and S. De Feyter, *Chem. Soc. Rev.*, 2009, **38**, 402–421.

19 X. Zhang, N. Li, L. Liu, G. Gu, C. Li, H. Tang, L. Peng, S. Hou and Y. Wang, *Chem. Commun.*, 2016, **52**, 10578–10581.

20 C.-A. Palma, A. Ciesielski, M. A. Oner, G. Schaeffer, J.-M. Lehn, J. V. Barth and P. Samori, *Chem. Commun.*, 2015, **51**, 17297–17300.

21 L. Piot, F. Silly, L. Tortech, Y. Nicolas, P. Blanchard, J. Roncali and D. Fichou, *J. Am. Chem. Soc.*, 2009, **131**, 12864–12865.

22 D. Grumelli, B. Wurster, S. Stepanow and K. Kern, *Nat. Commun.*, 2013, **4**, 2904.

23 D. den Boer, T. Habets, M. J. J. Coenen, M. van der Maas, T. P. J. Peters, M. J. Crossley, T. Khoury, A. E. Rowan, R. J. M. Nolte, S. Speller and J. A. A. W. Elemans, *Langmuir*, 2011, **27**, 2644–2651.

24 A. G. Slater, L. M. A. Perdigão, P. H. Beton and N. R. Champness, *Acc. Chem. Res.*, 2014, **47**, 3417–3427.

25 X. Bouju, C. Mattioli, G. Franc, A. Pujol and A. Gourdon, *Chem. Rev.*, 2017, **117**, 1407–1444.

26 R. Decker, U. Schlickum, F. Klappenberger, G. Zoppellaro, S. Klyatskaya, M. Ruben, J. V. Barth and H. Brune, *Appl. Phys. Lett.*, 2008, **93**, 243102.

27 S. Stepanow, N. Lin, J. V. Barth and K. Kern, *Chem. Commun.*, 2006, 2153–2155.

28 S. Stepanow, M. Lingenfelder, A. Dmitriev, H. Spillmann, E. Delvigne, N. Lin, X. B. Deng, C. Z. Cai, J. V. Barth and K. Kern, *Nat. Mat.*, 2004, **3**, 229–233.

29 S. Fabris, S. Stepanow, N. Lin, P. Gambardella, A. Dmitriev, J. Honolka, S. Baroni and K. Kern, *Nano Lett.*, 2011, **11**, 5414–5420.

30 W. Xiao, D. Passerone, P. Ruffieux, K. Aït-Mansour, O. Gröning, E. Tosatti, J. S. Siegel and R. Fasel, *J. Am. Chem. Soc.*, 2008, **130**, 4767–4771.

31 D. Bonifazi, A. Kiebele, M. Stöhr, F. Cheng, T. Jung, F. Diederich and H. Spillmann, *Adv. Funct. Mater.*, 2007, **17**, 1051–1062.

32 A. G. Phillips, L. M. A. Perdigão, P. H. Beton and N. R. Champness, *Chem. Commun.*, 2010, **46**, 2775–2777.

33 S. J. H. Griessl, M. Lackinger, F. Jamitzky, T. Markert, M. Hietschold and W. M. Heckl, *J. Phys. Chem. B*, 2004, **108**, 11556–11560.

34 J. A. Theobald, N. S. Oxtoby, M. A. Phillips, N. R. Champness and P. H. Beton, *Nature*, 2003, **424**, 1029–1031.

35 E. Mena-Osteritz and P. Bäuerle, *Adv. Mater.*, 2006, **18**, 447–451.

36 K. H. Chung, H. Kim, W. J. Jang, J. K. Yoon, S. J. Kahng, J. Lee and S. Han, *J. Phys. Chem. C*, 2013, **117**, 302–306.

37 J. Teyssandier, S. De Feyter and K. S. Mali, *Chem. Commun.*, 2016, **52**, 11465–11487.

38 S. Klyatskaya, F. Klappenberger, U. Schlickum, D. Kühne, M. Marschall, J. Reichert, R. Decker, W. Krenner, G. Zoppellaro, H. Brune, J. V. Barth and M. Ruben, *Adv. Funct. Mater.*, 2011, **21**, 1230–1240.

39 K. Cui, F. Schlütter, O. Ivasenko, M. Kivala, M. G. Schwab, S.-L. Lee, S. F. L. Mertens, K. Tahara, Y. Tobe, K. Müllen, K. S. Mali and S. De Feyter, *Chem.–Eur. J.*, 2015, **21**, 1652–1659.

40 S. Q. Zhang, J. Y. Zhang, K. Deng, J. L. Xie, W. B. Duan and Q. D. Zeng, *Phys. Chem. Chem. Phys.*, 2015, **17**, 24462–24467.

41 J. Adisoejoso, K. Tahara, S. Okuhata, S. Lei, Y. Tobe and S. De Feyter, *Angew. Chem., Int. Ed.*, 2009, **48**, 7353–7357.

42 R. Madueno, M. T. Räisänen, C. Silien and M. Buck, *Nature*, 2008, **454**, 618–621.

43 C. Silien, M. T. Räisänen and M. Buck, *Angew. Chem., Int. Ed.*, 2009, **48**, 3349–3352.

44 C. Silien, M. T. Räisänen and M. Buck, *Small*, 2010, **6**, 391–394.

45 M. T. Räisänen, A. G. Slater, N. R. Champness and M. Buck, *Chem. Sci.*, 2012, **3**, 84–92.

46 L. M. A. Perdigão, A. Saywell, G. N. Fontes, P. A. Staniec, G. Goretzki, A. G. Phillips, N. R. Champness and P. H. Beton, *Chem.–Eur. J.*, 2008, **14**, 7600–7607.

47 B. Karamzadeh, T. Eaton, I. Cebula, D. Muñoz Torres, M. Neuburger, M. Mayor and M. Buck, *Chem. Commun.*, 2014, **50**, 14175–14178.

48 D. Bléger, D. Kreher, F. Mathevet, A.-J. Attias, G. Schull, A. Huard, L. Douillard, C. Fiorini-Debuischert and F. Charra, *Angew. Chem., Int. Ed.*, 2007, **46**, 7404–7407.

49 G.-B. Pan, J.-M. Liu, H.-M. Zhang, L.-J. Wan, Q.-Y. Zheng and C.-L. Bai, *Angew. Chem., Int. Ed.*, 2003, **42**, 2747–2751.

50 M. Li, K. Deng, S.-B. Lei, Y.-L. Yang, T.-S. Wang, Y.-T. Shen, C.-R. Wang, Q.-D. Zeng and C. Wang, *Angew. Chem., Int. Ed.*, 2008, **47**, 6717–6721.

51 H. Spillmann, A. Kiebele, M. Stöhr, T. A. Jung, D. Bonifazi, F. Y. Cheng and F. Diederich, *Adv. Mater.*, 2006, **18**, 275–279.

52 L. M. A. Perdigão, P. A. Staniec, N. R. Champness and P. H. Beton, *Langmuir*, 2009, **25**, 2278–2281.

53 N. M. Jenny, M. Mayor and T. R. Eaton, *Eur. J. Org. Chem.*, 2011, 4965–4983.

54 E. A. Krasnokutskaya, N. I. Semenischeva, V. D. Filimonov and P. Knochel, *Synthesis*, 2007, 81–84.

55 T. Waldmann, R. Reichert and H. E. Hoster, *ChemPhysChem*, 2010, **11**, 1513–1517.

56 T. Waldmann, J. Klein, H. E. Hoster and R. J. Behm, *ChemPhysChem*, 2013, **14**, 162–169.

57 N. Wintjes, D. Bonifazi, F. Y. Cheng, A. Kiebele, M. Stöhr, T. Jung, H. Spillmann and F. Diederich, *Angew. Chem., Int. Ed.*, 2007, **46**, 4089–4092.

58 A. E. Baber, H. L. Tierney and E. C. H. Sykes, *ACS Nano*, 2008, **2**, 2385–2391.

59 D. Kühne, F. Klappenberger, W. Krenner, S. Klyatskaya, M. Ruben and J. V. Barth, *Proc. Natl. Acad. Sci. U. S. A.*, 2010, **107**, 21332–21336.

60 J. K. Gimzewski, C. Joachim, R. R. Schlittler, V. Langlais, H. Tang and I. Johannsen, *Science*, 1998, **281**, 531–533.
61 S. J. H. Griessl, M. Lackinger, F. Jamitzky, T. Markert, M. Hietschold and W. A. Heckl, *Langmuir*, 2004, **20**, 9403–9407.
62 B. Karamzadeh, T. Eaton, I. Cebula, D. M. Torres, M. Neuburger, M. Mayor and M. Buck, The Cambridge Structural Database, CCDC No. 1000630, 2014, DOI: 10.5517/cc12l7d9.
63 D. Écija, K. Seufert, D. Heim, W. Auwärter, C. Aurisicchio, C. Fabbro, D. Bonifazi and J. V. Barth, *ACS Nano*, 2010, **4**, 4936–4942.
64 D. Heim, D. Écija, K. Seufert, W. Auwärter, C. Aurisicchio, C. Fabbro, D. Bonifazi and J. V. Barth, *J. Am. Chem. Soc.*, 2010, **132**, 6783–6790.
65 T. Takeda and Y. Tobe, *Chem. Commun.*, 2012, **48**, 7841–7843.
66 G. Jeschke, M. Sajid, M. Schulte, N. Ramezanian, A. Volkov, H. Zimmermann and A. Godt, *J. Am. Chem. Soc.*, 2010, **132**, 10107–10117.
67 C. A. Palma, J. Bjork, F. Rao, D. Kühne, F. Klappenberger and J. V. Barth, *Nano Lett.*, 2014, **14**, 4461–4468.
68 I. Cebula, M. T. Raisanen, R. Madueno, B. Karamzadeh and M. Buck, *Phys. Chem. Chem. Phys.*, 2013, **15**, 14126.

Faraday Discussions

PAPER

Molecular self-assembly of substituted terephthalic acids at the liquid/solid interface: investigating the effect of solvent†

A. Della Pia,[a] D. Luo,[a] R. Blackwell,[b] G. Costantini*[a] and N. Martsinovich *[b]

Received 20th March 2017, Accepted 12th April 2017
DOI: 10.1039/c7fd00112f

Self-assembly of three related molecules – terephthalic acid and its hydroxylated analogues – at liquid/solid interfaces (graphite/heptanoic acid and graphite/1-phenyloctane) has been studied using a combination of scanning tunnelling microscopy and molecular mechanics and molecular dynamics calculations. Brickwork-like patterns typical for terephthalic acid self-assembly have been observed for all three molecules. However, several differences became apparent: (i) formation or lack of adsorbed monolayers (self-assembled monolayers formed in all systems, with one notable exception of terephthalic acid at the graphite/1-phenyloctane interface where no adsorption was observed), (ii) the size of adsorbate islands (large islands at the interface with heptanoic acid and smaller ones at the interface with 1-phenyloctane), and (iii) polymorphism of the hydroxylated terephthalic acids' monolayers, dependent on the molecular structure and/or solvent. To rationalise this behaviour, molecular mechanics and molecular dynamics calculations have been performed, to analyse the three key aspects of the energetics of self-assembly: intermolecular, substrate–adsorbate and solvent–solute interactions. These energetic characteristics of self-assembly were brought together in a Born–Haber cycle, to obtain the overall energy effects of formation of self-assembled monolayers at these liquid/solid interfaces.

1. Introduction

The ability of molecules to self-assemble into extended ordered structures thanks to specific intermolecular interactions opens many possibilities for applications in such diverse fields as biomedicine,[1,2] molecular electronics,[3–8] sensors[9] and

[a]Department of Chemistry, University of Warwick, Coventry, UK. E-mail: G.Costantini@warwick.ac.uk
[b]Department of Chemistry, University of Sheffield, Sheffield, UK. E-mail: n.martsinovich@sheffield.ac.uk

† Electronic supplementary information (ESI) available: Energies and structures for force field fitting, schematic of a 2D unit cell and lattice parameters, potential energy surfaces for 2D self-assembly of TPA, 2HTPA and 25DHTPA. See DOI: 10.1039/c7fd00112f

catalysis.[10] In particular, by confining the self-assembly process on solid substrates, two-dimensional (2D) structures can be formed[11,12] by exploiting a number of different intermolecular forces: from metal coordination[13,14] to hydrogen bonding,[14,15] to weaker dispersion interactions.[16] While the nature of the interactions between the molecular units is typically the key factor in determining the resulting assembly, other more subtle influences have also been reported to affect the final supramolecular structures: the chemistry and symmetry of the substrate (even for inert surfaces such as highly ordered pyrolytic graphite (HOPG) and Au(111)[17]), the temperature,[18–20] the ultra-high vacuum (UHV) or solution environment,[19,21,22] the nature of the solvent,[19,23–28] the concentration of the solute (the self-assembling molecule),[18,29–35] and any co-adsorption of solvent or guest molecules.[24,25,34,36,37] The possibility of controlling supramolecular polymorphism by weak intermolecular interactions, such as interactions with the solvent, is a new and fascinating approach to the ultimate goal of rationally programming molecular self-assembly. However, its fundamental mechanisms are still not clearly understood, and it is likely that multiple mechanisms may be simultaneously at play: from co-adsorption of solvent and guest molecules[25,31] to different solvation of small molecular aggregates – precursors to the extended self-assembly – in different solvents.[23]

In this work, we investigate the combined effects of the molecular structure and the nature of the solvent in the molecular self-assembly of benzene dicarboxylic acids at the liquid/solid (HOPG) interface. In particular, we study the self-assembly of a series of three molecules: terephthalic acid (TPA) and its hydroxyl-substituted analogues 2-hydroxyterephthalic acid (2HTPA) and 2,5-dihydroxyterephthalic acid (25DHTPA), shown in Fig. 1. Self-assembly of TPA has been widely studied on a variety of substrates (both inert, such as HOPG,[38–41] graphene,[42,43] Au(111),[44] Ag(111),[45] Pt(111)[46] and reactive, such as Cu(100)[47,48] and Cu(110),[49] Pd(111),[50] supported metallic multilayers,[51] doped Si(111),[45,52] TiO_2,[53,54] calcite[55]), and both in vacuum[42–45,47,49–55] and at the liquid/solid interface.[38–40] While on the

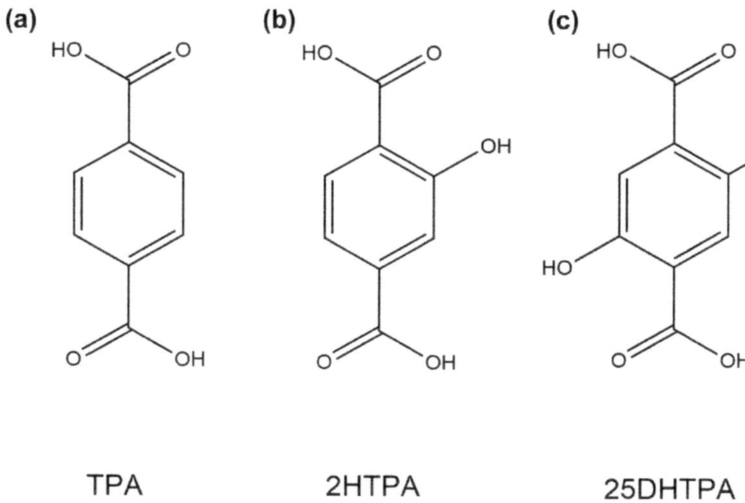

Fig. 1 Structures of (a) terephthalic acid (TPA), (b) 2-hydroxyterephthalic acid (2HTPA) and (c) 2,5-dihydroxyterephthalic acid (25DHTPA).

more reactive surfaces TPA can undergo different transformations that modify its chemical structure (*e.g.* deprotonation of the carboxylic moieties),[47–49,51,53,54] on inert substrates its self-assembly is characterised by the formation of molecular chains stacked in a brickwork fashion.[38–40,42–44] This supramolecular architecture is controlled by two types of interactions: intra-chain dimerisation of carboxylic groups to form strong hydrogen bonds (H-bonds) – as also observed for other carboxylic acid molecules: trimesic acid,[23,30,56] isophthalic acid,[40] 1,3,5-benzenetribenzoic acid[24] and stilbenedicarboxylic acid[57] – and secondary inter-chain dispersion interactions.

Here we introduce additional "lateral" OH moieties and vary their number to tune the inter-chain interactions and to study their effect on the self-assembly. We also use two different solvents: a nonpolar solvent, 1-phenyloctane (PO), and a polar solvent with an acid group, heptanoic acid (7A), to investigate the effect of solvent–solute interactions (π-stacking *vs.* H-bonds) on the assembly.

We use a combination of scanning tunnelling microscopy (STM) experiments and molecular mechanics (MM) and molecular dynamics (MD) calculations. The STM results reveal similarities in the 2D structures formed by these molecules, but also differences: (i) different surface coverage by adsorbates at the two liquid/solid interfaces, (ii) formation of two slightly different self-assembled structures of 25DHTPA depending on the solvent, and (iii) co-existence of several domains with different molecular orientations for the asymmetric 2HTPA molecule. Computational modelling is used to rationalise the observed 2D structures and the equilibrium between molecules in solution and self-assembled monolayers at the liquid/solid interface. Similarly to what was done in previous related work,[39,57] Born–Haber cycles are constructed to evaluate the energy gain upon formation of self-assembled monolayers from solution.

2. Methods

2.1. Experimental methods

A fresh graphite surface was obtained by cleaving a HOPG crystal (grade ZYB) with Scotch tape before each molecular deposition. A saturated solution was prepared by dissolving the molecules in the solvent (heptanoic acid or 1-phenyloctane) in a small glass vial; approximately 10–30 μL of the solution were then deposited on the HOPG substrate using a micropipette.

The adsorbed self-assembled 2D molecular structures were characterised using STM (Veeco with Nanoscope E controller and an A-type scanner) operating in ambient conditions at the solid–liquid interface, and using mechanically-sheared Pt/Ir (90/10) tips. For molecular imaging, the bias voltage (applied to the sample) ranged from −1.5 to 1.0 V, with typical currents between 70 and 100 pA. For atomic resolution imaging of the underlying HOPG surface, typical tunnelling parameters were −0.1/0.1 V and 100–800 pA. All STM images were processed using the WSxM software.[58] STM images which have been recalibrated by using half–half images containing both atomic resolution of the HOPG substrate and the molecular layer[56] are indicated as "rescaled STM images" in the figure captions.

2.2. Computational methods

2.2.1 Force field. The calculations of the 2D assembly of the TPA, 2HTPA and 25DHTPA molecules, adsorption of these molecules and solvent molecules on graphite, adsorption of the solvent on 2D molecular monolayers and solvation of these molecules by liquid solvent were carried out using molecular mechanics, with the Tinker software[59] and the MM3 force field.[60,61] The force field parameters for the H-bonding in the carboxylic acid dimer (interactions between carboxylic hydrogen, atom type 24, and double-bonded carboxylic oxygen, atom type 77) were taken from our previous work ref. 62: the energy parameter $\varepsilon_{24\cdots77} = 7.78$ kJ mol^{-1} and the distance parameter $r_{24\cdots77} = 1.75$ Å.

The 2HTPA and 25DHTPA molecules contain additional phenolic hydroxyl groups, and therefore different types of H-bonding interactions, both intra- and intermolecular, are expected between two hydroxyl groups and between hydroxyl and carboxylic groups (see Table 1). H-bonding parameters for these interactions are not available in MM3 (except for the interaction type 73⋯6: hydroxyl hydrogen – phenolic oxygen). Therefore, accurate quantum-chemistry calculations were performed using Møller–Plesset perturbation theory (MP2)[63] and were used to fit the missing H-bonding parameters. Four isomers of 2HTPA were considered, with different positions and conformations of the hydroxyl group relative to the carboxylic groups, as well as several 2HTPA and phenol dimers with a range of hydroxyl–hydroxyl and hydroxyl–carboxyl arrangements (see the ESI†, Section S1). MP2 calculations with the DZVP basis set were done using Gaussian09 (ref. 64) software; all binding energies were corrected for the basis set superposition error (BSSE). Some of the calculations were also done with the larger TZVP basis set, but the resulting binding energies and relative energies of isomers were similar to those obtained with the DZVP basis set within 2.5 kJ mol^{-1}. MM3 calculations were then done on the same systems, while varying the energy and distance parameters for each interaction, to achieve a good fit both in terms of energies (within 5.0 kJ mol^{-1}, see ESI Section S1†) and geometries (within 0.2 Å). The best parameters, shown in Table 1, were used for all the following MM calculations.

2.2.2 Calculations of 2D structures. 2D monolayer structures of 2HTPA and 25DHTPA were explored by scanning through combinations of their 2D lattice parameters. First, isolated 1D molecular chains were modelled: the lattice parameter a (along the molecular chain) was varied, with a step of 0.1 Å, to find the lowest-energy value of a. Then, while keeping a fixed at its optimum value, 2D arrangements of the molecular chains were modelled by simultaneously varying the parameters b_y (perpendicular distance between the chains) and b_x (the shift of the chains relative to each other along the chain direction, shown schematically in Fig. 2), with a step of 0.1 Å, similar to the procedure used in ref. 39. The parameters b_x and b_y are directly related to the parameters b ($b^2 = {b_x}^2 + {b_y}^2$) and γ ($\sin\gamma = b_y/b$) typically used to describe 2D lattices. The structures were kept

Table 1 Hydrogen bonding parameters for the MM3 force field, fitted in this work

Interaction (H⋯O)	Atom types	$r_{\mathrm{H\cdots O}}$, Å	$\varepsilon_{\mathrm{H\cdots O}}$, kcal mol^{-1}
$\mathrm{H_{(carboxyl)}\cdots O{=}_{(carboxyl)}}$	24⋯77 (ref. 62)	1.75	7.78
$\mathrm{H_{(phenol)}\cdots O{=}_{(carboxyl)}}$	73⋯77	1.75	7.78
$\mathrm{H_{(phenol)}\cdots OH_{(carboxyl)}}$	73⋯75	1.9	5.5
$\mathrm{H_{(phenol)}\cdots OH_{(phenol)}}$	73⋯6	2.3	3.2
$\mathrm{H_{(CH)}\cdots OH_{(phenol)}}$	5⋯6	2.6	1.0

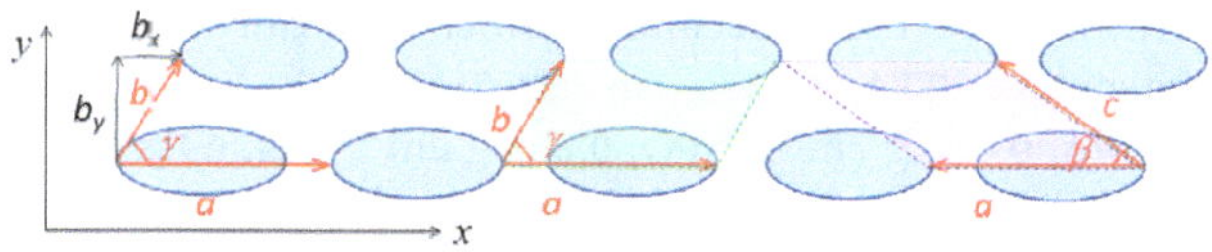

Fig. 2 Schematic showing (left) the definition of the lattice parameters a and b and the angle γ between them, and decomposition of b into the components b_y (inter-chain separation along the y axis) and b_x (shift along the direction of the chain, *i.e.* along the x axis); (middle and right) two choices of the inter-chain lattice parameter, b or c (and the corresponding angles γ and β) for the same lattice, leading to unit cells of different shapes but the same size (area).

planar by fixing the z coordinates of all atoms. The 2D potential energy surfaces (PES) obtained by varying b_x and b_y were analysed to identify the energy minima.

2.2.3 Calculations of adsorption. A large hydrogen-terminated graphene sheet (20 × 20 C atoms) was used to model the adsorption of the three terephthalic acid molecules, and of the solvent molecules, on HOPG. All atoms of the graphene sheet and the lateral coordinates (x and y) of the adsorbates were fixed, while the vertical coordinates of the adsorbates were allowed to optimise. A 2D grid of adsorption positions above the graphene sheet was considered, covering the rectangular-shaped graphene unit cell (2.46 × 4.26 Å, with step 0.2 Å). Adsorption of single solvent molecules above TPA and substituted TPA monolayers was modelled similarly: all atoms in the monolayer were fixed, and the lateral coordinates of only the first and last atoms of the solvent's alkyl chains were fixed, thus allowing the solvent molecules the flexibility to adjust their conformation (this flexibility was found to not be necessary on graphene, where the adsorbates' conformations remained essentially unchanged). As with the direct adsorption on graphene, a 2D grid of adsorption positions above the monolayers was considered, covering the whole area of each monolayer's unit cell, with step 0.2 Å.

2.2.4 Calculations of solvation. Molecular dynamics simulations were used to obtain solvation energies of TPA, 2HTPA and 25DHTPA in 7A and PO. To achieve good sampling of the solvent and solute-in-solvent systems, several 3D boxes of solvent were constructed, with periodic boundary conditions: a parallelepiped-shaped box containing 200 7A molecules and a roughly cubic box containing 192 7A molecules, and a parallelepiped-shaped PO box containing 192 PO molecules and a roughly cubic box containing 198 PO molecules. Cell volumes were chosen to reproduce the experimental densities of these solvents: 0.918 g cm^{-3} (7A) and 0.858 g cm^{-3} (PO). Solvent systems were first annealed from 1000 K to 298 K for 1 ns, then MD simulations using the canonical (NVT) ensemble were run until the change in energies (averaged every 0.5 ns) was less than 5 kcal mol^{-1} (this took 2–3 ns for PO and 4–6 ns for 7A, since hydrogen bonding of carboxylic groups took a longer time to equilibrate). The Nosé–Hoover thermostat was used; the integration time step was 1 fs, and the "rattle" algorithm was used to constrain all covalent bonds to H atoms to their ideal bond length. 6 simulations of PO solvent and 10 simulations of 7A solvent were run, and energies (collected over the last 1 ns) were averaged over these MD runs.

To create solvent–solute systems, one or two molecules of the solvent were removed and replaced by one or two molecules of the solute. The volume of the cell was adjusted, to account for the different molecular volume of the solute compared to the solvent (the molecular volumes were calculated from the molar masses and densities: TPA, 1.52 g cm^{-3};[65] 2HTPA, 1.61 g cm^{-3};[66] 25DHTPA, 1.779 g cm^{-3};[67] 7A, 0.918 g cm^{-3};[68] PO, 0.858 g cm^{-3} (ref. 69)). One solute molecule per 192–200 solvent molecules corresponds to the solute concentrations of ~0.035 mol dm^{-3} in 7A and ~0.023 mol dm^{-3} in PO. Several solvent–solute cells were built and simulated: 9 for TPA in 7A, 6 for 2HTPA and 25DHTPA in 7A and 6 for each solute molecule in PO. Solute-7A systems were initially annealed from 400 to 298 K for 1 ns; then all solute–solvent systems were simulated using MD (NVT ensemble) until the averaged energy change was below 5 kcal mol^{-1} in 0.5 ns. Similar to the pure solvents, solute in PO took less time to equilibrate (2–4 ns) than in 7A (3–7 ns). The last 1 ns of each MD simulation were used to determine the energies of solute in solvent.

3. Results and discussion

3.1. STM imaging

STM images of TPA, 2HTPA and 25DHTPA, obtained at the interface of HOPG with 7A and PO solutions, are shown in Fig. 3 and 4, demonstrating that in most cases the molecules formed ordered self-assembled layers. One notable exception is TPA in PO, where no self-assembled monolayer was observed, as discussed below.

The measured lattice parameters for all observed monolayers are summarised in Table 2. The images show many similarities: the molecules are imaged as bright spots corresponding to the benzene rings, sometimes with submolecular contrast; all observed monolayers have a brickwork-like pattern, indicative of the formation of chains held together by strong intra-chain interactions (dimeric hydrogen bonds) and weak inter-chain interactions.[40,62] The measured lattice parameters for TPA at the HOPG/7A interface (a = 10.0 Å, b = 7.7 Å, γ = 48°, relative error ± 5%) are in good agreement with previous studies of TPA self-assembled monolayers on a variety of substrates (HOPG, graphene, Au(111), Pt(111)), both at the liquid/solid interface and in UHV,[38–40,42,44,46] as shown in Table 3.

However, there are also notable differences between the monolayers formed in the two solvents. When deposited from 7A, all three molecules form ordered extended islands and completely cover the HOPG surface (Fig. 3). In contrast, with the PO solvent, the molecules tend to form isolated islands rather than a complete monolayer (Fig. 4). In the case of TPA in PO, the molecules do not adsorb at all: no molecules were observed on the HOPG surface despite extensive scanning.

Table 2 Experimental lattice parameters of TPA, 2HTPA (only regular brickwork assembly) and 25DHTPA monolayer structures from STM measurements (the relative uncertainty is 5% in all cases)

Molecule	Solvent	a/Å	b/Å	γ/°	Area/Å per molecule
TPA	7A	10.0	7.7	48	57.2
2HTPA	7A, PO	9.4	8.4	50	60.5
25DHTPA	7A	9.3	8.6	44	55.6
25DHTPA	PO	9.3	8.4	57	65.5

Table 3 Comparison of TPA lattice parameters obtained in this work with literature values for 2D monolayers of TPA on inert and weakly reactive substrates and for TPA bulk crystal

Source	Substrate	Solvent or UHV	a/Å	b/Å	γ/°
This work	HOPG	7A	10.0 ± 0.5	7.7 ± 0.4	48 ± 2
Ref. 38	HOPG	7A	10.0	7.5	60
Ref. 40	HOPG	7A	9.8	7.4	60
Ref. 41	HOPG	7A	9.6 ± 0.05	8.9 ± 0.05	70 ± 5
Ref. 39	HOPG	9A	9.6 ± 0.1	7.8 ± 0.1	50 ± 1
Ref. 70	Graphene	7A	9.5 ± 0.2	7.6 ± 0.6	53 ± 3
Ref. 42	Graphene	UHV	9.8 ± 0.6	7.4 ± 0.3	60
Ref. 44[a]	Au(111)	UHV	10.0 ± 0.3	7.3 ± 0.3	55 ± 3
Ref. 71[b]	Cu(111)	UHV	9.5 ± 0.1	N/A	N/A
Ref. 46	Pt(111)	UHV	9.6	7.3	49
Ref. 50[c]	Pd(111)	UHV	9.5 ± 0.6	N/A	N/A
Ref. 72	3D crystal	—	9.54	7.73	43

[a] Averaged over three distinct sets of a, b, γ for three non-equivalent directions on the reconstructed Au(111) surface. [b] Averaged over two main a values. [c] Value for 1D chains (2D monolayers of deprotonated molecules were also observed).

The overview of the TPA lattice parameters obtained in this study and reported in the literature (Table 3) shows that the structure of the monolayers formed by TPA is essentially unchanged on all inert substrates, both at the solid/liquid interface and in the UHV environment; therefore, the lack of adsorption at the HOPG/PO interface is unexpected. However, previous studies of TPA self-assembly were done only in UHV and in alkanoic acid (heptanoic acid[38,40] and nonanoic acid (9A)[39]) solvents; we are not aware of this molecule having been adsorbed from phenyloctane or other nonpolar solvents. There is, however, an example of a chemically similar system for which no adsorbed self-assembled layers were observed: phthalic acid at the HOPG/7A interface[40] – attributed to weak adsorption of this non-planar molecule on HOPG. In our case of TPA, the likely difference is the nature of solvation: hydrogen bonding interactions of TPA with 7A, against π-stacking in PO. The lack of TPA adsorption suggests strong π-stacking interaction with the PO solvent, which competes with the molecule–substrate interaction and limits the formation of an adsorbed layer.

Comparing the assembly of the symmetric molecules (TPA and 25DHTPA) to that of the asymmetric 2HTPA molecule, it can be observed that while the former arrange into extended islands with only one orientation with respect to the HOPG lattice (Fig. 3a, b, e, f and 4c, d), the latter forms several molecular domains with different orientations, both in 7A and PO (Fig. 3c, d and 4a, b). This can be attributed to the existence of different adsorption orientations for the 2HTPA molecule – with different domains containing molecules with the different orientations. The solvent affects the 2HTPA island size but not the molecular packing: the same brickwork-like structure with very similar lattice parameters (see Table 2) is seen for both solvent interfaces. The b parameter (describing the inter-chain distance) and the angle γ are slightly larger in 2HTPA than in TPA, indicating that 2HTPA chains are more widely spaced than TPA chains. This is clearly caused by the presence of the hydroxyl moiety in 2HTPA: the bulkier OH groups and the repulsion between oxygens in hydroxyl and carboxyl groups in

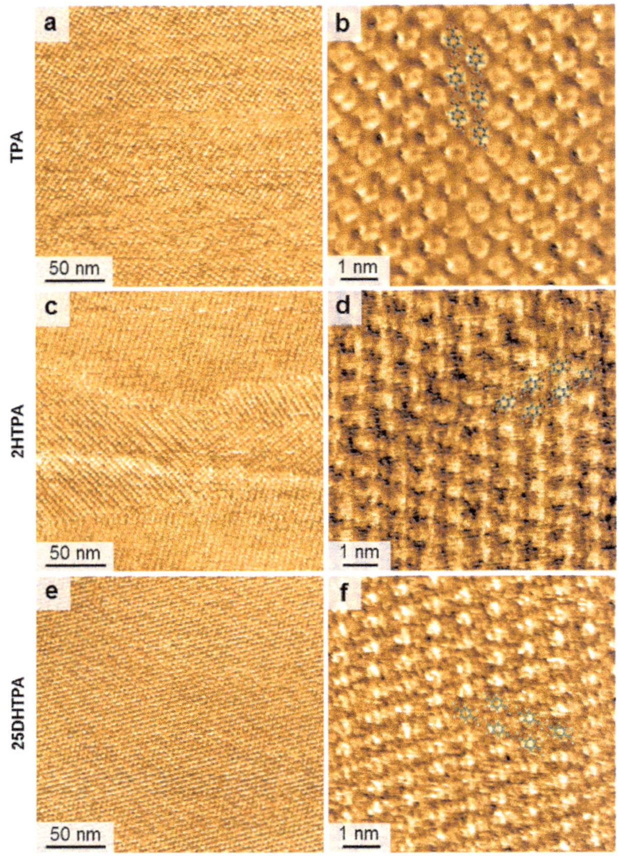

Fig. 3 STM images of self-assembled monolayers at the HOPG/7A interface: (a and b) TPA; (c and d) 2HTPA; (e and f) 25DHTPA. Overlays of molecular structures in (b, d and f) show proposed supramolecular arrangements in these 2D structures. (b, d and f) are rescaled STM images.

neighbouring chains are likely to both play a role here. Surprisingly, the distance along the chain, *i.e.* along the hydrogen-bonded carboxylic groups, is reduced compared to TPA, from 10.0 to 9.4 Å. A possible reason for this may be the effect of the substrate, *i.e.* the relationship between the substrate periodicity and the intra-chain periodicity,[71] and the possibility of inter-chain interactions (either weak or strong, depending on the presence of OH groups) modulating the substrate interactions.

A closer inspection of the 2HTPA images reveals that, besides regions characterised by a regular brickwork assembly (Fig. 3c, d and 4a, b), also other regions exist displaying an alternative assembly with a high variability in the inter-chain separation, noticeable as gaps between the chains (Fig. 5). This second type of assembly develops at the interface with both 7A and PO. While the inter-chain distance in the regular 2HTPA structure is 8.4 ± 0.4 Å, the other regions show a pairing of chains with alternating short (7.2–7.4 Å) and long (9.1–9.2 Å) separations and are therefore dubbed alternating 2HTPA assembly. The likely

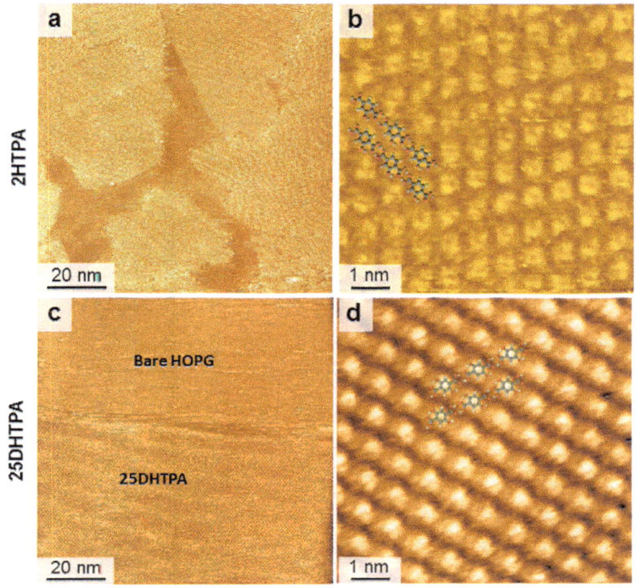

Fig. 4 STM images of self-assembled monolayers at the HOPG/PO interface: (a and b) 2HTPA; (c and d) 25DHTPA. Overlays of molecular structures in (b and d) show proposed supramolecular arrangements in these 2D structures. (b and d) are rescaled STM images.

explanation for these enlarged and shortened inter-chain distances is the repulsion between hydroxyl groups of adjacent 2HTPA molecules. Notably, the shorter inter-chain separation approaches the corresponding value in TPA (7.7 Å). It is thus likely that in the regular brickwork regions (Fig. 3c, d and 4a, b) the 2HTPA molecules have the OH groups all oriented in the same direction forming evenly spaced single chains (as shown schematically in Fig. 6a), while in the alternating assembly (Fig. 5), molecules with OH facing/opposing each other belong to chains with wider/smaller separations (Fig. 6b). Thus, 2HTPA displays polymorphism, which is not caused by the solvent but rather originates from the structure of the molecules themselves.

25DHTPA (Fig. 3e, f and 4c, d) also forms a brickwork structure, similar to TPA and 2HTPA. However, in this case there are quantitative differences between the structures formed at the interfaces with 7A and in PO (see Table 2): although the values of the a and b lattice parameters are very similar for both solvents, the angle γ between them is noticeably larger in PO (57°) than in 7A (44°). Thus, the structure formed at the interface with PO is 18% less densely packed than the structure formed at the interface with the 7A solvent, with the difference likely being caused by different orientations of hydroxyl groups. Thus, 25DHTPA monolayers display solvent-induced polymorphism.

To summarise, all three terephthalic acid molecules showed differences in their self-assembly behaviour at the two studied solid–liquid interfaces: the presence or absence of self-assembled monolayers at the solid–liquid interface (TPA); full or partial surface coverage of the molecular layers (2HTPA, 25DHTPA); singly-oriented (TPA, 25DHTPA) or multiply-oriented (2HTPA) molecular

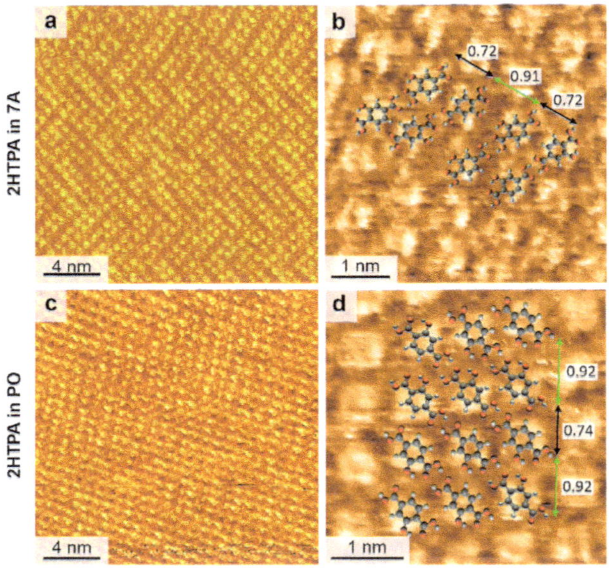

Fig. 5 STM images of the alternating 2HTPA assembly: (a and b) at the HOPG/7A interface; (c and d) at the HOPG/PO interface. Overlays of molecular structures in (b and d) show possible supramolecular arrangements in these 2D structures, while numbers show measured inter-chain separations in nm. Disorder in the arrangement of the double rows and a quite wide distribution of inter-row distances can be seen, which may be attributable to different positions of OH groups in 2HTPA molecules, as indicated by the tentative overlaid structures. (b and d) are rescaled STM images.

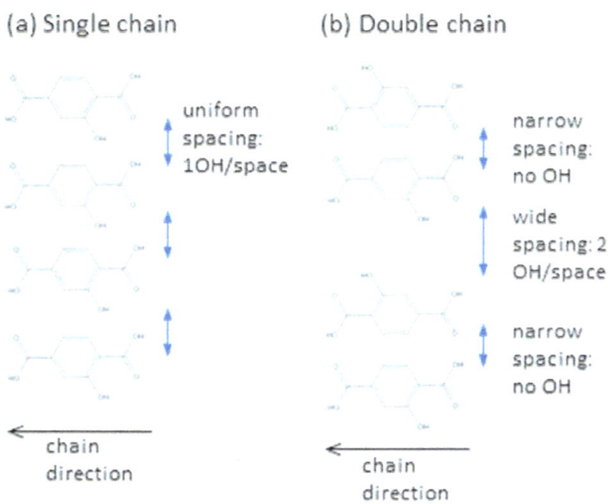

Fig. 6 Schematics of two possible types of arrangements of 2HTPA molecules in 2D periodic structures: (a) single chain structure: all 2HTPA molecules have the same orientation of the OH groups, resulting in uniform inter-chain spacing; (b) double chain structure: pairs of 2HTPA chains with alternating OH orientations, resulting in two different inter-chain spacings.

domains; co-existence of two polymorphs for both solvents (2HTPA); formation of two polymorphs depending on the solvent (25DHTPA). Theoretical insight is necessary in order to understand the origin of these differences and will be presented in the next section.

3.2. Calculations of 2D structures of 2HTPA and 25DHTPA

The whole "surface-adsorbate-solvent" system is too large to be modelled efficiently at once. However, it can be partitioned into key components: (i) 2D self-assembled monolayers (intermolecular interactions), (ii) individual molecules adsorbed on the graphite surface (molecule–substrate interactions) and (iii) solute molecules surrounded by solvent (solute–solvent interactions).

To understand the differences in the self-assembly and the polymorphism of 2HTPA and 25DHTPA molecules, calculations of their 2D periodic structures in isolation (*i.e.* without substrate and solvent) were done using MM, as described in the Computational methods section. To identify all possible stable 2D arrangements of these molecules, potential energy surfaces (PES) were obtained by scanning through combinations of the 2D lattice parameters. The monolayer structures (Table 4) were compared to TPA results published earlier[39,62] and to the experimental results found in this work.

3.2.1 2HTPA. 2HTPA is a non-symmetric molecule containing one hydroxyl group. Therefore, unlike the symmetric TPA, 2HTPA can adsorb on a surface in four different orientations: with the hydroxyl group in the top-right, top-left, bottom-right, and bottom-left positions. While 2HTPA molecules within a chain display all the same orientation, molecules in neighbouring chains can be in each of these four orientations. This gives rise to four different arrangements for the 2HTPA molecular chains: a single-chain structure (SC, Fig. 7a), where molecules have the same orientation over the entire monolayer, and three double-chain structures (DC, Fig. 7b–d), where the orientations of the 2HTPA molecules in two adjacent chains differ. The packing of the molecular chains is uniform only in the former case (SC in Fig. 7a), while different inter-chain distances arise for all other cases, depending on the number and position of OH groups in between neighbouring molecules: 1 OH per molecular pair (DC2 in Fig. 7c) or two OH between neighbouring chains followed by none in the successive pair (structures DC1, DC3 in Fig. 7b and d).

The potential energy surfaces for the 2D monolayers of the single-chain and double-chain 2HTPA structures are shown in the ESI,† the lowest-energy structures are displayed in Fig. 7, and the intermolecular distances in Table 4. While the unit cell of the SC structure contains only a single molecule, that of the DC structures comprises two molecules with two sets of inter-chain distances (described by b_1, c_1 and b_2, c_2) and two sets of angles (α_1, β_1, and α_2, β_2), reflecting the existence of two inter-chain arrangements.

A single minimum is found for the SC structure ($a = 9.4$ Å, $b = 8.0$ Å, $\gamma = 51°$, see Table 4), which is in good agreement with the experimentally observed regular 2HTPA monolayer ($a = 9.7$ Å, $b = 8.4$ Å, $\gamma = 50°$, see Table 2). In contrast, several minima are found for the double-chain structures. Notably, DC1-Min1, DC1-Min2 and DC2-Min1 are more stable than SC. This clearly shows that the 2HTPA molecule is capable of polymorphism. Moreover, the geometry of DC1-Min1 (the most stable calculated 2HTPA structure) is very similar to that of SC (distances

Table 4 Calculated lattice parameters, area per molecule and monolayer binding energies (relative to an isolated molecule) of low-energy 2D monolayers of TPA, 2HTPA and 25DHTPA. The calculated values for TPA from ref. 39 are included for comparison

Molecule	2D arrangement	E_{ML}/kJ mol^{-1}	a/Å	b_1; b_2/Å	c_1; c_2/Å	α_1, α_2/°	β_1; β_2/°	γ_1; γ_2/°	Area/Å^2 per molecule	Assignment to experimental structures
TPA	Ref. 39	−76.8	9.4	8.1; –	7.3; –	75; –	56; –	49; –	56.83	Regular TPA
2HTPA	SC-Min1	−79.0	9.4	7.5; –	8.0; –	74; –	51; –	55; –	58.28	Regular 2HTPA (?)
2HTPA	DC1-Min1	−82.6	9.4	7.3; 7.5	8.3; 7.9	77; 75	49; 51	58; 54	57.81	Regular 2HTPA
2HTPA	DC1-Min2	−82.3	9.4	7.5; 8.8	8.3; 7.2	73; 71	50; 62	58; 46	59.69	Alternating 2HTPA
2HTPA	DC1-Min3	−78.6	9.4	6.8; 8.1	11.1; 7.3	59; 75	39; 56	82; 49	60.16	—
2HTPA	DC2-Min1	−81.0	9.4	7.0; 6.9	11.5; 8.3	52; 76	36; 59	88; 59	60.63	—
2HTPA	DC2-Min2	−78.2	9.4	8.7; 7.0	7.1; 8.7	72; 73	62; 45	46; 62	58.28	Alternating 2HTPA (?)
2HTPA	DC3-Min1	−79.1	9.4	7.4; 7.8	9.5; 7.5	66; 76	46; 53	68; 51	60.63	—
25DHTPA	SC-Min1	−90.1	9.4	7.3; –	8.3; –	74; –	48; –	58; –	58.28	25DHTPA in 7A
25DHTPA	DC-Min1	−82.2	9.4	7.4; –	9.3; –	67; –	47; –	66; –	63.45	25DHTPA in PO
25DHTPA	DC-Min2	−81.9	9.4	9.5; 7.4	7.4; 9.6	66; 66	68; 46	46; 68	64.39	—

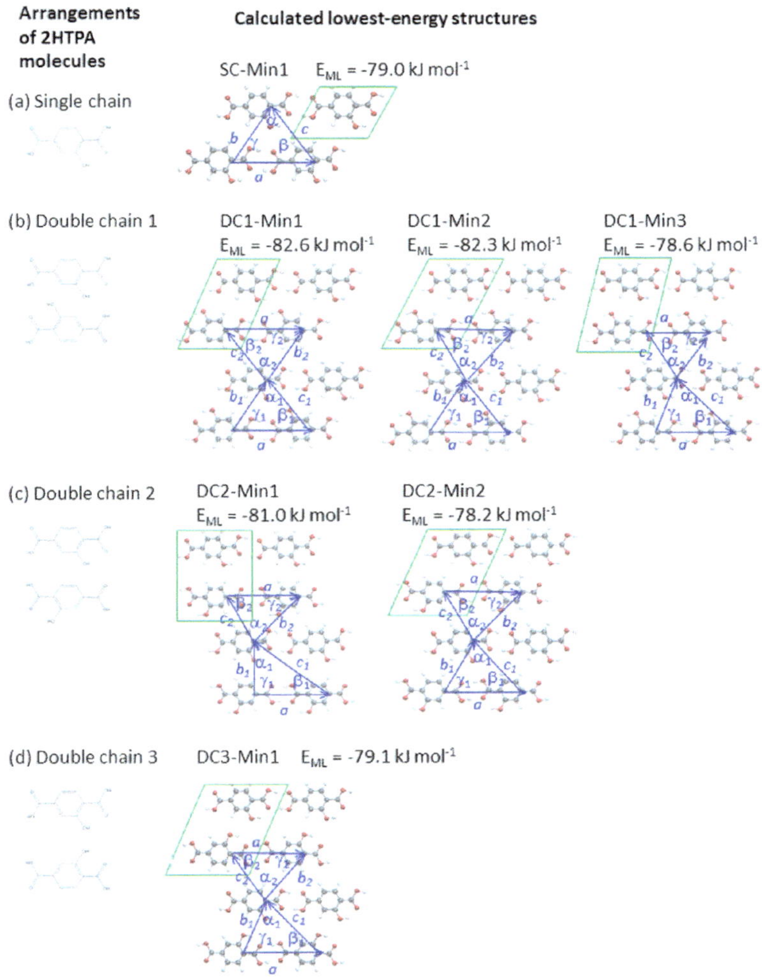

Fig. 7 Single-chain and double-chain arrangements of 2HTPA molecules (left) and calculated lowest-energy structures of 2HTPA 2D monolayers and their binding energies E_{ML} (right). Lattice parameters are shown in blue. Unit cells are highlighted in green.

within 0.3 Å, angles within 3°, *i.e.* differences below the ±5% accuracy of the experimental measurements). Thus DC1-Min1 is the most likely candidate structure for the experimentally observed regular 2HTPA monolayers.

The DC1-Min2 structure is only slightly less stable than DC1-Min1 (−82.3 *vs.* −82.6 kJ mol^{-1}), but has a different arrangement of chains, resulting in alternating large and small inter-chain distances (both b_1, b_2 and c_1, c_2). Therefore, this structure is the most likely candidate for the observed alternating 2HTPA assembly (Fig. 5). Among the other energy minima described in Table 4, one (DC2-Min2) also has the geometry similar to the alternating structure, but it is higher in energy, while the other minima are both high in energy and significantly different from the experimentally observed structures.

Therefore, two likely 2HTPA monolayer structures emerge: DC1-Min1 for the regular assembly, and DC1-Min2 for the alternating assembly. The very close similarity in energy of these two structures explains their experimentally observed coexistence. Moreover, the similarity in the monolayer binding energies also explains why this polymorphism of 2HTPA is not affected by the polar or apolar nature of the solvent. The specific pairing of 2HTPA molecular chains necessary for both DC1-Min1 and DC1-Min2 structures may also be the reason for the formation of molecular domains with different orientations (Fig. 3c, d and 4a, b): "wrong" molecular pairings may be encountered at grain boundaries.

3.2.2 25DHTPA. Since the 25DHTPA molecule has two OH groups, there are only two possible orientations it can take in adjacent chains: parallel and antiparallel, resulting in either single-chain or double-chain structures (Fig. 8). Because of its symmetry, the PES of 25DHTPA is also much simpler than that of 2HTPA: only one minimum is found for the SC structure, and two minima for the DC structure, as presented in Fig. 8 and Table 4. The most stable structure, SC-Min1 (lattice parameters a = 9.4 Å, b = 8.3 Å, γ = 48°), is in very good agreement with the experimentally observed 25DHTPA monolayers in 7A (a = 9.3 Å, b = 8.4 Å, γ = 44°, Table 2).

The two DC structures are less stable than SC by ~8 kJ mol^{-1}, and the agreement with the experimental monolayer geometries in either 7A or in PO is not very good. However, both DC structures have a larger area per molecule than the SC structure (63.5–64.4 Å^2 per molecule *vs.* 58.3 Å^2 per molecule), caused by the wider spacing between the chains. This sparser molecular packing is a characteristic of the experimental monolayers observed in PO, which have a larger area per molecule (65.5 Å^2 per molecule) than those observed in 7A (55.6 Å^2 per molecule). Thus, it is possible that the monolayers formed in PO are related to the calculated DC structures, in particular to DC-Min1, which matches better the

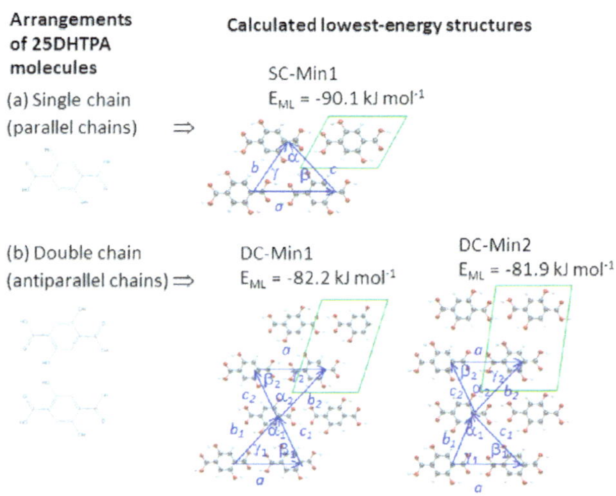

Fig. 8 Single-chain and double-chain arrangements of 25DHTPA molecules (left) and calculated lowest-energy structures of 25DHTPA 2D monolayers and their binding energies E_{ML} (right). Lattice parameters are shown in blue. Unit cells are highlighted in green.

experimentally observed uniform separation between the 25DHTPA chains. However, our force field was not able to fully reproduce the true structure of 25DHTPA chains in PO. The reason may be in the choice of the distance and energy parameters for the OH(phenol)···O(carboxylic) hydrogen bond: they were fitted to reproduce the strong intramolecular hydrogen bond in 2HTPA and 25DHTPA (see ESI Section S1†), but this may also lead to the intermolecular OH(phenol)···O(carboxylic) hydrogen bonds being artificially shortened.

Overall, the 25DHTPA molecule appears to be capable of polymorphism, similarly to 2HTPA, although its lowest energy monolayer structure, SC-Min1, is clearly significantly more stable than the alternatives.

3.3. Thermodynamic analysis of the self-assembly of substituted TPA

The calculations described above considered isolated monolayers, *i.e.* the effects of the substrate and the solvent were not explicitly included. To understand the nature of self-assembly at the solid–liquid interface, we need to take into account the fact that the molecules in a monolayer are adsorbed on a surface, are in contact with the solvent, and are in dynamic equilibrium with molecules dissolved in the solvent.

3.3.1 Born–Haber cycle. To achieve a quantitative description of the energetics of self-assembly at the solid–liquid interface and, in particular, of the effect of the solvent, we used the Born–Haber cycle (shown in Fig. 9 for TPA assembly at the HOPG/7A and HOPG/PO interfaces), similar to what was done in ref. 39 and 57.

The energy of a monolayer of solute molecules adsorbed at the solid–liquid interface is calculated as a sum of several contributions: (i) the monolayer binding energy E_{ML}, *i.e.* the difference between the energy of a single isolated solute molecule and that of the same molecule within a monolayer; (ii) the adsorption energy E_{ads}, calculated as the binding energy of a single solute molecule on the graphite substrate; (iii) the de-wetting energy $E_{dewet} = E_{desorb(solv)} = -E_{ads(solv)}$, which accounts for the fact that the solvent, initially covering the substrate, needs to be desorbed to make space for the adsorption of the solute molecules; (iv) the

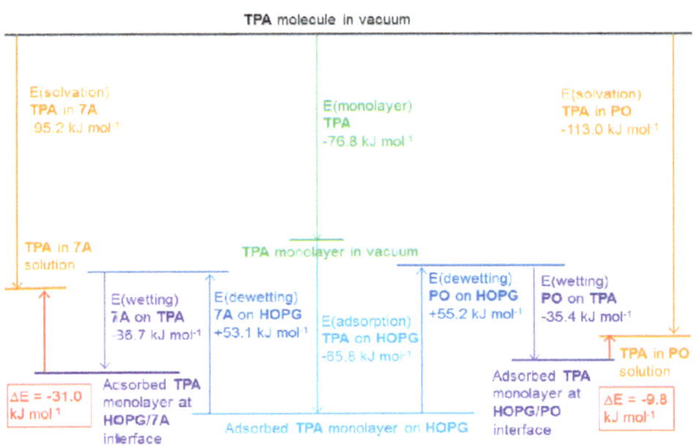

Fig. 9 Born–Haber cycle for the self-assembly of TPA at the HOPG/7A and HOPG/PO interfaces. The energy of the monolayer formation at the solid/liquid interface, with respect to molecules in solution, is highlighted in red.

wetting energy of the adsorbed monolayer $E_{\text{wetting}} = E_{\text{ads(solv-on-ML)}}$, that takes into consideration the fact that the monolayer of adsorbed solute molecules is in contact with a layer of solvent above it. Note that the latter two quantities, the energies of adsorption of the solvent on the substrate and on the monolayer, are calculated per 1 solvent molecule. On the other hand, the energetics of self-assembly is calculated per 1 molecule of solute. The solvent adsorption energies should therefore be re-scaled per area occupied by 1 solute molecule adsorbed on the substrate:[39]

$$E_{\text{ads(solv) scaled}} = E_{\text{ads(solv)}}/A_{\text{solv}} \times A_{\text{solute}}. \quad (1)$$

Thus, the energy of monolayer assembly at the solid–liquid interface, $E_{\text{ML@SLI}}$, relative to that of a solute molecule in vacuum, is:

$$E_{\text{ML@SLI}} = E_{\text{ML}} + E_{\text{ads}} - E_{\text{ads(solv) scaled}} + E_{\text{ads(solv-on-ML) scaled}}. \quad (2)$$

The energy of solvation, $E_{\text{solvation}}$, relative to a solute molecule in vacuum is simply calculated as the difference between the energy of the system composed of one solute molecule within the solvent and the energies of the pure solvent and of the isolated solute molecule.

Finally, the energy gain (or cost) for the monolayer formation at the solid–liquid interface is the difference between the energy of the monolayer at the solid–liquid interface and the energy of solvation:

$$E_{\text{monolayer formation}} = E_{\text{ML@SLI}} - E_{\text{solvation}}. \quad (3)$$

3.3.2 Energies. The energies of adsorption and solvation necessary for obtaining the monolayer formation energy have been calculated as described in the Computational methods section (mean values of solvation energies over several MD simulations, and mean values of adsorption energies for a grid of adsorption positions above the substrate), and are collected in Table 5. In particular, the calculated solvation energies are very similar between the three solute molecules, but vary with the solvent: −95.2 to −98.6 kJ mol^{-1} in 7A, −113.0 to −114.3 kJ mol^{-1} in PO. Interestingly, despite the possibility of strong hydrogen bond formation with the carboxylic groups of 7A, the solvation energies in PO are larger, showing that π-stacking in these systems is stronger than the hydrogen bonding. For comparison, the solvation energy of TPA in 9A calculated using the same method is −115.1 kJ mol^{-1}, and the experimental value is −114.4 kJ mol^{-1};[39] this is more than the solvation energies in 7A (−95.2 kJ mol^{-1}), showing that the dispersion interaction with the alkyl chains of the solvent is also non-negligible and is stronger for longer and more flexible alkyl chains. Note also that the variation (standard error of the mean) of the solvation energies is very large, up to ± 20.5 kJ mol^{-1} in PO and up to ±41.9 kJ mol^{-1} in 7A, representing the largest source of inaccuracy in our computational analysis.

Adsorption energies on HOPG progressively increase from TPA to 25DHTPA (from −65.8 to −75.4 kJ mol^{-1}, Table 5). Inspection of the potential energy surfaces of these molecules' adsorption shows that adsorption positions corresponding to AB stacking of the benzene ring above the underlying graphite are the most stable ones; however, the variation of energies between different adsorption

Table 5 Energies and areas per molecule involved in the Born–Haber cycle for TPA, 2HTPA and 25DHTPA at the HOPG/7A and HOPG/PO interfaces

Molecule	Solvent	Area (solute)/Å^2	Area (solvent)/Å^2	$E_{solvation}$/kJ mol^{-1}	E_{ML}/kJ mol^{-1}	E_{ads}/kJ mol^{-1}	$E_{ads(solv)}$/kJ mol^{-1}	$E_{ads(solv)\ scaled}$/kJ mol^{-1}	$E_{ads(solv\text{-}on\text{-}ML)}$/kJ mol^{-1}	$E_{ads(solv\text{-}on\text{-}ML)\ scaled}$/kJ mol^{-1}	$E_{ML@SLI}$/kJ mol^{-1}	$E_{monolayer\ formation}$/kJ mol^{-1}
TPA	7A	56.8	54.9	−95.2	−76.8	−65.8	−51.3	−53.1	−35.4	−36.7	−126.2	−31.0
TPA	PO	56.8	89.9	−113.0	−76.8	−65.8	−87.4	−55.2	−56.1	−35.4	−122.8	−9.8
2HTPA (DC Min1)	7A	57.8	54.9	−98.6	−82.6	−69.7	−51.3	−54.0	−41.8	−44.0	−142.3	−43.7
2HTPA (DC Min1)	PO	57.8	89.9	−114.3	−82.6	−69.7	−87.4	−56.2	−64.0	−41.1	−137.2	−22.9
25DHTPA (SC Min1)	7A	58.3	54.9	−93.8	−90.1	−75.4	−51.3	−54.5	−45.5	−48.3	−159.4	−65.6
25DHTPA (DC Min1)	PO	63.5	89.9	−114.1	−82.2	−75.4	−87.4	−61.7	−66.5	−46.9	−142.8	−28.7
25DHTPA (SC Min1)	PO	58.3	89.9	−114.1	−90.1	−75.4	−87.4	−56.6	−66.5	−43.1	−152.0	−37.9

positions is very small: the difference between the largest and smallest adsorption energy is only 1.1 kJ mol^{-1} for 25DHTPA, and 1.0 kJ mol^{-1} for 2HTPA (a similar difference of 0.8 kJ mol^{-1} between the adsorption minimum and maximum was found for TPA on HOPG previously[39]). This very flat potential energy surface for adsorption of these molecules on HOPG suggests that there is no strong preference towards specific adsorption positions.

The adsorption energy of PO on HOPG (-87.4 kJ mol^{-1}) is larger than that of 7A (-51.3 kJ mol^{-1}), in agreement with the area of the two solvent molecules and the presence/absence of phenyl rings. The difference in energies between adsorption maxima and minima is again small: 2.0 kJ mol^{-1} for PO, and 1.4 kJ mol^{-1} for 7A.

Adsorption of both solvents on monolayers is weaker than on HOPG (7A adsorption energies from -35.4 to -45.5 kJ mol^{-1}, PO adsorption energies from -56.1 to -66.5 kJ mol^{-1}, always strongest on 25DHTPA and weakest on TPA). This is as expected, because monolayers have a less dense structure than graphite and therefore have fewer atoms to interact with. Interestingly, the variation in these adsorption energies is larger than on HOPG (standard deviation up to 6.4 kJ mol^{-1} for 7A adsorption and up to 4.0 kJ mol^{-1} for PO adsorption). This can be rationalised, as there are preferential positions both for 7A (the carboxylic group of 7A pointing towards the carboxylic and hydroxyl groups of TPA and its analogues) and PO (the phenyl ring of PO above the phenyl rings of TPA).

3.3.3 Analysis of the energetics of self-assembly. The energies summarised in Table 5 can be combined according to eqn (1)–(3) to calculate the energy gain for monolayer formation at the solid–liquid interface, which is presented in the extreme right column of Table 5. TPA is the most interesting example. The experiments show that TPA forms adsorbed self-assembled monolayers at the HOPG/7A interface but not at the HOPG/PO interface. The breakdown of the overall monolayer formation energy into contributions according to eqn (1)–(3) is illustrated in Fig. 9. Two of the contributions (the binding energy of the TPA monolayer in vacuum and the adsorption energy of a single TPA molecule on HOPG) are independent of the solvent, while the solvent wetting–dewetting processes stabilise the structure at the HOPG/7A interface slightly more than at the HOPG/PO interface. However, the biggest difference is in the solvation energies: solvation of TPA in PO is much more favourable than in 7A. As a result, the energy gain in forming the monolayer from solution in PO is very small (-9.8 kJ mol^{-1}) compared to 7A (-31.0 kJ mol^{-1}).

Note that the energies described here are enthalpies, while Gibbs free energies would be needed for a definitive answer as to whether adsorption from solution is possible or not. Thus, although the self-assembly of TPA at the HOPG/PO interface still has a small enthalpic gain, this may be compensated by an entropic loss. The entropy of molecules in solution can be calculated[73] and in general depends on the structure of the molecule, concentration and temperature. For example, for TPA in 9A, the entropy term $-T\Delta S$ was estimated as +3.4 kJ mol^{-1},[39] and for a related slightly larger stilbenedicarboxylic molecule (SDA) as +12.5 kJ mol^{-1},[57] both values of similar magnitude to the enthalpy gain found here. Thus, the Gibbs free energy for this monolayer formation could be very close to zero, indicating that a stable adsorbed monolayer of TPA at the HOPG/PO interface should not form.

For TPA at the HOPG/7A interface and for all other 2HTPA and 25DHTPA systems considered here, the energy gain due to monolayer adsorption from solution (from −22.9 to −28.7 kJ mol^{-1} in PO and from −31.0 to −65.6 kJ mol^{-1} in 7A) is much larger than the entropy terms quoted above. Therefore, the Gibbs free energy for the self-assembly of these systems is always negative (favourable) – supported by the experimental observations of adsorbed monolayers. It can also be seen that the energy gain of self-assembly is always larger in 7A than in PO. This agrees with the experimentally observed full monolayer coverage in 7A and partial coverage in PO.

To summarise, the analysis of all energy contributions to the process of monolayer self-assembly at the solid–liquid interface enables us to explain the formation or absence of TPA monolayers in 7A and PO, respectively, and the differences in surface coverage of substituted TPA molecules at the interfaces between these solvents and HOPG.

4. Conclusions

Self-assembly of TPA and its hydroxylated analogues 2HTPA and 25DHTPA at liquid/solid interfaces (graphite/heptanoic acid and graphite/1-phenyloctane) was studied using a combination of STM measurements and molecular mechanics and molecular dynamics calculations. The aim was to investigate the effects of the polar and apolar solvents on the self-assembly, and their interplay with weak (dispersion) and strong (hydrogen-bonding) interactions. STM results show that all three molecules form brickwork structures, similar to what was previously reported for TPA. However, the coverage achieved is different: full surface coverage is observed for all three molecules in 7A, partial coverage for 2HTPA and 25DHTPA in PO, and no adsorption of TPA in PO. There are further differences related to the nature of the molecules: the symmetric TPA and 25DHTPA form domains with a single orientation, while the non-symmetric 2HTPA forms multiply-oriented domains. 2HTPA is also the only molecule that, besides the regular brickwork assembly, forms alternative structures characterised by the pairing of H-bonded molecular chains with alternating small and large inter-chain separations. 25DHTPA forms two different brickwork structures depending on the solvent: a dense structure in 7A and a ~18% less dense structure in PO. Thus, polymorphism was observed, both induced by the solvent (for 25DHTPA) and related to the molecular structure (2HTPA).

To rationalise these results, molecular mechanics investigations of 2D monolayers of 2HTPA and 25DHTPA were carried out. 2D arrangements for both molecules had multiple minima, showing that both molecules should be capable of polymorphism. In particular, two 2D structures, close in energy but slightly different in geometry, were identified for 2HTPA, which correspond well to the regular and the alternating structures observed in the experiments. Because of the close similarity in their energies (only 0.3 kJ mol^{-1} preference for the "regular" structure), these structures are expected to co-exist independent of the solvent. For 25DHTPA, one energetically favoured 2D structure is found (attributed to the structure experimentally observed in 7A), as well as two less favourable structures, which may be the candidates for the less dense structure experimentally observed in PO.

The energetics of self-assembly was explored by constructing the Born–Haber cycle and analysing the energy difference between adsorbed monolayers at the liquid–solid interface and molecules in solution. Solvation of all three molecules by PO was found to be more exothermic than solvation by 7A. For TPA at the HOPG/PO interface, the adsorbed monolayer and the solvated molecule were very close in energy, suggesting an equilibrium between molecular adsorption and molecules in solution, with no strong energetic preference for the TPA molecules to adsorb. By comparison, there is a strong preference for adsorption of TPA at the HOPG/7A interface, and for 2HTPA and 25DHTPA at both solid/liquid interfaces. The formation of an adsorbed monolayer is particularly favourable at the 7A interfaces, explaining why full monolayer coverage is achieved with this solvent but only partial coverage is observed in the PO solvent.

Thus, by studying the assembly of three very similar molecules, we obtained different outcomes: molecules self-assembling on a surface (forming a range of structures) or staying in solution. The outcome is controlled by a complex balance of solvent–solute, adsorbate–adsorbate and adsorbate–surface interactions. In the relatively simple model system studied here, the careful small changes in the molecules have allowed us to obtain a full insight into the causes behind the observed phenomenology, with an almost completely predictive model. This is a very important result, demonstrating the level of control that an integrated experiment-theory approach can achieve in the technologically relevant field of molecular functionalisation of surfaces by 2D self-assembly.

Acknowledgements

RB and NM acknowledge the use of high-performance computing facilities provided by the University of Sheffield (Sol and Iceberg clusters). ADP was funded through a WPRS scholarship and an IAS early career fellowship of the University of Warwick. GC acknowledges financial support from the EU through the ERC Grant “VISUAL-MS”.

References

1 M. Gimenez-Marques, T. Hidalgo, C. Serre and P. Horcajada, *Coord. Chem. Rev.*, 2016, **307**, 342–360.
2 P. Xing and Y. Zhao, *Adv. Mater.*, 2016, **28**, 7304–7339.
3 S. Casalini, C. A. Bortolotti, F. Leonardi and F. Biscarini, *Chem. Soc. Rev.*, 2017, **46**, 40–71.
4 K. W. Hipps, *Science*, 2001, **294**, 536.
5 S. R. Forrest, *Nature*, 2004, **428**, 911–918.
6 W. F. Smith, *Nat. Nanotechnol.*, 2007, **2**, 77–78.
7 N. Koch, *ChemPhysChem*, 2007, **8**, 1438–1455.
8 M. Fahlman, A. Crispin, X. Crispin, S. K. M. Henze, M. P. d. Jong, W. Osikowicz, C. Tengstedt and W. R. Salaneck, *J. Phys.: Condens. Matter*, 2007, **19**, 183202.
9 E. Busseron, Y. Ruff, E. Moulin and N. Giuseppone, *Nanoscale*, 2013, **5**, 7098–7140.
10 D. M. Vriezema, M. Comellas Aragonès, J. A. A. W. Elemans, J. J. L. M. Cornelissen, A. E. Rowan and R. J. M. Nolte, *Chem. Rev.*, 2005, **105**, 1445–1490.

11 J. V. Barth, G. Costantini and K. Kern, *Nature*, 2005, **437**, 671–679.
12 J. V. Barth, *Annu. Rev. Phys. Chem.*, 2007, **58**, 345–407.
13 L. Dong, Z. A. Gao and N. Lin, *Prog. Surf. Sci.*, 2016, **91**, 101–135.
14 J. A. A. W. Elemans, S. Lei and S. De Feyter, *Angew. Chem., Int. Ed.*, 2009, **48**, 7298–7332.
15 M. Lackinger and W. M. Heckl, *Langmuir*, 2009, **25**, 11307–11321.
16 S. De Feyter and F. C. De Schryver, *Chem. Soc. Rev.*, 2003, **32**, 139–150.
17 T. Balandina, K. Tahara, N. Sändig, M. O. Blunt, J. Adisoejoso, S. Lei, F. Zerbetto, Y. Tobe and S. De Feyter, *ACS Nano*, 2012, **6**, 8381–8389.
18 M. O. Blunt, J. Adisoejoso, K. Tahara, K. Katayama, M. Van der Auweraer, Y. Tobe and S. De Feyter, *J. Am. Chem. Soc.*, 2013, **135**, 12068–12075.
19 T. Sirtl, W. Song, G. Eder, S. Neogi, M. Schmittel, W. M. Heckl and M. Lackinger, *ACS Nano*, 2013, **7**, 6711–6718.
20 D. C. Y. Nguyen, L. Smykalla, T. N. H. Nguyen, T. Rüffer and M. Hietschold, *J. Phys. Chem. C*, 2016, **120**, 11027–11036.
21 M. N. Nair, C. Mattioli, M. Cranney, J.-P. Malval, F. Vonau, D. Aubel, J.-L. Bubendorff, A. Gourdon and L. Simon, *J. Phys. Chem. C*, 2015, **119**, 9334–9341.
22 G. M. Florio, B. Ilan, T. Müller, T. A. Baker, A. Rothman, T. L. Werblowsky, B. J. Berne and G. W. Flynn, *J. Phys. Chem. C*, 2009, **113**, 3631–3640.
23 M. Lackinger, S. Griessl, W. M. Heckl, M. Hietschold and G. W. Flynn, *Langmuir*, 2005, **21**, 4984–4988.
24 L. Kampschulte, M. Lackinger, A.-K. Maier, R. S. K. Kishore, S. Griessl, M. Schmittel and W. M. Heckl, *J. Phys. Chem. B*, 2006, **110**, 10829–10836.
25 W. Mamdouh, H. Uji-i, J. S. Ladislaw, A. E. Dulcey, V. Percec, F. C. De Schryver and S. De Feyter, *J. Am. Chem. Soc.*, 2006, **128**, 317–325.
26 S.-L. Lee, Y.-C. Chu, H.-J. Wu and C.-h. Chen, *Langmuir*, 2012, **28**, 382–388.
27 N. T. N. Ha, T. G. Gopakumar and M. Hietschold, *Surf. Sci.*, 2013, **607**, 68–73.
28 L. Cui, X. Miao, L. Xu, Y. Hu and W. Deng, *Phys. Chem. Chem. Phys.*, 2015, **17**, 3627–3636.
29 S. Lei, K. Tahara, F. C. De Schryver, M. Van der Auweraer, Y. Tobe and S. De Feyter, *Angew. Chem., Int. Ed.*, 2008, **47**, 2964–2968.
30 N. Thi Ngoc Ha, T. G. Gopakumar and M. Hietschold, *J. Phys. Chem. C*, 2011, **115**, 21743–21749.
31 X. Miao, L. Xu, Z. Li and W. Deng, *J. Phys. Chem. C*, 2011, **115**, 3358–3367.
32 A. Ciesielski, P. J. Szabelski, W. Rżysko, A. Cadeddu, T. R. Cook, P. J. Stang and P. Samorì, *J. Am. Chem. Soc.*, 2013, **135**, 6942–6950.
33 K. S. Mali, K. Lava, K. Binnemans and S. De Feyter, *Chem.–Eur. J.*, 2010, **16**, 14447–14458.
34 L. Xu, X. Miao, L. Cui, P. Liu, X. Chen and W. Deng, *Nanoscale*, 2015, **7**, 11734–11745.
35 X. Shen, X. Wei, P. Tan, Y. Yu, B. Yang, Z. Gong, H. Zhang, H. Lin, Y. Li, Q. Li, Y. Xie and L. Chi, *Small*, 2015, **11**, 2284–2290.
36 L. Cui, X. Miao, L. Xu and W. Deng, *Appl. Surf. Sci.*, 2014, **313**, 841–849.
37 J. M. MacLeod, O. Ivasenko, D. F. Perepichka and F. Rosei, *Nanotechnology*, 2007, **18**, 424031.
38 M. Lackinger, S. Griessl, L. Kampschulte, F. Jamitzky and W. M. Heckl, *Small*, 2005, **1**, 532–539.

39 W. T. Song, N. Martsinovich, W. M. Heckl and M. Lackinger, *J. Am. Chem. Soc.*, 2013, **135**, 14854–14862.
40 M. Lackinger, S. Griessl, T. Markert, F. Jamitzky and W. M. Heckl, *J. Phys. Chem. B*, 2004, **108**, 13652–13655.
41 S. Yasuda, A. Furuya and K. Murakoshi, *RSC Adv.*, 2014, **4**, 58567–58572.
42 R. Addou and M. Batzill, *Langmuir*, 2013, **29**, 6354–6360.
43 W. Zhang, A. Nefedov, M. Naboka, L. Cao and C. Woll, *Phys. Chem. Chem. Phys.*, 2012, **14**, 10125–10131.
44 S. Clair, S. Pons, A. P. Seitsonen, H. Brune, K. Kern and J. V. Barth, *J. Phys. Chem. B*, 2004, **108**, 14585–14590.
45 T. Suzuki, T. Lutz, D. Payer, N. Lin, S. L. Tait, G. Costantini and K. Kern, *Phys. Chem. Chem. Phys.*, 2009, **11**, 6498–6504.
46 Y.-G. Kim, S.-L. Yau and K. Itaya, *Langmuir*, 1999, **15**, 7810–7815.
47 S. L. Tait, Y. Wang, G. Costantini, N. Lin, A. Baraldi, F. Esch, L. Petaccia, S. Lizzit and K. Kern, *J. Am. Chem. Soc.*, 2008, **130**, 2108–2113.
48 S. Stepanow, T. Strunskus, M. Lingenfelder, A. Dmitriev, H. Spillmann, N. Lin, J. V. Barth, C. Woll and K. Kern, *J. Phys. Chem. B*, 2004, **108**, 19392–19397.
49 Y. L. Wang, S. Fabris, G. Costantini and K. Kern, *J. Phys. Chem. C*, 2010, **114**, 13020–13025.
50 M. E. Cañas-Ventura, F. Klappenberger, S. Clair, S. Pons, K. Kern, H. Brune, T. Strunskus, W. Ch, R. Fasel and J. V. Barth, *J. Chem. Phys.*, 2006, **125**, 184710.
51 H. Aitchison, H. Lu, S. W. L. Hogan, H. Fruchtl, I. Cebula, M. Zharnikov and M. Buck, *Langmuir*, 2016, **32**, 9397–9409.
52 T. Suzuki, T. Lutz, G. Costantini and K. Kern, *Surf. Sci.*, 2011, **605**, 1994–1998.
53 P. Rahe, M. Nimmrich, A. Nefedov, M. Naboka, C. Woll and A. Kuhnle, *J. Phys. Chem. C*, 2009, **113**, 17471–17478.
54 A. Tekiel, J. S. Prauzner-Bechcicki, S. Godlewski, J. Budzioch and M. Szymonski, *J. Phys. Chem. C*, 2008, **112**, 12606–12609.
55 P. Rahe, M. Nimmrich and A. Kuhnle, *Small*, 2012, **8**, 2969–2977.
56 J. M. MacLeod, J. A. Lipton-Duffin, D. Cui, S. De Feyter and F. Rosei, *Langmuir*, 2015, **31**, 7016–7024.
57 W. Song, N. Martsinovich, W. M. Heckl and M. Lackinger, *Phys. Chem. Chem. Phys.*, 2014, **16**, 13239–13247.
58 I. Horcas, R. Fernández, J. M. Gómez-Rodríguez, J. Colchero, J. Gómez-Herrero and A. M. Baro, *Rev. Sci. Instrum.*, 2007, **78**, 013705.
59 J. W. Ponder and F. M. Richards, *J. Comput. Chem.*, 1987, **8**, 1016–1024.
60 N. L. Allinger, Y. H. Yuh and J. H. Lii, *J. Am. Chem. Soc.*, 1989, **111**, 8551–8566.
61 J.-H. Lii and N. L. Allinger, *J. Comput. Chem.*, 1998, **19**, 1001–1016.
62 N. Martsinovich and A. Troisi, *J. Phys. Chem. C*, 2010, **114**, 4376–4388.
63 M. Head-Gordon, J. A. Pople and M. J. Frisch, *Chem. Phys. Lett.*, 1988, **153**, 503–506.
64 G. W. T. M. J. Frisch, H. B. Schlegel, G. E. Scuseria, M. A. Robb, J. R. Cheeseman, G. Scalmani, V. Barone, B. Mennucci, G. A. Petersson, H. Nakatsuji, M. Caricato, X. Li, H. P. Hratchian, A. F. Izmaylov, J. Bloino, G. Zheng, J. L. Sonnenberg, M. Hada, M. Ehara, K. Toyota, R. Fukuda, J. Hasegawa, M. Ishida, T. Nakajima, Y. Honda, O. Kitao, H. Nakai, T. Vreven, J. A. Montgomery Jr, J. E. Peralta, F. Ogliaro, M. Bearpark, J. J. Heyd, E. Brothers, K. N. Kudin, V. N. Staroverov, R. Kobayashi, J. Normand, K. Raghavachari, A. Rendell, J. C. Burant, S. S. Iyengar,

J. Tomasi, M. Cossi, N. Rega, J. M. Millam, M. Klene, J. E. Knox, J. B. Cross, V. Bakken, C. Adamo, J. Jaramillo, R. Gomperts, R. E. Stratmann, O. Yazyev, A. J. Austin, R. Cammi, C. Pomelli, J. W. Ochterski, R. L. Martin, K. Morokuma, V. G. Zakrzewski, G. A. Voth, P. Salvador, J. J. Dannenberg, S. Dapprich, A. D. Daniels, Ö. Farkas, J. B. Foresman, J. V. Ortiz, J. Cioslowski and D. J. Fox, Gaussian, Inc., Wallingford CT, 2009.

65 Pubchem – Open Chemistry Database: Terephthalic Acid, https://www.pubchem.ncbi.nlm.nih.gov/compound/7489#section=Density, accessed 02/03/2017.

66 Santa Cruz Biotechnology: 2-Hydroxyterephthalic Acid, https://www.scbt.com/scbt/product/2-hydroxyterephthalic-acid-636-94-2, accessed 02/03/2017.

67 ChemBK Database: 2,5-dihydroxyterephthalic Acid, http://www.chembk.com/en/chem/2,5-dihydroxyterephthalic%20acid, accessed 02/03/2017.

68 Pubchem – Open Chemistry Database: Heptanoic Acid, https://www.pubchem.ncbi.nlm.nih.gov/compound/8094#section=Density, 02/03/2017.

69 Sigma Aldrich: 1-Phenyloctane, http://www.sigmaaldrich.com/catalog/product/aldrich/113190?lang=en&region=GB, accessed 02/03/2017.

70 A. J. Marsden, Z. P. L. Laker, O. De Luca, A. Della Pia, L. M. A. Perdigao, G. Costantini and N. R. Wilson, *Nanoscale*, 2017, DOI: 10.1039/C7NR03588H.

71 T. W. White, N. Martsinovich, A. Troisi and G. Costantini, submitted.

72 M. Bailey and C. J. Brown, *Acta Crystallogr.*, 1967, **22**, 387–391.

73 M. Mammen, E. I. Shakhnovich, J. M. Deutch and G. M. Whitesides, *J. Org. Chem.*, 1998, **63**, 3821–3830.

Faraday Discussions

PAPER

Transfer of chiral information from a chiral solvent to a two-dimensional network†

Iris Destoop,‡[a] Andrea Minoia,‡[b] Oleksandr Ivasenko,[a] Aya Noguchi,[c] Kazukuni Tahara,[cd] Yoshito Tobe, *[c] Roberto Lazzaroni*[b] and Steven De Feyter *[a]

Received 10th March 2017, Accepted 7th April 2017
DOI: 10.1039/c7fd00103g

Chiral induction in self-assembled monolayers has garnered considerable attention in the recent past, not only due to its importance in chiral resolution and enantioselective heterogeneous catalysis but also because of its relevance to the origin of homochirality in life. Here, we demonstrate the emergence of homochirality in a supramolecular low-density network formed by achiral molecules at the interface of a chiral solvent and an atomically-flat achiral substrate. We focus on the impact of structure and functionality of the adsorbate and the chiral solvent on the chiral induction efficiency in self-assembled physisorbed monolayers, as revealed by scanning tunneling microscopy. Different induction mechanisms are proposed and evaluated, with the assistance of advanced molecular modeling simulations.

Introduction

Chiral surfaces have recently attracted considerable attention as they can be used for chiral separation, enantioselective heterogeneous catalysis and chiroptical materials.[1,2] Chiral surfaces can be obtained by different approaches. One way is to craft intrinsically chiral surfaces by cutting a crystal along one of its high Miller index directions, giving rise to a crystal plane that lacks mirror symmetry and therefore is chiral.[3,4] Another approach is to decorate surfaces by the irreversible adsorption of chiral molecules.[5,6] Similarly, organic building blocks physisorbed on achiral surfaces such as graphite or gold can create nanostructured chiral

[a]Division of Molecular Imaging and Photonics, Department of Chemistry, KU Leuven-University of Leuven, Celestijnenlaan 200F, B 3001, Leuven, Belgium. E-mail: steven.defeyter@kuleuven.be

[b]Laboratory for Chemistry of Novel Materials, University of Mons, 20, Place du Parc, B 7000 Mons, Belgium. E-mail: roberto.lazzaroni@umons.ac.be

[c]Division of Frontier Materials Science, Graduate School of Engineering Science, Osaka University, Toyonaka, Osaka 560-8531, Japan. E-mail: tobe@chem.es.osaka-u.ac.jp

[d]Department of Applied Chemistry, School of Science and Technology, Meiji University, 1-1-1 Higashimita, Tama-ku, Kawasaki, 214-8571, Japan

† Electronic supplementary information (ESI) available. See DOI: 10.1039/c7fd00103g

‡ I. D. and A. M. contributed equally.

surfaces. In this case, the presence of the surface prohibits the existence of symmetry elements such as inversion centers and mirror planes parallel to the surface. This inherent non-centrosymmetric nature of the surface ensures that not only chiral, but also prochiral molecules can become chiral upon surface confinement. Nevertheless, monolayers formed by prochiral building blocks will always express both mirror image motifs in equal abundances leaving the surface globally achiral. Enantiopure molecules, on the other hand, will form only one chiral motif (the mirror image motif is absent) on the surface, given that the stereogenic centre influences the molecular adsorption and self-assembly process effectively, creating a surface that displays global chirality.[7–10]

Nonetheless, achiral building blocks can be forced to form globally chiral surfaces. Usually this is done by adding a chiral dopant to the solution of the achiral building block.[11–17] The chiral dopant or sergeant is incorporated into the monolayer of the achiral building blocks (or soldiers) and transfers in this way its chirality, rendering a chiral surface. A quite similar approach consists of adding a small excess of one enantiomer to the racemate.[18–22] Here, the majority enantiomer will suppress the formation of the mirror image pattern. Typically, the addition of only a few percent of these sergeants or a small enantiomeric excess is sufficient to obtain global chirality.

Both these approaches were pioneered by Green *et al.* in polymer science and were named "sergeants-and-soldiers" and "majority-rules", respectively.[23,24] Over the last couple of years, it was shown that these principles also hold at the liquid/solid interface.

Alternatively, chiral solvents have recently shown great potential to induce homochirality in self-assembled two-dimensional organic networks.[20,25,26] The self-assembly of a *p*-phenylenevinylene oligomer, for instance, can be biased towards a specific rosette handedness by simply drop-casting it from enantiopure 1-phenyl-1-octanol.[27] Also, the high-density network formed by amidophenyl-substituted porphyrin derivatives has been revealed to be sensitive to the presence of chiral hydrogen-bonding solvents.[28] Another class of molecules that has been shown to be "sensitive" to the chirality of the solvent, and that will be the subject of this study, are alkoxylated dehydrobenzo[12]annulene (DBA) derivatives (Fig. 1A).[29–33] The porous networks formed by various DBAs are based on the interdigitation of alkoxy chains of adjacent molecules (Fig. 1). The relative alignment of the four interdigitated alkoxy chains per DBA pair ("+" or "−" type interdigitation) governs the chirality of the resultant honeycomb network. A virtual clockwise (CW) or counterclockwise (CCW) nanowell is obtained by combining six "−" type or "+" type interdigitation patterns, respectively. The labels CW and CCW refer to the sense of rotation of the six alkoxy chains making up the rim of the nanowell, and typically, the chirality of the nanowells is domain-specific.[34] Also, "mixed" nanowells (MIX) can be formed by combining both "−" and "+" type interdigitations in various ways. A relatively simple marker for the handedness of the nanowells is the relative shift of the neighboring DBA cores.

The tilt of the line joining the centers of neighboring DBA cores with respect to the rim of the honeycomb can be used to easily identify the type of interdigitation and the handedness of individual nanowells, as well of the domains they form (Fig. 1B and C). In the absence of any chiral influence, achiral DBA molecules (DBA-OC*n*) enantioseparate on highly oriented pyrolytic graphite (HOPG) into equal amounts of mirror image chiral domains, so CW and CCW.[11,35]

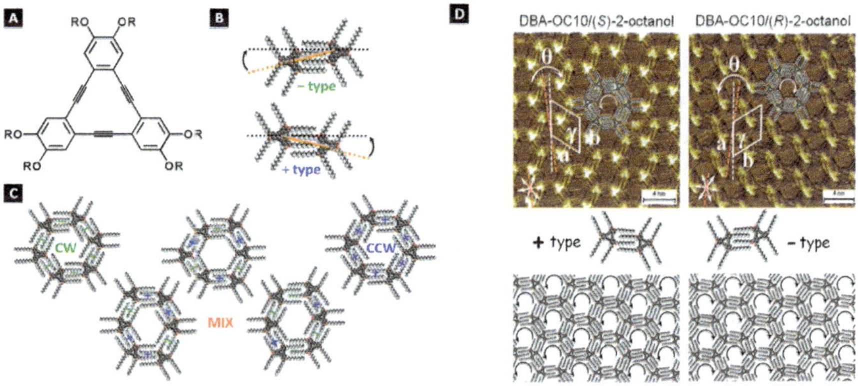

Fig. 1 (A) Molecular structure of an alkoxylated dehydrobenzo[12]annulene (R = C_nH_{2n+1}). (B) The interdigitation of the alkoxy chains can lead to the formation of a "−" or "+" type interdigitation depending on the relative alignment of the adjacent DBAs. (C) The combination of six interdigitation types leads to the formation of a hexagonal nanowell. If all "−" types are combined a CW pore is formed, all "+" types lead to a CCW pore and combinations of both lead to "mixed" pores. (D) Top: HR-STM images of the self-assembled networks of DBA-OC10 at the left (*S*)-2-octanol/HOPG and right (*R*)-2-octanol/HOPG interface (I_{set} = 280 pA, V_{bias} = 230 mV). Red dashed lines indicate the graphite reference axis. θ is the angle between the graphite reference axis and one of the unit cell vectors. Bottom: Molecular models for the self-assembled networks of DBA-OC10 in left (*S*)-2-octanol and right (*R*)-2-octanol. The unit cell parameters are identical in both solvents and they are: $a = 4.0 \pm 0.1$ nm, $b = 4.1 \pm 0.1$ nm, $\gamma = 61.0 \pm 1.0°$. Adapted with permission from the American Chemical Society.[26]

So, how is the self-assembly process of this system affected by the presence of chiral solvent? Previously, we showed that the self-assembly of DBAs with alkoxy chains containing an even number of carbon atoms (DBA-OCn_{even} with n = 6, 8, 10, 12, 14, 16) at the interface between HOPG and (*S*)-2-octanol gives rise to the preferential formation of CCW domains on the graphite surface, while deposition from (*R*)-2-octanol leads to preferential formation of CW domains (Fig. 1D).[26] Furthermore, a DBA alkoxy chain length-dependent induction was observed in these chiral solvents: derivatives with short alkoxy chains (*e.g.* DBA-OC6) only form domains of one particular chirality (and thus yield homochiral surfaces), while for the derivatives with longer alkoxy chains (*e.g.* DBA-OC14, DBA-OC16) the surface is not fully homochiral, or in other words, the chiral induction by the solvent seems less efficient.

In order to gain insight into the solvent-based chiral induction mechanism or mechanisms at play, we have carried out additional sets of experiments, and have engaged in extensive molecular modeling simulations. This manuscript is structured as follows. In the first part, we describe the results of a new series of scanning tunneling microscopy (STM) studies, and evaluate to what extent the new experimental results in combination with our previous data support or refute a number of induction mechanisms we propose. In the second part, we engage in advanced molecular modeling simulations to shine light on the most possible induction mechanism(s). Among the several mechanisms that we evaluate, those that involve on-surface transient complexation of the chiral solvent and the DBA molecules at the moment of nucleation turn out to be the most likely ones.

Results and discussion

The impact of DBA alkyl chain length and odd–even effects

To start with, the self-assembly behavior of DBA was further investigated by looking at three other DBA derivatives. For DBA-OC4, the shortest probed derivative, again CCW domains are obtained upon deposition from (*S*)-2-octanol, while the drop-casting from (*R*)-2-decanol led to only CW domains on the surface. The two other derivatives that were tested have an odd number of carbons in the alkoxy side chains (DBA-OCn_{odd}). Surprisingly, both DBA-OC11 and DBA-OC13 turned out to express equal amounts of CW and CCW pores on the surface and these derivatives are thus not in any way influenced by the presence of the chiral solvent. In addition, DBA-C10, the alkylated derivative of DBA-OC10, also displays chiral induction when self-assembled at the interface between HOPG and (*S*)-2-octanol.

In another set of experiments, we looked at the impact of the solvent. The effect of a series of enantiopure 2-alcohols, ranging from 2-hexanol up to 2-dodecanol, on the self-assembly of DBA-OC10 was probed (Fig. 2). The analysis of the STM data reveals that all the probed solvents do not yield a uniform result. Instead again an odd–even effect becomes apparent, in which the chiral solvents with an even chain length (*e.g.* 2-decanol) tend to form 90–95% induced domains and those with an odd chain length (*e.g.* 2-nonanol) only form about 70% induced domains (Fig. 2, Table 1). However, it should be noted that there are some exceptions. In the case of enantiopure (*S*)-2-dodecanol, only about 70% of the surface is covered with induced CCW domains. Note that (*S*)-2-dodecanol has a melting point close to room temperature and as a consequence the solution is quite viscous, which may hinder the system to reach equilibrium. Also, for commercial (*S*)-2-hexanol the chiral induction is lower than the anticipated value of 90% (79%). In this particular case, the deviation is believed to be caused by the

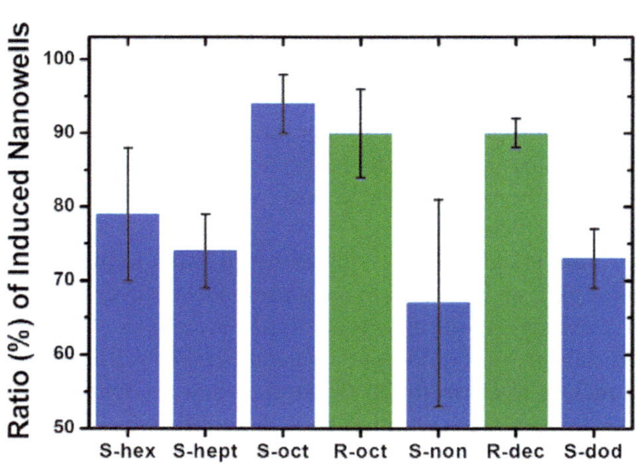

Fig. 2 An overview of the ratio (%) of induced nanowells for DBA-OC10 in a series of enantiopure 2-alcohol solvents. A clear odd–even solvent chain length effect is revealed. (*S*)-Stereogenic centers induce CCW nanowells (■), (*R*)-stereogenic centers induce CW nanowells (■). The error bars reflect the variation in the ratio (%) of induced nanowells in three sessions (about 10 images per session).

Table 1 Chirality expression of various DBA derivatives in chiral solvents

Compound	Solvent	CCW	CW	Mix	C (mol dm^{-3})	# pores	# domains
DBA-OC4	(*S*)-2-Octanol	**99%**	1%	0%	8.4×10^{-6}	4364	33
	(*R*)-2-Decanol	0%	**99%**	1%	8.4×10^{-6}	2691	23
DBA-OC6	(*S*)-2-Octanol	**98%**	0%	2%	4.2×10^{-6}	4045	36
	(*R*)-2-Octanol	0%	**100%**	0%	4.2×10^{-6}	4163	41
	(*S*)-2-Nonanol	**65%**	33%	4%	4.2×10^{-6}	6447	43
	(*R*)-2-Decanol	0%	**99%**	1%	2.2×10^{-6}	4007	33
	(*S*)-2-Dodecanol	**76%**	16%	8%	8.0×10^{-5}	4748	28
DBA-OC8	(*S*)-2-Octanol	**87%**	10%	3%	2.1×10^{-6}	2205	34
	(*R*)-2-Octanol	1%	**99%**	0%	2.1×10^{-6}	2644	37
	(*R*)-2-Decanol	3%	**97%**	0%	2.1×10^{-6}	3292	37
DBA-OC10	(*S*)-2-Hexanol	**79%**	18%	2%	1.4×10^{-6}	1897	74
	(*S*)-2-Heptanol	**74%**	25%	1%	2.1×10^{-6}	2447	64
	(*S*)-2-Octanol	**94%**	6%	0%	2.1×10^{-6}	2081	32
	(*R*)-2-Octanol	9%	**90%**	1%	2.1×10^{-6}	2596	33
	(*rac*)-2-Octanol	51%	49%	0%	2.1×10^{-6}	4439	58
	(*S*)-2-Nonanol	**69%**	27%	5%	2.1×10^{-6}	2614	39
	(*R*)-2-Decanol	5%	**90%**	5%	2.1×10^{-6}	3169	34
	(*S*)-2-Dodecanol	**73%**	18%	10%	8.0×10^{-5}	2081	29
DBA-C10	(*S*)-2-Octanol	**69%**	20%	11%	2.1×10^{-6}	2580	34
	(*R*)-2-Octanol	16%	**77%**	8%	2.1×10^{-6}	2259	47
DBA-OC11	(*S*)-2-Octanol	22%	48%	31%	2.1×10^{-6}	2669	34
	(*R*)-2-Octanol	27%	54%	18%	2.1×10^{-6}	2010	29
	(*S*)-2-Nonanol	21%	11%	**68%**	2.1×10^{-6}	3394	39
DBA-OC12	(*S*)-2-Octanol	**90%**	7%	3%	2.1×10^{-6}	2194	38
	(*R*)-2-Octanol	16%	**82%**	3%	2.1×10^{-6}	1832	34
DBA-OC13	(*S*)-2-Octanol	42%	44%	14%	2.7×10^{-6}	1800	34
	(*R*)-2-Octanol	44%	33%	23%	2.7×10^{-6}	1678	34
DBA-OC14	(*S*)-2-Octanol	**79%**	17%	3%	2.1×10^{-6}	1432	37
	(*R*)-2-Octanol	20%	**76%**	4%	2.1×10^{-6}	1124	32
	(*R*)-2-Decanol	8%	**84%**	7%	2.1×10^{-6}	1486	34
DBA-OC16	(*S*)-2-Octanol	**74%**	20%	6%	1.4×10^{-6}	1306	31
	(*R*)-2-Octanol	17%	**74%**	9%	1.4×10^{-6}	1098	38

presence of impurities (see Fig. S2†). The induction results for all combinations of DBAs and chiral solvents are summarized in Table 1.

Chiral induction in relation to the different stages of the self-assembled monolayer formation

Before we interpret these results, let's first consider the stage at which the chiral induction takes place. First of all, time-dependent STM experiments do not reveal any significant evolution of the handedness of the supramolecular network with time, indicating that the chiral induction takes place at the level of 2D nucleation, or in the very early stages of monolayer development. Thus, one can readily rule out the formation of racemic domains as a preceding step to the evolution of enantiopure domains.

The different modes of chiral induction

A first important question is where the induction takes place. Is the induction the result of (1) a phenomenon that takes place in the bulk solution, (2) a consequence of a specific interaction of solvent near the surface with adsorbed DBA molecules, the solvent molecules not being adsorbed on the surface, or (3) does the chiral induction require the chiral solvent molecules to be (co)adsorbed on the surface?

A. Chiral induction in the bulk (1). The "bulk solution process" (1) can be ruled out. This would require some interactions in solution between chiral solvent and DBA molecules, prior to molecular self-assembly on surfaces, beyond the formation of some transient solvent–DBA complexes in solution. UV-absorption, ^{1}H NMR and circular dichroism measurements do not reveal any sign of such pre-organization in solution.[26] While we don't rule out specific solvent–molecule interactions in the bulk, these do not lead to chiral supramolecular seeds in solution that would retain their structure upon adsorption on the surface.

B. Chiral induction orthogonal to the surface (2). Having ruled out the "bulk solution process", we consider the formation of a (transient) complex between one or more adsorbed or adsorbing DBA molecules and one or more chiral solvent molecules that are located in the supernatant solution just above the surface, but are not adsorbed on the surface (2). Irrespective of the mode of interaction, in a process where solvent molecules act from above the surface rather than on the surface, structural variations such as length should have a limited effect on the induction of chirality. However, the experimentally observed odd–even solvent and alkyl chain length behavior (Table 1, Fig. 2) cannot be reconciled with such a mechanism where solvent molecules act from the liquid phase on the adsorbed DBA molecules. While the experimental observations don't rule out such a mode of interaction, its importance, relative to other induction processes, should be moderate.

C. Chiral induction on and parallel to the surface (3). This brings us to the hypothesis where adsorbed solvent molecules are crucial in the chiral induction process. The outcomes of the experiments described in the next section hint indeed to the importance of (co)adsorbed solvent molecules in direct contact with the substrate. In this set of experiments, the chiral solvent is diluted with an achiral solvent, and competitive solvent adsorption is probed. DBA-OC10 was deposited from a mixture of 1-octanol and (*S*)-2-octanol at various volume ratios. A clear linear dependency can be observed between the chirality of the DBA network and the v/v% of the solvents (Fig. 3A). This seems to imply that the transfer efficiency of chiral information is rather poor; alternatively, this linear behavior might demonstrate the equal adsorption affinity of both solvents 1-octanol and (*S*)-2-octanol, which seems quite reasonable since they have similar alkyl chain lengths and will thus compete for the same adsorption sites on graphite. To put the latter interpretation to the test, the experiment was repeated with achiral and chiral solvents of different chain length, as several studies have indicated that upon self-assembly of bi-component systems, the molecule with the longest chain length is preferentially adsorbed.[36–40] Hence DBA-OC10 was deposited from a mixture of 1-hexanol and (*R*)-2-decanol at various v/v ratios: 50/50, 75/25 and 90/10. The analysis of the STM data now clearly shows a non-linear dependency between the v/v% and the induced CW nanowells (Fig. 3B). This indicates that in the previous case the linear dependency indeed reflected the similar adsorption affinity of both solvents. In this case, only about 50 v/v% of the enantiopure solvent is needed in order to gain the same monolayer chirality as for

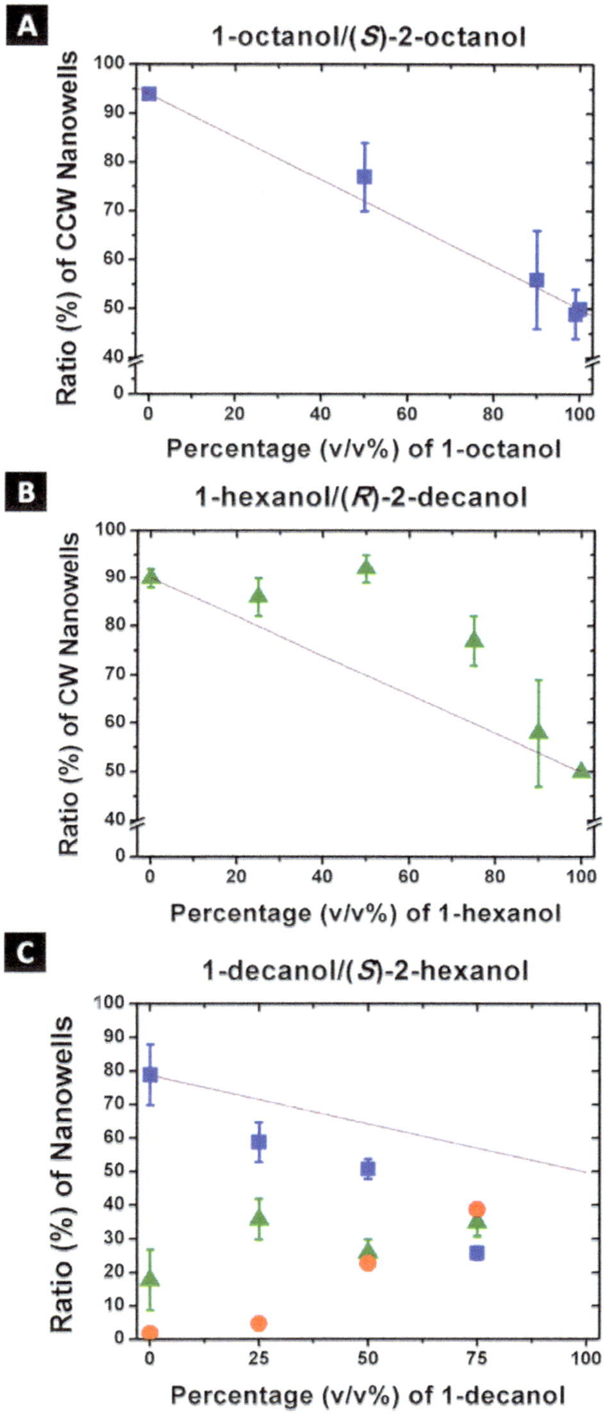

Fig. 3 Chirality expression of monolayers formed by DBA-OC10 in mixtures of an achiral and a chiral solvent, showing the population of the CW (▲), CCW (■) and MIX (●) nanowells. The line reflects the expected trend of the chiral induction when scaling linearly with the fraction of chiral solvent.

the pure enantiopure solvent. In addition, a control experiment was performed with 1-decanol and (*S*)-2-hexanol. From Fig. 3C, a negative non-linear dependency can be observed, although obscured by the presence of MIX pores. But the equal amounts of CW and CCW honeycombs present for 75/25 v/v% of 1-decanol/(*S*)-2-hexanol clearly illustrate the diminished chiral induction. The fact that the longer solvent always has the upper hand is a strong indication for the involvement of solvent adsorption as part of the mechanism for the transfer of chiral information to the two-dimensional network. All data are also presented in Table S1.†

This insight doesn't provide us with the mechanism yet, as several induction mechanisms can still be considered where (co)adsorbed solvent molecules are involved. We will now consider a few of them and evaluate their relative importance.

C1. Induction via a chiral self-assembled solvent layer. Indeed, there are quite some examples of self-assembled systems that are adsorbed on top of a static solvent buffer layer. Typically these layers are formed on top of a self-assembled layer of pentacontane or other long alkanes.[41,42] But also shorter hydrogen-bonding solvents like 1-decanol can act as a buffer layer.[43] One characteristic that all these solvents have in common is their ability to form well-ordered long-range domains which can be visualized by STM. However, the enantiopure 2-alcohols in this study show no ordered monolayer formation at room temperature and, even so, the stereogenic centre would probably create steric restraints for the formation of the DBA overlayer as either the hydroxyl or the methyl group (whichever is not adsorbed on the surface) must point in that direction. So, we regard this mechanism, based upon the formation of chiral self-assembled solvent layers on top of which the DBA molecules self-assemble, as unlikely.

C2. Induction via a dynamic interfacial solvent (mono)layer. Rather than adsorbing on top of those solvent molecules, DBA molecules may nucleate on the HOPG surface, thereby repelling solvent molecules from the surface. As those flexible solvent molecules get expelled, they contribute in entropy gain. But how would such a process lead to chiral induction? The DBA orientation with respect to HOPG symmetry lattices is different for odd and even DBAs and for *R*- and *S*-alcohols (see the ESI†), which means that the number of *S*-alcohol molecules replaced by CCW and CW nuclei would be different—hence chirality induction (for the details of the orientational aspects, we refer to the ESI†). Note that this mechanism does not require specific intermolecular interactions.

C3. Induction via co-adsorbed solvent with the DBA molecules on the surface. The formed DBA–solvent complex at the moment of nucleation or in the very early stages of monolayer development could potentially be stabilized by hydrogen bonds and van der Waals interactions between the adsorbed DBA and the adsorbed solvent molecules. In addition, solvent–substrate interactions could further enhance the stability of this DBA–solvent complex. Again, we can differentiate between a number of different cases.

C3.1. Stabilization of the nanowells. In the first case, solvent molecules could become co-adsorbed in the nanowells and stabilize nanoporous structures of a given handedness. Previous studies on DBAs in other solvents have revealed that solvent molecules can co-adsorb on the surface, in the nanopores. One DBA-OC16 pore for instance can host up to nineteen 1,2,4-trichlorobenzene molecules,[32] and also tetradecane molecules have been found to co-adsorb inside the DBA network.[39] One could expect to find the alcohol solvent molecules along the rim of the hexagonal pores to maximize the van der Waals contacts, supplemented by the ability to form hydrogen bonds with the DBA oxygen atoms. However, from the high resolution (HR) STM images obtained on the DBA derivatives in the two enantiomeric 2-octanol solvents, no 2-octanol molecules could be found inside the CW or CCW rotated nanowells of the low-density honeycomb phase. Even 2-dodecanol, which should be less dynamic when co-adsorbed in the pores, was not observed. So co-adsorbed solvent molecules could never be clearly visualized inside CW or CCW nanowells. While this observation does not rule out that the mechanism described above is active, it suggests that the solvent molecules must be mobile, which somehow could weaken if not their stabilizing effect, most likely their induction effect.

C3.2. Formation of on-surface DBA–solvent complex and integration in the monolayer. The second possible mechanism to consider is the chirality induction through the nucleation of a stable DBA–alcohol complex. Such a complex would act as a chiral "sergeant" defining the alkyl chain interdigitation of the other DBAs in the growing domain. An example of solvent-containing defects in DBA domains is shown in Fig. 4. Here, trapping solvent molecules between the alkoxy chains of DBA is reminiscent of the related co-assemblies of *n*-tetradecane and DBA-OC16, reported earlier.[44] There is, however, no evidence that: (1) such defects indeed form during domain nucleation and not at later stages (growth or ripening), and (2) that their formation is sufficiently enantioselective to promote chirality induction. Furthermore, alcohol inclusion was not observed if the difference between the carbon chain length in the alcohols and DBAs was larger than two

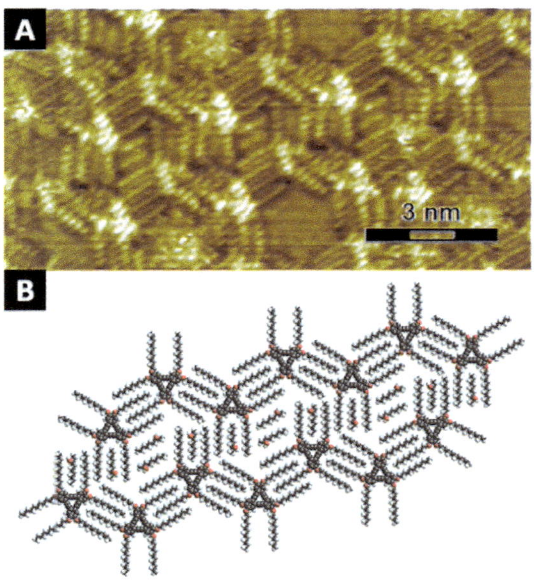

Fig. 4 (A) STM image of DBA-OC10 in 1-octanol/(*S*)-2-octanol (99 : 1) revealing the existence of defects involving co-adsorbed solvent molecules. The tunneling parameters are I_{set} = 750 pA; V_{bias} = −200 mV. (B) Model of the observed defect. In the model, 2-octanol molecules are adsorbed with an arbitrary orientation of the hydroxyl groups. There is no experimental evidence to justify the orientation. Moreover, in solvent mixtures with 1-octanol, most likely also 1-octanol is adsorbed, maybe even predominantly.

methylene groups. Therefore, even if valid, such a chirality induction pathway cannot account for the majority of the DBA–alcohol combinations studied in this work.

C3.3. Transiently formed DBA–solvent complexes at the stage of nucleation. As a third case, we envision that surface-confined solvent–DBA complexes can transiently be formed as part of the nucleation process, promoting one form of interdigitation ("+" or "−"), while penalizing the other form of interdigitation. Upon formation of the nanowell, the co-complexed solvent molecules are released in the supernatant solution. This mechanism will be discussed in more detail in the modeling section.

Discussion

Which of these "solvent-on-a-surface" induction models is most likely? The problem is that the time-resolution of STM is not fast enough to evaluate the existence of transient solvent–DBA complexes on the surface. Therefore the argumentation is rather qualitative or semi-quantitative at best. The fact that we can't image self-assembled monolayers of the solvent itself lowers the probability that self-assembled DBA monolayers are formed on top of a solvent monolayer (C1). The lack of any experimental evidence for structured solvent adsorption in nanowells disfavors also that pathway (C3.1). And if solvent co-adsorption is observed, it happens rather rarely, and doesn't give rise to the anticipated honeycomb lattice (C3.2). However, a solvent monolayer can be dynamic in nature and may still act at the moment of nucleation, not for DBA molecules adsorbing on top of the solvent layer, but for DBA molecules adsorbing in such a solvent layer, potentially making a transient DBA–solvent complex (C3.3.) – *i.e.*, a process that would be enthalpically-driven, or by replacing solvent molecules and expelling those to the supernatant solution in an enantiospecific way (C2) – *i.e.*, a process that would be entropically-driven. It is also possible that more than one mechanism is at play. While it is quite reasonable to assume that hydrogen bonding interactions involving the hydroxy group of the solvent and the oxygen atoms of the DBA alkoxy chains are at play, the simple fact that a DBA molecule with alkyl groups instead of alkoxy groups also undergoes chiral induction, though less efficiently, hints to the fact that several mechanisms act simultaneously or in a compound-specific fashion.

All these experiments were performed at room temperature. However, annealing to 60 °C leads, for many DBA derivatives, to quite different results. The high density phase, with no nanopores present, is favored for DBA-OC10, DBA-OC11 and DBA-OC14 (Fig. S5†). Only in the case of DBA-OC10, a honeycomb lattice is still observed. The same chiral induction, though, is observed for DBA-OC10, regardless of the chiral 2-alcohol used, and upon annealing, chiral induction is more efficient, *i.e.* more domains of a given handedness are observed compared to the corresponding layers at room temperature. So, the outcome of the monolayer formation at room temperature is, for a number of systems, probably not the thermodynamically most stable phase. However, the fact that the type of chiral induction (CW or CCW) remains upon annealing supports the fact that our discussion on the induction mechanism at room temperature is still valid.

Molecular modeling simulations

In order to gain insight into the chiral induction mechanisms involving interactions between DBA molecules and solvent molecules at the substrate surface, we have performed force-field-based simulations on model systems related to processes C3.1 (stabilization of the nanowells) and C3.3 (formation of a DBA–solvent complex at the nanopore nucleation stage).

All molecular modeling simulations have been conducted with the Biovia Materials Studio 7 molecular modeling package using the DREIDING force field. The choice of the force field was motivated by the fact that DREIDING comprises a specific H-bonding interaction term, which is essential for this study. The atomic charges were calculated for the DBA and solvent isolated molecules by using the Gasteiger method as implemented in Materials Studio. The graphite surface was modeled by using two flat, infinite (periodic), stacked graphene layers, which are kept rigid. No charges are assigned to the graphite carbon atoms; hence, the surface–molecule interactions are only van der Waals interactions. All molecular dynamics (MD) simulations were conducted in the NVT ensemble at a temperature of 250 K, kept constant by using the Andersen thermostat with a collision rate of 1. The time step used for the MD runs is 1 fs.

To test mechanism C3.1, we have modeled how (*S*)-2-octanol adsorbs in the nanopores of DBA-OC12. Those simulations examined the arrangement of the solvent molecules in the pore in contact with a solvent supernatant layer. In the data analysis, particular attention was paid to the orientation of the solvent molecules with respect to the pore rims and H-bonding interactions with the DBA molecules. Despite very extensive simulations involving a large number of different initial conditions, we have not found any indication for the stabilization of one type of nanopore (CW or CCW) by one solvent enantiomer. Concerning mechanism C3.3, most likely a (transient) complex can be formed between the oxygen atoms of the adsorbed DBA alkoxy chains and the hydroxyl group of the enantiopure solvent *via* stereospecific hydrogen-bonding.

In order to examine the relevance of the formation of such a complex to the DBA interdigitation process and the subsequent pore nucleation, we have simplified the system used in the simulations by making the following assumptions, based on the experimental findings:

• We assume that the DBA molecules interdigitate at the solvent/surface interface and do not aggregate in the liquid phase.

• We assume that the chiral induction occurs at the nucleation stage of the monolayer formation, where the presence of chiral solvent molecules will favor only one type of DBA interdigitation, *i.e.* + or −, leading to the formation of only one type of monolayer.

An iterative MD/quench scheme has been devised in order to locate preferential interaction sites between a DBA and a chiral solvent molecule. For this scheme, an initial 100 ps MD run is performed, and a structure for the system is extracted from the trajectory every 500th step. Those structures are then optimized and the lowest-energy optimized structure is used as a starting point for a new 100 ps-long MD run. The cycle is iterated until the energy of the optimized structures does not decrease further.

A second iterative scheme has been devised to obtain the energy profile for the interdigitation process between two DBA molecules, in the presence of chiral solvent molecules. This scheme consisted of an initial alignment between two DBA molecules, each one carrying a single chiral solvent molecule in its preferential adsorption site, which was previously obtained from the MD/quench scheme. Each DBA molecule is considered as a rigid object and the DBA–solvent complex is considered to be a group. At each iteration of the process, one group is moved towards the other by 1 Å, and the geometry of the system is optimized and then used as input for the next iteration.

DBA-OC6–2-alcohol complexation: 1-octanol, (*S*)-2-octanol, (*R*)-2-octanol

First we test our assumption that molecules of 1-octanol, (*S*)-2-octanol and (*R*)-2-octanol have preferred interaction sites with the DBA molecules. For this, we consider a single DBA-OC6 molecule adsorbed on graphite, and we observe how a solvent molecule, also adsorbed on the surface, interacts with the DBA core in two different places, called docking site A, DSA, and docking site B, DSB (Fig. 5).

In DSA, the solvent molecule can form hydrogen bonds with the oxygen atoms in the alkoxy groups, while in DSB an OH–π interaction can take place between the hydrogen of the hydroxyl group and the triple bond of the alkyne group of the annulene ring. When in DSA, the polar head of the solvent molecule strongly tends to interact *via* hydrogen bonding with the two oxygen atoms, labeled interaction site 1, IS1, and interaction site 2, IS2, of the DBA molecule. The space in DSA is large enough to allow for some conformational freedom of the solvent molecules. In particular, the alkyl part of the solvent molecule can align with one alkyl chain of the DBA molecule, either the one containing IS1 or the one containing IS2 (see Fig. 5C and D). Because of the presence of the stereogenic center in the solvent molecule, those two situations, which are called IS1 and IS2 below, are not equivalent.

To better assess the consequences of the IS1 *vs.* IS2 interaction, we modeled three systems, where a single molecule of 1-octanol, (*S*)-2-octanol or (*R*)-2-octanol is placed in the DSA site of a DBA-OC6 molecule, which is adsorbed on the graphite surface (Table 2). The MD/quench iterative scheme is used to find the most stable geometry for those systems when the solvent molecule is made to interact either with IS1 or IS2.

In the system with 1-octanol, the solvent molecule does not show a preferential interaction with one particular site in DSA, as the energies of the systems in the two situations differ only by about 0.7 kcal mol^{-1}, *i.e.*, about the thermal energy k_bT available at room temperature.

In contrast, when a chiral solvent molecule is considered, there is a clear indication of a preferential interaction site within DSA: (*S*)-2-octanol prefers to interact with IS1, while (*R*)-2-octanol prefers to interact with IS2. In both cases, the energy difference between the molecule interacting with the preferred site with respect to the other one is in the order of 3 kcal mol^{-1}, *i.e.*, 5 times k_bT. The difference in stability stems from the configuration around the polar head of the molecule.

Because the solvent molecule aligns with one of the alkoxy groups in the DSA cavity, the symmetry of DSA is broken and one kind of interdigitation may be hindered by the presence of the solvent molecule. In particular, when the solvent molecules interact with IS1, "+" interdigitation can be favored, while "−" interdigitation can be favored when the solvent molecules are interacting with IS2 (Fig. 6). This could be the reason explaining why a chiral solvent can promote chiral induction in the assembly of DBA molecules.

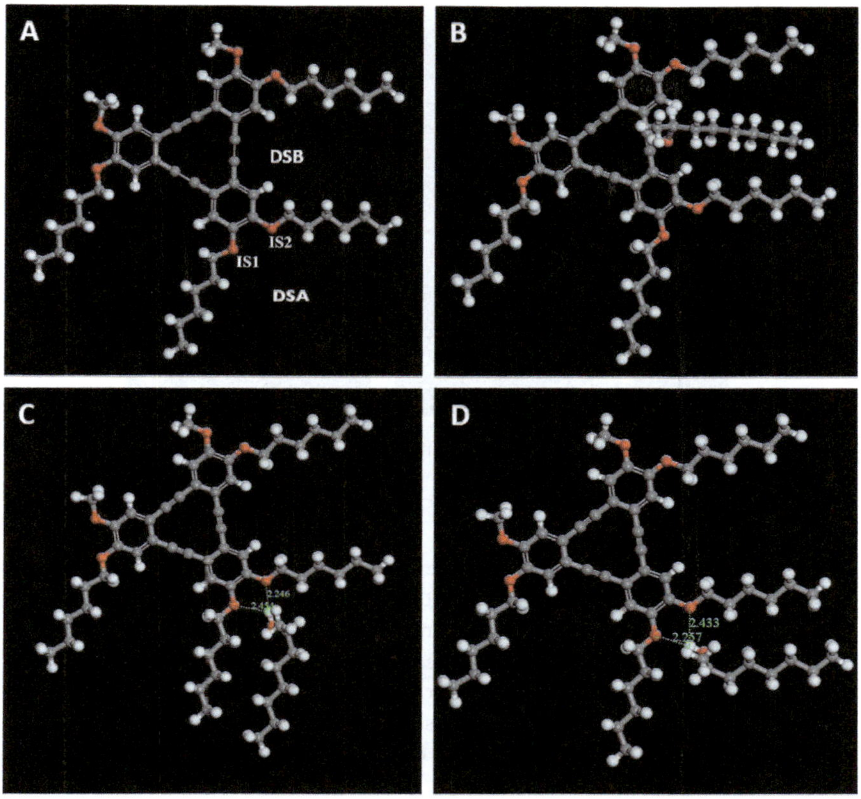

Fig. 5 (A) The molecule used in this work as a model for DBA-OC6 with indication of the different possible docking and interaction sites. (B) (*S*)-2-Octanol interacting in DSB, (C) (*S*)-2-octanol interacting in DSA with IS1 and (D) (*R*)-2-octanol interacting in DSA with IS2.

Table 2 Energies for the most stable structures obtained from the MD/quench iterative scheme for three different solvents in the DSA site of a DBA-OC6 molecule

Solvent	E_{IS1} (kcal mol^{-1})	E_{IS2} (kcal mol^{-1})	$\Delta E_{IS1/IS2}$ (kcal mol^{-1})
(*S*)-2-Octanol	−14.4	−11.4	−3.00
(*R*)-2-Octanol	−11.2	−14.1	2.90
1-Octanol	−10.4	−9.7	−0.70

In contrast, the smaller available space in DSB for the solvent molecule forces it to stay in the middle of the site and to align to both the alkoxy groups of the DBA molecule (Fig. 5B). So interaction at DSB is not expected to play a role in favoring "+" or "−" interdigitation.

The impact of the length of the alkoxy chains

It is experimentally known that at room temperature for a chiral solvent such as (*S*) or (*R*)-2-octanol, the chiral induction on DBA-OC6 self-assembly is 100%, while

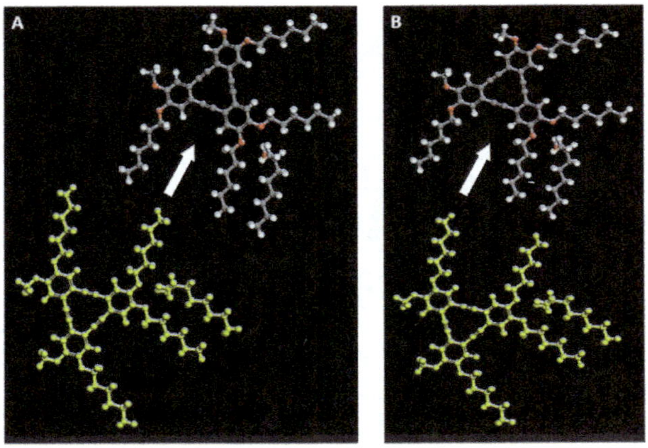

Fig. 6 Favorable (A) *versus* unfavorable (B) interdigitation for DBA-OC6 in the presence of a chiral solvent molecule in DSA.

for DBA-OC12 self-assembly, the degree of chiral induction is lowered to about 80%. The results from the simulations to determine the preferential interaction site for (*S*)-2-octanol and (*R*)-2-octanol interacting with DBA-OC12 show that the solvent selectivity of the interaction site drops from 3 kcal mol^{-1} in the case of DBA-OC6 to 2 kcal mol^{-1}, *i.e.*, to about $3k_bT$, for DBA-OC12. This is consistent with the experimental data. In both cases, (*S*)-2-octanol prefers to interact with IS1. The modeling of the influence on the chiral induction of the even or odd number of carbon atoms in the alkoxy chain is presented in the ESI (S6).†

The impact of the length of the chiral solvent molecule

Another aspect to consider, when trying to determine how much the selectivity of the solvent molecule to interact with IS1 or IS2 can hinder a particular type of interdigitation, is the length of the solvent molecule with respect to the length of the interdigitating alkoxy groups. As shown in Fig. 5, the length of the –OC6 alkoxy groups is comparable with that of a (*S*)-2-octanol molecule interacting with IS1.

This suggests that (*S*)-2-octanol favors "+" interdigitation whereas (*R*)-2-octanol, which preferentially interacts with IS2, favors "−" interdigitation, due to steric hindrance. When a (*S*)-2-octanol molecule is interacting with a DBA-OC12 molecule in IS1, the length of the alkoxy group will be double that of the alcohol. This suggests that, while "+" interdigitation is still favored, DBA molecules could begin to interdigitate, before being hindered by the presence of the solvent molecule (see Fig. 7, left panel). This hypothesis, together with the reduced selectivity for a specific interaction site, could explain why the chiral induction observed for the DBA-OC12 monolayers is less marked than for the DBA-OC6 monolayers. To test this hypothesis, we computed an interdigitation profile between two interdigitating DBA-OC12 molecules, as explained in the methodology section. Fig. 7 shows the model built to compute the interdigitation profile, forcing molecule DBA2 (at the bottom of the image) to move towards

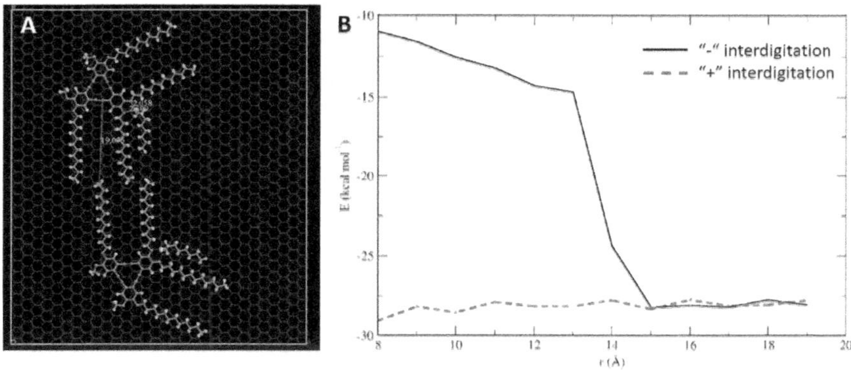

Fig. 7 (A) Model used to compute the interdigitation profile for DBA-OC12 in the presence of (*S*)-2-octanol ("−" interdigitation is shown). DBA1 is the molecule at the top of the image. (B) Interdigitation profiles for DBA-OC12 in the presence of (*S*)-2-octanol.

molecule DBA1 (on the top of the image) from a distance $r = 19$ Å (no interdigitation) down to $r = 8$ Å (maximum amount of interdigitation tested).

As expected, at the beginning of the interdigitation, there is no selectivity between the "+" and "−" interdigitation: the two energy profiles are superimposed. But from a certain degree of interdigitation ($r = 15$ Å), when the incoming alkoxy group starts to feel the presence of the solvent molecule, the "+" interdigitation becomes increasingly more favorable, with respect to the "−" counterpart: by up to 18 kcal mol^{-1} for the maximum amount of interdigitation tested ($r = 8$ Å).

It is important to note that at the end of the interdigitation process, the solvent molecule is expelled from the DSA site, suggesting that the strength of the interaction of the solvent molecule with the DBA is not sufficient to permanently keep the DBA–solvent complex together. This, in fact, would prevent DBA molecules from fully interdigitating to build the observed monolayers.

Conclusions

In conclusion, we have shown that chiral solvents can induce homochirality in surface-confined low-density networks. Moreover, odd–even chain length behavior was observed for both the chiral solvent and the DBA building block. An attempt to unravel the mechanism behind the chiral induction has led to the proposition of various mechanisms that could be responsible for the transfer of chiral information from a chiral solvent to a two-dimensional network. Albeit not fully conclusive, out of all the proposed mechanisms two are very compelling. These two mechanisms rely on the fact that solvent–substrate interactions are key. This means that the solvent molecules could form a DBA–solvent precursor complex that biases a specific handedness of the porous network, or that the solvent molecules could form a (metastable) layer from which solvent molecules have to be released upon DBA adsorption. Advanced modeling protocols provided valuable information on the DBA–solvent precursor induction mechanism, by identifying the enantiospecific DBA–solvent interaction favoring the occurrence of a biased interdigitation of the DBA molecules and by rationalizing the influence of the chain length. While chiral induction can be explained on the basis of thermodynamic considerations, kinetic aspects are also at play and may have an impact on the extent of chiral induction at a given temperature.

Acknowledgements

This work was supported by the Fund of Scientific Research – Flanders (FWO), Internal Funds of KU Leuven, the Belgian Federal Science Policy Office (IAP-7/05), and JSPS KAKENHI (15H02164). Research in Mons is also supported by the Région Wallonne OPTI2MAT program and FNRS-FRFC (CECI HPC facility). This research has also received funding from the European Research Council under the European Union's Seventh Framework Programme (FP7/2007–2013)/ERC grant agreement no. 340324. I. D. is grateful to the Agency for Innovation by Science and Technology in Flanders (IWT).

References

1 L. Perez-Garcia and D. B. Amabilino, *Chem. Soc. Rev.*, 2007, **36**, 941.
2 R. M. Hazen and D. S. Sholl, *Nat. Mater.*, 2003, **2**, 367.
3 J. D. Horvath, A. Koritnik, P. Kamakoti, D. S. Sholl and A. J. Gellman, *J. Am. Chem. Soc.*, 2004, **126**, 14988.
4 D. S. Sholl and A. J. Gellman, *AIChE J.*, 2009, **55**, 2484.
5 A. J. Baiker, *J. Mol. Catal. A: Chem.*, 1997, **115**, 473.
6 G. Y. Liu, S. Xu and Y. L. Qian, *Acc. Chem. Res.*, 2000, **33**, 457.
7 S. De Feyter, A. Gesquiere, P. C. M. Grim, F. C. De Schryver, S. Valiyaveettil, C. Meiners, M. Sieffert and K. Mullen, *Langmuir*, 1999, **15**, 2817.
8 J. Elemans, I. De Cat, H. Xu and S. De Feyter, *Chem. Soc. Rev.*, 2009, **38**, 722.
9 J. Zhang, A. Gesquiere, M. Sieffert, M. Klapper, K. Mullen, F. C. De Schryver and S. De Feyter, *Nano Lett.*, 2005, **5**, 1395.
10 A. Miura, P. Jonkheijm, S. De Feyter, A. Schenning, E. W. Meijer and F. C. De Schryver, *Small*, 2005, **1**, 131.
11 K. Tahara, H. Yamaga, E. Ghijsens, K. Inukai, J. Adisoejoso, M. O. Blunt, S. De Feyter and Y. Tobe, *Nat. Chem.*, 2011, **3**, 714.
12 A. Nuermaimaiti, C. Bombis, M. M. Knudsen, J. R. Cramer, E. Laegsgaard, F. Besenbacher, K. V. Gothelf and T. R. Linderoth, *ACS Nano*, 2014, **8**, 8074.
13 M. Parschau, S. Romer and K.-H. Ernst, *J. Am. Chem. Soc.*, 2004, **126**, 15398.
14 S.-Y. Li, T. Chen, J.-Y. Yue, L. Wang, H.-J. Yan, D. Wang and L.-J. Wan, *Chem. Commun.*, 2016, **52**, 12088.
15 T. Chen, S.-Y. Li, D. Wang and L.-J. Wan, *Angew. Chem., Int. Ed.*, 2015, **4**, 4309.
16 Y. Fang, E. Ghijsens, O. Ivasenko, H. Cao, A. Noguchi, K. S. Mali, K. Tahara, Y. Tobe and S. De Feyter, *Nat. Chem.*, 2016, **8**, 711.
17 H. Cao, I. Destoop, K. Tahara, Y. Tobe, K. S. Mali and S. De Feyter, *J. Phys. Chem. C*, 2016, **120**(31), 1744.
18 R. Fasel, M. Parschau and K.-H. Ernst, *Nature*, 2006, **439**, 449.
19 S. Haq, N. Liu, V. Humblot, A. P. J. Jansen and R. Raval, *Nat. Chem.*, 2009, **1**, 409.
20 T. Chen, W. H. Yang, D. Wang and L. J. Wan, *Nat. Commun.*, 2013, **4**, 8.
21 S.-Y. Li, T. Chen, L. Wang, D. Wang and L.-J. Wan, *Nanoscale*, 2016, **8**, 17861.

22 S.-Y. Li, T. Chen, L. Wang, B. Sun, D. Wang and L.-J. Wan, *Langmuir*, 2016, **32**, 6830.

23 M. M. Green, M. P. Reidy, R. J. Johnson, G. Darling, D. J. Oleary and G. Willson, *J. Am. Chem. Soc.*, 1989, **111**, 6452.

24 S. Lifson, C. E. Felder and M. M. Green, *Macromolecules*, 1992, **25**, 4142.

25 I. Destoop, H. Xu, C. Oliveras-Gonzalez, E. Ghijsens, D. B. Amabilino and S. De Feyter, *Chem. Commun.*, 2013, **49**, 7477.

26 I. Destoop, E. Ghijsens, K. Katayama, K. Tahara, K. S. Mali, Y. Tobe and S. De Feyter, *J. Am. Chem. Soc.*, 2012, **134**, 19568.

27 N. Katsonis, H. Xu, R. M. Haak, T. Kudernac, Z. Tomovic, S. George, M. Van der Auweraer, A. P. H. J. Schenning, E. W. Meijer, B. L. Feringa and S. De Feyter, *Angew. Chem., Int. Ed.*, 2008, **47**, 4997.

28 M. Linares, P. Iavicoli, K. Psychogyiopoulou, D. Beljonne, S. De Feyter, D. B. Amabilino and R. Lazzaroni, *Langmuir*, 2008, **24**, 9566.

29 K. Tahara, S. B. Lei, J. Adisoejoso, S. De Feyter and Y. Tobe, *Chem. Commun.*, 2010, **46**, 8507.

30 K. Tahara, S. Furukawa, H. Uji-i, T. Uchino, T. Ichikawa, J. Zhang, W. Mamdouh, M. Sonoda, F. C. De Schryver, S. De Feyter and Y. Tobe, *J. Am. Chem. Soc.*, 2006, **128**, 16613.

31 S. B. Lei, K. Tahara, F. C. De Schryver, M. Van der Auweraer, Y. Tobe and S. De Feyter, *Angew. Chem., Int. Ed.*, 2008, **47**, 2964.

32 M. O. Blunt, J. Adisoejoso, K. Tahara, K. Katayama, M. Van der Auweraer, Y. Tobe and S. De Feyter, *J. Am. Chem. Soc.*, 2013, **135**, 12068.

33 J. Adisoejoso, K. Tahara, S. Okuhata, S. B. Lei, Y. Tobe and S. De Feyter, *Angew. Chem., Int. Ed.*, 2009, **48**, 7353.

34 F. Masini, N. Kalashnyk, M. M. Knudsen, J. R. Cramer, E. Laegsgaard, F. Besenbacher, K. V. Gothelf and T. R. Linderoth, *J. Am. Chem. Soc.*, 2011, **133**, 13910.

35 S. Lei, M. Surin, K. Tahara, J. Adisoejoso, R. Lazzaroni, Y. Tobe and S. De Feyter, *Nano Lett.*, 2008, **8**, 2541.

36 G. Wang, S. Lei, S. De Feyter, R. Feldman, J. E. Parker and S. M. Clarke, *Langmuir*, 2008, **24**, 2501.

37 M. A. Castro, S. M. Clarke, A. Inaba, T. Arnold and R. K. Thomas, *J. Phys. Chem. B*, 1998, **102**, 10528.

38 N. Elbel, W. Roth, E. Gunther and H. Vonseggern, *Surf. Sci.*, 1994, **303**, 424.

39 A. P. Gunning, A. R. Kirby, X. Mallard and V. J. Morris, *J. Chem. Soc., Faraday Trans.*, 1994, **90**, 2551.

40 K. Kim, K. E. Plass and A. J. Matzger, *J. Am. Chem. Soc.*, 2005, **127**, 4879.

41 L. Piot, A. Marchenko, J. S. Wu, K. Mullen and D. Fichou, *J. Am. Chem. Soc.*, 2005, **127**, 16245.

42 B. Xu, S. X. Yin, C. Wang, X. H. Qiu, Q. D. Zeng and C. L. Bai, *J. Phys. Chem. B*, 2000, **104**, 10502.

43 A. Bellec, C. Arrigoni, L. Douillard, C. Fiorini-Debuisschert, F. Mathevet, D. Kreher, A.-J. Attias and F. Charra, *Nanotechnology*, 2014, **25**, 435604.

44 K. Tahara, K. Kaneko, K. Katayama, S. Itano, C. H. Nguyen, D. D. D. Amorim, S. De Feyter and Y. Tobe, *Langmuir*, 2015, **31**, 7032.

Faraday Discussions

PAPER

Peptide binding to metal oxide nanoparticles

S. P. Schwaminger, S. A. Blank-Shim, I. Scheifele, P. Fraga-García and S. Berensmeier

Received 13th March 2017, Accepted 28th March 2017
DOI: 10.1039/c7fd00105c

Magnetic metal oxide nanoparticles demonstrate great applicability in several fields such as biotechnology, medicine and catalysis. A stable, magnetic and low-cost material, nanoscale magnetite, is an interesting adsorbent for protein purification. Downstream processing can account for up to 80% of the total production costs in biotechnological production. As such, the development of new innovative separation methods can be regarded as highly profitable. While short peptide sequences can be used as specific affinity tags for functionalised adsorber surfaces, they need expensive affinity ligands on the particle surface for adsorption. In order to identify peptide tags for several non-functionalised inorganic surfaces, different binding conditions to iron oxide nanoparticles are evaluated. Therefore, magnetite nanoparticles in a range of 5–20 nm were synthesised with a co-precipitation method. Zeta potential measurements indicated an amphiphilic surface with an isoelectric point in the neutral pH region. Glutamic acid-based homo-peptides were used as affinity peptides for the magnetite nanoparticles. We demonstrate a dependence of the binding affinity of the peptides on pH and buffer ions in two different experimental set-ups. The nature of surface coordination for glutamic acid-based peptides can be demonstrated with different spectroscopic approaches such as infrared spectroscopy (IR), Raman spectroscopy and circular dichroism spectroscopy (CD). We want to emphasise the importance of physicochemical properties such as surface energy, polarity, morphology and charge. These parameters, which are dependent on the environmental conditions, play a crucial role in peptide interactions with iron oxide surfaces. The understanding of the adsorption of simple biomolecules on nanoscale metal oxide surfaces also represents the key to the even more complex interactions of proteins at the bio-nano interface. From the identification of interaction patterns and an understanding of the adsorption of these peptides, the up-scaling to tagged model proteins facilitates the possibility of an industrial magnetic separation process and might therefore reduce time and costs in purification processes.

Bioseparation Engineering Group, Department of Mechanical Engineering, Technical University of Munich, Boltzmannstraße 15, Garching, 85748, Germany. E-mail: S.Berensmeier@tum.de

Introduction

The adsorption of organic molecules and biomolecules on the solid/liquid interface of inorganic surfaces is a common phenomenon in nature and is said to have influenced the origin of life.[1] Many industries make use of this thermal/chemical process – including wastewater treatment,[2,3] mining,[4] food production[5] and the purification of pharmaceuticals.[6] Usually the formation of covalent bonds and hence a change of the adsorbate's chemical state can be differentiated from the physisorption, which is based on weaker binding mechanisms such as hydrogen bonding, dipole/dipole, induced dipole/dipole or induced dipole/induced dipole interactions.[7] The ability to design material surfaces which can prevent, facilitate or, generally, control the adsorption of biomolecules from an aqueous phase can be applied in multiple sectors of industry and research.[8–10] These different possibilities of application account for different surface requirements. A large specific surface area and a low-cost production and process operation are always recommended for materials employed in bioseparation processes,[11] while the following effects can be used to facilitate different adsorbent characteristics: the static binding capacity can be used as a descriptor for the control of adsorption processes. In this case, the equilibrium conditions of a system are utilised to control the process and thus the adsorption of the component with the lowest desorption rate is favoured. The thermodynamic binding enthalpy can be used as a tool to gain process control over displacement processes as well, especially in the case of an irreversible chemisorption. Furthermore, diffusion kinetics in porous or agglomerated systems can be used as a limitation criterion for the adsorption.[12] Typically, surface design is comprised of the synthesis of a novel adsorbent or the coating with a novel adsorbent, a change of the steric or porosity parameters by adding ligands or scaling the material in order to derive new properties.[9,10]

Instead of developing expensive new materials, our approach aims to reintroduce a familiar acquaintance, bare magnetite nanoparticles, as an innovative low-cost adsorbent for downstream processing of biomolecules. Magnetite nanoparticles possess several favourable characteristics to be used in industrial processes.[13,14] The co-precipitation of iron chlorides, as abundant resources, with sodium hydroxide without cooling represents a scalable minimum cost process with a high reproducibility.[15,16] The thermal and mechanical stability, a high specific surface area of $\sim$100 $m^2\ g^{-1}$ containing amphiphilic surface hydroxyl groups and especially the nanoparticles' superparamagnetic behaviour, which facilitates a sustainable recovery, are beneficial properties of these nanoscale materials.[17] Thus, the application of bare magnetite, and even of costly modified magnetite nanoparticles, ranges from catalysis, as a catalyst or carrier material,[18–20] to the use in media and energy storage,[21,22] medical implementations such as hyperthermia treatment,[23,24] contrast agents[16,25] or carrier materials in drug delivery,[25,26] and the use as reversible adsorbents in biomolecule purification.[27,28] For downstream processing in biotechnology, biomolecules containing peptide sequences with a specific binding affinity to the adsorbent, such as the His- or the Strep-tag, can be used for the separation of the target molecule from a complex suspension.[29] In the case of the His-tag,[30] the reversible covalent attachment of multiple histidine molecules to divalent ions such as Ni^{2+} is used as the binding mechanism, while the Strep-tag uses the high affinity of different noncovalent interactions between the peptide sequence Trp–Ser–His–Pro–Gln–Phe–Glu–Lys and streptavidin.[31]

Our main issue is to identify a peptide sequence specifically binding bare magnetite nanoparticles which can be easily fused to recombinant proteins like existing affinity tags.[31] However, the interaction of proteins at the solid/liquid interface of nanoscale surfaces is not yet wholly understood and difficult to analyse due to the complexity of these systems.[8,32–34] Which effects do the surface defects and chemistry of nanoparticles and the solute conditions have on the protein conformation and organisation on the surface? How are the colloidal stability and entropic effects of particle aggregation affected by the adsorption of peptides? Many different strategies exist that try to respond to these arising questions but they are often limited. The analytical methods are almost as diverse as the investigated interactions. From prediction simulation to screening methods, from microscopic to spectroscopic approaches, from kinetic studies to thermodynamic parameters, manifold techniques can be used to characterise molecular surface interactions.[32,35]

Here, we try to identify peptide sequences able to act as affinity tags towards bare magnetite nanoparticles and analyse the binding with a multi-analytical approach. To achieve this, we firstly screened for possible peptide sequences in different environments with homo-peptides anchored to a cellulose membrane. The best candidate, a glutamic acid octamer, was further investigated towards the binding mechanism and the binding affinity to magnetite nanoparticles. Therefore, the impact of adsorbed peptides on the electrophoretic behaviour of the suspended nanoparticles is investigated with zeta-potential measurements. Furthermore, the influence of the bound biomolecules on the magnetic properties and chemical oxidation state are verified with magnetometry and X-ray photoelectron spectroscopy (XPS), respectively. The binding mechanism and the conformation of the peptide are analysed with attenuated total reflection infrared spectroscopy (ATR-IR), circular dichroism spectroscopy (CD) and Raman spectroscopy. Additionally, we demonstrate the dependence of the binding on different pH conditions and in different environments and discuss the ability to control the interaction. Finally, we verify the transferability of the peptide binding to a full green fluorescent protein (GFP) containing a similar terminal peptide sequence.

Results and discussion

Particle synthesis and characterisation

Fig. 1 illustrates magnetite nanoparticles synthesised with an alkaline co-precipitation of ferrous and ferric chloride as described by Roth *et al.*[15] From the transmission electron microscopy (TEM) particle count, an average diameter of around 12 nm and a narrow size distribution can be determined for these nanoparticles, while the analysis of the 311 reflection as the most pronounced signal in the X-ray diffractogram (XRD) shown in Fig. 1 by use of the Scherrer equation yields a particle size of around 8.3 nm. The diffraction pattern of the synthesised nanoparticles is in good agreement with pure magnetite and does not show any indications for oxidation.[17] This can be verified *via* Raman spectroscopy of the nanoparticles. A band at around 670 cm^{-1} corresponding to an A_{1g} mode of

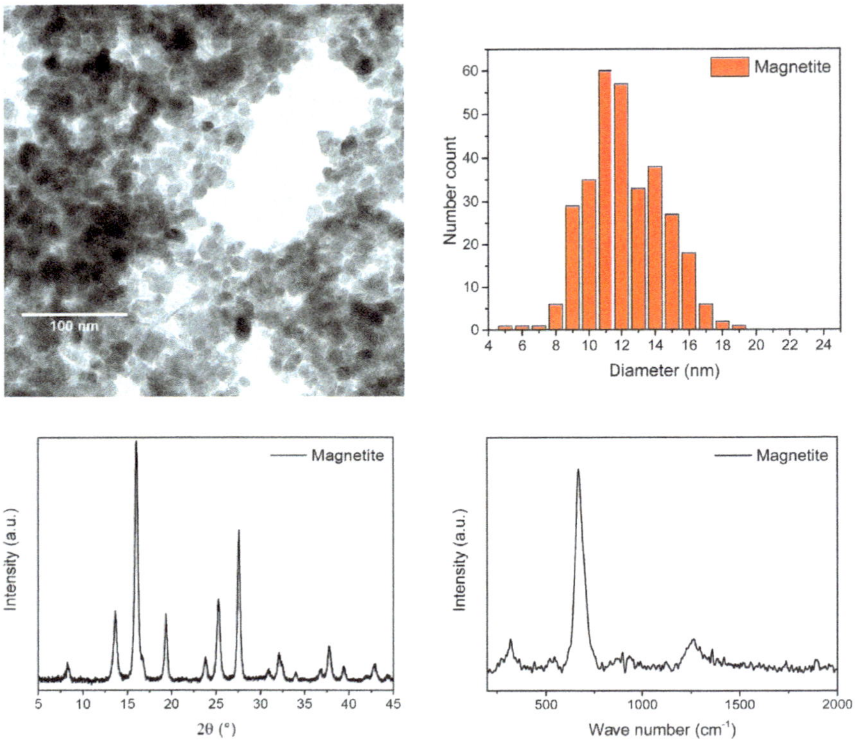

Fig. 1 TEM image of the as-synthesised nanoparticles (left top) and the corresponding particle size distribution (right top). XRD reflections (left bottom) and Raman spectrum (right bottom) of the freeze-dried magnetite nanoparticles.

Fe–O which is typical for magnetite can be identified, while the lack of a peak around 720 cm^{-1} excludes the presence of maghemite in the spectrum.[17] The specific surface area for these nanoparticles is determined to be 89 $m^2 g^{-1}$ with nitrogen adsorption isotherms at 77.15 K following the BET method.

Screening for peptides

The binding of these nanoparticles to homo-peptides at pH 6 was screened with a cellulose-based array where 5–10 nM hexa-peptides were anchored on spots of 2 mm in diameter.[36] The change of the membrane colour due to the adsorption of black magnetite nanoparticles was analysed and plotted as a binding score in Fig. 2. A negative binding score means that the spots are brighter than the background and hence no binding of magnetite can be assumed. Both hexamers which are negatively charged at pH 6, Glu_6 and Asp_6, demonstrated by far the highest binding score in this assay. The positively charged peptides (His_6, Lys_6 and Arg_6) all seemed to bind in a lower amount as well and exhibited a similar binding score. The nonpolar and polar amino acids behaved differently. While Ala_6, Leu_6, Met_6, Asn_6, Pro_6, Gln_6 and Thr_6 showed no binding at all, Cys_6, Gly_6, Ser_6 and Val_6 demonstrated a moderate tendency to bind magnetite nanoparticles. Interestingly, the aromatic peptides Phe_6, Trp_6 and Tyr_6 showed a significant amount of bound magnetite. From these results an electrostatic

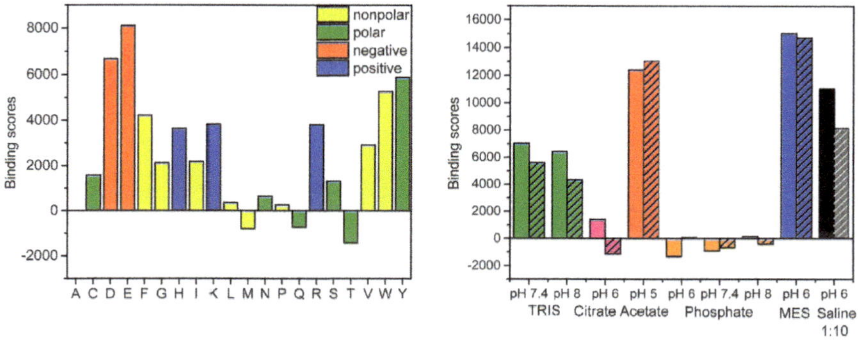

Fig. 2 Binding scores of magnetic nanoparticles (0.4 g L^{-1}) in saline (1 : 10) on different hexa-peptides anchored on a cellulose membrane at pH 6 (left) and binding scores of nanoparticles in different buffers on octa- (full colour) and hexa- (hatched) glutamic acid peptides (right).

component in the binding of peptides to magnetite can be assumed, taking into account that the nanoparticles are slightly positively charged at pH 6 as indicated by the zeta potential shown in Fig. 5. Furthermore, hydrophobic interactions seem to play a significant role in peptide adsorption towards an amphiphilic surface such as magnetite nanoparticles, which can be proven by this experiment. The amount of bound peptides on magnetite nanoparticles at pH 6 can be correlated to the adsorption of the respective single amino acid, as similar results were obtained here.[37,38] As glutamic acid demonstrated the highest binding score at pH 6 in saline (1 : 10), we decided to further investigate the influence of the buffer on the binding behaviour. In Fig. 2, the effects of varying the buffer system on the binding of magnetite nanoparticles to the peptides Glu_6 and Glu_8 were tested. Both peptides behaved similarly during all experiments, but the octamer demonstrated a slightly higher amount of bound nanoparticles under most conditions. Besides the unbuffered environment, we observed a high binding score for the glutamic acid-based peptides in the buffer Tris at pH 7.4 and 8 and an even further elevated score in MES at pH 6 and acetate at pH 5. On the other hand, no adsorption can be proven for citrate and phosphate buffer solutions. The buffer molecules seem to be loosely bound to the surface or the adjacent surrounding Helmholtz layer and affect the binding to the Glu_8 peptides. As reported in the literature, phosphate, nitrate and citrate have a high affinity to iron oxide surfaces and therefore shift the amphiphilic surface charge in a negative direction.[13,17,39,40] The Tris molecule contains an amine group which can be positively charged and therefore facilitates the binding of negatively charged peptides even at higher pH. However, why MES buffer and acetate favour binding compared to saline and Tris cannot be answered. As both peptides behave completely similarly, the octamer was chosen for further experiments as it demonstrates a better detection limit for photometric UV spectroscopy.

Effect of peptides on the adsorbent

Both particles demonstrate similar magnetisation curves at 300 K as shown in Fig. 3. No hysteresis or remanence can be observed for both materials, while

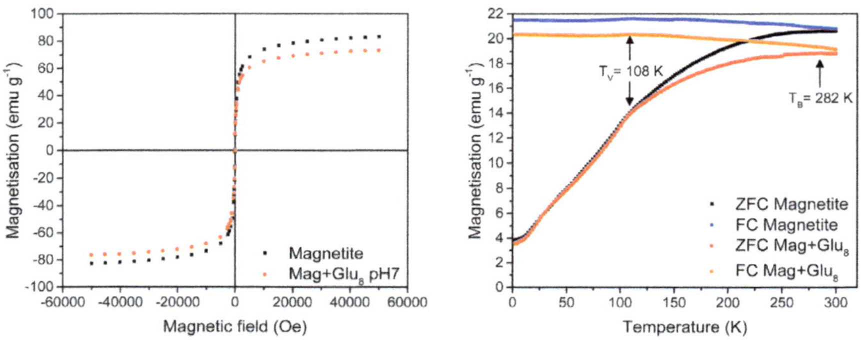

Fig. 3 Magnetisation loops of freeze-dried magnetite nanoparticles and particles with adsorbed peptides at 300 K (left). ZFC and FC curves of the same materials between 1.8 and 300 K with an applied field of 200 Oe.

saturation is not completely reached at 50 000 Oe. The decrease in maximal magnetisation from 83 emu g^{-1} for bare particles to 75 emu g^{-1} can be explained by the adsorbed mass of the peptide and is in good agreement with adsorption isotherms at pH 7, where a binding capacity of $\sim$0.1 g g^{-1} is reached as shown in Fig. 9. This behaviour means that the magnetic properties, which are important for a separation process, are not significantly affected by the adsorption of the peptide and the magnetite particles are stable against oxidation. The zero field cooling (ZFC) and field cooling (FC) curves shown in Fig. 3 do not differ significantly. In particular, the kink at around 108 K is visible for both ZFC curves, while a plateau for each FC curve can be observed at this temperature. This can be interpreted as the Verwey transition, which is at around 125 K (T_V) for bulk magnetite and marks the transition from a metal to an insulator material, and is therefore connected to abrupt changes in resistivity, heat capacity, magnetization and coercivity of the particles.[41–43] This transition does not occur in the oxidised iron oxide maghemite or in nanoscale magnetite nanoparticles with oxidised surface layers, and thus indicates pure magnetite nanoparticles, which do not oxidise upon adsorption of the Glu$_8$ peptide.[43] However, we cannot exclude a surface dead layer, as this measurement is not surface sensitive enough. The kinks are less pronounced than in the bulk material and the Verwey temperature is shifted to lower temperatures, which has also been reported for spherical magnetite nanoparticles.[44] The existence of a maximum in the ZFC curve can be connected to the blocking temperature (T_B) of magnetite nanoparticles. While the blocking temperature was not reached upon heating to 300 K for bare magnetite nanoparticles, the maximum in magnetisation for an applied field of 200 Oe was found to be at around 282 K for the material containing adsorbed Glu$_8$. This behaviour can be explained by a less ordered or aggregated state of the freeze-dried nanoparticles after adsorption of Glu$_8$, as the thermal relaxation (Brownian motion) is increased compared to that of bare nanoparticles. While individual nanoparticles change their alignment in a magnetic field with increasing temperatures fast, aggregated particles cannot change their position.[44] In this case this effect is very moderate, while stabilising materials, such as tetramethylammonium hydroxide, oleic acid or silica, tend to affect the blocking temperature in a more pronounced fashion.[44–46] For coatings, the blocking temperature and thus the paramagnetic behaviour of nanoparticles can be decreased to below 100 K.[45] Hence, the repulsion due to adsorbed Glu_8 and the thickness of the adsorbed layer either has to be significantly smaller than for coatings or not affect the agglomeration at all.

As the surface charge changes upon adsorption, we further verified the properties of magnetite nanoparticles *via* magnetometry and XPS before and after the adsorption of Glu_8 at pH 7. Fig. 4 illustrates the chemical composition of the particle surface in the region of the O1s, C1s, N1s and $Fe2p^{3/2}$ orbitals, respectively. In the O1s spectrum, a component corresponding to carboxylate groups occurs with adsorption of the peptide Glu_8 on the magnetite surface. It is also possible that adsorbed water and hydroxyl groups are replaced by adsorbed peptides, as a slight decrease of this species can be observed. However, the oxidation state is not affected by the adsorption, as there is no change in the $Fe2p^{3/2}$ spectrum. This is in good agreement with spectra reported for bulk and nanoscale magnetite.[17,47] The carbon content increases for the adsorbed peptide, especially for the chemically shifted C1s peaks to higher binding energies corresponding to amide and carboxylate groups, which proves a successful adsorption of Glu_8. This is furthermore emphasised by the spectrum of the N1s orbital. Here, only for the adsorbed species is a peak visible – located at around 400 eV. This means that the amide backbone does not interact with the surface through

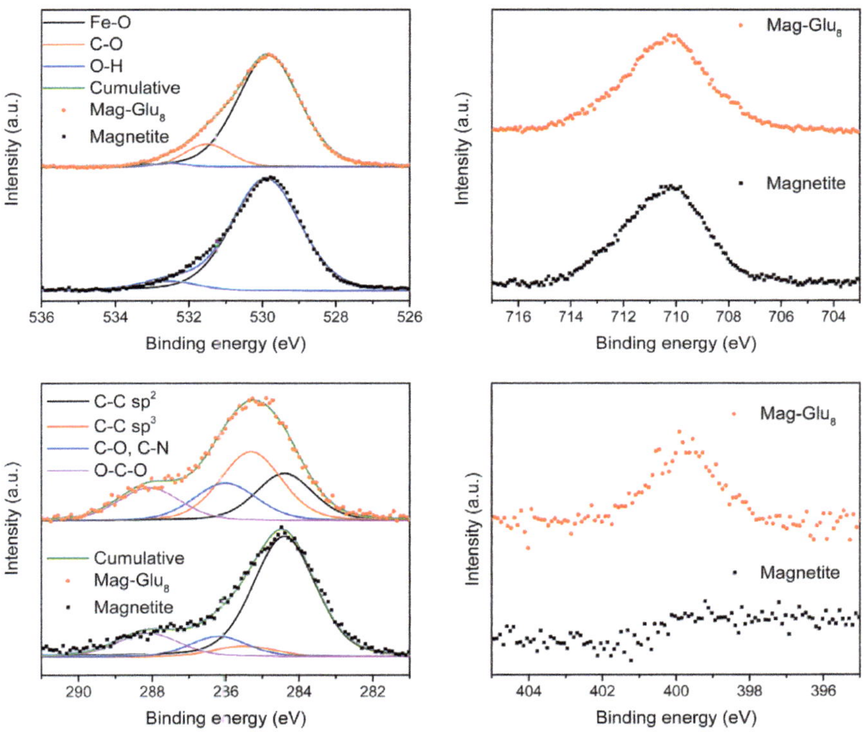

Fig. 4 XP spectra of freeze-dried magnetite nanoparticles and peptides bound to nanoparticles in the O1s (left top), the $Fe2p^{3/2}$ (right top), the C1s (left bottom) and the N1s region (right bottom).

hydrogen bonding, which is usually visible at 402 eV as observed for single amino acids.[37] Hence, for Glu_8 the adsorption takes place *via* the carboxylate groups only.

Interaction mechanisms between Glu_8 and magnetite nanoparticles

In order to further investigate the binding of Glu_8 and interpret the possibility of a predominance of electrostatic interactions, we measured the zeta potential of the peptide bound to magnetite, which is shown in Fig. 5. Bare magnetite nanoparticles present an amphiphilic surface with an isoelectric point (IEP) at around pH 6.5. Their net surface charge is higher than 30 mV below pH 4 and lower than −30 mV above pH 9, which are the regions where the particles are colloidally stable.[48] When Glu_8 is incubated with the magnetite nanoparticles at a low pH, a significant shift in the IEP to pH 3.5 can be observed in the titration curves. Furthermore, the minimum of the zeta potential is now in the region around −18 mV, which indicates a less colloidally stable system.[48] This change can be related to a different coordination of peptides to the nanoparticles. When most carboxylate groups are coordinated to the surface, the net charge on the shear plane is reduced as the negatively charged groups do not point outwards and fewer hydroxyl groups can be attached to the surface. The zeta potential behaves completely differently for Glu_8 adsorbed in an alkaline environment to the surface, especially at higher pH values. We interpret this effect as evidence for a different molecular conformation and coordination at the surface. Here, the zeta potential of magnetite containing Glu_8 adsorbed at pH 11 and titrated from pH 11 to 3 follows the trend of the bare nanoparticles at the beginning from −30 mV, but then demonstrates a low IEP at pH 3.2. Here, at high pH, not all carboxylate groups are directed to the already negatively charged surface, and therefore the charge at the shear plane is more pronounced in the negative region. Besides the different coordination at different pH values, the hysteresis demonstrates a distinct time dependence to reach the adsorption equilibrium.

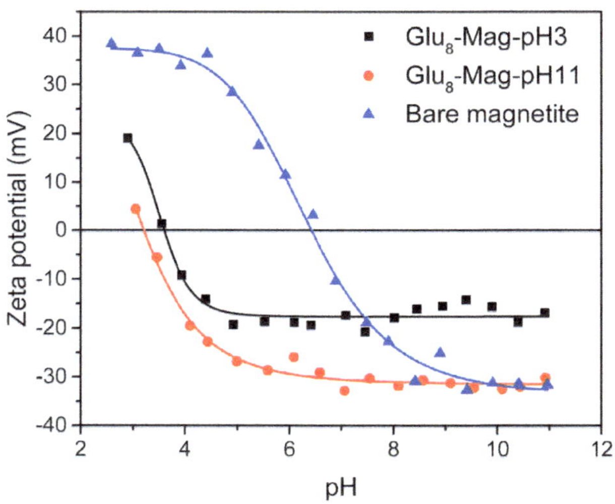

Fig. 5 Zeta potential of magnetite nanoparticles (1 g L^{-1}) and magnetite nanoparticles incubated with a Glu_8 peptide at high and low pH values, in the range of pH 3 to 11.

ATR-IR spectroscopy is a technique which facilitates visualisation of the bound state as well as the conformation of peptides and proteins. The change of pH influences the bands corresponding to the amide backbone, amide I (C=O stretch vibration) at around 1660 cm^{-1}, amide II (N–H bend and C–N stretch vibration) at around 1540 cm^{-1} and the amide III at around 1260 cm^{-1} (N–H bend and C–N stretch vibration).[49,50] Amide I and III are very sensitive to conformational changes of the peptide.[50] For the lowest pH, 5, a shift of the amide I to a maximum at around 1655 cm^{-1} can be observed. This can be linked to an increase in the α-helical content. Here, the peptides at pH 7 and 9 only differ slightly and show a maximum at 1680 cm^{-1}, which corresponds to the vibration of β-sheets. All peptides demonstrate a high amount of random coils corresponding to vibrations at around 1645 cm^{-1}, which is in good agreement with the CD spectra of the peptides shown in Fig. 6. At pH 5, a small blue shift of the amide I band can be observed upon adsorption, which can be related to an increase in either helical conformation or random coils. At the higher pH 7 and 9, the amide I shifts towards higher wavenumbers, indicating an increase in β-sheet content of the peptides. In the CD spectra, the absorbance of circularly polarised light is generally higher for the unbound peptides, as bound peptides might influence the spectrum to a lesser extent. The molar ellipticity angle change is more pronounced between 190 and 200 nm in the spectrum of Glu_8 at pH 5 in the presence of magnetite than in the rest of the spectrum. This behaviour can be related to a small increase in the α-helical or β-sheet content. Both have a maximum intensity at below 200 nm, compared to random coils, which usually exhibit a maximum at around 220 nm. In the region between 190 and 205 nm, the shift in the spectra of the peptides at pH 7 and 9 is similar but not as pronounced as for pH 5. In the region between 210 and 230 nm, the ellipticity angle decreases for peptides mixed with magnetite particles compared to pure peptide molecules. On the other hand, in the same wavelength range a slight increase in the ellipticity angle can be observed at pH 9. This suggests an increase in the random coil content at pH 9 with nanoparticles, while this content decreases with adsorption at pH 7. Thus, a negatively charged surface seems to promote a random coil, while an uncharged surface and a positively charged surface tend to influence the

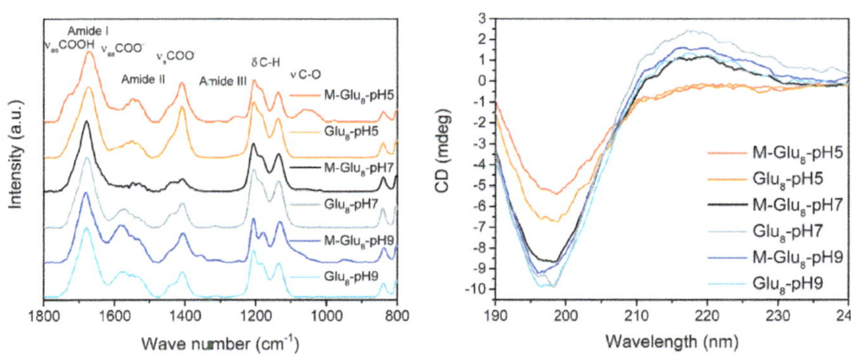

Fig. 6 ATR-IR spectra of peptides and peptide (0.25 g L^{-1}) magnetite (*M*) nanoparticle (0.1 g L^{-1}) mixtures at different pH values (left). CD spectra of peptides and peptide (0.25 g L^{-1}) nanoparticle (0.1 g L^{-1}) mixtures at different pH values (right).

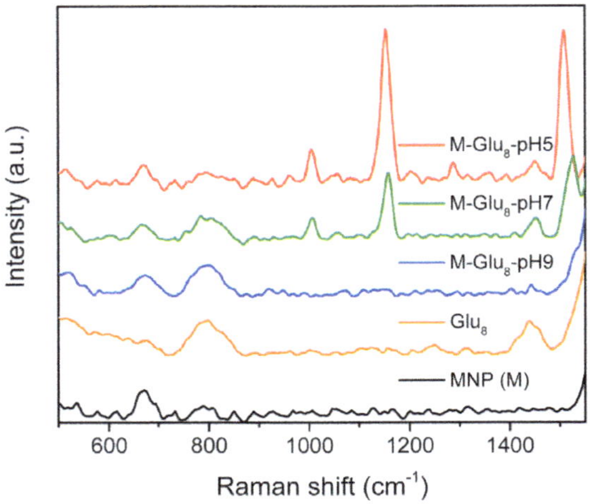

Fig. 7 Raman spectra of peptides (0.5 g L^{-1}) and peptide (0.25 g L^{-1}) magnetite (*M*) nanoparticle (0.1 g L^{-1}) mixtures at different pH values.

peptide conformation towards an α-helix or β-sheet structure. Beside the amide backbone, the vibrations corresponding to stretch vibrations of the γ-carboxylate groups in the peptide are strongly dependent on the pH and can be used for the determination of the binding mechanism. In the ionic state, the carboxylate groups demonstrate an asymmetric stretch vibration at around 1600 cm^{-1} and a symmetric stretch vibration at around 1400 cm^{-1} in the unbound state.[37,51] A significant increase of this distance Δ between the COO^- vibrations to 300–400 cm^{-1} usually points to a monodentate binding mechanism, while a slight decrease can represent bridged binding and a distinct decrease in Δ corresponds to bidentate binding. Typical binding mechanisms including monodentate and bridged bidentate binding of Glu_8 to magnetite surfaces as well as ionic interactions are illustrated in Fig. 8. We observe a decrease in the intensities of ν_{as} COO^- and ν_s COO^- at pH 5 and 7 with the presence of magnetite nanoparticles.

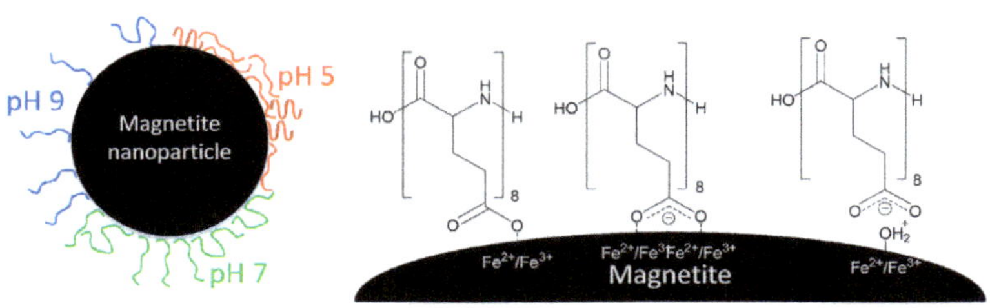

Fig. 8 Schematic of the peptide adsorption and the influence of the pH on the peptide conformation (left) and coordination (right) to the magnetite surface. The arrangement and the amount of bound functional groups as well as the secondary structure of peptides changes with pH, which is illustrated by different colours. The peptides can form different types of bonds. Here, the most probable bonding types, monodentate binding, bridged bidentate binding and electrostatic interactions of the oppositely charged surface and Glu_8 peptide, are illustrated from left to right.

For the pH 5, two distinct new bands at around 1720 and 1230 cm^{-1} occur in the presence of magnetite nanoparticles. This can be linked to a monodentate formation of carboxylate groups of the Glu_8 peptide with the surface. The significant shift of these bands also points to a participation of multiple carboxylate groups in the binding to magnetite. The further occurrence of a C–O vibration at around 1050 cm^{-1} is an additional hint to the formation of a bond between the peptide and the nanoparticles. At pH 7 the shift is less pronounced. This means that there can be several bound and unbound states of Glu_8. Only slight differences can be observed at pH 9. Here, the COO^- vibrations are still visible at around 1600 and 1400 cm^{-1}, while two small bands at around 1370 and 1080 cm^{-1} can be interpreted as evidence for bidentate and monodentate binding and most carboxylate groups are still negatively charged. The difference between these spectra can be explained on the basis of electrostatic interactions of the negatively charged peptides with a positively charged, an uncharged and a negatively charged surface.

Raman spectroscopy is another useful tool to obtain deeper insights towards the binding mechanism of peptides to the nanoparticle surface for different pH values. This is shown in Fig. 7. For the spectra containing magnetite nanoparticles, a small peak can be seen at around 670 cm^{-1} which corresponds to an A_{1g} phonon mode of magnetite and is similar to the freeze-dried particles shown in Fig. 1.[17,49] The unbound peptide and the peptide nanoparticle mixtures show a similar band at around 1450 cm^{-1} which corresponds to overlaying carboxylate anions and C–H deformation vibrations.[52] The spectrum of the magnetite peptide mixture is similar at a pH of 9 to that of the peptide at pH 7, while at lower pH values an enhancement effect of several bands is visible. The bands at around 1007, 1158, 1507 and 1526 cm^{-1} are significantly enhanced at pH 7. This effect is even stronger for the bands at around 1005, 1153 and 1507 cm^{-1} at pH 5. A similar surface enhancement effect for Raman spectroscopy with magnetite was observed for oxalic acid and catechol by Lee *et al.*[53] It is widely discussed that magnetite as a conductive material is able to enhance the spectra of chemically-bound components *via* a charge transfer mechanism.[54–56] For catechol, usually the vibration of the C–C binding between the bound groups is enhanced in Raman spectra.[53] Here, we assume a similar behaviour and an enhancement of the adjacent carbon atom and its C–C and C–H vibrations. As the peptide is visible with surface enhanced Raman spectroscopy (SERS), the type of binding should be a valid way (from a chemically point of view) to prove a charge-transfer effect. While the enhancement effect is often dependent on the pH, we assume that the strong enhancement at pH 5 is due to the formation of a chemical bond between Glu_8 and magnetite. Hence, the Raman spectra verify the ATR-IR spectra concerning the binding mechanism. However, in the case of SERS enhancement by magnetite, at least two atoms bind to the surface in order to establish a charge transfer from or to the molecule.[25,54,55] This behaviour was reported for oxalic acid and catechol but to the best of our knowledge, we are the first to report this SERS effect for peptides bound to magnetite. Such a mechanism is able to facilitate the possibility to use bare magnetite nanoparticles as sensor materials for tagged proteins.[53]

To sum up the findings derived from the multi-analytical approach, the binding mechanisms are briefly illustrated in Fig. 8. Both Raman and ATR-IR spectroscopy indicate the interplay of different binding mechanisms. While the adsorbed Glu_8 at pH 5 most explicitly indicates a monodentate coordination of carboxylate groups to the surface, here the highest amount of peptide is bound to the surface and crowding effects can influence the interpretation of the spectra. Moreover, the enhancement of peptide vibrations by the magnetite surface as demonstrated with Raman spectroscopy might hint to a charge transfer which might arise from a bidentate complexation. We emphasise the existence of charged carboxylate groups interacting electrostatically with the surface as well, as they can be observed with Raman and ATR-IR spectroscopy. ATR-IR, Raman and CD spectroscopy indicate a more pronounced interaction with the surface at lower pH values, which is illustrated in Fig. 8 and can be confirmed with zeta potential measurements. Here we also try to demonstrate the peptide conformation at the different pH values which is dominated by uncoiled peptides interacting with the surface.

Evaluation of binding parameters and transfer to protein interactions

The adsorption equilibrium of different Glu_8 peptide concentrations was determined in several binding experiments. Adsorption constants and the maximum load can be derived from Fig. 9. This allows for deeper insights into the thermodynamic driving force of the binding. The maximum load varies considerably for different pH values. For pH 4 and pH 5, no saturation could be achieved in the

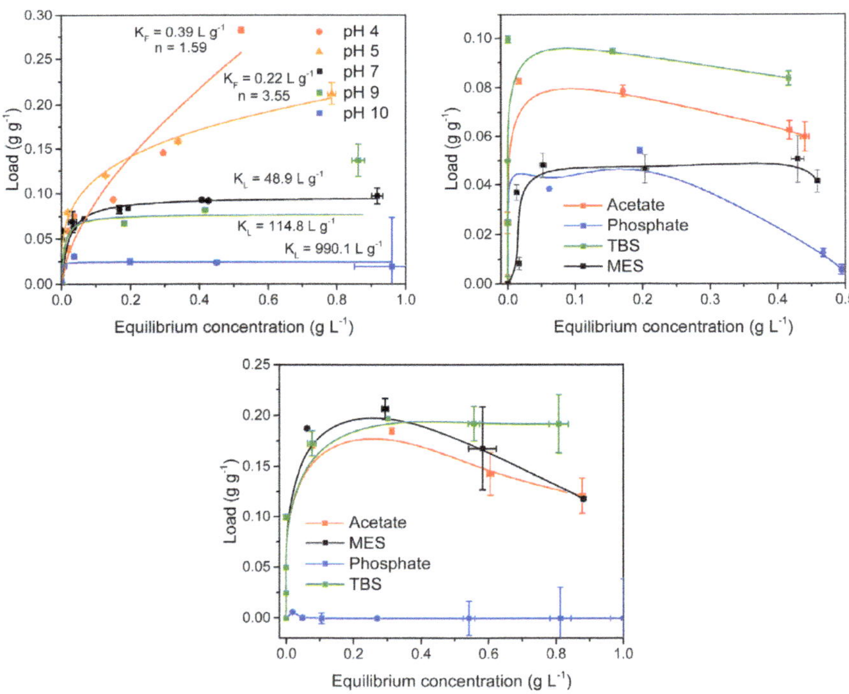

Fig. 9 Adsorption isotherms of octa-glutamic acid on magnetite nanoparticles (1 g L^{-1}) (top) at different pH (left) and buffer conditions (right) at pH 7 and of hexa-glutamic acid tagged GFP on magnetite nanoparticles with the similar buffers (bottom). Adsorption isotherms at pH 4, 5 and 7 are fitted with a Langmuir function, while the isotherms at pH 9 and 10 are fitted with a Freundlich function.

investigated concentration range and the highest loads obtained were 0.28 and 0.20 g g^{-1} (peptide per magnetite), respectively. For higher pH values, the maximum load of peptide further decreased from 0.10 to 0.08 and 0.02 g g^{-1} for pH values 7, 9 and 10. The amount of bound peptide and the interaction mechanism is strongly dependent on the pH, which is in good agreement with zeta potential measurements and the spectroscopic data. On the other hand, the affinity constants K_L, which are in the same range as those of peptides of comparable length to oxide surfaces, show an increase with increasing pH, indicating a higher binding affinity for higher pH values.[57,58] This trend is similar for the binding at low pH values. Here the adsorption isotherm at pH 5 demonstrates a steeper slope at the lower concentration region. This indicates a higher affinity than for Glu_8 at pH 4. The general trend of the adsorption isotherms points to an increased interaction between nanoparticles and peptides at lower pH values. This behaviour is expected for peptides, which are negatively charged from pH 4 to 10 as the amphiphilic surface is oppositely charged for pH values below pH 6.5. For all investigated buffered systems at pH 7, the maximum load is lower than in pure water. Here, a similar trend as in the peptide array experiments is visible. A high load can be obtained for Tris (0.10 g g^{-1}) and acetate buffered saline (0.08 g g^{-1}), while phosphate demonstrates the lowest adsorption capacity (0.04 g g^{-1}) for Glu_8 on magnetite. The discrepancy to the array results can be explained by the slightly different pH values and the lack of saline. The pH of the Tris solution is lower, while the other pH values are higher than in the array experiments. This might change the surface charge, as driving force of the adsorption. A decreasing load with higher equilibrium concentrations of peptide can be observed for all buffered solutions and is contrary to the behaviour in unbuffered solution. For phosphate buffered saline the peptide load even decreases to zero for an equilibrium concentration of 0.5 g g^{-1} Glu_8. The decreasing load for higher equilibrium concentration impedes the use of an adsorption model such as the Langmuir or Freundlich isotherm. Therefore, no affinity constants can be determined here. With the tests of these peptide adsorption experiments we analysed the binding of a terminally hexa-glutamic acid tagged green fluorescent protein (Glu_6-GFP) under the same conditions. The results of the protein are similar to those of the peptides, but show a higher maximum load of 0.20 g g^{-1} due to the greater mass of a 27 kDa protein compared to a 1 kDa peptide. Here, the load decreases for higher protein concentrations in acetate and MES buffer as well. In phosphate, no binding can be observed for the protein, suggesting a repulsion of the protein by the negatively charged phosphate ions around the magnetite nanoparticles. Hence, we are able to adapt the functionality of a peptide to a model protein.

Experimental

Materials

All reagents used are commercially available, have a purity of >95% (HPLC grade) and were used as received without further purification. The octa-glutamic acid peptide (101 mg) was obtained from metabion international AG (Germany) as a lyophilized powder with a purity of >95% and stored at +4 °C prior to use.

Magnetite was synthesised according to a co-precipitation method, stored and handled as described elsewhere.[15,17] A magnetite suspension (1 g L^{-1}) was incubated with a glutamic acid octamer (0.25 g L^{-1}) for 1 h at pH 7. After magnetic separation, the residue was washed twice and then freeze-dried for further analytics.

Green fluorescent protein (GFP) with a hexa-glutamate (Glu_6)-tag was produced by the *E. coli* strain BL21 (DE3) in Terrific broth media in a shaking flask. The media components are: 12 g L^{-1} tryptone, 24 g L^{-1} yeast extract, 9.4 g L^{-1} K_2HPO_4, 2.2 g L^{-1} KH_2PO_4 and 4 mL glycerol.[59] Glu_6-GFP was encoded on a pET21 vector containing a resistance gene against the antibiotic ampicillin. 100 mg L^{-1} of ampicillin was added to the medium as selection. For protein expression, the culture was incubated overnight and shaken at 200 rpm after induction with 50 μM IPTG (isopropyl-β-D-thiogalactopyranosid). Protein purities of 99% were achieved by preparative chromatography in two steps, namely anion exchange and size exclusion chromatography. The anion exchange resin was Unosphere Q (BioRad Laboratories, Hercules, USA) and the equilibration and elution buffers were 50 mM Tris pH 9.5 with 60 and 1000 mM NaCl, respectively. The size exclusion chromatography column HiLoad 16/60 Superdex 200 prep grade (GE Healthcare Life Sciences, Chicago, USA) was used for polishing with 20 mM Tris buffer pH 7.4 supplemented with 200 mM NaCl as a mobile phase. The elution fractions were pooled and the purity was assessed by SDS-PAGE and HPLC on an analytical column (Aeris 3.6 μm Widepore C4 200 Å 100 × 2.1 mm by Phenomenex, Torrance, USA).

Characterisation

The magnetite nanoparticles were furthermore characterised by X-ray diffraction (XRD) and Raman spectroscopy in the same manner as reported by Schwaminger *et al.*[17]

Particle size and shape were investigated by transmission electron microscopy (TEM) with a JEM 1400 Plus from JEOL GmbH, Germany. For the TEM measurements the colloidal samples were diluted in degassed and deionised water, ultrasonicated and placed on carbon-coated copper grids. In order to receive statistical particle distributions, at least 30 particles were counted per picture.

In order to identify binding candidates for bare magnetic nanoparticles (MNP), CelluSpot peptide arrays from Intavis with different homo hexa-peptides per spot anchored in a cellulose membrane were used. The membrane was conditioned with the buffer three times for ten minutes prior to the binding experiment. The buffers were based on 50 mM citrate (CBS), phosphate (PBS), tris(hydroxymethyl) aminomethane (TBS), 2-(*N*-morpholino)ethanesulfonic acid (MES-BS) containing 137 mM NaCl and 2.7 mM KCl, Tween 20 and double distilled water each. Saline solution contained one-tenth the NaCl and KCl concentration. Magnetite nanoparticles (0.4 g L^{-1}) were incubated for 60 minutes on the membrane and the unbound particles were removed by three washing steps under respective buffer conditions for 10 minutes. The different shades of black on the spots against the background were analysed with ImageJ and correlated to the amount of bound nanoparticles.

Magnetic properties of the precipitates were characterised with a superconducting quantum interference device (SQUID) magnetometer MPMS (Quantum Design Inc., USA) at a temperature of 300 K. The magnetic field varied from −50 kOe to +50 kOe. Zero field cooling/field cooling experiments were conducted in a range from 1.8 to 300 K with an applied field of 200 Oe.

Chemical analysis was accomplished by X-ray photoelectron spectroscopy (XPS) with a Leybold–Heraeus LHS 10 XPS system in ultrahigh vacuum (UHV) hosting a non-monochromatised Mg Kα source (1253.6 eV). The powder samples were compressed on hollow sample holders. The spectra were recorded at a constant pass energy mode set to 100 eV and a full width at half-maximum (FWHM) of ~1.1 eV. The C1s (284.5 eV) peak corresponding to adventitious carbon was used as energy reference to compensate energy shifts due to charging. Detail spectra were recorded by repeatedly scanning the same region 30 times in order to reduce statistical noise. All spectra were recorded in a UHV at a pressure below 1.5×10^{-8} mbar. The core level spectra were fitted by a mix of Gaussian and Lorentzian functions (Gaussian line width 0.7 eV and Lorentzian line width 0.3 eV). Shirley backgrounds were subtracted from all spectra.

For the zeta potential measurements, peptides (0.25 g L^{-1}) and magnetite nanoparticles (1 g L^{-1}) were incubated and titrated in the range of pH 3 to 11. The zeta potential was measured with a Delsa Nano C (Beckman Coulter Inc., USA) where 0.1 M NaOH and 0.1 M HCl were used as titrants.

For the binding experiments, different amounts of peptides were incubated with magnetite nanoparticles (1 g L^{-1}) at different pH and buffer conditions for at least one hour at 25 °C under vigorous shaking. The supernatant was separated from the particles with hand magnets, decanted and analysed with an Infinite M200 Microplate Reader (Tecan Deutschland, Germany) at 230 nm.

Raman spectroscopy, attenuated total reflection infrared spectroscopy (ATR-IR) and circular dichroism spectroscopy (CD) were measured of magnetite suspensions (0.1 g L^{-1}) incubated with peptides (0.25 g L^{-1}) for 1 h at different pH. Raman spectroscopy of suspensions was performed with a Raman Senterra spectrometer (Bruker Optics, Germany) at 40 mW with a 488 nm laser and a 60 s integration time. Measurements were performed in triplicates and the background was subtracted with a concave rubberband method. For the ATR-IR spectroscopy a Bruker Vertex 70 spectrometer (Bruker Optics, Germany) was used and the suspensions were precipitated on a diamond ATR prior to the measurement. The spectra were treated with a similar rubberband correction as the Raman spectra.

CD spectroscopy of the samples was performed with a Jasco 815 spectrometer (Jasco, Germany) in a sealed quartz cuvette (path length 0.5 mm).

Conclusions

This study describes the unique behaviour of a peptide confined at a magnetite nanoparticles water interface. We try to understand how a biomolecule binds to complex nanoparticle surfaces. With the combination of several different analytical approaches, the binding mechanism can be described, which is strongly dependent on the environment as it influences the interface. We identified oligomeric glutamic acid-based peptides as iron oxide binding peptides, and were able to follow the conformational changes of the peptides and their bonding mechanism to the surface. The binding of the Glu_8 peptide does not affect the chemical state and the physical properties of the adsorbent. This is the basis to recycle and reuse the nanoparticles, which is a nice feature for industrial applications such as protein purification. The conformation and orientation of adsorbed molecules was monitored with electrophoretic measurements as well as with spectroscopic methods. The carboxylate groups are more distinctly oriented to the surface at lower pH values and the peptide backbone is mostly randomly coiled. However, the peptide conformation is influenced by the surface which leads to a higher α-helix and β-sheet content of peptides. We assume that this change is influenced by the carboxylate groups which are directed and bonded to the nanoparticle surface as monodentate species. The existence of a chemical bond between a Glu_8 peptide and the surface can be proven with ATR-IR spectroscopy and Raman spectroscopy. The Raman spectra even demonstrated an unexpected enhancement effect of the peptide binding. Such an effect could form the basis of biomolecule sensing with bare magnetite nanoparticles, including all the advantages of a low-cost, environmentally-friendly system. Furthermore, the affinity of the peptide to the surface and the binding capacity was evaluated. Here, the pH and the buffer composition not only affects the binding, but can be used to control the adsorption process. An implementation of the peptide binding towards industrial application was investigated by addition of a terminal Glu_6 peptide to a model protein (GFP). Here, the demonstrated binding behaviour was similar to the peptide binding. Thus, we believe that it is possible to take advantage of the molecular properties of peptides at a complex interface for possible industrial applications and a new generation of affinity tags.

Acknowledgements

The authors would like to thank Dr Carsten Peters for support with TEM imaging, Prof. Sebastian Günther for support with XPS, Prof. Tom Nilges for the provision of the X-ray diffractometer, Prof. Bernhard Rieger for the provision of the IR spectrometer and Dr Kathrin Castiglione for the provision of the CD spectrometer. We thank the students Yasmin Kaveh Bagbadherani and Tobias Koch for the preparation of the tagged protein and Dominique Nolte Taboas for preliminary experiments on binding and zeta potential. We also thank Julia Schwaminger for language editing of the manuscript. Furthermore, we are particularly appreciative for the financial support of this work by the Federal Ministry of Education and Research (Grant number 031A173A).

References

1 J. D. Bernal, *Proc. Phys. Soc., London, Sect. A*, 1949, **62**, 537.

2 W. Weber, *Water Sci. Technol.*, 1999, **40**, 11–19.

3 E. Worch, *Adsorption technology in water treatment. Fundamentals, processes, and modeling*, de Gruyter, Berlin u.a., 2012.

4 G. A. Moldoveanu and V. G. Papangelakis, *Hydrometallurgy*, 2012, **117–118**, 71–78.

5 J. Kammerer, R. Carle and D. R. Kammerer, *J. Agric. Food Chem.*, 2011, **59**, 22–42.

6 R. L. Fahrner, H. L. Knudsen, C. D. Basey, W. Galan, D. Feuerhelm, M. Vanderlaan and G. S. Blank, *Biotechnol. Genet. Eng. Rev.*, 2001, **18**, 301–327.

7 J. N. Israelachvili, *Intermolecular and surface forces*, Elsevier, Amsterdam u.a., 3rd edn, 2011.

8 M. Penna, K. Ley, S. Maclaughlin and I. Yarovsky, *Faraday Discuss.*, 2016, **191**, 435–464.
9 B. R. Peelle, E. M. Krauland, K. D. Wittrup and A. M. Belcher, *Langmuir*, 2005, **21**, 6929–6933.
10 D. Ling and T. Hyeon, *Small*, 2013, **9**, 1450–1466.
11 R. G. Harrison, P. Todd, S. R. Rudge and D. P. Petrides, *Bioseparations science and engineering*, Oxford University Press, New York, NY, 2015.
12 D. Bathen and M. Breitbach, *Adsorptionstechnik*, Springer, Berlin u.a., 2001.
13 S. Laurent, D. Forge, M. Port, A. Roch, C. Robic, L. Vander Elst and R. N. Muller, *Chem. Rev.*, 2008, **108**, 2064–2110.
14 A.-H. Lu, E. L. Salabas and F. Schuth, *Angew. Chem., Int. Ed.*, 2007, **46**, 1222–1244.
15 H.-C. Roth, S. P. Schwaminger, M. Schindler, F. E. Wagner and S. Berensmeier, *J. Magn. Magn. Mater.*, 2015, **377**, 81–89.
16 D. Forge, A. Roch, S. Laurent, H. Tellez, Y. Gossuin, F. Renaux, L. Vander Elst and R. N. Muller, *J. Phys. Chem. C*, 2008, **112**, 19178–19185.
17 S. P. Schwaminger, D. Bauer, P. Fraga-García, F. E. Wagner and S. Berensmeier, *CrystEngComm*, 2017, **19**, 246–255.
18 H.-C. Roth, S. P. Schwaminger, F. Peng and S. Berensmeier, *ChemistryOpen*, 2016, **5**, 183–187.
19 V. Polshettiwar, R. Luque, A. Fihri, H. Zhu, M. Bouhrara and J.-M. Basset, *Chem. Rev.*, 2011, **111**, 3036–3075.
20 R. B. N. Baig, S. Verma, M. N. Nadagouda and R. S. Varma, *Aldrichimica Acta*, 2016, **49**, 35–41.
21 A. M. Bruck, C. A. Cama, C. N. Gannett, A. C. Marschilok, E. S. Takeuchi and K. J. Takeuchi, *Inorg. Chem. Front.*, 2016, **3**, 26–40.
22 B. Li, H. Cao, J. Shao, M. Qu and J. H. Warner, *J. Mater. Chem.*, 2011, **21**, 5069.
23 S. Laurent, S. Dutz, U. O. Hafeli and M. Mahmoudi, *Adv. Colloid Interface Sci.*, 2011, **166**, 8–23.
24 M. Levy, A. Quarta, A. Espinosa, A. Figuerola, C. Wilhelm, M. García-Hernández, A. Genovese, A. Falqui, D. Alloyeau, R. Buonsanti, P. D. Cozzoli, M. A. García, F. Gazeau and T. Pellegrino, *Chem. Mater.*, 2011, **23**, 4170–4180.
25 N. Lee, D. Yoo, D. Ling, M. H. Cho, T. Hyeon and J. Cheon, *Chem. Rev.*, 2015, **115**, 10637–10689.
26 T. Sun, Y. S. Zhang, B. Pang, D. C. Hyun, M. Yang and Y. Xia, *Angew. Chem., Int. Ed.*, 2014, **53**, 12320–12364.
27 M. Malmsten, *Curr. Opin. Colloid Interface Sci.*, 2013, **18**, 468–480.
28 T. Hawkins, *US Pat.* 5705628 A, 1994.
29 P. Fraga-García, M. Brammen, M. Wolf, S. Reinlein, M. Freiherr von Roman and S. Berensmeier, *Sep. Purif. Technol.*, 2015, **150**, 29–36.
30 E. Hochuli, W. Bannwarth, H. Döbeli, R. Gentz and D. Stüber, *Nat. Biotechnol.*, 1988, **6**, 1321–1325.
31 T. G. M. Schmidt and A. Skerra, *Nat. Protoc.*, 2007, **2**, 1528–1535.
32 A. E. Nel, L. Madler, D. Velegol, T. Xia, E. M. V. Hoek, P. Somasundaran, F. Klaessig, V. Castranova and M. Thompson, *Nat. Mater.*, 2009, **8**, 543–557.
33 M. Mahmoudi, I. Lynch, M. R. Ejtehadi, M. P. Monopoli, F. B. Bombelli and S. Laurent, *Chem. Rev.*, 2011, **111**, 5610–5637.
34 S. P. Schwaminger, S. A. Blank-Shim, M. Borkowska-Panek, P. Anand, P. Fraga-García, K. Fink, W. Wenzel and S. Berensmeier, *Eng. Life Sci.*, 2017, 1–17.

35 J. M. Slocik and R. R. Naik, *Chem. Soc. Rev.*, 2010, **39**, 3454–3463.
36 S. A. Blank-Shim, S. P. Schwaminger, M. Borkowska-Panek, P. Anand, P. Fraga-García, K. Fink, W. Wenzel and S. Berensmeier, *Sci. Rep.*, 2017, submitted.
37 S. P. Schwaminger, P. Fraga-García, G. K. Merck, F. A. Bodensteiner, S. Heissler, S. Günther and S. Berensmeier, *J. Phys. Chem. C*, 2015, **119**, 23032–23041.
38 K. Pušnik, M. Peterlin, I. K. Cigić, G. Marolt, K. Kogej, A. Mertelj, S. Gyergyek and D. Makovec, *J. Phys. Chem. C*, 2016, **120**, 14372–14381.
39 T. J. Daou, S. Begin-Colin, J. M. Grenèche, F. Thomas, A. Derory, P. Bernhardt, P. Legaré and G. Pourroy, *Chem. Mater.*, 2007, **19**, 4494–4505.
40 A. Situm, M. A. Rahman, S. Goldberg and H. A. Al-Abadleh, *Environ. Sci.: Nano*, 2016, **3**, 910–926.
41 E. Verwey and P. W. Haayman, *Physica*, 1941, **8**, 979–987.
42 J. P. Shepherd, J. W. Koenitzer, R. Aragón, J. Spal/ek and J. M. Honig, *Phys. Rev. B: Condens. Matter Mater. Phys.*, 1991, **43**, 8461–8471.
43 A. Mitra, J. Mohapatra, S. S. Meena, C. V. Tomy and M. Aslam, *J. Phys. Chem. C*, 2014, **118**, 19356–19362.
44 R. L. Rebodos and P. J. Vikesland, *Langmuir*, 2010, **26**, 16745–16753.
45 Y. A. Barnakov, M. H. Yu and Z. Rosenzweig, *Langmuir*, 2005, **21**, 7524–7527.
46 K. Chesnel, M. Trevino, Y. Cai, J. M. Hancock, S. J. Smith and R. G. Harrison, *J. Phys.: Conf. Ser.*, 2014, **521**, 12004.
47 A. P. Grosvenor, B. A. Kobe, M. C. Biesinger and N. S. McIntyre, *Surf. Interface Anal.*, 2004, **36**, 1564–1574.
48 R. J. Hunter, *Zeta potential in colloid science. Principles and applications*, Academic Press, London, San Diego, 1988.
49 S. P. Schwaminger, P. Fraga-García, F. Selbach, F. G. Hein, E. C. Fuß, R. Surya, H.-C. Roth, S. A. Blank-Shim, F. E. Wagner, S. Heissler and S. Berensmeier, *Adsorption*, 2017, **23**, 281–292.
50 P. Roach, D. Farrar and C. C. Perry, *J. Am. Chem. Soc.*, 2005, **127**, 8168–8173.
51 H.-C. Roth, S. Schwaminger, P. Fraga García, J. Ritscher and S. Berensmeier, *J. Nanopart. Res.*, 2016, **18**, 303.
52 P. K. Sengupta and S. Krimm, *Biopolymers*, 1985, **24**, 1479–1491.
53 N. Lee, P. J. Schuck, P. S. Nico and B. Gilbert, *J. Phys. Chem. Lett.*, 2015, **6**, 970–974.
54 S. J. Hurst, H. C. Fry, D. J. Gosztola and T. Rajh, *J. Phys. Chem. C*, 2011, **115**, 620–630.
55 P. Tarakeshwar, D. Finkelstein-Shapiro, S. J. Hurst, T. Rajh and V. Mujica, *J. Phys. Chem. C*, 2011, **115**, 8994–9004.
56 W. Ji, Y. Kitahama, X. Han, X. Xue, Y. Ozaki and B. Zhao, *J. Phys. Chem. C*, 2012, **116**, 24829–24836.
57 S. Joshi, I. Ghosh, S. Pokhrel, L. Madler and W. M. Nau, *ACS Nano*, 2012, **6**, 5668–5679.
58 A. D. Roddick-Lanzilotta, P. A. Connor and A. J. McQuillan, *Langmuir*, 1998, **14**, 6479–6484.
59 M. Kangwa, V. Yelemane, A. N. Polat, K. D. D. Gorrepati, M. Grasselli and M. Fernandez-Lahore, *AMB Express*, 2015, **5**, 70.

Faraday Discussions

PAPER

Elucidating "screw dislocation"-driven film formation of sodium thiosulphate with complex hierarchical molecular assembly†

Deepak Dwivedi,[a] Katerina Lepkova, *[a] Thomas Becker [b] and Matthew R. Rowles [c]

Received 6th March 2017, Accepted 7th April 2017

DOI: 10.1039/c7fd00092h

Sodium thiosulphate ($Na_2S_2O_3$) films were synthesized on carbon steel substrates through solution deposition, and a film formation growth mechanism is delineated in detail herein. Dislocation-driven film formation took place at the lower concentration of $Na_2S_2O_3$ (0.1 M) studied, where screw dislocation loops were identified. Interestingly, we observed the co-existence of screw dislocation spiral loops and hierarchically-ordered molecular assembly in the film, and showed the importance of hierarchical morphology in the origin of screw dislocation. The screw dislocation loops were, however, distorted at the higher studied concentration of $Na_2S_2O_3$ (0.5 M), and no hierarchical structures were formed. The mechanisms of film formation are discussed in detail and provide new insights into our understanding regarding morphology of the hierarchical molecular assembly, screw dislocation loop formation, and the role of chemical elements for their development. The main crystalline and amorphous phases in the surface films were identified as pyrite/mackinawite and magnetite. As sodium thiosulphate is widely used for energy, corrosion inhibition, nanoparticle synthesis and catalysis applications, the knowledge generated in this study is applicable to the fields of corrosion, materials science, materials chemistry and metallurgy.

1. Introduction

In the oil and gas industry, internal pipeline corrosion is one of the major concerns and can be mitigated through the use of corrosion inhibitor formulations containing chemical compounds that form protective films at the steel surface. Sodium thiosulphate ($Na_2S_2O_3$) is a known component in commercial corrosion-inhibitor formulations, where it acts synergistically with other

[a]Curtin Corrosion Engineering Industry Centre, Department of Chemical Engineering, Australia. E-mail: K.Lepkova@curtin.edu.au

[b]Nanochemistry Research Institute, Department of Chemistry, Australia

[c]Department of Physics and Astronomy, Fuels and Energy Technology Institute Faculty of Science and Engineering, Curtin University, GPO Box U1987, Perth, Australia

† Electronic supplementary information (ESI) available. See DOI: 10.1039/c7fd00092h

molecules, *e.g.* organic surfactants, and enhances the efficiency of these formulations against corrosion.[1] However, the exact role of sodium thiosulphate in corrosion film formation with organic molecules is yet to be elucidated. The possible mechanisms of the inhibition effect of $Na_2S_2O_3$-containing formulations could involve (1) interaction of $Na_2S_2O_3$ with organic surfactant molecules in solution, affecting micelle formations and their adsorption to the steel surface, or (2) competitive or synergistic adsorption of $Na_2S_2O_3$ and other components on the steel substrate.[2,3]

Besides these corrosion science applications, $Na_2S_2O_3$ also holds significant importance in catalysis, nanoscience, energy, and corrosion mitigation applications. It has recently been utilized in the synthesis of dodecanethiolate-stabilized magnetic iridium nanoparticles[4] and whisker-shaped MnO_2 nanoparticles,[5] and in the preparation of ω-carboxyl-*S*-hexanethiosulphate ligands.[6] Furthermore, a CdS photocatalyst was prepared by combining sodium thiosulphate and $CdSO_4$,[7] while CuS-containing TiO_2 fibres were prepared by mixing solutions of sodium thiosulphate and $CuSO_4$.[8] Sodium thiosulphate has also been used in hollow TiO_2 particle synthesis.[9]

This work examines the film formation, by $Na_2S_2O_3$, on carbon steel substrates under CO_2-saturated conditions that replicate the real time situation in oil pipelines, where the application of inhibitors is a widely used method for protecting carbon steels from corrosion.[10] It is envisaged that elucidating the role of sodium thiosulphate in film formation on metallic substrates will be applicable in various surface and interface-related applications.

We aim to investigate the complex phenomenon of the formation of the sodium thiosulphate film on carbon steel and to describe the mechanism for the dislocation-driven film formation by sodium thiosulphate at various concentrations. In particular, we are interested in describing the role of the interface between the hierarchical morphology of the sodium thiosulphate film[11] and the initially grown dislocation spiral, and the spiral propagation. The observed growth is expected to support Kossel's theory, where it has been universally accepted that the growth of non-Kossel crystals (consisting of different ions in their unit cell) depends on the degree of supersaturation and ion activity in solution.[12,13] To the best of our knowledge, screw dislocation-driven film formation has not been previously investigated on metallic substrates, such as carbon steel. Given the wide range of applications for sodium thiosulphate, knowledge on its film formation will benefit many areas of fundamental and applied science.

2. Experimental

2.1 Substrate preparation

1030 carbon steel was used as a substrate for the film formation. The steel samples were embedded in epoxy resin with an exposed surface area of 1.44 cm^2. The elemental composition of the carbon steel (weight%) was: C (0.37), Mn (0.80), Si (0.282), P (0.012), S (0.001), Cr (0.089), Ni (0.012), Mo (0.004), Sn (0.004), Al (0.01), and Fe (balance). The steel samples were ground and polished to a surface roughness of less than 1 μm. The polished samples were ultrasonically cleaned in ethanol for 2 min, rinsed with Milli-Q water (resistivity 18 MΩ cm), and dried with nitrogen gas before immersion in the test solution.

2.2 Solution preparation for film formation

The test solution consisted of standard brine, containing 30 g L^{-1} sodium chloride (NaCl; Ajax Finechem, analytical reagent, 99.9%) and 0.1 g L^{-1} sodium hydrogen carbonate ($NaHCO_3$; Merck, 99.5%) prepared in Milli-Q water (resistivity – 18.2 MΩ cm). Sodium thiosulphate pentahydrate ($Na_2S_2O_3 \cdot 5H_2O$; MW = 248.18 g mol^{-1}, Sigma Aldrich, >99.99%) was added to the brine solution and used for surface-film formation in concentrations of 0.1 M or 0.5 M. All experiments were conducted in a test solution saturated with CO_2 gas and no effort was made to maintain any specific pH tailoring. The test solution was pre-sparged with CO_2 gas for 2 hours, prior to immersion of the steel sample, to ensure low levels of dissolved oxygen (4–5 ppb). The saturated solutions were then pumped into the test cell using a low-oxygen penetration tube and the cell was continuously purged with CO_2 at ambient pressure (1 atm) during the experiment. All experiments were conducted at room temperature (25 ± 2 °C). Film formation duration was 24 hours. Samples were dried with nitrogen gas and stored in a vacuum desiccator after the test.

2.3 Microscopic and spectroscopic characterization

The microstructural characterization of films formed with 0.1 M and 0.5 M thiosulphate was carried out using a scanning electron microscope (SEM; Zeiss NEON 40EsB). Elemental depth distribution was confirmed by a focused ion beam-SEM time-of-flight secondary-ion mass spectrometer (FIB-SEM ToF-SIMS; Tescan Lyra). ToF-SIMS was conducted over various surface areas using an ion beam accelerating voltage of 30 kV and current of 35 pA. Infrared spectra were collected using a Fourier-transform infrared spectrometer (FTIR; Nicolet iS50) fitted with a dedicated single-bounce diamond attenuated total reflection (ATR) accessory purged with dry nitrogen. Sixty-four scans of the background and 64 sample scans were co-averaged and ratioed to produce a transmission spectrum. A spectral resolution of four wavenumbers, with two levels of post-zero filling, Norton–Beer strong apodization, and Mertz phase correction were used. The absorbance spectrum was ATR-corrected for the effect of wavelength on the depth of penetration, producing a spectrum with relative peak intensities that are comparable to transmission techniques (*i.e.* most library spectra). Chemical characterization of films was done through Raman spectroscopy (WITec Alpha 300 SAR, WITec, Ulm, Germany) with a 2ω NdYAG laser ($\lambda_{exc.}$ = 532 nm) using a 1800 g mm^{-1} grating. The instrument was controlled with the WITec Control FOUR software and data were processed and analysed with WITec Project FOUR. The objective used for these experiments was 100×, NA 0.9 (Zeiss).

Diffraction data were collected from steel samples which were subjected to treatment in standard brine containing 0.1 M $Na_2S_2O_3$ or 0.5 M $Na_2S_2O_3$. The specimens measured 1.2 × 1.2 × 0.5 cm. Data were collected using a PANalytical Empyrean diffractometer using Cu Kα radiation. The incident beam had a parabolic mirror with 2.3° Soller slits, and was fixed at 2° to the specimen surface. The diffracted beam had 2.3° Soller slits, a 0.18° parallel plate collimator, and a point detector with a 0.05 mm receiving slit. Diffraction data were taken over the range 10–100° 2θ in steps of 0.02°. The diffraction data were collected in this fixed-incident beam geometry in order to enhance the contribution from surface layers.

Phase identification was carried out using the software package EVA[14] with the ICDD PDF2 database. Crystal structures for the phases were taken from the ICDD PDF 4+ database. The phases were modelled using the Rietveld[15] method as implemented in TOPAS.[16] Diffraction geometry-specific corrections were implemented as previously outlined.[17,18] March–Dollase[19] preferred orientation corrections were applied for fixed-incident beam geometry.[20,21] Relative weight fractions and crystallite size were calculated using the matrix-flushing method[22] and double-Voigt approach,[23] respectively.

3. Results and discussion

3.1 Scanning electron microscopic analysis

Fig. 1(a and b) shows representative images of the formation of screw-dislocation-driven sodium thiosulphate ($Na_2S_2O_3$) films on low carbon steel. The concentration of $Na_2S_2O_3$ was 0.1 M. It is clearly noticeable that various loops have formed around the central core of the spiral (see also the enlarged area in Fig. 1(a)). It has been illustrated earlier that screw dislocation-driven spiral growth is more prominent at low supersaturation (*i.e.* $\sigma = \ln(C/C_0)$, where C is the precursor concentration and C_0 is the equilibrium concentration). This is because self-perpetuated spiral growth promotes crystal growth to occur below the supersaturation level, which is needed for layer-by-layer growth.[24,25] Fig. 1(b) shows an enlarged area of a spiral formed due to dislocations. This particular image reveals an interesting attribute about the inside structure of a spiral core, where we observed a hierarchical film formation inside the spiral.

The observation of both screw dislocation-driven spiral films and hierarchical-structured films in the presence of 0.1 M sodium thiosulphate is further demonstrated in Fig. 2, where areas denoted by 1 and 2 suggest the formation of screw dislocation-derived spiral formations (see also Fig. 1 for a detailed view), along with the hierarchical assembly denoted by 3 in the same figure and enlarged in Fig. 2(b).

The typical thickness of the hierarchical boundary, measured with SEM, was around 10 nm with significant porosity as exhibited by Fig. 3(a). We also

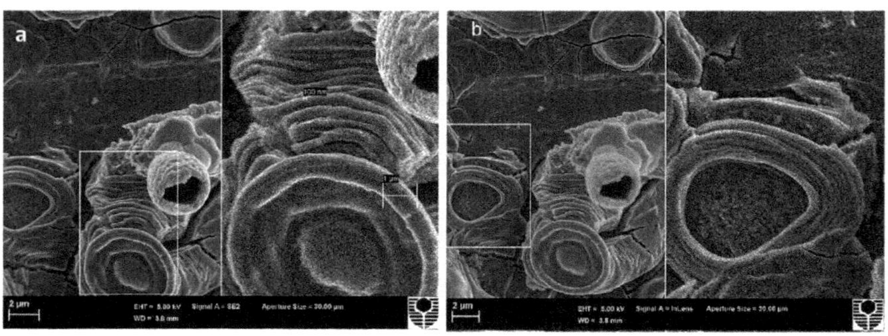

Fig. 1 Examples of screw dislocation spirals in film synthesized on a carbon steel substrate using 0.1 M sodium thiosulphate in CO_2-saturated brine at room temperature: (a) FESEM images showing enclosed and hollow loops, and (b) loops with hierarchically ordered assemblies inside the loop.

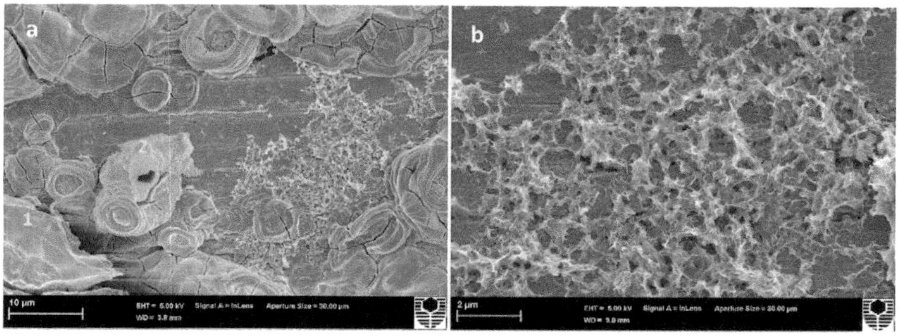

Fig. 2 FESEM images of surface film synthesized with 0.1 M sodium thiosulphate at room temperature in CO_2-saturated brine: (a) exhibiting the formation of dislocation loops and different shapes of screw dislocation, and (b) confirming the formation of hierarchical morphology that co-exists with dislocation spirals in the film; (b) is the magnified image of region 3 in (a).

measured the loop size in the spiral and noticed formation of different loop-sized spirals of about 100 nm and 200 nm in this particular example (Fig. 3(b)). It is interesting to observe that the spirals were grown with different core sizes too.

The above results highlight the fundamental aspect of dislocation-driven growth, which is to be considered as development of a low-index surface due to supersaturation. But the loops formed during spiral growth due to the screw dislocation are considered as high surface energy sites, which leads to the conclusion that certain sites were established as highly active sites.[26] Once these loops are created, it is assumed that the spiral receives a large portion of liquid molecules (containing precursor ions along with solvent ions) and a rapid growth proceeds, in the vertical direction.

Fig. 3(b) shows that during the vertical growth of loops, simultaneous growth of a hierarchically ordered film also took place. The proportion of hierarchical film not part of any loop formations was low in comparison to the loop formations that were the predominant features of the surface film. Hierarchically-ordered features were also found to form inside spiral dislocation loops in the presence

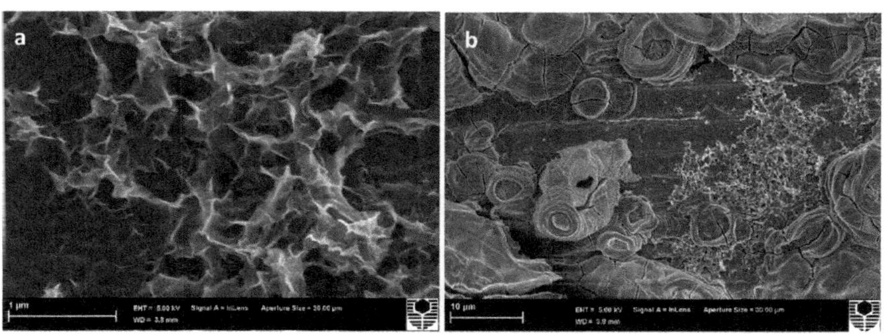

Fig. 3 FESEM images of surface film synthesized with 0.1 M sodium thiosulphate at room temperature in CO_2-saturated brine: (a) exhibiting the thickness of the hierarchical boundaries, and (b) elucidating the steps' thickness in the dislocation spirals.

of 0.1 M sodium thiosulphate (Fig. 1(b)) and at a low concentration of sodium thiosulphate,[11] which suggests the transformation of hierarchical structures into dislocation loops. Screw dislocation promotes the growth of spiral formations as discussed earlier, but the final shape of the spiral always depends on the competition between dislocation core and spiral velocities. We believe that the supply of ionic species was greater for large spirals than for smaller spirals, and therefore we noticed different shapes and sizes of spirals during the film growth with 0.1 M sodium thiosulphate.[27]

Interestingly, we have observed that thicker steps form near the dislocation core compared to the steps further away from the dislocation core, in each individual spiral loop. This again supports our belief that the ion supply was higher for the large spiral region and also established a core of dislocation at high energy sites. In other words, one could say that the steps generated near the dislocation core had higher energy than the steps away from the core. The same phenomenon was observed by Yang *et al.*[27] It should be mentioned that the surface energy of each individual step influences its growth. The observed growth of thick steps near the dislocation core leads us to speculate that a large amount of ion diffusion occurred near the dislocation core and established that site as a high energy site. Interestingly, it is worth pointing out that each step has a unique surface energy that defines its growth velocity and the length of the steps.

Another intriguing aspect in this study is the coexistence of two different shapes of the spiral that generated due to the screw dislocation-driven growth (Fig. 2(a)). The dislocation loops were identified in the hexagonal and pyramidal shapes, as was previously observed by Morin *et al.*[24] It is interesting to note that it has one dislocation core which was centred at the dislocation loop (Fig. 2(a), position 1). A hexagonal pyramidal shape can be observed in Fig. 2(a) at position 1, as well as at the adjacent position. It is evident that around the core, dislocation loops evolved and formed into a hexagonal shape. Few distorted hexagonal dislocation patterns were also noticed, which could further be related with the supply of ions and surface energy. These hexagonal dislocation patterns were rare at the surface compared to the spiral-loop-shaped dislocations, as represented in Fig. 2(a). The spiral formation occurred through one screw dislocation, whereas in a case of interaction of two closely spaced dislocations, a Frank–Read source prevented dislocation loop formation, resulting in a hexagonal pattern instead of a spiral loop.[28,29]

We further investigated whether the morphology and film formation remains the same at a higher sodium thiosulphate concentration. Fig. 4(a) and (b) shows the surface film formed with 0.5 M sodium thiosulphate. It can be seen that screw dislocation loops, such as those found at the lower concentration of sodium thiosulphate, and hierarchical assemblies were not observed in this case. We noticed bent and straight spirals as the predominant features of the surface film. We assume that these spirals could form due to generation of a strain field inside and around spirals. Nucleation and multiplication of any dislocation generates a strain field and it thus can be stated that screw dislocations also generate a strain field.[30] The bend dislocation spirals were seen in Fig. 1(a), and this bending has been speculated to be due to either the anisotropic contact angle made between the particle and its vapour, or to an asymmetric particle orientation, which leads to the torus geometry. We speculate that there may be regions

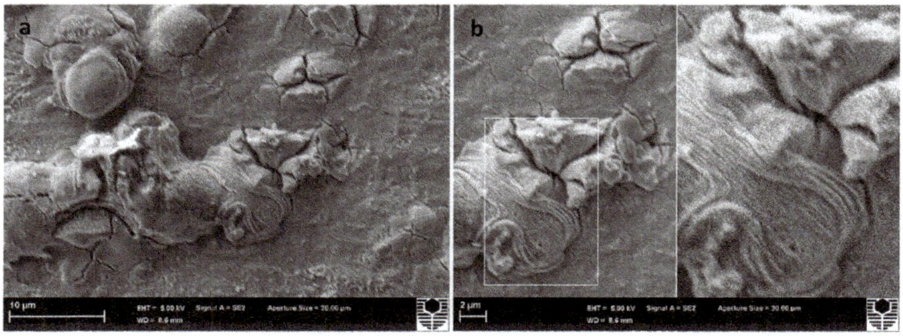

Fig. 4 (a) and (b) FESEM images of surface film synthesized with 0.5 M sodium thiosulphate at room temperature in CO_2-saturated brine. The images show thick film with few closed dislocation loops without any hierarchical structure.

with a higher concentration of ionic species that are considered as compressive, whereas regions with low concentrations of ionic species are responsible for tensile stress generation. This supports our hypothesis about the development of different spiral shapes due to screw dislocations.[31] We observed a few spirals that grew in a vertical direction and had a cylindrical structure, which happened due to the minimization of strain energy. These are further shown in Fig. 11. No hierarchical structures were observed in these cylindrical structures.

3.2 Focused ion beam scanning electron and secondary ion mass spectroscopy

In order to confirm the film composition and describe elemental depth distribution in the surface films, FIB-SEM and SIMS analysis were performed. Fig. 5(a), which is a top view of the depth profile of Na, confirms that the upper surface

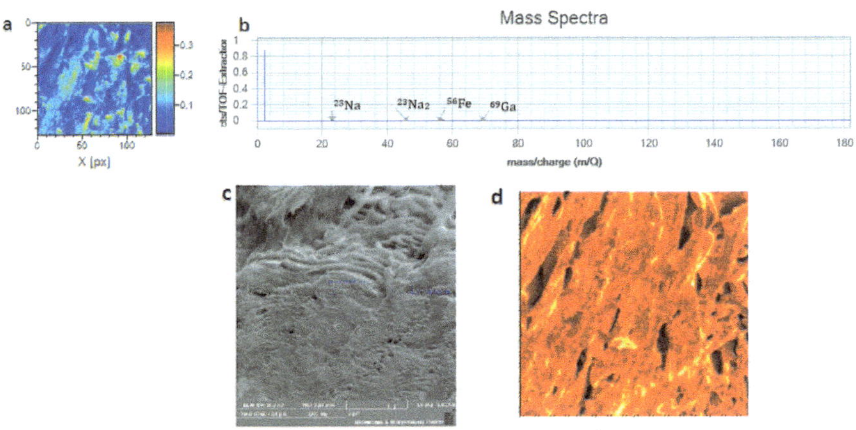

Fig. 5 Analysis of a film synthesized on a carbon steel substrate using 0.1 M sodium thiosulphate in CO_2-saturated brine at room temperature, where (a) represents the top view of the depth profile taken by ToF-SIMS for Na in the surface film, (b) a ToF SIMS mass spectrum confirming the presence of Na and Fe, and (c) and (d) are FIB SEM and FIB signal images, respectively, taken at the site of milling (region of interest).

layer mainly consists of Na, whereas lesser amounts of Na were diffused at a greater depth. Fig. 5(b) confirmed the presence of Na and Fe in the film formed with 0.1 M sodium thiosulphate. It should be noted that Ga^+ ions were used for FIB sectioning, and, therefore, appeared in the mass spectra. The corresponding images are the top and the front view of the elemental distribution. Fig. 5(c) shows the region where FIB milling was performed. It was a pearlitic region, which is part of the carbon steel structure. The corresponding FIB SEM image is shown in Fig. 5(d).

We aimed to mill the area known as pearlite (ferrite and cementite) as we noticed that the pearlitic region was completely covered with the surface film. Fig. 6(a) shows a dislocation loop with hierarchical morphology formed inside the loop; the same area after milling with Ga ions is shown in Fig. 6(b). It is interesting to note that after 50 frames of milling, the final milled depth varied across the studied area, *i.e.* across the loop. The depth values ranged from approximately 244 nm (close to the loop centre) to 395 nm outside the loop. These differences are attributed to the variable susceptibility to milling (hardness) at the various parts of the loop.

We wanted to further investigate the elemental depth distribution at the screw dislocation loop, and found that the amount of Na decreased significantly after a certain depth (approx. 36–55 nm, ~10–15 frames), followed by recordings of the Fe frame only in greater depths. This confirmed that a film has formed at the steel surface. As perceived through the SEM analysis, it should be noted here that screw dislocation with significant hierarchical assembly was also confirmed through the FIB SEM image (Fig. 6(a)), similarly to that shown earlier in Fig. 1(b).

Depth profile measurement of the surface films formed with 0.1 M and 0.5 M sodium thiosulphate confirmed the presence of O, Cl and S, which were also observed through energy-dispersive X-ray spectroscopy (EDS) analysis (see ESI

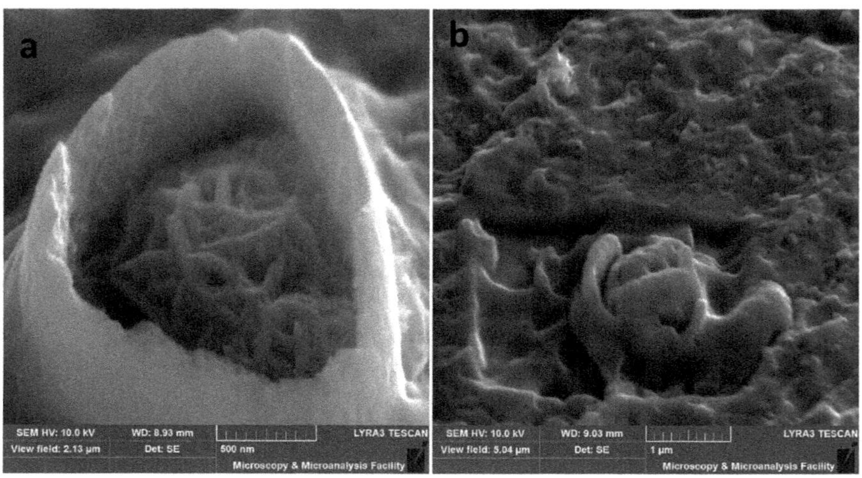

Fig. 6 Examples of screw dislocation spirals in film synthesized on a carbon steel substrate using 0.1 M sodium thiosulphate in CO_2-saturated brine at room temperature. (a) Confirming the association of hierarchical morphology with a dislocation loop for film formed with 0.1 M sodium thiosulphate; (b) exhibiting the same region after ion milling using FIB.

Fig. S1†). Depth distribution analysis done by ToF-SIMS suggested a uniform distribution of O and S, whereas the Cl signal dropped down after around 216 nm (~36 frames), while a high O level was observed at the top layer of the film (down to ~24 nm depth) (Fig. 7(a)). The relevant FIB inspector profile is depicted in Fig. 7(b) and corresponding top views for the O, S and Cl distributions are shown in Fig. 7(c), which elucidated the depth profiles of these elements and brought up an interesting aspect about their distribution within the surface film. The hierarchical morphology formed inside the dislocation loop consists of a significant amount of O and low amounts of Cl and S, while Cl was segregated at the boundary wall of the dislocation loop. This suggests the importance of O in the molecular assembly formation.[11] The formation of stacking faults offers a significant attraction from ions, and we believe that this can be the reason for the observed segregation of Cl at the boundary wall of the dislocation loop.

We have also investigated the effect of pearlite on the sodium thiosulphate film formation and a pearlite region was milled with the ion beam. Surprisingly, depth distribution analysis suggested a completely different phenomenon and confirmed the presence of high amounts of S at the top layer of the film along with O, while Cl remained almost unchanged, as shown in Fig. 8(a). It is worth noting that the opposite phenomenon occurred in the case of depth analysis at the dislocation loop, where high oxygen levels were found at the top of the film. Another noticeable aspect of the analysis was the appearance of a high amount of O at the upper portion of the film (around 18 nm in depth), but about 37 nm lower (approx. 10 frames), the O profile decreased, whereas S remained constant. The

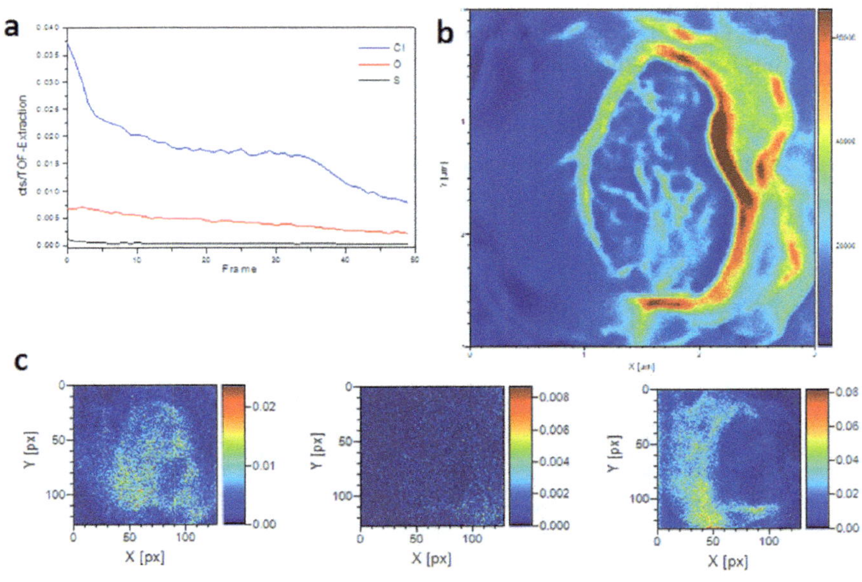

Fig. 7 Analysis of film (imaged in Fig. 6) synthesized on a carbon steel substrate using 0.1 M sodium thiosulphate in CO_2-saturated brine at room temperature. (a) ToF-SIMS depth profile for O, S and Cl in the film, (b) FIB inspector image for the region of interest where depth profiles were recorded, and (c) top views of ToF-SIMS depth profiles for O, S and Cl, respectively (left to right).

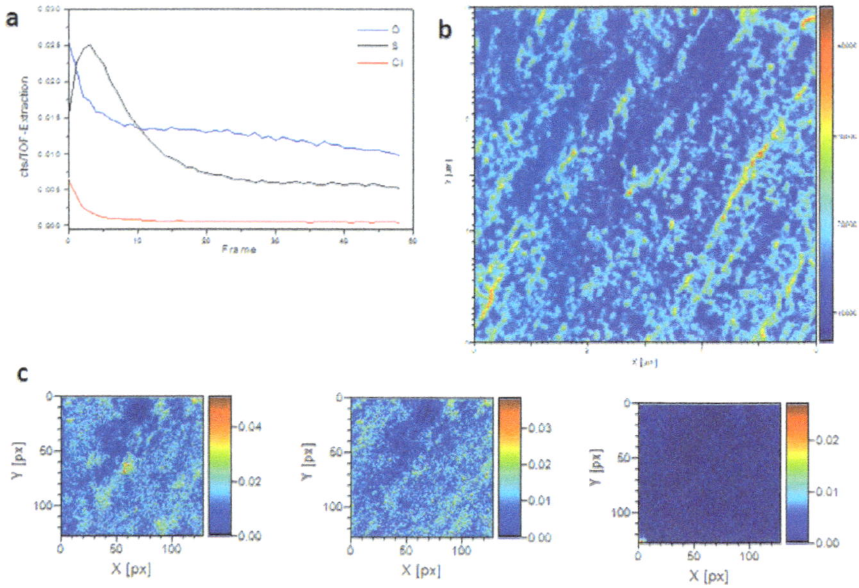

Fig. 8 Analysis of film synthesized on a carbon steel substrate, specifically at a pearlitic region at the substrate, using 0.1 M sodium thiosulphate in CO_2-saturated brine at room temperature. (a) ToF-SIMS depth profiles recorded for O, S and Cl in the film, (b) FIB inspector image for the region of interest where depth profiles were recorded, and (c) top views of ToF-SIMS depth profiles for O, S and Cl, respectively (from left to right).

FIB inspector image is shown in Fig. 8(b), along with depth profile top views for O, S and Cl in Fig. 8(c). It exhibits significant amounts of O and S, whereas not much signal has been obtained for Cl.

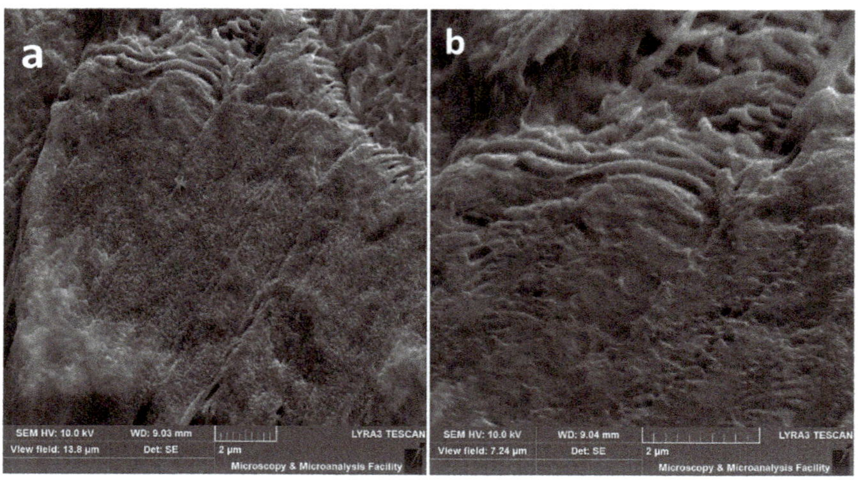

Fig. 9 Analysis of film synthesized on a carbon steel substrate using 0.1 M sodium thiosulphate in CO_2-saturated brine at room temperature. (a) Exhibiting the film formed at a pearlitic region of the steel substrate before ion milling, and (b) after ion milling.

FIB SEM images are shown in Fig. 9(a) for the pearlitic region where FIB milling was performed. Fig. 9(b) exhibits the FIB SEM morphology after milling and respective values for the depth of milling.

Analysis of film formation at pearlite with 0.5 M sodium thiosulphate on carbon steel confirmed the presence of Na, O, Cl and S, similarly to that observed in the case of the film formed with 0.1 M sodium thiosulphate. It was difficult to measure the exact depth of milling due to the lamellar structure of pearlite (Fig. 10(a) and (b)), but depth analysis revealed the accumulation of a higher amount of S on the top of the film compared to O and Cl (Fig. 10(c)). This is a completely different scenario compared to the film formed at pearlite (0.1 M sodium thiosulphate) where high O was measured. Both S and O diffused heavily at the top of the film, whereas S diffused to a greater depth in the film compared to O (Fig. 10(d)). It was portrayed that at around 166 nm (30 frames), the depth profile of S decreased (Fig. 10(c)) whereas the amounts of O and Cl were constant. Interestingly, Cl was segregated at the specific location and exhibited heavy diffusion within that domain (Fig. 10(d)).

Another interesting observation was the formation of dislocation loops without an associated hierarchical morphology. These loops were very few and this suggested that the increment of concentration of sodium thiosulphate in the test solution prevented the formation of screw dislocation and also hierarchical assembly formation (Fig. 11).

3.3 Raman spectroscopic analysis

Raman spectroscopy was used to characterize $Na_2S_2O_3$ films formed on carbon steel with 0.1 M and 0.5 M $Na_2S_2O_3$ brine solutions. Raman spectra acquired from different sites at the steel surfaces are presented in the ESI as Fig. S3 and S4† for $Na_2S_2O_3$ concentrations of 0.1 M and 0.5 M, respectively.

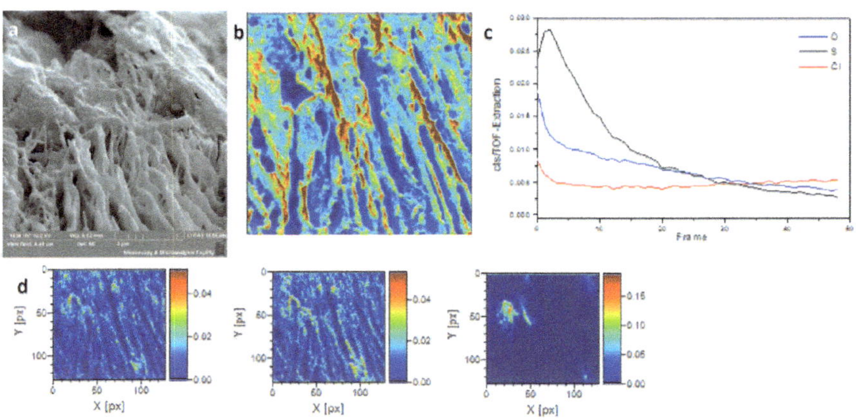

Fig. 10 Analysis of film synthesized on a carbon steel substrate using 0.5 M sodium thiosulphate in CO_2-saturated brine at room temperature. (a) FIB SEM image of a pearlitic region (region of interest), (b) FIB inspector image for the region of interest where depth profiles were recorded, (c) ToF-SIMS depth profiles for O, S and Cl in the synthesized film, and (d) top views of ToF-SIMS depth profiles for O, S and Cl. The relevant front views are presented in the ESI in Fig. S2.†

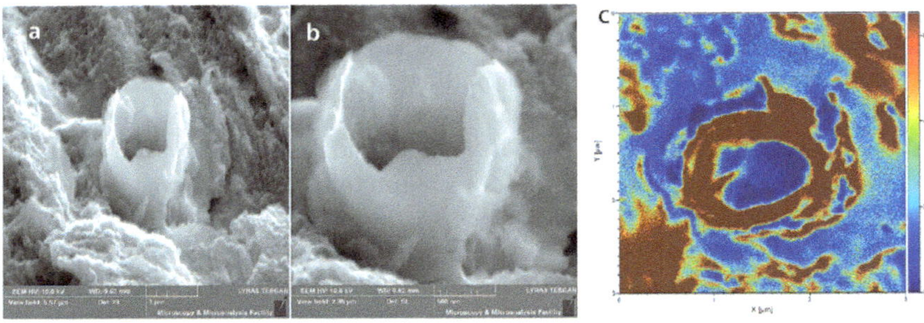

Fig. 11 Dislocation loop formed in film synthesized with 0.5 M sodium thiosulphate in CO_2-saturated brine at room temperature; (a) and (b) show dislocation loop embedded in the film. A limited number of dislocation loops were found in the film, with no hierarchical assembly; (c) shows a FIB inspector image of the dislocation loop shown in (a) and (b). The data confirm that there is no association of the loop formation with hierarchical assembly in the case of high sodium thiosulphate concentration (0.5 M).

For the 0.1 M solution, resonant bands were recorded at 339, 433, 666 and 1326 cm^{-1}. The resonant band at 339 cm^{-1} corresponds to S_2, libration (E_g mode) and can be attributed to pyrite (FeS_2) formation.[32] This band suggests the displacement of S atoms along an axis perpendicular to the S–S bond axis.[33–35] The absence of a resonant band at 373 cm^{-1} indicated the absence of S–S in-phase stretching. Resonant bands at 433 and 666 cm^{-1} confirmed the presence of sodium thiosulphate[36] and magnetite,[37,38] respectively. The appearance of the resonant band at 1326 cm^{-1} can be explained by a shifted band of 1306 cm^{-1} (assigned to Fe_2O_3). This shift could also occur due to two-magnon scattering that occurs due to the interaction of two magnons associated with the proximate anti-parallel spin site in Fe_2O_3.[39] Another assignment of the resonant band at 1326 cm^{-1} can be to hexagonal graphite layers and is termed as the D band. Appearance of the D band indicates the presence of defected and curved graphite sheets with lattice distortion in carbon structures.[40] We speculate that evolution of this band in the Raman spectrum could be due to the defected and curved Fe_3C.

Films synthesized with 0.5 M $Na_2S_2O_3$ showed Raman spectra with resonant bands at 337, 671 and 1357 cm^{-1}. Band formation at 337 cm^{-1} can be assigned to the characteristic peak of S–S (E_g), which is an indication of the formation of impurity-free pyrite.[41] Formation of magnetite or maghemite can be confirmed with the appearance of the band at 671 cm^{-1}.[42,43] The band formed at 1357 cm^{-1} is due to the D band of carbon, as described earlier in this section.[44] This band has also been assigned to polycrystalline graphite, including a C=C vibration indicator.[45] We speculate that this could emerge due to the cementite phase in the carbon steel. The curved lamellae of pearlite (ferrite and cementite) are evidenced from the SEM analysis (see Fig. 9), and confirm the defected phase structure.

In summary, the surface characterization of films assembled at low carbon steel through Raman spectroscopy confirmed the film formation along with the appearance of corrosion products, such as pyrite, magnetite, *etc.* It is worth mentioning that Raman spectroscopy is capable of identifying both amorphous and semicrystalline phases.

3.4 Fourier transform infrared spectroscopy analysis

Fourier transform infrared spectroscopy was used as an additional surface characterization tool to confirm the presence of various functional groups in the films synthesized at the various concentrations of sodium thiosulphate. An FTIR spectrum from a film synthesized with 0.1 M $Na_2S_2O_3$ (ESI, Fig. S5†) confirmed the presence of sodium thiosulphate at the surface, with the absorption bands at 670 cm^{-1} representing the C–S bond.[46,47] The absorption band at 998 cm^{-1} is associated with sulphide groups,[48] and also indicates the presence of $Na_2S_2O_3$. An FTIR spectrum of pure $Na_2S_2O_3$ exhibits bands at 1124 (ref. 49) and 1408 (ref. 50) cm^{-1}, assigned to the SO_2 (asymmetric) functional group. Along with this, bands at 869,[51] 1436,[52] 1635 (ref. 53) and 3353 (ref. 54) cm^{-1} are related to the presence of C–H out-of-plane stretching and O–H functional groups.

Similarly, absorption bands at 661 (ref. 55) and 1162 (ref. 56) cm^{-1} confirmed the presence of S-related groups, such as S–S (di-sulphides), C–S, sulphides and SO_2 group in the film formed with 0.5 M $Na_2S_2O_3$. A representative spectrum is shown in Fig. S6 (ESI†). At the same time, we noticed the appearance of bands at 771,[57] 839,[51,58] 994, 1114, 1616,[59] 1653, 1684 (ref. 60) and 3399 cm^{-1}, which are associated with different functional groups such as O–C–O (for 771 and 1616 cm^{-1}), C–H out-of-plane bending (for 839, 994 and 1114 cm^{-1}) and the OH functional group (for 3399 cm^{-1}). This analysis confirms the presence of a sodium thiosulphate film.

3.5 X-ray diffraction analysis

The only phase identified at the steel treated with 0.1 M $Na_2S_2O_3$ was iron. The phases identified in the presence of 0.5 M $Na_2S_2O_3$ were iron, halite (NaCl), mackinawite (FeS; PDF 015-0037), and sodium thiosulphate pentahydrate ($Na_2S_2O_3\cdot 5H_2O$; PDF 04-009-2374).

The presence of halite and sodium thiosulphate is assumed to be a result of crystallization of the respective phases from solution, whereas mackinawite is a corrosion product. Table 1 gives the calculated relative weight fractions. Mackinawite and sodium thiosulphate both exhibited some preferred orientation indicating that they were present as needles ($r = 1.58$) and plates ($r = 0.72$), respectively.

The widths of the diffraction peaks were also analysed to provide some insight into crystallite size.‡ In both specimens, the widths of the iron peaks indicated

Table 1 Relative weight fractions of the observed phases. The 0.5 M $Na_2S_2O_3$* column refers to relative weight fractions with the iron substrate contribution removed

	0.1 M $Na_2S_2O_3$	0.5 M $Na_2S_2O_3$	0.5 M $Na_2S_2O_3$*
Iron	100	14	—
Halite	—	8	9
Mackinawite	—	38	44
$Na_2S_2O_3\cdot 5H_2O$	—	41	47

‡ Where "crystallite size" means the volume-weighted mean diffraction column length.

a crystallite size of 44(8) nm. The halite peaks were at the instrumental resolution, indicating large, strain-free crystallites. Mackinawite and sodium thiosulphate had crystallite sizes of 19(1) and 32(6) nm, respectively. X-ray diffraction patterns for the films formed at 0.1 M and 0.5 M $Na_2S_2O_3 \cdot 5H_2O$ are presented in the ESI (Fig. S7 and S8,† respectively).

4. Proposed mechanism of screw dislocation-driven spiral growth

We have synthesized films of $Na_2S_2O_3$ with different concentrations which exhibited the spiral-driven growth, assisted by screw dislocation. Screw dislocation-driven growth has been noticed in many materials,[28,61–67] and especially in single crystal Ag,[68] but to the best of our knowledge, dislocation-driven film formation of sodium thiosulphate on low carbon steel under CO_2 corrosion is illustrated for the first time in this study. The screw dislocation was observed at a specific concentration (0.1 M $Na_2S_2O_3$). Increasing the concentration of $Na_2S_2O_3$ led to the formation of a film that did not grow by screw dislocation.

It has been described that the screw dislocation loop introduces stacking faults.[69] As discussed, this could be due to the generation of stress due to the origination of screw dislocation loops that develop to high energy sites. We speculate that the generation of stacking faults decides the growth of the dislocation loops. For example, intrinsic Frank-type stacking faults encourage overgrowth of the spiral loops as shown in Fig. 1, whereas Shockley stacking faults prevent the overgrowth of spirals. We believe that stacking fault generation and development of spirals could not have happened unless the low supersaturation condition is achieved. Generally, screw dislocation spirals form above stacking

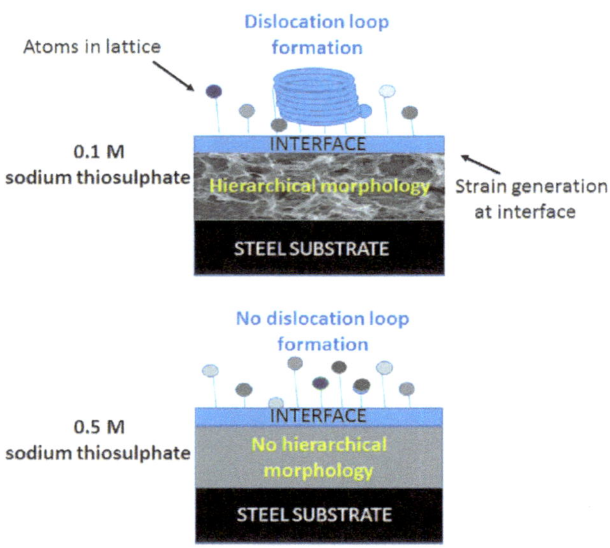

Fig. 12 Elucidating the plausible mechanism of screw dislocation-driven film formation by sodium thiosulphate in CO_2-saturated brine at room temperature.

faults and it is therefore hard to detect those in real scenarios. It has, however, been suggested that intrinsic Frank-type stacking faults are more common than the Shockley stacking faults.[69]

Based on the above facts, we hypothesize and develop a mechanism of growth of screw dislocation-assisted sodium thiosulphate films on carbon steel with significant hierarchical morphology. We speculate that in the early stages of nucleation and growth, a hierarchical assembly forms initially (see Fig. 3(a) and 6(a)). As soon as low supersaturation is attained, the formation condition for screw dislocation is established. The hierarchical morphological assembly has significant porosity (Fig. 3(a)), and we believe that further supply of ions from the bulk solution at a low supersaturation condition assists the generation of stacking faults. These surface sites are considered as high surface energy sites, which allows them to attract more ions from the solution. A strain field may develop by hierarchical assembly[70,71], due to strain associated with the hierarchical structure and the strain developed at the interface between a newly formed spiral and hierarchical assembly. This strain field may cause further development of the spirals. Since energy needs to be minimized, we believe that the minimization of the strain field is the driving force for closing the spiral loop. This phenomenon could be assisted by the generation of intrinsic Frank or Shockley stacking faults, which allows or prevents, respectively, the overgrowth of spirals.

On the contrary, we did not observe hierarchical assembly and dislocation spirals in 0.5 M sodium thiosulphate films. Some dislocation loops were observed at the surface, but those were without hierarchical assembly, which leads us to the assumption that the absence of hierarchical assembly prevents the growth of screw dislocation loops. We therefore propose that hierarchical assembly plays a major role in the dislocation-driven film growth on carbon steel under CO_2 condition at 0.1 M sodium thiosulphate concentration (Fig. 12).

5. Conclusions

We have shown the formation of sodium thiosulphate pentahydrate ($Na_2S_2O_3 \cdot 5H_2O$) dislocation loops with hierarchical assembly on 1030 carbon steel. The steel was exposed to CO_2-saturated brine (NaCl-containing solution) with $Na_2S_2O_3 \cdot 5H_2O$. The concentration of $Na_2S_2O_3 \cdot 5H_2O$ was varied (0.1 M or 0.5 M) in order to study the effect of diffusion flux of its ions on molecular assembling, while keeping constant other experimental parameters such as temperature, gas flow rate, *etc.* The screw dislocation-driven film formation took place by solution deposition and we have observed differences in complex molecular assemblies on the steel that are directly related to the concentration of the sulphur-containing inorganic molecule in the solution.

As an interesting outcome, the study elucidated dislocation-driven film formation at a low concentration of sodium thiosulphate (0.1 M), where "screw dislocation" loops were identified through field-emission scanning electron microscopy (FESEM) analysis. These loops were, however, found to be distorted at the higher concentration of sodium thiosulphate (0.5 M). Interestingly, we have observed complex molecular assembling inside the screw dislocation loop, which is known as "hierarchical assembly", at low concentration only. Furthermore, the surface films were characterized through focused ion beam-secondary ion mass spectroscopy (FIB-SIMS) and the elemental depth distributions were assessed using time-of-flight secondary ion mass spectroscopy (ToF-SIMS). The film analysis revealed significant differences within the dislocation loop structures with internal hierarchical assemblies as well as films formed at the two $Na_2S_2O_3$ concentrations. Besides $Na_2S_2O_3$, the surface films consisted of phases such as pyrite/mackinawite and magnetite, which result from iron dissolution (corrosion processes) under the acidic test conditions.

This study delineates the formation of a hierarchical assembly with screw dislocation spirals (loops). The role of hierarchical morphology in the development of screw dislocation spirals' generation has not been depicted in earlier studies, and we have proposed the plausible mechanism of the formation of screw dislocation loops and the role of hierarchical morphology in the emergence of dislocation loops. This development of a mechanistic understanding of the molecular assembly of $Na_2S_2O_3$ is of significant interest to scientific communities, such as crystal growth, catalysis, surface science, and specifically to corrosion scientists, as $Na_2S_2O_3$ serves as a performance enhancer in corrosion inhibitor formulations widely applied in the oil and gas industry.

Acknowledgements

We acknowledge the facilities, scientific, and technical assistance of the Curtin University Microscopy & Microanalysis Facility and the Australian Microscopy & Microanalysis Research Facility at the Centre for Microscopy, Characterisation & Analysis (CMCA), The University of Western Australia, both of which are partially funded by the University, State, and Commonwealth Governments, the WA X-ray Surface Analysis Facility, funded by the Australian Research Council LIEF grant (LE120100026), and the Department of Chemistry/Nanochemistry Research Institute, Curtin University. We specifically thank Dr Aaron Dodd (CMCA) and Dr William Rickard (Curtin University) for technical assistance. D. D. acknowledges Curtin University for a Curtin International Postgraduate Research Scholarship.

References

1 V. Jovancicevic, Y. S. Ahn, J. A. Dougherty and B. A. Alink, *CO_2 corrosion inhibition by sulfur-containing organic compounds*, NACE International, 2000.

2 M. Kappes, G. S. Frankel, N. Sridhar and R. M. Carranza, *J. Electrochem. Soc.*, 2012, **159**, C195–C204.

3 M. Kappes, G. S. Frankel, N. Sridhar and R. M. Carranza, *Corrosion*, 2012, **68**, 872–884.

4 D. J. Gavia, Y. Do, J. Gu and Y.-S. Shon, *J. Phys. Chem. C*, 2014, **118**, 14548–14554.

5 D. Jaganyi, M. Altaf and I. Wekesa, *Appl. Nanosci.*, 2013, **3**, 329–333.

6 V. Chen, H. Pan, R. Jacobs, S. Derakhshan and Y.-S. Shon, *New J. Chem.*, 2017, **41**, 177–183.

7 Q. Wang, J. Li, Y. Bai, J. Lian, H. Huang, Z. Li, Z. Lei and W. Shangguan, *Green Chem.*, 2014, **16**, 2728–2735.

8 G. Hou, Z. Cheng, L. Kang, X. Xu, F. Zhang and H. Yang, *CrystEngComm*, 2015, **17**, 5496–5501.

9 R. Ghosh Chaudhuri and S. Paria, *Dalton Trans.*, 2014, **43**, 5526–5534.

10 D. Dwivedi, K. Lepkova and T. Becker, *RSC Adv.*, 2017, **7**, 4580–4610.

11 D. Dwivedi, K. Lepkova, T. Becker and M. Rowles, 2017, manuscript in preparation.
12 M. Kucher, D. Babic and M. Kind, *Chem. Eng. Process.*, 2006, **45**, 900–907.
13 G. Nehrke, G. J. Reichart, P. Van Cappellen, C. Meile and J. Bijma, *Geochim. Cosmochim. Acta*, 2007, **71**, 2240–2249.
14 Bruker AXS, *EVA 3.2*, Karlsruhe, Germany, 2013.
15 H. M. Rietveld, *J. Appl. Crystallogr.*, 1969, **2**, 65–71.
16 Bruker AXS, *TOPAS 5*, Karlsruhe, Germany, 2014.
17 M. R. Rowles and I. C. Madsen, *J. Appl. Crystallogr.*, 2010, **43**, 632–634.
18 R. W. Cheary, A. A. Coelho and J. P. Cline, *J. Res. Natl. Inst. Stand. Technol.*, 2004, **109**, 1–25.
19 W. A. Dollase, *J. Appl. Crystallogr.*, 1986, **19**, 267–272.
20 T. Ida, *Effect of Preferred Orientation in Synchrotron X-ray Powder Diffraction, Nagoya Institute of Technology*, Annual Report of the Advanced Ceramics Research Center, 2013.
21 M. R. Rowles and C. E. Buckley, *J. Appl. Crystallogr.*, 2017, **50**, 240–251.
22 R. J. Hill and C. J. Howard, *J. Appl. Crystallogr.*, 1987, **20**, 467–474.
23 D. Balzar, in *Defect and Microstructure Analysis by Diffraction*, ed. R. L. Synder, J. Fiala and H. J. Bunge, Oxford University Press, Oxford, 1st edn, 1999, ch. 7, pp. 94–126.
24 S. A. Morin, A. Forticaux, M. J. Bierman and S. Jin, *Nano Lett.*, 2011, **11**, 4449–4455.
25 L. Chen, B. Liu, A. N. Abbas, Y. Ma, X. Fang, Y. Liu and C. Zhou, *ACS Nano*, 2014, **8**, 11543–11551.
26 Z. Ren and P.-X. Gao, *Nanoscale*, 2014, **6**, 9366–9400.
27 S. K. Yang, M. Y. Li, X. Zhu, G. Q. Xu and J. H. Wu, *J. Phys. Chem. C*, 2015, **119**, 14312–14318.
28 A. G. Shtukenberg, Z. Zhu, Z. An, M. Bhandari, P. Song, B. Kahr and M. D. Ward, *Proc. Natl. Acad. Sci. U. S. A.*, 2013, **110**, 17195–17198.
29 M. J. Shearer, L. Samad, Y. Zhang, Y. Zhao, A. Puretzky, K. W. Eliceiri, J. C. Wright, R. J. Hamers and S. Jin, *J. Am. Chem. Soc.*, 2017, **139**, 3496–3504.
30 J. J. Zhang, Z. F. Wang, Y. D. Yan and T. Sun, *RSC Adv.*, 2016, **6**, 59206–59217.
31 S. Amelinckx, X. B. Zhang, D. Bernaerts, X. F. Zhang, V. Ivanov and J. B. Nagy, *Science*, 1994, **265**, 635.
32 K. Sun, Z. Su, J. Yang, Z. Han, F. Liu, Y. Lai, J. Li and Y. Liu, *Thin Solid Films*, 2013, **542**, 123–128.
33 Y. Bi, Y. Yuan, C. L. Exstrom, S. A. Darveau and J. Huang, *Nano Lett.*, 2011, **11**, 4953–4957.
34 M. Y. C. Teo, S. A. Kulinich, O. A. Plaksin and A. L. Zhu, *J. Phys. Chem. A*, 2010, **114**, 4173–4180.
35 A. Dubey, S. K. Singh, B. Tulachan, M. Roy, G. Srivastava, D. Philip, S. Sarkar and M. Das, *RSC Adv.*, 2016, **6**, 16859–16867.
36 J.-L. Xia, Y. Yang, H. He, X.-J. Zhao, C.-L. Liang, L. Zheng, C.-Y. Ma, Y.-D. Zhao, Z.-Y. Nie and G.-Z. Qiu, *Hydrometallurgy*, 2010, **100**, 129–135.
37 A. Demoulin, C. Trigance, D. Neff, E. Foy, P. Dillmann and V. L'Hostis, *Corros. Sci.*, 2010, **52**, 3168–3179.
38 M. Etique, F. P. A. Jorand and C. Ruby, *Geobiology*, 2016, **14**, 237–254.
39 K. Rout, M. Mohapatra, S. Layek, A. Dash, H. C. Verma and S. Anand, *New J. Chem.*, 2014, **38**, 3492–3506.

40 X. Wang, P. Zhang, W. Wang, X. Lei, B. Zou and H. Yang, *RSC Adv.*, 2015, **5**, 27857–27861.

41 B. Yuan, W. Luan and S.-t. Tu, *Dalton Trans.*, 2012, **41**, 772–776.

42 G. Genchev and A. Erbe, *J. Electrochem. Soc.*, 2016, **163**, C333–C338.

43 L. M. Barge, I. J. Doloboff, E. Flores, S. S. S. Cardoso, J. H. E. Cartwright, E. Macias-Sanchez, C. I. Sainz-Diaz and P. Sobron, *Proc. R. Soc. A*, 2016, **472**, 20160466.

44 X. Huang, H. Yu, J. Chen, Z. Lu, R. Yazami and H. H. Hng, *Adv. Mater.*, 2014, **26**, 1296–1303.

45 Y. Wu, H. Yang, L. Zhu, A. Xie, S. Li, J. Song and Y. Shen, *Anal. Methods*, 2014, **6**, 7083–7087.

46 A. Mendez-Vilas, *Materials and Technologies for Energy Efficiency*, Brown Walker Press, 2015.

47 B. Saha, A. Singh and S. Sengupta, *RSC Adv.*, 2016, **6**, 76434–76447.

48 X. Guo, H.-S. Xiao, F. Wang and Y.-H. Zhang, *J. Phys. Chem. A*, 2010, **114**, 6480–6486.

49 L. Li, Z. M. Chen, Y. H. Zhang, T. Zhu, S. Li, H. J. Li, L. H. Zhu and B. Y. Xu, *J. Geophys. Res.: Atmos.*, 2007, **112**, D18301.

50 R. A. Nyquist, *Interpreting Infrared, Raman, and Nuclear Magnetic Resonance Spectra*, Elsevier Science, 2001.

51 N. M. Logacheva, V. E. Baulin, A. Y. Tsivadze, E. N. Pyatova, I. S. Ivanova, Y. A. Velikodny and V. V. Chernyshev, *Dalton Trans.*, 2009, 2482–2489, DOI: 10.1039/b819805e.

52 J. Ĉulić-Viskota, W. P. Dempsey, S. E. Fraser and P. Pantazis, *Nat. Protoc.*, 2012, **7**, 1618–1633.

53 F. C. Jentoft, *Advances in Catalysis*, Elsevier Science, 2014.

54 D. Y. Corredor and Kansas State University, *Pretreatment and Enzymatic Hydrolysis of Lignocellulosic Biomass*, Kansas State University, Department of Biological and Agricultural Engineering, 2008.

55 M. Area, A. Garau, F. Isaia and V. Lippolis, *Spectrosc. Lett.*, 1996, **29**, 697–709.

56 L. Boucher, U. Hettwer, E. Schleitzer-Rust, H. Demmer, M. Kotowski, K. Koeber and D. Tille, *Mn Manganese: Coordination Compounds 7*, Springer, Berlin, Heidelberg, 2013.

57 S. Gunasekaran and S. C. R. Academy, *Proceedings on International Conference on Recent Advances in Applied Sciences: ICRAAS 2016*, 2016.

58 A. Brandt, L. Chen, B. E. van Dongen, T. Welton and J. P. Hallett, *Green Chem.*, 2015, **17**, 5019–5034.

59 *Zeolites: A Refined Tool for Designing Catalytic Sites*, ed. L. Bonneviot and S. Kaliaguine, Elsevier, 1995, vol. 97.

60 F. Liu and H. He, *Catal. Today*, 2010, **153**, 70–76.

61 C. M. Pina, U. Becker, P. Risthaus, D. Bosbach and A. Putnis, *Nature*, 1998, **395**, 483–486.

62 S. Kodambaka, S. V. Khare, W. Swiech, K. Ohmori, I. Petrov and J. E. Greene, *Nature*, 2004, **429**, 49–52.

63 A. R. Verma, *Nature*, 1951, **167**, 939.

64 S. Amelinckx, *Nature*, 1951, **167**, 939–940.

65 H. H. Teng, P. M. Dove, C. A. Orme and J. J. De Yoreo, *Science*, 1998, **282**, 724.

66 S. R. Qiu, A. Wierzbicki, C. A. Orme, A. M. Cody, J. R. Hoyer, G. H. Nancollas, S. Zepeda and J. J. De Yoreo, *Proc. Natl. Acad. Sci. U. S. A.*, 2004, **101**, 1811–1815.

67 S. R. Qiu, A. Wierzbicki, E. A. Salter, S. Zepeda, C. A. Orme, J. R. Hoyer, G. H. Nancollas, A. M. Cody and J. J. De Yoreo, *J. Am. Chem. Soc.*, 2005, **127**, 9036–9044.
68 J. F. Wolf and H. Ibach, *Appl. Phys. A: Solids Surf.*, 1991, **52**, 218–221.
69 S. Harada, K. Seki, Y. Yamamoto, C. Zhu, Y. Yamamoto, S. Arai, J. Yamasaki, N. Tanaka and T. Ujihara, *Cryst. Growth Des.*, 2012, **12**, 3209–3214.
70 P. Kim, M. Abkarian and H. A. Stone, *Nat. Mater.*, 2011, **10**, 952–957.
71 Z.-C. Shao, Y. Zhao, W. Zhang, Y. Cao and X.-Q. Feng, *Soft Matter*, 2016, **12**, 7977–7982.

Faraday Discussions

DISCUSSIONS

Supramolecular systems at liquid–solid interfaces: general discussion†

David Amabilino, Ioan Bâldea, Pol Besenius, Peter Beton, Matthew Blunt, Manfred Buck, Neil R. Champness, Lifeng Chi, Stuart Clarke, Giovanni Costantini, Steven De Feyter, Yuri Diaz Fernandez, Deepak Dwivedi, Karl-Heinz Ernst, Amar Flood, Brandon Hirsch, Robert Jones, Angelika Kühnle, Markus Lackinger, Trolle R. Linderoth, Natalia Martsinovich, Andrew Mount, Martin Nalbach, Claire-Marie Pradier, Talat Rahman, Rasmita Raval, Neil Robinson, Marco Sacchi, Sebastian Schwaminger, Steven L. Tait, Phil Woodruff and Han Zuilhof

DOI: 10.1039/c7fd90074k

Ioan Bâldea opened a general discussion of the paper by Amar Flood: Could you please give details on how you model the interaction between the anion and the molecule?

Amar Flood answered: As noted in the paper (DOI: 10.1039/C7FD00104E), we geometry optimized the structure of the molecular receptor using DFT and kept this as fixed. We extracted point partial charges for the atoms. The charges were used to calculate the energy of the bromide ion in the proximity of the receptor. The locations of the atoms were used as the origins of the vdW radii of the molecule.

Rasmita Raval said: Yesterday we talked about not ignoring disorder. Today, you have demonstrated that we shouldn't ignore fuzzy STM images. Do you think that the natural focus of scientists using SPM techniques to image beautifully ordered regions is affecting our overview on complex molecule behavior at surfaces?

Amar Flood responded: Yes and no. You first have to prove your scientific credibility in correctly locating, identifying, and then interpreting the ordered regions. After that, you can tackle the harder, more fuzzy and less ordered regions of the datasets. The same dichotomy is seen with thermodynamics (easier and with permanence) and kinetics (harder and fleeting). "Flaunt it if you got it" is all I can say.

† Electronic supplementary information (ESI) available. See DOI: 10.1039/c7fd90074k

Steven De Feyter asked: If you kick out the anion, you go from one adsorbed phase to the other one? In other words, removing the anion destabilizes the adsorbed phase it was part of?

Amar Flood replied: Yes, with the correct molecular receptor on the surface, then ejection of the anions leads to the molecules transforming into another phase. By "correct", I mean to say that the receptors with C_{18} chains in the tails undergo the transformation. The factors (vdW and dipole) controlling this phase selection have been reported elsewhere.[1] To the second part of the question, removal of the anion does destabilize the adsorbed phase relative to another phase. Loss of the anion leads to larger molecular dipoles that introduce dipole–dipole repulsions that are relieved when the molecule forms the other ordered (and dipole stabilized) phase.

1 B. E. Hirsch *et al.*, *J. Chem. Phys.*, 2015, **142**, 101914.

Steven De Feyter enquired: What is the time scale of the shuttling of the anion, and of the change in adsorbate structure as a result of a "relocation" of the anion?

Amar Flood responded: The movement of the anion is many orders of magnitude faster than the change in adsorbate structure. From the model we created, the barrier to motion of the bromide anion when the tip is very far away from the anion is on the order of 10 s. As the tip gets closer, the barrier gets increasingly smaller until it disappears. For the adsorbates reordering on the surface (with presumed partial desorption into solution), it takes multiple minutes to see ordered regions by STM.

Steven L. Tait commented: It would be interesting to attempt these experiments using a constant-height scanning mode in STM so that the tip-surface distance could be controlled independent of the STM tip-surface bias. This may allow for more control of the dynamic behavior in the gap. It is difficult to interpret the dynamic behavior here because the size of the tip-surface gap is so small that there is room for only a few molecules in this region. We would like to consider the solution in the simplest picture of having uniform properties, but that model is obviously inadequate in the gap where there may be only one or two solvent molecules and they are likely to be somewhat restricted to a specific orientation, therefore not having much "liquid" character. Even though the STM tip-surface gap in ultra-high vacuum can be modeled well, we need more development of our theoretical description of that gap in solution/surface environments, as well as ambient environments. The work presented here by Prof. Flood highlights some of the complexity of that interface in the dynamic behavior of surface-bound anions, but there is much more to be explored here, including the dynamic behavior of molecules with dipole moments and the behavior of the solvent.

Andrew Mount said: Building on the discussion about the potential for carrying out experiments at constant height (known distance) above the surface, having performed a previous topological scan, could a small amplitude ac voltage oscillation superimposed on a chosen dc bias voltage between sample and tip as a function of frequency not then provide interesting experimental characterisation information? At high frequencies one might expect to see currents and hence impedances governed by the dielectric (probing solvent character), whilst at lower frequencies ion movement might be possible, and the differences between distinct surface states might become evident. Are there any technical challenges to making such measurements?

Amar Flood answered: Your suggested experiment is, in principle, exactly as one might imagine. As for the technical challenges and if they exist, those are for the instrument builders amongst us and that is not my area of expertise; rather, I am a molecule builder. So if you want a molecule to make this type of behavior "resonate", then I and others in my area of expertise can help.

Robert Jones asked: Due to the movement of electrons at a surface, a bromide ion on a metal conductor would have an image charge of opposite polarity located an equal distance below the surface as the bromide is above the surface. The bromide and its image form a relatively strong, attractive, Coulombic interaction holding the bromide on the surface. For the bromide on graphite case in this paper, does graphite have sufficient mobile electron density to allow such an image charge to form, and if so, was this taken into account when calculating the energetics?

Amar Flood responded: The assumption was made that the graphite can support image charges and there is nothing to suggest that it does not. Yes, the character of the graphite was taken into account in the form of the dielectric constant when including the image charge.

Robert Jones said: In answering the question about whether an image charge of the bromide could be formed in the graphite substrate, you mentioned that the STM tip was metallic, and that an image charge could therefore be formed in the tip. If such an image charge was formed in the STM tip, is it likely, given the distances involved, that the attraction of the bromide for its image in the STM tip could contribute to displacing it from the substrate to the tip?

Amar Flood replied: Perfection! Yes, that is exactly what we see. As the tip approaches the surface, the largest effect on the bromide's overall stability is the overlapping effects of the two image charges. However, the electric field and the receptor's stabilization of the bromide are what decide which location (tip or surface) has the lower energy for the anion.

Pol Besenius enquired: A supramolecular question from my side. On your first slide, you pointed out the importance of the hydrogen bond between the aryl amide and triazole units of the halogen receptor. The two amide carbonyl units do get fairly close, while the hydrogen bonds are locked up. Do you ever observe a change in the hydrogen bonded conformation of the physisorbed receptor unit on the surface in the absence of the bromide or iodide guests?

Amar Flood responded: Good question and insight on the molecular details and mutual differences in conformational energies that we need to consider. We have never seen the "unfolded" conformation on the surface. We do, however, have molecules in which some of the amides are missing and observe such conformations, giving us a little license to make these types of judgements.

Giovanni Costantini asked: How do your calculations scale with the effect of curvature of the tip? I guess it would be possible to have an essentially flat tip (extremely large radius of curvature) with a single atom protruding from it. This tip would be an excellent STM resolving tip but the electric field at the position where the image is generated and in a relatively large region around it would be essentially constant. How would this affect your results and main conclusions?

Amar Flood answered: We did not examine the effect of the tip's curvature on the electric field, and thus the relative stability of the bromide anion if the tip was essentially flat. However, if pressed, then I could speculate about this possibility. I imagine a tip some 100 nm in width with one atom protruding from it at the center of the 100-nm block (as suggested). At both very high positive or negative biases (*e.g.*, ± 1 V or greater), then all the bromide anions would either be attracted to or repelled, respectively, from the surface prior to the tunneling atom reaching the area being imaged. Correspondingly, one would expect to see all bromide ions sitting in the receptor pockets or the receptors empty. Thus, all the receptors in the field of view would be the same, which differs from the curved tip. With the curved tip, we saw loss of image quality when using negative surface biases of decreasing magnitude. In this case, the model indicates this effect arises from the bromide being stable in the surface receptor except when the tunneling region of the tip gets close to said bromide anions. With a 100-nm flat headed tip, all those anions will be expected to be ejected from the surface receptors at a distance far away from the tunneling region. In that case, any and all negative biases would show the empty receptors at the surface on account of the fact that the anions would have been ejected at times and locations far away from the tunnel region.

Brandon Hirsch commented: Pertaining to the nature of the effects that would impact the anion ejection, it is certainly necessary to consider local tip shape impacts, but one should also take into account the absolute slope of the surface when no fitting to the raster scan is done. In such a case we often see that the surface has a slight slope moving "uphill"/"downhill" depending on the trace/retrace. This subtle change in tip movement also acts in combination with tip shape effects to alter the voltage-dependent anion ejection.

Peter Beton addressed Amar Flood: I have a question about ion interaction on the surface, is it mostly bromine you're using or are there others? Is there any information about the binding energy?

Amar Flood responded: We have looked at bromide, iodide, and tetrafluoroborate (BF_4^-) anions. We have quantified their binding energies to the receptors in homogeneous solutions, which are all about 26 kJ mol^{-1} and with slight drop-offs $Br^- > I^- > BF_4^-$. So we know their relative binding energies in solution. We can see binding of bromide and iodide but not BF_4^- at the surface. Presumably binding in solution is allowed by expansion of the binding pocket to accommodate the BF_4^- anion's larger size. But this cannot occur in the crystalline (and fixed) receptors on the surface. If you really care to ask about the binding energies of the anion into the receptor pockets in the surface-arrayed system, then that is contained in the model. Therein, image charges, and electric fields add together. From the paper (DOI: 10.1039/C7FD00104E), at 0 V I can pull the number out at about 64 kJ mol^{-1} or ~0.66 eV or ~25 kT.

Peter Beton asked: For a bigger ion, would you get a smaller image charge?

Amar Flood answered: Focusing on the surface-bound anions, bigger ions are further away from the surface resulting in a smaller energy of interaction coming as a result of the image charge. I do not know if the spatial image would be bigger or smaller in size.

Peter Beton remarked: Is there a value for the adsorption energy of one of these ions on the surface, and does it vary between the different ions?

Amar Flood replied: As mentioned previously, the adsorption energy of the anions inside the receptors is close to 64 kJ mol^{-1} or ~0.66 eV or ~25 kT. Without the receptor, which confers around 26 kJ mol^{-1}, the adsorption energies are lower.

Peter Beton said: Regarding the counter ions, if it is an image charge holding the ions on the surface, would there be the same interaction for the positive ion?

Amar Flood replied: Yes, that is our expectation as well.

Steven L. Tait said: I have a comment following on from discussion about the tunnelling gap. The question of how to model the STM tip to surface gap is rather interesting and quite challenging. The gap is so small that with a molecular adsorbate present there may just be room for a single counter-ion (if a charged adsorbate) or a single solvent molecule. Therefore, it no longer seems reasonable to assume a mean field characterization for the solvent and one must question usual ideas about screening ions in solution and other bulk effects of the solution. In the calculations that Prof. Flood presented, we made a point to vary the size of the tip–surface gap to explore how this would impact the behaviors we observed. It would be very valuable to have other groups build on this work with more detailed and rigorous calculations of adsorbate (charged or neutral) interactions within the STM tip–surface gap. The study points out that the electric field off to the sides of the tunneling junction may affect the adsorbates. It would be useful to study this for different tip shapes.

Talat Rahman asked: Long back we did calculations on the effect of the shape of the tip in manipulation at surfaces and found that the shape of a tip could make a difference.[1] In our case, the tip was close to the surface, while in your case it is probably further away. However, a pointed tip could make a difference in the presence of the electric field that you have here. I wonder how difficult it would be to do a couple of calculations with pointed tips rather than the shapeless one here, or to do an experiment in which you use a pointed tip?

1 C. Ghosh, A. Kara and T. S. Rahman, *Surf. Sci.*, 2002, **502–503**, 519–526.

Amar Flood answered: This is nice work you cited. We have long considered the tip a black box. A mysterious entity constantly being shaped and reshaped during the actual experiments. We expect it to have a gross shape that is ill-defined and extremely difficult to alter or control at will. That said, my personal opinion is that we could examine all sorts of tip shapes. We may even collect data that corresponds to one shape over another. But, we will always have to face the reality that we will never know the tip's shape. The work on SERS of plasmonically active particles has the same problem. They got around it by TEM imaging of the actual particles they got SERS spectra from. Perhaps a similar method may be viable in this field. Until then, I worry about drawing too much (or perhaps too little) insight from such studies?

Talat Rahman commented: What we were doing was simpler, looking at the effect of the shape of the tip on the potential energy surface, adatoms diffusing, manipulation of atoms, and we found that electric field could concentrate at sharp tips. It might be something to concentrate on in the future.

David Amabilino asked: I'd like to come back to the counterion effect. I didn't notice the counterion thing taken into consideration in your model, could you comment?

Amar Flood replied: You are correct. We were not able to account for the counter cation. Our model could only account for the energy of a point charge in the tip-surface region. So, the counter cation's energy would largely be the inverse of the anion's. In addition, we were not able to include the dynamics for ion pair association and dissociation. That would require a much different approach.

David Amabilino enquired: What would happen if instead of putting an anion in your model, you put an associated dipole (because I guess ion pairing is tight)?

Amar Flood answered: If we put an associated dipole, let's call it a permanent electric dipole, into our model, then we would expect the dipole to orient rather than move within the electric field. In addition to that part of the model, we would need to include features for the dipole binding to the receptor (if the receptor was designed to bind the dipole), and for the image dipoles in the surface and tip to also stabilize the dipole. Presumably, the dipole–image dipole interactions would be weaker in magnitude than the charge–image charge interactions. In such cases, the receptor binding might play a bigger role than the other factors, which is not what we saw with the bromide.

David Amabilino remarked: I recognise the fact that this system is nice as you have a receptor for an anion which is not a good receptor for a cation, but nonetheless, the cation has to be around there somewhere, right? When you flip the system with the electric field, the anion is going to the tip as you showed, and therefore your cation is going to the surface presumably, but it doesn't bind well so is it mobile? Can you design a system where you can control the dipole orientation, *e.g.* the example that Giovanni showed (K docking on top of TCNQ), and can you think about flipping that around to try and control it in that way and switch orientation of the dipole? Could this be the basis for a device?

Amar Flood replied: Yes, the cation must be present for charge balance. We also assume that the cation goes to the surface at large negative surface biases but it cannot settle and order at the surface, and this situation offers no chance to image the cations by the trace and retrace needed for STM. As to the device idea you propose, yes, using an electric field orienting and reorienting dipoles could work and I believe it is the basis for ferroelectric systems. The dipole reorienting has also been reported in ref. 1.

1 S.-B. Lei *et al.*, *Nano Lett.*, 2008, **8**, 1836–1843.

David Amabilino asked: Does ion pairing change with electric field?

Amar Flood replied: I am not exactly sure on this one. We imagine that if a tight ion pair exists, that it will be a dipolar zwitterion. This dipole will orient in the field. However, all tight ion pairs have a finite lifetime. When the two ions dissociate then they will move in the field away from each other. So, it is likely that the electric field hastens the dissociation of tight ion pairs and destabilizes them.

Neil R. Champness opened a general discussion of the paper by Manfred Buck: In the systems that you have studied, what are the timescales associated with the rotation of the guest molecule?

Manfred Buck answered: While the temporal resolution of the STM is far too low to directly access the time scale of the rotation at the temperature of our experiments (room temperature), we estimated the range ($10–10^4\ s^{-1}$) based on literature work (the details are described on page 7 of our paper, DOI: 10.1039/C7FD00115K).

Neil R. Champness remarked: Is there any possibility that the guest molecule is not rotating, it's actually flipping perpendicular to the surface?

Manfred Buck responded: Flipping over should be much less likely compared to rotation, as the latter requires only one arm to be detached from the surface whereas the former would require a more substantial detachment and/or an energy-costly deformation of the molecule.

Markus Lackinger said: You mentioned that the preparation sequence is important, or in other words the system evolves under kinetic control. I can clearly see this for the system with thiol SAMs as guests, but for C_{60} as guests I would expect that the targeted structure is the thermodynamically most stable one. Could you comment on this?

Manfred Buck replied: For the honeycomb network only, adsorption of seven C_{60} molecules results in a thermodynamically stable state, in contrast to thiol adsorption which, in general, is kinetically controlled and, ultimately, displaces the network. For this reason the two types of molecules were chosen for the experiments. For the star-modified network it is, based on the existing data, difficult to differentiate between thermodynamic and kinetic control, *i.e.*, whether three fullerenes plus star are more stable than a cluster of seven fullerenes and no star. Our observation, that the immersion time for C_{60} is not critical in the case of the star-modified network, is consistent with a thermodynamically-controlled assembly process. However, it is also conceivable that once one or two subpores are filled, the activation barrier to displace the star molecules by C_{60} (*e.g.* due to suppression of the rotational motion with partial detachment) becomes too high to occur sufficiently fast. It would need *in situ* experiments to monitor individual pores in order to decide whether clusters of six or seven C_{60}, as seen sporadically in Fig. 6 of the paper (DOI: 10.1039/C7FD00115K), form only in those pores void of a star molecule or also in other pores by displacement of a star molecule.

Markus Lackinger asked: You were saying that lateral mobility is a necessity for replacing the excess network forming PTCDI molecules that are stably adsorbed within the pores by C_{60} guest molecules. Can you explain this in more detail?

Manfred Buck answered: PTCDI adsorbs strongly, so the desorption channel is negligible as concluded from our unsuccessful attempts to clean the pores by immersion of the network into the pure solvent at elevated temperature. That means an adsorption site occupied by a PTCDI molecule only becomes available for a fullerene molecule if the former diffuses laterally. Since the highly confined environment of a subpore eliminates this channel, C_{60} cannot adsorb in a subpore if already occupied by a PTCDI molecule.

Yuri Diaz Fernandez said: For the adsorption of the star molecules you have used two different solvents. Have you observed any effect of the solvent on the occurrence of defects, or on the occupancy of the pores? Is the choice of the solvent based on the solubility of the molecules?

Manfred Buck replied: While we have not studied this in any depth, no obvious difference between solvents has been observed. In general, solubility can indeed be an important criterion but that is dependent on the particular system. Boiling/freezing points are quite important as they define the temperature range available for the preparation. Another point is toxicity.

Martin Nalbach asked: If I understood correctly, you adsorb the molecules from solution and investigate the surface afterwards in air? How do you remove the solvent?

Manfred Buck replied: This is done by quickly blowing the sample dry in a stream of nitrogen when the sample is removed from the solution.

Martin Nalbach said: At the moment you take the sample out of the solution, how do you ensure that any solute in the solvent does not precipitate on the surface due to a local high supersaturation at the interface which is caused by the radical decrease of solvent volume? In the experimental part of your paper, you mention an adsorption time of about 2 min for the star molecules as well as for

C_{60}. How do you ensure that the molecular structures you observe on the surface are formed during the 2 min and not at the moment you locally supersaturate the molecule concentration at the interface?

Manfred Buck answered: Usually our solutions are rather dilute (typically in the range of 1 mM to 100 μM or even below). We exclude a significant contribution due to supersaturation as the coverage scales with the immersion time.

Marco Sacchi enquired: Could you say that C_{60} could jump from one nanopore to another?

Manfred Buck responded: We have no indication that C_{60} molecules jump spontaneously. However, even at the low currents used in the work we observed some tip-induced movement of C_{60} molecules, revealed by occasional streaks or only a fraction of the molecule being imaged.

Marco Sacchi remarked: Did you consider the possibility that the 3BPEB molecules could sit on top of one another inside the nanopores, or is this not energetically possible?

Manfred Buck responded: There is no indication that the molecules are systematically arranged on top of each other in a well-defined geometry. The possibility that they are exactly on top of each other is energetically unfavourable and should also give rise to a different tunneling contrast compared to a single molecule. Also, on uniform layers of 3BPEB (no network) the shapes of the molecules are clearly seen with no differences in contrast indicative of stacking. However, partial overlap (*e.g.* two arms of two different molecules) cannot be completely excluded to occur occasionally. It could give rise to irregular features such as the one highlighted by the dotted circle in Fig. 3 of the paper (DOI: 10.1039/C7FD00115K). It is noted that these features are different from the regular ones which arise from a stochastic rotational motion of a single molecule in a network pore.

Steven De Feyter asked: You image by STM under ambient conditions. Is there a specific reason why you don't do the imaging at the liquid–solid interface?

Manfred Buck responded: This comes from our motivation to build structures which are sufficiently stable to allow sequential assembly. The advantage is that very different types of molecules and different assembly schemes (*e.g.* hydrogen bonding molecules and thiols) can be combined. Furthermore, it also enables kinetic control, thus adding another dimension to the assembly process.

Steven De Feyter commented: The liquid–solid interface might promote solvation of the on-surface impurities and their desorption, favouring subsequently the adsorption of the guest species.

Manfred Buck answered: For some components this might work. For example, excess melamine present after the synthesis of a melamine-PTCDI network could be removed as its interaction with the surface is not that strong and its solubility sufficiently good at elevated temperature. However, for the more strongly adsorbing PTCDI this is not an option as the temperature required for solvation seems to lie above the temperature where the network is stable in a solution environment. Our attempts to clean the network pore by solvent treatment were unsuccessful.

Peter Beton asked: You mentioned that you tried some molecules that hadn't been successful. What are the essential ingredients for a molecule to be captured by the pores?

Manfred Buck responded: For the pores compartmentalised by the star molecules, the bottleneck is their adsorption strength. While C_{60} was not critical, thiols easily displace the star molecule as documented in Fig. 8 of our paper (DOI: 10.1039/C7FD00115K). So it is a balance between sufficiently strong adsorption (also to remove impurities) and a sufficient difference in the time scale of pore filling and displacement to allow for a kinetic control. Different thiols tried all had similar effects, *i.e.*, easily displaced the pore modifier.

Peter Beton said: Just as a related question, in your schematic, you've got the melamine being 'nudged out of the way'. Is the filling of the pores completely random? Is there evidence that what's going on in a neighbouring pore affects this?

Manfred Buck replied: We do not have any evidence that processes in adjacent network pores affect each other. For example, the orientation of the stationary star molecules is random. With regard to C_{60} adsorption in subpores, our present data base does not allow an analysis as any correlation effects and blocking of sites by impurities would be hard to disentangle.

Brandon Hirsch enquired: Have you attempted to position the tip over the pore and monitor the tunnelling current fluctuations?

Manfred Buck responded: Our experiments were performed at room temperature in an ambient environment. The drift under these conditions did not allow this type of experiments.

Neil R. Champness opened a general discussion of the paper by Natalia Martsinovich: In your model, when you have introduced hydroxy groups to the terephthalic acid core, the hydroxy groups are involved in intramolecular hydrogen bonding. Are you confident that this is the most stable arrangement?

Natalia Martsinovich replied: Yes, we are confident that the structure involving intramolecular hydrogen bonding is the most stable one. We have calculated four isomers of 2HTPA, with different positions of the OH group: next to carboxylic OH or carbonyl oxygen, with or without intramolecular hydrogen bonding (the last four structures shown in Table S1 of our paper, DOI: 10.1039/C7FD00112F). The structures with intramolecular hydrogen bonding were 6–10 kcal mol^{-1} more stable (the most stable isomer has the hydrogen bonding to the carbonyl oxygen). These energies are shown in the Supporting Information in Table S1.

Neil R. Champness said: If the intramolecular hydrogen bonds involving the hydroxy groups represent the more stable arrangement, can you rule out intermolecular hydrogen bonds to adjacent chains?

Natalia Martsinovich replied: We cannot firmly rule out intermolecular hydrogen bonding, but we have not observed them in our search of 2D structures. In all our 2D arrangements, the starting structures involved intramolecular hydrogen bonding, and all the resulting minimum-energy structures preserved this bonding. There was no flipping of OH to form intramolecular bonds. However, we did not use alternative starting 2D structures without intramolecular hydrogen bonding – these structures would be more likely to lead to intermolecular hydrogen bonds to adjacent chains. I would expect such 2D arrangements to be less stable than with intramolecular bonds: our calculations of various isolated dimers (Table S1) showed the 2HTPA dimer with flipped OH groups as unfavourable; intramolecular hydrogen bonds are favourable here because they lead to formation of stable 6-membered rings.

Brandon Hirsch asked: In Fig. 5b of your paper (DOI: 10.1039/C7FD00112F), it appears that you have placed the OH group in a random orientation. Is there a substantial basis for this? From the STM image it appears that there are double rows, which have epitaxial turns possibly due to kinetic trapping between two relatively degenerate states. Is it not then likely that the double rows result from the interactions between OH groups from molecules in neighbouring rows?

Giovanni Costantini answered: The molecular structures proposed to interpret Fig. 5b are only tentative but the orientation of the OH groups has not been assigned in a random way. It is rather based on the measured intermolecular distances (as shown for three specific molecular pairs in the same figure) which we believe to be indeed determined by the interaction between the lateral OH groups. What should be noted is that there is much disorder in the arrangement of the double rows, which are extremely short and have quite a wide distribution of inter-row distances.

Matthew Blunt addressed Natalia Martsinovich: For terephthalic acid (TPA), no self-assembly was observed at the graphite–phenyloctane interface for the experimental section of your presented work. However, the results of your molecular mechanics/dynamics calculations on this system, when analysed as part of the Born–Haber cycle, suggest a small enthalpic gain associated with two-dimensional self-assembly by this system (-9.8 kJ mol^{-1}). In the paper and in your presentation, you state that as the enthalpic gain is small it may be offset by the entropic losses associated with self-assembly. Do you have any idea what size of enthalpy gain as obtained from your calculations would provide a unambiguous prediction for self-assembly?

Natalia Martsinovich responded: We cannot at this stage make an exact prediction of the enthalpy gain that is sufficient to ensure self-assembly, because of the difficulty and complexity of calculating the entropy term. We can make only qualitative estimates. First, based on comparison with entropy calculations for similar systems ($-T\Delta S = 3$–13 kJ mol^{-1} for related dicarboxylic acids[1,2]), the enthalpy gain should be at least 5–10 kJ mol^{-1}, dependent on the system. Second, by comparing the calculated enthalpy gains in the systems where self-assembly was observed ($\Delta H = -22$ kJ mol^{-1} and larger) and the system where self-assembly was not observed ($\Delta H = -9.8$ kJ mol^{-1}), an enthalpy gain of at least -10 kJ mol^{-1} is necessary and -20 kJ mol^{-1} is sufficient. However, this value would also depend on the nature and concentration of the solvent and solute and, in particular, on the number of solvent molecules that are associated with one solute molecule in solution.

1 W. Song *et al.*, *J. Am. Chem. Soc.*, 2013, **135**, 14854–14862.
2 W. Song *et al.*, *Phys. Chem. Chem. Phys.*, 2014, **16**, 13239–13247.

Matthew Blunt asked: Do you have any idea how accurate the calculations are that are used to obtain values for the entropic losses associated with 2D self-assembly (*i.e.* $T\Delta S$ terms)?

Natalia Martsinovich replied: They are not very accurate. First, the calculated entropy term depends on the structure and concentration of the solute and on the assumption of the free volume of the solvent,[1] leading to perhaps a few kJ mol^{-1} uncertainty in $T\Delta S$. The second and larger source of uncertainty is the assumption on the number of solvent molecules that are associated with one solute molecule and are released during adsorption and self-assembly – here the uncertainty may be as large as several tens of kJ mol^{-1} (*e.g.* ref. 2).

1 M. Mammen *et al.*, *J. Org. Chem.*, 1998, **63**, 3821–3830.
2 W. Song *et al.*, *J. Am. Chem. Soc.*, 2013, **135**, 14854–14862.

Markus Lackinger responded: I believe the accuracy and reliability of $T\Delta S$ estimates is rather poor. For the calculation several assumptions have to be made, where the most crucial ones relate to the solvation shell. Here we do not even know the number of solvent molecules involved and also the corresponding structure is not known. However, both factors significantly affect the outcome of the simulations. In this respect, it appears more reasonable to refrain from an *ab initio* calculation of $T\Delta S$, and rather benchmark theoretical against experimental results.

Steven De Feyter said: It's tricky to conclude something based on the fact you don't see anything, especially using STM. I'm sure there were many attempts to visualize this system. However, I was wondering if you looked, for instance, into the impact of bias polarity and TPA concentration? Bias polarity might have a strong influence on the probability to "see" a monolayer. Furthermore, to what extent is the relative surface coverage of the different polymorphs dependent on annealing or TPA concentration?

Giovanni Costantini communicated in response: Several attempts were done to image TPA in PO by varying the tunnelling conditions (both current and bias voltage, at positive and negative values), but none of them produced an appearance of any ordered molecular layer. Concerning the effect of polymorphism (observed for 2HTPA but not for TPA), we did not explore the effect of concentration, although we expect this to potentially play a role.

Markus Lackinger addressed Natalia Martsinovich: Do the solvation energies you quote for heptanoic acid *vs.* phenyloctane refer to vacuum, and are those experimental or theoretical values? I'm surprised to see an even slightly higher value for phenyloctane, given the fact that TPA can form strong solute–solvent hydrogen bonds in heptanoic acid. Can you explain what accounts for the relatively high solvation energy in phenyloctane?

Natalia Martsinovich responded: The solvation energies reported in the paper are with respect to an isolated molecule in vacuum. These are theoretical values (experimental determination of solvation energies was not attempted in this work). I believe that the high solvation energy in phenyloctane is caused by strong π-stacking interactions of the phenyl rings of the solvent and solute molecules. A solute molecule can have two solvent molecules stacked above and below it. As an upper limit of the π-stacking energy, we can consider the energy of adsorption of one phenyloctane molecule on a TPA monolayer, -87.4 kJ mol^{-1} (Table 5 of the manuscript, DOI: 10.1039/C7FD00112F) – larger than the carboxylic acid dimer hydrogen bond energy (-60– -68 kJ mol^{-1}), although this adsorption energy also contains the interaction of the alkyl chain. The contribution of the dispersion interaction with the solvent's alkyl chain (which is longer in phenyloctane than in heptanoic acid) is also non-negligible: for comparison, the solvation energy of TPA in a longer alkanoic acid, nonanoic acid, was calculated as -115.1 kJ mol^{-1} (also with respect to vacuum),[1] larger than the value obtained here for heptanoic acid (-93.8– -98.6 kJ mol^{-1}). Thus a longer and, therefore, more flexible alkyl chain will also improve solvation. I also note that there is a significant error bar in the energies obtained in these molecular dynamics simulations: the standard error of the mean of the solvation energies is up to 20.5 kJ mol^{-1} in phenyloctane and up to 41.9 kJ mol^{-1} in heptanoic acid (manuscript section 3.3.2), therefore the calculated solvation energies in heptanoic acid are not very accurate.

1 W. Song *et al.*, *J. Am. Chem. Soc.*, 2013, **135**, 14854–14862.

Steven L. Tait opened a general discussion of the topics raised at the session: What analytical techniques can be used at liquid/solid interfaces to complement scanning tunneling microscopy (STM)? We have discussed the complexity of self-assembly and have seen several examples of effectively combining structural analysis, spectroscopic analysis, and computational modeling to better understand these systems. Please respond with your ideas for effective analysis tools for molecular assembly in ambient or at liquid/solid interfaces. Please give a brief statement about what type of information can be obtained and what are the essential system requirements and key features of the technique.

Han Zuilhof communicated in reply: Laser Ablation Electrospray Ionization (LAESI) MS would be useful, as long as the liquid is water/protic. In this technique the surface is irradiated with a pulsed red/IR laser. This heats up the water very rapidly, leading to an 'explosion' at the surface and a plume bursting upwards. This (vertically moving) plume with the materials of interest can be crossed by a (horizontal) ESI spray, which transfers desorbed materials from the surface to the MS. This technique currently offers spatial resolution in the 10's of micrometers, and of course coupling it to a high-resolution mass spectrometer yields very specific molecular information. For ambient conditions this is also fine. In those cases DART-MS might also work well and complement STM and other techniques.

Deepak Dwivedi communicated in response: I think small angle neutron scattering can be an alternative technique for characterizing solid/liquid interfaces. It has been widely used for this purpose.[1] Our recently published book chapter describes the role of small angle neutron scattering for film characterization and we discussed the application of this technique in corrosion science.[2]

1 W. C. Forsman and B. E. Latshaw, *Polym. Eng. Sci.*, 1996, **36**, 1114–1124.
2 D. Dwivedi and K. Lepková, *SAXS and SANS Techniques for Surfactant Characterization: Application in Corrosion Science*, in *Application and Characterization of Surfactants*, ed. R. Najjar, InTech, 2017, DOI: 10.5772/intechopen.69290.

David Amabilino said: Rachel Barnard and Adam Matzger used thermal analysis to study heat of adsorption.[1] Simple alkane derivatives on graphite, but in principle you could use that technique more generally.

1 R. A. Barnard and A. J. Matzger, *Langmuir*, 2014, **30**, 7388–7394.

Stuart Clarke communicated: There are a number of experimental methods that can be used for the study of particular aspects of molecular species adsorbed from liquids to solid surfaces. In general, a combination of these methods is recommended to give a clear picture of these complex and challenging systems. The ESI† of this discussion section presents an outline of several of these and some references for the interested reader. It is not intended as a full review of all the methods one might consider, nor a comprehensive literature review, but a suggestion of some methods that might be used to complement STM measurements, as prompted by the question raised by Steven L. Tait at the Faraday meeting. There are also other more established techniques that colleagues from a UHV background will be familiar with, such as XPS, SIMS, EDX, BSED, FIB-TEM, EELS, EXAFS *etc.* which are generally available through national and local services that are not included, that should also be employed to characterise surfaces before and/or after adsorption from solution. Other methods not discussed in detail include QCM and SPR, which are helpful in identifying that something has adsorbed, but do not identify the species. In simple terms these approaches may be considered in several broad groups: a) structural methods, 2D – in plane diffraction and structure normal to the surface; b) spectroscopic methods, *e.g.* surface-specific methods (such as SFG) and surface sensitive approaches (such as RAIRS, ATR and PM-IRRAS) and related methods such as AFM-IR; c) thermodynamics/calorimetric methods, such as Differential Scanning Calorimetry (DSC), solution depletion adsorption isotherms and surface titration; d) methods to investigate dynamics of adsorbed species, which include low and high field NMR and incoherent neutron scattering (IQNS), exploiting 'dynamic contrast'; e) and others, such as the Surface Force Balance.

Angelika Kühnle commented: Regarding further techniques for investigating the solid–liquid interface, I would like to add that scanning force microscopy is not limited to 2D images. It allows for acquiring 3D maps of the interfacial region.

From these maps, one can deduce conventional images parallel to the surface, but also slices perpendicular to the surface. In these vertical slices, you see the position of the solvent molecules above the surface.[1,2] With this technique, we can study the impact of the solvation structure on molecular self-assembly. For many systems, the solvation structure will be crucial to understand molecular self-assembly at the solid–liquid interface.

1 H. Söngen *et al.*, *Rev. Sci. Instrum.*, 2016, **87**, 063704.
2 H. Söngen *et al.*, *Langmuir*, 2017, **33**, 125.

Deepak Dwivedi communicated in response: Thank you, Professor Kühnle, for sharing these nice papers. I would like to ask if you could comment about the image processing technique which gets used for making 2D and 3D images? Is it possible to figure out phase segmentation by scanning force microscopy as was done with X-ray nano-computed tomography?[1]

1 T. M. M. Heenan *et al.*, *Fuel Cells*, 2017, **17**, 75–82.

Angelika Kühnle communicated in reply: The AFM imaging technique is based on probing the interface with a sharp tip mounted on a mechanical resonator. As the tip is scanned directly in real space, 2D images and 3D data are directly recorded. Thus, no elaborate image processing is necessary. Both 2D and 3D AFM allow various pieces of information about the tip–sample force to be obtained (see ref. 1). With AFM, a phase boundary could therefore only be identified based on the force measurement.

1 H. Söngen, R. Bechstein and A. Kühnle, *J. Phys.: Condens. Matter*, 2017, **29**, 274001.

Neil Robinson opened the discussion of the paper by Steven De Feyter: Could you tell us a little more about the possible applications of the transfer of chiral information from solvent to the solid–liquid interface? In your opinion, could such processes be applicable to the separation of miscible liquids?

Steven De Feyter responded: In my view, this approach has no direct applications. While I don't think it can be used for the separation of miscible liquids, there are other possibilities. For instance, as we have shown some years ago in a collaboration with David Amabilino,[1] chiral physisorbed monolayers can be used to trap analytes in an enantiospecific fashion, akin to processes that happen in chromatography.

1 H. Xu, W. J. Saletra, P. Iavicoli, B. Van Averbeke, A. P. H. J. Schenning, D. Beljonne, R. Lazzaroni, D. B. Amabilino and S. De Feyter, *Angew. Chem. Int. Ed.*, 2012, **51**, 11981–11985.

Claire-Marie Pradier asked: What happens if you change the solvent? Are these arrangements reversible? What happens if you dry the system?

Steven De Feyter responded: Changing the solvent has an important impact on the appearance and the stability of certain phases. In short, the outcome of the self-assembly process at the liquid–solid interface can be quite different, depending on the solvent used. This certainly also relates to the impact of the solvent used on the dynamics at the liquid–solid interface. For instance, the time scale of out-of-plane dynamics, *i.e.* adsorption and re-adsorption, is system- and solvent-dependent. Under conditions where the dynamics of individual molecules can be traced, the residence time is shown to range from sub-milliseconds to several hours, depending on the molecular system and solvent probed. Obviously, the dynamics and the time scale of the dynamics are influenced by the presence of the solvent. Solvent evaporation or removal may lead to "freezing" and stabilizing the molecular pattern. However, one should note that the appearance and stability of a pattern often depends on the concentration of the molecule in the solution phase. A change in concentration, *e.g.* as a result of solvent evaporation, may induce the formation of a new, often higher-density phase. However, for some systems, it is possible to kinetically trap a metastable pattern.

Talat Rahman asked: What are the essential elements in terms of where theory is at this point? In a sense, MD is the obvious answer, since it has the ability to include all the effects of temperature and pressure in a natural way. The problems with MD, of course, are the length and time scales, particularly for *ab initio* MD simulations. The good news is that we are not too far away from getting there. Also, the combination of DFT and kinetic Monte Carlo simulations, for which you need to have prior knowledge of all mitigating processes, is very promising and some efforts are being made in that direction. To get back to your experiments, do you get an indication of whether nucleation is important? I assume it is. Could we then get some idea about nucleation sites from experiments? This would save a lot of computational time.

Steven De Feyter replied: For sure, nucleation is important. And indeed, there are efforts to use MD to gain insight into the nucleation process itself, but as you mention, it's time consuming. Getting experimental information on the nucleation sites is possible, though not trivial, at least not at the liquid–solid interface, where stable STM imaging often goes hand in hand with the presence of a monolayer. We are looking now into new experimental designs that could overcome that problem, and allow us to directly visualize the nucleation process itself.

Talat Rahman noted: Regarding these systems containing molecules with 50–100 atoms, it may be possible to carry out *ab initio* MD simulations for them. For the record, I think I can give you a reference of where people are doing this kind of work, free energy calculations.[1]

1 M. A. Caro, T. Laurila and O. Lopez-Acevedo, *J. Chem. Phys.*, 2016, **145**, 244504.

Trolle R. Linderoth remarked: Your detailed analysis shows that the chiral induction by the solvent is likely to occur in the nucleation stage on the surface. Could UHV-STM experiments at low temperatures help to pinpoint and identify the nucleation complexes by co-adsorbing the DBA molecules and the chiral octanol solvent?

Steven De Feyter responded: I believe that this is indeed possible. In particular, the aspect of temperature control in a UHV environment is attractive. It may allow the nucleation step to be recorded or visualized and provide meaningful insight into the intermolecular interactions that are involved.

Karl-Heinz Ernst commented: We have heard several times about nucleation. But the most stable nuclei are often not the ones that are viable to grow. This can be due to the fact that with increasing coverage the self-assembled layer does not fit perfectly to the substrate, and strain in the overlayer is increased. Moreover, grain boundaries play a very important role. If they can't accommodate the right-handedness, certain domains may be unfavorable and become rearranged into the more favored one. A comment to Trolle R. Linderoth: Concerning experimental methodology, I wouldn't focus always on STM here. These questions may be addressed by high-resolution electron or photoelectron microscopy (LEEM/PEEM) working under ambient pressure (as for ambient pressure XPS today). It would be difficult to build this, but not impossible.

Lifeng Chi remarked: We tried vacuum deposition with molecules which have a rigid core and three arms (alkyl chains). The molecules form very complex structures, not only coverage- but also process-dependent. I mean that with different deposition speed and substrate temperature, the structure could be different. We observed at least ten phases. In contrast, when liquid preparation was used for the same molecule, much fewer phases were observed. For this reason, this kind of molecules may be suited for liquid preparation rather than vacuum preparation.

Phil Woodruff commented: It would surely be nice if it were possible to adapt the LEEM and PEEM instruments to operate at 'near ambient' pressures, but there are likely to be insurmountable technical problems. These instruments achieve their high spatial resolution by operating the electron optics at ~20–30 keV, and this is achieved with electrons emerging from the surface at only a few eV by applying voltages of 20–30 kV across a very narrow gap between the sample and the last optical element. Clearly at higher pressures this will lead to breakdown. By contrast, in 'near ambient pressure' XPS, the 'only' challenge is to have differential pumping such that the pressure in the electron energy analyser is much lower than that in the narrow space in front of the sample.

Manfred Buck addressed Steven De Feyter: An odd-even effect in the chiral selectivity has been observed for chain lengths between 10 and 14. What is the explanation and is this effect limited to this narrow range of chain lengths?

Steven De Feyter replied: We don't have a clear view on the reasons behind this "odd-even" effect. Only a few molecules with an odd number of carbon atoms in the alkyl chains were tested. In all cases, a clear difference with the response of the "even" analogues was obtained.

Karl-Heinz Ernst commented: We have environmental scanning electron microscopy (SEM) that works with higher voltages. The same applies to transition electron microscopy. If there would be higher interest in having systems (PEEM, LEEM) that would probe self-assembly at the solid interface, they would probably exist. There is no fundamental reason not to be able to build something that I just proposed. The design, of course, would be completely different from what we have for UHV machines.

David Amabilino opened the discussion of the paper by Sebastian Schwaminger: I really enjoyed your presentation and your paper. I have some questions about the magnetite nanoparticles that you prepare. You prepared them with a covering, what was that covering? You show a binding with your peptide immobilised on paper, what's happening at the interfaces there? How do you immobilise the peptide in this way? What's the change in the covering of your nanoparticle in the process of adsorption to the peptide? Finally, how do you get the nanoparticle off the paper to study it?

Sebastian Schwaminger responded: Thank you. The nanoparticles are bare particles which are not coated but are terminated with hydroxy groups, which can be proven with XPS and ATR-IR. Furthermore, in most experiments, buffers containing different ion species are used, which means that the surface should contain bound species of sodium ions, chlorine ions and other buffer components as well. It is a good question regarding what is really happening on the interfaces between peptides and nanoparticles. We assume that there is a mixture of electrostatic interaction and coordinative binding between the peptides and particles. However, this experiment is not able to tell us much about the local interactions between single peptide molecules and particle surfaces, but can be used as a screening method which hints global interactions and, to some extent, the affinity of the used peptides to the nanoparticles. The cellulose membrane (basically paper) is amine-terminated, where the peptide can be linked to *via* its C-terminus by peptide synthesis methods. The membrane we used is commercially available and the peptide composition can be designed by the user. We think that a lot of molecules are playing a role in the binding process. We think it is possible to have a direct coordination between peptides and metal ions (here the surface composition is changed and one bound molecule is replaced by a peptide) as well as a binding mediated by H-bonds and electrostatic interactions between buffer components bound to the particles and peptides (here the surface composition will not change). We are able to elute the particles from the Glu8 and Glu6 peptides with citrate and phosphate-based buffers.

David Amabilino asked: How do you think the peptides are binding to the magnetite; what's the chemical process? What are the functional groups involved (amino acid residues) and chemical interactions? You looked at XPS; what are the main interactions with the surface? Where is the optimum point on your surface for binding?

Sebastian Schwaminger answered: I think there is a mixture of electrostatic interactions and coordination chemistry possible. For example, in the paper (DOI: 10.1039/C7FD00105C) we were able to show a monodentate coordination between a Glu8 peptide and magnetite nanoparticles. So it is definitely reasonable that hydroxy groups can be replaced by carboxy groups. However, it is also possible that H-bonds between the hydroxy groups and carboxy groups are formed. An electrostatic interaction between negatively-charged peptides and the nanoparticles is possible as well. I think it is really difficult to exactly define the interactions for such complex systems. Usually we are only able to observe interactions between charged amino or acid groups and magnetite nanoparticles. The amount of bound peptide type is strongly dependent on pH and buffer. However, the results shown in the article do demonstrate interactions between magnetite and aromatic peptides as well. This behavior was only observed in unbuffered saline. In most buffers used, usually either negatively- or positively-charged peptides are favored for the binding with magnetite particles. From XPS, it is difficult to say what the main interactions with the surface are, as Fe 2p spectra are quite difficult to interpret. Furthermore, it is always difficult to describe an ambient system *ex situ* with an *in vacuo* technique. However, a reduction of hydroxy groups can be interpreted as an at least partial replacement by carboxy groups. The optimum point for binding can only be answered globally and not locally for this system. We observed the best binding conditions for negatively-charged peptides in MES (pH 6) and acetate buffer (pH 5).

Claire-Marie Pradier enquired: If you do XPS on a layer of your peptide on your nanoparticles, you claim there is no change in the Fe peak, but would you detect a change on the Fe peak, coming from the surface layer?

Sebastian Schwaminger replied: This is indeed a really good question. The aim of our statement in the paper was to show that magnetite does not oxidise and change its physical properties upon adsorption and processing. As XPS is a very surface-sensitive technique, it should be possible to detect changes in the outermost layer. However, the high-spin multiplets of Fe^{3+} and Fe^{2+} compounds and their resulting Fe 2p 3/2 peaks are quite difficult to analyse. Therefore, I think for these nanoparticles, I am not able to distinguish between OH-terminated magnetite surfaces and surfaces coordinated by carboxy groups with the XPS device used for this study. Further experiments with Mössbauer spectroscopy, which is less surface-sensitive, were also not able to detect a change of the electron density and the magnetic field at the iron ions upon adsorption of amino acids.

Claire-Marie Pradier said: Then you do IR spectroscopy and you look at the shift of the amide bond. Is it with the same coverage of peptide as the one used for XPS measurements? You deduce there is a change in the conformation, but does this not assume that you have a thick enough layer, of at least several peptides?

Sebastian Schwaminger replied: Yes, the IR measurement at pH 7 is of the same sample as the one investigated with XPS. We also see changes of the amide band with different pH values. However, the peptides are quite short and therefore even small changes like adsorption or orientation of some side chains to the surface can induce a conformational change. Furthermore, the pH is critical for the peptide conformation as well, and there is an overlay of amide bands and carboxy groups. Here, glutamic-based homopeptides were investigated and therefore the carboxy group is very dominant in the IR spectrum. The change in conformations does not necessarily result in layers of several peptides but, especially for lower pH values, lateral interactions between peptides will definitely influence the binding.

Yuri Diaz Fernandez asked: You mentioned the formation of peptide multilayers around the particles, but for a wide pH range poly-glutamate molecules are negatively charged and hence repulsive interactions will dominate, preventing formation of a multilayer. How could you explain, then, the relatively strong circular dichroism signals observed, that seem too large for a monolayer?

Sebastian Schwaminger replied: In the case of circular dichroism, we do not only measure the adsorbed peptides but a defined amount of peptides mixed with a small number of nanoparticles. So what we see in the paper (DOI: 10.1039/C7FD00105C) is how some of the peptides are affected and not a multilayer, but otherwise you would be right. The peptides are negatively charged for all pH regions which were investigated.

Yuri Diaz Fernandez remarked: In the experiments of attachment of the particles to cellulose, you mentioned that depending on the pH, the nanoparticles are detached, suggesting that the interactions with the peptides are reversible with the pH, as expected due to the protonation/deprotonation equilibrium of the peptides and the change in the charge of the iron oxide nanoparticles. When you discuss your model, even at acidic pH at which you may expect glutamates to be fully protonated, you still consider the peptides to be bound to the surface of the particle. What evidence do you have that the peptides are still attached?

Sebastian Schwaminger replied: Yes, we were able to detach the particles from the peptides by changing the buffer, which is not shown in the article. We were able to release the particles by adding citrate-buffered saline. Thereby we change the charge of the particles to a negative charge and coordinated peptides can be replaced by citrate ions which are added in excess. At lower pH values the magnetite surface is charged more positively, while the Glu8 peptide is still negatively charged at pH 4. We measured adsorption isotherms at pH 4, 5, 7, 9 and 10 and by analyzing the supernatant, we were able to determine the amount of peptide. Hence, at all pH values an adsorption can be observed, while the highest loading capacity can be monitored at low pH values.

Claire-Marie Pradier asked: In the last part of your paper, the attachment of the protein, you said that proteins have a behavior similar to that of peptides. Does that mean that the interaction of the protein with the nanoparticles is dominated by electrostatic interactions?

Sebastian Schwaminger replied: In our case, using green fluorescent protein (GFP) with a Glu6 tag, electrostatic interaction plays an important role in the binding to the magnetite nanoparticles. Furthermore, from experiments where we used mixtures of Glu6- and Gly6-tagged GFP, we can conclude that the binding is *via* the tag and that the affinity is definitely influenced by the tag. However, this can be completely different when other particles or other proteins are used. In this case, electrostatics are most probable to dominate the interaction.

Claire-Marie Pradier said: Can you detect by IR spectroscopy some change in the conformation of the protein upon adsorption?

Sebastian Schwaminger replied: We have not analysed the protein binding with IR spectroscopy, yet. However, this would be very interesting, and as GFP is dominated by a β-barrel structure, such measurements might prove or disprove the interaction of magnetite and GFP *via* a glutamic acid-based tag.

Neil R. Champness asked: What are your opinions on the comparison between nanoparticles and flat surfaces? What have we learned from studies on flat surfaces and how do these inform nanoparticle studies?

Sebastian Schwaminger responded: I think it is really important to think about similarities of flat surfaces and nanoparticle surfaces. There is a reason why I like to address a nanoparticle problem to a surface science community. In my opinion, it is always better to know what happens on a flat defined surface before changing to a more complex system. It is always easier to simplify a system and understand the basics of what is possible (chemically) here. Many ways of analyzing and understanding surface/interface properties have their roots in surface science and are based on flat surfaces. However, it will always be difficult to make the transfer from flat to curved surfaces.

Claire-Marie Pradier commented: Comparing results on NPs and on planar surfaces enables us to apply complementary techniques, like NMR on powders and "surface science" techniques on planar surfaces. As another example, IR in various modes can be applied to nanoparticles or to flat surfaces. It's very interesting to compare results from those various techniques for a similar system (same molecule, same substrate in the form of either powder or of a single crystal).

Sebastian Schwaminger replied: Yes, this might be very valuable to compare single crystal behaviour to nanoparticles with orthogonal techniques. We thought about this and actually wanted to start with single crystal surfaces. Actually, techniques like IRRAS or AFM using a tip with attached peptides would definitely improve the understanding of interactions and only work on flat surfaces. The problem hereby is that the behaviour is very different, as small spherical nanoparticles do not possess defined regular surfaces, and more manpower, knowledge transfer and interdisciplinarity is necessary to understand two systems. I think that after more than four years, I am still far away from wholly understanding the behaviour of magnetite nanoparticles and I do not believe that it will be so much easier to understand the interactions of peptides and proteins on a single crystal.

Rasmita Raval opened the discussion of the paper by Deepak Dwivedi: Corrosion in pipes is a big problem and can involve bacteria and marine species in the corrosion process. Do we know how corrosion inhibitors work with respect to such living species?

Deepak Dwivedi responded: Thanks, Prof. Raval, for raising this question. Corrosion of pipelines is a big issue and has been addressed in our recently published articles.[1,2] I have not worked with microbiological corrosion, but I am familiar with the concept that inhibitor selection depends on the type of microorganism present in the pipeline. If it contains S-reducing bacteria, we should try to avoid S-containing inhibitor molecules. In another article which is on the corrosion of ancient artifacts, we have described the effect of microbiological corrosion on ancient ship wrecks, buried inside the sea.[3] I must say that this problem is a big issue for oil and gas companies and still the corrosion mechanism is not yet well understood, as described in ref. 4. The microorganisms alter the interfacial chemistry between corrosive fluid and substrate, and anaerobic sulfate-reducing bacteria (SRB) is the most common type of bacteria which we encounter in pipelines. Various inhibitors such as imidazolines, amino-amines, fatty acids, *p*-dodecyl benzene sulfonic acids *etc.* are inhibitor compounds (a complete list is mentioned in a review article, ref. 1) and organisms present degrade the inhibitive film formed on the steel (as organisms cause enzymatic attack on specific functional groups of inhibitors such as benzene rings, CN groups *etc.*) As far as characterization is concerned, environmental microscopy, Raman spectroscopy and X-ray CT are various techniques which are getting used currently by researchers. Techniques for inhibitor characterization are still a challenge, and we have discussed other techniques such as neutron scattering and neutron reflectivity in our articles.[1,2,5]

1 D. Dwivedi, K. Lepková and T. Becker, *RSC Adv.*, 2017, **7**, 4580–4610.
2 D. Dwivedi, K. Lepková and T. Becker, *Proc. R. Soc. A*, 2017, **473**, 20160852.
3 D. Dwivedi, K. Lepková and J. Mata, Archaeometallurgical investigation of ancient artefacts' corrosion phenomenon: A critical analysis for paving the future path, submitted.
4 K. M. Usher *et al.*, *International Biodeterioration & Biodegradation*, 2014, **93**, 84–106.
5 D. Dwivedi and K. Lepková, *SAXS and SANS Techniques for Surfactant Characterization: Application in Corrosion Science. Application and Characterization of Surfactants*, InTech, 2017.

Martin Nalbach asked: How do you saturate your test solution experimentally with CO_2 and how do you quantify or measure that it is really saturated?

Deepak Dwivedi answered: Thanks, Martin, for asking this question. For CO_2 corrosion, we usually try to keep the oxygen ppm level as low as possible, and by observing the turbidity and color of the solution, we can judge about the saturation.

Manfred Buck remarked: Most of us at this meeting work on rather simple systems involving well-defined substrates like single crystals, which allow studies down to the level of atomic/molecular resolutions. For your much more complex steel surfaces, how much information do you have on the initial stages of film formation? To what extent can you relate nucleation to topography such as screw dislocations or surface heterogeneities and/or chemical effects?

Deepak Dwivedi answered: We are thankful to Prof. Buck for pointing out the complexity associated with film formation on carbon steel surfaces. There are various factors such as the texture of carbon steel, the surface energy of carbon steel, phases such as ferrite and pearlite, and microstructure (grain size and grain boundaries *etc.*). We have characterized these factors with the help of various characterization techniques. Film formation has been characterized with *in situ* synchrotron XRD and *in situ* Raman spectroscopy, which illustrated the phases formed at the initial stages of the film formation. From our point of view, the initial hierarchical film formation promoted the screw dislocation formation, which was further supported with the stacking fault energy, inclusion *etc.* Surface heterogeneities were characterized with FESEM and SIMS, and we have noticed the screw dislocation loop with hierarchical morphology. *In situ* AFM experiments are in our future plans.

Steven De Feyter commented: In your conclusions, you write that sodium thiosulphate serves as a performance enhancer in corrosion inhibitor formulations. What is the impact of other ingredients in the formulation on the process that you describe here?

Deepak Dwivedi replied: Thanks, Prof. De Feyter, for raising this issue. As discussed during the meeting, we have investigated the interaction of sodium thiosulphate with other organic inhibitors (surfactants) under similar CO_2 corrosion conditions and we have noticed that the corrosion inhibition efficiency was improved. Various techniques such as *in situ* synchrotron XRD, *in situ* Raman, *ex situ* Raman *etc.* have been utilized for investigation.

David Amabilino said: Very nice presentation and nice work. Out of curiosity, I was wondering if you'd tried looking at the first steps of film formation? Those initial steps are presumably what gets the film going. To try and see that nucleation, maybe it's a common theme in this session, did you try using atomic force microscopy to see the initial stages of the formation of the films?

Deepak Dwivedi responded: Thank you, Prof. Amabilino, for this nice suggestion. We believe that *in situ* AFM could be a nice technique to investigate about the initial steps formation. It is worth trying and, as mentioned earlier, it is in our future experiment list which is going to be conducted.

David Amabilino said: To make the surface smoother, maybe you could polish it. I seem to recall seeing corrosion studies where they look at it using electrochemical STM, though roughness is of course a big issue. Systems like yours, it's those rough points where nucleation is starting, and I suppose you want to look there. STM and AFM in liquid could be useful, could you try? To make controlled defects and see what happens there in terms of nucleation is a really interesting opportunity. Can you comment on this?

Deepak Dwivedi responded: It is a nice suggestion. As discussed, we plan to perform *in situ* AFM with the electrochemical cell. The application of STM, as you suggested, could be another interesting technique and this has been highlighted by our review article published recently.[1] We will follow this nice advice to get further insights about a dislocation formation process.

1 D. Dwivedi, K. Lepková and T. Becker, *RSC Adv.*, 2017, 7, 4580–4610.

Andrew Mount asked: Following on from the discussion about applying electrochemical measurements (*e.g.* in SECM), there is the potential in this corrosion system for $S_2O_3^{2-}$ to undergo oxidation to form $S_4O_6^{2-}$ (and there may also be other oxidised species). Might these be implicated in nucleation and film formation, and which of the characterisation techniques could distinguish such species? Electrochemical measurement, or control, of the substrate potential should be able to probe the importance of such effects. Have such measurements been carried out, and if so, what do they show?

Deepak Dwivedi answered: Prof. Mount, your suggestion to use scanning electrochemical microscopy for investigating the ionic species formation and its impact on screw dislocation-derived film formation is nice. I agree that these species may affect nucleation and growth. It is also worth mentioning that surface roughness of the substrate as well as inherent defects and/or inclusions may also act as nucleating sites. We have not used SECM, but we have discussed SECM and its application in corrosion science in our review article[1] (including other techniques[2]); we have not used this technique yet and it is a nice suggestion to include this technique in our project.

1 D. Dwivedi, K. Lepková and T. Becker, *RSC Adv.*, 2017, **7**, 4580–4610.
2 D. Dwivedi, K. Lepková and T. Becker, *Proc. R. Soc. A*, 2017, **473**, 20160852.

Neil R. Champness opened a general discussion of the topics raised at the session: A general theme that is coming out of the discussion is nucleation; what is the real question that needs to be answered and how do we answer that particular question?

Karl-Heinz Ernst commented: I'd like to emphasize the importance of better understanding nucleation in the liquid phase, which is relevant for polymorphism, and therefore for crystallization in general, but in particular also for pharmaceutical drugs. Because it requires lower supersaturation, nucleation basically occurs only on surfaces, hardly ever as homogeneous nucleation. So surface structure plays an important role. If one thinks about the role of impurities at the stage of nucleation, the ripening of the colloidal nucleus into more stable crystallites under the exclusion of solvent: this is not understood at all. Recent progress in environmental microscopy has revealed or proved new forms of ripening, like Viedma ripening, where stereoselective fusion of small crystallites into larger ones occurs. So the importance and impact to materials science demand new approaches and techniques towards better insight.

Talat Rahman remarked: In terms of theory, *ab initio* MD is the best way to go as this captures the dynamics of the system and you don't have to put the forces from empirical potentials. How to do time- and length-scales large enough to capture what's going on? There have been promising recent advances. Someone has been able to show, for molecules in solvents, at the liquid–solid interface, that you could do MD *ab initio* at timescales large enough to be able to capture the free energies of these systems. Promising steps are being taken by a number of groups.

Talat Rahman commented: There are some new experimental setups in which they can measure the vibrational density of states while reactions are happening, using inelastic Mössbauer scattering (Prof. Roldan's group at Ruhr University Bochum). These are very promising as they allow investigation of the surface *in operando*. They also provide measures that can relate directly to theory.

Neil R. Champness asked: The crystal engineering community talks about similar issues relating to nucleation. What can we learn from those who work on crystal structure prediction?

Deepak Dwivedi communicated in response: Thanks, Prof. Champness, for raising this nice question. I would like to highlight a paper by C.-Y. Wu *et al.* who have used synchrotron-radiation-based infrared nanospectroscopy (SINS).[1] The application of such *in situ* synchrotron techniques may enable the nucleation and growth phenomenon to be described in great detail.

1 C.-Y. Wu *et al.*, *Nature*, 2017, **541**, 511–515.

Talat Rahman commented: I am more interested here in the nucleation phenomenon than the crystal structure – how are things happening in operando. A lot of factors are involved, and we need to understand under what kinds of conditions you could stabilise structures. In this context, I agree that DFT calculations may not be the optimal way to proceed. A lot can be learned from molecular dynamics and kinetic Monte Carlo simulations using empirical and semi-empirical potentials, which include the effects of temperature either directly through the lattice (molecular dynamics) or indirectly (kinetic Monte Carlo). While molecular dynamics simulations can be achieved to microseconds, kinetic Monte Carlo simulations can be run to seconds and minutes, as in the lab. For accuracy checks, DFT calculations can then be performed for selected cases and rate-limiting parameters validated. Of course, it is imperative that contact be made with relevant experimental data at all stages possible.

Amar Flood responded: To address these types of phenomena, we have been thinking about using multi-scale simulations. However, after some involvement with molecular dynamics and multi-scaling from that starting point, it is clear that 10-fold speed-ups are what are expected (and are applauded) from within that community. However, and as noted by many others already, we do not need to go from one to ten microseconds but to get into seconds and minutes.

Talat Rahman said: Minutes is when we see something that's beyond nucleation stage.

Amar Flood answered: The only solution we have for the use of MD and multiscale simulations to understand the self-assembled structures is the same strategy used with DFT, and that is to start as close as possible to the thermodynamic equilibrium point.

Talat Rahman remarked: Relying on DFT for everything may not be the best way to go. Classical MD with empirical or semi-empirical force fields are good enough now to give us a starting point, but I think you still need that accuracy that you get from DFT to check things. In the *ab initio* MD simulations that I mentioned earlier, the point is not to carry out simulations for nanoseconds or microseconds, but rather to use analytical methods to obtain information from the trajectories you get for short time scales ($\sim$10 ps, or 1 ns). It appears that you can get reliable values of free energies from such procedures. That beats looking at the trajectory of each species involved; that could take forever. The point is, we are addressing these issues in the computational modelling of materials.

Faraday Discussions

PAPER

A new on-surface synthetic pathway to 5-armchair graphene nanoribbons on Cu(111) surfaces†

Kewei Sun,[a] Penghui Ji,[a] Haiming Zhang,*[a] Kaifeng Niu,[a] Ling Li,[a] Aixi Chen,[a] Qing Li,[a] Klaus Müllen*[b] and Lifeng Chi *[a]

Received 11th April 2017, Accepted 10th May 2017
DOI: 10.1039/c7fd00129k

We report a new pathway to fabricate armchair graphene nanoribbons with five carbon atoms in the cross section (5-AGNRs) on Cu(111) surfaces. Instead of using haloaromatics as precursors, the 5-AGNRs are synthesized *via* a surface assisted decarboxylation reaction of perylene-3,4,9,10-tetracarboxylic dianhydride (PTCDA). The on-surface decarboxylation of PTCDA can produce extended copper–perylene chains on Cu(111) that are able to transform into graphene nanoribbons after annealing at higher temperatures (*ca.* 630 K). Due to the low yield (*ca.* 20%) of GNRs upon copper extrusion, various gases are introduced to assist the transformation of the copper–perylene chains into the GNRs. Typical reducing gases (H_2 and CO) and oxidizing gas (O_2) are evaluated for their performance in breaking aryl–Cu bonds. This method enriches on-surface protocols for the synthesis of AGNRs using non-halogen containing precursors.

1. Introduction

In recent years, the bottom-up synthesis of graphene nanoribbons (GNRs) has attracted growing interest in view of their potential applications in next-generation carbon-based electronics. One main advantage of GNRs over other carbon-based materials lies in the adjustable band-gap. As predicted by theory,[1,2] the electronic structures of GNRs sensitively depend upon their geometric structure, specifically the width of the GNRs. Although top-down fabrication can hardly obtain atomically precise edges,[3–5] by choosing proper precursor molecules the on-surface synthesis of GNRs provides attractive and promising pathways to fabricate GNRs with controlled width and edges at an atomic level. Both 7-AGNRs

[a]Institute of Functional Nano and Soft Materials (FUNSOM), Jiangsu Key Laboratory for Carbon-Based Functional Materials &devices, Soochow University, Suzhou 215123, P. R. China. E-mail: hmzhang@suda.edu.cn; chilf@suda.edu.cn

[b]Institute of Physical Chemistry, Johannes Gutenberg University Mainz, Duesbergweg 10-14, D-55128 Mainz, Germany. E-mail: muellen@mpip-mainz.mpg.de

† Electronic supplementary information (ESI) available. See DOI: 10.1039/c7fd00129k

and 13-AGNRs (where the numbers 7 and 13 represent the carbon number across the width of the GNR) have been successfully prepared in a multi-step approach: the dehalogenation of suitable precursors, intermolecular coupling of the intermediate radicals, and subsequent intramolecular dehydrogenation.[6–11] Different from the above protocol, a novel route toward GNRs has been realized in the synthesis of 5-AGNRs on Au(111) using 1,4,5,8-tetrabromonaphthalene as the precursor molecule. To date, various types of GNRs, such as zigzag GNRs, cove-GNRs, and doped GNRs, as well as the heterojunctions of GNRs have been fabricated[12–20] *via* on-surface synthesis. In addition to the synthesis of GNRs on Au(111), the preparation of GNRs on Ag(111) and Cu(111) has also been reported.[21–25] In these cases, the precursor molecules are always bromo-substituted aromatics, which allow for polymerization to occur after homolytic dissociation of the C–Br bonds. However, chemisorbed bromine ions/atoms on metal surfaces are contaminants in further measurements. An example is the investigation of the electronic structures of GNRs using scanning tunnelling spectroscopy (STS), which requires further annealing at higher temperatures to remove bromine species. It is therefore challenging to explore a non-halogen containing process towards the synthesis of GNRs on metal surfaces.

Here we report a new pathway to 5-AGNRs by employing perylene-3,4,9,10-tetracarboxylic dianhydride (PTCDA) as a precursor molecule on Cu(111) surfaces under ultra-high vacuum (UHV). PTCDA is a well-known colorant and is commonly used as the semiconductor component of organic field effect transistors. Its self-assembled structures on metal single crystal surfaces are well characterized in previous studies.[26–29] Although the copper-substrate catalyzed decarboxylation of aromatic acids has been reported recently,[30] the removal of acid anhydride groups and the following connection of aromatic rings toward the formation of GNRs have not yet been described. In light of this, we select PTCDA as a non-halogen containing precursor molecule for the on-surface synthesis of GNRs. STM investigations reveal that copper-organic chains are the initial product after the decarboxylation of PTCDA by thermal annealing at 470 K on a Cu(111) surface. Annealing at a higher temperature ($\sim$630 K) can lead to the dissociation of aryl–Cu bonds, which generates 5-ANGRs on the Cu(111) surface. To the best of our knowledge, this new synthetic route is the first "halogen-free" process enabling the on-surface synthesis of GNRs, as displayed in Scheme 1. Due to the low yield (*ca.* 20%) of the thermal annealing, some gases (H_2, CO, and O_2) have been introduced during thermal annealing in order to evaluate their role in the transformation of copper-organic chains to 5-AGNRs.

2. Experimental

All samples were prepared under ultra-high vacuum conditions with a base pressure of 5×10^{-10} mbar. A Cu(111) single crystal was cleaned by repeated Ar^+ sputtering and annealing at 750 K. The temperature of the samples was measured using a thermocouple. PTCDA molecules were deposited on a Cu(111) surface at 300 K using organic molecular beam epitaxy (OMBE). After deposition, the sample was first annealed at 470 K (15 min) and then at 630 K (20 min) for the formation of 5-AGNRs. Low temperature STM, using an instrument from Omicron Nanotechnology, was used to characterize the samples at the temperature of 77 K under ultra-high vacuum conditions (base vacuum $< 1.0 \times 10^{-10}$ mbar). High purity

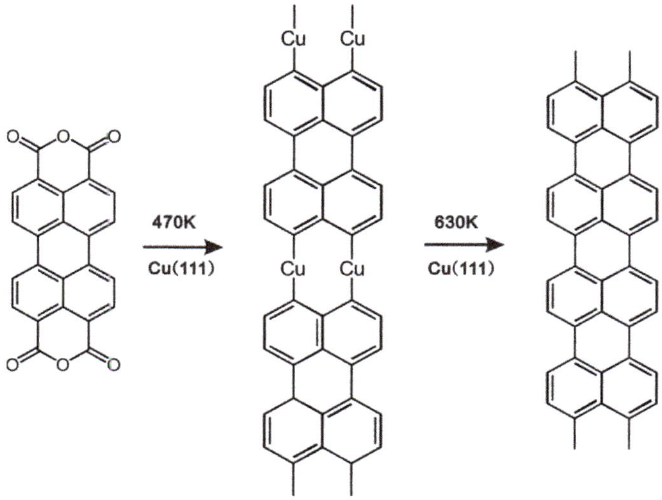

Scheme 1 Reaction processes of PTCDA towards the formation of 5-AGNRs.

(99.999%) reducing gases (H_2 and CO) and an oxidizing gas (O_2) were dosed into the preparation chamber *via* a leak valve.

3. Results and discussion

PTCDA molecules are firstly deposited on Cu(111) surfaces at room temperature. Displayed in Fig. 1a is a STM image of PTCDA self-assembled monolayers (SAMs), in which each building block represents a PTCDA molecule. According to the measured parameters of the unit cell ($a = 1.23 \pm 0.01$ nm, $b = 1.94 \pm 0.03$ nm), a structural model is proposed (inset, Fig. 1a) which is consistent with previous results from LEED and STM investigations.[27,31] When PTCDA molecules are deposited on the hot Cu(111) substrate (held at 520 K), newly formed linear chains

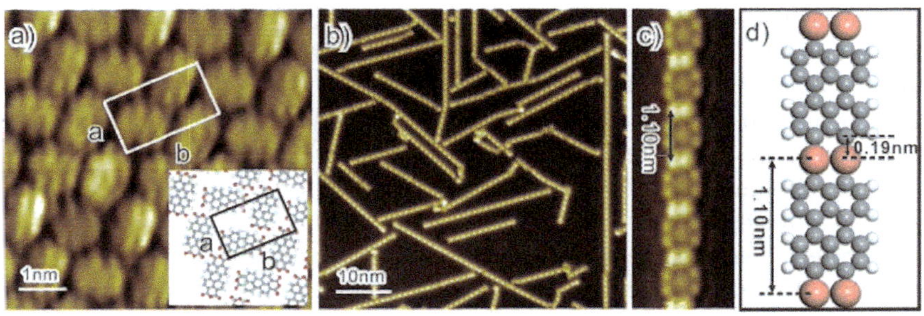

Fig. 1 (a) A representat ve STM image of the PTCDA SAMs prepared at room temperature on Cu(111). Scanning parameters: voltage $V_b = 100$ mV and current $I_t = 1$ nA. (b) A STM image ($V_b = -1$ V, $I_t = 50$ pA) of the sample prepared by depositing PTCDA on the Cu(111) surface held at 520 K. (c) A high resolution STM image ($V_b = 2$ mV, $I_t = 2$ nA) of a copper–perylene chain. (d) Optimized geometric structures of a copper–perylene chain. (Gray atoms are carbon, white atoms are hydrogen, and orange atoms are copper).

are frequently observed by STM on the surface. Fig. 1b illustrates the chain-like structure obtained by STM. Detailed information can be collected from the high resolution STM image (see Fig. 1c), in which the periodic distance along the chain is measured to be 1.10 ± 0.03 nm. The structure proposed in Fig. 1d is based on a natural assignment. The measured distance (1.10 nm) fits well with the periodic length obtained from the optimized geometric structure of an alternating copper–perylene arrangement. The two bright bumps in the chain are attributed to copper atoms where each copper atom is connected to two carbon atoms, giving a Cu–C distance of about 2.0 Å.

Copper atoms should play a critical role in the decarboxylation of PTCDA, which is similar to the copper-catalyzed aryl–aryl coupling of aromatic acids.[30,32,33] Although no metal–organic structures exist on surfaces after annealing at temperatures higher than 470 K,[30,33] the stable copper–perylene chains observed in our case are thought to result from the bonding geometry of multiple aryl–Cu bonds, which is similar to the case of multiple aryl–Au bonds.[34] The copper–perylene chains show robust stability, as observed by STM on samples prepared by depositing PTCDA molecules on hot Cu(111) surfaces held at temperatures ranging from 470 K to 600 K. Since copper–organic intermediates are often observed in a surface-supported debromination and the subsequent C–C coupling reaction, such as the reaction of 4,4″-dibromo-*p*-terphenyl on Cu(111), the initial formation of copper–perylene chains is not surprising in the surface assisted decarboxylation of PTCDA.[33] A single aryl–Cu covalent bond breaks at 470 K,[33,35,36] whereas the multiple aryl–Cu bonds in the copper–perylene chains appear to be more stable. This might be caused by the interaction between the two adjacent Cu atoms. In addition, PTCDA molecules have also been deposited on other metal substrates to explore the probability of the subsequent reactions. STM investigations reveal that surface assisted decarboxylation of PTCDA cannot occur on Au(111) and Ag(111) surfaces. This is reminiscent of previous studies on the catalytic differences of coin metal substrates in decarboxylation.[30]

The decarboxylation of PTCDA and other aromatic acids should proceed in a different way on Cu(111). In the decarboxylation of 2,6-naphthalenedicarboxylic acid (NDCA) on a copper substrate, a metal–organic coordinated complex (MOCC) has been observed as a result of dehydrogenation of adsorbed NDCA.[30] The formation of a MOCC with a higher thermal stability than that of NDCA may ensure the subsequent decarboxylation on the surfaces. However, in the case of PTCDA, there are not any intermediate states observed by STM before decarboxylation. When PTCDA molecules are deposited on Cu(111) held at 500 K, both self-assembled molecules and copper–perylene chains can be simultaneously observed by STM (see ESI Fig. S1†). The transformation of the anhydride group and the subsequent decarboxylation reaction must be too fast to be observed by *ex situ* STM investigations.

Further annealing the sample with copper–perylene chains at 630 K for 20 minutes gives rise to the breaking of aryl–Cu bonds and subsequent aryl–aryl couplings, producing 5-AGNRs on the surface, as depicted in Fig. 2a. Some copper–perylene chains are still observable, featuring periodic bright bumps in the chain. In contrast, a few ribbons, exhibiting identical tunnelling probability, can be recognized from the image. Typical and high resolution STM images of one ribbon are displayed in Fig. 2b and c respectively. The periodic distance along the

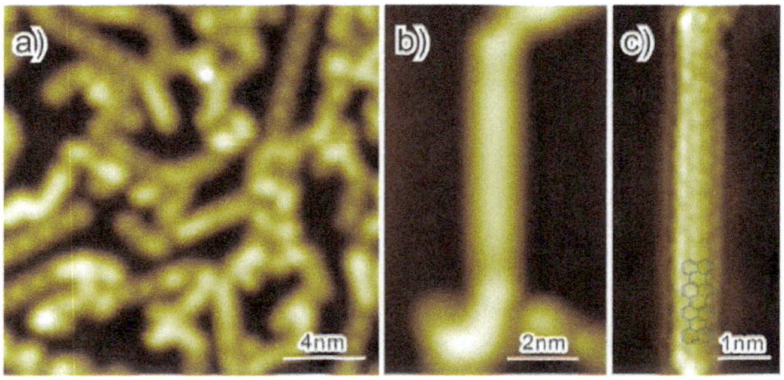

Fig. 2 (a) A typical STM image of a Cu(111) surface containing 5-AGNRs after annealing at 630 K ($V_b = 1$ V, $I_t = 20$ pA). (b) STM image of one 5-AGNR on Cu(111) ($V_b = 1$ V, $I_t = 10$ pA). (c) High resolution image of the 5-AGNR ($V_b = 20$ mV, $I_t = 1.2$ nA).

ribbon is measured to be 0.42 nm, which perfectly coincides with the periodic structural features of 5-AGNRs. The ribbons are therefore attributed to 5-AGNRs.

Although the annealing temperature applied for the transformation from copper–perylene chains to 5-AGNRs is as high as 630 K, the reaction efficiency toward 5-AGNRs is quite low (*ca.* 20%). Some copper–perylene chains are still observable from the STM image (Fig. 2a), implying insufficient annealing. Considering the fact that the temperature to initiate dehydrogenation on Cu(111) is as low as ~500 K,[37,38] many distorted structures (see Fig. 2a) resulting from side reactions on the surface are not unexpected. Due to the robustness of the copper–perylene chains and the increasing probability of byproducts at higher temperatures, thermal annealing alone is not efficient enough and alternative routes need to be explored for the breaking of aryl–Cu bonds under milder conditions. In tentative experiments, several gases have been introduced into the UHV chamber during substrate annealing of copper–perylene/Cu(111) at 570 K. The effect of each gas in promoting the transformation of copper–perylene chains has been monitored by STM investigations. Displayed in Fig. 3 are the STM images of the copper–perylene/Cu(111) annealed at 570 K in the presence of H_2 (a), CO (b), and O_2 (c). No significant changes are observable for the samples annealed with H_2

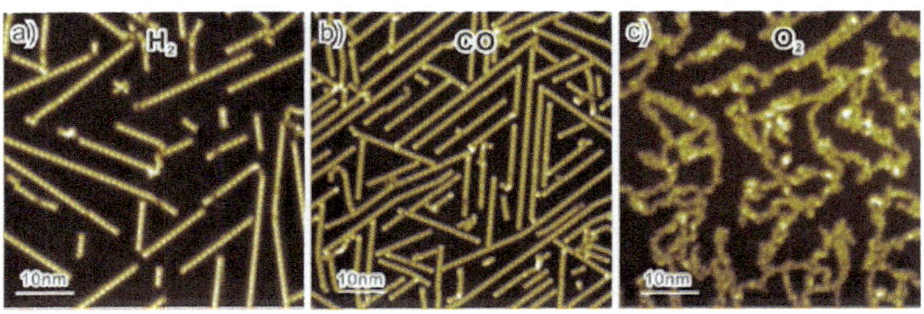

Fig. 3 Representative STM images of the copper–perylene/Cu(111) after substrate annealing at 570 K in the presence of (a) hydrogen (chamber pressure: 7×10^{-7} mbar, scanning parameters: $V_b = -1$ V, $I_t = 10$ pA), (b) carbon monoxide (chamber pressure: 1×10^{-7} mbar, scanning parameters: $V_b = 1$ V, $I_t = 100$ pA), and (c) oxygen (chamber pressure: 1×10^{-8} mbar, scanning parameters: $V_b = 1$ V, $I_t = 100$ pA).

(7×10^{-7} mbar) and CO (1×10^{-7} mbar), but the presence of O_2 (1×10^{-8} mbar) causes a remarkable change in the structure of the copper–perylene chains. Mostly disordered structures are observed, showing the participation of O_2 in assisting the breaking of aryl–Cu bonds.

In order to better understand the influence of oxygen on the copper–perylene chains, temperature-dependent structural changes have been explored by annealing the copper–perylene/Cu(111) sample at various temperatures. For substrate temperatures ranging from 300–470 K, only growing patches (see Fig. 4, panel a to c, highlighted by red arrows) are recognizable. STS measurements on the patches show a prominent feature at 0.75 V (ESI Fig. S2†) that is consistent with the electronic structure reported previously from STS investigations of Cu_2O.[39–42] Since no obvious changes of the copper–perylene chains are recognizable from the STM images, the substrate Cu atoms must be more susceptible to oxidation than the Cu atoms bridged between perylenes. When the sample is heated to 520 K with the same O_2 exposing pressure, the aryl–Cu bonds demonstrate an initial state of oxidation. On further increasing the temperature to 550 K, more of the aryl–Cu bonds are oxidized, leading to the disordered structures displayed in Fig. 4d. In addition to the distorted chains, segments of the chains

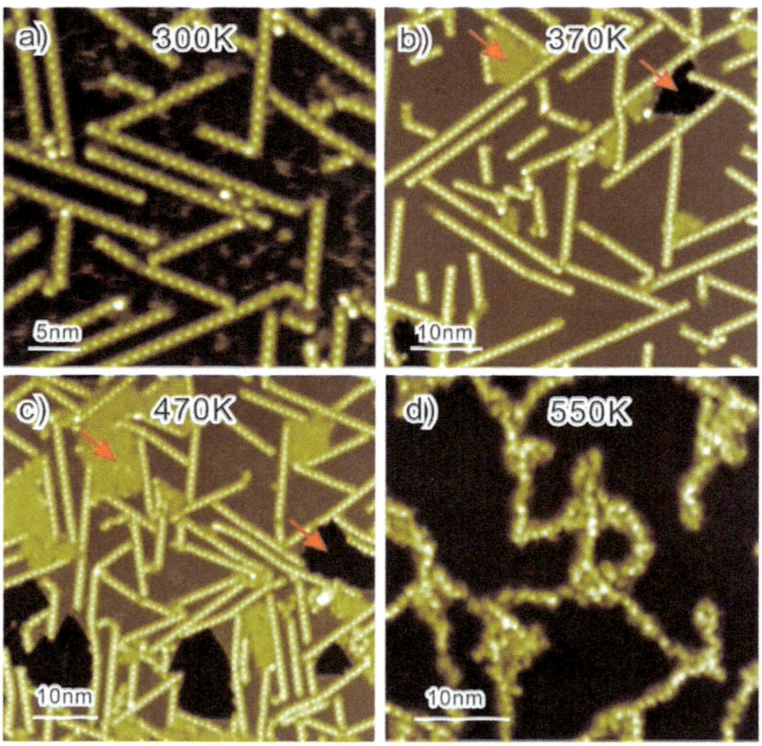

Fig. 4 Typical STM images of the copper–perylene chains on Cu(111) after exposure to oxygen (pressure: 2–3 × 10^{-8} mbar) for thermal annealing at (a) RT, (b) 370 K, (c) 420 K, and (d) 550 K. All images are observed with the scanning parameters of $V_b = 1$ V and $I_t = 100$ pA.

are also frequently observed in the STM images (see ESI Fig. S3†), suggesting that oxygen may promote the breaking of aryl–Cu bonds. Unfortunately, although the aryl–Cu bonds are susceptible to oxidation, this protocol fails to assist the transformation to 5-AGNRs. Obviously, the temperature needed to affect the aryl–Cu bond under O_2 exposure is still too high.

4. Conclusions

In summary, PTCDA molecules are selected as non-halogen containing precursors to synthesize 5-AGNRs on Cu(111) surfaces. A surface assisted decarboxylation reaction occurs upon annealing of the PTCDA SAMs at 470 K for 15 minutes, producing firstly copper–perylene chains with the formation of aryl–Cu bonds. High-temperature annealing at 630 K for 20 minutes results in the breaking of the aryl–Cu bonds and the subsequent aryl–aryl coupling towards forming 5-AGNRs on the surface. Due to the low yield of 5-AGNRs (20%) with high temperature annealing, the copper–perylene chains have been exposed to several gases during annealing in order to promote the formation of 5-AGNRs. STM investigations reveal that the copper–perylene chains are resistant to the reducing gases H_2 and CO and more susceptible to the oxidizing gas O_2. This work provides an alternative way to generate graphene nanoribbons. Further experiments and concepts are needed to reduce the activation energy for breaking the aryl–Cu bonds and to promote the high quality preparation of 5-AGNRs from non-halogen containing precursors.

Acknowledgements

This work was supported by the National Natural Science Foundation of China (NSFC, Grant No. 21527805 and 21673154). We also thank the Collaborative Innovation Center of Suzhou Nano Science & Technology, and the Priority Academic Program Development of Jiangsu Higher Education Institutions. K. M. acknowledges support from the Johannes Gutenberg University *via* a Gutenberg Research Fellowship. Open Access funding provided by the Max Planck Society.

Notes and references

1 Y. W. Son, M. L. Cohen and S. G. Louie, *Phys. Rev. Lett.*, 2006, **97**, 216803.
2 L. Yang, C. H. Park, Y. W. Son, M. L. Cohen and S. G. Louie, *Phys. Rev. Lett.*, 2007, **99**, 186801.
3 D. V. Kosynkin, A. L. Higginbotham, A. Sinitskii, J. R. Lomeda, A. Dimiev, B. K. Price and J. M. Tour, *Nature*, 2009, **458**, 872–876.
4 X. L. Li, X. R. Wang, L. Zhang, S. W. Lee and H. J. Dai, *Science*, 2008, **319**, 1229–1232.
5 X. R. Wang, Y. J. Ouyang, L. Y. Jiao, H. L. Wang, L. M. Xie, J. Wu, J. Guo and H. J. Dai, *Nat. Nanotechnol.*, 2011, **6**, 563–567.
6 J. M. Cai, P. Ruffieux, R. Jaafar, M. Bieri, T. Braun, S. Blankenburg, M. Muoth, A. P. Seitsonen, M. Saleh, X. L. Feng, K. Müllen and R. Fasel, *Nature*, 2010, **466**, 470–473.
7 Y. C. Chen, D. G. de Oteyza, Z. Pedramrazi, C. Chen, F. R. Fischer and M. F. Crommie, *ACS Nano*, 2013, **7**, 6123–6128.

8 H. M. Zhang, H. P. Lin, K. W. Sun, L. Chen, Y. Zagranyarski, N. Aghdassi, S. Duhm, Q. Li, D. Y. Zhong, Y. Y. Li, K. Müllen, H. Fuchs and L. F. Chi, *J. Am. Chem. Soc.*, 2015, **137**, 4022–4025.

9 A. Basagni, F. Sedona, C. A. Pignedoli, M. Cattelan, L. Nicolas, M. Casarin and M. Sambi, *J. Am. Chem. Soc.*, 2015, **137**, 1802–1808.

10 L. Talirz, H. Söde, T. Dumslaff, S. Y. Wang, J. R. Sanchez-Valencia, J. Liu, P. Shinde, C. A. Pignedoli, L. B. Liang, V. Meunier, N. C. Plumb, M. Shi, X. L. Feng, A. Narita, K. Müllen, R. Fasel and P. Ruffieux, *ACS Nano*, 2017, **11**, 1380–1388.

11 H. Sakaguchi, Y. Kawagoe, Y. Hirano, T. Iruka, M. Yano and T. Nakae, *Adv. Mater.*, 2014, **26**, 4134–4138.

12 J. M. Cai, C. A. Pignedoli, L. Talirz, P. Ruffieux, H. Söde, L. B. Liang, V. Meunier, R. Berger, R. J. Li, X. L. Feng, K. Müllen and R. Fasel, *Nat. Nanotechnol.*, 2014, **9**, 896–900.

13 Y. C. Chen, T. Cao, C. Chen, Z. Pedramrazi, D. Haberer, D. G. de Oteyza, F. R. Fischer, S. G. Louie and M. F. Crommie, *Nat. Nanotechnol.*, 2015, **10**, 156–160.

14 Y. Zhang, Y. F. Zhang, G. Li, J. C. Lu, X. Lin, S. X. Du, R. Berger, X. L. Feng, K. Müllen and H. J. Gao, *Appl. Phys. Lett.*, 2014, **105**, 023101.

15 S. Blankenburg, J. M. Cai, P. Ruffieux, R. Jaafar, D. Passerone, X. L. Feng, K. Müllen, R. Fasel and C. A. Pignedoli, *ACS Nano*, 2012, **6**, 2020–2025.

16 J. Z. Liu, B. W. Li, Y. Z. Tan, A. Giannakopoulos, C. Sanchez-Sanchez, D. Beljonne, P. Ruffieux, R. Fasel, X. L. Feng and K. Müllen, *J. Am. Chem. Soc.*, 2015, **137**, 6097–6103.

17 R. R. Cloke, T. Marangoni, G. D. Nguyen, T. Joshi, D. J. Rizzo, C. Bronner, T. Cao, S. G. Louie, M. F. Crommie and F. R. Fischer, *J. Am. Chem. Soc.*, 2015, **137**, 8872–8875.

18 S. Kawai, S. Saito, S. Osumi, S. Yamaguchi, A. S. Foster, P. Spijker and E. Meyer, *Nat. Commun.*, 2015, **6**, 8098.

19 G. D. Nguyen, F. M. Toma, T. Cao, Z. Pedramrazi, C. Chen, D. J. Rizzo, T. Joshi, C. Bronner, Y. C. Chen, M. Favaro, S. G. Louie, F. R. Fischer and M. F. Crommie, *J. Phys. Chem. C*, 2016, **120**, 2684–2687.

20 P. Ruffieux, S. Y. Wang, B. Yang, C. Sánchez-Sánchez, J. Liu, T. Diene, L. Talirz, P. Shinde, C. A. Pignedoli, D. Passerone, T. Dumslaff, X. L. Feng, K. Müllen and R. Fasel, *Nature*, 2016, **531**, 489–492.

21 H. Huang, D. C. Wei, J. T. Sun, S. L. Wong, Y. P. Feng, A. H. C. Neto and A. T. S. Wee, *Sci. Rep.*, 2012, **2**, 983.

22 K. A. Simonov, N. A. Vinogradov, A. S. Vinogradov, A. V. Generalov, E. M. Zagrebina, N. Mårtensson, A. A. Cafolla, T. Carpy, J. P. Cunniffe and A. B. Preobrajenski, *J. Phys. Chem. C*, 2014, **118**, 12532–12540.

23 P. Han, K. Akagi, F. F. Canova, H. Mutoh, S. Shiraki, K. Iwaya, P. S. Weiss, N. Asao and T. Hitosugi, *ACS Nano*, 2014, **8**, 9181–9187.

24 P. Han, K. Akagi, F. F. Canova, R. Shimizu, H. Oguchi, S. Shiraki, P. S. Weiss, N. Asao and T. Hitosugi, *ACS Nano*, 2015, **9**, 12035–12044.

25 K. A. Simonov, N. A. Vinogradov, A. S. Vinogradov, A. V. Generalov, E. M. Zagrebina, G. I. Svirskiy, A. A. Cafolla, T. Carpy, J. P. Cunniffe, T. Taketsugu, A. Lyalin, N. Mårtensson and A. B. Preobrajenski, *ACS Nano*, 2015, **9**, 8997–9011.

26 K. Glöckler, C. Seidel, A. Soukopp, M. Sokolowski, E. Umbach, M. Böhringer, R. Berndt and W. D. Schneider, *Surf. Sci.*, 1998, **405**, 1–20.

27 T. Wagner, A. Bannani, C. Bobisch, H. Karacuban and R. Möller, *J. Phys.: Condens. Matter*, 2007, **19**, 056009.

28 Y. Zou, L. Kilian, A. Schöll, Th. Schmidt, R. Fink and E. Umbach, *Surf. Sci.*, 2006, **600**, 1240–1251.

29 Q. H. Wang and M. C. Hersam, *Nat. Chem.*, 2009, **1**, 206–211.

30 H. Y. Gao, P. A. Held, M. Knor, C. Mück-Lichtenfeld, J. Neugebauer, A. Studer and H. Fuchs, *J. Am. Chem. Soc.*, 2014, **136**, 9658–9663.

31 Th. Wagner, A. Bannani, C. Bobisch, H. Karacuban, M. Stöhr, M. Gabriel and R. Möller, *Org. Electron.*, 2004, **5**, 35–43.

32 W. Zhao, L. Dong, C. Huang, Z. M. Win and N. Lin, *Chem. Commun.*, 2016, **52**, 13225–13228.

33 W. H. Wang, X. Q. Shi, S. Y. Wang, M. A. Van Hove and N. Lin, *J. Am. Chem. Soc.*, 2011, **133**, 13264–13267.

34 H. M. Zhang and L. F. Chi, *Adv. Mater.*, 2016, **28**, 10492–10498.

35 T. A. Pham, F. Song, M. T. Nguyen, Z. S. Li, F. Studener and M. Stöhr, *Chem. - Eur. J.*, 2016, **22**, 5937–5944.

36 L. Dong, P. N. Liu and N. Lin, *Acc. Chem. Res.*, 2015, **48**, 2765–2774.

37 F. Schulz, P. H. Jacobse, F. F. Canova, J. van der Lit, D. Z. Gao, A. van den Hoogenband, P. Han, R. K. Gebbink, M. Moret, P. M. Joensuu, I. Swart and P. Liljeroth, *J. Phys. Chem. C*, 2017, **121**, 2896–2904.

38 J. J. Zhang, C. R. Chang, B. Yang, N. Cao, C. C. Peng, H. M. Zhang, D. D. Tang, F. Glorius, G. Erker, H. Fuchs, Q. Li and L. F. Chi, *Chem.–Eur. J.*, 2017, **23**, 1–6.

39 A. Gharachorlou, M. D. Detwiler, X. K. Gu, L. Mayr, B. Klötzer, J. Greeley, R. G. Reifenberger, W. N. Delgass, F. H. Ribeiro and D. Y. Zemlyanov, *ACS Appl. Mater. Interfaces*, 2015, **7**, 16428–16439.

40 C. X. Kronawitter, C. Riplinger, X. B. He, P. Zahl, E. A. Carter, P. Sutter and B. E. Koel, *J. Am. Chem. Soc.*, 2014, **136**, 13283–13288.

41 N. Nilius, H. Fedderwitz, B. Groβ, C. Noguera and J. Goniakowski, *Phys. Chem. Chem. Phys.*, 2016, **18**, 6729–6733.

42 H. Sträter, H. Fedderwitz, B. Groβ and N. Nilius, *J. Phys. Chem. C*, 2015, **119**, 5975–5981.

Faraday Discussions

PAPER

Surface induced crystallization of polymeric nano-particles: effect of surface roughness

Ahmad Jabbarzadeh * and Xin Chen

Received 20th February 2017, Accepted 23rd March 2017
DOI: 10.1039/c7fd00071e

Molecular dynamics simulations are conducted to study the crystallization of a polymeric system as a drop in an isolated state and on a surface. It is shown that crystallization kinetics for the polymeric system as a particle on a smooth surface is much faster than in the isolated form. We show however that as the surface becomes rough the crystallization rate of the polymeric particle decreases. The effect of roughness was compared for two cases of a polymer drop, partially (Wenzel state) and fully (fully confined) wetting the cavities on a rough surface. In both cases it was observed that crystallization was slower than that on a smooth surface, and crystal growth rate was decreased by increasing the characteristic roughness ratio. The crystallization on rough surfaces was still faster than that of the isolated polymer drop.

1 Introduction

Surface induced crystallization (SIC) is an important phenomenon which is encountered when liquid phase material comes into contact with a solid surface. Understanding SIC is important for many scientific and technological processes such as polymer coating of surfaces and glass forming, and for the production of advanced materials such as fiber reinforced polymers, polymer composites, polymer/quantum dot conjugates, and polymeric nano-particles.[1] Surfaces affect crystallization kinetics as well as the morphology of the crystals that form on the surface. Surface induced phenomena also have important effects on the formation of glassy states, and vapour deposition on surfaces is shown to result in denser glasses.[2] Here we will focus on the effect of surfaces on crystallization kinetics. Depending on the polymer and the film thickness, the surface may enhance or impede the crystallization rate and nucleation process. For example, poly (methyl methacrylate) (PMMA) shows increased glass transition temperature and much faster crystallization on polyethylene surfaces.[9] For films thinner than 60 nm however, PMMA shows slower crystallization and decreased glass

School of Aerospace, Mechanical and Mechatronic Engineering, The University of Sydney, NSW 2006, Australia. E-mail: ahmad.jabbarzadeh@sydney.edu.au

transition.[3] Molecular dynamics (MD)[4] and Monte Carlo methods[5,6] have been used to study the process on smooth surfaces. Interesting experimental studies[7,8] have been conducted on the nucleation process of polymeric droplets of various sizes, as small as containing only a few molecular chains, on rough and smooth surfaces. They have demonstrated that the nucleation kinetic time constant depends on whether the nucleation process is driven by the edge, surface, or within the droplet by homogenous nucleation. In those experiments the length scale of the roughness was much smaller than the average radius of gyration of the polymer. For small molecular systems such as water, neo-pentanol and tetrabromomethane[10] deposited on a mica surface, it has been shown that roughness induced by scratching the surface enhances the nucleation process. Another recent work[11] on surface topology effect on crystallization showed that the crystallization of aspirin molecules from solution was much easier in the vicinity of angular nanopores whereas spherical nanopores hindered crystallization. The behaviour of polymers in the presence of roughness in the form of regular patterns or irregular features on the surfaces has been correlated with confinement effects. Therefore, in recent years, many experiments have been conducted to find the effects of confinement on polymer crystallization processes, including crystallization under thin films,[12,13] nanopores[14] and inside spheres,[15] which can be considered as one-, two-, and three-dimensional confinement, respectively. The effect of the size of confinement on crystallization processes was experimentally studied by Shin *et al.*[16] for linear PE in cylindrical nanopores. The experimental results showed that crystallization was impeded inside the nanopores and the amount of crystallite decreased with a decrease in the size of nanopores. Despite these studies, the effect of surface roughness on surface induced crystallization kinetics of polymers and droplets has received little attention. Here we are interested in the crystallization process of polymeric nanodroplets which come into contact with a surface which has wetting properties that allow for fast diffusion of the polymeric material on the surface and subsequent penetration inside the grooves made by rough features on the surface. Here, the crystallization process is considered for cases where the polymer has partially or fully penetrated the grooves. This is compared with the crystallization on a smooth surface of the same material and also the crystallization of the same polymer in the absence of any surface effects. This should reveal how the crystallization is affected by the presence of a surface in comparison to crystallization in isolation, and how roughening the surface may affect the results.

While it is difficult to control all of the parameters experimentally, such as impurities in the system and the creation of an ideal smooth surface, computational modelling allows for a controlled environment and conditions. This helps to obtain valuable information on the effects of important parameters on the observed crystallization process. Monte Carlo simulations of polymeric nanodroplet crystallization have revealed the effect of droplet size on crystallization on a smooth surface.[6] Here we will study surface induced crystallization on both smooth and rough surfaces by using molecular dynamics simulations. A linear alkane, namely *n*-hexacontane ($C_{60}H_{122}$, referred to as C60) is simulated. This molecule is long enough to show features relevant to polymeric materials and can be considered as a shorter version of polyethylene (PE), and its crystallization time is short enough to be captured using MD simulations. In the case of a rough surface, the crystallization will take place within rectangular gaps on the surface.

The roughness characteristic size will be varied for values larger than the radius of gyration of the molecule, and the effect will be investigated. This will lead to situations where the polymer will be partially or fully confined by the roughness features of the surface. The crystallization of C60 is simulated by gradual cooling to a target temperature for four cases, including for a drop in isolated state, a drop in Wenzel state, a drop fully wetting the rough surface (fully confined state), and crystallization on a smooth surface. We have fitted the data into Avrami's model of crystallization and have compared the results for these four cases. The results provide a full picture of crystallization, revealing the effects of surface induced crystallization, and how the surface roughness affects the kinetics of crystallization.

2 Methodology and results

The simulations are separated into four stages, consisting of generating C60 in bulk and as a drop, the diffusion of C60 drops on smooth crystalline substrates and subsequent crystallization after wetting, and finally diffusion of C60 on rough crystalline substrates and subsequent crystallization. For the case of rough surfaces, the wetting is conducted at two different temperatures. This has allowed us to study the effect of roughness on crystallization for two cases, (a) for a drop partially penetrating the rough features on the surface and forming a so called Wenzel state and; (b) for a drop fully spreading into the grooves formed by roughness, we call this the fully confined state. For each case the roughness characteristics are varied and the effect of roughness is examined.

2.1 Molecular model

The molecular model is based on our earlier studies.[17,18] The CH_2 and CH_3 groups are treated as united atoms for the linear C60 molecules. Intra-molecular interactions are governed through bond stretching, torsional and angle bending potentials. For non-bonded van der Waals interactions, a 6–12 Lennard-Jones potential is used with a cut-off radius of $r_c = 2.5\sigma = 0.9825$ nm, where σ is the LJ length parameter of the methyl–methyl interaction. Lorentz–Berthelot's combining rules are used to calculate the LJ interactions for unlike groups so the $\varepsilon_{ij} = (\varepsilon_i\varepsilon_j)^{1/2}$ and $\sigma_{ij} = (\sigma_i + \sigma_j)/2$. For the simulations here the surface atoms have values for the length and energy parameters of $\sigma_w = 0.26$ nm and $\varepsilon_w = \varepsilon_{CH_2} = 47$ K. All potential parameters for the C60 molecules can be found in ref. 17 and 18. A bulk C60 system was initially equilibrated under NVT conditions (constant number of molecules, volume and temperature) at a density of 750 kg m^{-3} respectively. For the bulk state, C60 is simulated using periodic boundary conditions in all three directions representing an infinite system. The temperature was kept constant at 500 K throughout the equilibration stage. There were 1024 C60 molecules (61 440 united atoms) in a cubic simulation box of 12.286 nm size. The resulting configuration was further equilibrated at constant atmospheric pressure ($P = 0.101$ MPa) under NPT conditions. The SLLOD[19] equations of motion (eqn (1)) are used to track the motion of united atoms.

$$\begin{aligned}\dot{\boldsymbol{r}}_i &= \frac{\boldsymbol{p}_i}{m_i} + \dot{\varepsilon}\boldsymbol{r}_i \\ \dot{\boldsymbol{p}}_i &= \boldsymbol{F}_i - \zeta\boldsymbol{p}_i - \dot{\varepsilon}\boldsymbol{p}_i \\ \dot{V} &= 3\dot{\varepsilon}V\end{aligned} \tag{1}$$

Here $\boldsymbol{r}_i$, $\boldsymbol{p}_i$, and m_i are the position vector, peculiar momenta, and mass of the united atom i, respectively. $\boldsymbol{F}_i$ is the total force applied by all other atoms in the system on atom i. Coefficient ζ is the Gaussian thermostat given by eqn (2)

$$\zeta = \frac{\sum_{i=1}^{N} \boldsymbol{F}_i \cdot \boldsymbol{p}_i}{\sum_{i=1}^{N} \boldsymbol{p}_i^{\,2}} \tag{2}$$

V is the volume of the cubic simulation box. During NPT simulation the volume (V) is allowed to vary according to eqn (1) whose dynamics is governed by the dilation rate $\dot{\varepsilon}$, which itself is calculated by integrating eqn (3).

$$\ddot{\varepsilon} = \frac{(P - P_0)V}{QNk_\mathrm{B}T} \tag{3}$$

Here P_0, P, Q, N, k_B and T are the target pressure, instantaneous pressure, damping constant, total number of atoms, Boltzmann constant and temperature, respectively. The equilibration under NPT conditions at 1 atm resulted in a C60 system whose density was about 771 kg m$^{-3.}$ The resulting configuration of the C60 system was subsequently used in the isolated drop crystallization of C60, and as a droplet which was released on various surfaces for wetting and crystallization studies.

For the isolated drop the configuration resulting from the bulk was used and the polymer system was released into a cubic box of 65.9 nm size which was more than 4 times the diameter of the drop in spherical form. The large size of the box ensured there were no interactions between the periodic images of the drop. The simulations were then carried out using an NVT ensemble.

2.2 Wetting studies

2.2.1 Diffusion on smooth surfaces. For the diffusion stage the C60 droplet configuration at 500 K was placed near the surface and allowed to be captured by weak van der Waals forces and diffuse and wet the surface. The C60 drop forms into a spherical shape of approximately ~15.24 nm diameter once released near the surface. After equilibration for 9.42 ns the contact angle is measured and the wettability property of the surface is determined. At this stage the goal is to identify a substrate whose wetting properties is such that allows for the diffusion of the polymer inside the gaps that would be formed by the square posts when the surface is roughened. Here, the Lennard-Jones interaction energy parameter of the surface-polymer is the same as the polymer–polymer, and surface energy is varied by changing the lattice constant of the substrate. Four different crystalline materials were used as the substrate, all with FCC structure whose [100] plane was exposed to the C60 droplet. The lattice constants of these four surfaces were chosen to be of common materials including silicon, copper, magnesium and iron. The lattice constants for the simulated substrates are given in Table 1. We

Table 1 Nominal smooth surfaces and their lattice constants

Nominal surface material	Lattice constant (a, pm)
'Silicon'	543.07, 543.07, 543.07
'Iron'	286.65, 286.65, 286.65
'Copper'	361.49, 361.49, 361.49
'Magnesium'	320.94, 320.94, 521.08

emphasize that the surface energies of these substrates are not parameterized to represent actual silicon, copper, magnesium and iron, and a varying lattice constant was used to simply create surfaces of varying intrinsic wettability. To signify this, the names of these materials are used in quotation marks when referring to them.

To measure the contact angle a similar procedure was used as described in our earlier work.[20] As demonstrated in Fig. 1a, the density contour of the C60 drop is extracted from the simulation and is used to determine the contact angle, θ, as

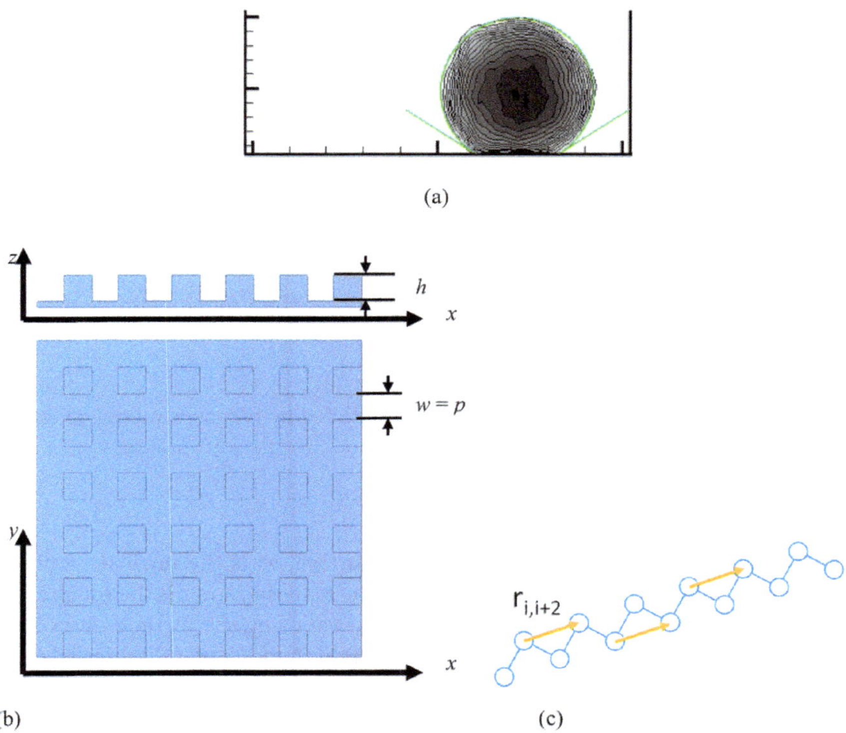

Fig. 1 (a) Contact angle measurement using ImageJ, the density contour of C60 is fitted to the green curve, and the contact angle is determined by the software; (b) roughness profile created on the surface using a periodic square wave. The roughness ratio is controlled by the gap width (w), pitch (p) and height (h) of the square pillar; (c) schematic view of chord vector between united atoms i and $i + 2$, which is used for the calculation of the degree of crystallinity.

a measure of the wettability of the surface. Here, the Low Bond Axisymmetric Drop Shape Analysis plug-in, also known as LBADSA, of ImageJ software is used to measure the contact angle.[21] An example of the contact angle measurement is shown in Fig. 1a. This step is critical since the subsequent diffusion process on the rough surface should allow the C60 droplet to wet the surface, therefore, the contact angle on the smooth surface should be less than 90°, representing an oleophilic substrate.

2.3 Diffusion on rough surfaces

The smooth surfaces are roughened by creating square pillars,[20] thus between the pillars there are rectangular grooves through which the liquid polymer diffuses and is subsequently crystallized. A rough surface topology is shown in Fig. 1b. The dimensions of the groove, including width (w), and separation (p) can be controlled by the number of pillars and wall dimensions. The height (h) can be changed independently, where a larger h requires carving from thicker substrates. Here, in order to create confinements with different dimensions, the number of pillars is set to 6, 7, 8, 9, 10 and 12 in both x and y directions, resulting in gap widths of 3.15 nm, 2.72 nm, 2.44 nm, 2.15 nm, 1.86 nm and 1.58 nm, respectively. The calculated radius of gyration from the bulk simulation for C60 at 500 K is $R_g = 1.27$ nm. Therefore the gap size used in the simulations is always larger than the radius of gyration. This arrangement makes it possible to investigate the effect of gap size on the subsequent crystallization. Here we also have used the process to study the effect of roughness on the wetting properties for one of the surfaces at a temperature of 500 K.

The roughness ratio (r) is often used to get an insight into the effect of surface roughness on the contact angle. It is defined as the ratio of the actual geometric area of a rough surface to the projected area.[22] Based on this definition, the roughness ratio r of the patterned surface shown in Fig. 1b is given by eqn (4).

$$r = 1 + \frac{4wh}{(p+w)^2} \tag{4}$$

The contact angle on a rough surface is predicted by Wenzel's model given by eqn (5):[23]

$$\cos\theta_W = r\cos\theta^* \tag{5}$$

where θ_W is the contact angle on the rough surface, θ^* is the intrinsic contact angle on an ideal flat smooth surface of the same material, and r is the roughness ratio. We will determine θ^* for our smooth surface using the method described earlier. Since the roughness ratio r is always larger than 1, eqn (5) indicates that the oleophilicity of oleophilic surfaces will be enhanced by surface roughness and the oleophobicity of oleophobic ones will be enhanced by surface roughness as well.

2.4 Crystallization of C60

The melting point of C60 is ~368 K,[24] and a crystallization temperature of 325 K is used for all simulations in the crystallization stage. This supercooling should speed up the process of crystallization allowing observation in timescales accessible to MD simulations. The crystallization processes of four states have been studied using gradually cooling down instead of quenching. Crystallization of a C60 drop in the Wenzel state on rough surfaces begins at an initial temperature of $T_i = 500$ K, and the cooling rate is set to 0.106 K per picosecond to the target crystallization temperature of $T_c = 325$ K. The crystallization simulation for the Wenzel state runs for 9.42 nanoseconds. For the fully confined state on rough surfaces, $T_i = 600$ K and the cooling rate was doubled to 0.212 K ps^{-1}. The crystallization duration was the same as that of the Wenzel state. For the isolated drop the initial temperature was set to 500 K with a cooling rate of 0.106 K ps^{-1}, and the crystallization duration was 37.68 nanoseconds, which is four times that of the Wenzel state crystallization. This larger simulation time was necessary as the crystallization rate was much slower for this isolated drop system. Crystallization on the smooth surface was simulated under the same conditions of $T_i =$ 500 K and a cooling rate of 0.106 K ps^{-1}. Details of these four simulations are summarized in Table 2.

2.5 Kinetics of crystallization and degree of crystallinity

Generally, the kinetics of crystallization can be described by the Avrami equation, eqn (6):

$$1 - \chi_c = \exp(-K(T)t^n) \tag{6}$$

here χ_c is the relative crystallinity at time t, $K(T)$ and the exponent n are referred to as the Avrami time constant, and Avrami exponent. $K(T)$ is a function of overall crystallization rate while n, the Avrami exponent, characterizes nucleation and the geometry of the growing crystallites.[25]

It is reported that all changes in n from 1 to 3 reflect the superposition of the homogeneous (in the bulk) and heterogeneous (near the interface) components of crystallization.[26] $n = 0$ indicates that the crystallization has stopped. If the phase separation rate exceeds the crystallization rates, n is close to 3; otherwise if the crystallization rate is much higher than phase separation n approaches 1.

In this simulation, to quantify crystallinity, we are using a method which is described in our earlier studies.[18,27] Here the relative orientation of the chord vectors within a molecule and neighboring molecules is used to detect the crystallinity. Parallel orientation of these vectors is indicative of crystallization (Fig. 1c). This method is applied to quantify the degree of crystallinity. Between the united atom i and $i + 2$, the chord vector is positioned at atom i, pointing to

Table 2 Simulation set up for four different crystallization simulations

Model	Wenzel state	Fully confined state	Isolated drop state	Smooth surface
Initial temperature (K)	500	600	500	500
Target temperature (K)	325			
Cooling rate (K ps^{-1})	0.106	0.212	0.106	0.106
Duration (ns)	9.42	9.42	37.68	4.9

atom $i + 2$. This orientational order is monitored by second and fourth rank correlation functions g_2 and g_4 defined by eqn (7):

$$g_2(\Gamma) = \langle \cos 2[\theta_i - \theta_j] \rangle; \; g_4(\Gamma) = \langle \cos 4[\theta_i - \theta_j] \rangle \tag{7}$$

where $\theta_i - \theta_j$ is the angle between vector i and j; these pair-wise correlations are calculated within the spatial domain Γ; $\langle \, \rangle$ bracket indicates an ensemble average. g_2 and g_4 have values in the range of 0 to 1, and $g_2 \sim g_4 \sim 1$ indicates that the chord vectors are fully aligned parallel and the polymer is totally crystallized.[18] Here only g_2 is shown in the results as the primary parameter for degree of crystallinity.

3 Results and discussion

3.1 Wetting and diffusion of C60 droplet

The first stage of this simulation is to create bulk C60. The simulation started at a density of 750 kg m^{-3} for C60,[18] under constant volume and temperature (NVT)

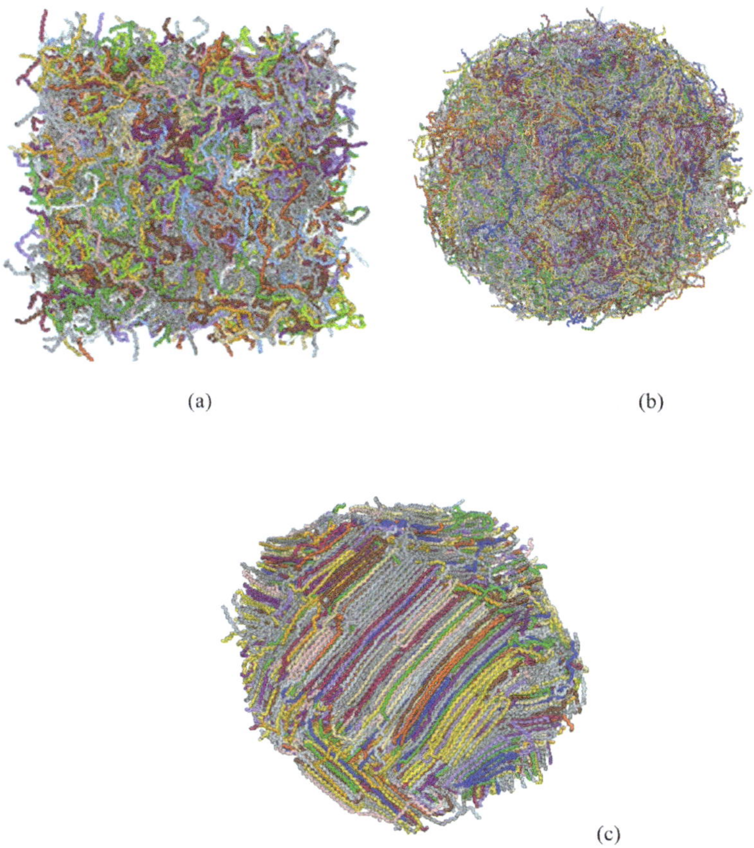

Fig. 2 Equilibrated C60 (a) in bulk form in the amorphous state at 500 K, (b) after leaving the vacuum state to form a spherical shape of an isolated drop at 500 K, and (c) after crystallization at 325 K. Molecules are shown in various colours to improve clarity.

and generated the initial system configuration. The subsequent production simulation was conducted based on this configuration, and was equilibrated for 4.7 ns at a constant pressure of 0.101 MPa under NPT conditions. A snapshot of the isotropic system of *n*-hexacontane obtained from the simulation of bulk material is shown in Fig. 2a. This configuration was used in the simulation for the isolated drop and also as a droplet in the wetting and crystallization simulations on various surfaces. For the isolated drop state, the bulk configuration equilibrated under NPT conditions is released into a cubic simulation box several times the diameter of the drop. The drop forms a spherical shape (Fig. 2b) after releasing it into this vacuum chamber.

3.1.1 The effect of surface lattice constant on the contact angle. Diffusion on a smooth crystalline surface is a critical step to ensure the contact angle is less than 90°, hence the C60 drop will successfully penetrate into the gaps on a patterned surface. Therefore, here we measured the contact angle when the droplet is equilibrated on four different crystalline smooth substrates. Fig. 3

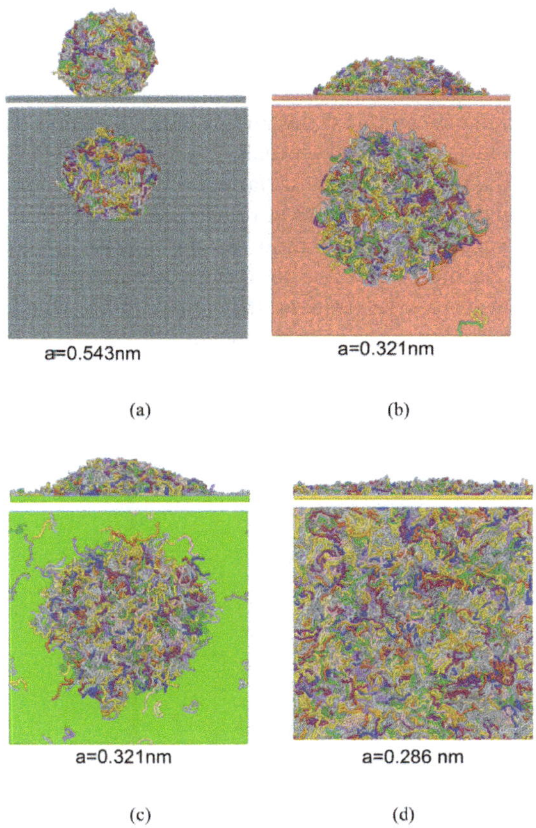

Fig. 3 Equilibrated C60 drop at 500 K on step free smooth crystalline surfaces shown as top and side views. All substrates have an FCC(100) structure with different lattice constants as shown in the figure. Oleophobic behavior can be seen when the substrate has the same lattice size as silicon (a), while surfaces with the same lattice size as copper (b), magnesium (c) and iron (d) show oleophilic properties.

shows the top and side views of C60 drop wetting flat smooth substrates whose lattice constants are given in Table 1. From the snapshots in Fig. 3 we can see that the contact angle is larger than 90° when the substrate has the same lattice constant as silicon. By contrast, the contact angle is less than 90° when the substrate has the same lattice constant as copper, magnesium or iron, showing oleophilic properties. The average contact angle is quantitatively measured using the density contours and the results are plotted as a function of lattice constant in Fig. 4. We can see from this plot that the contact angle increases by increasing the lattice constant. A good review on the effects of surface crystallographic direction on wetting can be found in ref. 28. The trend observed here is in general agreement with the experimental evidence in the literature[28] that suggest that in single component metallic melt-surface systems where weak van der Waals forces are dominant, surface facets with higher atomic density will lead to better wettability. Furthermore, experiments on metal–carbon interfaces have shown the contact angle of liquid metal on more compact facets of diamond are smaller, demonstrating increasing atomic density (decreasing nearest neighbour) results with decreasing contact angle.[29]

A trend line is plotted in Fig. 4 and a dashed line intersecting this line is drawn at $\theta = 90°$, which shows the oleophobic to oleophilic transition threshold. The corresponding value of the surface lattice constant a_{cr} is ~420 pm. This indicates an oleophilic surface will result when $a < 420$ pm and an oleophobic surface when $a > 420$ pm. This trend shows that if the surface lattice constant reduces further to ~280 pm, the contact angle will eventually reach 9°. Therefore, this particular surface ('iron') is the most oleophilic surface of the four tested here, and we used it as the surface to examine the effect of roughness.

3.1.2 The effect of roughness ratio on the contact angle. Based on the simulation results in the previous section, the crystalline surface with the same lattice constant as iron is selected as the substrate on which 2D roughness features are created. The temperature is kept constant at 500 K. Fig. 5a shows the density contour of C60 when it is equilibrated on a rough 'iron' surface. Compared with that on the smooth 'iron' surface, the contact angle is

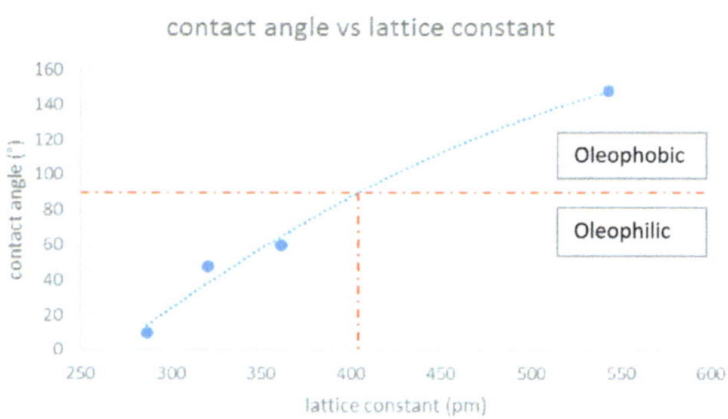

Fig. 4 The intrinsic contact angle (θ^*) as a function of the lattice constant of the smooth crystalline surface. The dotted blue line shows the trend line, which is intersected by a dash-dotted red line, which separates oleophilic and oleophobic regions. The critical lattice constant, $a_{cr} \sim 420$ pm, is identified as the threshold for oleophobic to oleophilic transition.

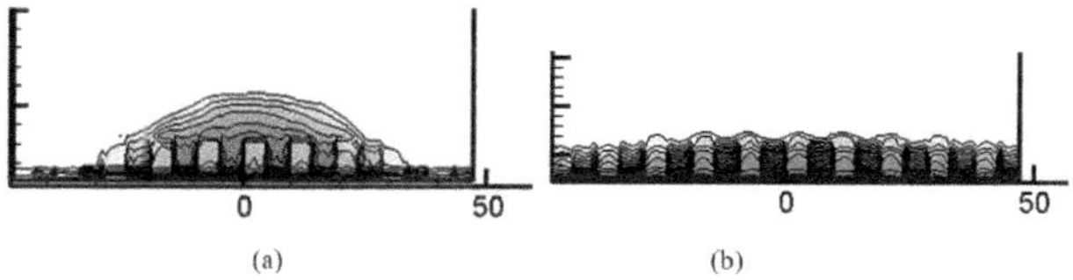

Fig. 5 Density contours of the C60 drop on rough substrates (a) in the Wenzel state at $T = 500$ K, and (b) in the fully confined state when the temperature is raised to 600 K. For both surfaces, the gap between the pillars, w, is 1.862 nm.

significantly larger on the rough surface. The wetting on this rough surface represents a Wenzel state where the drop wets cavities but does not spread beyond its boundaries. The molecules have penetrated in the gaps in the case of the rough 'iron' substrate. We also examined other rough surfaces including that with magnesium's lattice size and, as expected, found that there was less penetration by the polymer inside the gaps. Therefore, in order to investigate the effect of roughness on the degree of crystallinity, 'iron' was selected as the substrate for further simulations. Various values of gap width (w) were used to generate different roughness ratios. In Table 3, the contact angle is listed for rough "iron" surfaces with different roughness ratios, together with the geometrical dimensions of the roughness features.

In order to compare the simulation results with Wenzel's prediction of contact angles on rough surfaces, we used the intrinsic contact angle of $\theta^* = 9°$, calculated for the flat smooth 'iron' surface in Section 3.1.1. The Wenzel model basically predicts that a surface which is oleophilic will become even more oleophilic if roughened. Based on Wenzel's model for the case of rough 'iron' surfaces simulated here, the predicted contact angle for all the roughness ratios examined here would be $\theta = 0°$. Fig. 6 shows the contact angle on rough surfaces obtained from simulations (shown by filled symbols) and those predicted by the Wenzel model (shown by open symbols), *versus* the roughness ratio (r). Wenzel's model, predicts that the oleophilicity will be enhanced if the intrinsic contact angle is less than 90°, and if the roughness ratio is large enough, the contact angle will eventually be 0°. However, the simulation results demonstrate a completely opposite trend where the contact angle mostly increases with the roughness ratio. Here, the results show that Wenzel's predictions for contact angles on rough surfaces do not work well at the nanoscale. Similar deviations from the Wenzel model have been reported[20] for decane and hexadecane nano-droplets. As the

Table 3 Contact angle measured on crystalline 'iron' surfaces with different roughness ratios. Geometrical data for roughness pillars are also given

w, p separation/ width (nm)	h, height (nm)	r, roughness ratio	θ, average contact angle (°)
3.153	3.153	2.00	44.421 ± 1.011
2.723		2.16	34.945 ± 2.483
2.437		2.29	40.489 ± 1.375
2.150		2.47	43.248 ± 0.708
1.863		2.69	44.873 ± 0.842
1.577		3.00	46.705 ± 1.650

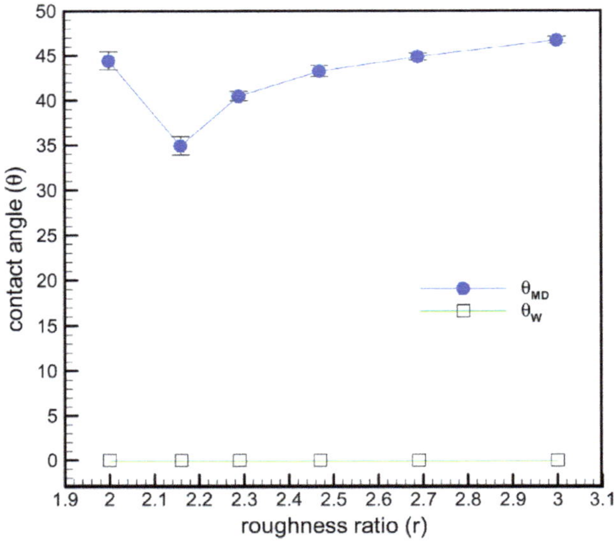

Fig. 6 Contact angle *versus* roughness ratio, the filled blue symbols show the simulation results for the contact angle on rough 'iron' surfaces, and the green open symbols show the theoretical contact angle on rough surfaces predicted using Wenzel's model with the intrinsic contact angle, $\theta^* = 9°$, of the smooth 'iron' surface.

wettability is not the main part of this study, we now turn our focus to the crystallization.

3.2 Crystallization analysis

3.2.1 Crystallization on a smooth surface. In order to compare the effect of surfaces with and without roughness on the degree of crystallinity, initially, a crystallization simulation on a smooth 'iron' surface is conducted and the degree of crystallinity as a function of time is plotted in Fig. 7. The duration of this crystallization simulation on the smooth surface was 4.9 ns after which a maximum degree of crystallinity of 0.582 was achieved. Snapshots in Fig. 8 show top and side views of the morphology of the polymer film before and after crystallization. These snapshots illustrate that molecules are not ordered at the beginning, but after the crystallization process ends, most molecules stretched or folded and aligned with each other, presenting an ordered structure. The resulting crystallized polymer is a thin film of roughly ~1.6 nm.

3.2.2 Crystallization on rough surfaces. Here we will focus on the crystallization on a rough 'iron' surface for the C60 drop in both Wenzel and fully confined states. The density contours of C60 at 500 K shown in Fig. 5a indicate the Wenzel state where molecules have not completely penetrated inside the cavities of the entire area of the surface. Therefore, in order to obtain a fully confined system, the temperature is raised to 600 K to ease the flow of the C60 melt into the cavities. This results in the C60 polymer fully penetrating the grooves and submerging the pillars. The density contour for this fully confined state is shown in Fig. 5b clearly demonstrating that the polymer has filled in the gaps.

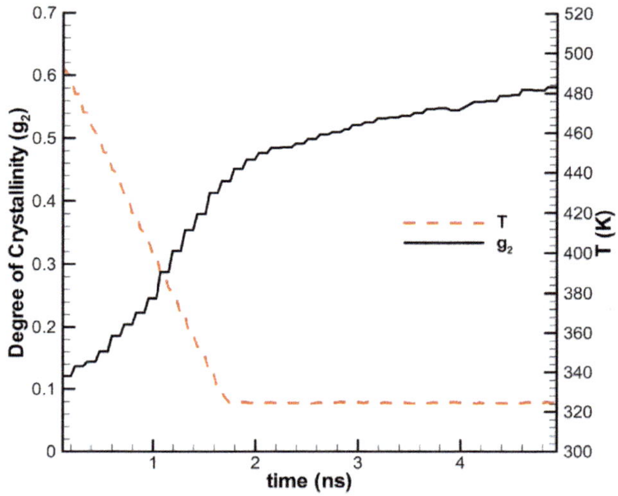

Fig. 7 Degree of crystallinity of C60 *versus* time on the smooth 'iron' surface, $T_i = 500$ K and $T_c = 325$ K with a cooling rate of 0.106 K ps^{-1}. The total simulation duration was 4.9 ns.

A. Crystallization in the Wenzel state. For the drop in the Wenzel state, the growth in the degree of crystallinity as a function of time is plotted in Fig. 9 for several rough surfaces of various gap widths (w). In all cases at the beginning, as the temperature drops the crystallization of C60 is accelerated. When the

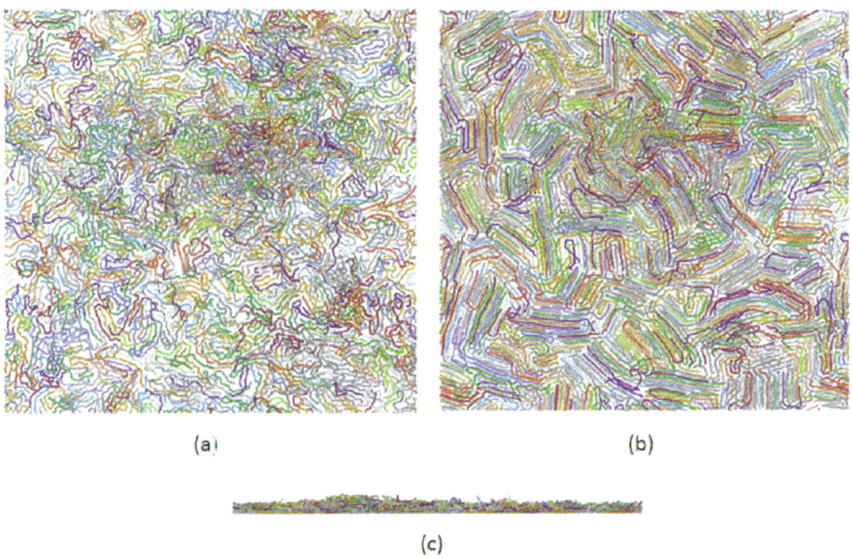

Fig. 8 Crystallization stages on the smooth 'iron' surface. Top views of the molecular configuration for the C60 drop: (a) the initially diffused polymer at $T = 500$ K, and (b) after the 4.9 ns crystallization simulation at $T_c = 325$ K. (c) Shows the side view of the final state. The substrate is not shown for clarity.

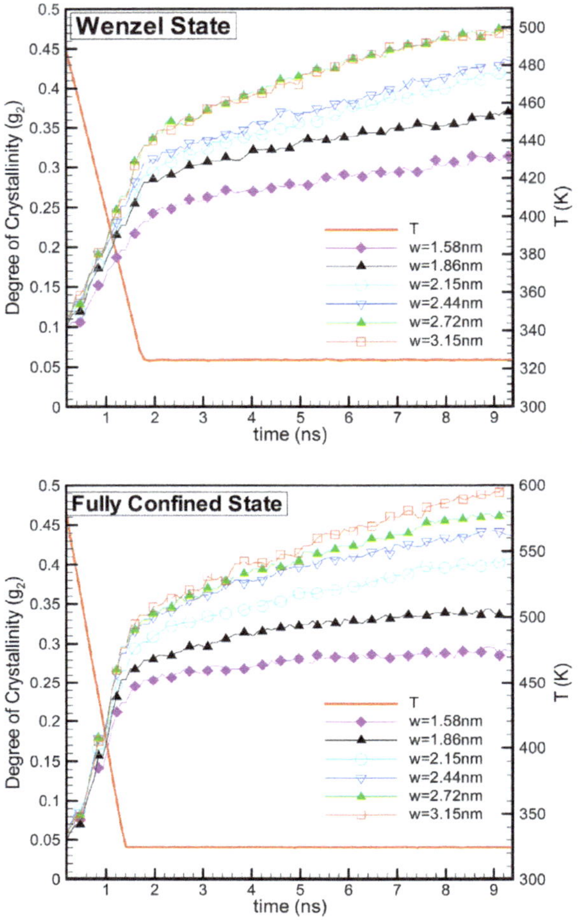

(a)

(b)

Fig. 9 Degree of crystallinity (g_2) as a function of time for the crystallization simulations of C60 on rough 'iron' surfaces patterned with rectangular grooves of various widths. The polymer drop was cooled to the target temperature $T_c = 325$ K. (a) Wenzel state and (b) fully confined state.

temperature reaches the target and is kept at 325 K, the crystal growth slows down. We can also see from these results that with the decrease in gap width (w), the maximum crystallinity declines. The degree of crystallinity for $w = 3.15$ nm has an almost overlapped trend line with that of $w = 2.72$ nm. This indicates that $w = 2.72$ nm could be a threshold in terms of impeding the crystal growth. The mean-square radius of gyration for polymer molecules is defined by eqn (8):[27]

$$\left\langle R_g{}^2 \right\rangle = \left\langle \frac{\sum_i m_i |\boldsymbol{r}_i - \boldsymbol{r}_c|^2}{\sum_i m_i} \right\rangle \tag{8}$$

where $\boldsymbol{r}_i$ is the position vector of the ith site on the molecule, and $\boldsymbol{r}_c$ is the position of the center of mass of the molecule. The $\langle\,\rangle$ denotes the ensemble average of this

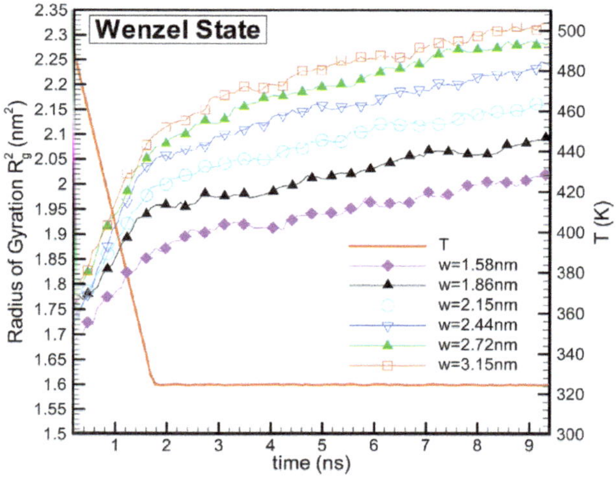

Fig. 10 Mean-square radius of gyration (R_g^2) as a function of time during crystallization of the drop in the Wenzel state for various rough 'iron' surfaces.

quantity. The mean-square radius of gyration against time is also plotted in Fig. 10 which shows a similar trend to that observed for the degree of crystallinity. Radius of gyration is an indicator of the stretching of molecules in the crystallization process. With the growth of crystals, the number of stretched molecules or segments will increase and hence R_g increases. There is also a strong correlation between the gap size and R_g which increases with increasing gap size.

The Avrami model, given by eqn (6), is used to describe the kinetics of crystallization. This equation can be rearranged[30] by applying a double natural logarithm on both sides resulting in eqn (9):

$$\ln[-\ln(1-\chi_c)] = \ln K(T) + n\ln(t) \tag{9}$$

The Avrami constants ln $K(T)$ and n are the crystal growth function and Avrami exponent, respectively. $\ln[-\ln(1-\chi_c)]$ is plotted as a function of $\ln(t)$ in Fig. 11a for the case of the Wenzel state crystallization. Thus, n and ln $K(T)$ are determined from the slope and intercept of the fitted lines for various gap size values. From this plot it can be detected that there are two linear regimes for similar segments of these curves, the first one starts from $\ln(t) \approx -21.5$ to -20, here defined as stage 1, and stage 2 which starts from $\ln(t) \approx -20$ to -18.5. One can find out the Avrami constants by using a linear fit to each crystallization stage, as shown in Fig. 12. In these plots, the dash-dot lines are the linear fit to the data. The Avrami constants are found for stages 1 and 2 and listed in Table 4. The Avrami exponent (n) is plotted as a function of the gap width (w) on the surface in Fig. 13 for stages 1 and 2. In the first stage, both the time constant ln $K(T)$ and the Avrami exponent n have larger values than those at the second stage due to the fast growth of crystals. Studies in the literature suggest that polymers in confined spaces exhibit first order or lower nucleation kinetics corresponding to an Avrami exponent of 1 or lower.[14] In this case, the nucleation dominates crystallization resulting in n being close to 1 or lower. These studies also suggest that the Avrami exponent decreases

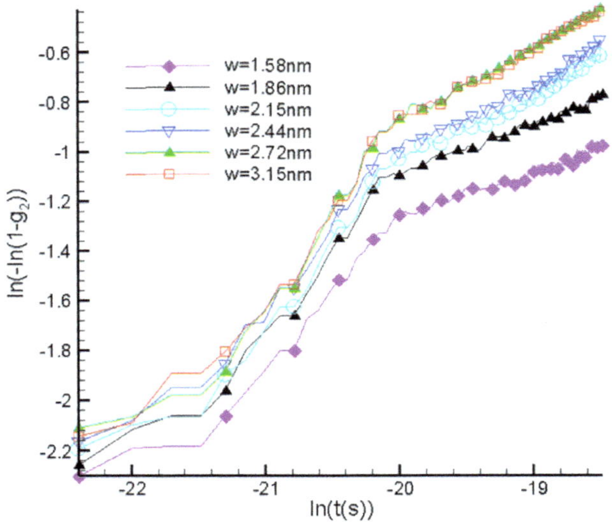

(a)

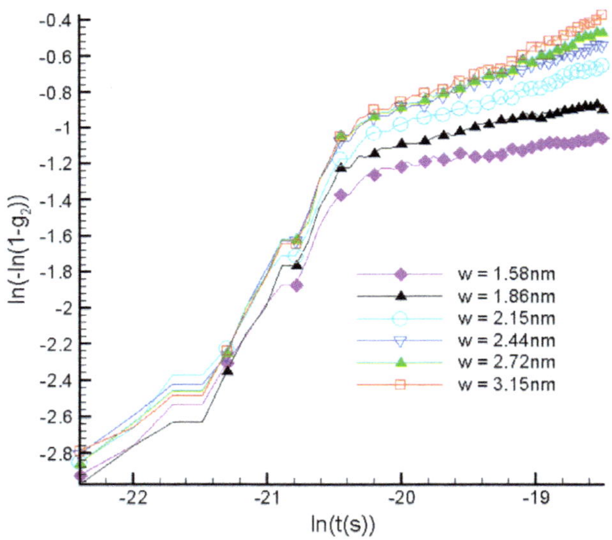

(b)

Fig. 11 Avrami plot: $\ln(-\ln(1-g_2))$ (g_2 represents the degree of crystallinity χ_c) as a function of $\ln(t(s))$ for the crystallization of a C60 drop in the (a) Wenzel state, and (b) fully confined state on rough 'iron' substrates with different gap widths. The time unit is seconds.

with increased confinement. This is consistent with the results in Table 4 for Avrami constants $\ln K(T)$ and n which show a decreasing trend with decreasing gap width (w). For the Wenzel state, the maximum values for Avrami constants are found for $w = 2.72$ nm in the first stage, which indicates that the crystal growth rate reaches a maximum under this condition.

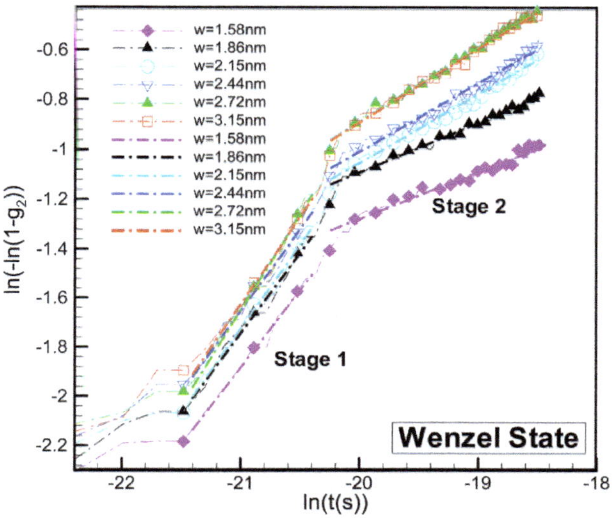

Fig. 12 Avrami constants are found by applying a linear fit (shown with thick dash-dot lines) to the Avrami plot. The slope represents n and the intercept represents $\ln K(T)$. Avrami constants are analyzed for stages 1 and 2. The results here are shown for the crystallization of C60 in the Wenzel state for different rough 'iron' surfaces.

The maximum degree of crystallinity and mean-square radius of gyration are plotted as a function of gap width, shown in Fig. 14a (black lines and open symbols for the Wenzel state). The results for the smooth 'iron' surface are also shown in this plot as filled blue symbols. For the rough surfaces generally, both crystallinity and ${R_g}^2$ increase as the size of the gaps increases, and the maximum degree of crystallinity starts to plateau at a gap width of 2.72 nm. We can see that both g_2 and ${R_g}^2$ are much larger when the surface is smooth. The results indicate that the pillars act as barriers that impede the stretching and folding of molecules and hence the crystal growth. There is however a gap width of ~2.72 nm beyond which it does not make much of a difference in the overall maximum crystallinity.

Table 4 Avrami constants extracted for the crystallization of C60 on rough surfaces of various gap widths for the Wenzel and fully confined states crystallized at $T_c = 325$ K. The results are shown for two stages of crystallization

Drop state	Wenzel				Fully confined			
	Stage 1		Stage 2		Stage 1		Stage 2	
Configuration	$\ln K(T)$ (s^{-n})	n	$\ln K(T)$ (s^{-n})	n	$\ln K(T)$ (s^{-n})	n	$\ln K(T)$ (s^{-n})	n
$w = 1.58$ nm	11.40	0.632	2.36	0.181	20.39	1.066	0.87	0.104
$w = 1.86$ nm	12.81	0.693	2.88	0.199	24.76	1.273	2.01	0.155
$w = 2.15$ nm	13.32	0.717	4.48	0.277	22.03	1.138	3.45	0.222
$w = 2.44$ nm	12.99	0.697	4.58	0.280	24.71	1.264	3.95	0.242
$w = 2.72$ nm	14.99	0.792	5.04	0.296	25.51	1.302	4.69	0.279
$w = 3.15$ nm	13.95	0.741	5.01	0.295	26.00	1.326	5.40	0.313

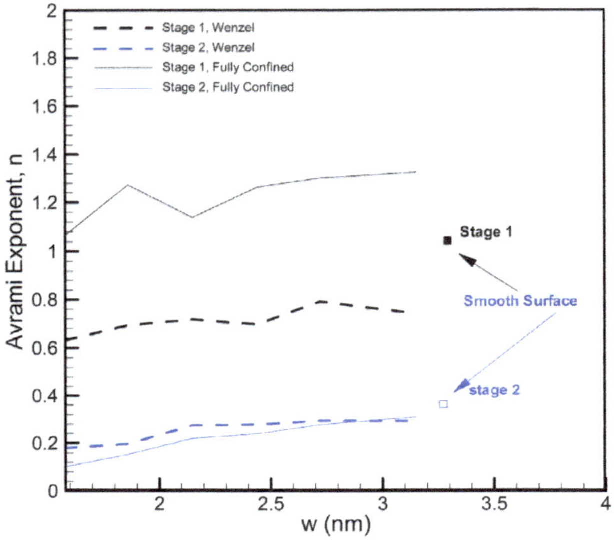

Fig. 13 The Avrami exponent (*n*) as a function of gap width on rough surfaces. Here, the results for the Wenzel (dashed lines) and fully confined (solid lines) states are compared for stage 1 (black line) and stage 2 (blue lines) of the crystallization. The trend shows that *n* increases as the gap width increases. A similar trend is observed for ln *K*(*T*) (see Table 4). The square symbols are exponents for crystallization on the smooth surface.

The maximum crystallinity and radius of gyration are also plotted as functions of the roughness ratio in Fig. 14b which clearly show that g_2 and ${R_g}^2$ are inversely proportional to the roughness ratio. That is, as the roughness increases the crystallinity decreases. The extrapolation of the ${R_g}^2$ data to $r{\sim}1$ approximately coincides with the result obtained for crystallization on a smooth surface.

B. Crystallization in the fully confined state. The same method was applied when analyzing the crystallization process on rough surfaces where the C60 drop fully penetrated the gaps between the pillars and made them fully confined. In this case the initial temperature was 600 K and therefore the cooling rate was doubled to 0.212 K ps^{-1} to reach the target temperature of 325 K in the same time as before. The trend of crystal growth for the fully confined state (shown in Fig. 9b) was almost the same as that of the Wenzel state. Here, the degree of crystallinity monotonically increases with the gap width. The Avrami plots for crystallization of the submerged state are shown in Fig. 11b using the same method described earlier.

By comparison of the two plots in Fig. 9a and b, one can see that in stage 1 the slope of crystallization curve for the fully confined state is sharper than that for the Wenzel state, whereas its slope in stage 2 of crystallization is gentler than that for the Wenzel state. We believe this is due to surface effects. Since, for the fully confined state the polymer is in contact with a much larger surface area than that in the Wenzel state, crystallization is faster during this first stage. The linear fit for stages 1 and 2 are plotted and the Avrami constants are derived from the plots by applying the same method used for the Wenzel state. The growth function ln *K*(*T*) and Avrami exponent *n* are listed in Table 4. The Avrami exponent is plotted as

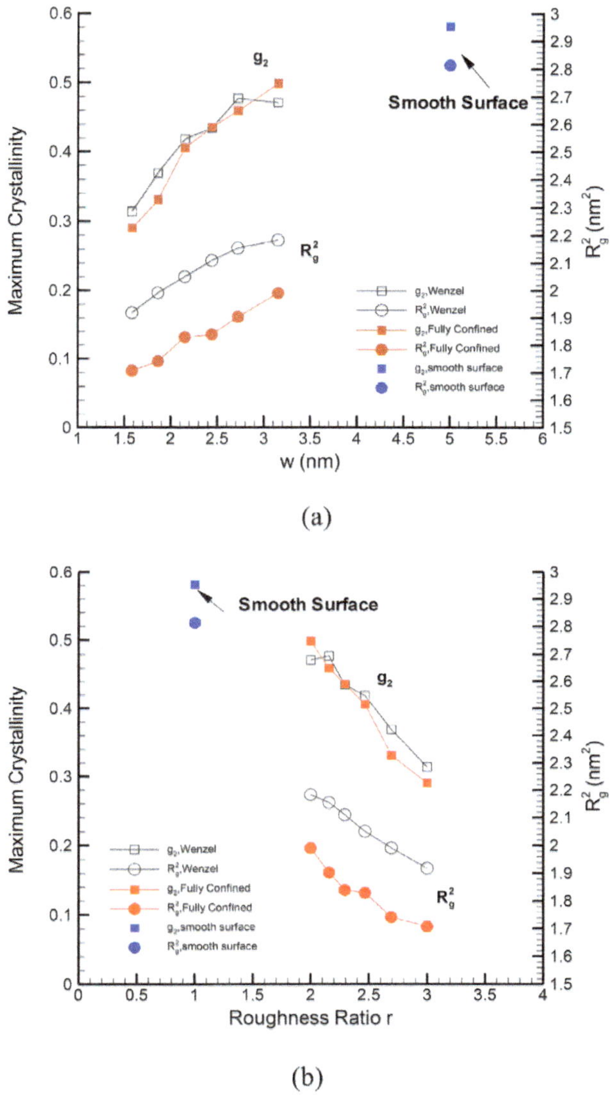

Fig. 14 Comparison of the maximum degree of crystallinity (square symbols) and mean-square radius of gyration (circle symbols) for C60 in the Wenzel (open symbols) and fully confined (red filled symbols) states at $T = 325$ K. The results on the smooth surface are shown by filled blue symbols. The results are shown against (a) the gap width (w) and (b) against the roughness ratio (r).

a function of gap width for the fully confined state in Fig. 13. By comparing Avrami constants, it is found that generally in the first stage, ln $K(T)$ and n are much larger for the fully confined state than those of the Wenzel state. In contrast, for the second stage, both these parameters (n and ln $K(T)$) are very close for the fully confined and Wenzel states. On the same in plot in Fig. 13 we have included the Avrami exponent for the smooth surface and it can be seen that the exponent in stage 1 is somewhere between the Wenzel and fully confined states.

However, in stage 2 the exponent for the rough surface converges at high values of w to that of the smooth surface. At this stage, the crystal growth rate under different confinement sizes varies slightly but both growth function and Avrami exponent grow gradually as the gap size increases. Despite different crystallization kinetics in the first stage for the Wenzel and fully confined states, the difference in final crystallinity at the end is not significant. This can be seen from Fig. 14 which shows the maximum degree of crystallinity and the mean-square radius of gyration as a function of gap width and also as a function of roughness ratio for both the Wenzel and fully confined states. The largest g_2 and R_g^2 are found when the gap width equals 3.15 nm, however, both g_2 and R_g^2 are much smaller than those for crystallization on a smooth 'iron' substrate. We can see that for the crystallization of C60 in the fully confined and Wenzel states, as the roughness ratio increases, the crystallinity and radius of gyration decrease. The results for the smooth surface fall approximately near the extrapolation of the rough surface results at $r = 1$. Comparing the results for the Wenzel and fully confined states, there is little difference in the final crystallinity, however the Wenzel state shows a larger radius of gyration. This is due to the fact that for the Wenzel state, a larger portion of the polymer drop crystallizes outside of the grooves, where the polymer molecules have room to extend as they crystallize. For the fully confined state the molecules are fully confined in the grooves and they are geometrically restricted and therefore, the radius of gyration is smaller on average.

3.2.3 A comprehensive crystallization picture of C60. Fig. 2b and c show snapshots of the beginning and final state of the isolated C60 crystallization after 37 ns of crystallization time. It takes almost six times longer for this isolated drop to achieve the same final crystallinity that is obtained on the smooth surface. The crystal morphology for this isolated polymer drop is well developed showing the

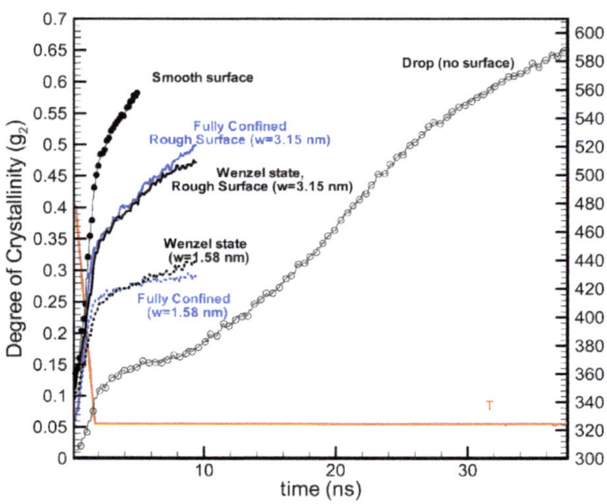

Fig. 15 Comparison of the crystallization kinetics of C60 as an isolated drop, as a drop on the smooth surface, and on the rough surface in the Wenzel and fully confined states. The results for the Wenzel and fully confined states are shown for two rough surfaces, one with a large gap size of $w = 3.15$ nm (solid lines), and one with smaller gap size of $w = 1.58$ nm (dashed lines).

formation of several crystal lamellae. To gain a comprehensive understanding of the crystallization of C60, we have put the results for the isolated C60 drop, C60 drop on the smooth surface, and C60 drop on rough surfaces (w = 1.58 nm and 3.15 nm) in the Wenzel and submerged states in one plot shown in Fig. 15. In this plot, the degree of crystallinity as a function of time is plotted for comparison with these four crystallization states.

This is a remarkable plot, which shows at one extreme we have the isolated drop crystallization which has the slowest growth rate. In the other extreme we have the droplet crystallization on the smooth surface which shows the fastest growth. Between these two states we have the crystallization on rough surfaces where in one case the polymer drop has fully wetted the surfaces submerging the pillars, and in another case forming a Wenzel state, just filling the gaps under the droplet. Here, the results are shown for two rough surfaces with gap sizes of w = 3.15 and 1.58 nm. For both gap separation values, we can see the crystal growth trend is very similar for the Wenzel and fully confined cases. The fully confined case shows slightly higher final crystallinity when the gap size is the largest (3.15 nm), however for other gap sizes including the smallest, w = 1.58 nm, there is an opposite effect where the final crystallinity is slightly larger for the Wenzel state. The crystallinity of the fully confined case is slightly larger than that of the Wenzel state for the largest gap as there is a larger contact surface between the molecules and the substrate, and the gaps are large enough for the crystallization to be enhanced by surface induced crystallization. For the smaller gaps however, the confined polymers face more restricted diffusion, and therefore the surface effects become less dominant.

For the isolated drop the final degree of crystallinity at a given time is the least among the four cases. By comparison, crystallization on the smooth surface has the largest degree of crystallinity within the shortest duration. For a drop on a rough surface the final crystallinity can be anything between the isolated drop and fully spread drop on the smooth surface.

By using a linear fit of the Avrami plots, the Avrami constants are obtained from these four different crystallization configurations and are listed in Table 5. It shows that at the first stage, the isolated drop has the largest ln $K(T)$ and n, while the Wenzel drop on a rough surface has the smallest values. In the second stage, ln $K(T)$ and n for all three surfaces are close and much smaller than those in the first stage. However, at the second stage, the isolated drop shows much larger ln $K(T)$ and n than any other states suggesting that significant crystallization still

Table 5 Comparison of the Avrami constants for crystallization of the isolated C60 drop, and its drop on smooth and rough surfaces in different states

	Stage 1		Stage 2	
Configuration	ln $K(T)$ (S^{-n})	n	ln $K(T)$ (S^{-n})	n
Rough 'iron', w = 3.15 nm, Wenzel state	13.95	0.741	5.01	0.295
Rough 'iron', w = 3.15 nm, fully confined state	26.00	1.326	5.40	0.313
Smooth 'iron' surface	20.02	1.022	6.49	0.347
Isolated drop in vacuum	27.00	1.463	15.61	0.920

continues. This suggests that for cases when the polymer comes into contact with a surface, the crystal growth rate is initially higher due to surface induced crystallization, however further development of crystallinity depends on the geometrical constraints.

4 Conclusions

In this study, the diffusion and crystallization of a linear C60 alkane as an isolated drop and as a drop on smooth and rough surfaces were studied using molecular dynamics simulations. By adjusting the lattice size of the underlying crystal structure, we produced substrates of varying degrees of oleophilicity and created a surface that allowed penetration of the polymer droplet into cavities. This surface was subsequently roughened with square pillars creating surfaces with rectangular grooves with different gap widths and roughness ratios. The effect of surface roughness on the crystallinity of the polymer was investigated by varying the size of the surface features on this strongly oleophilic surface. The crystallinity was measured for a drop which was partially penetrated inside the grooves (Wenzel state) and for a drop which fully penetrated the grooves. It was found that in both cases the rough features impeded the crystal growth. For the same crystallization time, the maximum degree of crystallinity was reduced when the gap width and size of the surface features were reduced (roughness ratio increased). Such results were consistent with experiments with polyethylene crystallization on silicon surfaces where decreased crystallinity was observed as the size of the nano-pores decreased.[16] The crystal growth could be divided into two stages, with a much larger growth rate observed in the first stage. We found that the overall crystal growth rate became smaller when the roughness ratio increased.

Crystallization of the C60 drop in four different states was compared, including an isolated drop in vacuum, on a smooth oleophilic surface, and on rough surfaces for the Wenzel and fully confined states. Significant differences were found for these modes of crystallization. The isolated drop in the absence of any surface showed the slowest crystallization and the crystal growth rate was the smallest. By contrast, drops crystallized on a surface had a larger degree of crystallinity demonstrating surface-induced crystallization. It was demonstrated that introducing roughness in the form of square pillars on the surface impeded the stretching and folding of molecules. Therefore, the degree of crystallinity for drops on rough surfaces either in the Wenzel or fully confined states was less than that of C60 on the smooth surface.

The Avrami parameters $\ln K(T)$ and n were obtained for the first and second stages of the crystallization process in all cases. For crystallization on rough surfaces the Avrami constants decreased by decreasing the size of the gaps (increasing the roughness ratio), indicating lower dimensional growth due to the presence of roughness which induced confinement effects. This finding was in agreement with existing experimental observations that reported the Avrami exponent decreased with increased confinement.[14] The isolated drop showed higher dimensional growth (n) in stages 1 and 2, whereas the polymer on the surface showed higher growth exponent only in stage 1. This was expected, as for the polymer in contact with a surface, the crystallinity develops due to surface induced crystallization as soon as it comes into contact with the surface. For the isolated drop crystallization, nucleation happens within the polymer or its external surface and crystallization accelerates once critical nucleation sites are generated.

Acknowledgements

The authors gratefully acknowledge the support of this study by the Australian Centre of Advanced Computing and Communications and the University of Sydney's high performance computational facility (Artemis).

References

1 T. Yamamoto, Molecular dynamics of polymer crystallization revisited: crystallization from the melt and the glass in longer polyethylene, *J. Chem. Phys.*, 2013, **139**(5), 054903.
2 L. Berthier and M. D. Ediger, "Facets of glass physics", *Phys. Today*, 2016, **69**, 40.
3 J. L. Keddie, R. A. Jones and R. A. Cory, *Faraday Discuss.*, 1994, **98**, 219–230.
4 J. Liu, J. Wang, H. Li, D. Shen, J. Zhang, Y. Ozaki and S. Yan, *J. Phys. Chem. B*, 2006, **110**, 738.
5 J. A. Torres, P. F. Nealey and J. J. De Pablo, Molecular simulation of ultrathin polymeric films near the glass transition, *Phys. Rev. Lett.*, 2000, **85**(15), 3221.
6 G. Reiter and J. U. Sommer, Crystallization of adsorbed polymer monolayers, *Phys. Rev. Lett.*, 1998, **80**(17), 3771.
7 W. Hu, T. Cai, Y. Ma, J. K. Hobbs, O. Farrance and G. Reiter, Polymer crystallization under nano-confinement of droplets studied by molecular simulations, *Faraday Discuss.*, 2009, **143**, 129–141.
8 J. L. Carvalho and K. Dalnoki-Veress, Homogeneous bulk, surface, and edge nucleation in crystalline nanodroplets, *Phys. Rev. Lett.*, 2010, **105**(23), 237801.
9 M. V. Massa, J. L. Carvalho and K. Dalnoki-Veress, Confinement effects in polymer crystal nucleation from the bulk to few-chain systems, *Phys. Rev. Lett.*, 2006, **97**(24), 247802.
10 J. L. Holbrough, J. M. Campbell, F. C. Meldrum and H. K. Christenson, Topographical control of crystal nucleation, *Cryst. Growth Des.*, 2012, **12**(2), 750–755.
11 Y. Diao, T. Harada, A. S. Myerson, T. A. Hatton and B. L. Trout, The role of nanopore shape in surface-induced crystallization, *Nat. Mater.*, 2011, **10**(11), 867–871.
12 J. M. Carr, D. S. Langhe, M. T. Ponting, A. Hiltner and E. Baer, Confined crystallization in polymer nanolayered films: a review, *J. Mater. Res.*, 2012, **27**(10), 1326–1350.
13 F. Zhang, J. Liu, H. Huang, B. Du and T. He, Branched crystal morphology of linear polyethylene crystallized in a two-dimensional diffusion-controlled growth field, *Eur. Phys. J. E: Soft Matter Biol. Phys.*, 2002, **8**(3), 289–297.
14 M. R. Mary, *et al.*, Confinement effects on polymer crystallization: from droplets to alumina nanopores, *Polymer*, 2013, **54.16**, 4059–4077.
15 Y. Su, G. Liu, B. Xie, D. Fu and D. Wang, Crystallization features of normal alkanes in confined geometry, *Acc. Chem. Res.*, 2013, **47**(1), 192–201.
16 K. Shin, E. Woo, Y. G. Jeong, C. Kim, J. Huh and K.-W. Kim, Crystalline structures, melting, and crystallization of linear polyethylene in cylindrical nanopores, *Macromolecules*, 2007, **40**(18), 6617–6623.
17 A. Jabbarzadeh, P. Harrowell and R. I. Tanner, Crystal bridges, tetratic order, and elusive equilibria: the role of structure in lubrication films, *J. Phys. Chem. B*, 2007, 11354–11365.

18 A. Jabbarzadeh and R. Tanner, Crystallization of alkanes under quiescent and shearing conditions, *J. Non-Newtonian Fluid Mech.*, 2009, **160**(1), 11–21.

19 P. J. Daivis and D. Evans, *J. Chem. Phys.*, 1994, **100**, 541.

20 A. Jabbarzadeh, Effect of nano-patterning on oleophobic properties of a surface, *Soft Matter*, 2013, **9**(48), 11598–11608.

21 M. D. Abràmoff, P. J. Magalhães and S. J. Ram, Image processing with ImageJ, *Biophotonics Int.*, 2004, **11**(7), 36–42.

22 M. Miwa, A. Nakajima, A. Fujishima, K. Hashimoto and T. Watanabe, Effects of the surface roughness on sliding angles of water droplets on superhydrophobic surfaces, *Langmuir*, 2000, **16**(13), 5754–5760.

23 R. N. Wenzel, *Ind. Eng. Chem.*, 1936, **28**, 988–994.

24 P. Shashikanth and P. Prasad, Intrachain defects in n-$C_{60}H_{122}$ hydrocarbon: a low angle powder XRD study, *Curr. Sci.*, 2001, **80**(12), 1578–1581.

25 M. Avrami, Kinetics of phase change. I General theory, *J. Chem. Phys.*, 1939, **7**(12), 1103–1112.

26 T. Irzhak, S. Mezhikovskii and V. Irzhak, The physical meaning of the Avrami equation in oligomer curing reactions, *Polym. Sci., Ser. B*, 2008, **50**(7–8), 201–203.

27 A. Jabbarzadeh and R. I. Tanner, Flow-induced crystallization: unravelling the effects of shear rate and strain, *Macromolecules*, 2010, **43**(19), 8136–8142.

28 P. Shen, H. Fujii and K. Nogi, Effect of substrate crystallographic orientation on wettability and adhesion in several representative systems, *J. Mater. Process. Technol.*, 2004, **155**, 1256–1260.

29 O. Dezellus and N. Eustathopoulos, The role of van der Waals interactions on wetting and adhesion in metal/carbon systems, *Scr. Mater.*, 1999, **40**, 1283–1288.

30 A. Genovese, G. Amarasinghe, M. Glewis, D. Mainwaring and R. A. Shanks, Crystallisation, melting, recrystallisation and polymorphism of n-eicosane for application as a phase change material, *Thermochim. Acta*, 2006, **443**(2), 235–244.

Faraday Discussions

PAPER

What can be inferred from moiré patterns? A case study of trimesic acid monolayers on graphite†

Saskia Spitzer,[abc] Oliver Helmle,[ab] Oliver Ochs,[abc] Joshua Horsley,[d] Natalia Martsinovich,[d] Wolfgang M. Heckl[abc] and Markus Lackinger*[abc]

Received 20th March 2017, Accepted 9th May 2017
DOI: 10.1039/c7fd00113d

Self-assembly of benzene-1,3,5-tricarboxylic acid (trimesic acid – TMA) monolayers at the alkanoic acid–graphite interface is revisited. Even though this archetypal model system for hydrogen bonded porous networks is particularly well studied, the analysis of routinely observed superperiodic contrast modulations known as moiré patterns lags significantly behind. Fundamental questions remain unanswered such as, are moiré periodicity and orientation always the same, *i.e.* is exclusively only one specific moiré pattern observed? What are the geometric relationships (superstructure matrices) between moiré, TMA, and graphite lattices? What affects the moiré pattern formation? Is there any influence from solvent, concentration, or thermal treatment? These basic questions are addressed *via* scanning tunneling microscopy experiments at the liquid–solid interface, revealing a variety of different moiré patterns. Interestingly, TMA and graphite lattices were always found to be ~5° rotated with respect to each other. Consequently, the observed variation in the moiré patterns is attributed to minute deviations (<2°) from this preferred orientation. Quantitative analysis of moiré periods and orientations facilitates the determination of the TMA lattice parameter with picometer precision.

Introduction

Moiré patterns, *i.e.* the expression of a superperiodicity upon the superposition of two periodic structures, are abundant and can even be observed in daily life. Particularly in surface science, moiré patterns play an important role for epitaxy when a periodic adsorbate structure grows on a crystalline substrate. True epitaxy

[a] Department of Physics, Technische Universität München, James-Franck-Str. 1, 85748 Garching, Germany. E-mail: markus@lackinger.org
[b] Nanosystems-Initiative-Munich, Center for Nanoscience, Schellingstrasse 4, 80799 München, Germany
[c] Deutsches Museum, Museumsinsel 1, 80538 München, Germany
[d] Department of Chemistry, University of Sheffield, Sheffield S3 7HF, UK

† Electronic supplementary information (ESI) available: Additional STM data and simulation results. See DOI: 10.1039/c7fd00113d

implies congruence of adsorbate and substrate lattices at the unit cell level. However, often this strict criterion cannot be met, and one possible alternative is the formation of a higher order superstructure that can be accompanied by a moiré pattern. Experimentally, these moiré patterns can be observed in reciprocal space using low energy electron diffraction (LEED) and other diffraction techniques, or more directly in real space using high resolution scanning tunneling microscopy (STM). Moiré patterns occur in diverse epitaxial systems such as sulfur, noble gas or halogen monolayers,[1–3] for graphene and hexagonal boron-nitride (h-BN) monolayers on crystalline supports,[4–8] and also for self-assembled molecular monolayers.[9–12] Even the surface of pristine graphite can already show moiré patterns caused by rotational misalignment of the topmost layer.[13]

An important distinction of moiré patterns is between commensurate and incommensurate patterns. Commensurate moiré patterns are strictly periodic and originate from the coincidence of the adsorbate and substrate lattices at some point. However, even in the absence of any coincidence incommensurate moiré patterns can emerge. Albeit not strictly periodic, incommensurate moiré patterns exhibit defined spatial beating frequencies. Although this allows the definition of a moiré lattice, the corresponding unit cells are not equivalent. Graphene grown by chemical vapour deposition on transition metal surfaces is a versatile example where both types of moiré patterns are observed, for instance incommensurate on Cu(111)[8] and commensurate on Ru(0001).[4] Especially for very large moiré unit cells, the experimental distinction between commensurate and incommensurate moirés can become challenging, if not impossible. More complex epitaxial systems can also show a combination, with commensurate domains separated by incommensurate soliton walls.[7,14]

Commensurate moiré patterns can be viewed as actual superstructures and accordingly be described by a unit cell and a lattice defined through base translation vectors $\vec{A}_{\text{moiré}}$ and $\vec{B}_{\text{moiré}}$. These are simultaneously lattice vectors of both the adsorbate and substrate lattices:

$$\vec{A}_{\text{moiré}} = a_{11}\cdot\vec{a}_{\text{surf}} + a_{12}\cdot\vec{b}_{\text{surf}} \tag{1}$$

$$\vec{B}_{\text{moiré}} = a_{21}\cdot\vec{a}_{\text{surf}} + a_{22}\cdot\vec{b}_{\text{surf}} \tag{2}$$

as well as:

$$\vec{A}_{\text{moiré}} = b_{11}\cdot\vec{a}_{\text{ads}} + b_{12}\cdot\vec{b}_{\text{ads}} \tag{3}$$

$$\vec{B}_{\text{moiré}} = b_{21}\cdot\vec{a}_{\text{ads}} + b_{22}\cdot\vec{b}_{\text{ads}} \tag{4}$$

with integer coefficients a_{ij} and b_{ij}, respectively, and $\vec{a}_{\text{surf}}$, $\vec{b}_{\text{surf}}$ and $\vec{a}_{\text{ads}}$, $\vec{b}_{\text{ads}}$ denote the base translation vectors of the surface and adsorbate lattice, respectively. Conversely, if $\vec{A}_{\text{moiré}}$ and $\vec{B}_{\text{moiré}}$ are lattice vectors of the adsorbate or substrate lattice, the reciprocal lattice vectors of the adsorbate or substrate are lattice vectors of the reciprocal moiré lattice.

Moiré patterns were the subject of many experimental and theoretical studies over recent decades, and their emergence in the context of graphene and other 2D materials has led to a renaissance. An interesting, but understudied model system for moiré patterns is self-assembled molecular monolayers on crystalline surfaces. The present contribution focuses on weakly interacting graphite surfaces, where moiré patterns are more commonly observed than on metal surfaces. In addition, straightforward experiments can be carried out at liquid–solid interfaces – molecules are dissolved in appropriate dielectric solvents (*e.g.* phenyloctane, 1,2,4-trichlorobenzene, alkanoic acids, *etc.*) and the solution is then applied to inert substrates, mostly graphite. In most cases, STM is used for characterization in facile experiments, where the tip is directly immersed into the solution.[15–18]

However, moiré patterns are more than just a peculiar phenomenon, and profound insights can be obtained from their detailed analysis. As a result of the subtle balance between molecule–molecule and molecule–surface interactions, moiré patterns can provide evidence for the relative strength of intermolecular and surface potentials. Moreover, the moiré structure parameters can be utilized to determine adsorbate lattice parameters with unprecedented precision, exceeding that of standard STM measurements by orders of magnitude.[12,19] These precise and reliable data can serve as pivotal benchmarks for structure simulations.

The moiré patterns of molecular adlayers exhibit important differences in comparison to those of graphene or h-BN; the lattice parameter differences between molecular adlayers and the surface are typically large, whereas graphene ($a_{\text{graphene}} = 0.246$ nm) and its supports feature similar lattice parameters (*e.g.* $a_{\text{Ir}(111)} = 0.271$ nm). Molecular adlayers are stabilized by relatively weak non-covalent intermolecular bonds such as hydrogen or van der Waals bonds. Consequently, molecular lattices are more flexible and adaptive as compared to their covalently interlinked counterparts. Lastly, graphene and h-BN are grown at high temperatures, typically up to 1000 °C, and in most cases studied at room temperature or even below after cooling down. Owing to differences in the thermal expansion between graphene and the underlying metal surface, the lattice parameter ratios at growth *vs.* characterization temperature can differ, with implications for the moiré patterns. In contrast, the molecular monolayers at the liquid–solid interface are normally studied at room temperature, *i.e.* the same temperature at which they self-assembled.

In the following, the routinely observed moiré patterns in hydrogen bonded monolayers of benzene-1,3,5-tricarboxylic acid (trimesic acid – TMA) on graphite are analysed and discussed in more detail.

Results and discussion

TMA on graphite was chosen as a model system for this study, because it reliably self-assembles into long-range ordered structures with clearly discernible moiré patterns in STM. Despite its simple chemical structure, rigidity, and high symmetry, self-assembly of TMA is complex and shows rich behaviour with different surface polymorphs.[20–28] Yet, the so-called chickenwire structure – a hexagonal porous honeycomb network with a lattice parameter of ~1.7 nm and a pore diameter of ~1.0 nm – is the most abundant structure (*cf.* inset in Fig. 5). Other, typically more densely packed, polymorphs were observed for various conditions and preparation procedures. For instance, at the liquid–solid interface TMA self-assembly is known for solvent-induced polymorphism, where a polymorph called flower structure occurred when shorter chain length alkanoic acids are used as solvents.[28] The chickenwire polymorph exclusively features two-fold cyclic $R_2^2(8)$ hydrogen bonds in a straight geometry between all of the carboxylic acid groups of each TMA. Owing to the high bond strength of these resonance enhanced hydrogen bonds,[29] the network is also relatively strong; for a free-standing TMA monolayer a binding energy of 1.28 eV per molecule (corresponding to 0.86 eV per double hydrogen bond) is obtained from dispersion-corrected density functional theory (DFT) simulations. In addition, TMA molecules also adsorb strongly on graphite with an adsorption energy of ~0.86 eV for isolated molecules according to molecular mechanics simulations. Both contributions total an overall binding energy in excess of ~2 eV per TMA molecule in the adsorbed monolayer. Interestingly, in accordance with a previous study,[25] the TMA adsorption energy on graphite did not vary significantly between different sites, indicating a shallow surface potential.

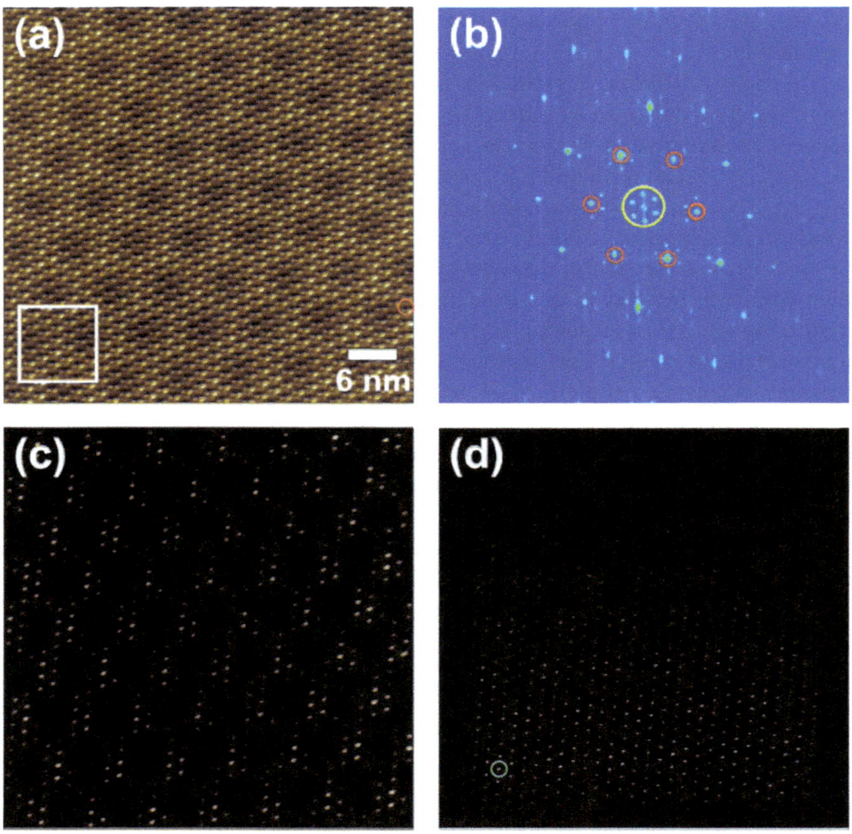

Fig. 1 (a) STM image of the TMA chickenwire structure acquired at the heptanoic acid–graphite(0001) interface (−588 mV, 52 pA) and (b) the corresponding FFT (Hannig filter, square root magnitude). First order spots of the moiré and TMA are marked by the inner yellow and the outer red circles, respectively. (c) The same image as in (a), but low-pass filtered with a cut off just above the first order TMA. The colour scale was adjusted to highlight differences between the topographic maxima. (d) Cross-correlation of the main image (a) with the close-up marked by the rectangle. The green circle indicates the original position of the close-up with the highest intensity.

Over the years TMA has become a fairly well-studied model system. Yet, its moiré patterns were not analysed in greater detail. Fig. 1 shows a large scale STM image of the self-assembled TMA chickenwire structure at the heptanoic acid–graphite interface. The apparent height of the honeycomb network is periodically modulated with hexagonal symmetry. The structural origin of this moiré pattern lies in the mismatch between TMA and the underlying graphite lattice. The moiré pattern is also clearly recognizable in reciprocal space as illustrated by the corresponding two-dimensional Fast Fourier Transform (FFT) in Fig. 1(b). First order TMA spots are marked by the red circles, and higher orders exhibit significant intensity. The larger spatial periodicity of the moiré gives rise to the inner Fourier components marked by the yellow circle. The hexagonally arranged satellites around the TMA spots are a further manifestation of the moiré.

The emergence of a moiré with large period clearly indicates the incommensurability of TMA and graphite at the level of a single or a few unit cells. This implies that the TMA dimers – the basic structural motif – are not adsorbed on the equivalent sites. As a further consequence, the pores defined by the six surrounding TMA molecules feature different positions with respect to graphite, with important implications for host–guest chemistry. An illustrative example, therefore, is coronene (COR) guest molecules. Their snug fit in the TMA pores prohibits any lateral degrees of freedom, hence the COR adsorption sites are rigorously defined by the TMA lattice.[30] In STM images the spatial variation of adsorption sites within the pores manifest themselves as pronounced intramolecular contrast modulation of the COR guests.[31]

Although the TMA moiré on graphite was already reported some time ago,[32] a detailed analysis is still outstanding. Even the fundamental question of commensurability has not been addressed so far. A possible first step is expressing $\vec{A}_{\text{moiré}}$ as a linear combination of $\vec{a}_{\text{ads}}$ and $\vec{b}_{\text{ads}}$. In principle, this could be done simply by counting in the STM images. However one may encounter ambiguities, so a reciprocal space analysis provides spatially averaged, and hence potentially more accurate information. According to crystallographic notation, base translation vectors with 120° angles are used for all hexagonal lattices in the following. The subsequent analysis is exemplified for hexagonal lattices, but could similarly be adapted to other lattices. A practical way to find the coefficients b_{ij} of eqn (3) and (4) is to first determine the length ratio of the lattice vectors, Φ, by direct measurement in the FFT:

$$\Phi = \frac{|\vec{A}_{\text{moiré}}|}{|\vec{a}_{\text{ads}}|} = \frac{|\vec{a}^{*}_{\text{ads}}|}{|\vec{A}^{*}_{\text{moiré}}|}. \tag{5}$$

Asterisks denote the reciprocal lattice vectors. Similarly the rotation angle α (defined as the smallest angle between $\vec{A}^{*}_{\text{ads}}$ and $\vec{A}^{*}_{\text{moiré}}$) can be directly inferred from the FFT. Then the coefficients can be determined according to:

$$b_{11} = \Phi\left(\cos\alpha + \frac{1}{\sqrt{3}}\sin\alpha\right) \tag{6}$$

$$b_{12} = \frac{2}{\sqrt{3}}\Phi\sin\alpha \tag{7}$$

$$b_{21} = -b_{12} \tag{8}$$

$$b_{22} = b_{11} - b_{12}. \tag{9}$$

Unless all coefficients are integers, the moiré is incommensurate. However, how large are the acceptable tolerances for these experimental values? STM imaging, and even more so experiments under ambient conditions, are prone to image distortions caused by thermal drift as well as piezo creep and hysteresis. Consequently, "real space crystallography" requires as undistorted as possible images. In any case, the coefficients should be evaluated with subsequent up and down scans to judge the reproducibility. Further inaccuracies can result from the widths of the FFT peaks that are inversely proportional to the number of periods. Hence it is important that the STM images capture a sufficient number of moiré periods, while the pixel resolution of the molecular lattice should be high enough to avoid aliasing effects. This condition becomes increasingly challenging to fulfil with an increasing moiré period.

Applying this procedure to the image in Fig. 1(a) results in $b_{11} = 4.77 \pm 0.06$ (4.78 ± 0.04) and $b_{12} = 1.64 \pm 0.06$ (1.65 ± 0.06), clearly indicating incommensurability.‡ The values in parentheses refer to the subsequently acquired down scan, indicating a quite satisfying reproducibility, even though some residual drift remained during STM imaging. Comparing two lattices in the same image means that any possible distortion similarly affects both lattices, resulting in beneficial error compensation. So it is less surprising that determination of the coefficients from a distortion-corrected image with an enforced hexagonal TMA lattice resulted in similar coefficients.

The non-integer coefficients indicate incommensurability of the moiré with respect to the TMA lattice. This, not necessarily, but reasonably also implies incommensurability of the TMA and graphite lattices. In theory, true incommensurability requires irrational coefficients, because any rational coefficients have integer multiples. Nevertheless, we conclude practical incommensurability of the TMA lattice for the following reason: even under favourable assumptions, *i.e.*

$$b_{11} = 4\frac{3}{4} \text{ and } b_{12} = 1\frac{2}{3}$$

commensurability is only achieved after 12 moiré periods, a length that possibly exceeds the domain size. More importantly, a pronounced and specific energetic preference for such exceedingly large unit cells appears highly unlikely. So this lattice relation is presumably not related to the presence of few energetically favourable adsorbate positions on the surface.

Concluding incommensurability also means that the moiré unit cell is not an actual repeat unit. Even though this should be directly visible in the STM image, it is relatively difficult to recognize. Hence the STM image in Fig. 1(a) was further processed by low-pass filtering in Fourier space with a cut-off frequency just above first order TMA. The so processed image is shown in Fig. 1(c) with an adjusted

‡ Lengths and angles were measured in the FFTs and averaged over the three symmetrically equivalent directions. Maximum deviations are stated as errors.

colour scale to highlight the topographic maxima. While largely arranged on the moiré lattice, these groups of maxima appear clearly different at different moiré lattice points. This violation of translational symmetry further confirms the incommensurability of the TMA moiré on graphite. This is further substantiated by the cross-correlation shown in Fig. 1(d), evaluated for the close-up region marked in Fig. 1(a) with its main image. This cross-correlation shows maxima with a distribution corresponding to the TMA lattice, because relative shifts with the TMA periodicity result in a certain level of coincidence. However, slightly higher intensities result for the simultaneous coincidence of both the TMA and moiré lattices. Accordingly, for a fully commensurate moiré equally high intensity maxima should appear with the moiré periodicity. This is clearly not the case in Fig. 1(d), where the colour scale was adjusted to highlight differences across the cross-correlation. Some higher intensities, *i.e.* more favourable coincidences, are still observed in the vicinity of the original position of the close-up, yet the intensities decay for increasing distance. This indicates dephasing of the TMA and moiré for increasing distance, and hence incommensurability. For further illustration, the same image processing procedures were applied to simulated commensurate *vs.* incommensurate moiré patterns (*cf.* ESI†).

Kinetics *vs.* thermodynamics

The unexpected incommensurability also gives rise to the question of whether the experimentally observed structure is thermodynamically most stable or whether it is a metastable intermediate. Intuitively, one would expect a higher order commensurate structure as the energetic optimum. The intermolecular arrangement of TMA in the chickenwire polymorph exhibits only energetically ideal hydrogen bonds, hence it is the energetically most stable structure per molecule. However, as recently worked out, the free energy per unit area is decisive for the relative thermodynamic stability of competing structures.[33] This might also lead to a thermodynamic preference for polymorphs with energetically inferior intermolecular bonds, but a higher packing density.[34] Furthermore, the thermodynamically most stable structure is not only defined by the intermolecular arrangement, but also by its epitaxy with respect to the underlying surface. Here optimization of molecule–surface interactions should be most important, with only minor influences of entropy variations. In other terms, the thermodynamically most stable adsorbed TMA monolayer structure is also characterized by a unique epitaxial relationship with respect to graphite. Accordingly, in thermodynamic equilibrium either no moiré or exclusively one distinct moiré should be observed. Again a comparison to covalently linked networks, *i.e.* graphene epitaxially grown on transition metal surfaces, is instructive; on Ir(111) different moiré patterns with 0°, 14°, 18.5°, and 30° orientation are observed,[5,35] whereas on Rh(111) only one distinct moiré pattern is known.[36] Moreover, on Ir(111) growth temperature and rate influence the graphene moiré patterns, indicating the importance of nucleation and growth kinetics even for networks that feature the ultimate bond strength.

The large scale STM image obtained in heptanoic acid in Fig. 2 shows three domains with moiré patterns, but each with a different period and orientation. Interestingly, an earlier study reported just one distinct moiré for the TMA network itself, and another one for the TMA network with incorporated COR guest

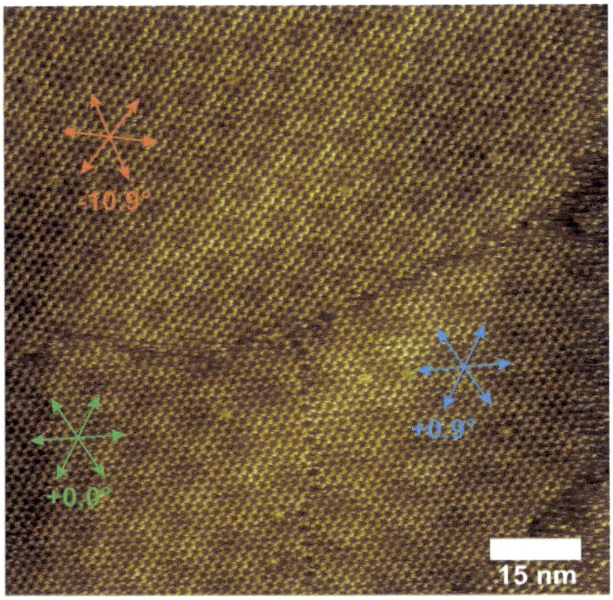

Fig. 2 Overview STM image of the TMA chickenwire structure obtained at the heptanoic acid–graphite interface (−500 mV, 40 pA). Each of the three larger domains shows an individual moiré pattern with a different period and orientation. The orientation of the TMA lattice in the green marked domain in the lower left corner was defined as 0°, the other angles are stated relative to this reference. Arrows indicate equivalent directions of the TMA lattices.

molecules.[32] With reference to the discussion above, our experimental findings provide evidence for a kinetically controlled and trapped system. Although self-assembly favours the chickenwire polymorph, the molecule–surface interactions do not attain the energetic optimum.

Closer inspection of the relative domain orientations provides further information. While the orientations of the lower left and right domain differ by only 0.9°, the upper domain has a relative orientation of ~11°. In all of the STM data either approximately aligned or ~10° rotated domains were consistently observed, suggesting a relative rotation between the TMA and graphite lattices of approximately ±5°. This finding is further corroborated by split-images (*vide infra*, ESI†), and in accordance with an earlier study[32] as well as recent work by the groups of Rosei and De Feyter.[25] The invariant rotational orientation of the TMA with respect to the graphite in combination with the fact that moiré parameters respond extremely sensitively to rotation angle changes,[37] suggest that minute deviations from this preferred rotational orientation account for the variety of experimentally observed moirés.

Solvent dependence and heating experiments

Similar observations, *i.e.* various different moiré patterns, either ~0° or ~10° relative domain orientations, and ±5° orientation between the TMA and graphite, were also made using nonanoic acid as the solvent, suggesting no or a very minor solvent influence.

The general reason for kinetic trapping is insufficient thermal energy to overcome barriers. Consequently, a transition into a thermodynamically more stable or most stable state could be promoted by heating. Since these transitions are irreversible, characterization at elevated temperatures is not required, and *ex situ* studies after cooling down to room temperature are sufficient. However, care has to be taken because of unwanted chemical reactions between the solute and solvent. Heating TMA in nonanoic acid solution to only ~50 °C already results in intense yellow discolouration, indicating chemical changes of the TMA, most probably anhydride formation with solvent molecules. This also affects the TMA self-assembly, where an overall chickenwire structure is still observed but with many defects. While these experiments with defective TMA molecules can provide further insights into the dynamic exchange and self-healing properties, extensive heating experiments to promote thermodynamic equilibration are ruled out for TMA with alkanoic acids. Improvements for homologous alkanoic acid solvents or lower concentrations cannot be expected. In principle, adding water should shift the equilibrium for the reversible anhydride formation, but might also induce competitive hydrogen bonding. Carrying out experiments without solvent under ultra-high vacuum (UHV) conditions would be a feasible alternative, but this sacrifices the possibility of studying the important role of dynamic exchange with solution for equilibration.

Nevertheless, heating experiments in solution were performed in a slightly different manner. First the pristine graphite sample (without any solution applied) was heated up to ~100 °C on a hot plate. Then the graphite was removed from the hot plate and room temperature solution was immediately applied to the hot surface. The sample was allowed to cool down to room temperature and subsequently characterized using STM. This type of experiment offers less control than a slow heating experiment, because the exact temperature profile during cooling depends on heat capacities and transport. However, this approach facilitates self-assembly studies at relatively high temperatures, while minimizing the risk of chemical alterations. Longer exposure times are still not possible, but according to the Arrhenius law the influence of temperature is significantly more severe than that of time. For the proposed experiments at least, TMA self-assembly takes place at elevated temperatures with increased chances to avoid kinetic trapping. Subsequently acquired STM images show the chickenwire polymorph without any defects. The still present moiré patterns were likewise incommensurate and similar to those observed for room temperature samples. Only an increase in domain size was observed, which was not further studied, indicating thermal activation of Ostwald ripening.

Reciprocal space analysis

The period of commensurate moiré patterns is unambiguously defined by the smallest translation between two coincident points of the adsorbate and substrate lattices. Once the (integer) coefficients of the moiré lattice vectors with respect to the adlayer and surface are known, the adsorbate lattice parameter can be determined with the precision of the surface lattice parameters. An intriguing question is, whether the incommensurate moiré patterns of TMA on graphite can similarly be utilized to precisely determine the adsorbate lattice parameters. For incommensurate moiré patterns it is less obvious how the moiré parameters are related to the relative rotation and lattice parameters of the adsorbate and surface. Recently Günther *et al.* reported an intuitively understandable method for the geometric construction of moiré beating frequencies.[37] Their approach aimed to explain the multitude of moiré patterns observed for graphene on transition metal surfaces. There it was exemplified for hexagonal lattices with comparable lattice parameters, but the approach is generally applicable. In an analogy to optics, the moiré was obtained as the product of two lattice-periodic functions, representing the adlayer and surface, respectively:

$$f_{\text{moiré}}(x,y) = f_1(x,y) \times f_2(x,y). \tag{10}$$

According to the convolution theorem, the Fourier transform of the moiré $\mathscr{F}\{f_{\text{moiré}}\}$ is equal to the convolution of the Fourier transforms of f_1 and f_2:

$$\mathscr{F}\{f_{\text{moiré}}\} = \mathscr{F}\{f_1\} \otimes \mathscr{F}\{f_2\}. \tag{11}$$

In $\mathscr{F}\{f_{\text{moiré}}\}$ additional Fourier components with sizable intensities occur for connecting vectors between strong Fourier components in $\mathscr{F}\{f_1\}$ and $\mathscr{F}\{f_2\}$. The orientation and period of the moiré lattice are defined by the shortest of these vectors.

In the following, this approach is applied to the TMA monolayers on graphite, whose reciprocal lattices are both hexagonal and 30° rotated with respect to the corresponding real space lattices. In order to draw both the TMA and graphite reciprocal lattices together, the ratio of the real space lattice parameters and relative rotation angle is required. This information can best be obtained from STM in so-called "split-images", where the tunnelling parameters are changed to subsequently image the adsorbate and substrate within the same scan frame. From split-images a rotation angle of $\alpha = 5°$ and a TMA-to-graphite lattice parameter ratio of $\sim$6.75 can be acquired, corresponding to $a_{\text{TMA}} = 1.66$ nm. Interestingly, the rotation angles in all of the split-images agreed within ±1° (ESI†), in accordance with the observed relative domain orientations as discussed above, as well as with the literature.[25,32] This experimental TMA lattice parameter also agrees well with our own and previously published DFT simulations.[25,32]

Fig. 3 shows a sketch of both the TMA and graphite reciprocal lattices according to the experimental data. The closest proximity of lattice points of both of the reciprocal lattices occurs between (7,1) TMA and (1,0) graphite. This finding already highlights an important difference to graphene on transition metals: both have comparable lattice parameters, so the closest proximity appears between Fourier components of comparable order, *e.g.* (2,0) and (1,1). Owing to the almost seven-fold difference in the lattice parameter between TMA and graphite, this is simply not possible, and inevitably requires the involvement of higher order TMA Fourier components. In contrast, a contribution from higher order graphite Fourier components (*e.g.* (2,0) graphite) appears rather unlikely, because the intensity of the required TMA Fourier components at $\sim$14th order is extremely low. Otherwise higher order moiré Fourier components should also be experimentally observed, which is not the case. According to this geometric

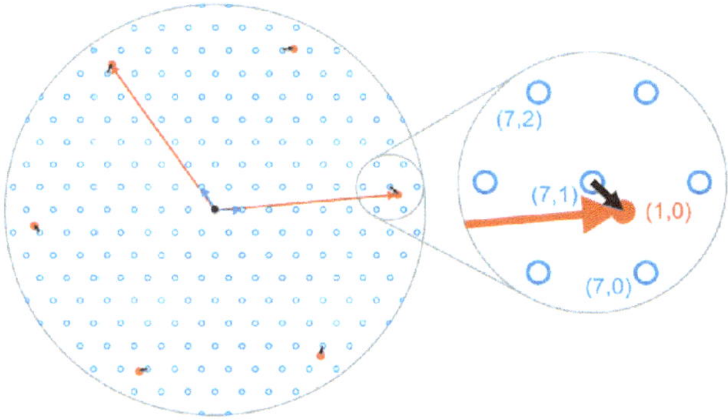

Fig. 3 Reciprocal lattices of TMA (blue open circles) and graphite (red closed circles) drawn for a relative rotation of $\alpha = 5°$ and $a_{TMA} = 1.66$ nm. Red and blue arrows indicate the base translation vectors of the respective reciprocal lattices. The close-up on the right hand side shows the vicinity of the (7,1) TMA Fourier component, which is closest to the (1,0) graphite Fourier component. The black rather short arrows indicate the resulting base translation vectors of the moiré reciprocal lattice.

construction, the base translation vector of the reciprocal moiré lattice $\vec{A}^*_{\text{moiré}}$ (short black arrow in Fig. 3) is given by:

$$\vec{A}^*_{\text{moiré}} = \left(1 \cdot \vec{a}^*_{\text{graphite}} + 0 \cdot \vec{b}^*_{\text{graphite}}\right) - \left(7 \cdot \vec{a}^*_{\text{TMA}} + 1 \cdot \vec{b}^*_{\text{TMA}}\right) . \tag{12}$$

From here the moiré period can be calculated as a function of the rotation angle α between the TMA and graphite lattices using simple trigonometry. The results are shown in Fig. 4 for rotation angles in the vicinity of the experimental value of $\alpha = 5°$ and for three slightly different TMA lattice parameters around the

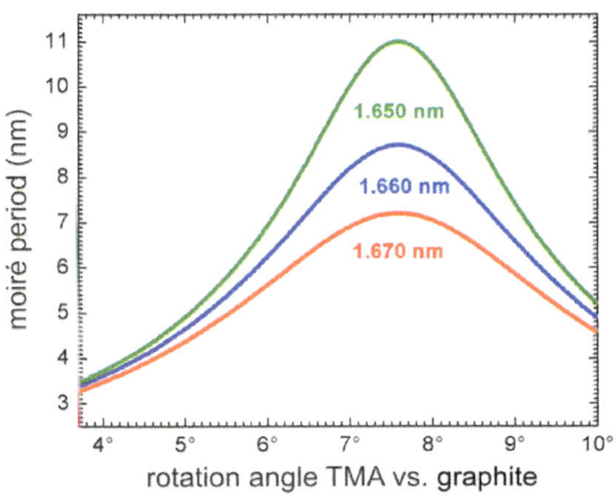

Fig. 4 The moiré period as a function of the relative rotation between the TMA and graphite lattices around the experimental value of $\alpha = 5°$. The plot assumes the closest proximity between the (1,0) graphite and (7,1) TMA Fourier components. The moiré period also depends on the TMA lattice parameter, as illustrated by the different curves for specific a_{TMA} values.

experimental value. In all cases, the largest moiré lattice parameter is obtained for $\alpha \approx 7.6°$, *i.e.* when the (1,0) graphite and (7,1) TMA reciprocal lattice vectors are aligned, resulting in the shortest $\vec{A}^*_{\text{moiré}}$. The moiré period also increases with decreasing a_{TMA}, because the corresponding enlargement of a^*_{TMA} further decreases the length of $\vec{A}^*_{\text{moiré}}$ as evident from Fig. 3. It is worth noting that this plot is only valid for $3.7° < \alpha < 11.7°$. Outside this range the distance between (1,0) graphite and (7,0) or (7,2) TMA becomes shorter than to (7,1) TMA, and hence defines $\vec{A}^*_{\text{moiré}}$.

The graphs in Fig. 4 demonstrate how slight variations of α result in large variations of the moiré period $A_{\text{moiré}}$. However, the same is true for modest variations of a_{TMA}. For instance, for fixed $\alpha = 6°$, $a_{\text{TMA}} = 1.650$ nm results in $A_{\text{moiré}} = 6.9$ nm, whereas $a_{\text{TMA}} = 1.670$ nm results in a notably smaller value of $A_{\text{moiré}} = 5.6$ nm.

In order to judge the adaptability of the TMA lattice, DFT calculations of the energy *vs.* a_{TMA} were performed. The results shown in Fig. 5 indicate a relatively shallow energy minimum at $a_{\text{TMA}} = 1.650$–1.660 nm, allowing slight variations of the TMA lattice parameters at modest energy costs.

Since the moiré period is sensitive to both α and a_{TMA}, the origin of the variety of moiré patterns is not clear *a priori*. However images at domain boundaries provide further insights. The two lower domains in Fig. 2 show only a small, but detectable relative rotation of $\sim 1°$, but different moiré patterns, suggesting a major influence of α.

So far only moiré periods were analysed. However, even overview STM images provide further easily accessible information, namely the rotation angle φ between the moiré and TMA lattices. Theoretical values for φ can similarly be derived from the geometrical construction in Fig. 3 as a function of α with a_{TMA} as an additional parameter (*cf.* ESI†). For further analysis, however, a combined plot of moiré orientation *vs.* periodicity as shown in Fig. 6 is more useful. Constant

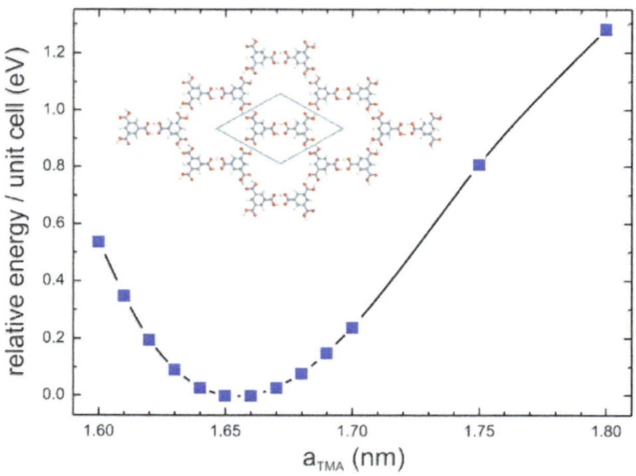

Fig. 5 DFT-derived energies per unit cell for a free-standing TMA honeycomb network (*i.e.* surface neglected) as a function of the lattice parameter of a hexagonal lattice. The black line serves as a guide to the eye. The curve shows a relatively shallow minimum at 1.650 nm–1.660 nm. The inset depicts the optimized structure (unit cell marked, grey shows carbon, red shows oxygen, and white shows hydrogen).

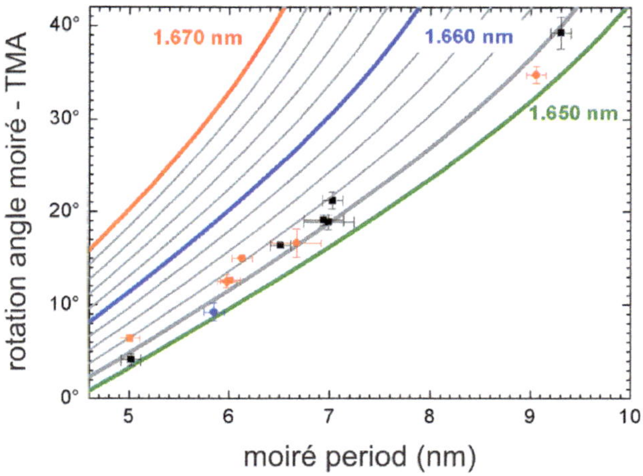

Fig. 6 Theoretical curves for the moiré orientation with respect to graphite *vs.* the moiré period. The curves were evaluated for different a_{TMA} as indicated. Along each curve the rotation angle α between the TMA and graphite lattices varies (approximately from 4° to 8°). The gray lines represent iso-lines for increments of $\Delta a_{TMA} = 0.002$ nm. Each data point corresponds to one moiré pattern observed in an independent experimental run in heptanoic acid (black squares) or nonanoic acid (red circles: room temperature deposition, red squares: deposition onto a hot surface, red diamond: image acquired at 50 °C after room temperature deposition) (*cf.* ESI†). The blue circle represents a literature result.[32]

values of a_{TMA} result in iso-lines that uniquely correlate the moiré orientation and period. This diagram highlights once more the extreme sensitivity of both the moiré period and orientation to slight variations in a_{TMA}. For further analysis, experimentally observed moiré patterns (*cf.* ESI†) were evaluated in large scale STM images§ and included in Fig. 6 as one data point for each observed moiré.

Despite the fact that data were acquired using different solvents (heptanoic *vs.* nonanoic acid) and for different preparation protocols (deposition onto room temperature surfaces *vs.* heated surfaces) all data points fall onto the same iso-line, (thicker grey line) corresponding to $a_{TMA} = (1.652 \pm 0.002)$ nm. Moreover, the literature value from ref. 32 (blue full circle) also falls on the same iso-line. Based on the small deviations, we propose that the comparison of easily and reliably measurable moiré parameters to the reciprocal space model facilitates the determination of a_{TMA} with picometer precision.

Materials and methods

STM data were acquired at the solid–liquid interface using home-built instruments driven by ASC 500 and SPM 100 control electronics from attocube systems AG and RHK, respectively. All voltages refer to the sample. A droplet of TMA in heptanoic or nonanoic acid solution was applied to the freshly cleaved surface of highly oriented pyrolytic graphite. Images were recorded with mechanically cut PtIr tips directly immersed into solution. For any measurement, the instruments

§ These values were obtained from distortion-corrected images with hexagonal TMA layers and averaged over symmetrically equivalent pairs of Fourier components. Maximum deviations are stated as errors.

were allowed to equilibrate until the similarity of subsequent up and down scans indicated low drift.

The DFT calculations were performed for free-standing TMA monolayers with the CP2K software[38] using the PBE functional[39] and empirical dispersion correction.[40] DZVP basis sets were used for all atoms. Periodic boundary conditions were applied, using a hexagonal unit cell with variable $a = b$ and fixed $c = 1.00$ nm. No further constraints were applied.

Summary and conclusions

In summary, the moiré patterns routinely observed for the TMA chickenwire polymorph on graphite were studied in greater detail. The precise evaluation of the moiré lattice parameters with respect to the TMA lattice reveals incommensurability of both lattices. This is further corroborated by the observed aperiodicity in the STM images, related to the dissimilarity of the moiré unit cells. This surprising finding indicates a weak preference of TMA–graphite interactions for specific sites (*i.e.* modest spatial variation of the surface potential) as compared to the strong intermolecular TMA–TMA hydrogen bonds.

Moreover, various moiré patterns with different periods and orientations were found. This indicates that the samples are not in thermodynamic equilibrium, but are kinetically stabilized. Yet it was not possible to influence the moiré pattern formation by providing thermal energy in experiments with deposition of solution onto heated surfaces.

The TMA lattice is not statistically oriented with respect to graphite, but was always found with a relative rotation of $\sim 5°$. This was inferred either directly from split-images or indirectly from the $\sim 10°$ angle found between differently oriented TMA domains. Consequently, the variety of different moiré patterns is explained by slight deviations from this preferred orientation. The non-random orientation between the TMA and graphite lattices implies that the reciprocal moiré lattice vector always occurs between the same Fourier components, *i.e.* (7,1) TMA and (1,0) graphite. This provides the basis for a quantitative analysis, where both the moiré period and orientation are expressed as functions of the TMA–graphite rotation angle α. In addition, the moiré parameters sensitively depend on the exact value of the TMA lattice parameter, where even picometer changes in a_{TMA} have a notable effect on $A_{\mathrm{moir\acute{e}}}$. Accordingly, the TMA lattice parameter can be derived from the moiré period and orientation, quantities that are both easily and accurately accessible from STM data. Interestingly, for all moiré patterns observed, irrespective of solvent or heat treatment, a TMA lattice parameter of $a_{\mathrm{TMA}} = (1.652 \pm 0.002)$ nm is obtained. This not only suggests picometer accuracy, but also corroborates the quite rigid nature of the TMA lattice, also in the presence of an evidently weak surface potential.

A clear advantage of this approach is that the input data, *i.e.* moiré period and orientation, can be measured in relatively large scale STM images in the order of 50–100 nm, where drift is less of a problem than for atomically resolved images. Moreover, image distortions simultaneously affect both the moiré and TMA lattices and thus errors partly compensate. The preferred rotational alignment of TMA with respect to graphite of $\sim 5°$ is convenient, as it implies that always the same Fourier components are involved in the moiré formation. If this was not the case, knowledge of the adsorbate–substrate rotation angle would be required for each recorded moiré. Yet, approximate values for α with STM accuracy are sufficient in order to determine the involved Fourier components. In practice, a split-image has to be acquired along with each moiré.

Finding an incommensurate moiré for TMA on graphite leads to many further interesting research questions. What drives the preferred orientation of TMA with respect to graphite? What controls the formation of incommensurate domains and what stabilizes them? Finally, what is the thermodynamically most favourable structure of TMA on graphite?

As for the stabilization, the absence of Ostwald ripening in the TMA monolayers at room temperature, *i.e.* the coalescence of domains in order to minimize energetically unfavourable domain boundaries, plays an important role. At room temperature, coalescence is only observed for comparatively small TMA domains, whereas larger domains remain stable.[41] Most domain boundaries are straight and exhibit defined crystallographic orientations, *i.e.* <10> or <12> with respect to the TMA lattice. This implies that TMA molecules at domain boundaries are still hydrogen bonded to two adjacent TMA molecules, and are consequently stabilized by two pairs of resonance enhanced cyclic hydrogen bonds. According to molecular mechanics simulations,¶ this gives rise to a high binding energy of 1.81 eV per TMA molecule, explaining the high stability and persistence of the domain boundaries.[41]

A further interesting observation – in particular for an incommensurate structure – is the preferred rotational orientation of $\sim 5^\circ$ with respect to graphite. Most likely, this orientation preference is already predetermined at an early stage of growth, since a concerted reorientation of domains will most likely not take place beyond a certain size. In this respect, a small size of the critical nucleus for the TMA monolayers can be presumed. Accordingly, the growing TMA domains become thermodynamically stable at a relatively small size, *i.e.* the preferred orientation of a few TMA molecules (maybe as little as two) with respect to graphite already determines the domain orientation that is then kinetically trapped. Of course, this does not yet explain the energetic preference for this specific orientation. Relating to this, an important contribution might arise from the preference of more strongly interacting oxygen atoms for specific sites.[42] In any case, farther-reaching quantum chemical simulations of TMA on graphite are likely to aid in elucidating the prominent preference for the $\sim 5^\circ$ orientation. In this respect, it might be necessary to consider at least TMA dimers, *i.e.* the two non-equivalent molecules within the unit cell, or even larger aggregates as cyclic hexamers, rather than just individual TMA molecules. Further insights could be expected from a full evaluation of the adsorption energy as a function of the rotational orientation of TMA with respect to graphite. A relatively shallow energy minimum could explain the small orientation deviations that eventually account for the variety of observed moiré patterns.

Finding an incommensurate moiré for TMA on graphite is still a surprising result. One would expect that reasonably strong spatial variations of TMA–graphite interactions should result in preferred adsorption sites and orientations, and hence promote commensurability. This may not be achievable at the level of

¶ This binding energy also includes molecule–surface interactions and was obtained using the standard Dreiding force field that features an explicit hydrogen bonding term, but does not account for the resonance enhancement of two-fold cyclic hydrogen bonds.

a single unit cell, but doubling or tripling the unit cell already increases the number of possibilities of favourable adsorption sites substantially. Thereby, the almost seven-fold difference between the TMA and graphite lattice parameters fosters a large number of possibilities. Moreover, the relatively small strain energy of the chickenwire structure (*cf.* Fig. 5) implies some adaptability of the TMA lattice parameter. Both factors should promote the lock-in into a high order commensurate structure. However it can be shown that for a sufficiently small corrugation of the surface potential, incommensurate superstructures can become energetically favourable.[14] In this respect, evaluation of monolayer energies in reciprocal space based on the Fourier components of the surface potential, as previously proposed for commensurate superstructures,[43] is a very promising approach. Without any doubt, the situation with reduced-symmetry adsorbates such as molecules requires adaptations and can become quite challenging, but avoids the notorious problems with applying periodic boundary conditions to incommensurate structures.

Our comparative experiments in heptanoic *vs.* nonanoic acid suggest a negligible solvent influence on both the preferred rotational orientation and the precise TMA lattice parameter. In this regard further studies in the absence of solvent, *i.e.* under UHV conditions, might be interesting. While the nucleation kinetics at the liquid–solid interface is determined by adsorption rates that cannot be directly controlled, the deposition rate under UHV is an experimental parameter, and typically many orders of magnitude lower. Moreover, growth of TMA domains could also be studied at lower temperature with reduced influence of thermal fluctuations.

In summary, we have demonstrated a straightforward method to extract precise values for the lattice parameters of adsorbed molecular monolayers from the analysis of incommensurate moiré patterns as routinely observed for TMA monolayers on graphite. In this respect, it would be interesting to apply this method to comparable systems in order to see whether a similar level of precision is attainable.

Acknowledgements

O. O. gratefully acknowledges generous financial support by the Helmut-Fischer-Stiftung. J. H. and N. M. acknowledge access to the high-performance computing cluster Iceberg at the University of Sheffield.

Notes and references

1 D. Wang, Q. M. Xu, L. J. Wan, C. Wang and C. L. Bai, *Surf. Sci.*, 2002, **499**, L159–L163.

2 N. S. Komarov, T. V. Pavlova and B. V. Andryushechkin, *Surf. Sci.*, 2016, **651**, 112–119.

3 F. Brunet, R. Schaub, S. Fedrigo, R. Monot, J. Buttet and W. Harbich, *Surf. Sci.*, 2002, **512**, 201–220.

4 W. Moritz, B. Wang, M. L. Bocquet, T. Brugger, T. Greber, J. Wintterlin and S. Günther, *Phys. Rev. Lett.*, 2010, **104**, 136102.

5 J. Coraux, A. T. N'Diaye, C. Busse and T. Michely, *Nano Lett.*, 2008, **8**, 565–570.

6 S. Joshi, D. Ecija, R. Koitz, M. Iannuzzi, A. P. Seitsonen, J. Hutter, H. Sachdev, S. Vijayaraghavan, F. Bischoff, K. Seufert, J. V. Barth and W. Auwärter, *Nano Lett.*, 2012, **12**, 5821–5828.
7 C. R. Woods, L. Britnell, A. Eckmann, R. S. Ma, J. C. Lu, H. M. Guo, X. Lin, G. L. Yu, Y. Cao, R. V. Gorbachev, A. V. Kretinin, J. Park, L. A. Ponomarenko, M. I. Katsnelson, Y. N. Gornostyrev, K. Watanabe, T. Taniguchi, C. Casiraghi, H. J. Gao, A. K. Geim and K. S. Novoselov, *Nat. Phys.*, 2014, **10**, 451–456.
8 L. Gao, J. R. Guest and N. P. Guisinger, *Nano Lett.*, 2010, **10**, 3512–3516.
9 J. P. Rabe and S. Buchholz, *Science*, 1991, **253**, 424–427.
10 F. Silly, *Nanotechnology*, 2012, **23**, 225603.
11 L. K. Thomas, A. Kühnle, S. Rode, U. Beginn and M. Reichling, *J. Phys. Chem. C*, 2010, **114**, 18919–18924.
12 K. G. Nath, O. Ivasenko, J. M. MacLeod, J. A. Miwa, J. D. Wuest, A. Nanci, D. F. Perepichka and F. Rosei, *J. Phys. Chem. C*, 2007, **111**, 16996–17007.
13 M. Kuwabara, D. R. Clarke and D. A. Smith, *Appl. Phys. Lett.*, 1990, **56**, 2396–2398.
14 P. Bak, *Rep. Prog. Phys.*, 1982, **45**, 587–629.
15 S. De Feyter and F. C. De Schryver, *Chem. Soc. Rev.*, 2003, **32**, 139–150.
16 S. De Feyter and F. C. De Schryver, *J. Phys. Chem. B*, 2005, **109**, 4290–4302.
17 M. Lackinger and W. M. Heckl, *Langmuir*, 2009, **25**, 11307–11321.
18 K. S. Mali, J. Adisoejoso, E. Ghijsens, I. De Cat and S. De Feyter, *Acc. Chem. Res.*, 2012, **45**, 1309–1320.
19 A. Stabel, R. Heinz, J. P. Rabe, G. Wegner, F. C. Deschryver, D. Corens, W. Dehaen and C. Siiling, *J. Phys. Chem.*, 1995, **99**, 8690–8697.
20 N. T. N. Ha, T. G. Gopakumar, R. Gutzler, M. Lackinger, H. Tang and M. Hietschold, *J. Phys. Chem. C*, 2010, **114**, 3531–3536.
21 Y. C. Ye, W. Sun, Y. F. Wang, X. Shao, X. G. Xu, F. Cheng, J. L. Li and K. Wu, *J. Phys. Chem. C*, 2007, **111**, 10138–10141.
22 S. Griessl, M. Lackinger, M. Edelwirth, M. Hietschold and W. M. Heckl, *Single Mol.*, 2002, **3**, 25–31.
23 Y. Ishikawa, A. Ohira, M. Sakata, C. Hirayama and M. Kunitake, *Chem. Commun.*, 2002, 2652–2653.
24 G. J. Su, H. M. Zhang, L. J. Wan, C. L. Bai and T. Wandlowski, *J. Phys. Chem. B*, 2004, **108**, 1931–1937.
25 J. M. MacLeod, J. A. Lipton-Duffin, D. Cui, S. De Feyter and F. Rosei, *Langmuir*, 2015, **31**, 7016–7024.
26 V. Iancu, K. F. Braun, K. Schouteden and C. Van Haesendonck, *Langmuir*, 2013, **29**, 11593–11599.
27 M. S. Baviloliaei and L. Diekhöner, *Phys. Chem. Chem. Phys.*, 2014, **16**, 11265–11269.
28 M. Lackinger, S. Griessl, W. M. Heckl, M. Hietschold and G. W. Flynn, *Langmuir*, 2005, **21**, 4984–4988.
29 P. Gilli, V. Bertolasi, V. Ferretti and G. Gilli, *J. Am. Chem. Soc.*, 1994, **116**, 909–915.
30 G. Eder, S. Kloft, N. Martsinovich, K. Mahata, M. Schmittel, W. M. Heckl and M. Lackinger, *Langmuir*, 2011, **27**, 13563–13571.
31 S. J. H. Griessl, M. Lackinger, F. Jamitzky, T. Markert, M. Hietschold and W. M. Heckl, *Langmuir*, 2004, **20**, 9403–9407.

32 M. Li, K. Deng, Y. L. Yang, Q. D. Zeng, M. He and C. Wang, *Phys. Rev. B: Condens. Matter Mater. Phys.*, 2007, **76**, 155438.
33 S. Conti and M. Cecchini, *Phys. Chem. Chem. Phys.*, 2016, **18**, 31480–31493.
34 J. F. Dienstmaier, K. Mahata, H. Walch, W. M. Heckl, M. Schmittel and M. Lackinger, *Langmuir*, 2010, **26**, 10708–10716.
35 E. Loginova, S. Nie, K. Thurmer, N. C. Bartelt and K. F. McCarty, *Phys. Rev. B: Condens. Matter Mater. Phys.*, 2009, **80**, 085430.
36 B. Wang, M. Caffio, C. Bromley, H. Fruchtl and R. Schaub, *ACS Nano*, 2010, **4**, 5773–5782.
37 P. Zeller and S. Günther, *New J. Phys.*, 2014, **16**, 083028.
38 J. Hutter, M. Iannuzzi, F. Schiffmann and J. VandeVondele, *Wiley Interdiscip. Rev.: Comput. Mol. Sci.*, 2014, **4**, 15–25.
39 J. P. Perdew, K. Burke and M. Ernzerhof, *Phys. Rev. Lett.*, 1996, **77**, 3865–3868.
40 S. Grimme, *J. Comput. Chem.*, 2006, **27**, 1787–1799.
41 M. Lackinger, S. Griessl, L. Kampschulte, F. Jamitzky and W. M. Heckl, *Small*, 2005, **1**, 532–539.
42 F. Shayeganfar and A. Rochefort, *Langmuir*, 2014, **30**, 9707–9716.
43 A. Tkatchenko, *Phys. Rev. B: Condens. Matter Mater. Phys.*, 2007, **75**, 235411.

Faraday Discussions

PAPER

The influence of nearest-neighbour interactions and assembly dynamics on the transport properties of porphyrin supramolecular assemblies on Au(111)†

Alison A. Pawlicki,‡[a] Ayelet Vilan,[b] Matthew Jurow,§[d] Charles Michael Drain[d] and James D. Batteas*[ac]

Received 27th March 2017, Accepted 23rd May 2017
DOI: 10.1039/c7fd00118e

Here we report on the effect of local molecular organization or "tertiary structure" on the charge transport properties of thiol-tethered tetraphenylporphyrin ($ZnTPPF_4$-SC_5SH) nanoscale clusters of *ca.* 5 nm in lateral dimension embedded within a dodecanethiol (C12) monolayer on Au(111). The structure of the clusters in the mixed monolayers and their resulting transport properties were monitored by Atomic Force Microscopy (AFM), Scanning Tunneling Microscopy (STM) and Spectroscopy (STS). The mixed films were deposited on Au(111) for a period of one to five days, during which the lateral dimensions of the $ZnTPPF_4$-SC_5SH islands that were formed after one day reduced by nearly 35% on average by five days, accompanied by a noticeable depletion of the surrounding C12 monolayer. These subtle changes in mixed monolayer morphology were accompanied by drastic differences in conductance. The $ZnTPPF_4$-SC_5SH clusters assembled for one day exhibited highly reproducible *I*–*V* spectra with simple tunneling behavior. By three days, this evolved into bias-induced switching of conductance, with a ~100–1000 fold increase. Furthermore, current fluctuations started to become significant, and then dominated transport across the $ZnTPPF_4$-SC_5SH clusters assembled over five days. Our data suggests that this evolution can be understood by slow surface diffusion, enabling the $ZnTPPF_4$-SC_5SH molecules to overcome initial steric hindrance in the early stages of island formation in the C12 monolayer (at day one), to reach a more energetically-favored, close-packed organization, as noted by the

[a]Department of Materials Science and Engineering, Texas A&M University, PO Box 3003, College Station, TX 77842, USA. E-mail: batteas@chem.tamu.edu

[b]Department of Chemical Physics, Weizmann Institute of Science, Herzl St. 234, Rehovot 76100, Israel

[c]Department of Chemistry, Texas A&M University, PO Box 30012, College Station, TX 77842, USA

[d]Department of Chemistry and Biochemistry, Hunter College of the City University of New York, 695 Park Avenue, New York, NY 10021, USA

† Electronic supplementary information (ESI) available. See DOI: 10.1039/c7fd00118e

‡ Present address: Oak Ridge National Labs and the University of Tennessee, Knoxville, TN 37830 USA.

§ Present address: Lawrence Berkeley National Labs, One Cyclotron Road, MS: 67R6110, Bldg. 67, Rm. 6113, Berkeley, CA 94720 USA.

decrease in island size (by day three). However, when desorption of the supporting matrix of C12 became pronounced (by day five), the $ZnTPPF_4$-SC_5SH clusters began to lose stabilization, and stochastic switching was then observed to dominate transport in the clusters, illustrating the critical nature of the local organization on these transport properties.

Introduction

In molecular electronics, molecules are the fundamental, electro-active materials used to modulate charge transport in devices.[1,2] The main attractions for using molecules over, or in conjunction with, other conventional materials for electronics is their inherent nano-metric dimensions, coupled with the wide versatility of their electronic properties that can be brought to bear through directed chemical syntheses. In addition to their dimensions and readily variable chemical composition, the structural flexibility of molecules can also play a key role in dictating the net electrical performance of a device or assembled architecture,[3–9] whereby additional tertiary structure can be driven by nearest neighbour interactions[10] and control of the assembly process,[11,12] as considered in this paper.

Molecular organization and structural flexibility are predicted to affect charge transport in molecular junctions *via* polaronic effects,[13–16] which has been suggested to greatly impact hopping transport in organic electronics (20–100 nm thick organic films).[17] Polaron formation is strongly related to the structural organization of molecules in the film, and is facilitated by aggregation, especially pi-stacking, which delocalizes molecular orbitals between non-covalently bonded pi-systems, yielding effects such as spectral shifts in dye-aggregates[18,19] or enhanced field-effect mobilities in densely packed conductive oligomers.[20]

In the context of molecular electronics, the principles of molecular self-assembly can be used to control such local organization and molecular structure.[12] Chemical synthesis offers a rich toolbox for tuning chemical interactions with the substrate[21–24] along with inter-molecular forces[25–27] from (relatively) weak van der Waals attractions, to hydrogen bonds, to pi-stacking, to coordinative bonds and to even cross-links, that can be used to form pre-designed structures based on these specific interactions. Moreover, the thermodynamically stable structures that form through non-covalent interactions such as pi-stacking offer the opportunity for self-repair.[28]

A prominent factor dictating monolayer structures formed *via* self-assembly is the assembly time, especially for thiol-based monolayers on Au. Even for simple systems, the self-assembly duration can have a drastic effect on the phases of alkanethiol self-assembled monolayers (SAMs) and their internal order.[29] However, the evolution of supramolecular order is less understood for more complex monolayers with pi-stacked systems and the question remains as to how subtle organizational variations due to prolonged self-assembly time may impact the electronic properties of such molecular systems. To this end, we have been exploring these ideas by examining few-molecule clusters of thiol-tethered porphyrins within a host alkanethiol SAM that subtly self-organize, to examine how cluster size, internal molecular organization, and nearest neighbor interactions within the host monolayer vary with assembly time. In our previous studies, we have explored the assembly of porphyrin-thiol islands of 5,10,15-tri(4-pyridyl)-20-(4-(1′,5′-dithiopentyl)-2,3,5,6-tetraflourophenyl)porphyrin ($H_2TPy_3PF_4SC_5SH$)

and its Zn(II) containing analog, $ZnTPy_3PF_4SC_5SH$, whereby the local interactions between the molecules were driven by combinations of pi-stacking as well as metal-pi nearest-neighbor interactions, where the presence of the Zn in the porphyrin macrocycle enabled the formation of larger islands than in the corresponding free-base. Here, the transport properties were found to vary specifically with island size, and in the case of the Zn(II) containing porphyrin, upon reaching an island size of *ca.* 5–10 nm, the transport properties were seen to change from simple tunneling behavior (for single molecules) to charge hopping.[25]

In the work presented here, the porphyrin-thiol chosen – 5,10,15-tri(4-phenyl)-20-(4-(1′,5′-dithiopentyl)-2,3,5,6-tetraflourophenyl) ($ZnTPPF_4$-SC_5SH), is analogous to the molecule studied by Schuckman *et al.*[25] with one variation; the pyridyl groups are replaced with phenyl groups (Fig. 1). This variation eliminates any possibility for axial coordination of Zn(II) with pyridyl groups on neighboring porphyrins and favors exclusively pi-stacking of the porphyrin macrocycles during assembly. As in our prior studies, the porphyrin-thiols were inserted into a pre-formed dodecanethiol (C12) SAM thereby providing stabilization of a standing-up orientation,[25,30] and allowed to self-assemble for one, three, or five days. The physical properties were examined using AFM and STM imaging. The electronic properties were measured through comparative STM imaging and STS, to evaluate relative changes in conductance to the background C12 monolayer and to measure the *I*–*V* characteristics for assemblies formed after different immersion times. As STS data are known to suffer from large spreads in recorded data, we were able to gain statistically-significant quantification on the evolution of the electronic properties of the $ZnTPPF_4$-SC_5SH clusters with assembly time by defining new parameters to quantify bias-induced switching and stochastic switching of conductance. These methods were applied to compare clusters of approximately similar sizes formed over different assembly durations to eliminate the effect of cluster size.

Experimental methods

Synthesis of $ZnTPPF_4$-SC_5SH

The 5,10,15-tri(4-phenyl)-20-(4-(1′,5′-dithiopentyl)-2,3,5,6-tetraflourophenyl) porphyrin was synthesized in two steps similar to a previously published procedure.[30] This was

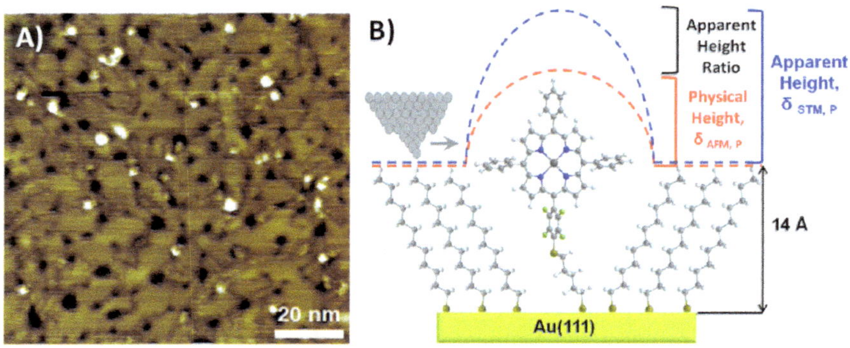

Fig. 1 $ZnTPPF_4$-SC_5SH within the C12 matrix, showing (A) an STM image with height contrast at a set point of 20 pA at 1.4 V and (B) a schematic side view illustrating that the measured heights are relative to the surrounding matrix, with a height of 14 Å.

followed by chelation of Zn(II) ions using the same procedure previously described to form [(5,10,15-tri(4-phenyl)-20-(4-(5-mercaptopentyl)thio]-2,3,5,6-tetrafluorophenyl) porphyrinato)zinc(II) ($ZnTPPF_4$-SC_5SH).[25]

Mixed self-assembled monolayer (SAM) preparation

The mixed monolayers were adsorbed onto 200 nm thick, flame annealed Au(111) films on mica (Phasis, Geneva, Switzerland). Before mixed monolayer absorption, the Au(111) films were cleaned using UV/ozone for 10 minutes, rinsed with deionized (18.2 MΩ cm) water (NANOpure Diamond, Barnstead), rinsed with ethanol, dried under streaming N_2, and then exposed to UV/ozone for another 5 minutes and rinsed and dried again with the same sequence described above. The substrates were subsequently immersed in 1 mM dodecanethiol (C12) (Aldrich, 98%, used as received) in ethanol solution for one day. Following formation of the C12 matrix SAM, the substrates were rinsed with ethanol and immersed in 0.05 mM of $ZnTPPF_4$-SC_5SH in dichloromethane for one, three, or five days. The concentration of $ZnTPPF_4$-SC_5SH was smaller than what has been previously used[25,30] to avoid aggregation in solution, that would result in highly crowded clusters on the Au(111) surface. After one, three, or five days, the substrates were removed from the $ZnTPPF_4$-SC_5SH solution, rinsed with dichloromethane, and dried under streaming N_2.

Scanning tunneling microscopy (STM)

STM images and I–V spectroscopy were collected with an Omicron UHV-XA STM system ($<3 \times 10^{-10}$ Torr) using mechanically cut Pt/Ir (80/20) tips and analyzed with Scanning Probe Image Processor (SPIP) software (version 6.0.2, Image Metrology, Lyngby, Denmark). Images were collected at 1.4 V and 20 pA in which 100–350 $ZnTPPF_4$-SC_5SH clusters were imaged for each immersion time (one, three, and five days) and their dimensions were extracted using grain analysis. I–V spectroscopy was collected over several clusters on different samples with different tips. 100 consecutive I–V spectra were collected above selected $ZnTPPF_4$-SC_5SH clusters, by sweeping the bias from −2 to +2 V (300 data points, 0.1 ms per point, see Fig. S1 in the ESI† for detailed bias-scan timing). The feedback loop was activated before each scan to maintain the tip height according to a set point of 20 pA at 1.4 V. To avoid spurious tip effects, 100 scans were also collected on a nearby C12 matrix immediately before and after the $ZnTPPF_4$-SC_5SH spectroscopy. If these reference spectra were of wide distribution or highly asymmetric, the results were discarded (Fig. S2 in the ESI†). I–V data was also discarded in cases where re-imaging the insertion area showed that tip changes had occurred. Further data statistics were carried out using MatLab.

Atomic force microscopy (AFM)

AFM images were obtained with an Agilent 5500 AFM using commercial Si_3N_4 tips (Bruker AFM Probes, MSCT, Sunnyvale, CA) with nominal 10 nm radii and 0.03 N m^{-1} spring constants. Samples were imaged inside an ethanol liquid cell using contact mode with forces between 0.015 and 0.050 nN. Low imaging forces were used to minimize compression of the $ZnTPPF_4$-SC_5SH molecules. Between 100 and 350 $ZnTPPF_4$-SC_5SH clusters were imaged using AFM and for each immersion time (one, three, and five days) and their dimensions were extracted using grain analysis.

Results and discussion

Changes in mixed monolayer morphology with immersion time

A typical STM topography image is shown in Fig. 1(A) wherein the bright features are $ZnTPPF_4$-SC_5SH and the dark features are etch pits, characteristic of alkanethiols on Au(111). Similar images were obtained by AFM, though with poorer lateral resolution. A comparison between AFM and STM images is given in the ESI.† The height measured by either AFM or STM is the height difference relative to the surrounding C12 matrix, as schematically shown in Fig. 1(B). The AFM height, $\delta_{AFM,P}$, is the difference in physical height. Full histograms and statistical analysis is provided in the ESI.† The mean and standard deviation are not adequate descriptors for asymmetric distributions, thus Table 1 summarizes the location and spread of the distribution using the median and half of the interquartile range (IQR) of all imaged clusters. The time evolution of the clusters is statistical in nature and the large spread in values implies that not all have evolved in the same manner.

We can further use $\delta_{AFM,P}$ to estimate the orientation of the $ZnTPPF_4$-SC_5SH clusters, expressed as the tilt angle, α, of the $ZnTPPF_4$-SC_5SH long axis with respect to the surface normal:

$$\cos\alpha = \frac{\delta_{AFM,P} + 1.4}{2.3} \tag{1}$$

where 1.4 nm and 2.3 nm are the heights of the C12 and the nominal length of the $ZnTPPF_4$-SC_5SH thiol derivative, respectively.[25,31] The tilt angles derived from eqn (1) are given in Table 1. The $ZnTPPF_4$-SC_5SH clusters are at an angle of approximately 41–45° with respect to the surface normal, which is reasonable compared to a tilt angle of approximately 30° for alkyl thiols on Au(111).[31] For one to five days, the measured relative height difference was largely seen to be unchanged at ~0.3 nm.

The lateral dimensions of the $ZnTPPF_4$-SC_5SH clusters were determined from the STM data due to its superior lateral resolution, as compared to AFM. While the distribution was rather broad, a 30–60% reduction in the lateral cluster dimensions after prolonged immersion time was observed in all cases (Fig. S7 in the ESI†). Based on these dimensions, a rough estimate for the number of $ZnTPPF_4$-SC_5SH molecules per cluster could be made using the 1.6 nm diameter of a single

Table 1 Statistics on cluster morphology where the reported values are the median of the distribution and error is half of the interquartile range (25 to 75% of the population). The tilt angle is relative to the surface normal and computed per eqn (1). The apparent height ratio is computed per eqn (2) with median values of physical and apparent heights

Immersion time (days)	1	3	5
AFM height (nm)	0.24 ± 0.04	0.30 ± 0.06	0.31 ± 0.05
Tilt angle (degree)	44.6 ± 1.5	42.6 ± 1.9	41.9 ± 1.7
Lateral dimension (nm)	5.2 ± 1.9	3.5 ± 1.6	3.3 ± 1.7
STM app. height (nm)	0.62 ± 0.3	1.04 ± 0.6	0.45 ± 0.2
Apparent height ratio (ratio)	0.38	0.74	0.14

$ZnTPPF_4$-SC_5SH and 0.6 nm as the distance between pi-stacked porphyrins.[25] Given this, smaller cluster sizes of 1–2 nm are most likely from a single $ZnTPPF_4$-SC_5SH molecule and large clusters, 10 nm in lateral dimension (approximately 100 nm^2), could contain a range of 30 loosely packed to 120 tightly packed $ZnTPPF_4$-SC_5SH molecules. The trend of reducing cluster size with immersion time from approximately 5 nm for one day to 3 nm on average for three and five days is the opposite of what we observed with the pyridyl Zn(II) porphyrin derivative, where the size of the inserted clusters increased considerably with time.[25] The difference in self-organization can likely be attributed to the lack of Zn(II) metalloporphyrin coordination by the pyridyl groups on neighboring molecules in the clusters.[25] Considering the large number of $ZnTPPF_4$-SC_5SH molecules in the immersion solution, one would expect continuous exchange between the solution phase and adsorbed molecules, resulting in slow desorption of the C12 and additionally, adsorption of $ZnTPPF_4$-SC_5SH that would cause a gradual increase in the cluster size with immersion time, as opposed to the observed size reduction.

Given this, here we suggest that the observed size reduction originates from reorganization of $ZnTPPF_4$-SC_5SH molecules within the individual clusters. Despite their strong surface binding, adsorbed thiols diffuse on Au surfaces quite easily and the Au surface atoms diffuse to relax lattice strain.[32] Reducing the intermolecular distance might be driven by pi-stacking between the aromatic macrocycles, which results in $\leq$0.5 nm distances between the mean planes. Considering that it takes approximately one day for flexible alkyl-thiols to achieve their optimal surface organization, one can speculate that it would take much longer for the rigid and more bulky $ZnTPPF_4$-SC_5SH to reach organizational equilibrium.

In addition to the changes in the porphyrin cluster size, subtle changes were also observed in the C12 SAM in the vicinity of the $ZnTPPF_4$-SC_5SH clusters with immersion time (Fig. 2). Any morphological change in the C12 is important because the dense packing of the C12 aids in stabilizing the standing-up orientation[30] of the $ZnTPPF_4$-SC_5SH molecules, and any change in the C12 is likely to affect the $ZnTPPF_4$-SC_5SH clusters through a change in nearest-neighbor interactions.[33] The subtle changes observed here illustrate that the C12 desorbs from the sample surface, and since the mixed monolayer is being immersed in a solution with a lack of C12, there is insufficient exchange with solution phase C12 as is typical in a solution with a sufficient concentration,[34] but a depletion of the C12 SAM from the surface. An indication of C12 depletion is shown in the STM image featuring a particular $ZnTPPF_4$-SC_5SH cluster formed after immersion for five days (Fig. 2). With increased immersion time, a lying-down, stripe phase emerges in the C12 surrounding the $ZnTPPF_4$-SC_5SH clusters. The lying-down, striped phase is known to form with a lower density of alkanethiols on the surface compared to the high density standing-up phase.[35]

Variation in transport properties with immersion time

Overall, Table 1 indicates that increasing the immersion time has little effect on the physical dimensions of the $ZnTPPF_4$-SC_5SH clusters beyond the local decrease in average cluster size. In contrast, the STM apparent height changes markedly by ~0.4 nm upon increasing the immersion time from one to three days, as we have

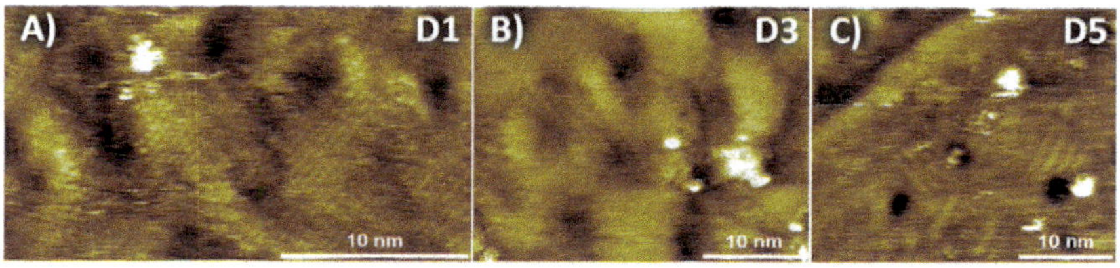

Fig. 2 Typical STM topography images of the C12 SAM quality surrounding the ZnTPPF$_4$-SC$_5$SH clusters. STS on these particular ZnTPPF$_4$-SC$_5$SH clusters has been included in the intensity maps shown in Fig. 4 as well as in the statistical data treatment. (A) The C12 monolayer after immersion for one day in the ZnTPPF$_4$-SC$_5$SH solution appears to be well-ordered in the vicinity of the ZnTPPF$_4$-SC$_5$SH clusters. However evidence of SAM degradation appears after (B) immersion for three days and even more so after (C) immersion for five days with the emergence of a striped or laying down phase.

observed previously for porphyrin-thiol clusters.[25] The scheme in Fig. 1(B) illustrates that the apparent STM height, $\delta h_{STM,P}$, is the sum of the physical height difference, $\delta h_{AFM,C12}$, and a variation in the vacuum gap to compensate for changes in the conductivity of ZnTPPF$_4$-SC$_5$SH clusters compared to that of the surrounding C12 matrix, to maintain the set point current. The range in apparent height values is the greatest for the clusters after immersion for three days, and the range narrows after immersion for five days. After five days, the apparent height is the smallest, indicating the conductance is "switched-off" in these clusters. The change in apparent height can be used to quantify the relative conductance *via* the apparent height ratio of the ZnTPPF$_4$-SC$_5$SH clusters, Δh_P, to the surrounding matrix, Δh_{C12}:

$$\frac{\Delta h_P}{\Delta h_{C12}} = \frac{\delta h_{STM,P} - \delta h_{AFM,P}}{\delta h_{STM,C12} - \delta h_{AFM,C12}} \approx \delta h_{STM,P} - \delta h_{AFM,P} \tag{2}$$

where $\delta h_{STM,P}$ is the apparent ZnTPPF$_4$-SC$_5$SH height extracted from STM measured with respect to the C12 SAM and $\delta h_{AFM,P}$ is the physical ZnTPPF$_4$-SC$_5$SH height extracted from AFM measured with respect to the C12 SAM. The denominator $\delta h_{STM,C12} - \delta h_{AFM,C12}$ is normalized to one since the porphyrin heights are always measured with respect to the C12 SAM (Table 1). The full data set of the apparent height ratio as a function of immersion time is shown in Fig. S5 in the ESI.† In general, the apparent height ratio increases for increasing lateral cluster dimension, but the spread in the apparent height ratio with lateral cluster dimension is large and becomes larger as the lateral cluster dimension increases. This suggests that the lateral dimension is not the solitary factor impacting the electronic properties. As the lateral cluster dimension increases, the possibilities of internal ZnTPPF$_4$-SC$_5$SH organization within a particular cluster increases as well, which is likely to be the origin of the large spread in height for larger lateral cluster dimensions. To investigate this possibility, the electronic properties of clusters formed over different immersion times with approximately similar lateral dimensions were investigated.

Current–voltage intensity maps

Three ZnTPPF$_4$-SC$_5$SH clusters of approximately similar lateral dimension, formed after one, three, and five days (designated as D1, D3 and D5), and one on a neat C12 monolayer, were closely examined. As summarized in Table 1 the $ZnTPPF_4$-SC_5SH clusters appear similar however, their current–voltage (*I*–*V*) spectra differ markedly. The lateral dimensions of the clusters are not *a priori* representative of the general population, and the $ZnTPPF_4$-SC_5SH clusters chosen for spectroscopy were amongst the wide-end of the population. To manage the large variance in the measured *I*–*V*, we show intensity maps in Fig. 3 of 100 successive scans recorded over $ZnTPPF_4$-SC_5SH clusters. The intensity *I*–*V* map over a neat C12 monolayer is also shown for comparison. The pure C12 matrix also included a low density of apparent low-contrast C12 islands. *I*–*V* scans were both taken over background and C12 "islands" and were identical. Fig. 3 reveals a prominent change in the transport characteristic solely as a function of immersion time and consequently, reorganization of the mixed monolayer.

Transport across a $ZnTPPF_4$-SC_5SH cluster grown for one day is qualitatively similar to the neat C12. Both *I*–*V* intensity maps have a narrow data distribution and are continuous across zero bias. In contrast, transport across a $ZnTPPF_4$-SC_5SH cluster, grown for three and five days, results in drastically different *I*–*V* intensity maps: the distribution becomes wider (poorer reproducibility) and a gap of low-conductivity near 0 V emerges. Panels D3 and D5 of Fig. 3 show current variations of orders of magnitude (note the semi log-scale of Fig. 3). Such random fluctuations are attributed to stochastic switching. The low-conductivity gap is clearer for the three day cluster (panel D3) where sharp current onsets occur at *ca.* ±1 V. This effect is termed bias-induced switching as the sharp increase in current occurs at a well-defined bias. We argue that the two mechanisms: bias-induced switching and stochastic switching evolve with the immersion time.

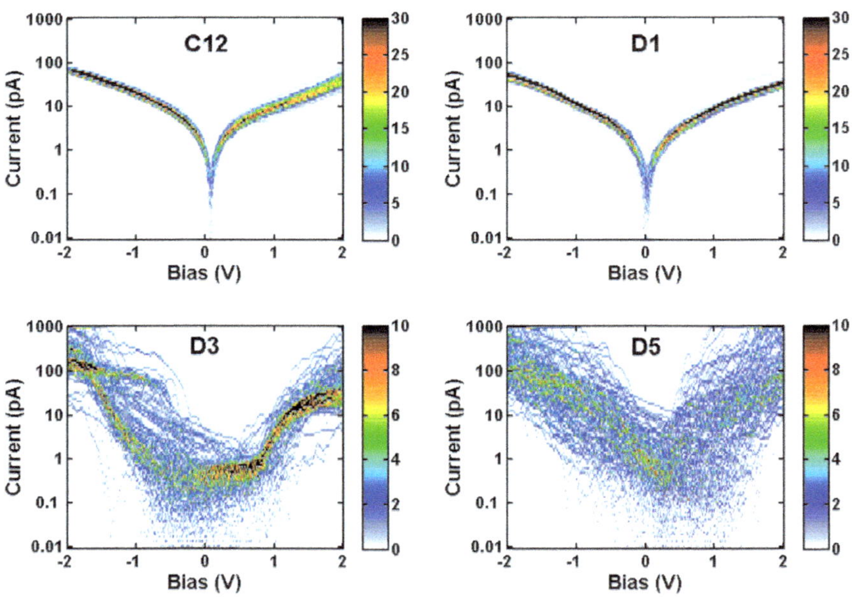

Fig. 3 Current intensity maps for neat C12 matrix (C12) and $ZnTPPF_4$-SC_5SH clusters after one day (D1), three days (D3) and five days (D5) on a semi-log scale against the applied bias. The color scale represents the recorded counts out of 100 replicas (note the different color scale bars for top and bottom plots).

However, the stochastic switching blurs the bias-switching, as can be seen for the five day cluster. While the overall shape of the *I*–*V* response in panel D5 of Fig. 3 is similar to that of three days (non-conducting near 0 V) the wider data distribution (stochastic switching) masks a clear observation of bias-induced switching.

There is a gradual increase in maximum current between the four samples. The maximum current does not exceed 100 pA for the neat C12 and the one day cluster, while there is a considerable increase in the maximum current reaching 1 nA for the three day cluster, and an even further increase for the five day cluster, often reaching the compliance limit of the current amplifier (3 nA). Rather than the current maximum simply increasing for the three and five day clusters, the number of current data counts becomes more intense in the low current range (<1 pA) suggesting conductivity switching.

Characteristic *I*–*V* curves

To illustrate the observed trend in Fig. 3, typical *I*–*V* curves are shown in Fig. 4. It is challenging to select a single, "representative" curve because of the large variance within consecutively measured *I*–*V* curves, especially for longer immersion times. Averaging over all data smears-out details; therefore, we select the *I*–*V* curve nearest to the median to be the "representative" one. The median (*cf.* mean) eliminates the effect of outliers and answers the need to average both over log-normal distribution at high bias[36] and noisy, varying polarity near low bias. Different approaches to define a "representative" curve are described in the ESI.† The single cubic polynomial fits of the clusters for the one day cluster and the neat C12 (continuous lines in Fig. 4) indicate a generic, single tunneling mechanism[37,38] over the entire ± 2 V bias range. In contrast, the *I*–*V* curves recorded over the three and five day clusters could not be generally fitted to a polynomial relation, except over a limited range near zero bias (see green and blue lines in Fig. 4). Thus, what appears to be a conductance gap in the *I*–*V* intensity maps shown in Fig. 4 is actually an onset of an additional transport channel.

The additional high-bias conductance channel behaves differently for the three and five day clusters. For the three day cluster, the second channel is characterized by a huge conductance onset, above approximately ± 1 V, or a bias-

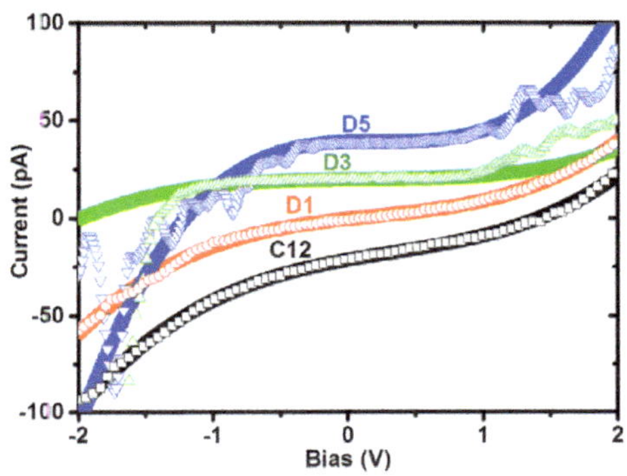

Fig. 4 *I*–*V* curves for the different equilibration times and the C12 matrix that is nearest to the median. The symbols are the data and the solid lines are fits to a third order polynomial relation. Curves are shifted vertically (by 20 pA each) for clarity.

induced switching of conductivity. For the five day cluster, the onset occurs at a lower bias (<1 V) but leads to a rather noisy conductance of moderate magnitude. This noisy *I–V* curve in Fig. 4 qualitatively agrees with the wide intensity spread in Fig. 3. In particular, this noise is not simply between consecutive scans but also within each individual scan. Both observations point to stochastic switching of conductivity. Stochastic switching was also observed as a function of imaging time and STM set-point (Fig. S14, Table S3, and S4 in the ESI†).

Statistics quantifying bias-induced and stochastic switching

Overall, Fig. 3 and 4 indicate that prolonged assembly of $ZnTPPF_4$-SC_5SH clusters within the C12 matrix promotes both bias-induced switching (conductance onset) and stochastic switching (noise). To corroborate our claims, we have identified three quantitative criteria that allow us first to quantify our former qualitative observation, and more importantly to examine a much larger data set including 7 to 14 different clusters for each immersion time. Formulating robust and meaningful criteria was not trivial and they may be found useful for future studies.

Quantifying bias-induced switching

The concept of bias-induced switching is represented by two possible parameters: (1) the bias window of noisy current below 0.5 pA and (2) the onset of high-conductance. The term V_{noise} measures the width of the bias window with absolute current values below 0.5 pA (*i.e.* 2.5% of the set-point current and 0.02% of the amplifier saturation). Its extraction is demonstrated in Fig. S8(A) of the ESI.† Fig. 5(A) presents the distribution in V_{noise} for all measured *I–V* sets over all measured clusters (*cf.* specific clusters in Fig. 3). Fig. 5(A) confirms the evolution of the low-conductance gap (bias-induced switching) with assembly time. The mean noise-region is 0.09 V for C12 and doubles for one day $ZnTPPF_4$-SC_5SH clusters. Further increasing the immersion time considerably increases the noise region up to 0.54 V on average for the five day clusters. The width of the distribution also increases considerably with immersion time. The extension of the low current region for five day clusters is in accordance with the trend of smaller STM height for the longest immersion time (Table 1).

The advantages of the V_{noise} criterion are directness and simplicity; however, it is sensitive to stochastic switching and STM set-point settings (affecting the dynamic range). Identifying the bias-induced switching with an onset of a second transport channel (see Fig. 4) suggests a voltage-gap criterion based on a relative increase in conductance. Therefore, the negative, V_N, and positive, V_P, onsets of high conductance (see Fig. S8(B) in the ESI†) were arbitrarily set to the voltage where the conductance reached 10-fold its value at 0 V (G_{Eq}, see Fig. S8 in ESI† for histograms of G_{Eq}). Within this view, the tunneling range (V_{tunn}) extends from V_N to V_P. In cases where none of the $\frac{dI}{dV}$ values exceed $10G_{Eq}$, the value of $V_{N(P)}$ is set to ± 2.5 V making it distinctly different than the measured range of ± 2 V.

As above, the values of V_{tunn} were extracted for all measured *I–V* sets and shown as histograms in Fig. 5(B). It shows that 78% of the neat C12 *I–V* curves had no onset of high conductance at either positive or negative bias (*i.e.* $V_{tunn} = 5$ V) and only 2% of the curves reached $10G_{Eq}$ on both polarities. For both C12 and D1 clusters, V_{tunn} covers the majority of the bias range, implying no bias-induced

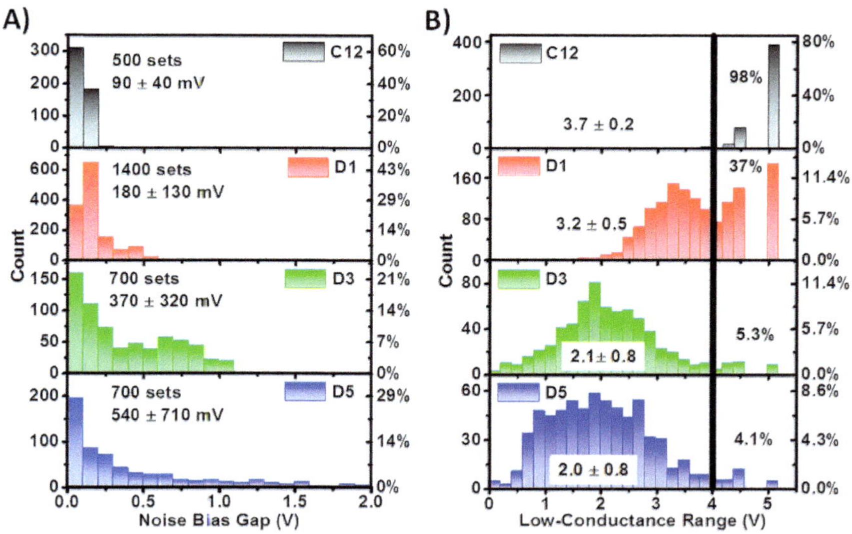

Fig. 5 Histograms of the two parameters used to quantify bias-induced switching for all measured I–V sets and clusters. (A) Noise voltage window using an absolute threshold value of $|I| \leq 0.5$ pA. (B) Histograms of the low-conductance range, determined as the difference between the positive and negative onset bias, where the numerical conductance exceeds 10 times its value near 0 V. A range of 4 V (vertical line) is the full bias scan; any value larger than this indicates that at least over one polarity all conductance points are $<10G_{Eq}$ and an arbitrary onset value of ±2.5 V is designated. In both panels, the left y-axis is the net count of I–V sets and the right y-axis is the percentage.

switching at all (only natural evolution of conductance with bias that is typical to transport by tunneling, as discussed in Fig. 4). Onset of high conductance was much more frequent for three and five day ZnTPPF$_4$-SC$_5$SH clusters, where the gap is reduced to approximately 2 V or onsets at *ca.* ±1 V. The mean of the distribution was similar for both three and a five day clusters, however, the distribution of the five day clusters' low-conductance range is flatter (less Gaussian) and wider than that of the three day clusters', due to enhanced stochastic switching for longer immersion times.

Quantifying stochastic switching

The concept of stochastic switching is quantified by a third parameter describing the ratio between the measured current and nominal set-point current. The magnitude of STM current depends on its tip height which is set arbitrarily to yield a certain set-point value. For example, in our measurement procedure the height of the STM tip was re-adjusted before each I–V scan to yield a current of 20 pA at a bias of +1.4 V, and then the bias was swept from −2 V to +2 V; approximately 35 ms elapsed between turning-off the feedback loop and scanning the bias back to 1.4 V (Fig. S1 in the ESI†); if any stochastic switching occurred during this time interval the current re-measured at 1.4 V would differ from the set-point of 20 pA. Therefore, we define the set-point ratio, P_{sp} as (see Fig. S8(A) in the ESI†):

$$P_{sp} = \frac{I(1.4\ \text{V})}{20\ \text{pA}} \tag{3}$$

As above, we apply this analysis over all measured I–V sets, and summarize the results as a histogram in Fig. 6, where the x-axis is the $\log_{10}(P_{sp})$. In all four cases the histograms are centered at zero which is the stable or non-switching case ($P_{sp} = 1$). However, the distribution in P_{sp} becomes much wider for three day clusters and increases by three orders of magnitude for five day clusters. The reasonable uniformity in P_{sp} of the neat C12 and the one day clusters confirms that the noise is a true molecular feature and not a measurement artifact. In addition, the P_{sp} values of a single STS set (100 traces) are completely random (Fig. S11 in the ESI†) verifying the stochastic nature of these fluctuations rather than aging or deformation. The trend in Fig. 6 agrees well with the qualitative observation of the large spread in the I–V intensity maps (Fig. 3) for the three and five day clusters.

Finally, the observed distribution in the extracted parameters (Fig. 5 and 6) is not due to cluster to cluster variation, but occurs within each individual cluster or measurement set. This is demonstrated by averaging the 100 values V_{tunn} extracted for each cluster and plotting these cluster-averages against the lateral dimension of the cluster (Fig. 7(A)) extracted from STM images. Fig. 7(A) shows a clear difference in the tunneling gap between the one day clusters and the three–five day clusters. However, the standard deviation within each cluster (error bars in Fig. 7(A)) is similar or larger than the difference between clusters of the same immersion time.

Fig. 7(B) tests this concept for the variance in the stochastic current ratio, P_{sp} within each location. Also here, there is a clear trend in increasing stochastic variance, but no trend with the lateral cluster dimension. The lateral cluster dimension used in Fig. 7 is specifically determined for the clusters where the I–V spectroscopy was performed, in contrast to Table 1 which refers to a much larger

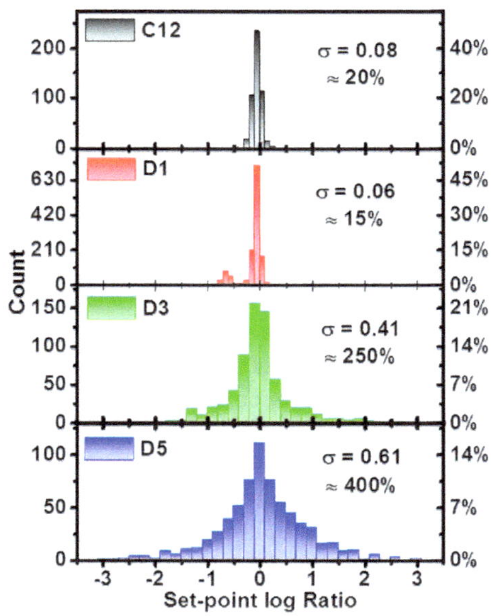

Fig. 6 Histograms of the parameters used to quantify stochastic switching; the ratio between measured current and nominal set-point current for all measured P_{sp} data, see eqn (3). $\log_{10}$ of this ratio is used as the bins axis.

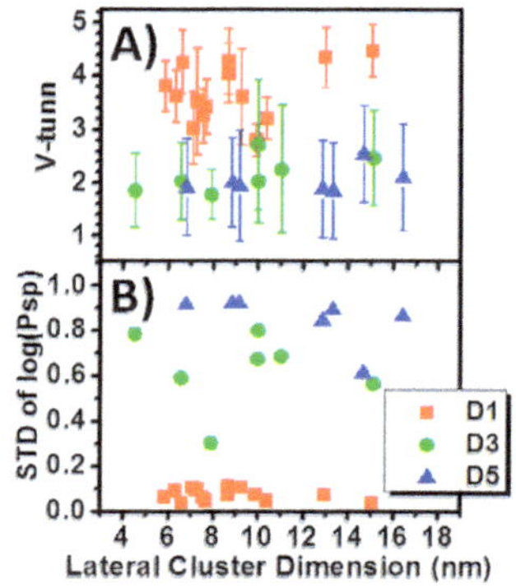

Fig. 7 Averaged parameters as a function of lateral cluster dimension, showing (A) the low-conductance range, V_{tunn} (eqn (S4) in the ESI† and Fig. 5) and (B) the standard deviation of stochastic switching, $\log_{10}(P_{sp})$ (eqn (3) and Fig. 6). The measured lateral cluster dimension is larger than the average (Table 1) because we measured tunneling spectroscopy on larger clusters to minimize the tip drifting away from the $ZnTPPF_4$-SC_5SH clusters.

set of imaged clusters. The fact that both parameters (Fig. 7) show no dependence on the lateral cluster dimension provides strong evidence that the change in transport mechanism cannot be induced by a geometrical effect, and must originate in a subtler structural change.

Discussion

Our statistical analyses show that by allowing the $ZnTPPF_4$-SC_5SH/C12 SAM mixed monolayer an extended period of time to rearrange, the net electrical conductivity of the $ZnTPPF_4$-SC_5SH molecules drastically changes (Table 1 and Fig. 3) with the emergence of both bias-induced (D3) and stochastic switching (D5). Bias-induced switching is clear after immersion for three days and then overruled by stochastic switching (Fig. 3) after immersion for five days. On the other hand, the structural changes in the mixed monolayer (Table 1 and Fig. 2) are subtle and not easily discerned with room-temperature STM. The apparent trend is that the clusters laterally squeeze to reduce in size and the C12 matrix depletes with immersion time. A model to summarize our findings is presented in Fig. 8.

Lateral cluster reduction and the effect on net conductance

Increasing the immersion time from one day to three days reduced the clusters by ~1.7 nm on average (estimated 55% area reduction) and between three and five days there is less reduction in size; 0.2 nm on average (estimated 12% area reduction) further than the original (one day) cluster size. This is comparable to studies of two-dimensional aggregation of porphyrins using a Langmuir–Blodgett (LB) trough in which nearest neighbor interactions altered the area per monomer (*i.e.*, a 66–71% area difference) accompanied by an impact on photo-physical properties.[19,39] Optical effects differ from charge transport since they are electrically neutral excited states. Still, they involve similar energy levels and charge propagation mechanisms. Therefore, the most probable mechanism that can

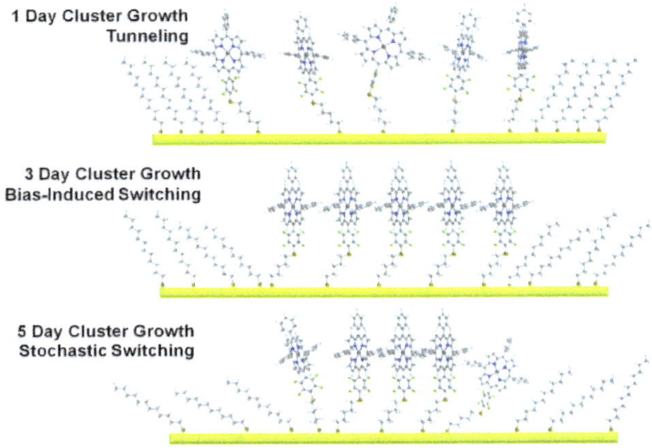

Fig. 8 Model of the reorganization of the $ZnTPPF_4$-SC_5SH clusters formed after one, three, and five days. After one day, the clusters are fairly isolated from their neighbors while the clusters formed after three days represent strong to medium inter-molecular interactions. Finally, after five days, due to the depletion of the surrounding C12, the clusters become more disordered and still interacting but not as uniformly as in the three day clusters.

account for both reduction in cluster size and increased conductivity is improved pi-stacking that stabilizes the molecular energy levels or reduces the polaron formation energy.

We suggest that the one day clusters are composed of $ZnTPPF_4$-SC_5SH molecules that are fairly isolated from their neighbors while the three and five day clusters represent medium to strong inter-molecular interactions (Fig. 8). Based on the change in lateral cluster dimension (Table 1) and assuming that the $ZnTPPF_4$-SC_5SH molecules in the clusters formed after three days are pi-stacked with a separation of 0.6 nm,[25] the $ZnTPPF_4$-SC_5SH separation after one day immersion is estimated to be 0.9 nm which is larger than the typical van der Waals interaction length, supporting that after one day the $ZnTPPF_4$-SC_5SH molecules are fairly isolated. After five days of immersion, the $ZnTPPF_4$-SC_5SH separation is estimated to be 0.5 nm which is within the typical van der Waals interaction length and slightly smaller than the $ZnTPPF_4$-SC_5SH separation after immersion for three days supporting that after immersion for three and five days, the $ZnTPPF_4$-SC_5SH clusters have medium to strong inter-molecular interactions.

We suggest that these moderate structural changes have a critical effect on the strength of the pi-interactions and therefore reduce the polarization energy. Namely, the richness of conductance behavior of $ZnTPPF_4$-SC_5SH clusters formed after three and five days is facilitated by their lower polarization energy and ease of polaron formation. Polaron mediated transport was used to explain negative differential resistance (NDR) or hystersis.[14,40,41] The bias- and stochastic-switching observed here are conceptually related to NDR and hysteresis. Hysteresis requires sharp conductance onsets as observed for the three day clusters. The multiple peak characteristics of the five day clusters can be viewed as ill-developed NDR. The absolute reproducibility of these noisy peaks in terms of bias (energy) and intensity was rather poor, which is at least partially because the measurements were done at room temperature. It was suggested that both NDR and hysteresis can originate from a redox event such as charging of the molecular wire. If the charged state is long lived, *e.g.*, for an isolated molecule, it will appear as stable switching or hysteresis, while the presence of an electrolyte will reduce the reorganization energy and increase the fluctuations of the charged state, which appears as NDR.[42] In analogy with this concept, we can attribute the evolution of the sharp conductance onset (D3) into multiple small peaks (D5) as a further reduction in the polaron energy for the five day clusters (*e.g.*, by closer pi-stacking). The medium polaron energy of the three day clusters is accessible through a bias of up to 2 V, but discharging is restricted. The even lower polaron energy of the five day clusters requires smaller onset bias (Fig. 4 and 5(A)) and is also less stable. The fluctuations in the charged state appear as multiple peaks in the *I–V* curves (Fig. 4), large spreading of the intensity maps (Fig. 3) and large distribution of the onset current (Fig. 5).

Depletion of C12 SAM and the effect on net conductance

Increasing the immersion time from one to five days influences the entire mixed monolayer, not just the electro-active, guest molecule. Accordingly, the surrounding C12 monolayer depletes with immersion time and is consistent with other SAM morphology studies in organic solvents over an extended period of time.[34,43] This depletion causes the C12 monolayer morphology to evolve from a densely-packed standing-up organization after one day, to the development of some less-dense lying down domains after three days, and to more prevalent lying down domains after five days (Fig. 2). In this mixed monolayer system, the role of the C12 monolayer is to stabilize the $ZnTPPF_4$-SC_5SH clusters and since the morphology is evolving with immersion time, it is likely that the interaction between the surrounding C12 and $ZnTPPF_4$-SC_5SH clusters is also evolving; providing less stabilization with immersion time leading to more $ZnTPPF_4$-SC_5SH conformational variations[44] within a particular cluster. Conformational and supramolecular variations in molecules,[44,45] nanoclusters,[46] and morphological changes of the background SAM in mixed monolayers[47,48] are known to impact the electronic characteristics. In particular, guest molecules within a host SAM[33] similar to the $ZnTPPF_4$-SC_5SH clusters presented herein, were found to switch stochastically when the host SAM was disordered. Therefore, it is conjectured that the stochastic switching dominant in the five day clusters (Fig. 3) originates from further local packing of the $ZnTPPF_4$-SC_5SH clusters as discussed above accompanied by the depletion of the surrounding C12 monolayer after immersion for five days.

Conclusions

Within the context of investigating the consequences of molecular structural flexibility and interactions with surrounding molecules, $ZnTPPF_4$-SC_5SH molecules were examined with a detailed focus on the impact of prolonged solution phase self-assembly on the resulting net electrical conductance of the structures formed. The $ZnTPPF_4$-SC_5SH molecules were incorporated within a host C12 SAM and assembled over one, three, or five days to self-organize into small clusters (5.2 to 3.3 nm on average, respectively). The charge transport was measured with STM and STS across the $ZnTPPF_4$-SC_5SH clusters and found to evolve over the prolonged self-assembly time. The *I–V* curves were characterized by an increasing current maximum and data spread for increasing assembly duration along with an opening of a new transport channel, expressed as a sharp conductance onset. For the one day clusters, the *I–V* curves are comparable to that of alkanethiols and similar to single porphyrin-thiol molecules,[30] indicating simple tunneling. For the three day clusters, the *I–V* curves show a conductance gap at low bias indicating bias-induced opening of a new conductance channel, known as bias-switching. Finally, for the five day assemblies, the conductance gap narrows and the *I–V* traces are highly spread leading to an unclear conductance onset, indicating stochastic switching. Although there are marked electronic changes, the accompanying structural changes are rather minor over the self-assembly time. Some depletion of the host C12 SAM was observed in the proximity of the $ZnTPPF_4$-SC_5SH clusters after an assembly time of three days and to a greater extent after five days.

The direct, though qualitative evidence for evolution of charge transport with prolonged self-assembly time, is further confirmed by statistical approaches to quantify the broad variation in STS. The bias-induced switching of conductance was quantified in two ways: the fraction of noise-level current and the onset voltage of high-conductance. The stochastic switching of conductance was quantified by the ratio between the measured current and nominal set-point current. Quantization was done on all STS data over several different clusters of approximately similar sizes, and indeed no correlation was found between the cluster size and statistical parameters.

We propose that the drastic transformation in electronic properties can be explained by slow aggregation of the absorbed $ZnTPPF_4$-SC_5SH molecules accounting for the reduction in the lateral cluster dimensions, thereby resulting in varying degrees of pi-stacking between $ZnTPPF_4$-SC_5SH within the assembled clusters. In this view, molecules within the clusters formed after one day weakly interact and can be treated as single, isolated molecules. Then, the $ZnTPPF_4$-SC_5SH molecules in clusters formed after three days strongly interact after slow aggregation into a closely-packed, and likely pi-stacked arrangement. However, the $ZnTPPF_4$-SC_5SH clusters formed after five days have more conformational variations because of the local loss of the stabilizing C12 SAM.

Closer-packing within the $ZnTPPF_4$-SC_5SH clusters and the aforementioned diverse switching was qualitatively considered in terms of polaron meditated transport. We conjecture an association between the internal cluster organization and observed electron transport, wherein closer-packing reduces the polaron reorganization energy, enabling bias-driven charging for clusters assembled after three days and stochastic charging for clusters assembled after five days.

Acknowledgements

The authors would like to acknowledge financial support for this work from the US National Science Foundation grants CHE-1213082 and CHE-1611119 to J. D. B. and CHE-1213962 and CHE-161-755 to C. M. D., as well as the US National Science Foundation Graduate Research Fellowship (NSF-GRF) to A. A. P.

References

1 M. Ratner, *Nat. Nanotechnol.*, 2013, **8**, 378.
2 A. Vilan, D. Aswal and D. Cahen, *Chem. Rev.*, 2017, **117**(5), 4248–4286.
3 Y. Selzer, L. Cai, M. A. Cabassi, Y. Yao, J. M. Tour, T. S. Mayer and D. L. Allara, *Nano Lett.*, 2005, **5**, 61.
4 W. Haiss, H. van Zalinge, D. Bethell, J. Ulstrup, D. J. Schiffrin and R. J. Nichols, *Faraday Discuss.*, 2006, **131**, 253.
5 Z. Donhauser, B. Mantooth, K. Kelly, L. Bumm, J. Monnell, J. Stapleton, D. Price, A. Rawlett, D. Allara and J. Tour, *Science*, 2001, **292**, 2303.
6 A. M. Moore, B. A. Mantooth, Z. J. Donhauser, Y. Yao, J. M. Tour and P. S. Weiss, *J. Am. Chem. Soc.*, 2007, **129**, 10352.
7 G. D. Harzmann, R. Frisenda, H. S. van der Zant and M. Mayor, *Angew. Chem., Int. Ed.*, 2015, **54**, 13425.
8 Y. Kitaguchi, S. Habuka, H. Okuyama, S. Hatta, T. Aruga, T. Frederiksen, M. Paulsson and H. Ueba, *Beilstein J. Nanotechnol.*, 2015, **6**, 2088.
9 A. Donarini, A. Yar and M. Grifoni, *Eur. Phys. J. B*, 2012, **85**, 1.
10 M. L. Perrin, F. Prins, C. A. Martin, A. J. Shaikh, R. Eelkema, J. H. van Esch, T. Briza, R. Kaplanek, V. Kral and J. M. van Ruitenbeek, *Angew. Chem., Int. Ed.*, 2011, **50**, 11223.
11 Z. J. Donhauser, B. A. Mantooth, T. P. Pearl, K. F. Kelly, S. U. Nanayakkara and P. S. Weiss, *Jpn. J. Appl. Phys.*, 2002, **41**, 4871.
12 A. Pawlicki, E. Avery, M. Jurow, B. Ewers, A. Vilan, C. M. Drain and J. J. Batteas, *Phys. Condens. Matter*, 2016, **28**, 094013.
13 M. Galperin, M. A. Ratner, A. Nitzan and A. Troisi, *Science*, 2008, **319**, 1056.
14 M. Galperin, M. A. Ratner and A. Nitzan, *Nano Lett.*, 2005, **5**, 125.
15 M. Galperin, A. Nitzan and M. A. J. Ratner, *Phys. Condens. Matter*, 2008, **20**, 374107.
16 M. Galperin, M. A. Ratner and A. J. Nitzan, *Phys. Condens. Matter*, 2007, **19**, 103201.
17 J. He, E. S. Forzani, L. A. Nagahara, N. Tao and S. J. Lindsay, *Phys. Condens. Matter*, 2008, **20**, 374120.
18 D.-J. Qian, C. Nakamura and J. Miyake, *Langmuir*, 2000, **16**, 9615.
19 B. Choudhury, A. C. Weedon and J. R. Bolton, *Langmuir*, 1998, **14**, 6192.
20 R. Österbacka, C. P. An, X. Jiang and Z. V. Vardeny, *Science*, 2000, **287**, 839.
21 K. Suto, S. Yoshimoto and K. Itaya, *J. Am. Chem. Soc.*, 2003, **125**, 14976.
22 A. J. Wang, W. Yu, Z. P. Huang, F. Zhou, J. B. Song, Y. L. Song, L. L. Long, M. P. Cifuentes, M. G. Humphrey, L. Zhang, J. D. Shao and C. Zhang, *Sci. Rep.*, 2016, **6**, 12.
23 S. Yoshimoto, N. Higa and K. Itaya, *J. Am. Chem. Soc.*, 2004, **126**, 8540.
24 S. Biswas, H.-Y. Ahn, M. V. Bondar and K. D. Belfield, *Langmuir*, 2012, **28**, 1515.
25 A. E. Schuckman, B. W. Ewers, L. H. Yu, J. P. C. Tome, L. M. Perez, C. M. Drain, J. G. Kushmerick and J. D. Batteas, *J. Phys. Chem. C*, 2015, **119**, 13569.
26 N. Nerngchamnong, L. Yuan, D.-C. Qi, J. Li, D. Thompson and C. A. Nijhuis, *Nat. Nanotechnol.*, 2013, **8**, 113.
27 Y. Li, T. Liu, H. Liu, M.-Z. Tian and Y. Li, *Acc. Chem. Res.*, 2014, **47**, 1186.
28 T. Yokoyama, S. Yokoyama, T. Kamikado, Y. Okuno and S. Mashiko, *Nature*, 2001, **413**, 619.

29 J. C. Love, L. A. Estroff, J. K. Kriebel, R. G. Nuzzo and G. M. Whitesides, *Chem. Rev.*, 2005, **105**, 1103.
30 Y.-H. Chan, A. E. Schuckman, L. M. Pérez, M. Vinodu, C. M. Drain and J. D. Batteas, *J. Phys. Chem. C*, 2008, **112**, 6110.
31 J. D. Monnell, J. J. Stapleton, S. M. Dirk, W. A. Reinerth, J. M. Tour, D. L. Allara and P. S. Weiss, *J. Phys. Chem. B*, 2005, **109**, 20343.
32 P. Maksymovych, O. Voznyy, D. B. Dougherty, D. C. Sorescu and J. T. Yates, *Prog. Surf. Sci.*, 2010, **85**, 206.
33 Z. J. Donhauser, B. A. Mantooth, K. F. Kelly, L. A. Bumm, J. D. Monnell, J. J. Stapleton, D. W. Price, A. M. Rawlett, D. L. Allara, J. M. Tour and P. S. Weiss, *Science*, 2001, **292**, 2303.
34 J. B. Schlenoff, M. Li and H. Ly, *J. Am. Chem. Soc.*, 1995, **117**, 12528.
35 C. Vericat, M. E. Vela and R. C. Salvarezza, *Phys. Chem. Chem. Phys.*, 2005, **7**, 3258.
36 W. F. Reus, C. A. Nijhuis, J. R. Barber, M. M. Thuo, S. Tricard and G. M. Whitesides, *J. Phys. Chem. C*, 2012, **116**, 6714.
37 A. Vilan, D. Cahen and E. Kraisler, *ACS Nano*, 2013, **7**, 695.
38 J. C. Cuevas and E. Scheer, *Molecular electronics: an introduction to theory and experiment*, World Scientific Publishing Company Incorporated, 2010, vol. 1.
39 G. Y. Liu, S. Xu and Y. L. Qian, *Acc. Chem. Res.*, 2000, **33**, 457.
40 J. H. Wei, S. J. Xie, L. M. Mei, J. Berakdar and Y. J. Yan, *New J. Phys.*, 2006, **8**, 9.
41 A. Bandyopadhyay and Y. Wakayama, *Appl. Phys. Lett.*, 2007, **90**, 023512.
42 J. He, Q. Fu, S. Lindsay, J. W. Ciszek and J. M. Tour, *J. Am. Chem. Soc.*, 2006, **128**, 14828.
43 J. Noh and M. Hara, *Langmuir*, 2001, **17**, 7280.
44 H. Wu, K. Sotthewes, P. M. Schon, G. J. Vancso and H. J. W. Zandvliet, *RSC Adv.*, 2015, **5**, 42069.
45 T. A. Su, H. X. Li, M. L. Steigerwald, L. Venkataraman and C. Nuckolls, *Nat. Chem.*, 2015, **7**, 215.
46 X. G. Li, X. G. Zhang and H. P. Cheng, *Nano Lett.*, 2014, **14**, 4476.
47 M. Leijnse, *Phys. Rev. B: Condens. Matter Mater. Phys.*, 2013, **87**, 9.
48 L. Jiang, C. S. S. Sangeeth, L. Yuan, D. Thompson and C. A. Nijhuis, *Nano Lett.*, 2015, **15**, 6643.

Faraday Discussions

PAPER

The culture of HaCaT cells on liquid substrates is mediated by a mechanically strong liquid–liquid interface

D. Kong,[ab] K. D. Q. Nguyen,[ab] W. Megone,[ab] L. Peng[ab] and J. E. Gautrot *[ab]

Received 5th March 2017, Accepted 31st March 2017
DOI: 10.1039/c7fd00091j

The mechanical properties of naturally-derived matrices and biomaterials are thought to play an important role in directing cell adhesion, spreading, motility, proliferation and differentiation. However, recent reports have indicated that cells may respond to local nanoscale physical cues, rather than bulk mechanical properties. We had previously reported that primary keratinocytes and mesenchymal stem cells did not seem to respond to the bulk mechanical properties of poly(dimethyl siloxane) (PDMS) substrates. In this study, we examine the mechanical properties of weakly crosslinked PDMS substrates and observe a liquid-like behaviour, with complete stress relaxation. We then report the observation that HaCaT cells, an epidermal cell line, proliferate readily at the surface of uncrosslinked liquid PDMS, as well as on low viscosity (0.77 cSt) fluorinated oil. These results are surprising, considering current views in the field of mechanotransduction on the importance of bulk mechanical properties, but we find that strong mechanical interfaces, presumably resulting from protein assembly, are formed at liquid–liquid interfaces for which cell adhesion and proliferation are observed. Hence our results suggest that cells sense the nanoscale mechanical properties of liquid–liquid interfaces and that such physical cues are sufficient to sustain the proliferation of adherent cells.

Cell adhesion, spreading, proliferation and differentiation are regulated by a complex collection of factors, including growth factors, cytokines, extracellular matrix (ECM) composition and the presence of neighbouring cells.[1,2] In addition to biochemical cues, physical properties of this microenvironment have been shown to modulate cell phenotype.[3,4] For example, the nanoscale topography of biomaterials was found to regulate the adhesion, spreading and fate of stem cells.[5–7] Similarly, mechanical properties of the ECM have been shown to affect cell adhesion, spreading,[8,9] motility,[10,11] proliferation,[12,13] differentiation[14,15] and

[a]School of Engineering and Materials Science, Queen Mary, University of London, Mile End Road, London, E1 4NS, UK. E-mail: j.gautrot@qmul.ac.uk
[b]Institute of Bioengineering, Queen Mary, University of London, Mile End Road, London, E1 4NS, UK

stem cell self-renewal.[16] Changes in ECM mechanics are also often correlated with pathologies such as cancer, where they are thought to contribute to the development of an abnormal and potentially invasive phenotype.[17–19] Such interactions not only have important implications for our understanding of biological processes and signalling underlying the control of cell phenotype, but also for the design of biomaterials supporting the long term expansion of stem cells as well as acting as scaffolds for the regeneration of damaged or diseased tissues.

In many cases, stem cell phenotype has been correlated with the bulk mechanical properties of substrates designed to study mechanotransduction, such as ECM protein-functionalised poly(acrylamide) or poly(ethylene glycol) hydrogels.[14,15] In the vast majority of these works, cells are most sensitive to stiffness in the range of 0.1–100 kPa, a very broad range considering it refers to forces that are thought to be generated *via* one single main contractile machinery (*e.g.* mediated by integrin, focal adhesion and the actin cytoskeleton). However, we previously reported that primary keratinocytes and mesenchymal stem cells (MSCs) did not apparently respond to the mechanical properties of hydrophobic substrates (crosslinked poly(dimethyl siloxane), PDMS): their spreading and differentiation remained unchanged on PDMS matrices with a range of Young's moduli of 0.1 kPa to 2 MPa.[20] This suggested that cells respond to nanoscale mechanical properties of their microenvironment, at least at the single cell level, rather than bulk mechanical properties. More recently, a number of other reports also evidenced that direct correlations between bulk mechanical properties and cell phenotypes are perhaps misleading. For example, whilst MSCs displayed maximum spreading and differentiated into osteoblasts on stiff poly(acrylamide) hydrogels,[14] their spreading remained unchanged on soft quasi-2D nanofibrous mats compared to stiffer mats.[21] This was attributed to the increased remodelling of soft fibres and their clustering into bundles supporting the formation of focal adhesions. Similarly, whilst MSCs encapsulated into stiff 3D covalently crosslinked alginate hydrogels committed to osteogenic lineages,[22] they differentiated into osteoblasts when seeded within soft viscoelastic alginate gels or degradable hyaluronic acid matrices.[23] In such cases, it was again the ability of cells to remodel their softer environment and allow the clustering of ligands that was attributed to the seemingly contradicting phenotypes observed.

These results indicate that cells sense other physical cues, in addition to bulk mechanics, such as nanoscale ligand clustering and matrix remodelling, and that these cues can dominate. This concept has already been proposed by Sheetz and Ladoux, who proposed that mechanosensing arises from nanoscale (near 100 nm) strain sensing events.[24–26] Recently, del Rio and co-workers confirmed the occurrence of such a nanoscale strain sensing process by studying the spreading of cells on self-assembled monolayers displaying tethers of varying length. Single molecule force spectroscopy identified that a critical ligand displacement exists (falling between 40 and 200 nm) beyond which integrin receptors presumably disengage, and that prevents cell spreading.[27]

However, the absence of cell response we previously reported on PDMS substrates[20] was not fully explained and the precise origin of the divergence between bulk and nanoscale mechanical properties in such substrates remained unclear. Could such divergence arise from the rates at which cells probe their mechanical environment, compared to the rates at which mechanical characterisation is typically carried out? If not, what is the molecular origin of this divergence and can we quantify the corresponding nanoscale and bulk mechanical properties? Our work in this area led us to investigate further the mechanical properties of very weakly crosslinked Sylgard 184 PDMS substrates and to study the behaviour of cells spreading at the surface of non-crosslinked oils.

A. Materials and methods

1. Materials

Trimethylamine, triethoxy(octyl)silane, trichloro (1*H*,1*H*,2*H*,2*H*-perfluorooctyl) silane, 2,3,4,5,6-pentafluorobenzoyl chloride, and poly(dimethyl siloxane) (PDMS, trimethyl siloxane terminated) were purchased from Sigma-Aldrich. The fluorinated oil (Novec 7500) was from ACOTA. Ethanol was purchased from VWR chemicals. The DMEM cell culture medium, Hoechst 33 342, LIVE/DEAD assay kit and versene were from Thermofisher Scientific. SYLGARD 184 PDMS was from Ellsworth. Reagents were used as received unless specified otherwise. Double distilled water (ddH_2O) was obtained using a Purelab® Chorus system.

2. Formation of liquid–liquid interfaces for cell culture

24-well plates were plasma oxidized using a plasma coater (Diener, 100% intensity) for 10 min. 500 μL ethanol, 10 μL trimethylamine and 10 μL of the desired silane (triethoxy(octyl)silane to prepare interfaces with liquid PDMS or trichloro (1*H*,1*H*,2*H*,2*H*-perfluorooctyl) silane for the fluorinated oil) were added into each well. Ethanol was added in-between wells to slow down evaporation and Parafilm was used to seal the well plate cover. After incubating for 24 h, the wells were washed in a sterile environment with 70% ethanol (twice) and ddH_2O (three times). 500 μL fluorinated oil with fluorinated surfactant (2,3,4,5,6-pentafluorobenzoyl chloride) at final concentrations of 0.01, 0.005 and 0 mg mL^{-1} was added in the fluorophilic 24 well plate to form the bottom liquid layer. Cell culture medium (2 mL, containing foetal bovine serum) was directly added into each well to generate corresponding liquid–liquid interfaces and cells were added shortly after the formation of these interfaces.

3. HaCaT culture and seeding

Human keratinocyte HaCaT cells were cultured in DMEM containing 10% foetal bovine serum (FBS, Labtech), 1% L-glutamine (200 mM), and 1% penicillin–streptomycin (5000 U mL^{-1}). For proliferation assays, HaCaT cells were harvested with trypsin (0.25%) and versene solutions (0.2 g L^{-1} EDTA in phosphate buffered saline) in a ratio of 1/9, centrifuged, counted and resuspended in DMEM at the desired density before seeding onto substrates (conditioned as stated above, in a 24-well plate) at a density of 2000 cells per well (1000 cells per cm^2). Cells were left to adhere and proliferate in an incubator (37 °C and 5% CO_2) for different periods of time, prior to staining and imaging. For passaging, cells were reseeded in a T75 at a density of 250k cells per flask.

4. Immunofluorescence staining and antibodies

After washing once with PBS, the samples were fixed with 4% paraformaldehyde (Sigma-Aldrich) for 10 min and permeabilized with 0.2% Triton X-100 (Sigma-

Aldrich) for 5 min at room temperature. Following blocking for 1 h in 10% foetal bovine serum and 0.25% gelatine (from cold water fish skin, Sigma-Aldrich), substrates were incubated with primary antibodies for 1 h at room temperature, and with Alexa Fluor 488-conjugated secondary antibody (Thermofisher Scientific) for 1 h at room temperature (1 : 1000). When relevant, tetramethyl rhodamine isothiocyanate phalloidin (1 : 500, Sigma-Aldrich) was included in the blocking solution and DAPI (1 : 1000, Sigma-Aldrich) in the secondary antibody solution. Samples of cells adhering to oils were imaged directly without mounting.

5. Hoechst staining and LIVE/DEAD cell viability assay

Cell proliferation was assessed *via* Hoechst (nuclear) staining. Cells were incubated in DMEM containing 5 μL Hoechst 33 342 (1 mg mL^{-1} stock solution) for 30 min before imaging by epifluorescence microscopy (see below). Viability of HaCaT cells on fluorinated oil interfaces was quantified by LIVE/DEAD viability/cytotoxicity assay using a kit. In brief, HaCaT cells were incubated in DMEM with 2 μM calcein AM and 4 μM ethidium homodimer for 30 min, stained cells were imaged using a Leica DMI4000 fluorescence microscope (see below). The percentage of viable cells was calculated by counting the number of green (live) cells and dividing by the total number of cells.

6. Fluorescence microscopy and data analysis

Fluorescence microscopy images were acquired with a Leica DMI4000B fluorescence microscope (CTR4000 lamp; 63 × 1.25 NA oil lens; 10 × 0.3 NA lens; 2.5 × 0.07 NA lens; DFC300FX camera). To determine cell densities (per area), cell counting was carried out by thresholding and watershedding nuclear images using ImageJ. In the case of cell clumps, for which this protocol did not allow the isolation of individual nuclei, cells were counted manually. Statistical analysis was carried out using Origin 8 through one-way ANOVA with Tukey test for *post hoc* analysis. Significance was determined by $*P < 0.05$, $**P < 0.01$, $***P < 0.001$ and n.s., non-significant.

7. Rheology of cured PDMS samples

Sylgard 184 PDMS samples were prepared at a base/cross linker ratio of 100/1. The base and crosslinker were mixed and degassed until no bubbles were present under vacuum. The sample was then tested in a TA Discovery HR3 Rheometer. The PDMS was initially cured *in situ* at 60 °C for 3 h, using a 40 mm parallel plate, under oscillating time sweep at a frequency of 1 Hz and an oscillating amplitude of 1×10^{-4} rad. Once the sample had cured it was cooled to room temperature (25 °C) and left to settle for 300 s before a series of stress relaxation tests were performed. All relaxation tests were 300 s long with a strain rate of 1% per s and the samples were strained to 0.01%, 0.1%, 1% and 10%. After the stress relaxation tests, a frequency sweep was performed at a displacement of 10^{-4} rad from 0.1 to 100 Hz on a logarithmic scale getting 10 points per decade. This was followed by a stress sweep performed at 1 Hz, from 0.1 to 100 Pa on a logarithmic scale with 10 points per decade.

8. Interfacial rheology

Rheological measurements were carried out on a hybrid rheometer (DHR-3) from TA Instruments fitted with a double wall ring (DWR) geometry and a Delrin trough with a circular channel. The double wall ring used for this geometry had a radius of 34.5 mm and the thickness of the Platinum–Iridium wire was 1 mm. The diamond-shaped cross-section of the geometry's ring provided the capability to pin directly onto the interface between two liquids and monitor the interface properties without complicated sub-phase correction. 19 mL of the fluorinated oil pre-mixed with surfactant was placed in the Delrin trough and the ring was lowered, ensuring contact with the surface, *via* an axial force procedure. The measuring position was set 500 µm lower than the contact point of the ring with the oil-phase surface. Thereafter, 15 mL of the PBS buffer were carefully syringed on top of the oil phase. Time sweeps were performed at a constant frequency of 0.1 Hz and a temperature of 25 °C, with a displacement of 1.0×10^{-3} rad to follow the formation of the protein layers at the interface. The concentration of BSA used for all rheology experiments was 1 mg mL^{-1} (with respect to aqueous phase volume). Before and after each time sweep, frequency sweeps (with a constant displacement of 1.0×10^{-3} rad) were conducted to examine the frequency-dependant characteristics of the interface whilst amplitude sweeps (with constant frequencies of 0.1 Hz) were carried out to ensure that the chosen displacement was within the linear viscoelastic region.

B. Results and discussion

1. Stress relaxation properties of weakly crosslinked PDMS

Having previously established that human primary keratinocytes and mesenchymal stem cells adhere well even at the surface of ultra-soft PDMS (100 : 1 crosslinker ratio of Sylgard 184 PDMS), we further examined the mechanical properties of this specific type of materials. In particular, we aimed to determine whether the rate of deformation applied by cells could account for the behaviours observed, as weakly crosslinked viscoelastic materials are known to display strong frequency dependent profiles.[28] We crosslinked Sylgard 184 (base : crosslinker 100 : 1) *in situ* in a rheometer and carried out amplitude and frequency sweeps (Fig. 1A and B). We observed a strong frequency dependent profile (quasi-linear log–log relationship) of the shear moduli as a function of the oscillation frequency (Fig. 1B), with the storage modulus varying by nearly 3 orders of magnitude over the frequency range probed. Interestingly, the highest moduli, measured at a frequency of 100 Hz, was only slightly above 3.4 kPa (corresponding to a Young's modulus of 10 kPa). However, considering the 2 µm displacement (for the periphery of the 40 mm plate geometry) used for these measurements, this would correspond to a deformation rate of 200 µm per second.

Cells adhere to 2D materials and interfaces *via* the formation of focal adhesions thought to be the main elements responsible for the transmission of contractile forces to the extra-cellular matrix.[29–32] These adhesions, typically in the range of 0.5–10 µm^2, apply nN forces to the matrix (typically 1–20 nN),[31,33] mediated by actin polymerisation and actin flow at rates of 100–200 nm s^{-1}.[24] Such forces are thought to be oriented mainly in the plane of the substrate (although some out-of-plane contribution has been observed).[34] Considering

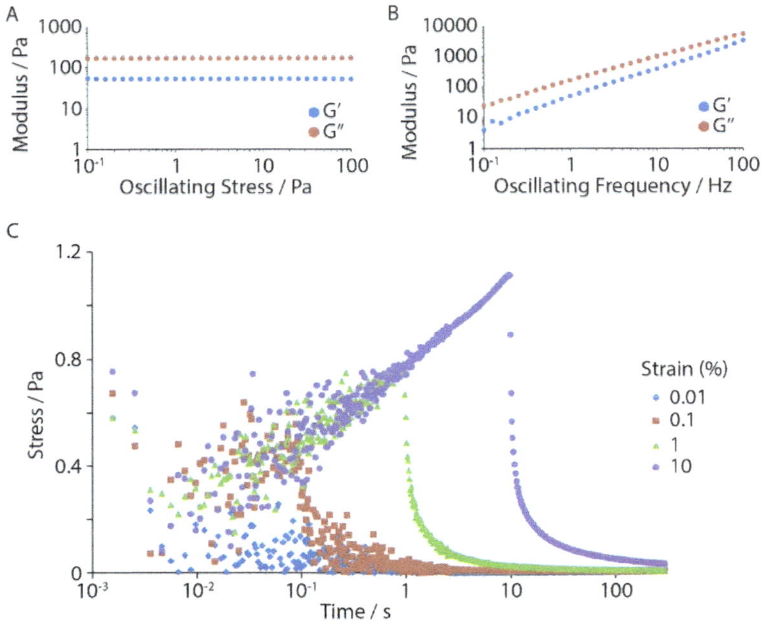

Fig. 1 Mechanical characterisation of Sylgard 184 crosslinked 100 : 1. Samples were cured *in situ* and characterised *via* stress sweep (A) (at 1 Hz), frequency sweep (B) (oscillating amplitude of 10^{-4} rad) and stress relaxation (C) (strain rate of 1% per s).

these deformations and their rates, cells should sense shear moduli corresponding to a range of 1–10 Pa for the weakly crosslinked 100 : 1 PDMS studied, significantly below the range typically thought to be sensed (for example by cells spreading on hydrogels).[14,24] Hence our results indicate that the apparent absence of response of cell spreading on weakly crosslinked Sylgard 184 PDMS cannot simply be explained by the strong frequency dependent mechanical behaviour of this material.

However, the modulus of 100 : 1 PDMS materials also displayed hallmarks of viscous materials, rather than a viscoelastic behaviour, with the loss modulus slightly higher than the storage modulus of the samples tested over the entire frequency range (Fig. 1B). In addition, stress relaxation experiments demonstrated full relaxation of these samples, indicating a classical viscous liquid behaviour (Fig. 1C). Cell adhesion to viscous fluids with short relaxation times (a few seconds) is surprising as it should not allow ligand and protein clustering for sufficient periods of time to sustain actomyosin-based cell contractility. Although recent reports gave evidence that increased stress relaxation in viscoelastic hydrogels results in ligand clustering and cell spreading,[23,35] such a process still requires residual stress to persist for prolonged times (min to h). This suggested that bulk mechanical properties are perhaps not representative of nanoscale interfacial mechanical properties of the PDMS substrates we previously tested and led us to ask whether any bulk crosslinking of PDMS substrates was actually required in order to sustain cell spreading, proliferation and to regulate stem cell fate decision.

2. HaCaT cells proliferate on uncrosslinked liquid Sylgard PDMS

In order to test to which extent crosslinking of PDMS was required to sustain cell proliferation, we simply omitted to crosslink the base Sylgard 184 PDMS mixture (no addition of crosslinker) and seeded HaCaT cells onto the resulting oil interfaces. The tissue culture polystyrene (TPS) surface and the PDMS oil–water surface were treated with a collagen solution (20 μg mL^{-1}) in order to sustain cell adhesion, but without any chemical treatment to promote coupling to the corresponding interfaces. After 3 and 7 days of culture, we stained cells with the nuclear staining agent Hoechst 33 342 and imaged the resulting samples *via* epifluorescence microscopy to determine any increase in cell density (Fig. 2). We clearly observed the formation of cell colonies on TPS as well as Sylgard 184 liquid PDMS, with comparable cell densities 7 days after culture.

Recently, two articles have reported the impact of matrix viscosity and stress relaxation on the clustering of ligand receptors and the ability of cells to spread at the surface of or within the corresponding materials.[23,35] It was proposed that the stress relaxation of viscoelastic materials enables the reorganisation of ligands, upon loading under cell-mediated tension, enabling the clustering of integrins and the generation of cell-mediated traction forces supporting the assembly of the actin cytoskeleton and cell spreading. To test further whether Sylgard PDMS could display sufficiently strong viscoelastic properties (despite full stress relaxation even in lightly cured samples, as discussed above), we cultured cells at the surface of liquid PDMS with a defined composition (fully methylated PDMS) and controlled viscosity (up to 5000 cSt, comparable to the viscosity of Sylgard 184 PDMS). However, no significant cell proliferation was observed (Fig. 2), although some small, non-spread colonies were observed in some of the experiments (but not all) carried out on the most viscous PDMS oil tested.

This highlighted that the proliferation of HaCaT cells observed on liquid Sylgard 184 PDMS is not resulting purely from its high viscosity. However, Sylgard PDMS is a relatively complex mixture of alkene functionalised PDMS and alkene–silane functionalised materials (full composition not disclosed) and our results suggested that perhaps some of the components present in Sylgard PDMS may alter the interfacial properties of the resulting materials or oils, when in contact with cell culture medium. In fact, the culture of cells (fibroblasts) on liquid substrates (PDMS and fluorinated oils with low viscosities) has been reported

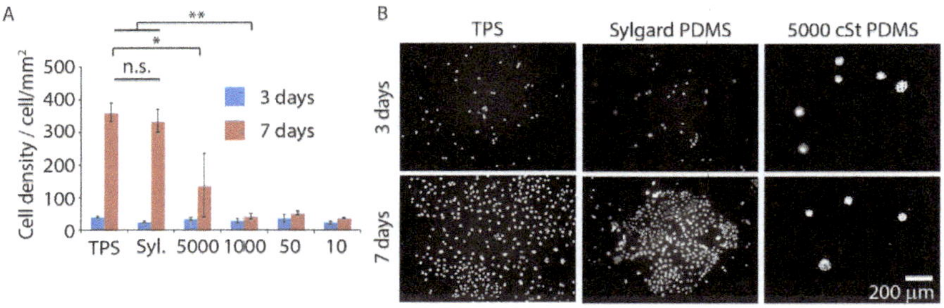

Fig. 2 (A) HaCaT cell proliferation on silicone-based liquids. TPS, tissue culture polystyrene; Syl, Sylgard 184; 5000/1000/50/10 describes the viscosity (cSt) of the PDMS oils used. (B) Corresponding nuclear staining images.

previously by Keese and Giaever in the 80s.[36,37] In such cases, the authors showed that fibroblasts proliferate at relatively high rates on liquid substrates. They also highlighted that the presence of surfactants in the oil phase was required to mediate such phenomenon.[38] Hence, our experiments and studies indicate that bulk mechanical properties are not necessarily required to mediate cell adhesion and proliferation and that liquids can support such processes.

3. HaCaT cells proliferate on a low viscosity fluorinated oil

In order to probe further the phenomenon of cell culture at liquid–liquid interfaces, we repeated the experiment of Keese and Giaever with a fluorinated carrier commonly used in the field of droplet microfluidics, Novec 7500 (viscosity 0.77 cSt). We used a surfactant that was reported to mediate the growth of fibroblasts, pentafluorobenzoyl chloride (PFBC). In the absence of surfactant, no significant cell culture was observed. However when PFBC was added at a concentration of 0.005 or 0.01 mg mL^{-1}, a significant amount of cell proliferation was observed, with relatively large and spread colonies apparent after 7 days of culture, although smaller than those formed on tissue culture polystyrene (TPS, Fig. 3). This is in good agreement with the reports of Keese and Giaever, which showed that fibroblasts proliferated at high rates, although direct comparisons with tissue culture polystyrene was not made. Our results indicate that such behaviour is also applicable to the case of epithelial cells such as HaCaTs.

We examined whether differences between proliferative behaviours observed on TPS and at oil interfaces were due to toxicity of some of the oil/surfactant mixtures used to generate liquid–liquid interfaces (Fig. 4). We observed that cell viability was high (comparable to TPS) when cells were cultured on oil interfaces generated in the presence of PFBC. When no surfactant was used, few rounded colonies formed and displayed increased toxicity, although viabilities remained close to 70%. Hence our viability assay suggests that rather than being due to the toxicity of components of the Novec 7500/PFBC mixtures, the decreased proliferation of HaCaT cells compared to TPS is due to other signals, potentially mechanical.

However, HaCaT cells being adherent cells requiring integrin binding for their proliferation (as for primary keratinocytes),[39,40] the fact that cell proliferation and the formation of spread colonies was possible on liquid substrates suggested that the formation of a mechanically strong interface was occurring. This would be likely mediated by the adsorption of proteins from the medium and serum (FBS). Protein adsorption is known to strongly impact cell adhesion and proliferation to solid substrates, including hydrophobic materials,[41,42] although it is typically not thought to impact the mechanical properties of the matrix that cells sense when exerting forces to sustain spreading. However, weak anchorage has been reported to result in cell detachment, despite initial adhesion, due to ECM proteins being peeled off the corresponding substrate.[43] Protein adsorption also occurs at liquid–liquid interfaces and is known to mediate important processes such as the stabilisation of emulsions or foams.[44] Hence we proposed that protein adsorption to the PDMS and fluorinated liquid–liquid interfaces studied, modulated by the presence of surfactants or additives, mediated cell adhesion and proliferation.

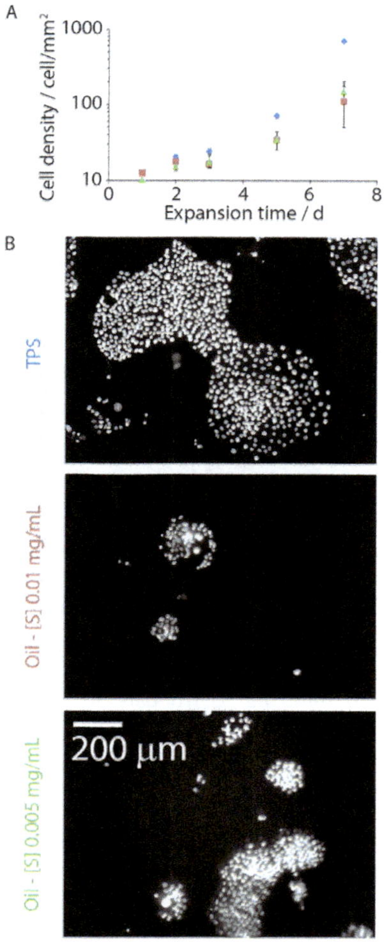

Fig. 3 (A) HaCaT cell proliferation profile on interfaces conditioned with cell culture medium (containing 10% FBS); blue diamonds, TPS; red square, Novec 7500 +0.01 mg mL^{-1} PFBC; green triangles, Novec 7500 +0.005 mg mL^{-1} PFBC). (B) Corresponding nuclear staining images (at day 7).

4. Cell proliferation on fluorinated oil is mediated by a strong mechanical interface

Keese and Giaever also came to the conclusion that protein films must form at liquid–liquid interfaces and be sufficiently mechanically strong to the sustain contractile forces generated.[45] Using a modified viscometer, they estimated the mechanical strength of the interfaces generated, in the range of 0.005 to 0.07 N m^{-1} and found that fibroblast proliferation reached a plateau at PFBC concentrations of above 10 μg mL^{-1}. In their measurements, they specifically investigated the mechanical properties of interfaces generated from albumin solutions, rather than formed in the presence of a cell culture medium. To quantify the mechanical properties of the interfaces we generated, we used interfacial shear rheology and a du Noüy ring with a diamond-shape cross section, limiting the

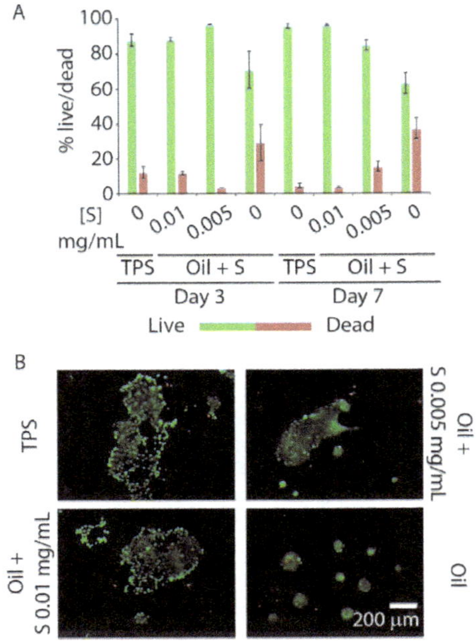

Fig. 4 (A) HaCaT cell viability (live, green; dead, red) at days 3 and 7 when cultured on interfaces conditioned with cell culture medium. TPS, tissue culture polystyrene; Novec 7500 + PFBC (0, 0.005 and 0.01 mg mL^{-1}). Error bars are s.e.m.; $n = 3$. (B) Corresponding images at day 7.

shear contribution from the fluid phases.[46] Upon injection of 10% FBS to the aqueous phase (PBS) of an interface formed with the fluorinated oil Novec 7500, the storage shear modulus increased by nearly two orders of magnitude (Fig. 5). This increase was relatively fast and complete within 30 min. Similarly, when this experiment was repeated in the presence of 10 μg mL^{-1} PFBC, a fast increase of

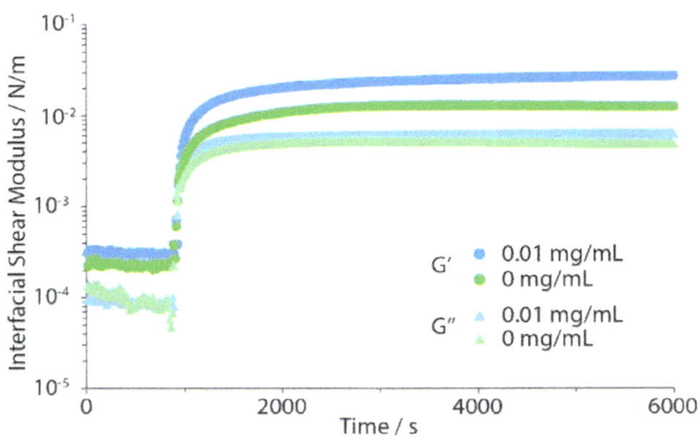

Fig. 5 Time sweep interfacial rheology experiments at interfaces between Novec 7500 oil containing PFBC (0 and 0.01 mg mL^{-1}) and PBS containing 10% FBS. 0.1 Hz, oscillating amplitude of 10^{-3} rad.

the interfacial storage modulus was observed, with comparable rate, but reaching slightly higher plateau values, above 10^{-2} N m^{-1}.

Further investigation of the frequency dependency of the interfacial storage and loss moduli after full equilibration of the protein sheets assembled was carried out (Fig. 6). We measured storage moduli significantly higher in the presence of 10 μg mL^{-1} PFBC compared to the naked oil (43.6 ± 7.5 mN m^{-1} and 18.9 ± 6.8 mN m^{-1}, respectively). This is in good agreement with the measurements previously made by Keese and Giaever, who had measured a modulus of *ca.* 20 mN m^{-1} for an albumin functionalised interface in the presence of 10 μg mL^{-1} PFBC,[45] a concentration that was found to support fibroblast growth and spreading well.[38] However, despite the general agreement of the scale of the interfacial moduli measured by Giaever *et al.* in our study, we found that the interfacial modulus required to sustain cell spreading had to be 2 fold higher (43 mN m^{-1} *vs.* 20 mN m^{-1}). This could be a result of the greater sensitivity of our experimental set up (weaker contribution of the shear properties of the liquid carriers), and the methodology used (paddle of the viscometer applying a mono-directional deformation in the case of the work of Giaever *et al.* and oscillating strain in our experiment). These values are also in good agreement with those measured for globular proteins such as lysozyme and albumin associating at oil-water interfaces.[47,48] Hence, we propose that proteins contained in the serum of cell culture medium (such as albumin, which is typically present at mg mL^{-1} concentrations in media supplemented with 10% serum) form a stiff sheet at the interface of the two liquids, enabling cell-mediated contractile forces to be sustained.

However, the striking difference observed between cell behaviour on interfaces generated in the absence of PFBC or in the presence of 10 μg mL^{-1} of this molecule are surprising as they only arise from a 2-fold increase in the interfacial shear modulus. Interestingly, we observed that the frequency-dependent profile of the two types of interfaces was significantly different (Fig. 6). A weaker variation over the range of frequencies probed for interfaces generated was observed in the presence of PFBC. At low frequencies (close to 10^{-2} Hz), the interfacial storage and loss shear moduli measured in the absence of surfactant were found to be

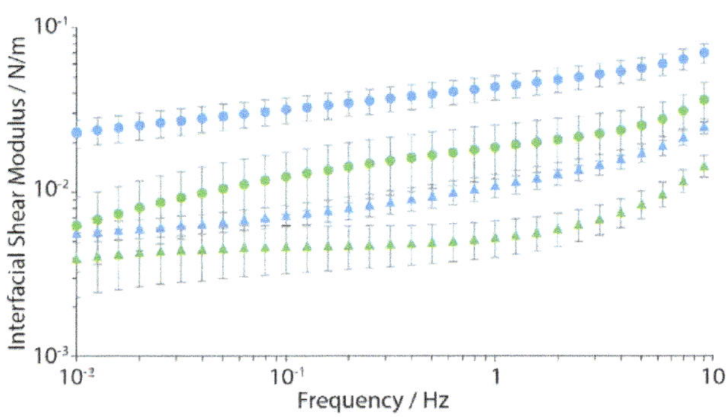

Fig. 6 Oscillatory shear response of interfaces generated from PBS solutions containing 10% 5000 oil containing no PFBC (green) and 0.01 mg mL^{-1} PFBC (blue). Circles, G'; triangles, G''. Oscillating amplitude of 10^{-3} rad.

very close (only 2-fold higher in the case of the storage modulus), indicating significant changes in the viscoelastic behaviour of these interfaces. Hence it could be proposed that changes in viscoelastic properties regulate cell phenotype at liquid–liquid interfaces, similarly to what was observed for cells interacting at the surface of, or within hydrogels.[23,35] In addition, changes in the type of biomolecules adsorbed at liquid–liquid interfaces, depending on the nature and concentration of surfactants present at corresponding interfaces, are likely to impact cell adhesion and spreading, further contributing to alter cell response at such interfaces.

C. Conclusions

Overall, our results imply that cell adhesion, proliferation and differentiation at the surface of hydrophobic materials such as PDMS may be regulated by nanoscale mechanical properties of protein interfaces assembled at the surface of such materials, rather than their bulk mechanical properties. We find that the culture of epithelial cells such as HaCaT cells is possible at the surface of very low viscosity liquids, providing a mechanically strong interface is assembled. However several questions remain with such phenomenon. The precise nature of the protein aggregates generated and their interaction with surfactants has not been investigated. The impact of such chemistry on the viscoelastic properties of such interfacial layers could be important and key to the design of a broader range of interfaces supporting the culture of more demanding adherent cell types such as stem cells. In addition, the precise mechanisms *via* which cell adhesion and phenotype are mediated and regulated at liquid–liquid interfaces are not known and have never been studied: are they integrin mediated, as in the vast majority of adherent cells spreading on solid 2D surfaces? What are the precise mechanical, morphological and biochemical parameters controlling such interactions? Is stem cell phenotype regulated by such liquid–liquid interfaces? The implications of these questions are high for the design of novel stem cell technologies, for example for the design of emulsion-based 3D bioreactors allowing stem cell expansion or simple delivery.

Acknowledgements

We thank Dr Britta Trappmann for insightful discussions and for pointing out Giaever's work. K. D. Q. N. thanks FormFormForm Ltd. for an industrial studentship. W. M. thanks the Engineering and Physical Sciences Research Council (EP/L505602/1) for a case studentship.

References

1 M. P. Lutolf and J. A. Hubbell, Synthetic biomaterials as instructive extracellular microenvironments for morphogenesis in tissue engineering, *Nat. Biotechnol.*, 2005, **23**, 47–55.

2 D. E. Discher, D. J. Mooney and P. W. Zandstra, Growth factors, matrices, and forces combine and control stem cells, *Science*, 2009, **324**, 1673–1677.

3 F. Guilak, *et al.*, Control of stem cell fate by physical interactions with the extracellular matrix, *Cell Stem Cell*, 2009, **5**, 17–26.

4 S. Di Cio and J. E. Gautrot, Cell sensing of physical properties at the nanoscale: mechanisms and control of cell adhesion and phenotype, *Acta Biomater.*, 2016, **30**, 26–48.
5 M. J. Dalby, *et al.*, The control of human mesenchymal cell differentiation using nanoscale symmetry and disorder, *Nat. Mater.*, 2007, **6**, 997–1003.
6 R. J. McMurray, *et al.*, Nanoscale surfaces for the long term maintenance of mesenchymal stem cell phenotype and multipotency, *Nat. Mater.*, 2011, **10**, 637–644.
7 J. E. Gautrot, *et al.*, The nanoscale geometrical maturation of focal adhesions controls stem cell differentiation and mechanotransduction, *Nano Lett.*, 2014, **14**, 3945–3952, DOI: 10.1021/nl501248y.
8 R. J. Pelham and Y.-L. Wang, Cell locomotion and focal adhesions are regulated by substrate flexibility, *Proc. Natl. Acad. Sci. U. S. A.*, 1997, **94**, 13661–13665.
9 T. Yeung, *et al.*, Effect of substrate stiffness on cell morphology, cytoskeletal structure, and adhesion, *Cell Motil. Cytoskeleton*, 2005, **60**, 24–34.
10 S. R. Peyton and A. J. Putnam, Extracellular matrix rigidity governs smooth muscle cell motility in a biphasic fashion, *J. Cell. Physiol.*, 2005, **204**, 198–209.
11 T. Tzvetkova-Chevolleau, *et al.*, The motility of normal and cancer cells in response to the combined influence of the substrate rigidity and anisotropic microstructure, *Biomaterials*, 2008, **29**, 1541–1551.
12 T. A. Ulrich, E. M. de Juan Pardo and S. Kumar, The mechanical rigidity of the extracellular matrix regulates the structure, motility, and proliferation of glioma cells, *Cancer Res.*, 2009, **69**, 4167–4174.
13 K. Bott, *et al.*, The effect of matrix characteristics on fibroblasts proliferation in 3D gels, *Biomaterials*, 2010, **31**, 8454–8464.
14 A. J. Engler, S. Sen, H. L. Sweeney and D. E. Discher, Matrix elasticity directs stem cell lineage specification, *Cell*, 2006, **126**, 677–689.
15 K. Saha, *et al.*, Substrate modulus directs neural stem cell behavior, *Biophys. J.*, 2008, **95**, 4426–4438.
16 P. M. Gilbert, *et al.*, Substrate elasticity regulates skeletal muscle stem cell self-renewal in culture, *Science*, 2010, **329**, 1078–1081.
17 K. R. Levental, *et al.*, Matrix crosslinking forces tumor progression by enhancing integrin signalling, *Cell*, 2009, **139**, 891–906.
18 R. W. Tilghman, *et al.*, Matrix rigidity regulates cancer cell growth and cellular phenotype, *PLoS One*, 2010, **5**, e12905.
19 S. C. Wei, *et al.*, Matrix stiffness drives epithelial–mesenchymal transition and tumour metastasis through a TWIST1-G3BP2 mechanotransduction pathway, *Nat. Cell Biol.*, 2015, **17**, 678–688.
20 B. Trappmann, *et al.*, Extracellular matrix tethering regulates stem cell fate, *Nat. Mater.*, 2012, **11**, 642–649.
21 B. M. Baker, *et al.*, Cell-mediated fibre recruitment drives extracellular matrix mechanosensing in engineered fibrillar microenvironments, *Nat. Mater.*, 2015, **14**, 1262–1268.
22 N. Huebsch, *et al.*, Harnessing traction-mediated manipulation of the cell/matrix interface to control stem-cell fate, *Nat. Mater.*, 2010, **9**, 518–526.
23 O. Chaudhuri, *et al.*, Hydrogels with tunable stress relaxation regulate stem cell fate and activity, *Nat. Mater.*, 2016, **15**, 326–334.

24 S. W. Moore, P. Roca-Cusachs and M. P. Sheetz, Stretchy proteins on stretchy substrates: the important elements of integrin-mediated rigidity sensing, *Cell*, 2010, **19**, 194–206.

25 G. Jiang, A. H. Huang, Y. Cai, M. Tanase and M. P. Sheetz, Rigidity sensing at the leading edge through αvβ3 integrins and RPTPα, *Biophys. J.*, 2006, **90**, 1804–1809.

26 A. Saez, A. Buguin, P. Silberzan and B. Ladoux, Is the mechanical activity of epithelial cells controlled by deformations or forces?, *Biophys. J.*, 2005, **89**, L52–L54.

27 S. J. Attwood, *et al.*, Adhesive ligand tether length affects the size and length of focal adhesions and influences cell spreading and attachment, *Sci. Rep.*, 2016, **6**, 34334.

28 J. C. Scanlan and H. H. Winter, Composition dependence of the viscoelasticity of end-linked poly(dimethylsiloxane) at the gel point, *Macromolecules*, 1991, **24**, 47–54.

29 C. G. Galbraith, K. M. Yamada and M. P. Sheetz, The relationship between force and focal complex development, *J. Cell Biol.*, 2002, **159**, 695–705.

30 P. Roca-Cusachs, N. C. Gauthier, A. del Rio and M. P. Sheetz, Clustering of α5β1 integrins determines adhesion strength whereas αvβ3 and talin enable mechanotransduction, *Proc. Natl. Acad. Sci. U. S. A.*, 2009, **106**, 16245–16250.

31 N. Q. Balaban, *et al.*, Force and focal adhesion assembly: a close relationship studied using elastic micropatterned substrates, *Nat. Cell Biol.*, 2001, **3**, 466–472.

32 B. Geiger, J. P. Spatz and A. D. Bershadsky, Environmental sensing through focal adhesions, *Nat. Rev. Mol. Cell Biol.*, 2009, **10**, 21–33.

33 H. van Hoorn, *et al.*, The nanoscale architecture of force-bearing focal adhesions, *Nano Lett.*, 2014, **14**, 4257–4262.

34 W. R. Legant, *et al.*, Multidimensional traction force microscopy reveals out-of-plane rotational moments about focal adhesions, *Proc. Natl. Acad. Sci. U. S. A.*, 2013, **110**, 881–886.

35 O. Chaudhuri, *et al.*, Substrate stress relaxation regulates cell spreading, *Nat. Commun.*, 2015, **6**, 6364.

36 I. Giaever and C. R. Keese, Behavior of cells at fluid interfaces, *Proc. Natl. Acad. Sci. U. S. A.*, 1983, **80**, 219–222.

37 C. R. Keese and I. Giaever, Cell growth on liquid microcarriers, *Science*, 1983, **219**, 1448–1449.

38 C. R. Keese and I. Giaever, Cell growth on liquid interfaces: role of surface active compounds, *Proc. Natl. Acad. Sci. U. S. A.*, 1983, **80**, 5622–5626.

39 S. Rahavan, C. Bauer, G. Mundschau, Q. Li and E. Fuchs, Conditional ablation of β1 integrin in skin: severe defects in epidermal proliferation, basement membrane formation and hair follicle invagination, *J. Cell Biol.*, 2000, **150**, 1149–1160.

40 P. Boukamp, *et al.*, Normal keratinization in a spontaneously immortalized aneuploid human keratinocyte cell line, *J. Cell Biol.*, 1988, **106**, 761–771.

41 F. Bathawab, *et al.*, Lateral chain length in polyalkyl acrylates determines the mobility of fibronectin at the cell/material interface, *Langmuir*, 2016, **32**, 800–809.

42 P. Rico, *et al.*, Substrate-induced assembly of fibronectin into networks: influence of surface chemistry and effect on osteoblast adhesion, *Tissue Eng., Part A*, 2009, **15**, 3271–3281.

43 N. B. Guerra, *et al.*, Subtle variations in polymer chemistry modulate substrate stiffness and fibronectin activity, *Soft Matter*, 2010, **6**, 4748–4755.

44 M. A. Bos and T. van Vliet, Interfacial rheological properties of adsorbed protein layers and surfactants: a review, *Adv. Colloid Interface Sci.*, 2001, **91**, 437–471.

45 C. R. Keese and I. Giaever, Substrate mechanics and cell spreading, *Exp. Cell Res.*, 1991, **195**, 528–532.

46 G. G. Fuller and J. Vermant, Complex fluid-fluid interfaces: rheology and structure, *Annu. Rev. Chem. Biomol. Eng.*, 2012, **3**, 519–543.

47 E. M. Freer, K. S. Yim, G. G. Fuller and C. J. Radke, Shear and dilatational relaxation mechanisms of globular and flexible proteins at the hexadecane/water interface, *Langmuir*, 2004, **20**, 10159–10167.

48 S. G. Baldursdottir, M. S. Fullerton, S. H. Nielsen and L. Jorgensen, Adsorption of proteins at the oil/water interface-Observation of protein adsorption by interfacial shear stress measurements, *Colloids Surf., B*, 2010, **79**, 41–46.

Faraday Discussions

PAPER

Surface-bound quadruple H-bonded dimers: formation and exchange kinetics†

Digvijay Gahtory, ‡[a] Rickdeb Sen,‡[a] Maarten M. J. Smulders *[a] and Han Zuilhof *[ab]

Received 16th February 2017, Accepted 11th May 2017

DOI: 10.1039/c7fd00068e

While the mechanistic details of the dimerization of the self-complementary 2-ureido-4(1*H*)-pyrimidinone (UPy) motif are well studied in solution, no such investigation is available on a surface. Here we report an extensive study of hydrogen binding kinetics for quadruply H-bonded UPy arrays on aluminum surfaces and explore the ON/OFF capability of such arrays under externally controllable conditions. Also, we investigate the dynamic nature of this system whereby the interfacially H-bonded UPy is displaced by another UPy derivative in solution, and reveal the kinetics of the exchange process.

Introduction

Supramolecular chemistry, often referred to as 'chemistry beyond the molecule', utilizes reversible, non-covalent interactions for the formation of large molecular assemblies.[1,2] While the advance of supramolecular chemistry has spurred great development in various research areas, such as functional supramolecular polymers,[3] supramolecular catalysts[4] and supramolecular DNA assemblies,[5] the potential of such stable, yet dynamic interactions for surface applications is still underexplored.

Current surface attachment strategies largely rely on strong covalent bonding of a molecule of interest to self-assembled monolayers (SAMs) on the surface.[6] In this regard, several reactions have been utilized for various applications, *e.g.* preparation of protein-repellent surfaces,[7] attachment of biomolecules such as DNA,[8] preparation of carbohydrate and protein arrays[9] and nanoparticle functionalization.[10] This approach, although providing a robust attachment strategy, precludes a reversible attachment strategy that offers control with external stimuli such as solvents, light, temperature, pH and chemical influences. While supramolecular surface modification strategies allow easy manipulation of the

[a]*Laboratory of Organic Chemistry, Wageningen University and Research, Stippeneng 4, 6708 WE Wageningen, The Netherlands. E-mail: han.zuilhof@wur.nl; maarten.smulders@wur.nl*

[b]*Department of Chemical and Materials Engineering, King Abdulaziz University, Jeddah, Saudi Arabia*

† Electronic supplementary information (ESI) available. See DOI: 10.1039/c7fd00068e

‡ These authors contributed equally.

interfacial chemical composition, architecture, topology, *etc.*, to yield responsive and reusable surfaces, so far only a limited number of reports dealing with dynamic covalent[11,12] or non-covalent interactions[13,14] has been reported. A major advantage of any dynamic interfacial interaction is its ability to confer a stimulus-dependent ON/OFF handle for the surface. The utility of such rewritable substrates is beneficial for the development of (regeneratable) bio-sensors, electro-(de)activation of surfaces, nanofabrication platforms, *etc.*[15,16]

Non-covalent interactions such as hydrogen bonding and π–π interactions offer a distinct advantage in this regard as they provide an easily addressable handle for quick attachment and removal of the bound moiety, which thus enables a surface-bound ON/OFF switch. As a virtue of their strength, specificity and dynamic nature, arrays of multiple hydrogen bonds have garnered significant interest for the assembly of complex supramolecular structures with useful properties in a predictable and reliable fashion.[17–19] Despite a ubiquitous presence of such motifs in supramolecular applications, their role in surface science is still underexamined. Reports by Ward *et al.* and Rotello *et al.* have demonstrated the utility of simple two- and three-fold hydrogen bonding arrays such as adenine–thymine[20] and thymine–diamidopyridine.[21] Although explored for different applications such as polymer attachment[22,23] and attachment of electroactive groups,[15] these motifs are quite labile and several of these motifs need to be applied in concert.[24] In contrast, increasing the number of available hydrogen bonding groups in a motif such as for quadruple[25] or hexafold[26] bonding arrays should produce much more stable dimeric structures.

An interesting candidate among such hydrogen bonding motifs is 2-ureido-4(1*H*)-pyrimidinone (UPy), a self-complementary quadruple hydrogen bonding array (Scheme 1).[27,28] Easily synthesized and capable of strong self-dimerization[29] ($K_{\mathrm{dim}} = 6 \times 10^8\ \mathrm{M}^{-1}$ in toluene) through a self-complementary donor–donor–acceptor–acceptor H-bonding motif, this system has been widely utilized for various applications such as polymeric materials with self-healing properties[30–32] and responsive nanoparticles.[33,34]

Despite numerous reports on aforementioned H-bonding motifs, the exploration of the UPy motif at a surface has received surprisingly little attention. As shown by single-molecule force spectroscopy (SMFS), the UPy motif forms a reversible, yet strongly bonded, UPy dimer on the surface.[35,36] Interestingly, the dimer between surface-bound UPy and soluble UPy can also be displaced, as

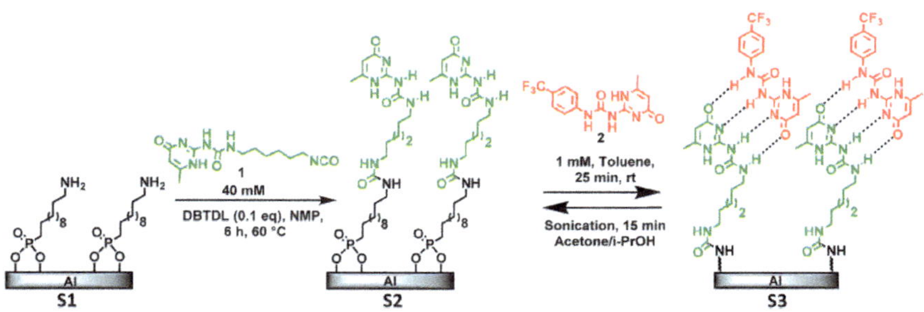

Scheme 1 Preparation of Link–UPy and Link–UPy···UPy–CF_3 surfaces, S2 and S3, respectively.

shown by Tolbert *et al.* for glass substrates.[37] However, their study was significantly hampered by non-specific interactions of UPy molecules with the substrate itself, in which no significant preference was observed for the UPy dimer during the exchange process.

Notwithstanding this early work, the potential for reversibility and exchangeability of the UPy motif to yield dynamic surfaces with properties that can be easily manipulated has not been fully investigated. Moreover, while the mechanistic details of UPy dimerization and other complementary hydrogen bonding counterparts have been studied comprehensively in solution,[38–40] kinetics inquiries on a surface have not been conducted to the best of our knowledge. We believe that insight into the extent and kinetics of surface binding and exchange by hydrogen bonding can facilitate the design and development of functional surfaces that rely on dynamic surface modification (*e.g.* regeneratable sensor surfaces), in an analogous manner to how insight into solution-based binding has contributed to the development of tunable responsive polymers.[41] In this study, we examine the quadruply H-bonded UPy dimer formation on aluminum surfaces and exploit its reversible attachment by ON/OFF and exchange experiments. For the characterization we primarily use X-ray photoelectron spectroscopy (XPS), grazing angle total reflection-Fourier transform infrared spectroscopy (GATR-FTIR), and direct analysis in real time-high resolution mass spectrometry (DART-HRMS).[42,43] Recently, we have demonstrated the potential of DART-HRMS for quantitative surface analysis by elucidating interfacial kinetics and activation parameters of strain-promoted click reactions.[44] In addition, using a combination of aforementioned techniques, we further explore the reconfigurable and dynamic nature of the UPy H-bonding system to obtain rewritable and exchangeable dynamic surfaces.

Results and discussion

Monolayer formation and Link–UPy attachment

Our experimental design involved the attachment of a ureido-[2-(4-pyrimidone)] functionalized with 1-hexylisocyanate (**UPy–Link**, **1**) onto amine-terminated aluminum (Al) substrates using isocyanate coupling (Scheme 1). The choice of substrate was based on previous experience with interfacial kinetics determination: oxide-coated Al samples are easily modified with highly stable phosphonate-based monolayers and offer a distinct advantage during DART measurements at higher temperatures (see below).[45] Amine-terminated monolayers **S1** were prepared by linking ω-NH_2-functionalized alkyl phosphonic acids (PAs) onto hydroxylated aluminum oxide surfaces. The C/P ratio (13.0 $\pm$ 0.5) and N/P ratio (1.0 $\pm$ 0.3) in XPS wide scans (Fig. 1a and ESI, Fig. S1†) confirmed the formation of C_{12}–NH_2 monolayers. GATR-FTIR analysis of the monolayers showed peaks at 2925 cm^{-1} and 2854 cm^{-1}, corresponding to antisymmetric and symmetric methylene stretching, respectively (ESI, Fig. S2†). Simulation studies with varying degrees of monolayer attachment using Material Studio (see ESI,† Section 7) revealed that ideally about half the surface sites are bound to a phosphonate moiety (in terms of obtaining the lowest packing energy per molecule).[46]

S1 substrates were reacted with a 40 mM solution of **1** in NMP for 6 h at 60 °C (Scheme 1). The successful coupling of **UPy–Link** became evident from the increase in the XPS N1s signal (Fig. 1b). This signal, after sonication, only results

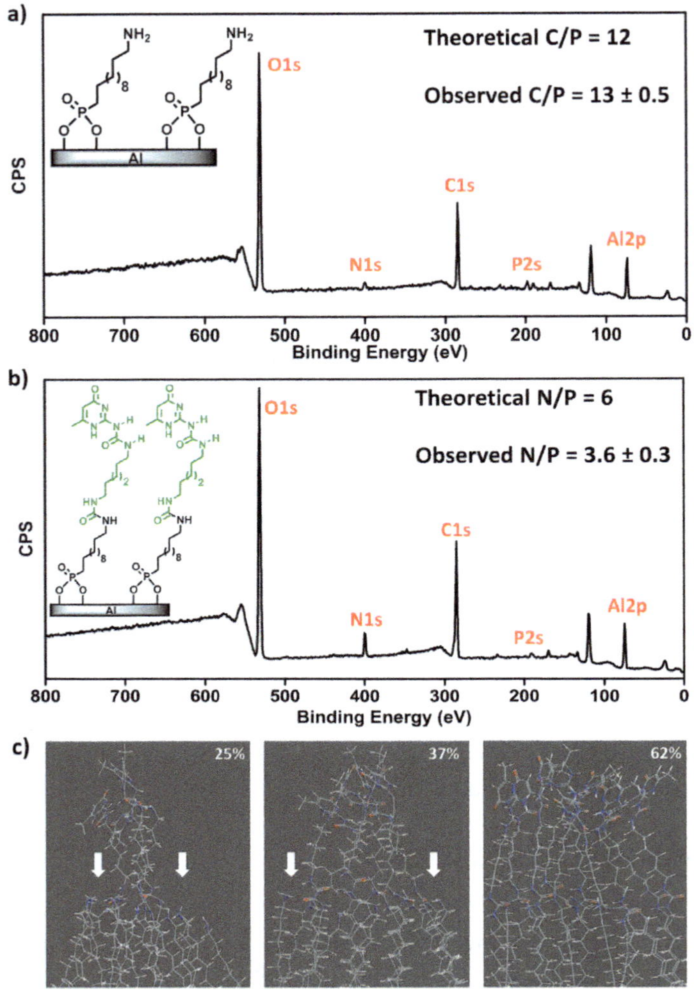

Fig. 1 (a and b) XPS wide range spectra of surfaces **S1** (a) and **S2** (b). (c) Front view of the simulation cell after **UPy–Hex** attachment on the aluminum surface with increasing coverage (25, 37 and 62%). After approx. 60% coverage with **UPy–Hex**, the amine groups (shown by white arrows) are completely buried and inaccessible for incoming 1-hexylisocyanate UPy molecules.

from covalently bound UPy: GATR-FTIR spectra of **S3** (see ESI, Fig. S2†) showed no peak around 2270 cm^{-1}, ascribable to isocyanate groups, indicating that after sonication no **Link–UPy···UPy–Link** dimer is present on the surface.

Further confirmation of UPy attachment was obtained by the appearance of a signal corresponding to C=O in the C1s narrow scan (Fig. S5†). While the presence of UPy was thus confirmed, it was also observed that the conversion tended to an asymptotic limit of 52 ± 5%, as determined from the N/P ratio in the XPS spectra (Fig. 1b and ESI, Fig. S3 and S4†). Attempts to enhance the surface yield by increased reaction times or temperatures rendered no improvement. A less than 100% conversion can be rationalized by the significantly larger footprint of a UPy moiety compared to an amine. To further rationalize this asymptotic limit, the availability of surface amine groups was compared at different yields of the surface-bound isocyanate coupling by theoretical 3D models (Fig. 1c, 2 and ESI, Section 7 and ESI Movie 1†). We found that approaching 60% coverage, the surface becomes increasingly and highly crowded. Hereby, the remaining amine groups are rendered inaccessible for any further functionalization with the incoming **UPy–Link** molecules. Although evidently only qualitatively, this confirms that for steric reasons the reaction of **UPy–Link** to the amine-terminated surface cannot reach full conversion, but is limited to around 50%.

Indicative 3D structures were obtained by constructing surfaces **S2** with a surface coverage of 60% in a molecular mechanics simulation program using the PCFF force field.[46] First, such a surface (with 320 UPy moieties on 512 chains) was constructed with repeating boundary conditions to eliminate edge effects. Second, a series of randomized structures was obtained by running molecular dynamics structures at temperatures of 500 °C or higher, which we subsequently optimized with geometry optimization routines. Typical structures of **S2** are presented in Fig. 2 and in the ESI (Section 7.8)†, and such structures invariably show the presence of surface-bound fourfold H-bonded UPy dimers, amidst a range of not ideally H-bonded UPy moieties. For the latter, the optimal fourfold H-bonding is apparently hampered by the increased steric hindrance that this would cause. In summary, surface-bound UPy is partially internally H-bonded which will influence the equilibrium with UPy moieties in solution.

Interfacial UPy dimer H-bonding and kinetics determination

In order to study the H-bonding phenomena quantitatively by XPS and DART-HRMS, we required a UPy tag that could fulfil specific roles: (1) when surface-bound it should yield a characteristic signal in XPS, and (2) it should be sufficiently volatile and ionizable to be detected in DART-HRMS. 4-(Trifluoromethyl) phenyl UPy (**UPy–CF$_3$, 2**) was found to be a good candidate that meets both criteria (Fig. 3a and b). For kinetics determination, **UPy–Link**-terminated samples (**S2**) were immersed in a 1 mM solution of **2** in toluene at 25 °C for different time intervals, followed by sonication for 5 min in toluene. The sonication step was necessary to ensure that all **UPy–CF$_3$** binding to the surface was specifically bound through strong (likely fourfold) hydrogen bonding, and that all weakly and non-

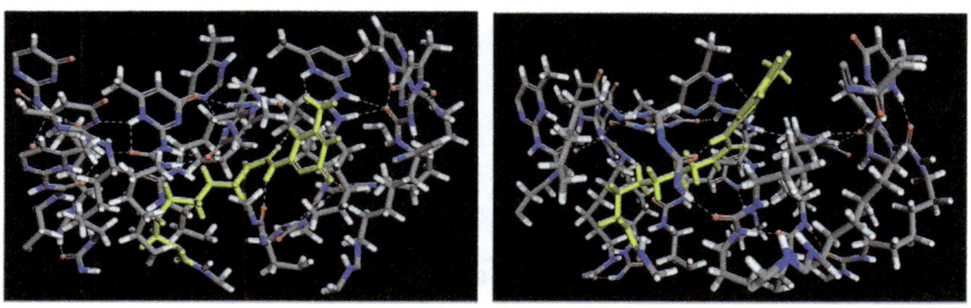

Fig. 2 Different views of **UPy–Link** on surfaces obtained after molecular dynamics simulations (one of the **UPy–Link** molecules has been highlighted in yellow). The intermolecular H-bonding interactions with adjacent UPy molecules can be clearly seen and possibly compete with the incoming soluble **UPy–CF$_3$** molecules.

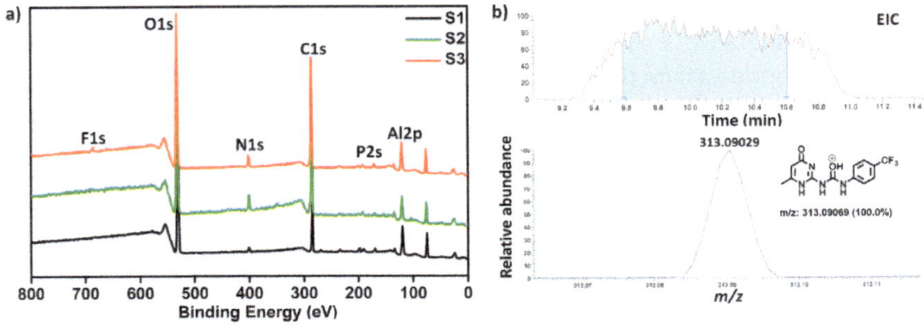

Fig. 3 Analysis of surfaces **S1**, **S2** and **S3**: (a) XPS wide scans and (b) DART ion intensity EIC (top panel) for **Link–UPy···UPy–CF$_3$** dimers and peak of ion of interest (bottom panel).

specifically physisorbed **UPy–CF$_3$** was removed. The extent of surface immobilization was subsequently determined by XPS and DART-HRMS. Monitoring the F/P ratio in XPS *versus* time (ESI, Fig. S9†) and the DART-HRMS intensity of protonated **2** (Fig. 4a) both showed that the extent of H-bonding converged to a limiting value around 20 min. Careful analysis of the data revealed (pseudo-)first-order reaction kinetics, with a first-order rate constant for UPy binding to the surface, at 25 °C in toluene at [UPy] = 1 mM, of $(22 \pm 1) \times 10^{-4}\ s^{-1}$ (ESI, Section 4.5†). The limiting F/P ratio corresponded to about 40% coverage of all surface-bound UPy moieties with **2** (Fig. 4b and ESI, Fig. S10†).

Overall, this means that the monolayer covers 50% of the surface sites with amines; approximately 52% of those amine groups are reacted with **1**; and 35% of the resulting surface-bound **UPy–Link** form UPy···UPy dimers. This yields a maximum overall dimer coverage of ~9%. This maximum value hinges on the formation of H-bonded dimers which induce additional steric constraints within the monolayer. Molecular dynamics indeed revealed that crowding at the surface by surface-bound UPy can hamper further binding (Fig. 2 and ESI, Section 7; see also ESI Movie 2†). As a result, when the steric hindrance becomes too large, binding of a surface-bound UPy with a soluble UPy will be outcompeted by the

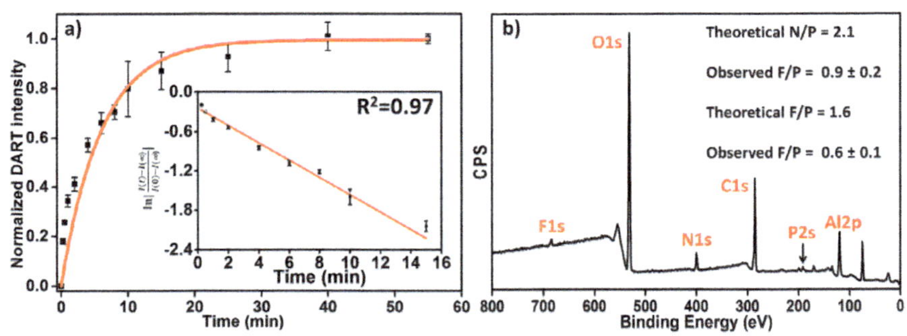

Fig. 4 (a) Normalized EIC intensity of protonated **2** (positive mode) for the dimerization process at 25 °C (each data point is from 6-fold repetition). Inset: plot of $\ln|(I_t - I_\infty)/(I_0 - I_\infty)|$ *versus* time used to obtain the first-order rate constant. (b) XPS wide scan of **S3**, showing a ~40% dimerization by **UPy–CF$_3$**.

competitive equilibrium between the dimers in solution. In addition, GATR-FTIR spectra of these surfaces (Fig. 3b and ESI, Fig. S2†) displayed distinct peaks at 1582 cm^{-1} and 1653 cm^{-1}, attributable to amide I and II bands for dimeric UPy as previously reported (at ~1580 cm^{-1} and ~1660 cm^{-1} respectively).[47]

Control H-bonding experiments with NH_2-terminated surfaces (**S1**) under similar conditions were also performed, and these showed the absence of any **UPy–CF$_3$** bonded non-specifically to the amino monolayers. This combination of XPS and DART analysis of control samples (ESI, Fig. S12 and S13†) thus confirmed that immobilization of **2** onto **S2** surfaces occurred through specific, likely four-fold, UPy···UPy hydrogen bonding. The coverage of all surface-bound UPy moieties with **2** was about 38% (Fig. 4b and ESI, Fig. S10†).

Surface reusability and exchange kinetics

We observed that this interfacial dimer (ON state) could be disrupted upon sonication in acetone or i-PrOH for 15 min, yielding the UPy monomer-terminated surface (OFF state). The OFF surface thus obtained could be changed back into the ON state simply by stirring it for 20 min in a solution of **2** (in toluene). The ON/OFF process was easily monitored by following the XPS F1s signal (Fig. 5 and ESI, Fig. S14†).

Interestingly, the disassembly of the UPy dimer showed a strong solvent dependence. Sonication in chloroform, dichloromethane and NMP was found to be ineffective in disrupting the hydrogen bonding assembly. Water, like acetone and i-PrOH, a competing solvent for H-bonding was expected to disrupt the interfacial H-bonds, but this was surprisingly not found to be true. This can possibly be explained by the hydrophobic 4-(trifluoromethyl)phenyl groups that hamper interaction of the solvent with surface-bound dimers. It should be noted that the observed solvent dependence may not apply to other interfacial H-bonded species. However, a useful implication is that in our case of **CF$_3$–UPy···UPy–Link** dimers, the system is stable under specific, easily obtainable conditions, while it can be easily dismantled in others.

Akin to their ON/OFF characteristics, dynamic systems also offer the additional advantage of exchangeability. We hypothesized that it would be possible to

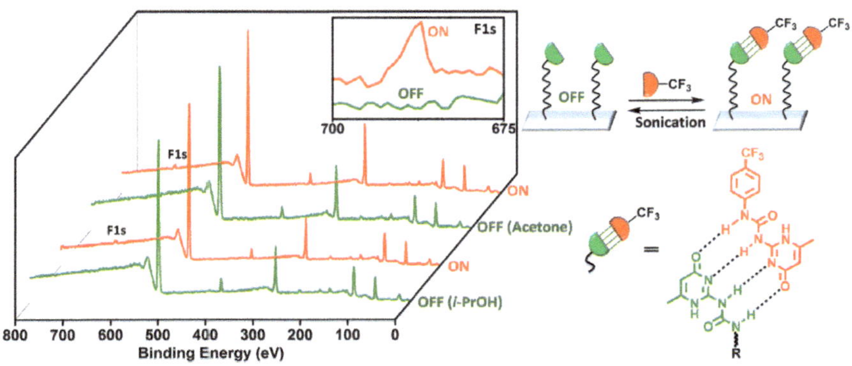

Fig. 5 XPS wide scans for solvent-mediated removal and re-attachment of **UPy–CF$_3$**, inset: F1s region. ON refers to the dimeric state and OFF refers to monomeric UPy.

exchange the **Link–UPy···UPy–CF$_3$** complex using a different UPy motif. Hence, we synthesized 4-(iodo)phenyl UPy (**UPy–I, 3**), which is structurally similar to **2**, but possesses a different substituent, and exchanged this for **2** on surface **S2**. The formation of **Link–UPy···UPy–I** dimers on a surface was confirmed by XPS (ESI, Fig. S15 and S16†), and compound **3** easily ionized in DART-HRMS, as was seen upon simple dipping of a glass slide in a solution of **3** and observing subsequent strong signals of protonated **3** (m/z = 370.9986) in DART-HRMS. This allowed us to follow the growth of the signal of its protonated species over time, in order to quantify the rate of exchange.

Samples with **Link–UPy···UPy–CF$_3$** dimers were allowed to react with a 1 mM solution of **3** in toluene. DART-HRMS analysis of these substrates showed the presence of protonated species for both **2** and **3** (ESI, Fig. S17†). The investigation of intensities thus obtained over different time intervals revealed that the decrease in DART intensity for **2** nicely correlated with the increase in signal intensity for **3** (ESI,† Fig. 6a and c). After 3 h the two signals acquired a limiting

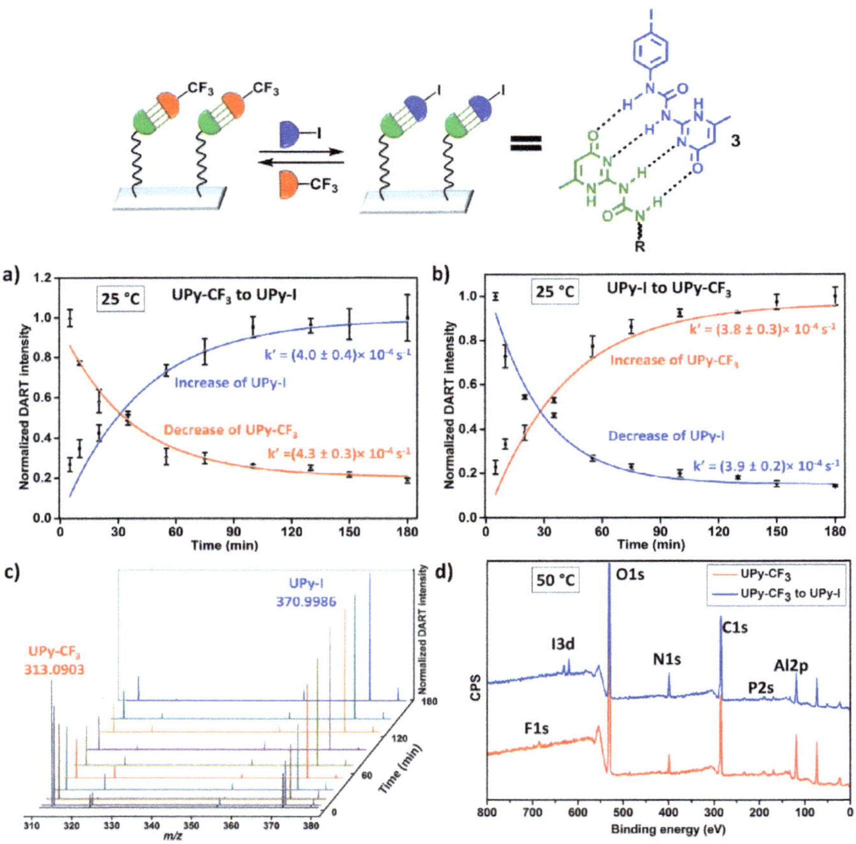

Fig. 6 Normalized EIC intensity of protonated **2** (m/z = 313.0903) and **3** (m/z = 370.9986) during (a) the **UPy–CF$_3$** to **UPy–I** exchange, and (b) the **UPy–I** to **UPy–CF$_3$** exchange. (c) EIC intensity plot *vs.* time for m/z of **2** and **3** during **UPy–CF$_3$** to **UPy–I** exchange. (d) Stacked XPS wide spectra of the **UPy–CF$_3$** to **UPy–I** exchange at 50 °C.

value indicating that an equilibrium between the two motifs had been achieved. Monitoring the decrease of F/P ratio for the exchange process in XPS (ESI, Fig. S18 and S19†) also confirmed that at room temperature the exchange was slow, requiring several hours to reach equilibrium. Also, at room temperature, displacement of **UPy–CF$_3$** for **UPy–I** was not complete: 15–20% of **UPy–CF$_3$** remained surface-bound, and can likely only gradually be lost *via* an exchange process with a higher activation barrier.

Schematically, one might tentatively think about H-bonded **UPy–CF$_3$** that is bound *via* vertically oriented H-bonds at the top of the monolayer without further embedding (so that loss of **UPy–CF$_3$** only means breaking the four H-bonds) *versus* a smaller fraction of **UPy–CF$_3$** that is fully embedded in the monolayer, so that the loss of **UPy–CF$_3$** not only requires the loss of H-bonds, but also of π–π stacking.[48] In reality, a range of dimer complexes with different degrees of embedding within the monolayer will likely be present, the ratios of which, of course, also depend on the precise composition of the monolayer and degree of H-bonding. However, at 50 °C the exchange was not only faster (completed within 2 h), but also complete as evidenced by the disappearance of the F1s signal in XPS (Fig. 6d and ESI, Fig. S20†) and the *m*/*z* intensity for **2** in DART (ESI, Fig. S21†).

A similar exchange process was observed when the system was reversed: *i.e.* the initial dimeric state as **Link–UPy···UPy–I** and displacement with **UPy–CF$_3$** in solution (Fig. 6b and ESI, Fig. S22†). The pseudo-first order rates (at [**UPy–CF$_3$**] = 1 mM) for both exchange reactions were found to be similar, 4×10^{-4} s^{-1} (Fig. 6a and b), but approximately five times slower than the initial **Link–UPy···UPy–CF$_3$** dimer (**S3**) formation process on a monomeric UPy surface. In addition, during the initial H-bond forming process the total number of H-bonded complexes is increased, which drives the reaction, while the number of hydrogen bonds stays constant during the exchange process. Thirdly, we can ascertain that exchange/displacement of UPy at a surface is indeed independent of the UPy substituent and takes place with precisely determinable, slower kinetics than simple dimer formation.

Finally, we were interested in a comparison between the equilibrium constant (K) for dimer formation in solution and interfacial dimer formation. To this aim, we studied the rate of loss of dimers from the **Link–UPy···UPy–CF$_3$** system using XPS at 80 °C (ESI, Fig. S23–S26†). Extrapolating the observed rate constant to 25 °C (see ESI, Fig. S26†), we found that the dissociation rate (k_{off}) was much slower than in the dimer-forming reaction, and estimated k_{off} to be in the order of 10^{-7} s^{-1}.

Based on the experimentally determined pseudo-first-order rate constant for dimer formation (k'_{on}) of 22×10^{-4} s^{-1} and based on the known (and constant) UPy concentration of 1.0 mM, we could calculate a second-order rate constant of $k_{on} = 2.2$ M^{-1} s^{-1} (see ESI, Section 4.5 and Fig. S26†). This fixes the equilibrium constant (k_{on}/k_{off}) for interfacial H-bonded dimer formation in toluene to be in the order of 10^7 M^{-1}. This value is about an order of magnitude lower than that observed for soluble dimers in toluene (6×10^8 M^{-1}).[29] Three aspects may contribute to this observed difference between solution and surface-bound data.

First, the difference may be caused by already present H-bonds within the monolayer and the restrictions in orientation between surface-bound H-bond forming UPy moieties and soluble UPy moieties. This likely lowers the value for surface-bound dimerization. Second, while the solution data refers to observations at low concentrations, the surface-bound data reflect highly interacting systems with varying entropy and enthalpy contributions. Given *e.g.* increased sterics on the one hand, but increased π–π interactions on the other hand, it is difficult to foresee the direction of this argument on K prior to more detailed experiments. Finally, while k_{on} is accurately determined, k_{off} is characterized with a lower precision, which simply adds uncertainty to the eventual outcome for K. Most importantly, we would therefore argue that the value of K is comparable to the high value observed in solution, and that this formation of stable dimers on a surface provides the potential for further studies on H-bonded responsive surfaces.

Conclusions

In summary, we have demonstrated that the fourfold hydrogen bonding UPy motif can be used for reversible surface modification. The inherently dynamic nature of this highly stable dimer (on-surface dimerization: $K \sim 10^7\ M^{-1}$) allows straightforward switching between a surface-bound (ON state) and a surface-free (OFF state) by control over the solvent. In addition, exchange of a hydrogen-bonded UPy moiety for a competing hydrogen-bonding UPy motif was possible, which yields further routes for exchangeable surface modification. By detailed kinetic analysis we have established the kinetics of binding, release and exchange. These studies not only revealed the typical time scales of these processes, but also enabled us to confirm that the strength of the quadruple hydrogen bond motif on the surface is of comparable strength to the solution-based binding.

Our study provides a rational approach towards the formation of dynamically controllable complex H-bonding arrays at a surface. This also opens up new avenues for the preparation of dynamic and regeneratable surface platforms with ON/OFF capability that may find various applications in molecular electronics, bio-sensing and other applications.

Acknowledgements

The authors thank NanoNextNL (program 5A), a micro and nanotechnology consortium of the Government of The Netherlands and 130 partners, and NWO (ECHO project 712.012.006) for funding of this project. The authors also thank Dr Sidharam Pujari for technical assistance.

Notes and references

1 F. Huang and E. V. Anslyn, *Chem. Rev.*, 2015, **115**, 6999–7000.
2 G. R. Desiraju, *Nature*, 2001, **412**, 397–400.
3 T. Aida, E. W. Meijer and S. I. Stupp, *Science*, 2012, **335**, 813–817.
4 M. Schulze, V. Kunz, P. D. Frischmann and F. Würthner, *Nat. Chem.*, 2016, **8**, 576–583.
5 C. K. McLaughlin, G. D. Hamblin and H. F. Sleiman, *Chem. Soc. Rev.*, 2011, **40**, 5647–5656.
6 S. P. Pujari, L. Scheres, A. T. M. Marcelis and H. Zuilhof, *Angew. Chem., Int. Ed.*, 2014, **53**, 6322–6356.
7 Z. Wang and H. Zuilhof, *Langmuir*, 2016, **32**, 6310–6318.

8 A. Abi and E. E. Ferapontova, *J. Am. Chem. Soc.*, 2012, **134**, 14499–14507.
9 X.-L. Sun, C. L. Stabler, C. S. Cazalis and E. L. Chaikof, *Bioconjugate Chem.*, 2006, **17**, 52–57.
10 C. Tassa, M. Liong, S. Hilderbrand, J. E. Sandler, T. Reiner, E. J. Keliher, R. Weissleder and S. Y. Shaw, *Lab Chip*, 2012, **12**, 3103–3110.
11 F. della Sala and E. R. Kay, *Angew. Chem., Int. Ed.*, 2015, **54**, 4187–4191.
12 D. I. Rozkiewicz, B. J. Ravoo and D. N. Reinhoudt, *Langmuir*, 2005, **21**, 6337–6343.
13 F. Arias, L. A. Godínez, S. R. Wilson, A. E. Kaifer and L. Echegoyen, *J. Am. Chem. Soc.*, 1996, **118**, 6086–6087.
14 K. Jo, T. Lee, H. J. Choi, J. H. Park, D. J. Lee, D. W. Lee and B.-S. Kim, *Langmuir*, 2011, **27**, 2014–2018.
15 G. M. Credo, A. K. Boal, K. Das, T. H. Galow, V. M. Rotello, D. L. Feldheim and C. B. Gorman, *J. Am. Chem. Soc.*, 2002, **124**, 9036–9037.
16 X. Du, J. Li, A. Welle, L. Li, W. Feng and P. A. Levkin, *Adv. Mater.*, 2015, **27**, 4997–5001.
17 L. Brunsveld, B. J. B. Folmer, E. W. Meijer and R. P. Sijbesma, *Chem. Rev.*, 2001, **101**, 4071–4098.
18 S. K. Yang and S. C. Zimmerman, *Isr. J. Chem.*, 2013, **53**, 511–520.
19 A. J. Wilson, *Soft Matter*, 2007, **3**, 409–425.
20 K. Viswanathan, T. E. Long and T. C. Ward, *Langmuir*, 2009, **25**, 6808–6812.
21 E. Jeoung, J. B. Carroll and V. M. Rotello, *Chem. Commun.*, 2002, 1510–1511.
22 K. Viswanathan, H. Ozhalici, C. L. Elkins, C. Heisey, T. C. Ward and T. E. Long, *Langmuir*, 2006, **22**, 1099–1105.
23 T. B. Norsten, E. Jeoung, R. J. Thibault and V. M. Rotello, *Langmuir*, 2003, **19**, 7089–7093.
24 A. Sanyal, T. B. Norsten, O. Uzun and V. M. Rotello, *Langmuir*, 2004, **20**, 5958–5964.
25 A. Embrechts, A. H. Velders, H. Schönherr and G. J. Vancso, *Langmuir*, 2011, **27**, 14272–14278.
26 N. Wedler-Jasinski, N. Delbosc, M.-A. Virolleaud, D. Montarnal, A. Welle, L. Barner, A. Walther, J. Bernard and C. Barner-Kowollik, *Chem. Commun.*, 2016, **52**, 8753–8756.
27 R. P. Sijbesma, F. H. Beijer, L. Brunsveld, B. J. B. Folmer, J. H. K. K. Hirschberg, R. F. M. Lange, J. K. L. Lowe and E. W. Meijer, *Science*, 1997, **278**, 1601–1604.
28 F. H. Beijer, R. P. Sijbesma, H. Kooijman, A. L. Spek and E. W. Meijer, *J. Am. Chem. Soc.*, 1998, **120**, 6761–6769.
29 A. W. Bosman, R. P. Sijbesma and E. W. Meijer, *Mater. Today*, 2004, **7**, 34–39.
30 A. M. Kushner, V. Gabuchian, E. G. Johnson and Z. Guan, *J. Am. Chem. Soc.*, 2007, **129**, 14110–14111.
31 J. Hentschel, A. M. Kushner, J. Ziller and Z. Guan, *Angew. Chem., Int. Ed.*, 2012, **51**, 10561–10565.
32 K. E. Feldman, M. J. Kade, T. F. A. de Greef, E. W. Meijer, E. J. Kramer and C. J. Hawker, *Macromolecules*, 2008, **41**, 4694–4700.
33 E. J. Foster, E. B. Berda and E. W. Meijer, *J. Am. Chem. Soc.*, 2009, **131**, 6964–6966.
34 E. B. Berda, E. J. Foster and E. W. Meijer, *Macromolecules*, 2010, **43**, 1430–1437.
35 S. Zou, H. Schönherr and G. J. Vancso, *J. Am. Chem. Soc.*, 2005, **127**, 11230–11231.

36 S. Zou, Z. Zhang, R. Förch, W. Knoll, H. Schönherr and G. J. Vancso, *Langmuir*, 2003, **19**, 8618–8621.
37 L. L. Crowe, K. M. Solntsev and L. M. Tolbert, *Langmuir*, 2007, **23**, 6227–6232.
38 A. T. ten Cate, H. Kooijman, A. L. Spek, R. P. Sijbesma and E. W. Meijer, *J. Am. Chem. Soc.*, 2004, **126**, 3801–3808.
39 G. B. W. L. Ligthart, H. Ohkawa, R. P. Sijbesma and E. W. Meijer, *J. Am. Chem. Soc.*, 2005, **127**, 810–811.
40 T. F. A. de Greef, G. Ercolani, G. B. W. L. Ligthart, E. W. Meijer and R. P. Sijbesma, *J. Am. Chem. Soc.*, 2008, **130**, 13755–13764.
41 X. Yan, F. Wang, B. Zheng and F. Huang, *Chem. Soc. Rev.*, 2012, **41**, 6042–6065.
42 J. H. Gross, *Anal. Bioanal. Chem.*, 2014, **406**, 63–80.
43 R. K. Manova, S. Joshi, A. Debrassi, N. S. Bhairamadgi, E. Roeven, J. Gagnon, M. N. Tahir, F. W. Claassen, L. M. W. Scheres, T. Wennekes, K. Schroën, T. A. van Beek, H. Zuilhof and M. W. F. Nielen, *Anal. Chem.*, 2014, **86**, 2403–2411.
44 R. Sen, J. Escorihuela, F. L. van Delft and H. Zuilhof, *Angew. Chem., Int. Ed.*, 2017, **56**, 3299–3303.
45 R. Sen, J. Escorihuela, M. M. J. Smulders and H. Zuilhof, *Langmuir*, 2016, **32**, 3412–3419.
46 B. Rijksen, S. P. Pujari, L. Scheres, C. J. M. van Rijn, J. E. Baio, T. Weidner and H. Zuilhof, *Langmuir*, 2012, **28**, 6577–6588.
47 T. F. A. de Greef, M. M. L. Nieuwenhuizen, P. J. M. Stals, C. F. C. Fitié, A. R. A. Palmans, R. P. Sijbesma and E. W. Meijer, *Chem. Commun.*, 2008, 4306–4308.
48 D. Guo, R. P. Sijbesma and H. Zuilhof, *Org. Lett.*, 2004, **6**, 3667–3670.

Faraday Discussions

DISCUSSIONS

Preparing macromolecular systems on surfaces: general discussion

David Amabilino, Ioan Bâldea, Olga Barykina, James Batteas, Pol Besenius, Peter Beton, Nerea Bilbao, Manfred Buck, Lifeng Chi, Stuart Clarke, Giovanni Costantini, Jonathan Davidson, Philip Davies, Steven De Feyter, Yuri Diaz Fernandez, Deepak Dwivedi, Karl-Heinz Ernst, Amar Flood, Julien Gautrot, Ahmad Jabbarzadeh, Vladimir Korolkov, Angelika Kühnle, Markus Lackinger, Claire-Marie Pradier, Talat Rahman, Rasmita Raval, Sebastian Schwaminger, Johannes Seibel, Steven L. Tait, Joan Teyssandier and Han Zuilhof

DOI: 10.1039/C7FD90C76G

Karl-Heinz Ernst opened a general discussion of the paper by Lifeng Chi: I was a little surprised at how high in temperature you have to go for this deoxygenation. Ullmann coupling is normally around 200 °C. Why do you need a higher temperature? Is it because of dehalogenation?

Lifeng Chi replied: Decarboxygenation seems to be more difficult than dehalogenation.

Karl-Heinz Ernst asked: Does this reaction (decarboxygenation) also work on gold?

Lifeng Chi answered: No, it does not work on gold, also not on silver.

Claire-Marie Pradier commented: At high temperature, don't you have dehydrogenation of the molecules?

Lifeng Chi responded: From our experience, I believe that hydrogenation should take place. This might be the reason why it is not easy to increase the yield for obtaining graphene nanoribbons.

Peter Beton commented: Following on from the last question, have you tried with perylene itself (rather than PTCDA)?

Lifeng Chi replied: We have not tried perylene yet. It could be an interesting system to study.

Philip Davies asked: In figure 4, after you've exposed it to oxygen at 470 K, there are patches on the surface apparently bound by perylene chains. Are those areas where the surface has oxidised? Is it that the perylene chains are confining the area of oxidation, or are they organising themselves so that they bind to the edges of the areas where there's oxidation?

Lifeng Chi answered: The patches are oxidized areas. Although other empty areas (not occupied by the perylene chains) could be oxidized too (see the ESI for our paper, DOI: 10.1039/C7FD00129K), those patches confined by the perylene chains are apparently higher. This may hint that the confined areas are easier to oxidize. This phenomenon needs to be further investigated.

Joan Teyssandier asked: What happens in terms of structures formed and reaction efficiency if you anneal at a higher temperature than 630 K? In particular, have you ever observed any lateral fusion of graphene nanoribbons on Cu like it has been seen on Au and Ag?

Lifeng Chi replied: Because we already have a more disordered structure at 630 K, probably due to the dehydrogenation, we did not try to further increase the annealing temperature.

Claire-Marie Pradier queried: In the absence of oxygen, what is the effect of possible defects, steps, *etc.* on the pristine surface with which you're working, on the alignment of your ribbons?

Lifeng Chi responded: Step edges surely help with the alignment of the ribbons. From our experience, high index crystal surfaces are even better.

Rasmita Raval commented: The organometallic linked species you show all have good registry with the underlying surface. However, this registry is lost in the C–C coupled systems you showed. We see the same type of behaviour.[1] Do you think the C–C coupled polymers have a much weaker interaction with the surface?

1 S. Haq *et al.*, *ACS Nano*, 2014, **8**, 8856.

Lifeng Chi responded: Compared to the metal coordinated organic complexes, I think that the C–C polymers have weaker interactions with the surfaces. But for aromatic polymers, the interaction cannot be too weak.

Steven L. Tait asked: At the high temperatures used in your studies, it seems likely that there would be some dehydrogenation, as has been observed in other systems on copper surfaces.[1] Do you think it is possible that some of the disorder in the systems after high temperature treatment is due to dehydrogenation and C–Cu bond formation (chemisorption) that would prevent the adsorbates from diffusion and re-ordering and thereby lead to a more disordered structure?

1 C. G. Williams, M. Wang, D. Skomski, C. D. Tempas, L. L. Kesmodel and S. L. Tait, *Surf. Sci.*, 2016, **653**, 130–137.

Lifeng Chi replied: Absolutely. We think that the dehydrogenation and the uncontrolled C–C coupling is the main reason for the disorder. In fact this is the reason we tried the oxidation, in order to lower the temperature for the C–Cu bond in the metal-organo chains.

Nerea Bilbao queried: If I understood correctly, you are looking for ways to weaken this C–Cu bond in the reaction intermediate. I was wondering if you had tried to decouple the molecular structure from the surface, by intercalation of atoms or small molecules? For instance, iodine atoms can be intercalated between the metal surface and the organic product, as shown by the group of Prof. Markus Lackinger in their recently published work.[1] They report the successful post-synthetic decoupling of covalent polyphenylene networks from a Au(111) surface by intercalation of a chemisorbed iodine monolayer. Another relevant example was also published this year by Deniz *et al.* In this work, they intercalate *in situ* Si to suppress the strong contribution of the Au(111) surface state. Their first-principles calculations provide insight into substrate induced screening effects, which result in a width-dependent band gap reduction for substrate-supported AGNRs.

1 A. Rastgoo-Lahrood *et al.*, *Nanoscale*, 2017, **9**, 4995.
2 O. Deniz *et al.*, *Nano Lett.*, 2017, **17**, 2197–2203.

Lifeng Chi answered: Thanks for the suggestion. We have not yet tried decoupling by introducing other atoms or small molecules (except oxygen). We may try this in future work.

Steven De Feyter asked: I'm not so familiar with this copper oxidation process that can take place, but it seems to be confined in space by areas defined by the adsorbed polymers. Is it of interest to learn more about the copper oxidation process itself?

Lifeng Chi answered: It should be an interesting issue to follow. It could be related to oxygen diffusion on the surface, but this is not confirmed yet.

Claire-Marie Pradier asked: If I understood correctly, the oxygen is added after the formation? What is this for? And what were you expecting from using reducing gases?

Lifeng Chi replied: Yes, oxygen is added after the formation of the perylene–Cu complex. The aim is to oxidize the Cu which may help to break the Cu–C bond more easily. Reducing gas is used mainly as the control experiment.

David Amabilino commented: Very nice talk. Following on from the last point, from the work we did with Rasmita Raval and others, when you have this cuprate bound to the surface it is tremendously stable. It is interesting that your surface reacts faster than the cuprate (in solution, these compounds are quite reactive). Are we missing something here? Chemically, that's really interesting. You'd think they would be more reactive. Could that chemical architecture be interesting for something, what possibilities do you see?

Lifeng Chi responded: We are also very interested in such unique cuprate structures and their stability against temperature, even in oxygen. We are going to explore the catalytic activities of such structure, both from experimental and theoretical sides. Electrochemical and spectroscopic detection could be suitable methods to prove their catalytic activities.

Steven De Feyter asked: Probably a more general question and maybe a bit provocative. You synthesize a new material on a copper surface in ultrahigh vacuum. What are the steps that must be taken to really use such material, for instance by removing it from copper and transferring it to another substrate? What are the important steps and how feasible are they?

Lifeng Chi responded: Directly using the structured materials on surfaces might be a more effective way than to remove them from the surfaces. For instance, the copper–perylene complex is quite stable against annealing. The single (or bi-) metal atom (not necessarily copper) might be catalytically active. If one can directly apply the materials for catalytic use, it will become very attractive.

Steven De Feyter opened a general discussion of the paper by Ahmad Jabbarzadeh: Is the surface that you used structured?

Ahmad Jabbarzadeh replied: As described in the manuscript, all surfaces are atomically structured. The lattice size of various surfaces can be found in Table 1 in our paper (DOI: 10.1039/C7FD00071E).

Steven De Feyter continued: What is the relation between the orientation of the alkane chains in the crystallizing droplets and the symmetry axes of the substrate underneath?

Ahmad Jabbarzadeh responded: Fig. 8b in our paper (DOI: 10.1039/C7FD00071E) shows the molecular configuration for the entire film, and the attached Fig. 1 shows the molecules in the first layer immediately next to the

Fig. 1 Molecules in the first later immediately next to the smooth surface.

smooth surface. We can see from both figures, the orientation of the molecules is random and does not follow any particular lattice symmetry of the underlying substrate. However, the molecules seem to be mostly parallel with other molecules in a given domain, where they share a common orientation. Even the smooth surface has some atomic level corrugation however, since the lattice size is 0.286 nm and the LJ length parameter of the polymers is $s = 0.393$ nm, it is less likely for them to give rise to a directional order to template the underlying crystal. The smoothness of the surface, on the other hand, is expected to help with creating a planar order. We know, from previous work with shorter confined alkanes, that for confinement induced order, the planar order depends on both the smoothness and degree of the corrugation of the underlying substrate.[1–5]

1 A. Jabbarzadeh, P. Harrowell and R. I. Tanner, Crystal bridges, tetratic order, and elusive equilibria: The role of structure in lubrication films, *J. Phys. Chem. B*, 2007, **111**, 11354–11365.
2 A. Jabbarzadeh, P. Harrowell and R. I. Tanner, The structural origin of the complex rheology in thin dodecane films: Three routes to low friction, *Tribol. Int.*, 2007, **40**, 1574.
3 A. Jabbarzadeh, P. Harrowell and R.I. Tanner, Crystal bridge formation marks the transition to rigidity in a thin lubrication film, *Phys. Rev. Lett.*, 2006, **96**, 206102.
4 A. Jabbarzadeh, P. Harrowell and R.I. Tanner, Low friction lubrication between amorphous walls: unravelling the contributions of surface roughness and in-plane disorder, *J. Chem. Phys.*, 2006, **125**, 034703.
5 A. Jabbarzadeh, P. Harrowell and R. I. Tanner, The very low friction state of a dodecane film confined between mica surfaces, *Phys. Rev. Lett.*, 2005, **94**, 126103.

Steven De Feyter queried: Does the surface, without the nanostructured features, template the crystallization in any way?

Ahmad Jabbarzadeh responded: Even the smooth surface has some atomic level corrugation, however since the lattice size is 0.286 nm and the LJ length parameter of the polymers is $s = 0.393$ nm, it is less likely for them to give rise to a directional order to template the crystal. The smoothness of the surface on the other hand is expected to help with creating a planar order.

Steven De Feyter commented: What is the impact of the initial size of the alkane droplet on the crystallization process?

Ahmad Jabbarzadeh responded: We have done preliminary studies with an isolated drop (without surfaces) on the effect of droplet size on crystallization kinetics. This shows great dependence, and we hope to publish the full results in the near future. For the droplet on the surface, the droplet size was chosen as such to fill in the cavities formed by the pillars when the surface is fully wetted by the droplet. As described in the paper, for such a system, a variety of roughness sizes are simulated to explore the effect of the roughness size.

Sebastian Schwaminger said: You investigated the surface wetting and the surface structure for the Wenzel case. For the Cassie-Baxter case, would different wetting behaviour affect the crystallinity of the polymers? And would these results be different from flat surfaces?

Ahmad Jabbarzadeh replied: We have not simulated the crystallization of a drop in the Cassie-Baxter state. We would expect different behavior as there will be less contact with the underlying surface, therefore the crystallization kinetics will be closer to that of an isolated drop under vacuum.

Sebastian Schwaminger asked: Would it be difficult to simulate the Cassie-Baxter wetting, considering the surface roughness and the lower interfacial area compared to flat surfaces?

Ahmad Jabbarzadeh responded: We already have the configurations for the Cassie-Baxter state and it will be relatively easy to conduct crystallization simulations for this case.

Sebastian Schwaminger continued: Another question, could you also think about changing the crystallinity with surface curvature? How would a convex or concave surface affect the crystallinity of your polymers? Here, the degree of curvature and the size of the droplet should play a critical role as well.

Ahmad Jabbarzadeh answered: We have not simulated such cases, however we believe that it will be relevant in the context of surface-liquid interactions and only important if the radius of curvature of both the droplet and the surface are very close. In other words, it is only important if the radius of curvature of the surface is such that the interaction of the liquid-surface is affected.

Talat Rahman commented: I'm confused. You start with the nanoparticle that's composed of these polymer molecules. There's some interaction that holds these molecules together and they form a nanoparticle of $\sim$15 nm in size, then you drop it on a surface. First, you drop it on a smooth surface, then you have some Lennard-Jones interaction which couples the molecules in the nanoparticle with the surface. Then you do your simulation for various temperatures, and at some point you find that the nanoparticle breaks apart?

Ahmad Jabbarzadeh replied: The initial wetting simulations on the smooth surfaces are all conducted at the same temperature. What makes the nano-drop spread and break-up on one of the surfaces (a = 0.286 nm), is the stronger wettability of the surface. The higher wettability is due to the smaller lattice size (packing more atoms on the same area).

Talat Rahman continued: So the role of the surface is through this Lennard-Jones potential? It has a corrugation. Next you replace that surface with a rough surface? That roughness is nothing but a vicinal surface of that crystal instead of Cu(100), is that correct? Or are you saying that you create a modulation of the surface, producing confined structures which are not related to vicinal surfaces?

Ahmad Jabbarzadeh answered: The surface atoms, and polymer molecules interact only through this Lennard-Jones potential. The smooth surface has corrugation only at the atomic length scale. The underlying atomic crystal structure for all the rough surfaces is Fe(100). The roughness is created by cutting patterns through a crystal Fe slab on the (100) plane. That is akin to machining a crystalline surface *via* some precision machining or ablation method. The rough surfaces create deep confinement space for the polymer molecules. I believe this structure is somehow different to vicinal surfaces where surface steps are rather shallow.

Giovanni Costantini said: The system you are simulating seems essentially to be a very concentrated, liquid drop of polymers, something that I might imagine could occur at quite high temperatures. How representative is this of the actual conditions used in the growth and crystallization of polymeric thin films?

Ahmad Jabbarzadeh replied: A representative condition relevant to the simulated system occurs in the recent development of a variation of PVD (Physical Vapour Deposition) using matrix assisted pulsed laser evaporation (MAPLE).[1] The MAPLE method is based on additive deposition of nanodroplets of the polymer on a substrate. Polymeric thin films are often deposited by casting from solutions. Despite this being an easy method, due to the short processing time, there is limited control of the nucleation and crystallization and eventual morphology of the polymeric films. The development of other deposition methods, such MAPLE, offers more control of the crystallinity and morphology of the deposited films.

1 J. Hyuncheol, K. B. Shepard, G. E. Purdum, Y. Guo, Y.-L. Loo, C. B. Arnold and R. D. Priestley, Additive growth and crystallization of polymer films, *Macromolecules*, 2016, **49**, 2860–2867.

Giovanni Costantini commented: A filed of research where it's absolutely essential to be able to control and modify the crystallinity of polymeric thin films, is that of conjugated polymers used in organic electronics and photovoltaic devices. Would you be able to include the solvent and the conditions used in spin coating in your simulations? I guess that this might have a high relevance.

Ahmad Jabbarzadeh answered: It is true that polymeric thin films are often deposited from solutions. Despite this being an easy method, due to short processing time, there is limited control of the nucleation and crystallization and eventual morphology of the polymeric films. We have not simulated such systems, however, it is possible to extend our methodologies to study crystallization from solutions. In fact, we expect this to be more feasible when simulating such processes due to the shorter times scales for molecular ordering.

Rasmita Raval remarked: Your paper is really interesting. A major comment that can be made is that the roughness of the surface determines the physical state of the adsorbed layer, *i.e.* you have shown that surfaces are hydrophobic, hydrophilic, molecule-phobic, molecule-philic depending on their roughness. This leads to the question, is there sufficient knowledge at present to tune the state of the surface layer required simply by engineering surfaces accordingly?

Ahmad Jabbarzadeh responded: Thank you for your comments. I believe there are methodologies that allow for the control of surface roughness at various length scales using surface engineering methods.

Claire-Marie Pradier asked: Your model predicts that the contact angle varies linearly with the lattice parameter; are you talking about the periodicity of the pillar or the lattice parameter of the substrate surface?

Ahmad Jabbarzadeh replied: This refers to the dependence of wettability on the lattice size of the smooth surfaces only, as shown in Fig. 4 of our manuscript (DOI:10.1039/C7FD00071E). We do not claim the dependence is linear, however the trend line shown in this figure shows that the contact angle increases with an increase in the lattice parameter. The figure should be considered as per the conditions described in section 3.1.1.

Claire-Marie Pradier continued: Do you have experimental evidence of this linear variation of the contact angle with the lattice parameter?

Ahmad Jabbarzadeh responded: The trend observed here is in general agreement with the experimental evidence in ref. 1 that suggest that in single component metallic melt-surface systems where weak van der Waals forces are dominant, surface facets with higher atomic density will lead to better wettability. Furthermore, experiments described in ref. 2 on metal–carbon interfaces have shown that the contact angle of liquid metal on more compact facets of diamond are smaller, demonstrating increasing atomic density (decreasing nearest neighbour) results with decreasing contact angle.

1 P. Shen, H. Fujii and K. Nogi, Effect of substrate crystallographic orientation on wettability and adhesion in several representative systems,*J. Mater. Process. Technol.*, 2004, **155**, 1256–1260.
2 O. Dezellus and N. Eustathopoulos, The role of van der Waals interactions on wetting and adhesion in metal/carbon systems, *Scr. Mater.*, 1999, **40**, 1283–1288.

Johannes Seibel asked: Did you also increase the surface roughness to the point where there is no difference between the crystallization on the nano-patterned surface and the crystallization on the smooth surface?

Ahmad Jabbarzadeh answered: No, but as can be seen from figures 14a and 14b in our paper (DOI: 10.1039/C7FD00071E), within the limits of roughly $w = \sim 5–6$ nm, or roughness ratio $\rightarrow 1$ the effect will vanish.

Angelika Kühnle opened a general discussion of the paper by Markus Lackinger: Markus, I really enjoyed your detailed description of the moiré pattern. I wonder, though, why these slight rotations are observed. You can envision that the energy landscape is very flat. However, the layers are large and a small energy penalty would add up to a large number. Wouldn't it?

Markus Lackinger responded: Thank you very much Angelika. In principle, I would agree, however, this argument implicitly assumes that the experimentally observed TMA structures represent the energetically most favorable epitaxial relation to the graphite substrate. I believe that optimizing the adsorption configuration of a single TMA molecule could result in a slight energetic advantage. However, the network of strong hydrogen bonds in the quite rigid chickenwire structure imposes constraints and dictates the mutual arrangement of further TMA molecules. Hence, it appears unlikely that the other TMA molecules can simultaneously optimize their interactions with the surface. Accordingly, the corrugation of the surface potential averages out, even more so for increasing domain size.

In fact, the emergence of moiré patterns could be interpreted as an indication of relatively weak molecule–surface interactions as compared to molecule–molecule interactions. This argument even gains strength for incommensurate structures, as proposed for the TMA chickenwire structure on graphite.

I would like to put an alternative explanation forward: the orientation of the TMA domains is already determined at an early stage of growth, *i.e.* during the nucleation. We propose that the critical nucleus has an energetic preference for the observed 5° orientation. However, the potential well around this energetically ideal orientation is rather shallow, and thermal fluctuations facilitate small deviations from the optimal orientation. Subsequent growth kinetically stabilizes this domain orientation close but not necessarily equal to the ideal angle.

To test this hypothesis, I propose to grow TMA monolayers at low temperatures in order to suppress the thermal fluctuations. This can only be done in a UHV experiment, where LEED would facilitate a direct and global measurement of the respective orientation, or distribution of orientations, between TMA and graphite.

Steven De Feyter queried: Do you expect that the precision with which you determined distances based on the appearance of moiré patterns can help to investigate other phenomena? For instance, would it be possible to probe the impact of the thermal expansion of the graphite substrate on the structure of the isophthalic acid monolayer? Could you use it to probe local stress caused by host–guest chemistry (isophthalic acid–coronene)?

Markus Lackinger replied: Studying thermal expansion of the TMA network would definitely make an interesting case. For one data point we have already carried out the imaging at 50 °C, but we did not find any significant deviations of the lattice parameter. At the liquid–solid interface, the accessible temperature range is rather limited at the lower end by freezing of the solvent. For elevated temperatures, we have observed a discoloration of the solution, pointing towards chemical reactions between solute and/or solvent molecules. This sets the upper limit for the temperature. However, carrying out experiments in UHV would allow us to study a wider range of temperatures.

A further idea would be to study static isotope effects with a deuterated version of TMA. It is known that this affects the hydrogen bond length. However, the effect is quite small, probably even below the accuracy of our method.

We have done some experiments with coronene guest molecules. If we add the coronene molecules after formation of the TMA network, we believe it just fills the pores without causing a rearrangement of the TMA network. Thereby, the intramolecular contrast of the coronene guest molecules becomes likewise modulated, and the guests even enhance and amplify the moiré pattern as previously observed,[1] but the moiré patterns are effectively similar to the pure TMA network. Yet, we found differences when the network is formed in the presence of coronene molecules, resulting in small domains. This clearly shows an influence of the coronene guests on the TMA network formation process, however, we did not observe moiré patterns with a defined periodicity for these small domains.

1 S. J. H. Griessl, M. Lackinger, F. Jamitzky, T Markert, M. Hietschold and W. M. Heck, *Langmuir*, 2004, **20**, 9403–9407.

Vladimir Korolkov commented: Often people look at the moiré patterns, some people look at them in great detail and some people just report them, you observed quite a few moiré patterns. However, in dry films we have observed only one pattern and never seen any other moiré patterns. You quite clearly see variations in moiré patterns, is this possibly due to the presence of a solvent layer and possible diffusion of the molecules within it? And a further comment; the reality is with AFM, corrugation of the moire pattern, depending on the molecular system, could be as small as a few picometres. So it is limited by base noise levels in AFM and could be missed.

Markus Lackinger responded: This is a very interesting, and for me quite surprising, result. I would indeed anticipate a distinct influence of the different preparation conditions. The occurrence of one unique moiré pattern is a necessary condition for thermodynamic equilibrium. However, finding equilibrium structures in dried samples as opposed to the liquid–solid interface is counter-intuitive for me. I would have expected that equilibrium structures are more likely to be observed at liquid–solid interfaces due to possibilities of dynamic exchange between molecules in solution and on the surface. So it would be definitely worthwhile to look more into the details of your preparation protocol, *i.e.* the type of solvent, concentration, method and speed of drying.

Peter Beton asked: Could small changes in the lattice constant be an alternative explanation of the variation in angle and period of the moiré patterns? The energy variation for small changes in lattice constant appear very small.

Markus Lackinger answered: In principle yes, yet the experimental observation of a unique relative rotation between TMA and the graphite lattice of approximately 5° with only small deviations implies that the reciprocal lattice vectors of the moiré always occurs between the (7,1) TMA and the (1,0) graphite reciprocal lattice points. In this respect, the theoretical curve of the moiré orientation (with respect to the TMA lattice) *vs.* the moiré period in Fig. 6 of the manuscript (DOI: 10.1039/C7FD00113D) with the precise TMA lattice parameter as an additional parameter can be considered exact. So for each moiré pattern one data point can be plotted which allows us to deduce a unique and exact value of the TMA lattice parameter. Interestingly, all experimentally observed moiré patterns result in a similar TMA lattice parameter of 1.652 nm $\pm$ 0.002 nm, suggesting that the main reason for the different moiré patterns is indeed small variations in the relative rotational orientation between TMA and the graphite lattice. However, small variations in the TMA lattice parameter within the experimental uncertainty of $\pm$ 0.002 nm cannot be excluded.

Fig. 2 of the manuscript (DOI: 10.1039/C7FD00113D) provides additional and more direct evidence for this hypothesis: the two TMA domains in the lower half show a very small, but safely detectable relative rotation of $\sim$1° and different moiré patterns. This very justified question could, and should, be experimentally settled by precisely measuring the TMA lattice parameter either by LEED or powder X-ray diffraction.

Jonathan Davidson said: There has been some discussion of using LEED to further study these systems, but have you considered using powder X-ray diffraction? This technique can achieve even higher levels of precision and is ideally suited to the study of monolayers on HOPG substrates.

Markus Lackinger responded: I was not aware that monolayer sensitivity can be accomplished with powder X-ray diffraction. I fully agree that the higher levels of precision in the lattice parameter could be particularly useful for further studies of TMA monolayers on graphite. Thank you very much for pointing this out to me, we will definitely look into this type of experiment for further studies, *e.g.* of the thermal expansion coefficient or isotope effects.

Deepak Dwivedi said: Thanks for the nice presentation. Can this model be applicable to X-ray diffraction? As for reciprocal space lattice calculations, we follow the Ewald sphere calculation which allows us to predict the possible diffraction that could occur for the system. Can I apply your model to monolayers or polycrystalline films? I encounter monolayer or multilayer films on steel surfaces and am trying to explore the possibility of applying your model to my system. If you could comment on this, it would be helpful.

Markus Lackinger responded: Emergence of these very defined moiré patterns of highly ordered molecular monolayers is presumably only observable on very defined surfaces prepared according to surface science standards, *i.e.* atomically flat and clean. Hence, I would doubt that this will work on more technological and presumably rough steel surfaces. I'm not sure if I understand the point about X-ray diffraction correctly. The main idea here is to use real space microscopy (*i.e.* STM) images to acquire information about the moiré pattern directly. Since the moiré pattern arises from the superposition of the surface and monolayer lattice, additional information about the monolayer lattice parameters can be obtained from the (generally known) surface lattice parameters.

Deepak Dwivedi asked: When we measure the lattice parameter with X-ray diffraction, it involves various errors as diffraction depends on atomic scattering, structure factors, thermal factors *etc.* If you measure the lattice parameter with those, it may involve some error. What would you recommend to mitigate this issue? In your model you have taken a lattice parameter, was it a modified one and free from these errors? How do you see your model applied to solve this problem? Can we use this model to answer this question either directly or indirectly?

Markus Lackinger answered: First of all I would like to state that using real space measurements of moiré patterns for the precise determination of lattice parameters for one of the two involved lattices is by no means a general method and remains limited to special model systems, where TMA monolayers on graphite appear particularly suitable. The input data, *i.e.* moiré period and orientation with respect to either graphite or the TMA lattice, are obtained from high resolution STM images. So the accuracy of the method depends on the accuracy with which these geometric parameters can be measured in real space images. For geometric measurements in STM images, the limiting factors are thermal drift, piezo hysteresis and piezo creep. Even though the proposed method features relatively high precision as compared to directly measuring lattice parameters from STM images, the high precision of an appropriately carried out diffraction experiment cannot be attained.

Deepak Dwivedi continued: Out of curiosity, I wanted to ask about the application of this model for predicting the effects of doping or substitutional diffusion (either in single crystal monolayers or polycrystalline multilayer films), as we know that a change in the lattice parameter (Vegard's law) can suggest doping (insertion of some elements inside your unit cell changes the lattice parameters). Do you think your model could be applied in this way?

Markus Lackinger responded: Vegard's law proposes a linear relation between the lattice parameter and composition for alloys, *i.e.* completely miscible systems. Since we are looking at pure TMA monolayers on (also chemically pure) graphite surfaces, I do not understand what is possibly meant by doping or substitutional diffusion in this context. So I do not see a feasible way in which the model could be applied for this.

Talat Rahman remarked: Whenever you have these moiré patterns arising from slightly incommensurate structures, a vibrational mode should exist, a very low frequency mode, connected to the slushing of one structure over the other. It is the so-called Goldstone mode. It should be possible to measure it, and it gives you an idea about the strength of the interaction. Have you thought about that? Another question, Would you see the kind of effect you're seeing here (multiple moire patterns) on any other substrate?

Markus Lackinger answered: As far as I know, vibrational spectroscopy data for TMA monolayers on graphite are not available. Admittedly, I have not thought about the Goldstone mode so far, but this is definitely an interesting suggestion and it is worthwhile to look deeper into the details. These moiré patterns are easily observed by STM in particular, thus limiting the choice of possible alternative substrates to electrical conductors. In general, the molecule–surface interactions are stronger on metals. This fact bears two implications: (1) commensurate structures with defined equivalent adsorption sites become increasingly more likely, and (2) on even more reactive surfaces, such as Cu(111), deprotonation of the carboxylic acid groups and subsequent formation of metal-coordination bonds with surface adatoms inhibits self-assembly of the hydrogen bond network. With Au(111), on the other hand, the chickenwire structure was observed, yet with an additional contrast modulation through the herringbone reconstruction. So one could think of other weakly interacting surfaces, such as MoS_2 or appropriately passivated metal surfaces.

Rasmita Raval asked: Do you think entropic effects and the inherent anharmonicity of hydrogen bonds are important in stabilizing your incommensurate structures?

Markus Lackinger answered: Your question already implies that the observed structures are the thermodynamically most stable structures. However, the observation of different moiré patterns indicates that the observed TMA structures have slightly different epitaxial relations to graphite.

Even though we cannot rigorously prove it, we propose that the precise final orientations of the TMA domains are already determined in the nucleation phase, where only an approximate orientation with respect to graphite is preferred, but small deviations are still possible. The absence of domain ripening also suggest that the orientation of larger domains is fixed by kinetic entrapment.

Models suggest that the occurrence of incommensurate superstructures is a question of the relative strength of adsorbate–adsorbate *vs.* adsorbate–substrate interactions, where significant insights can already be obtained from relatively simple 1D models.[1] These conditions appear to be met by TMA monolayers on graphite, where the two-fold cyclic hydrogen bond between TMA molecules is relative strong, whereas DFT calculations suggest a relatively small corrugation of the surface potential. So for this type of explanation the anharmonicity is not necessarily needed, but of course more subtle effects cannot be ruled out.

1 P. Bak, *Rep. Prog. Phys.*, 1982, **45**, 587.

Claire-Marie Pradier queried: You never observed an influence of the solvent, is this not surprising?

Markus Lackinger response: For TMA, solvent-induced polymorphism is well documented, resulting in the so called flower-structure for shorter chain length fatty acids as solvents.[1] So we have deliberately chosen two solvents (heptanoic acid and nonanoic acid) for our studies that both afford the same chickenwire structure. For these two solvents, we find a similar orientation of the TMA lattice with respect to the graphite lattice, resulting in comparable moiré patterns and in experimentally indistinguishable TMA lattice parameters. This may be taken as indication that the specific solvent does not affect nucleation and growth of the TMA chickenwire polymorph, and does not affect the intermolecular hydrogen bond strength and length, *e.g.* by screening effects that should depend on the dielectric constant of the solvent.

Overall, this can be taken as evidence that the molecule–molecule interaction in the TMA chickenwire structure is by far dominating over any other possible influence.

1 M. Lackinger *et al.*, *Langmuir*, 2005, **21**, 4984–4988.

Steven De Feyter commented: What is the origin of the appearance of moiré patterns in the STM images? Is it an electronic effect or is it the result of slight topographical changes in the monolayer?

Markus Lackinger replied: In principle both are possible, but I'm inclined to believe that electronic effects dominate. Offhand, I could not imagine any strongly site-specific TMA–surface interactions that would possibly cause the tilting of TMA molecules and hence buckling of the monolayer, even though simulations propose that the oxygen atoms of the carboxylic acid groups might interact more strongly with the graphite surface.[1] On the other hand, we also have tried to study the moiré corrugation as a function of tunneling voltage. However, in contrast to what would be expected for an electronic origin, we could not find any systematic dependency. Maybe complementary NC-AFM experiments that are more, but not exclusively, sensitive to topographic effects could shed more light on the origin of the moiré patterns. Appropriate calculations would also have the potential to address this question, however, these would have to deal with the incommensurability of the TMA monolayer on graphite.

1 F. Shayeganfar and A. Rochefort, *Langmuir*, 2014, **30**, 9707–9716.

Angelika Kühnle said: I want to comment on the issue of whether or not one can differentiate topography and/or electronic effects as an origin of the moiré patterns using AFM. It was kind of suggested that you can use AFM to measure topography. However, like any other SPM technique, AFM does not reveal a signal that exclusively originates from topography. In fact, AFM measures forces. Of course, by combining information from STM and AFM, one gains complementary information, which is often helpful.

Ioan Bâldea opened a general discussion of the paper by James Batteas: I am confused. In your experiments, you see that the *I–V* curves measured on the one day clusters are qualitatively different from the *I–V* curves recorded over the three day and five day clusters. However, the argumentation put forward in your paper on why the one day clusters behave differently from the three and five day clusters has nothing to do with that of your five minute presentation.

In your paper, you based your argument against charge transport *via* tunneling in the three and five day samples on the impossibility of fitting the measured *I–V* curves with cubic polynomials. In two recent papers,[1–2] I discussed in detail the shortcomings of cubic polynomial approaches and the fact that they are unable to correctly describe charge transport *via* tunneling, therefore I will not refer to this issue here.

In your five minute presentation you claimed that, in contrast to the one day samples wherein charge transport proceeds *via* tunneling, in your three and five day samples the charge transport proceeds *via* hopping, and you referred to a Coulomb blockade, a notion that cannot be found in your paper. I would note that strong *I–V* nonlinearities like those you have measured (*cf.* your Fig. 4, DOI: 10.1039/C7FD00118E) are quite unusual for a hopping mechanism; typical *I–V* curves do not so strongly depart from linearity. As a constructive suggestion (the above remarks are not intended as polemics), in order to get more insight into the conduction mechanism you may consider performing transport measurements at variable temperatures and see, for example, whether the transport is thermally activated or not.

1 I. Bâldea, *Phys. Chem. Chem. Phys.*, 2015, **17**, 20217.
2 I. Bâldea, *Phys. Chem. Chem. Phys.*, 2015, **17**, 31260.

James Batteas replied: Thank you for your comments. With regards to the criticism about the charge transport for 3 and 5 days. The truth is that we are fully clear on how to describe the charge transport for the aged clusters. This is a common problem with all molecular electronics, and other groups (*e.g.* Whitesides, McCreery) have also observed such phenomena that they cannot explain (*e.g.* transport over huge distances, >30 Å, without temperature dependence). It is true that describing this as hopping or Coulomb blockade may be more confusing than helping. But we have certainly seen this in our prior low temperature (~4 K) measurements on such clusters made in cross-wire junctions, and expect a similar effect here.[1]

The differences between "hopping" and "tunneling" are much smaller than normally assumed (there is an excellent review by X. Y. Zhu[2]). It mostly refers to the residence time of the passing charge or its degree of occupation of a molecular level (this is also explained in the papers by Galperin, Ratner and Nitzan noted in our paper – see ref. 16, DOI: 10.1039/C7FD00118E). As the molecular level gets closer in energy to the Fermi-level the transport gains more "hopping character". The important effect of the aggregation is to change the energy of the molecular levels, either by narrowing the HOMO–LUMO gap or by stabilizing the excited state (*i.e.* it is easier to charge the molecule within a hopping view).

It is also correct that for the 3 and 5 day clusters it is possible that the charge propagates through the molecule *via* tunneling. In this view, the 3 and 5 day clusters would have many energy levels within the ±2 eV that we measure, and these would produce the sharp steps in conductance (or resonances). What we are really suggesting is that a new conduction channel is opened, which can be understood as either near-Fermi DOS or a completely different effect, like charging of the molecule.

With regards to the fitting, while we recognize the challenges with the cubic model, here we have only used the fitting to the cubic model in a qualitative manner. The argument that the one day clusters can be fitted to a single tunneling channel while the 3 and 5 day clusters cannot, holds regardless of the choice of model (cubic, single model, Simmons, super-exchange – those tunneling models which are simplistic enough to have an analytical form), and perhaps a better distinction is between "off-resonance-tunneling" for the one day and either "resonance tunneling" or "hopping" for the aged clusters.

In summary, our analysis is not perfect and we continue to seek better solutions. We are also actively seeking to carry-out additional low temperature STM measurements to continue to evaluate on a cluster-by-cluster basis the thermal effects, but first we are trying to create well defined cluster structures whereby the local organization can be well controlled, since this will be the only way to then definitively address these questions.

1 A. E. Schuckman *et al.*, *J. Phys. Chem. C.*, 2015, **119**, 13569–13579.
2 X.-Y Zhu, *J. Phys. Chem. B.*, 2004, **108**, 8788–8793.

Ioan Bâldea remarked: I think that extra characterization of your samples would be highly desirable in order to get more insight into the differences between the one day clusters and the three and five day clusters. It would be useful if you can measure frequency-dependent transport, possibly supplemented by measurements of vibrational properties, which may provide information on the adsorption of your porphyrin-based clusters on the Au(111) substrate.

James Batteas replied: This is an excellent suggestion and we are very keen to carry-out AC tunneling spectroscopy on these. In our previous work, we have examined the low temperature inelastic tunneling spectra of similar porphyrin islands, and that is described in detail in the paper noted here.[1]

1 A. E. Schuckman *et al.*, *J. Phys. Chem. C.*, 2015, **119**, 13569–13579.

Manfred Buck said: Referring to the Coulomb barrier observed for the porphyrin thiols, you mentioned the contribution of molecular levels and aggregation. Could you explain their contribution in more detail. Is this effect dependent on the type of porphyrin, *e.g.* whether the porphyrin is metallated or not?

James Batteas answered: The systems we have seen where this occurs have only thus far been in the metallated porphyrins. We have seen this here with the Zn tri-phenyl porphyrin thiol version as well as in our earlier studies of tri-pyridyl porphyrin thiols with Zn.[1] Our hypothesis for this is that the charging energies of the individual molecules are too large, but are reduced to a suitable level (*i.e.* <0.2 eV) when an aggregate of >5–10 nm in lateral dimension is formed *via* pi-stacking.

1 A. E. Schuckman *et al.*, *J. Phys. Chem. C.*, 2015, **119**, 13569–13579.

Steven De Feyter asked: To what extent are the time-dependent effects, *i.e.* the changes in the monolayer structure, affecting any "practical" applications and how do you deal with them?

James Batteas responded: This is actually a very important point. The time-dependent fluctuations in the structures will presumably impact their local order and in that regard also their local electronic properties. Only in having well-defined and ordered structures with minimal variations would you one be able to predictably control the transport if the focus was, say, specifically electronic switching for molecular electronics. For systems such as dye sensitized solar cells however, I would expect reduced sensitivity to these local effects so long as the cluster size was sufficiently large.

Claire-Marie Pradier remarked: Surprisingly, when the time of immersion increases, the thiols tend to desorb and the porphyrin islands shrink; can you comment on that?

James Batteas replied: We believe that the change in cluster size is from a re-organization of the porphyrins in the islands to adopt a more pi-stacked (more closely packed) configuration. The loss of thiols that is observed is due to the desorption of the surrounding dodecanethiol matrix back into solution. This has been seen routinely for alkanethiols on Au when the samples are soaked in solvents such as dichloromethane or ethanol.

Han Zuilhof asked: Firstly, have you considered using multi-sulfur moieties (such as tri-S-adamantane-like systems or even simple di-thiol systems) which can couple your tetra-phenyl system to stabilize interactions with the surface? Secondly, regarding the tetra-phenyl system, why did you pick that in the first place, rather than a *para*-di-phenyl system in which the in-plane steric hindrance (with either the alkyl monolayer or adjacent porphyrin system) would be much (somewhat?) smaller?

James Batteas responded: Yes we have, and in fact we have molecular systems with two alkylthiol linkers to the surface, but we have yet to explore them in detail. We picked this system (actually originally a tetra-pyridyl system) with the vision in mind of building larger hierarchical 3D structures through additional metal–ligand binding. This phenyl ring system was chosen to simply follow along the same lines, but to eliminate possible metal–ligand binding between neighboring molecules. Certainly a *para*-di-phenyl system (as you suggest) would also be a reasonable choice.

David Amabilino commented: Regarding the last point, the tripyridyl system you talked about. There's a system out there, I'm sure you're aware of, where they grow a ferrocene dendrimer off the surface.[1] Could you say what you think about that system with respect to yours, given the results you've seen, especially thinking about the stability of alkyl or aromatic thiolate monolayers on gold?

1 R. Sakamoto, S. Katagiri, H. Maeda, Y. Nishimori, S. Miyashita and H. Nishihara, *J. Am. Chem. Soc.*, 2015, **137**, 734.

James Batteas responded: The system they have shown is very interesting. Stable redox centers, such as Fe, in their paper provide a nice route toward directing hopping in molecular systems. We have considered creating similar assemblies with our pyridyl system to create hierarchical structures that could direct energy flow *via* vectoral energy transfer, but it is clear from their work, that directing electron flow would also be possible. With regards to your comment on alkyl *vs.* aromatic thiol subunits, the stability of both attachment groups is similar, and both suffer long term oxidation effects to sulfonates, which limits their applications in some devices.

David Amabilino said: I was a little surprised from looking at Fig. 1 in your article (DOI:10.1039/C7FD00118E), which shows different sized clusters. Why is it that the large ones you see just after immersion have a completely different behaviour to the ones where you have a long immersion time and then you start to get bigger clusters? How is it different to the initial immersion stages?

James Batteas responded: Our proposed model is that the large aggregates formed in the initial immersion stages are more disordered than those seen following longer immersion times. Here, we suggest that the initially disordered islands slowly organize into more closely pi-stacked structures that are then also more electronically coupled, allowing them to act almost like an "organic quantum dot". This local reorganization is suggested based on the observed reduction in island size as a function of time as the porphyrins in the islands form more closely pi-stacked structures. This results in the reduced charging energies, and contributes to a small HOMO–LUMO gap.

Amar Flood commented: Regarding the parameters that you have available to dial in the polaron stability, you spoke about extending the size, and increasing pi-stacking. Are there other extraneous experimental parameters you might be able to introduce, *e.g.* solvent vapors, to stabilize the build-up of charge that constitutes a polaron? What features of the system (chemical or physical attributes) could support a polaron besides pi-stacking?

James Batteas responded: That is an exceptionally interesting suggestion! Carrying out these studies in a solvent that could further help stabilize the charge on the islands would be quite interesting. Here, we could explore how something like phenyloctane *vs.* octanol would impact this.

Rasmita Raval said: It would appear that a critical number of your functional porphyrin-thiol units have to self-assemble locally in order to create the requisite function. Do you know what "magic" combination is required for a working unit? Presumably the surrounding sea of C_{12} molecules is critical in achieving this, can you comment on this please?

James Batteas replied: What our proposed model suggests is that for the voltage range we are measuring ($\pm$ 2 V), the island dimensions must be on the order of 5–10 nm in order for the charging energy to be sufficiently reduced to observe controlled switching. It is our intention to carry forward a number of patterning experiments to design well organized clusters of specific dimensions to create a clear correlation between the specific sizes, local molecular organization (*i.e.* degree of order) and the observed electronic properties.

Here we are also seeking to address the instabilities of the surrounding C_{12} molecules. Since we have seen their desorption and transition to lower density structures for the longer assembly times, we are now planning to replace them with diacetylene containing thiols of the same chain length where we can photo-crosslink the matrix molecules to increase their stability and hopefully further stabilize the resulting porphyrin-thiol clusters.

Peter Beton asked: Looking at your models of the cluster, is there a possibility of (or have you already been) looking at the optical properties? Porphyrin clusters should change their fluorescence peak positions, are any shifts observed?

James Batteas responded: We have attempted to look at the optical spectra on the surface, but the very low densities have been prohibitive in obtaining sufficient signals (of either absorbance or fluorescence) for us to evaluate the expected changes in optical spectra that would accompany aggregation. To do this, we would likely need to carry-out near-field spectroscopy measurements, which we currently do not have the capabilities to do.

Peter Beton continued: You give a figure in your paper of about 0.6 nm for the pi–pi stack, that seems a bit bigger than the figure that's normally quoted?

James Batteas responded: Yes, this is a bit larger than normally reported. We have found in our prior STM studies a pi-stacking spacing of 0.5–0.6 nm,[1] which is slightly larger than the value of 0.3 nm seen for many other pi–pi systems. We attribute this larger spacing to the steric hindrance associated with the canted phenyl rings on the porphyrin macrocycle. We have also carried out DFT calculations for dimers of these types of molecules and find a similar 0.5–0.6 nm spacing for the energy optimized structure.

1 A. E. Schuckman *et al.*, *J. Phys. Chem. C.*, 2015, **119**, 13569–13579.

Rasmita Raval opened the discussion of the paper by Julien Gautrot: Really nice work. How much is understood about the nano-protein layer that is formed, which appears to be the key to how the cell subsequently reacts and behaves?

Julien Gautrot responded: Very little is known of the structure and properties of such protein layers. In the case of the interfaces generated between the fluorinated oil studied and the culture medium used for HaCaT cells, we characterised the mechanical properties of the interfaces generated and found differences in interfacial shear properties. Further studies will characterise the interfaces generated as well as other designs fully.

Claire-Marie Pradier commented: You describe the role of protein adsorption on cell adhesion; can it be considered similar to the mechanics of biofilm formation known to start with proteins and extracellular polymer adsorption?

Julien Gautrot replied: There are similarities in the fact that proteins and other biopolymers are assembled at interfaces by bacteria to help stabilise biofilms. Differences are in the importance of the local, nanoscale, mechanical environment of the adsorbed layers. Biofilms do not exert contractile forces, but the extracellular matrix may help biofilms resist macroscopic mechanical challenges.

Yuri Diaz Fernandez asked: Regarding the rheology data in Fig. 5 of the paper (DOI: 10.1039/C7FD00091J), if I interpret this correctly, the protein film is formed irrespectively of the presence of the surfactant. Could you comment on the function of the surfactant on the proliferation of the cells on the interface?

Julien Gautrot responded: This is correct. In this respect, PFBC is perhaps better described as a co-surfactant and proteins contained in the tissue culture medium are thought to act as the primary surfactant. What we propose is that the co-surfactant is strengthening the mechanics of the protein film formed at the interface.

Claire-Marie Pradier remarked: There's an ongoing controversy about the scale (size of the surface defect or roughness), which can influence cell adhesion. It is even claimed that defects on the nanoscale may influence cell adhesion. Could this be related to, *e.g.*, protein adsorption?

Julien Gautrot responded: Integrins are important surface receptors mediating cellular sensing of the extra-cellular matrix. Spacing between such integrins in the range of 60–70 nm has been proposed as a critical distance to sustain the formation of focal adhesion and the establishment of a stable cytoskeleton (see the work of Spatz). That was proposed to correspond to the distance between integrins in clusters preceding the formation of focal adhesions, and the distance between ligand sites on ECM fibres such as collagen fibres. Hence, a direct sensing of ligand assembly at scales below this critical distance would require other mechanisms rather than simple geometrical considerations. For example, nanoscale changes in surface chemistry (roughness, slight changes in the chemical composition of the surface such as in monolayers or in polymers or ceramics with very slight differences in chemistry) can have an important impact on the adsorption of proteins and the structure/morphology of the assemblies formed (see the work of Salmeron-Sanchez). This in turn may significantly impact cell phenotype, so indirect effects. That being said, integrins have been shown to be sensitive to synergistic ligand binding (RGD/PHSRN for example), with spacing below this critical distance.

Claire-Marie Pradier asked: In your paper, you mention a PDMS material surface whose exact composition was not given, do you have any ideas about this? And could it influence protein adsorption?

Julien Gautrot replied: That is correct, we do not know the precise composition of the interface formed between Sylgard 184 (a silicone material routinely used in microfabrication). It is reasonable to expect albumin molecules to adsorb at the surface of liquid silicone as they do to many other hydrophobic surfaces and liquids, but the molecule(s) that may be involved in the further maturation of the associated interfaces (from a physico-chemical and mechanical point of view) are not known. Vinyl functions present in Sylgard 184 would not be expected to directly result in protein conjugation under the experimental conditions used, but they are abundant in Sylgard 184. Other molecules that may be present are residual monomers from the synthesis of the silicone polymers.

Ahmad Jabbarzadeh asked: Have you measured, or do you have an estimate of the thickness of the protein interface between PDMS and the cell? Do you expect the results to be dependent on the thickness?

Julien Gautrot replied: We have measurements of thicknesses in the case of other systems, but not the particular case of interfaces conditioned with medium or serum. Typically, thicknesses measured for protein adsorption at the surface of hydrophobic oils are in the range of 30 to 50 nm.

Stuart Clarke opened a general discussion of the paper by Han Zuilhof: It's nice to see other techniques being developed in this area. As mentioned previously, there is a suite of experimental methods that can now be used for the study of these complex solid/liquid interfaces, each providing part of the story. In this case, I might suggest neutron reflectivity for consideration.

The neutron reflection method is *in situ* (the adsorbed layers remain in the liquid and can be studied without removal or drying) and the approach is non-destructive.

One can determine the absolute composition (mol m^{-2}) and some structural information, particularly the structure normal to the surface (often represented as layers). However, usually there is limited in-plane structural information and other approaches are preferred.

Although not an inherently fast method, due to the relatively low flux of neutrons, in favourable cases timescales of second/minutes should be accessible, which should be suitable for the particular problem of interest here. There are some much faster neutron reflection methods (such as stroboscopic approaches) which can assess much faster changes.

Importantly, isotopic variation enables one to highlight the behaviour and location of different components in a complex system (particularly hydrogen *vs.* deuterium). In this case, for example, the binding of the hydrogen bonded guest should be readily accessible.

See the studies of Bob Thomas and Jeff Penfold.[1,2]

Question: We have a little experience with MS and have struggled to get quantitative data (in absolute units). Partly this arises due the competition of the different components for ionisation. Hence, the same amount of a component of interest, in the presence of different amounts of other species, can give different ion intensities at a fixed *m*/*z*. Hence, although we can readily follow changes in a component, the concentration is not generally absolute. Do you have any helpful comments on this matter?

1 G. Fragneto, R. K. Thomas, A. R. Rennie and J. Penfold, Neutron Reflection from Hexadecyltrimethylammonium Bromide Adsorbed on Smooth and Rough Silicon Surfaces, *Langmuir*, 1996, **12**, 6036–6043.
2 A. R. Rennie, E. M. Lee, E. A. Simister and R. K. Thomas, Structure of a Cationic Surfactant Layer at the Silica-Water Interface, *Langmuir*, 1990, **6**, 1031–1034.

Han Zuilhof replied: In response to your comment, we have simply never considered neutron scattering as a tool – it may indeed be interesting to study the supramolecular binding of a well-studied case (for now: our UPy system) to investigate the potential thereof. I will undertake further interactions with Dr Clarke to see what might be feasible.

In response to your question, there are 2 facets to be considered here. First, the intrinsic noise in a technique like DART-MS, that is intrinsic to it being an atmospheric MS technique. The standard deviation can, in principle, be reduced to <10% by repetitive measurements. This is the reason why all our MS measurements are done in hexaplet. In addition, we look at the development of an MS signal over time – this can be benchmarked in individual cases by, *e.g.*, independent XPS measurements. However, if the question is: can we, from a single measurement (even if performed in *e.g.* hexaplet), determine how much material is on the surface? Then the answer is: no, since the ionizability of compounds and issues like ion suppression might play a role. So we have highly accurate, but relative amounts. Great for kinetics, not (at least not *via* a trivial method) for quantification on a surface. In the latter case, XPS is to be preferred.

Steven L. Tait commented: The system you studied here is covalently linked to the surface with a quadruple hydrogen bond to the upper part of the adsorbate. Is this type of configuration necessary for the DART method? Would you please comment on what types of samples the DART method is able to analyze? Can the charge state and other chemical characteristics of the sample be analyzed accurately by DART?

Han Zuilhof responded: We analysed a great variety of monolayers in the past; see some of our papers for examples.[1–3] From these studies we learned which groups cleave easily under these DART conditions, and which groups do not (or do not yield recognizable fragments). The structure of the UPy dimer system was chosen in such a manner that the most easily cleavable moiety would be the H-bonds. In other studies, we have built-in an easily cleavable reporter moiety in the groups, so as to be able to monitor the formation of even highly stable groups, as in a Diels Alder reaction. See, *e.g.*, ref. 4 and 5, especially Fig. 3. Regarding monitoring the charge state (*e.g.* of inorganic or hybrid complexes), I am afraid to have to say I do not know, but think I am in the company of about 7 billion others. Likely, the ionizability will be determined partially by the oxidation state, but I am not aware of studies of this facet. This is certainly something of interest for further work.

1 R. K. Manova *et al.*, Ambient Mass Spectrometry of Covalently Bound Organic Monolayers, *Chem. Commun.*, 2013, **49**, 922–924.
2 R. K. Manova *et al.*, Ambient Surface Analysis of Organic Monolayers using Direct Analysis in Real Time Orbitrap Mass Spectrometry, *Anal. Chem.*, 2014, **86**, 2403–2411.
3 R. Sen, J. Escorihuela, F. van Delft and H. Zuilhof, *Angew. Chem., Int. Ed.*, 2017, **56**, 3299.
4 R. Sen *et al.*, *Angew. Chem., Int. Ed.*, 2017, **56**, 4130–4134.
5 R. Sen, D. Gahtory, J. Escorihuela, J. Firet, S. P. Pujari and H. Zuilhof, *Chem.-Eur. J.*, 2017, **23**, 13015.

Olga Barykina asked: Is there any restriction on the surface of the analyzed material? Can it be an organic pigment surface or any other surface made of small organic molecules? Is DART analysis suitable for this? How do you subtract signals from the surface itself? Do you run the background?

Han Zuilhof responded: There are no restrictions. We analysed plastics, inorganic materials – flat and micro/nanostructured surfaces, *etc.* DART is, as far as we have seen, limited to the analysis of "relatively small" molecules, such as m/z < 2 000. Subtraction is not required, as the substrate typically does not display these signals (namely, masses measured with a precision of 4 digits, *e.g.* 234.5678 amu). In addition, the m/z ratio should be > *ca.* 100, as otherwise the atmospheric background is too high.

Giovanni Costantini said: Could you please comment on differences or similarities between DART and DESI?

Han Zuilhof replied: Both DART and DESI have their own positives and negatives in the study of surface-bound materials. Given their different modes of ionization, each also yields slightly different fragmentation patterns.

We have both types (DART and DESI) in our lab, and used both. In our hands, DART turned out to be much more stable (it works all of the time...) than DESI, and by now we have almost turned to DART exclusively.

There are again other, more recent, ionization sources that we are also using, such as LAESI. This offers a clearly better lateral resolution than DART for imaging MS, but requires sufficient water to ionize a sufficient amount of material. Our organic monolayers on surfaces are not so suitable for that, but, for example, coatings on fibres that we recently studied with that technique worked beautifully.[1]

1 F. A. M. G. van Geenen, M. C. R. Franssen, A. H. M. Schotman, H. Zuilhof and M. W. F. Nielen, *Anal. Chem.*, 2017, **89**, 4031–4037.

Karl-Heinz Ernst asked: I would also like you to compare DART to different techniques like ToF-SIMS, for example. You could use these techniques, freeze the sample, go into vacuum, and then you have depth profiling through thicker layers, or chemical imaging of cells that were cut in slices with a microtome, then you have lateral resolution plus depth information. Do you think you can install lateral resolution in your system by scanning the beam in X and Y?

Han Zuilhof answered: Yes, lateral resolution is available (we have it routinely on our system), and limited by the width of the DART beam. While this can be optimized to the 100 micron scale (in the literature, not by us), ToF-SIMS allows for a much higher lateral resolution: with DART this is typically on the scale of hundreds of microns. However, DART is much milder, and really limited to the top nanometers of a sample. In addition, DART-HRMS can be used on a Tuesday for DART-HRMS measurements of a monolayer, be decoupled (in 5 min) to turn into an LC-MS on Tuesday night and Wednesday, and be transformed (within 30 min) back into a DART-HRMS system again by Thursday morning. In other words, the versatility of use is much bigger, and the costs very low (you probably already have a suitable MS in your department, and a DART source costs only ~25k $).

Pol Besenius commented: I enjoyed your talk and paper, and I think your analytical approach is very elegant for people that are interested in the preparation of dynamic and functionalised surfaces. With respect to your exchange kinetics of the UPy units, your schematic picture is a bit misleading given that both the surface-bound and solution based quadruple hydrogen bond units do self-associate into dimers. Have you looked into this using spectroscopic tools and is the self-association important for the kinetics of the formation of the heterodimers? In addition, would you expect the exchange kinetics to be faster if you were to apply a hydrogen bonding partner in solution that has a much lower driving force for the formation of self-complementary dimers?

Han Zuilhof replied: First of all: yes, the ChemDraw based cartoon simplifies the situation a lot (but is hopefully not misleading). As we demonstrate and discuss extensively in the paper, there are eventually free (unreacted) NH_2 groups, undimerized UPy moeities, surface-bound UPy moieties dimerized with other surface-bound moieties, and surface-bound UPy moieties dimerized with UPy units coming from solution. The latter take up only about 10% of all surface sites.

The self-interaction of surface-bound UPy units is important. Perhaps the clearest picture thereof comes from our modelling studies, which show extensive self-dimerization of surface-bound UPy units as long as no soluble UPy units are around, and still significant self-dimerization with the soluble UPy units present.

Your final question is an intriguing one (as well as the suggestion for NaPy derivates that you made). We are not fully sure of the answer, but/and think that this is indeed a great topic for further studies.

Rasmita Raval remarked: I can see the value of your DART-HRMS technique in identifying molecular species at an interface and indicating the strength of their interaction at that interface. However, I am unclear about whether it gives us mechanistic information on the specific nature of the self-assembly or supramolecular chemistry that is involved. Can you please elaborate on this? Would it be possible, using highly defined interfaces, to create a database that would allow specific binding events to be recognised using DART-HRMS?

Han Zuilhof responded: The issue is what "specific nature" means here. As shown in the paper, DART-HRMS has provided rate information for supramolecular processes in a very facile manner. This is a big plus, as such rate information for unlabelled systems is rare to absent up to now. In addition, by measuring forward and backward rates, it allows us to derive equilibrium data. That is also a big step forward for supramolecular processes at a surface. Currently, I would argue that binding events that are in the range of $k = 10^2 - 10^8\ M^{-1}$ should be measurable, if the components are in the mass range of DART-HRMS. Such a range would, indeed, allow for the formation of a database, but with the current range of processes truly studied in detail (only one, namely our UPy case, with a few more in the pipeline), this needs to be shown in the coming 5 years.

Steven De Feyter asked: A question on sample preparation. To what extent does sonication affect the results (at the level of dimerization)? Does sonication have an impact on the final structure?

Han Zuilhof answered: The results of sonication depend on the solvent. In solvents like toluene, DCM, *etc.*, we only find removal of the physisorbed, non-specifically attached material. In other words, after the fast removal of some fraction of UPy-based material, we observe a plateau intensity. We attribute this plateau to only H-bonded UPy molecules. This plateau is fairly stable, as the energy involved in sonication is not so large as to break up the H-bonded UPy dimers. Prolonged sonication very slowly removes material from the surface, but we did not study to which degree this loss of H-bonded dimers is aided by the sonication (rather than just being an unassisted reverse process). In good solvents the dimer is breaking up, and sonication speeds this up slightly.

Claire-Marie Pradier asked: To measure the amount of attached UPy dimers, you calculate the F over P ratio, obtained by XPS, right? Did you consider attenuation by the organic layer, plus the difference in the sensitivity between F and P?

Han Zuilhof responded: Yes, we measure F/P. Also, N/P is informative. We do correct routinely for differences in the sensitivity using response factor corrections – for the first (taking account of the attenuation) we did not, given the relatively thin layers (typically 1.5–2 nm) – for such scales the attenuation is measurable, but not so big.

Claire-Marie Pradier opened a general discussion on the topics raised during the session: Would anyone like to comment on the techniques one should/could use to characterize surfaces under vacuum? I would like to mention IR spectroscopy, applied in the reflection mode, at grazing angle. It's a very sensitive technique which gives information about the nature of the chemical groups that are present, even at the monolayer scale, in addition to XPS, as well as on the molecule orientation/conformation.

Deepak Dwivedi responded: We have used Raman spectroscopy to characterize the film formed on carbon steel substrate and it was an *in situ* characterization, but at the same time, I would like to mention that it was not under vacuum.

I think it is possible to make cells which can work under vacuum. Making cells similar to *in situ* TEM cells can be helpful in this regard.

Faraday Discussions

PAPER

Generic nature of long-range repulsion mechanism on a bulk insulator?†

J. L. Neff,[a] A. Richter,[a] H. Söngen,[ab] C. Venturini,[c] A. Gourdon,[c] R. Bechstein[a] and A. Kühnle *[a]

Received 3rd March 2017, Accepted 24th March 2017
DOI: 10.1039/c7fd00089h

Dynamic atomic force microscopy measurements are reported that provide evidence for the presence of long-range repulsion in molecular self-assembly on a bulk insulator surface. We present the structures formed from four different benzoic acid derivatives on the (10.4) cleavage plane of calcite kept in ultra-high vacuum. These molecules have in common that they self-assemble into molecular stripes when deposited onto the surface held at room temperature. For all molecules tested, a detailed analysis of the stripe-to-stripe distance distribution reveals a clear deviation from what would be expected for randomly placed, non-interacting stripes (*i.e.*, geometric distribution). When excluding kinetic effects during growth, this result gives evidence for a long-range repulsion mechanism acting during the assembly of these stripes. The fact that this finding is robust against changes in the molecular structure indicates a generic nature of the observed mechanism, implying a ubiquitous origin such as electrostatic repulsion. Finally, we discuss parameters that might affect the unambiguous observation of this generic repulsion under specific experimental conditions.

1 Introduction

Molecular self-assembly is recognized as a most versatile tool to create molecular structures at surfaces.[1–3] Tuning the subtle balance between intermolecular as well as molecule–surface interactions is known to allow for tailoring the resulting supramolecular network in a rational fashion.[4] So far, the majority of the systems investigated make use of short-range attraction between the molecular building blocks. Besides van der Waals and attractive electrostatic forces, hydrogen bonds constitute a popular binding motif for molecular self-assembly due to their selectivity and directionality.[5,6] Exploiting intermolecular attraction has resulted

[a] *Institute of Physical Chemistry, Johannes Gutenberg University Mainz, 55099 Mainz, Germany. E-mail: kuehnle@uni-mainz.de*

[b] *Graduate School Materials Science in Mainz, Staudinger Weg 9, 55128 Mainz, Germany*

[c] *CNRS, CEMES, Nanoscience Group, BP 94347, 29 Rue J. Marvig, 31005 Toulouse, France*

† Electronic supplementary information (ESI) available: Overview over all molecules tested in this study. See DOI: 10.1039/c7fd00089h

in an impressive control over the order of the resulting structures, which range from clusters,[7,8] one-dimensional chains,[6,7] two-dimensional overlayers to most complex assemblies such as host–guest networks.[9] In contrast, intermolecular repulsion has been comparatively little explored for inducing order during molecular structure formation on surfaces in a rational fashion.[10–16] In most of the reported examples, electrostatic interaction has been identified as the origin of the repulsion.[17] While polar molecules carry an intrinsic dipole moment that can directly affect the ordering,[18,19] adsorption-induced dipoles of apolar molecules have also been shown to result in long-range repulsion.[13,14] An example for the latter is tetrathiafulvalene deposited onto Au(111), which has been shown to arrange in an ordered fashion with a range of up to 4 nm.[13] In a recent study, we have shown that 3-hydroxybenzoic acid (3-HBA) on calcite (10.4) self-assembles into molecular stripes.[16] These stripes exhibit an ordered arrangement induced by long-range repulsion that can be observed even at distances as large as 16 nm.

It should be noted, however, that the observation of an ordered arrangement of molecular structures is not necessarily a proof of the existence of repulsive interaction, but it can be due to kinetic effects during growth.[20–22] Such correlated growth kinetics have been demonstrated experimentally, *e.g.*, during the growth of pentacene islands on silicon oxide surfaces.[22,23] A possible way to exclude such deposition-dependent effects is comparing the structures obtained from different deposition protocols. For the structures formed from 3-HBA we have shown that the specific deposition protocol appears to be irrelevant,[16] indicating that the observed order is not determined by the specific diffusion and desorption kinetics,[22] but is, indeed, caused by long-range repulsion.

Here, we address the question of whether this example of long-range repulsion is special for 3-HBA or represents a general mechanism that might be used for rationally controlling molecular self-assembly on a bulk insulator surface. To this end, we have investigated twenty-eight different benzoic acid derivatives (for a full list, see the ESI†). We select those four molecules (including 3-HBA) that self-assemble into stable, uni-directional stripes when deposited on calcite (10.4) held at room temperature. A detailed analysis of the next-neighbour stripe distance distribution reveals that none of the molecules tested exhibits a geometric distance distribution,‡ which would be the distribution expected for randomly placed, non-interacting stripes. When excluding correlated growth kinetics, our results provide experimental evidence for the presence of a long-range repulsive interaction, regardless of the specific molecular structure. This robustness against variations in the molecular structure and the inner structure of the resulting stripes points towards a ubiquitous origin of the long-range repulsion, supporting the interpretation of an electrostatic repulsion mechanism. We discuss possible scenarios that might affect the unambiguous observation of this long-range repulsion under specific experimental conditions, such as elevated temperatures or kinetic trapping that might prevent to arrive at the thermodynamically favoured spacing. Keeping these parameters in mind might allow for a controlled use of long-range repulsion in future studies, enriching the available parameter space for tuning molecular self-assembly.

‡ For a full derivation of the geometric distribution and the validity to apply this to randomly placed, non-interacting stripes see the supporting information in ref. 16.

2 Experimental details

2.1 Dynamic atomic force microscopy measurements

Dynamic atomic force microscopy (AFM) measurements were performed with a variable-temperature AFM from ScientaOmicron (Taunusstein, Germany) operated under ultra-high vacuum (UHV) conditions. We used doped silicon cantilevers from Nanosensors (Neuchâtel, Switzerland) with a typical spring constant of 40 N m^{-1} and eigenfrequency of 300 kHz in UHV. A phase-locked loop and amplitude controller from NanoSurf (Liestal, Switzerland) was used for signal demodulation and amplitude stabilization. A typical oscillation amplitude of 10 nm was used. All measurements shown here were taken with the sample kept at room temperature.

2.2 Sample preparation

The calcite samples used were purchased from Korth GmbH (Altenholz, Germany). Prior to each experiment, the sample was cleaved *in situ* and annealed at 640 K for one hour to obtain a clean (10.4) cleavage plane (Fig. 1a). The molecules were obtained from Sigma-Aldrich (Munich, Germany), Alfa Aesar (Karlsruhe, Germany) or, in the case of 4-ethynylbenzoic acid (4-EBA), were synthesized from commercial 3-iodobenzoic acid by analogy with literature procedures.[24,25] A home-built Knudsen cell was used for depositing the molecules onto the freshly prepared calcite surface held at room temperature. Twenty-eight different benzoic acid derivatives were tested in this study in total (for a full list and molecular

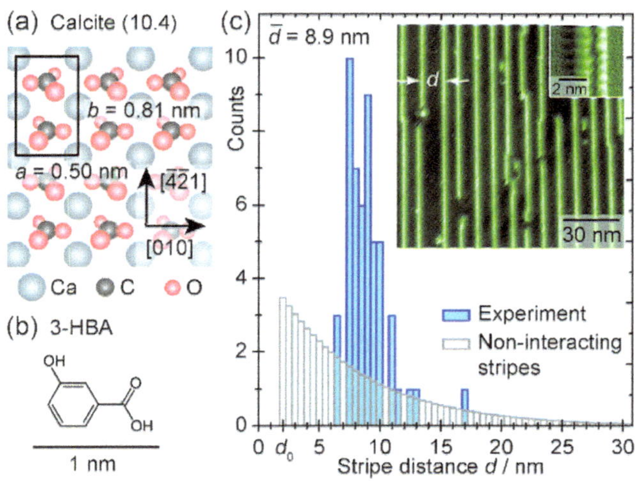

Fig. 1 (a) Model of the calcite (10.4) cleavage plane. The surface unit cell consists of two calcium cations and two carbonate groups. (b) Structural formula of 3-hydroxybenzoic acid (3-HBA). (c) Molecular double rows formed from 3-HBA on calcite (10.4). As can be seen, these stripes align along the [$\bar{4}\bar{2}1$] substrate direction. From the AFM image, an equidistant appearance of the stripes is evident. Inset: zoom onto a stripe showing the internal zig-zag structure of the stripe. The experimentally obtained next-neighbour stripe distance distribution (blue bars) clearly deviates from the distance distribution that would be obtained for randomly placed, non-interacting stripes (grey bars, geometric distribution). The minimum or nearest possible stripe distance is denoted by d_0.

structures, see the ESI†). Here, we focus on four molecules, namely 3-hydroxybenzoic acid (3-HBA, Fig. 1b), 3-aminobenzoic acid (3-ABA), 3,5-dihydroxybenzoic acid (3,5-DHBA), and 4-EBA. The sublimation parameters to arrive at submonolayer coverages were obtained from a different setup equipped with a quartz crystal microbalance.

3 Results and discussion

3.1 Proof-of-principle: molecular stripes from 3-HBA

As reported previously, when deposited onto calcite (10.4) kept at room temperature, 3-HBA self-assembles into molecular zig-zag rows (see inset in Fig. 1c) aligned along the $[\bar{4}\bar{2}1]$ substrate direction, indicating that short-range attraction dominates along the row direction.[16] The specific nature of this attraction remains speculative at the moment but appears irrelevant for the generic repulsion discussed in this work.§ The molecular stripes exhibit a strikingly ordered arrangement on the surface with an equidistant appearance, as seen in the AFM image shown in Fig. 1c. To obtain further insights into this ordering, we analyse the stripe distance distribution (blue bars in Fig. 1c) and use its standard deviation to quantify the degree of order. For randomly placed, non-interacting stripes, a geometric distribution is obtained (indicated by the grey bars in Fig. 1c). As can be seen in Fig. 1c, the distance distribution for 3-HBA at the given coverage (quantified using the mean stripe distance $\bar{d}$ of 8.9 nm) is clearly different from a geometric distribution. Moreover, the standard deviation of the distance distribution for 3-HBA (1.8 nm) is significantly smaller than the standard deviation of the corresponding geometric distribution (7.2 nm). Clearly, the stripes exhibit an ordering that corresponds to a non-random distribution.

For molecules deposited on a surface, such a non-random distribution can have two different origins. It is known that correlated growth kinetics can result in an ordered arrangement even in the absence of intermolecular interactions.[20,21] Thus, only when such growth kinetic effects can be excluded can the deviation from the geometric distribution be unambiguously assigned to the presence of a long-range repulsion mechanism. To exclude correlated growth effects for 3-HBA, we have previously performed experiments with varying deposition protocols and confirmed that the same stripe distance distribution is obtained regardless of the specific deposition sequence.[16] In a deposition sequence, we have demonstrated that the stripe distribution is adjusted to changes in the molecular coverage. This is clear evidence for the existence of a repulsive driving force causing this re-arrangement. Here, we further discuss a sequence of AFM images collected at the same sample position to assess the molecule's mobility at a fixed coverage (see the ESI for a complete 13 hours video†). In Fig. 2a, an AFM image with the characteristic 3-HBA stripe structure is shown. Fig. 2b shows the changes to another AFM image obtained at the same position about three hours later. We mark vanishing structures by blue colour in the initial image (Fig. 2c), while red colour is associated with newly-formed structures in the final image (Fig. 2d). As can be seen, the stripes do not only shrink and grow in size at the stripe ends, but entire stripes appear and vanish. This indicates sufficient

§ Analysing the stripe length distribution should shed light on the strength of this short-range attraction. However, for simplicity we ignore the finite stripe length in this work.

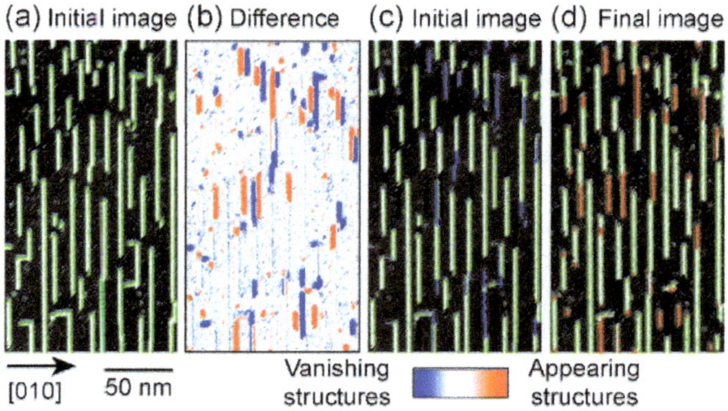

Fig. 2 Two AFM images from a series of images (see the ESI for the complete series†) illustrating the mobility of 3-HBA. (a) Initial image showing the characteristic 3-HBA stripe structure. (b) Changes to an image taken at the same sample position about three hours later, calculated according to: difference equals final image minus initial image. Blue colour marks vanishing structures in the initial image (c) and red colour marks appearing structures in the final image (d).

mobility for adopting the thermodynamic equilibrium structure within the time span given by the experiment, *i.e.*, any initial processes such as nucleation and ripening are completed and do, therefore, not affect the discussion made here. Even if an initial arrangement might be affected by correlated growth kinetics, the experimentally proven mobility would eventually result in a geometric distribution if long-range repulsion would be absent. The fact that the observed distance distribution deviates from the geometric distribution is, thus, clear experimental evidence for a long-range repulsion between the stripes. In our previous work,[16] we speculated that this repulsion arises from electrostatic repulsion of adsorption-induced dipoles. While we cannot prove that the observed repulsion is of electrostatic origin, we can calculate from a simple model – ignoring details such as diffusing single molecules and finite stripe lengths – that electrostatic repulsion of adsorption-induced dipoles with a reasonable dipole moment is of sufficient strength to explain the observed ordering.[16] In this previous work, we made an estimate as to how large an adsorption-induced dipole moment is needed for observing the standard deviation as we did in our experiments. For this, we performed MC simulations that gave a necessary dipole moment of 30 D/$\sqrt{(n)}$, where n is the number of molecules in a critical stripe interacting with infinitely long neighbouring stripes. When we assume that a single (neutral) molecule creates an adsorption-induced dipole normal to the surface of about 5 D, then only critical stripes of at least 7 nm length would reveal the observed order (to arrive at the total dipole moment of 30 D). If we assume the molecules to deprotonate, then a single molecule would have a dipole moment of about 15 D, which then reduces the necessary number of molecules in a critical stripe to four.

3.2 Generic nature: further evidence

Here, we address the question of whether the observed long-range repulsion is specific for 3-HBA or might be of a general nature. When considering electrostatic repulsion of adsorption-induced dipoles, we should expect the effect to be present for other molecules forming molecular stripes as well. To this end, we have investigated the self-assembly of twenty-eight different benzoic acid derivatives and focused on those that form uni-directional rows. We found three further molecules, 3-ABA, 3,5-DHBA and 4-EBA, that form stable molecular stripes orienting along the $[\bar{4}\bar{2}1]$ direction. Representative images of the stripes along with the corresponding stripe distance distributions are given in Fig. 3. In each distance histogram, the experimentally obtained distance distribution (blue bars) is compared with the geometric distribution for the respective coverage (grey bars).

As can be seen from the clear deviation of the experimental results from the geometric distribution, none of these experimentally obtained distributions can be explained by randomly placed, non-interacting stripes. Thus, when excluding correlated growth kinetics, a long-range repulsion mechanism must be present for all the molecules shown here.

It is interesting to note that this finding is not limited to the specific zig-zag structure of the stripes, which we identify not only for 3-HBA (see inset in Fig. 1c), but also for 3-ABA and 3,5-DHBA. To demonstrate this, we refer to the stripes that are formed by 4-EBA. These stripes are qualitatively different from the zig-zag structure of the three other molecules discussed before as they exhibit a (1×1) internal structure (see inset in Fig. 3c). Still, the distribution of these stripes shows a clear deviation from the geometric distribution. Excluding correlated growth kinetics during the assembly, these results again give evidence for the presence of a long-range repulsion. From this finding we can clearly deduce that the long-range repulsion is not linked to a specific stripe structure, indicating the generic nature of this mechanism.

3.3 Which parameters affect the degree of manifestation?

In the following, we want to discuss parameters that might affect a pronounced manifestation of long-range repulsion during molecular self-assembly in an equidistant arrangement. Obviously, at elevated temperatures the manifestation will become less pronounced. Also, short-range attractions can disturb an equidistant arrangement. Furthermore, we address the impact of kinetic trapping. These parameters need to be considered when aiming for taking advantage of

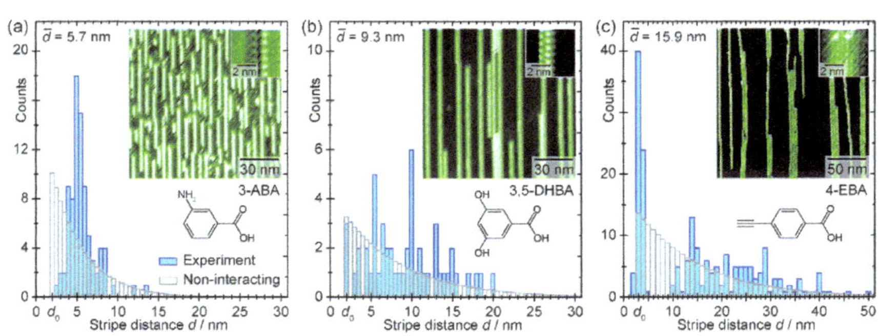

Fig. 3 Representative images and corresponding next-neighbour stripe distance distributions (blue bars) for (a) 3-ABA, (b) 3,5-DHBA and (c) 4-EBA. The geometric distribution calculated for the same coverage is shown with grey bars.

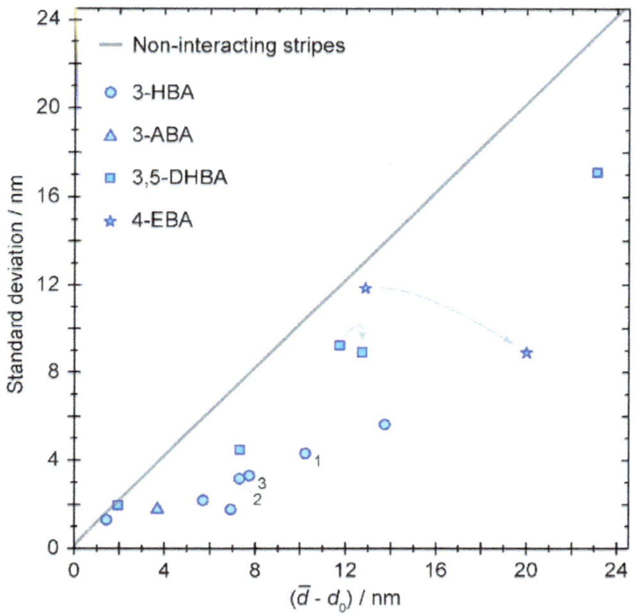

Fig. 4 Experimentally obtained standard deviation as a function of coverage for the four molecules discussed here. The grey line shows the standard deviation that is expected for randomly placed, non-interacting stripes (*i.e.*, a geometric distribution). Two distance distributions, namely one for 3,5-DHBA and one for 4-EBA, were analysed in two ways: first taking all stripe distances into account (origin of arrows) and, second, using all stripe distances except for the peak at d_0 (end of arrows). To exclude correlated growth effects, sequential deposition experiments can be carried out. We tested this for 3-HBA by comparing an experiment obtained after sequential deposition (corresponding standard deviations are marked by (1) for the first and (2) for the additional second deposition) with the situation obtained after a single deposition arriving at a similar coverage (marked by (3)). The similarity between the obtained values for (2) and (3) indicates sufficient stripe mobility.

a rational design of attractive short-range and repulsive long-range interactions in molecular self-assembly.

Elevated temperatures. When increasing the temperature, excited states become increasingly populated, which will inevitably result in a deviation from the ground state, *i.e.*, a perfect equidistant arrangement. Eventually, at a given temperature, the order might become difficult to detect in the experimental data.

Short-range attraction. The fact that stripes can be found directly next to each other for 4-EBA even for the low coverage shown in Fig. 3c sheds light onto a further interesting aspect: for 4-EBA, the short-range intermolecular attraction along the [010] direction (perpendicular to the long island axis) appears to favour broader islands composed of up to four stripes to be formed as compared to single stripes. Obviously, the above discussed long-range repulsion must also be present within the islands, but if the short-range attraction is large enough, islands can form. In fact, the observed island width is governed by the interplay between short-range attraction and long-range repulsion. This short-range attraction is evident in the displayed histogram for 4-EBA from a clear additional peak at d_0, the nearest possible neighbour distance. Such a peak is also present, but less dominant, in the displayed histogram for 3,5-DHBA in Fig. 3b. For high molecular coverages, the peak at the nearest neighbour distance gets more and more pronounced for all molecules. To quantify the degree of order, we plot the experimentally obtained standard deviation as a function of coverage and compare this standard deviation with what would be expected for randomly placed, non-interacting stripes (Fig. 4). As can be seen, all experimentally obtained standard deviations are clearly smaller than the value for a geometric distribution with the same coverage. We attribute the peak at d_0 to nearest neighbour attraction. In such a case, the standard deviation is clearly no longer a good measure for the induced order. Thus, in this case, we need to discard this first peak from the analysis of the distance distribution. We have done that for two cases of 3,5-DHBA and 4-EBA. The resulting two values for the standard deviation (square for 3,5-DHBA and star for 4-EBA) move towards even lower values as indicated by the two arrows in Fig. 4.

Kinetic trapping. Kinetics can have two different effects on the self-assembly. While correlated growth kinetics – as discussed above – can induce an ordered arrangement even in the absence of long-range repulsion, kinetic effects can also hamper the manifestation of long-range repulsion in the obtained structures. This situation occurs when the stripe mobility is not sufficient to arrive at the thermodynamic equilibrium. In such a case, a possibly existing repulsion might not manifest itself in the corresponding distance distribution.

4 Conclusions

In conclusion, we have presented four molecules that self-assemble into molecular stripes due to short-range attractive interactions. A detailed analysis of the next-neighbour stripe distance distributions reveals that all these arrangements clearly deviate from randomly placed, non-interacting stripes. When excluding correlated growth kinetics, this deviation provides unambiguous evidence for a long-range repulsion being present during the assembly of the stripes. The fact that the repulsion is observed regardless of the molecular structure and also regardless of the inner structure of the molecular stripes indicates a generic mechanism. We propose electrostatic interaction as a possible origin for the repulsion because of its ubiquitous nature. We discuss parameters that might hamper an experimental manifestation of the long-range repulsion during self-assembly in an equidistant arrangement. Obviously, elevated temperatures result in a deviation from the ground state, which counteracts an equidistant ordering. Short-range attraction can result in island formation which gives rise to an additional peak in the distance histogram at the nearest neighbour distance. Additionally, high diffusion barriers might result in a kinetically trapped structure that cannot develop into an ordered arrangement. With these insights, adjusting the balance between (anisotropic) short-range attraction and long-range repulsion appears to be a promising strategy for a rational design of the resulting structures in terms of the distance distribution.

Acknowledgements

We gratefully acknowledge financial support from the EU through grant PAMS (seventh framework program GA 610446). H. S. is a recipient of a fellowship through the Excellence Initiative (DFG/GSC 266). We thank Philipp Maass (Osnabrück University, Germany) and Oliver Hofmann (Graz University, Austria) for most stimulating discussions. We acknowledge Philipp Rahe, Markus Kittelmann, Christopher Hauke, Robert Lindner, Felix Kling, Simon Aeschlimann, René Scheid and Julia Enns for AFM imaging of other benzoic acid derivatives. We are also grateful to the group of Diego Peña (University of Santiago de Compostela, Spain) for supplying further benzoic acid derivatives.

Notes and references

1 M. De Wild, S. Berner, H. Suzuki, L. Ramoino, A. Baratoff and T. A. Jung, *Ann. N. Y. Acad. Sci.*, 2003, **1006**, 291–305.

2 J. V. Barth, *Annu. Rev. Phys. Chem.*, 2007, **58**, 375–407.

3 F. Klappenberger, *Prog. Surf. Sci.*, 2014, **89**, 1–55.

4 F. Kling, R. Bechstein, P. Rahe and A. Kühnle, *Self-Assembly of Organic Molecules on Insulating Surfaces*, Springer Series, Heidelberg, 2015.

5 M. Yu, N. Kalashnyk, W. Xu, R. Barattin, Y. Benjalal, E. Lægsgaard, I. Stensgaard, M. Hliwa, X. Bouju, A. Gourdon, C. Joachim, F. Besenbacher and T. R. Linderoth, *ACS Nano*, 2010, **4**, 4097–4109.

6 J. V. Barth, J. Weckesser, C. Cai, P. Günter, L. Bürgi, O. Jeandupeux and K. Kern, *Angew. Chem., Int. Ed.*, 2000, **39**, 1230–1234.

7 M. Böhringer, K. Morgenstern, W.-D. Schneider, R. Berndt, F. Mauri, A. D. Vita and R. Car, *Phys. Rev. Lett.*, 1999, **83**, 324–327.

8 A. Kühnle, T. R. Linderoth and F. Besenbacher, *J. Am. Chem. Soc.*, 2003, **125**, 14680–14681.

9 J. A. Theobald, N. S. Oxtoby, M. A. Phillips, N. R. Champness and P. H. Beton, *Nature*, 2003, **424**, 1029–1031.

10 G. Pawin, K. L. Wong, K. Y. Kwon and L. Bartels, *Science*, 2006, **313**, 961–962.

11 S. Lukas, G. Witte and C. Wöll, *Phys. Rev. Lett.*, 2002, **88**, 028301.

12 C. Stadler, S. Hansen, I. Kroeger, C. Kumpf and E. Umbach, *Nat. Phys.*, 2009, **5**, 153–158.

13 I. Fernandez-Torrente, S. Monturet, K. J. Franke, J. Fraxedas, N. Lorente and J. I. Pascual, *Phys. Rev. Lett.*, 2007, **99**, 176103.

14 J. Fraxedas, S. Garcia-Gil, S. Monturet, N. Lorente, I. Fernandez-Torrente, K. J. Franke, J. I. Pascual, A. Vollmer, R. P. Blum, N. Koch and P. Ordejon, *J. Phys. Chem. C*, 2011, **115**, 18640–18648.

15 G. Tomba, M. Stengel, W. D. Schneider, A. Baldereschi and A. De Vita, *ACS Nano*, 2010, **4**, 7545–7551.

16 J. L. Neff, H. Söngen, R. Bechstein, P. Maass and A. Kühnle, *J. Phys. Chem. C*, 2015, **119**, 24927–24931.

17 A. M. Gabovich and A. I. Voitenko, *Low Temp. Phys.*, 2016, **42**, 661–671.

18 C. Iacovita, P. Fesser, S. Vijayaraghavan, M. Enache, M. Stohr, F. Diederich and T. A. Jung, *Chem.–Eur. J.*, 2012, **18**, 14610–14613.

19 T. Yokoyama, T. Takahashi, K. Shinozaki and M. Okamoto, *Phys. Rev. Lett.*, 2007, **98**, 206102.

20 P. A. Mulheran and J. A. Blackman, *Philos. Mag. Lett.*, 1995, **72**, 55–60.

21 J. A. Blackman and P. A. Mulheran, *Phys. Rev. B: Condens. Matter Mater. Phys.*, 1996, **54**, 11681–11692.

22 M. Brinkmann, S. Pratontep and C. Contal, *Surf. Sci.*, 2006, **600**, 4712–4716.

23 S. Pratontep, M. Brinkmann, F. Nuesch and L. Zuppiroli, *Phys. Rev. B: Condens. Matter Mater. Phys.*, 2004, **69**, 165201.
24 E. Merkul, D. Urselmann and T. J. J. Müller, *Eur. J. Org. Chem.*, 2011, 238–242.
25 W. B. Austin, N. Bilow, W. J. Kelleghan and K. S. Y. Lau, *J. Org. Chem.*, 1981, **46**, 2280–2286.

Faraday Discussions

PAPER

Erecting buckybowls onto their edge: 2D self-assembly of terphenylcorannulene on the Cu(111) surface

Quirin S. Stöckl, †[a] Tsun-Cheng Wu,[b] Anaïs Mairena, [a]
Yao-Ting Wu*[b] and Karl-Heinz Ernst *[ac]

Received 16th March 2017, Accepted 25th April 2017
DOI: 10.1039/c7fd00109f

A 2D self-assembly of a $C_{32}H_{12}$ buckybowl on the Cu(111) surface has been studied by means of scanning tunnelling microscopy. Additional aromatic rings at the rim of the corannulene core cause the bowl-shaped molecule to stand on its edge. This adsorption mode allows distinct π–π and C–H⋯π interactions between the convex bowl surfaces as well as between the hydrogen-terminated rim and the convex bowl faces.

Introduction

Controlling the alignment of molecules at surfaces and the way they self-assemble through targeted functionalization is a modern way to prepare new advanced materials. However, this approach requires a good understanding of the non-covalent interactions in such materials.[1] Bowl-shaped fragments of fullerenes, so-called buckybowls, exhibit a pronounced difference in chemical activity between their concave and convex sides, which makes them very interesting as carbon-based materials for organic electronics, such as components of photovoltaic devices, organic field effect transistors (OFETs) or organic light emitting devices (OLEDs). Buckybowls also show rich metal coordination chemistry.[2,3]

Adsorbed on surfaces, buckybowls have been found to show very interesting results in phenomena like reversible phase transitions,[4,5] chiral surface reconstructions,[6] bowl-in-bowl complexation,[7,8] pentagonal molecular tiling of the plane,[9–12] and giant adsorbate dipole moments.[13] Due to the strong interaction of the convex bowl side with metal surfaces, single buckybowls or their aggregates are usually oriented such that the bowl opening points away from the surface,[4–13]

[a]Empa – Swiss Federal Laboratories for Materials Science and Technology, Überlandstrasse 129, 8600 Dübendorf, Switzerland. E-mail: karl-heinz.ernst@empa.ch

[b]Department of Chemistry, National Cheng Kung University, 70101 Tainan, Taiwan. E-mail: ytwuchem@mail.ncku.edu.tw

[c]Department of Chemistry, University of Zurich, Zürich, Switzerland

† Present address of QS: Berry College, Mount Berry, USA.

while partial bowl-opening-down configurations due to lateral π–π interactions remain the exception.[14,15]

Using scanning tunnelling microscopy (STM), we show here that additional annulated benzo groups at the rim of the parent buckybowl corannulene (Fig. 1) align the bowl on a copper(111) surface such that its opening is pointing parallel to the surface. In this adsorbate mode π–π and C–H···π interactions dominate as an intermolecular binding motif.

Experimental

1,1′:4′,1″-Terpenyl[3,2:2′,3′:2″,3″-*a*,*d*,*g*]corannulene (TPC) was synthesized as reported previously.[16] The Cu(111) single crystal surface (MaTecK, Jülich) was cleaned by cycles of argon ion sputtering and subsequent annealing to 900 K for 2 min. TPC was sublimed at 493 K *in vacuo* from a crucible onto the Cu(111) surface held at room temperature during deposition. After cooling the Cu crystal to ≈60 K in the STM apparatus, images were acquired in constant current mode with a variable temperature instrument (Omicron Nanotechnology GmbH, SCALA PRO 4.1 software). For molecular modelling, single molecules were prepared with HyperChem™ Release 7.1 for Windows and the intermolecular interaction was modelled with AMBER force fields.

Results and discussions

At very low coverage (≪10% of a complete monolayer) STM shows single adsorbed TPC molecules (Fig. 2) as well as small aggregates. The STM appearance of a single molecule is dominated by a very bright protrusion on one side, a dark protrusion in the centre and two bright lobes on each side of the dark protrusion (Fig. 2b). Such an appearance indicates a substantial tilt of the molecule. Moreover, the adsorbate appears to have C_S symmetry, which suggests in combination with the tilt that the molecule stands such that the middle benzo ring of the terphenyl

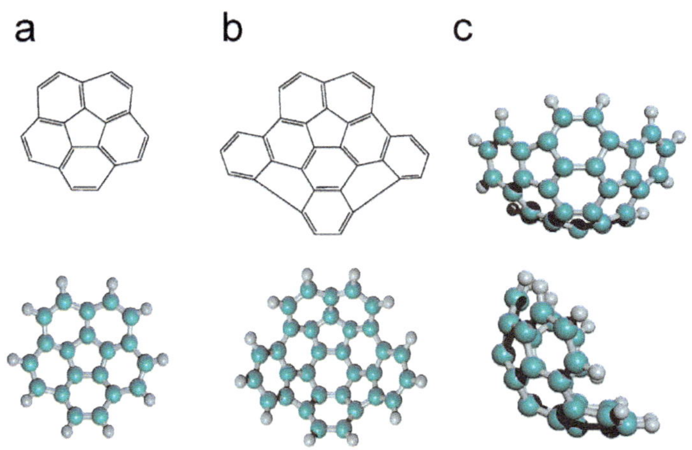

Fig. 1 Molecular structures and models of corannulene (a) and terphenylcorannulene (TPC) (b). Different side views of TPC illustrate its bowl shape (c).

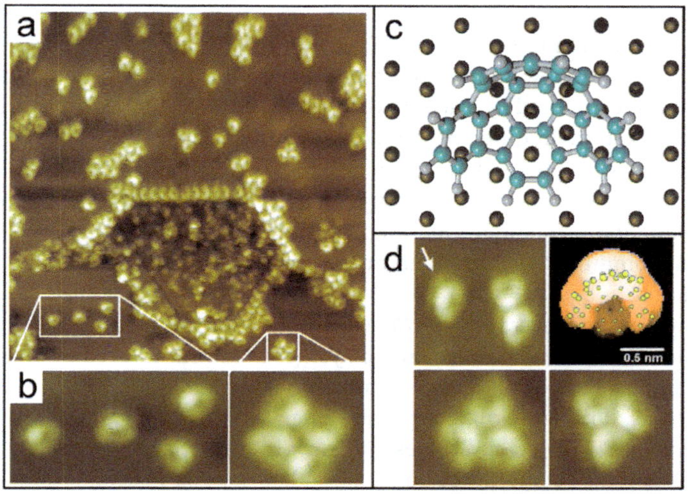

Fig. 2 (a) STM-image at very low coverage ($\theta < 0.1$, STM-image parameters: 40×40 nm^2, $U = -2.365$ V, $I = 61$ pA, $T = 60$ K) of TPC on Cu(111). (b) Single molecules and small aggregates indicate a substantial tilt of the bowl. (c) Ball-and-stick molecular model with the middle benzo group of the terphenyl group aligned parallel to the surface. (d) Back-to-back and back-to-edge assemblies and the electron density of the HOMO, as calculated using the extended Hückel theory (scale bar: 5 Å).

group is aligned parallel to the surface. In the model shown in Fig. 2c the benzo ring has been placed arbitrarily on top of a Cu atom. On Cu(111), however, aromatic rings are usually oriented above a threefold hollow-site.[17–20] This has been especially confirmed for corannulene.[18] Due to the steep decline on the convex side of the bowl molecule, there is some intensity near the bright protrusion in the STM image (arrow in Fig. 2d), which is due to a much larger tip apex that cannot image such steep lateral change perfectly. The tilt is further confirmed by comparison of the STM appearance with the expected appearance, *i.e.* the spatial electron density in the adsorbate. For the experimental bias voltages used, this would be essentially the HOMO orbital. Calculated using the extended Hückel theory, the HOMO charge density for the tilted species, sitting on the middle terphenyl benzo group, reproduces the experimental STM appearance well (Fig. 2d).

Small ensembles show in principle two interaction motifs. One is a 'convex–convex' alignment, and the second a 'rim–convex' alignment (Fig. 2d). This suggests that non-covalent $\pi\cdots\pi$ and C–H$\cdots\pi$ interactions govern the initial stages of the supramolecular assembly.

In order to verify in part this assumption we modelled the interaction of two molecules in that particular surface alignment with AMBER force field calculations. A slab of four fixed Cu(111) layers, with periodic boundary conditions, was used as a template. The central terphenyl C_6 ring at the rim of each $C_{32}H_{12}$ was constrained to the surface. The other atoms of the molecule were allowed to relax. Constraint of the central terphenyl C_6 ring is required to achieve the experimentally observed tilting of the molecule. A fully relaxed adsorbate preferred to align with its convex side, but with a central C_6 ring parallel to the surface.

Consequently, the constraint did not allow the molecule to move on the surface any more. Compared to the free gas phase molecule the fully and partly relaxed adsorbates did not show notable differences in conformation. The binding energy of a dimer was calculated by placing two constrained adsorbates in a manual trial-and-error procedure in different relative configurations until a minimum energy value was achieved. Several configurations (>16) with different distances and angles between the two molecules were considered. The dimer with the lowest energy configuration showed a binding energy of 7.6 kcal mol^{-1} (Fig. 3). Note that the surface binding sites are different for the monomers in this lowest-energy configuration.

With increasing coverage ($\approx$0.75 ML) ribbons with long-range order appear within the otherwise disordered monolayer (Fig. 4a). All molecules are equally oriented into a (4 0, −1 4)[21] phase, meaning that one molecule covers 16 Cu(111) surface atoms. Within the ribbon the molecules are still in the same configuration as observed for isolated molecules. Judging by the relative alignment of the molecules one may come to the conclusion that the dominating intermolecular binding motif is a C–H$\cdots\pi$ interaction between the hydrogen-terminated rim of the molecules in one row and the convex face of the molecules in adjacent rows. However, in densely packed crystals or self-assemblies one has to be cautious with such conclusions. Often short distances between parts of different molecules are taken as attractive interactions. However, short distances may rather emerge from the packing constrains of all molecules in the vicinity.[22] This applies in particular to molecular assemblies at surfaces, where favoured surface adsites may substantially contribute to the outcome of the relative alignment of molecules in extended aggregates.

With increasing coverage a denser, long-range ordered 2D phase forms, which stays up to monolayer saturation. This phase is shown in the STM-image in Fig. 5. Judged again from the STM appearance of the single adsorbate within an extended domain, the relative alignment with respect to the surface is identical with the one concluded for the isolated adsorbate. All molecules show a bright protrusion, and a C-like appearance as observed for single molecule adsorption

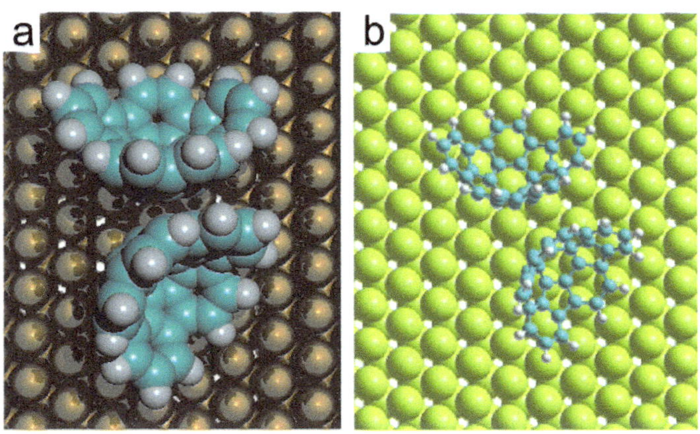

Fig. 3 Two views of the lowest energy configuration of a TPC dimer. The binding energy was calculated as 7.6 kcal mol^{-1}.

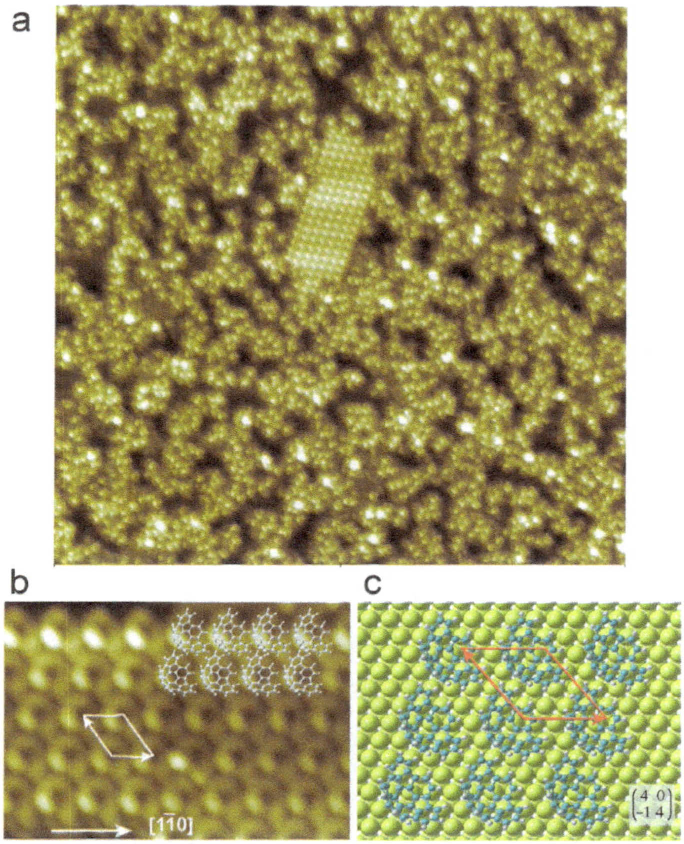

Fig. 4 (a) STM-image at intermediate coverage of TPC on Cu(111). Ribbons with ordered molecules appear. (80 × 80 nm^2, $U = 857$ mV, $I = 135$ pA, gray scale z-range: 0.6 nm, $T = 60$ K). (b) STM-image of a ribbon structure superimposed in part with molecular models. All molecules of this domain are equally aligned. The top row presents the edge of the domain. (c) Model of alignment in the ribbon.

(Fig. 5b). Only in every second row do the molecules show the same azimuthal orientation. The molecules of adjacent rows are rotated by 60° with respect to each other. The unit cell of this phase was determined as (4 0, −1 7). A model of this 'stripe phase', based on the STM-image, is shown in Fig. 5c. The binding site of all of the molecules is again the middle benzo ring of the additional terphenyl unit on top of a threefold hollow surface site. The azimuthal angle of 60° is in perfect agreement with the surface symmetry of a (111) surface. With two molecules per (4 0, −1 7) unit cell the coverage is one molecule per 14 Cu surface atoms.

A third phase was found to coexist beside the stripe phase at coverages close to the saturated monolayer. Annealing of the sample did not lead to any preference of one of the two phases. Fig. 6 displays an STM-image showing domains of this structure. A domain of the stripe phase is identified at the top of the image. The other structure, however, appears in double rows. Such a motif is in part also

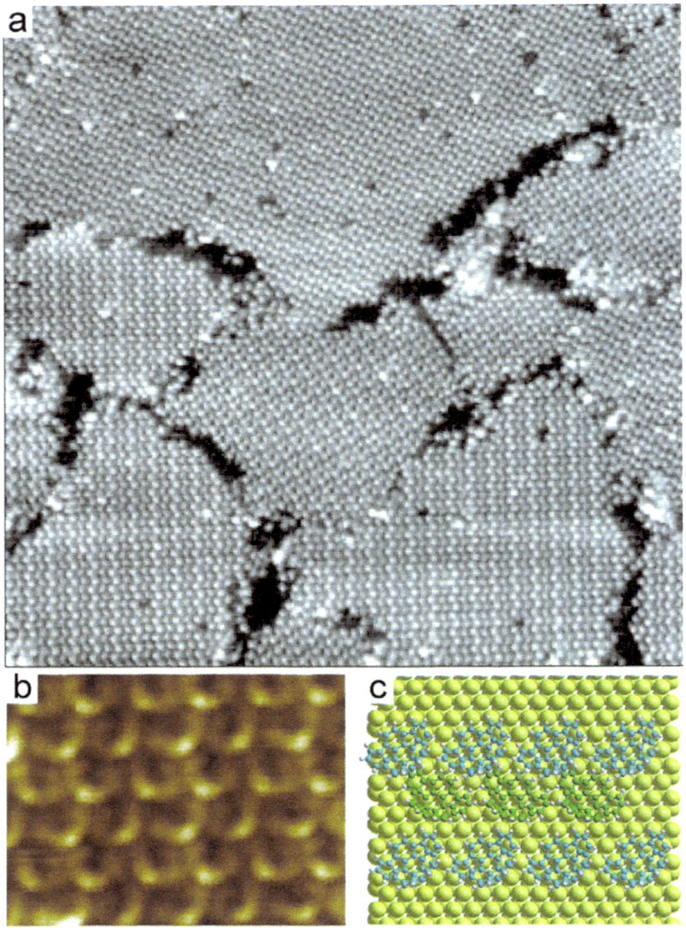

Fig. 5 (a) STM-image (80 × 80 nm^2, $U = -1.21$ V, $I = 140$ pA) of a 2D TPC phase. (b) High magnification STM-image revealing the relative orientation of the molecules. (c) Structure model of this 'stripe phase'.

observed in small assemblies (Fig. 6b). There the molecules are aligned with bowls pointing in opposite directions, having a convex–convex face alignment. Based on this observation a tentative model of the structure is presented (Fig. 6c). Due to the co-existence with the stripe phase, an identical coverage is assumed. The unit cell of this phase was determined as (4 2, 0 7), again with a lateral density of one molecule per 14 Cu(111) surface atoms. This model proposes C–H⋯π rim-to-convex face bonding within a row, but π–π bonding between rows. We note that temperature-dependent polymorphism, including co-existence of the two polymorphs with identical lattice densities within a certain temperature interval, has previously been reported for corannulene.[5] The co-existence observed here at this low temperature must be then due to limited diffusion for this species during the phase transition upon cooling.

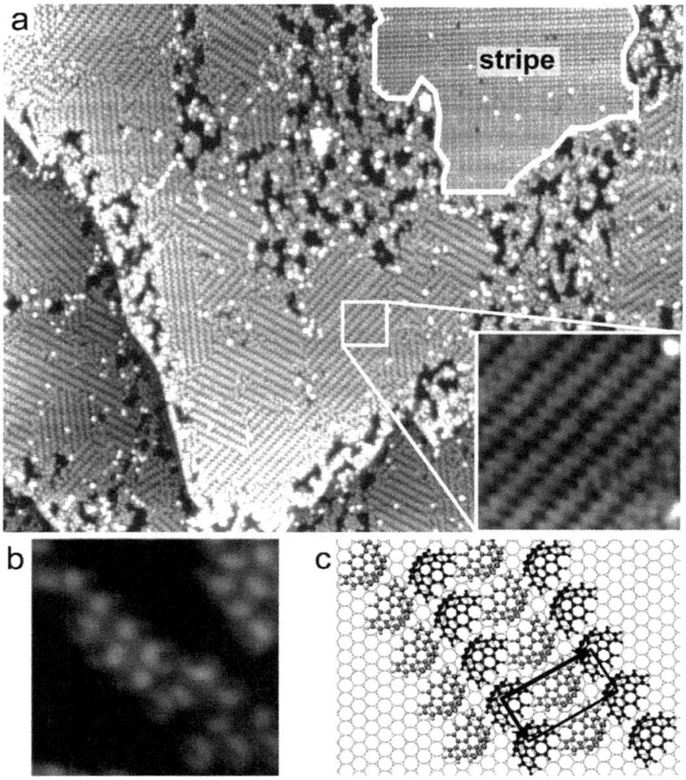

Fig. 6 (a) STM-image (200 × 160 nm^2, $U = 1.162$ V, $I = 137$ pA) of a phase in co-existence with the stripe phase (bordered by a white line). The molecules align in close-packed double rows (inset). (b) STM-image (7 × 7 nm^2, $U = 671$ mV, $I = 79$ pA) of a short double row segment without adjacent molecules as in an extended domain. (c) Tentative model for the 'double-row' structure with a coverage identical to the stripe phase. The unit cell periodicity is (4 2, 0 7).

Due to strong interactions with the metal surface, buckybowls usually adsorb such that their bowl opening points away from the surface, orienting the bowl axis more or less perpendicular. Additional functionalization of the rim with aromatic rings, as shown here for corannulene with a terphenylene group, causes the bowl to stand on its rim, having the bowl axis pointing more or less parallel to the surface. For TPC, interacting mainly with the middle benzo ring of the terphenyl group, this alignment does not change with coverage, indicating a strong molecule–surface interaction. The fact that single molecules can be imaged at 60 K, a temperature at which similar sized molecules diffuse rapidly on Cu(111),[23,24] supports this conclusion. However, the contact area with the surface is too small for a substantial van der Waals interaction that would explain such low mobility. As observed for corannulene,[13] most likely a strong electrostatic interaction with the substrate exists. It arises from the intrinsic dipole of the bowl molecule with its higher electron density at the convex face causing substantial Pauli repulsion in the top metal layers. This strong bond avoids a partial bowl-upside-down arrangement in close-packed layers, as observed for the somewhat larger $C_{36}H_{14}$ bowl. As judged by observations made for diluted, low coverage scenarios, lateral interactions of TPC molecules include intermolecular π–π and C–H$\cdots\pi$ bonding.

Conclusions

Functionalizing the buckybowl corannulene with further aromatic rings at the rim leads to a substantial change in the adsorption mode, *i.e.* the molecule stands on its rim. This single adsorption motif is quite stable due to the strong electrostatic interactions with the copper surface and does not change upon increasing the lateral interactions at higher coverages. Three different extended 2D structures were observed; two of them coexisting at high coverage. Upon better availability, it is anticipated that buckybowls will play a more important role in organic electronics due to their interesting properties at metal and electrode surfaces.

Acknowledgements

Financial support by the Schweizerischer Nationalfonds (SNF) is gratefully acknowledged. We thank Jay Siegel for fruitful discussions.

References

1 D. Umadevi, S. Panigrahi and G. N. Sastry, *Acc. Chem. Res.*, 2014, **47**, 2574–2581.

2 Y.-T. Wu and J. S. Siegel, *Chem. Rev.*, 2006, **106**, 4843–4867.

3 M. A. Petrukhina and L. T. Scott, *Dalton Trans.*, 2005, 2969–2975.

4 L. Merz, T. Bauert, M. Parschau, G. Koller, J. S. Siegel and K.-H. Ernst, *Chem. Commun.*, 2009, 5871–5873.

5 L. Merz, M. Parschau, L. Zoppi, K. K. Baldridge, J. S. Siegel and K.-H. Ernst, *Angew. Chem., Int. Ed.*, 2009, **48**, 1966–1969.

6 W. Xiao, K.-H. Ernst, K. Palotás, Y. Zhang, E. Bruyer, L. Peng, T. Greber, W. A. Hofer, L. T. Scott and R. Fasel, *Nat. Chem.*, 2016, **8**, 326–330.

7 T. Bauert, K. K. Baldridge, J. S. Siegel and K.-H. Ernst, *Chem. Commun.*, 2011, **47**, 7995–7997.

8 W. Xiao, D. Passerone, P. Ruffieux, K. Aït Mansour, O. Gröning, E. Tosatti, J. S. Siegel and R. Fasel, *J. Am. Chem. Soc.*, 2008, **130**, 4767–4771.

9 L. Zoppi, T. Bauert, J. S. Siegel, K. K. Baldridge and K. H. Ernst, *Phys. Chem. Chem. Phys.*, 2012, **14**, 13365–13369.

10 Q. Stöckl, D. Bandera, C. S. Kaplan, K.-H. Ernst and J. S. Siegel, *J. Am. Chem. Soc.*, 2014, **136**, 606–609.

11 T. Bauert, L. Merz, D. Bandera, M. Parschau, J. S. Siegel and K.-H. Ernst, *J. Am. Chem. Soc.*, 2009, **131**, 3460–3461.

12 M. Parschau, R. Fasel, K.-H. Ernst, O. Gröning, L. Brandenberger, R. Schillinger, T. Greber, A. P. Seitsonen, Y.-T. Wu and J. S. Siegel, *Angew. Chem., Int. Ed.*, 2007, **46**, 8258–8261.

13 T. Bauert, L. Zoppi, G. Koller, A. Garcia, K. K. Baldridge and K.-H. Ernst, *J. Phys. Chem. Lett.*, 2011, **2**, 2805–2809.

14 Q. S. Stöckl, Y.-C. Hsieh, A. Mairena, Y.-T. Wu and K.-H. Ernst, *J. Am. Chem. Soc.*, 2016, **138**, 6111–6114.

15 R. Jaafar, C. A. Pignedoli, G. Bussi, K. Aït Mansour, O. Groening, T. Amaya, T. Hirao, R. Fasel and P. Ruffieux, *J. Am. Chem. Soc.*, 2014, **136**, 13666–13671.

16 T.-C. Wu, H.-J. Hsin, M.-Y. Kuo, C.-H. Li and Y.-T. Wu, *J. Am. Chem. Soc.*, 2011, **133**, 16319–16321.

17 K. Toyoda, Y. Nakano, I. Hamada, K. Lee, S. Yanagisawa and Y. Morikawa, *J. Electron Spectrosc. Relat. Phenom.*, 2009, **174**, 78–84.

18 L. Merz, M. Parschau, L. Zoppi, K. K. Baldridge, J. S. Siegel and K.-H. Ernst, *Angew. Chem., Int. Ed.*, 2009, **48**, 1966–1969.

19 J. Larsson, S. Elliott, J. Greer, J. Repp, G. Meyer and R. Allenspach, *Phys. Rev. B: Condens. Matter Mater. Phys.*, 2008, **77**, 115434.

20 J. A. Smerdon, M. Bode, N. P. Guisinger and J. R. Guest, *Phys. Rev. B: Condens. Matter Mater. Phys.*, 2011, **84**, 165436.

21 The (2 × 2) transformation matrix, linking the adsorbate lattice vectors (b1, b2) to the substrate lattice vectors (a1, a2) via b1 = m11a1 + m12a2 and b2 = m21a1 + m22a2, is written here in the form (m11 m12, m21 m22). See: L. Merz and K.-H. Ernst, *Surf. Sci.*, 2010, **604**, 1049–1054.

22 J. D. Dunitz, *IUCrJ*, 2015, **2**, 157–158.

23 M. Parschau, R. Fasel and K.-H. Ernst, *Cryst. Growth Des.*, 2008, **8**, 1890–1896.

24 A. Mairena, L. Zoppi, J. Seibel, A. F. Tröster, K. Grenader, M. Parschau, A. Terfort and K.-H. Ernst, *ACS Nano*, 2017, **11**, 865–871.

Faraday Discussions

PAPER

Exploring catalyst passivation with NMR relaxation†

Neil Robinson, Lynn F. Gladden and Carmine D'Agostino *

Received 9th March 2017, Accepted 6th April 2017
DOI: 10.1039/c7fd00098g

NMR relaxation has recently emerged as a novel and non-invasive tool for probing the surface dynamics of adsorbate molecules within liquid-saturated mesoporous catalysts. The elucidation of such dynamics is of particular relevance to the study and development of solvated green catalytic processes, such as the production of chemicals and fuels from bio-resources. In this paper we develop and implement a protocol using high field ^{1}H NMR spin–lattice relaxation as a probe of the reorientational dynamics of liquids imbibed within mesoporous oxide materials. The observed relaxation of liquids within mesoporous materials is highly sensitive to the adsorbed surface layer, giving insight into tumbling behaviour of spin-bearing chemical environments at the pore surface. As a prototypical example of relevance to liquid-phase catalytic systems, we examine the mobility of liquid methanol within a range of common catalyst supports. In particular, through the calculation and comparison of a suitable interaction parameter, we assess and quantify changes to these surface dynamics upon replacing surface hydroxyl groups with hydrophobic alkyl chains. Our results indicate that the molecular tumbling of adsorbed methanol is enhanced upon surface passivation due to the suppression of surface-adsorbate hydrogen bonding interactions, and tends towards that of the unrestricted bulk liquid. A complex analysis in which we account for the influence of changing pore structure and surface chemistry upon passivation is discussed. The results presented highlight the use of NMR spin–lattice relaxation measurements as a non-invasive probe of molecular dynamics at surfaces of interest to liquid-phase heterogeneous catalysis.

1. Introduction

While many of the current experimental and theoretical developments across the field of heterogeneous catalysis concern themselves with the design, synthesis and optimisation of supported active species,[1–3] the nature of the catalyst support itself can significantly influence catalytic reactions. For example, supports may directly influence the performance of supported metal species through strong

Department of Chemical Engineering and Biotechnology, University of Cambridge, Philippa Fawcett Drive, West Cambridge Site, Cambridge, CB3 0AS, UK. E-mail: cd419@cam.ac.uk

† Electronic supplementary information (ESI) available. See DOI: 10.1039/c7fd00098g

metal–support interactions,[4–6] including metal cluster anchoring[7] and nanoparticle encapsulation,[8,9] as well as direct participation in reaction pathways.[10] Away from the active regions of the catalyst surface, the highly porous nature of typical catalyst supports also dictates mass transport properties through the existence of highly tortuous pore networks.[11] Liquid-phase catalytic systems are typically of relevance to solvated green chemical processes. Here, the high molecular density of molecules containing polar functional groups may result in further mass transport limitations due to adsorption interactions at the support surface.[11] While such surface interactions have been observed to enhance catalyst performance,[12,13] this effect is also highly dependent on the nature of competitive adsorption between reagents and any solvent molecules present.[14] One approach to limit unfavourable interactions between the support and mobile phase is to passivate the support surface *via* the covalent modification of surface hydroxyl species. Indeed, the passivation of silica-based supports has been observed to enhance the performance of epoxidation,[15,16] oxidation[17] and hydrogenation catalysts.[18–21] The influence of passivation also extends to other oxide materials; for example, Ellis *et al.* recently reported that passivation of γ-alumina leads to a 50% increase in activity for the dehydration of 1,2-propanediol.[22]

In this work we wish to establish the use of liquid-phase nuclear magnetic resonance (NMR) relaxation as a surface sensitive method with which changes to adsorbate mobility may be explored upon support passivation. The elucidation of such dynamics is of importance to the development of a range of liquid-phase catalytic processes. The successful functionalisation of catalyst surfaces has been extensively quantified using magic angle spinning solid-state NMR (see, for example ref. 15, 17 and 20). However, the dynamic behaviour (diffusion, rotation) of liquid molecules imbibed within the pore network of functionalised catalysts has attracted far less attention. Pulsed-field gradient (PFG) NMR diffusometry has been used by Hansen[23] and Courivaud[24] to explore the influence of surface hydrophobicity on the self-diffusion of *n*-hexane imbibed within MCM-41. Elsewhere Weber *et al.* used PFG NMR analysis of Pd/θ-alumina saturated with 1-octene to illustrate how molecular self-diffusion with the adsorbed surface layer is inhibited through passivation of the pore surface.[25]

Herein, we describe the use of high field NMR spin–lattice relaxation measurements to probe changes in surface dynamics experienced by methanol – used here to represent a prototypical polar molecule of relevance to green chemical processes – within liquid-saturated catalyst supports which have been functionalised with alkane groups.

2. Surface relaxation theory

Dipole–dipole interactions provide the primary contribution to NMR relaxation in the liquid phase. The dipolar spin–lattice relaxation rate for a collection of homonuclear spins, I, within a static magnetic field of strength B_0 is given by the well-known equation[26]

$$\frac{1}{T_1} = \left(\frac{\mu_0}{4\pi}\right)^2 \gamma_I{}^2 \hbar^2 \frac{3}{2} I(I+1)[J(\omega_0) + 4J(2\omega_0)] \tag{1}$$

where μ_0 is the permeability of free space, γ_I is the magnetogyric ratio of spin I, $\omega_0 = \gamma_I B_0$ is the corresponding Larmor frequency and T_1 is the characteristic time constant for spin–lattice relaxation. $J(\omega)$ is the normalised spectral density function typically given by[27]

$$J(\omega) = \frac{2\tau_c}{1 + \omega^2 {\tau_c}^2} \tag{2}$$

where the rotational correlation time τ_c describes the average time taken for spin-bearing molecules to rotate by 1 radian.

As a result of this relation, T_1 measurements may be used as an indicator of the rates of molecular tumbling within a given system; such an approach is often referred to as 'BPP Theory' after the seminal work in deriving this relation by Bloembergen, Purcell and Pound.[28]

In liquid-saturated mesoporous solids the observed relaxation rates are determined by the different liquid environments present. In this work we will assume the conditions of biphasic fast molecular exchange between an adsorbed surface layer of thickness δ and bulk liquid towards the centre of the pores, as described by Brownstein and Tarr.[29,30] The observed relaxation rate $1/T_{1,\text{obs}}$ is a weighted average of bulk and surface contributions

$$\frac{1}{T_{1,\text{obs}}} = \frac{1 - P_{\text{surf}}}{T_{1,\text{bulk}}} + \frac{P_{\text{surf}}}{T_{1,\text{surf}}}. \tag{3}$$

Here, P_{surf} represents the proportion of spins (molecules) in the adsorbed surface layer and takes the form

$$P_{\text{surf}} = \frac{\delta S}{V} \tag{4}$$

where S/V is the surface-to-volume ratio of the porous solid under study. Following eqn (1) and (2) the reduced molecular rotational mobility experienced by adsorbed species at the pore surface leads to $T_{1,\text{bulk}} \gg T_{1,\text{surf}}$. In combination with the high S/V of liquid-saturated mesoporous media, the bulk term in eqn (3) becomes negligible, rendering relaxation measurements of such systems inherently surface sensitive. The observed relaxation rate is therefore often approximated to

$$\frac{1}{T_{1,\text{obs}}} \approx \frac{S}{V}\rho_1 \tag{5}$$

where $\rho_1 = \delta/T_{1,\text{surf}}$ is known as the spin–lattice relaxivity. As a result, T_1 measurements performed on liquid-saturated mesoporous solids may be used to probe molecular mobility within the adsorbed surface layer; spin–lattice relaxation is a particularly useful probe of such dynamics as T_1 is independent of any internal magnetic field gradients which may occur at the solid–liquid interface under the influence of an applied magnetic field.[31] Herein, we use this approach to compare the observed changes in methanol mobility across the surfaces of four common oxide support materials functionalised with octyl groups.

3. Experimental

3.1 Materials and surface passivation

γ-Alumina (γ-Al_2O_3), θ-alumina (θ-Al_2O_3) and silica (SiO_2) were obtained from Johnson Matthey, while anatase-titania (A-TiO_2) was obtained from Evonik-

Degussa. Each oxide was functionalised with surface octyl groups through liquid-phase treatment with triethoxy(octyl)silane (TEOS, Sigma Aldrich, ≥96%). Oxide pellets were first dried at 105 °C for 2 hours before being soaked in excess TEOS for 12 hours. The pellets were then removed and washed several times in cyclohexane to remove any unreacted TEOS, before being dried at 105 °C for a further 12 hours.

3.2 Materials characterisation

Diffuse Reflectance Infrared Fourier Transform Spectroscopy (DRIFTS) measurements were performed to confirm successful passivation of the oxide supports. Measurements were performed on a ThermoFischer Nicolet iS50 FT-IR spectrometer equipped with a Praying Mantis diffuse reflectance cell and high temperature reaction chamber. Samples were ground by hand using a pestle and mortar and were analysed without dilution. Approximately 50 mg of each catalyst support was loaded into the reaction cell supported by a small amount of quartz wool. To remove the influence of physisorbed water, all samples were heated to 150 °C at a rate of 10 °C min^{-1} under a low flow of helium; this temperature was maintained for 30 minutes after which samples were cooled to 25 °C at the same rate. Spectra were acquired with 64 repeat scans with a resolution of 4 cm^{-1}, relative to a KBr background acquired under the same conditions. Nitrogen isotherm measurements were performed using a Micromeritics TriStar 3000 automated gas adsorption analyser. Specific surface areas were obtained by fitting to the Brunauer–Emmett–Teller (BET) equation, while pore volumes were calculated using the Barrett–Joyner–Halenda (BJH) method. All nitrogen adsorption measurements were carried out at −195 °C.

3.3 NMR measurements

Unfunctionalised oxides were dried for at least 12 hours at 105 °C before use. In order to ensure full saturation of the mesoporous structures each material was soaked in an excess of methanol (Sigma Aldrich, ≥99.8%) for at least 24 hours. Separately, each material was also saturated with cyclohexane (Sigma Aldrich, ≥99.5%) in order to provide a weakly-interacting reference. The oxide materials were then removed from the liquid and rolled across a pre-soaked filter paper in order to remove any surface excess. Finally, the samples were transferred to sealed 5 mm NMR tubes for analysis. Each sample consisted of between 5 and 10 saturated catalyst pellets so as to provide a well-averaged measurement of the surface-adsorbate interactions present between the imbibed liquids and the pore surfaces present throughout each oxide support.

^{1}H NMR spin–lattice relaxation measurements were performed using a Bruker DMX 300 spectrometer with a ^{1}H operational frequency of 300.13 MHz. All measurements were performed under ambient pressure at 20 ± 0.2 °C. T_1 values were acquired using the inversion-recovery method,[27] during which 16 recovery delays were employed ranging logarithmically from 1 ms to ~5T_1. For bulk liquids, T_1 values were obtained by fitting the acquired data to

$$\frac{M_z(t)}{M_0} = 1 - 2\,\exp\left(\frac{-t}{T_1}\right) \tag{6}$$

where $M_z(t)$ is the time-dependent longitudinal magnetisation and M_0 is the magnetisation at equilibrium.

From eqn (5) it is clear that for liquid-saturated mesoporous media, T_1 must form a distribution related to the spread of pore sizes throughout the solid. The acquired data may therefore be described by a Fredholm integral equation of the first kind.[32]

$$\frac{M_z(t)}{M_0} = \int_{-\infty}^{\infty} K(t, T_1)F(T_1)d\log(T_1) + \varepsilon \tag{7}$$

The kernel function $K(t,T_1) = 1 - 2\exp(-t/T_1)$ describes the predicted form of T_1 relaxation and ε represents the experimental noise. For each sample analysed the required T_1 distribution $F(T_1)$ was obtained *via* a numerical inversion of eqn (7).[32] Stability of the inverted distributions in the presence of noise was achieved through the use of Tikhonov regularisation,[33] with the degree of smoothing chosen using the generalised cross validation method.[34] As the shape and width of the resulting distributions are highly susceptible to both experimental noise and the signal-to-noise ratio of the acquired data, we take only the modal values $\langle T_{1,\mathrm{obs}}\rangle$ from these distributions in the following analysis. Further details are provided in the ESI.†

4. Results and discussion

4.1 Surface chemistry and pore structure

We first examine the influence of passivation on the surface chemistry and pore structure of the mesoporous oxides studied. Fig. 1 shows the difference in infrared adsorption spectra between passivated and unpassivated oxides. It is clearly apparent that functionalisation with TEOS results in a loss of surface hydroxyl groups, characterised by the negative peaks present above 3500 cm^{-1}. Indicated are the frequencies and corresponding assignments for the peaks which appear upon passivation. Four stretches are observed around 2900 cm^{-1} corresponding to $-CH_2$ (2926 cm^{-1} asymmetric and 2855 cm^{-1} symmetric) and

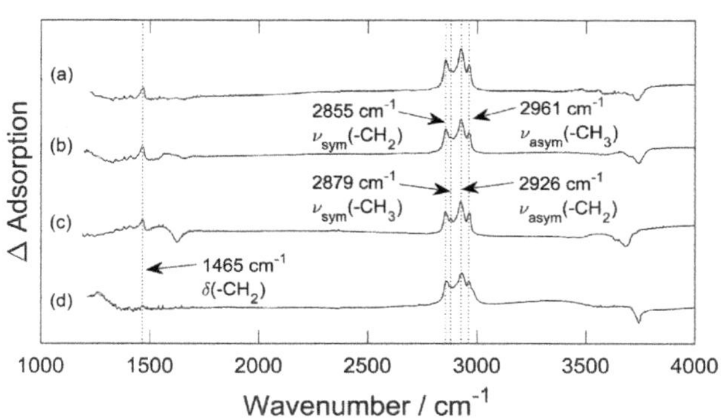

Fig. 1 DRIFTS spectra for (a) γ-alumina, (b) θ-alumina, (c) A-titania and (d) silica. Spectra are reported as the change in absorption upon passivation of the oxide surface with TEOS. Negative peaks therefore represent a loss of signal upon surface functionalisation while positive peaks indicate a gain in signal upon functionalisation.

$-CH_3$ (2961 cm^{-1} asymmetric and 2879 cm^{-1} symmetric) stretches.[35] An additional peak is observed at 1465 cm^{-1} which we assign to a $-CH_2$ bending mode.[35] The presence of these peaks confirms qualitatively the successful passivation of surface hydroxyls with alkyl groups upon treatment of our oxides with TEOS.

To examine the effect of passivation on the pore structure of our oxides we have performed specific surface area and pore volume measurements pre- and post-functionalisation. Results are provided in Table 1 and indicate a reduction in both surface area and pore volume upon passivation. The surface-to-volume ratio (S/V) of each material calculated from these results is also given, indicating an increase in S/V upon passivation.

4.2 NMR spin–lattice relaxation

The acquired $\langle T_{1,\text{obs}} \rangle$ values characterising the observed rates of longitudinal proton relaxation within our methanol-saturated oxides are presented in the top two sections of Fig. 2. Fig. 2a displays the relaxation time constants within unpassivated oxides. It is evident from these data that the methyl group presents a longer $\langle T_{1,\text{obs}} \rangle$ time than that exhibited by the methanol hydroxyl groups imbibed within the same support. This is consistent across the range of oxides investigated and may be explained in terms of the relative motional freedom of the two proton environments. According to the BPP theory of nuclear spin relaxation, as $\tau_c \rightarrow 0$, *i.e.* as tumbling becomes more rapid, $T_1 \rightarrow \infty$. A longer relative T_1 may therefore be recognised as an increase in observed tumbling rate, and hence motional freedom at the pore surface. The reduced relative motional freedom exhibited by hydroxyl groups within the adsorbed surface layer is easily explained through their ability to form hydrogen bonds with other polar moieties, resulting in favourable adsorption interactions with the hydroxyl groups which decorate the pore surfaces within unpassivated oxides. Hydroxyl protons therefore experience reduced mobility at the pore surface relative to the apolar methyl group, leading to more rapid $\langle T_{1,\text{obs}} \rangle$ times. Other surface interactions may also cause a bias towards a reduced T_1 of the hydroxyl group, such as the reversible dissociation of labile hydroxyl protons from adsorbed methanol molecules onto the pore surface, and interactions with acidic metal sites.

Fig. 2b displays the same data for methanol-saturated oxides which have been passivated. We observe through a comparison of Fig. 2a and b that the difference in T_1 experienced by the two proton environments is noticeably reduced upon

Table 1 BET and BJH analysis for passivated (oxide + TEOS) and unpassivated mesoporous oxide supports

	BET surface area ($m^2\ g^{-1}$)		BJH pore volume ($cm^3\ g^{-1}$)		S/V ($m^2\ cm^{-3}$)	
Oxide support	Oxide	Oxide + TEOS	Oxide	Oxide + TEOS	Oxide	Oxide + TEOS
γ-Al_2O_3	90	82	0.49	0.34	184	244
θ-Al_2O_3	108	99	0.57	0.52	188	189
A-TiO_2	40	37	0.28	0.19	142	196
SiO_2	272	236	1.29	1.01	211	235

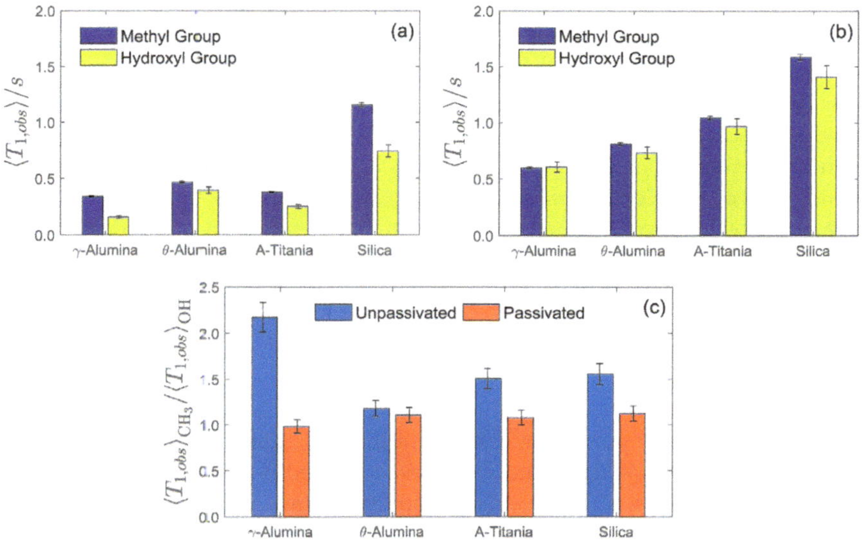

Fig. 2 $\langle T_{1,obs} \rangle$ values for methanol imbibed within (a) unpassivated and (b) passivated oxides, together with (c) the ratio of $\langle T_{1,obs} \rangle$ exhibited by the methyl (CH_3) and hydroxyl (OH) proton environments within each material.

passivation. We may quantify this change by taking the ratio of observed relaxation time constants, $\langle T_{1,obs} \rangle_{Me} / \langle T_{1,obs} \rangle_{OH}$. This is a simple analysis, and may be performed without any consideration of the changes to pore structure or surface chemistry upon passivation. This is because the terms S/V and δ within eqn (5) cancel exactly when considering molecules within the same pore structure. Values obtained by taking this ratio are displayed in Fig. 2c. In the case of bulk methanol (ESI Fig. S2†), the methyl group was found to have $T_{1,Me} = 4.22 \pm 0.01$ s while the hydroxyl group was found to have $T_{1,OH} = 3.92 \pm 0.01$ s. The resulting ratio of relaxation time constants in the bulk is therefore $T_{1,Me}/T_{1,OH} \approx 1$. From consideration of the ratios presented in Fig. 2c it appears that, within our passivated oxides in general, the ratio of relaxation time constants tends towards that of the bulk liquid. In turn, this indicates that bulk-like liquid dynamics occur near the surface of passivated pores, whereas the increased ratio of relaxation time constants observed within unpassivated supports indicates a more significant difference in the mobility of the two functional groups. Such a difference is of course expected following the above discussion on hydrogen bonding interactions with unpassivated surfaces.

4.3 Influence of passivation on relaxation time interpretation

It is also apparent from Fig. 2 that $\langle T_{1,obs} \rangle$ increases upon passivation regardless of the proton environment considered. From the above discussion we may be tempted to suggest that this is an obvious indication of enhanced molecular freedom within the adsorbed surface layer of passivated oxides. However, from eqn (5) it is clear that in order to compare the observed tumbling behaviour of molecules within different materials, we must also take into account any changes to the surface-to-volume (S/V) ratio of the pores within each material, as well as the influence of passivation on the thickness of the adsorbed surface layer, δ.

In this section, we therefore wish to provide an alternative method of analysing our relaxation data, in which the influence of passivation on pore structure and oxide surface chemistry is taken into account. We also aim to provide a more general method of analysing changes to molecular mobility at pore surfaces, in which the presence of multiple spin-bearing environments is not a prerequisite. We begin by defining a dimensionless 'spin–lattice interaction parameter' $\hat{\eta}(\delta)$, given by the ratio of T_1 in the bulk liquid to that of the same liquid at the pore surface.

$$\hat{\eta}(\delta) = \frac{T_{1,\text{bulk}}}{T_{1,\text{surf}}(\delta)} \tag{8}$$

Such a parameter provides a convenient limiting value of $\hat{\eta}(\delta) = 1$, in which case molecular rotational mobility within the adsorbed surface layer is considered identical to that within the bulk. Adsorption interactions leading to a reduction in surface mobility will therefore be expressed as an increase in $\hat{\eta}(\delta)$ from this limit. In previous work we have made use of a similar metric $\eta = T_{1,\text{bulk}}/T_{1,\text{obs}}$ to compare the surface dynamics of polar liquids within a range of unfunctionalised catalyst supports.[11] However, such an approach would be inappropriate here as it must be assumed that no change occurs to the thickness of the adsorbed surface layer upon passivation. In the current analysis we avoid this assumption by replacing the observed relaxation time constant with that associated with relaxation directly at the surface; $T_{1,\text{surf}}(\delta)$ is obtained through a simple rearrangement of eqn (3) such that

$$T_{1,\text{surf}}(\delta) = P_{\text{surf}}\left[\frac{1}{\langle T_{1,\text{obs}}\rangle} - \frac{(1 - P_{\text{surf}})}{T_{1,\text{bulk}}}\right]^{-1}. \tag{9}$$

A dependence on the unknown surface layer thickness δ arises from the presence of the surface population term P_{surf}. S/V – also present within P_{surf} – is estimated through the nitrogen absorption analysis discussed previously. Values are given in Table 1, from which it is clear that treatment of the oxide supports with TEOS causes an increase in S/V across the range of oxides considered. From examination of the approximate relation between $T_{1,\text{obs}}$ and S/V provided by eqn (5), and by assuming a constant surface layer thickness and $T_{1,\text{surf}}$ across both functionalised and unfunctionalised oxides, it is instructive to note that such a change should lead to a reduction in $\langle T_{1,\text{obs}}\rangle$ upon functionalisation, rather than the observed increase. Our results, therefore, cannot be explained through the inclusion of S/V alone. Indeed, it is necessary that the change in surface relaxivity ($\delta/T_{1,\text{surf}}$) which occurs as a result of surface passivation outweighs the influence of increasing S/V.

4.4 Analysis of methyl group dynamics

We chose to analyse our spin–lattice interaction parameter across an adsorbed surface thickness range of $\delta = 1$–5 molecular layers. This is an appropriate range for the analysis of alcohols interacting with oxide surfaces. For example, molecular dynamics simulations of methanol adsorbing at a model silica surface suggest that a surface layer thickness of 2 molecular layers is readily formed.[36] Experimental studies of the adsorption of ethanol at silica suggest a similar coverage.[37] We define the thickness of a single adsorbed layer within a given liquid-saturated pore to be that of the kinetic diameter of the molecule under study, d_k. For methanol, it is assumed that $d_k \approx 3.6$ Å during our relaxation experiments.[38]

To simplify the analysis of methanol dynamics, and to generate a method dependent on the presence of only a single relaxation environment, the T_1 relaxation of the methyl group is considered an appropriate representation of molecular mobility within our oxides. The spin–lattice relaxation of alkyl protons has previously been applied as a measure of overall molecular mobility in a number of liquid-saturated mesoporous oxides.[11] Indeed, in comparison to the relaxation characteristics of the hydroxyl group, relaxation within the methyl environment presents an attractive measure of molecular dynamics. Not only does it contain a significantly higher fraction of the total spins present within a given methanol-saturated pore, but its relaxation is also independent of any proton hopping, or other dissociative dynamics exhibited by the hydroxyl group. Indeed, we note that while the rotational mobility of methyl protons about the C–O axis of a given methanol molecule will be relatively unaffected by the interaction of the hydroxyl group with the pore surface, the rates of tumbling about axes parallel to the pore surface will still be significantly reduced, leading to sensitivity in changes to the overall molecular tumbling motion upon adsorption.

The calculated rotational interaction parameters for each of the methanol-saturated oxides investigated are presented in Fig. 3. On examination of these results it is clear that functionalisation of the pore surface has a significant impact on $\hat{\eta}(\delta)$. More specifically, passivation of the surface hydroxyl groups results in a notable decrease in $\hat{\eta}(\delta)$, irrespective of oxide material or surface layer thickness. From eqn (8) it is clear that this is indicative of an increase in the tumbling mobility of methanol molecules within the adsorbed surface layer. In particular, our results indicate that the motional freedom of methanol molecules within the adsorbed surface layer tends towards that of the unrestricted bulk liquid upon interaction with passivated pore surface. As this change is accompanied by a decrease in surface hydroxyl density upon functionalisation, we may attribute this observation to a reduction in the hydrogen bonding ability of the oxide surfaces upon passivation. Importantly, it should be noted that this result is qualitatively identical to that proposed by the simple analysis in Section 4.2, however, here we have performed a far more extensive assessment, involving the evaluation and elimination of the effects of changing S/V and δ values.

4.5 Derivation of a surface layer independent interaction parameter

It is interesting to observe that the data presented in Fig. 3 fit a power law of the form $\hat{\eta}_i(\delta) = a_i\delta^{-b_i}$. Here a_i (m^{b_i}) and b_i (dimensionless) are constants and the index i indicates the oxide under study. Moreover, this simple relationship suggests an approximate method to remove the dependency of $\hat{\eta}(\delta)$ on δ. We attempt to achieve this through the normalisation of $\hat{\eta}(\delta)$ against a weakly-interacting reference liquid. We first define a corresponding 'normalised' spin–lattice interaction parameter as

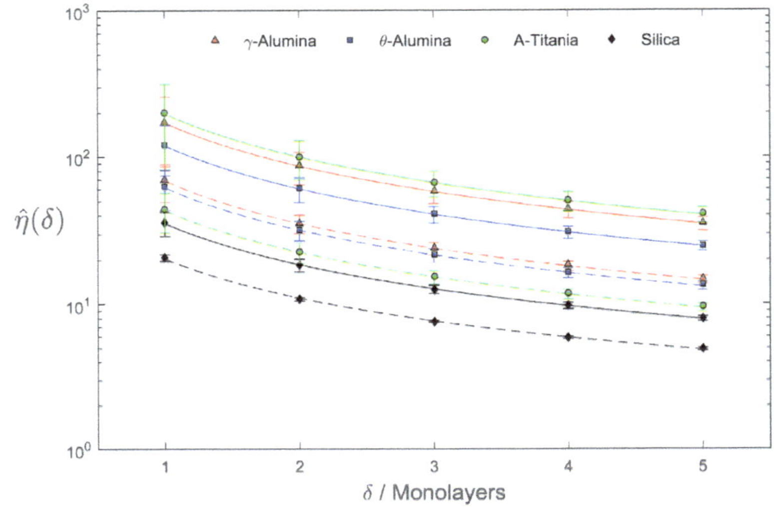

Fig. 3 Rotational surface interaction parameter as a function of surface layer thickness for methanol saturated mesoporous oxides. A fit to the data according to the power law discussed in the text is shown in each case, where solid lines indicate unpassivated oxides and dashed lines indicate those passivated with TEOS.

$$\hat{\eta}_i^N = \frac{\hat{\eta}_i^{Me}(\delta_j)}{\hat{\eta}_i^{CHX}(\delta_j)} \tag{10}$$

where $\hat{\eta}_i^{CHX}(\delta_j)$ is the spin–lattice interaction parameter acquired from cyclohexane-saturated oxides. Cyclohexane ($d_k \approx 6.2$ Å)[38] is often chosen as a weakly-interacting reference species for NMR relaxation and diffusion studies involving heterogeneous catalysts due to its single proton environment, slow relaxation characteristics and low affinity for oxide surfaces.[11–13,39,40] Within the oxide materials investigated here, cyclohexane also exhibits the same power law relationship between $\hat{\eta}(\delta)$ and δ as observed for methanol. We note from eqn (10) and from our empirical power law relation, that in order for the dependency on δ to be removed, b values acquired from the fitting of methanol and cyclohexane interaction parameter data must satisfy the condition $b_i^{Me} - b_i^{CHX} \approx 0$. In turn, the respective magnitudes of $\hat{\eta}_i^N$ will be proportional to the ratio a_i^{Me}/a_i^{CHX}. The power law constants a and b are listed in Table 2 for both methanol- and cyclohexane-saturated oxides. It is clear from these data, and by consulting the above condition on the magnitude of relative b values, that in order to remove the dependency on δ, the methanol relaxation data pertaining to both passivated and unpassivated oxides must be normalised by the relaxation behaviour of cyclohexane in unpassivated oxides alone. Indeed, the combination of these b values provides the closest match to the above condition on b values for the four oxides considered, while b values obtained from cyclohexane-saturated oxides passivated with TEOS are significantly different to those obtained for methanol in either passivated or unpassivated materials.

The index j within eqn (10) highlights the fact that the values of δ considered must be equal in units of length. Despite the differing values of d_k for methanol

Table 2 *a* and *b* values for passivated (oxide + TEOS) and unpassivated mesoporous oxide supports saturated with either methanol and cyclohexane. These constants correspond to the power law fit $\hat{\eta}_i(\delta) = a_i\delta^{-b_i}$ where i indicates the identity of the oxide

	Methanol				Cyclohexane			
	Oxide		Oxide + TEOS		Oxide		Oxide + TEOS	
Oxide support	a/m^b	b	a/m^b	b	a/m^b	b	a/m^b	b
γ-Al_2O_3	8.42×10^{-8}	0.99	5.23×10^{-8}	0.97	6.11×10^{-8}	0.95	6.06×10^{-7}	0.76
θ-Al_2O_3	6.66×10^{-8}	0.98	5.08×10^{-8}	0.96	7.58×10^{-8}	0.91	3.11×10^{-7}	0.80
A-TiO_2	9.32×10^{-8}	0.99	4.96×10^{-8}	0.95	8.14×10^{-8}	0.90	1.51×10^{-6}	0.70
SiO_2	5.18×10^{-8}	0.94	7.64×10^{-8}	0.89	4.10×10^{-7}	0.78	6.79×10^{-7}	0.75

and cyclohexane, the correct values of $\hat{\eta}_i^{CHX}(\delta)$ are readily obtained by consulting the appropriate power law constants governing the behaviour of cyclohexane (ESI Fig. S20†) and solving for the corresponding methanol δ values.

The normalised spin–lattice interaction parameters for methanol imbibed within unpassivated and passivated oxides are shown in Fig. 4a and b, respectively. From these data we may deduce that, within the experimental error indicated for each oxide, the dependency on the surface layer thickness has indeed been removed through the normalisation process detailed above. As a result we obtain a single, δ independent value of η^N for each material by averaging the acquired η^N values across the range of δ values investigated. In unpassivated oxides these are 3.1, 3.7, 6.7 and 3.0 for γ-alumina, θ-alumina, anatase-titania and silica, respectively. Upon passivation these values reduced to 1.3, 2.0, 1.5 and 1.8, respectively. It is again apparent that passivation of the surface hydroxyls within each mesoporous oxide has significantly enhanced the molecular tumbling of liquid methanol at the pore surface, indicating a reduction in surface–adsorbate interactions upon removal of surface hydroxyl groups. Furthermore, it is significant to note that while our previous methods of evaluating changes to the motional freedom of imbibed methanol concentrate on the comparison of the imbibed liquid behaviour relative to that of the unrestricted bulk, this analysis

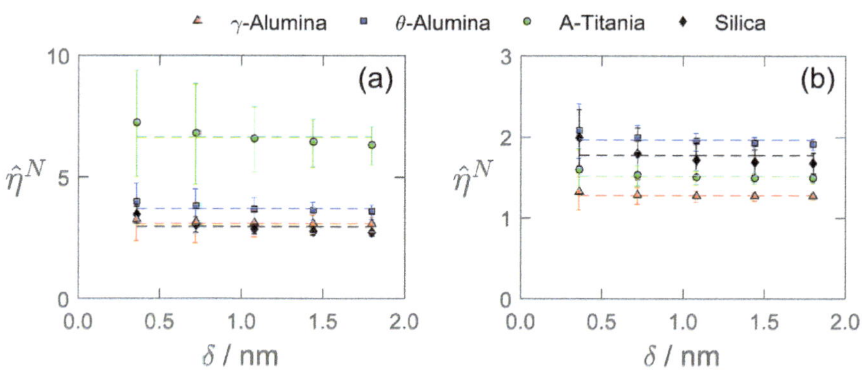

Fig. 4 Normalised rotational interaction parameters for methanol within (a) unpassivated and (b) passivated oxides, as a function of surface layer thickness.

compares the dynamics of imbibed methanol to that of an imbibed reference. More specifically, as $\hat{\eta}_i^N \to 1$ the molecular tumbling rates of methanol tend towards that of our cyclohexane reference, when imbibed within unpassivated oxide materials. This is deemed to be an adequate, non-interacting reference system, as interactions between cyclohexane and hydroxylated oxide surfaces are known to be negligible.[11]

5. Summary and conclusions

In this work we have demonstrated the use of NMR spin–lattice relaxation measurements as a non-invasive probe of molecular dynamics within liquid-saturated mesoporous oxides. Such measurements are particularly sensitive to the tumbling motion of molecules within the adsorbed surface layer. These measurements have been applied in order to assess changes in the surface dynamics of methanol upon passivation of surface hydroxyl groups. Our results show that passivation significantly increases the motional freedom of adsorbed liquid methanol. We have presented an extensive protocol designed to analyse in detail changes to the tumbling dynamics of adsorbed methanol upon passivation. By taking the ratio of observed spin–lattice relaxation time constants of the two proton environments present, and by comparing the results to that obtained from unrestricted bulk liquid, we have shown that passivation of the oxide supports cause methanol mobility in the adsorbed surface layer to tend towards that of the bulk. A more complex analysis based on the rotational dynamics of the methyl group alone leads to the same conclusion. This method is based on the calculation of a dimensionless interaction parameter and considers the influence of changes in surface chemistry and pore structure upon passivation. As such, this method not only provides a rigorous assessment of changes to alcohol dynamics within these oxides, but may also be extended, in future work, to more general cases of surface functionalisation. Finally, we have proposed an approximate method to remove any influence of the surface layer thickness on our results, based on the normalisation of methanol dynamics by a weakly-interacting reference system. Overall, this work illustrates NMR spin–lattice relaxation measurements as a useful probe of adsorbate dynamics within systems of relevance to liquid-phase catalysis. In particular, the calculation of a relevant interaction parameter and its comparison to simpler methods of mobility analysis enhances our ability to interpret the results of spin–lattice relaxation measurements in terms of modified surface affinity within functionalised mesoporous materials.

Acknowledgements

N. R. would like to thank the Catalysis@Cambridge initiative for the studentship. C. D. would like to acknowledge the support of Wolfson College Cambridge. Mr Zlatko Saracevic is also gratefully acknowledged for performing the pore size measurements.

References

1 C. H. Christensen and J. K. Nørskov, *J. Chem. Phys.*, 2008, **128**, 182503.

2 F. Calle-Vallejo, J. Tymoczko, V. Colic, Q. H. Vu, M. D. Pohl, K. Morgenstern, D. Loffreda, P. Sautet, W. Schuhmann and A. S. Bandarenka, *Science*, 2015, **350**, 185–189.
3 A. A. Latimer, A. R. Kulkarni, H. Aljama, J. H. Montoya, J. S. Yoo, C. Tsai, F. Abild-Pedersen, F. Studt and J. K. Nørskov, *Nat. Mater.*, 2017, **16**, 225–229.
4 J. C. Matsubu, S. Zhang, L. DeRita, N. S. Marinkovic, J. G. Chen, G. W. Graham, X. Pan and P. Christopher, *Nat. Chem.*, 2017, **9**, 120–127.
5 J. A. Farmer and C. T. Campbell, *Science*, 2010, **329**, 933–936.
6 S. J. Tauster, *Acc. Chem. Res.*, 1987, **20**, 389–394.
7 R. Burgess, C. Buono, P. R. Davies, R. J. Davies, T. Legge, A. Lai, R. Lewis, D. J. Morgan, N. Robinson and D. J. Willock, *J. Catal.*, 2015, **323**, 10–18.
8 M. Bowker and R. Sharpe, *Catal., Struct. React.*, 2015, **1**, 140–145.
9 O. Dulub, W. Hebenstreit and U. Diebold, *Phys. Rev. Lett.*, 2000, **84**, 3646–3649.
10 J. Graciani, K. Mudiyanselage, F. Xu, A. E. Baber, J. Evans, S. D. Senanayake, D. J. Stacchiola, P. Liu, J. Hrbek, J. F. Sanz and J. A. Rodriguez, *Science*, 2014, **345**, 546–550.
11 C. D'Agostino, J. Mitchell, L. F. Gladden and M. D. Mantle, *J. Phys. Chem. C*, 2012, **116**, 8975–8982.
12 C. D'Agostino, G. L. Brett, P. J. Miedziak, D. W. Knight, G. J. Hutchings, L. F. Gladden and M. D. Mantle, *Chem.–Eur. J.*, 2012, **18**, 14426–14433.
13 C. D'Agostino, T. Kotionova, J. Mitchell, P. J. Miedziak, D. W. Knight, S. H. Taylor, G. J. Hutchings, L. F. Gladden and M. D. Mantle, *Chem.–Eur. J.*, 2013, **19**, 11725–11732.
14 C. D'Agostino, M. R. Feaviour, G. L. Brett, J. Mitchell, A. P. E. York, G. J. Hutchings, M. D. Mantle and L. F. Gladden, *Catal. Sci. Technol.*, 2016, **6**, 7896–7901.
15 K. Lin, L. Wang, F. Meng, Z. Sun, Q. Yang, Y. Cui, D. Jiang and F.-S. Xiao, *J. Catal.*, 2005, **235**, 423–427.
16 M. Guidotti, I. Batonneau-Gener, E. Gianotti, L. Marchese, S. Mignard, R. Psaro, M. Sgobba and N. Ravasio, *Microporous Mesoporous Mater.*, 2008, **111**, 39–47.
17 N. Igarashi, K. Hashimoto and T. Tatsumi, *Microporous Mesoporous Mater.*, 2007, **104**, 269–280.
18 M. Ojeda, F. J. Pérez-Alonso, P. Terreros, S. Rojas, T. Herranz, M. López Granados and J. L. G. Fierro, *Langmuir*, 2006, **22**, 3131–3137.
19 L. Jia, L. Jia, D. Li, B. Hou, J. Wang and Y. Sun, *J. Solid State Chem.*, 2011, **184**, 488–493.
20 A. S. Zola, L. S. da Silva, A. L. Moretti, A. do C. Fraga, E. F. Sousa-Aguiar and P. A. Arroyo, *Top. Catal.*, 2016, **59**, 219–229.
21 N. A. Fellenz, J. F. Bengoa, M. V. Cagnoli and S. G. Marchetti, *J. Porous Mater.*, 2016, 1–12.
22 L. D. Ellis, S. Pylypenko, S. R. Ayotte, D. K. Schwartz and J. W. Medlin, *Catal. Sci. Technol.*, 2016, **6**, 5721–5728.
23 E. W. Hansen, F. Courivaud, A. Karlsson, S. Kolboe and M. Stöcker, *Microporous Mesoporous Mater.*, 1998, **22**, 309–320.
24 F. Courivaud, E. W. Hansen, A. Karlsson, S. Kolboe and M. Stöcker, *Microporous Mesoporous Mater.*, 2000, **35–36**, 327–339.
25 D. Weber, A. J. Sederman, M. D. Mantle, J. Mitchell and L. F. Gladden, *Phys. Chem. Chem. Phys.*, 2010, **12**, 2619–2624.

26 A. Abragam, *The Principles of Nuclear Magnetism*, Clarendon Press, 1961.
27 M. H. Levitt, *Spin Dynamics: Basics of Nuclear Magnetic Resonance*, Wiley-Blackwell, Chichester, England, Hoboken, NJ, 2nd edn, 2008.
28 N. Bloembergen, E. M. Purcell and R. V. Pound, *Phys. Rev.*, 1948, **73**, 679–712.
29 K. R. Brownstein and C. E. Tarr, *J. Magn. Reson.*, 1977, **26**, 17–24.
30 K. R. Brownstein and C. E. Tarr, *Phys. Rev. A*, 1979, **19**, 2446–2453.
31 K. E. Washburn, C. D. Eccles and P. T. Callaghan, *J. Magn. Reson.*, 2008, **194**, 33–40.
32 J. D. Wilson, *J. Mater. Sci.*, 1992, **27**, 3911–3924.
33 A. N. Tikhonov and V. I. Arsenin, *Solutions of ill-posed problems*, Winston, 1977.
34 G. Wahba and Y. Wang, *Comm. Stat. Theor. Meth.*, 1990, **19**, 1685–1700.
35 F. Scheinmann, *An Introduction to Spectroscopic Methods for the Identification of Organic Compounds: Nuclear Magnetic Resonance and Infrared Spectroscopy*, Elsevier, 2013.
36 D. Roy, S. Liu, B. L. Woods, A. R. Siler, J. T. Fourkas, J. D. Weeks and R. A. Walker, *J. Phys. Chem. C*, 2013, **117**, 27052–27061.
37 D. Wu, X. Guo, H. Sun and A. Navrotsky, *J. Phys. Chem. C*, 2015, **119**, 15428–15433.
38 H. Wu, Q. Gong, D. H. Olson and J. Li, *Chem. Rev.*, 2012, **112**, 836–868.
39 C. D'Agostino, M. R. Feaviour, G. L. Brett, J. Mitchell, A. P. E. York, G. J. Hutchings, M. D. Mantle and L. F. Gladden, *Catal. Sci. Technol.*, 2016, **6**, 7896–7901.
40 S. N. Lanin, E. V. Vlasenko, N. V. Kovaleva and F. T. Zung, *Russ. J. Phys. Chem. A*, 2008, **82**, 2152–2155.

Faraday Discussions

PAPER

The role of halogens in on-surface Ullmann polymerization†

Gianluca Galeotti,[a] Marco Di Giovannantonio,‡[b] Josh Lipton-Duffin,[c] Maryam Ebrahimi,*[a] Stefano Tebi,[b] Alberto Verdini,[d] Luca Floreano,[d] Yannick Fagot-Revurat,[e] Dmitrii F. Perepichka,[f] Federico Rosei*[a] and Giorgio Contini*[bg]

Received 9th March 2017, Accepted 30th March 2017

DOI: 10.1039/c7fd00099e

Ullmann coupling is the most common approach to form surface-confined one- and two-dimensional conjugated structures from haloaryl derivatives. The dimensions of the formed nanostructures can be controlled by the number and location of halogens within the molecular precursors. Our study illustrates that the type of halogen plays an essential role in the design, orientation, and extent of the surface-confined organometallic and polymeric nanostructures. We performed a comparative analysis of five 1,4-dihalobenzene molecules containing chlorine, bromine, and iodine on Cu(110) using scanning tunneling microscopy, fast-X-ray photoelectron and near edge X-ray absorption fine structure spectroscopies. Our experimental data identify different molecular structures, reaction temperatures and kinetics depending on the halogen type. Climbing image nudged elastic band simulations further clarify these observations by providing distinct diffusion paths for each halogen species. We show that in addition to the structure of the building blocks, the halogen type has a direct influence on the morphology of surface-confined polymeric structures based on Ullmann coupling.

[a]*Centre Énergie, Matériaux et Télécommunications, Institut National de la Recherche Scientifique, 1650 Boulevard Lionel-Boulet, Varennes, QC, J3X 1S2, Canada. E-mail: rosei@emt.inrs.ca; maryam.ebrahimi@emt.inrs.ca*

[b]*Istituto di Struttura della Materia, CNR, Via Fosso del Cavaliere 100, 00133 Roma, Italy. E-mail: giorgio.contini@cnr.it*

[c]*School of Chemistry, Physics and Mechanical Engineering, Institute for Future Environments, Queensland University of Technology (QUT), 2 George Street, Brisbane, QLD 4001, Australia*

[d]*Istituto Officina dei Materiali CNR, Laboratorio TASC, s.s. 14 km 163.5, 34149 Trieste, Italy*

[e]*Institut Jean Lamour, UMR 7198, Université Lorraine/CNRS, B.P. 239 FE-54506, Vandoeuvre-les-Nancy, France*

[f]*Department of Chemistry, McGill University, 801 Sherbrooke Str. West, Montreal, QC H3A 0B8, Canada*

[g]*Department of Physics, University of Rome Tor Vergata, Via della Ricerca Scientifica 1, 00133 Roma, Italy*

† Electronic supplementary information (ESI) available. See DOI: 10.1039/c7fd00099e

‡ Present address: Empa – Swiss Federal Laboratories for Materials Science and Technology, Überlandstrasse 129, 8600 Dübendorf, Switzerland.

Introduction

The discovery of graphene, an atomically thin layer of carbon which exhibits remarkable thermal, mechanical, and optoelectronic properties, highlights the greatest breakthrough in materials science of the past decade.[1–4] However, graphene's zero bandgap limits its application as a semiconductor in electronic devices, which has stimulated a growing interest to synthesize organic analogues of graphene.[5–9] One of the most compelling strategies in this arena is to design surface-confined fully-conjugated polymers with tunable bandgaps based on a rational choice of monomer.[10–13] The flexibility of organic synthesis offers a broad playground for creating two-dimensional (2D) conjugated polymers by exploring various surface reactions using different building blocks.[14–17] Surface-confined 2D polymers may be synthesized through bottom-up methodologies which yield large molecular structures resulting from small monomers.[10] The surface acts as template, creating ordered 1D[18] or 2D[19] polymeric structures by suppressing the entropically-driven disorder of solution synthesis.

In addition to passive templating, the surface can also act as a catalyst, accelerating the reaction and guiding the orientation of the growing polymer structure.[20] Although it is relatively easy to obtain long-range order in self-assembled molecular networks (SAMNs) which are driven by weak intermolecular interactions, this is rarely the case for covalently bound structures, due to the irreversible nature of covalent coupling. A possible approach to forming long-range ordered covalent networks is to use a multi-step reaction, in which the monomers first form a dynamic intermediate which arranges into a 2D monolayer. The neighboring molecules can subsequently undergo a coupling reaction forming a polymer made of covalent C–C bonds.[21,22] Ullmann coupling, the most popular reaction used in on-surface chemistry, expresses these features, it is a two-step process in which the transition metal supporting substrate catalyzes[23] the dehalogenation of aryl halide precursors, forming an organometallic (OM) intermediate[18] which subsequently couples into a covalent structure *via* new C–C bonds.[24] The activation energy of the reaction depends on both the particular molecule[25] and the surface.[26,27] On Cu and Ag surfaces, the deposited molecules can dehalogenate at room temperature (RT) and form an OM intermediate phase that is stable up to the polymerization temperature.[28–30]

While the majority of recent studies on surface-confined Ullmann coupling focused on understanding the structure and order of the polymers on different substrates[27] and the density of defects,[31] the effects of the halogen by-product have not been explored in detail. A few theoretical[20] and experimental[32,33] studies have reported the dissociation of halobenzenes on metallic surfaces, addressing the differences in the adsorption energies. However, the effects of halogens on the reaction outcome have not been explored. The derivatives of 1,3,5-tris(4-halophenyl)benzene functionalized with bromine (TBB),[26,34] iodine (TIB),[25] and both halogens (TBIB)[30] have been studied on various surfaces, yet a coherent comparison to reveal the role played by the halogens was not reported. Our recent work described the contribution of halogens in the formation of 1D polymers *via* Ullmann coupling,[28,29] showing that the halogens are part of the OM unit cell after the deposition of 1,4-dibromobenzene (dBB) on Cu(110). These studies also illustrate that the structures obtained from dBB[28] and 1,4-

diiodobenzene (dIB)[18] on Cu(110) are different, which suggests the importance of studying the influence of halogens on the processes that transform the intermediate phase into a polymeric structure.

Since the first report describing on-surface Ullmann coupling under ultra-high vacuum (UHV) conditions,[35] several studies have provided insights into its reaction mechanisms and kinetics.[23,24,36] A kinetic model based on a combined experimental and theoretical approach was applied to the polymerization of dBB on Cu(110), showing that the reaction follows a nucleation-and-growth mechanism.[37] However, there are still open questions regarding both the fundamental understanding and the practical realization of on-surface Ullmann coupling.[38] In line with this perspective, we address how the type of halogen can influence the design of the OM and polymer structures.

Here, we report a systematic study of five different 1,4-dihalobenzene molecules, a phenyl ring containing two halogens (Cl, Br or I) located at the *para* positions, on Cu(110). Our findings illustrate the role of the metal halide by-product in determining the rate of C–C coupling step and the order of the formed π-conjugated polymers based on Ullmann coupling.

Materials and methods

All experiments were performed under UHV conditions with base pressures below 2×10^{-10} mbar. The precursors dBB (98% purity), dIB (99%), 1,4-dichlorobenzene (dCB, ≥99%), 1-bromo-4-chlorobenzene (BCB, 99%) and 1-bromo-4-iodobenzene (BIB, 98%), shown in Fig. 4, were purchased from Sigma-Aldrich and were deposited through a leak valve onto Cu(110) (MaTecK GmbH), with the substrate held at RT. All experiments were performed starting from a saturated monolayer coverage. The substrate was prepared by repeated cycles of Ar^+ sputtering (1.2 keV) and annealing (500 °C). Scanning tunneling microscopy (STM) was performed at RT in constant-current mode using an Omicron VT-STM. Quoted bias voltages are measured from the tip to the sample. STM images were analyzed using WSxM,[39] and were treated for plane subtraction, line-by-line flattening, contrast enhancement, and were corrected based on the known lattice parameters of the substrate. X-ray photoelectron spectroscopy (XPS), fast-XPS and near edge X-ray absorption fine structure (NEXAFS) spectroscopy measurements were performed at the ALOISA beamline of the Elettra synchrotron-radiation facility in Trieste (Italy). XPS experiments were performed in normal emission geometry, with a 4° grazing incidence of linearly polarized radiation, using a pass energy of 18 eV and a home-built hemispherical electron analyzer equipped with a multichannel plate (MCP) detector. The C 1s, Br 3d and Cl 2p core levels were measured at a photon energy of 390 eV with an overall energy resolution of 200 meV. The I 3d peak was measured at a photon energy of 750 eV with an overall energy resolution of 300 meV. Unless stated otherwise, the spectral peak fitting based on residual minimization with Gaussian–Lorentzian line shape functions (30% Lorentzian) and linear backgrounds was performed using Casa XPS[40] and CONTUR[41] software.

The fast-XPS maps of C 1s were acquired by increasing the sample temperature from RT to 230 °C with a heating rate of 0.2 °C s^{-1}, using a radiative heating element behind the sample. Every line on each map is a snapshot of the C 1s peak, taken at the rate of one spectrum per second at a photon energy of 390 eV and pass energy set to 30 eV (overall energy resolution of ~350 meV). The NEXAFS C K-edge spectra were recorded in partial electron yield by means of a channeltron equipped with a filtering grid biased at 240 V, to reject the tail of secondary electrons. The photon energy resolution was set to 100 meV. To calibrate the photon energy scale, the drain current on the last refocusing mirror of the beamline was measured along with the C K-edge, and subsequently calibrated back to reference data. The C K-edge spectrum acquired from the clean Cu(110) surface was used to normalize the spectra from surfaces containing the molecular layers. Polarization dependent measurements were performed by rotating the sample about the beam axis, changing the angle (θ) of the surface with respect to the polarization vector from transverse magnetic (TM, almost p-polarization) to transverse electric (TE, s-polarization) geometry, while keeping the grazing photon incidence angle fixed at 6° (see Floreano *et al.*[42] for details about NEXAFS geometry and calibration).

Density functional theory (DFT) calculations were performed with the Vienna Ab-initio Simulation Package (VASP)[43,44] installed at the SciNet[45] supercomputer clusters of Compute Canada. DFT calculations were executed using the Perdew–Burke–Ernzerhof approximation (PBE)[46] of the exchange–correlation potential, the projector augmented wave (PAW)[47,48] method, and a plane-wave basis set with an energy cut off of 450 eV. Final calculations were performed using the zero-damping DFT-D3 [49] methods of Grimme, in which vdW correction for potential energy and dispersion effects are taken into account *via* a semi-empirical approach. The k-point sampling was restricted to the Γ point to cope with the large size of the supercells. The energy barriers for the diffusion of the halogens (Cl, Br, I) were determined by calculating the minimum energy pathways (MEPs) along the [001], [1−10], [1−12], and [1−11] surface directions using the climbing image nudged elastic band (CI-NEB) method.[50,51] For the initial states, the most stable geometry/location of the halogen was found to be short-bridge for Cl and Br, and hollow for I. The path of diffusion from the initial state to the nearest similar adsorption site (as final state) along four directions of the surface was calculated for each halogen. First, the structures of the initial and final states were optimized, and then the intermediate structures (16 structures) were calculated using DFT-D3, so that, at each iteration, the NEB method nudges the reaction path (atomic coordinates) towards the minimum energy geometry until convergence is reached. The 216-atom supercell used for calculating the diffusion barrier includes a nine-layer thick 4 × 6 Cu(110) slab and a 20 Å thick vacuum region. The bottom five layers in the Cu slab were fixed throughout the NEB calculations, while all of the other atomic coordinates were relaxed until all forces were smaller than 0.05 eV Å^{-1}.

Results and discussion

(1) Fast-XPS and reaction overview

On-surface Ullmann coupling is a two-step reaction,[28,52,53] with the formation of an OM intermediate and metal halide by-product in the first step, followed by the formation of a C–C bond in the second, rate-limiting step. As such, the nature of the halide is not generally considered to affect the overall rate of reaction. However, we show below that the C–C coupling step is in fact halogen-dependent (Fig. 1 and 4). The role of the halogen is manifested in the OM chain termination

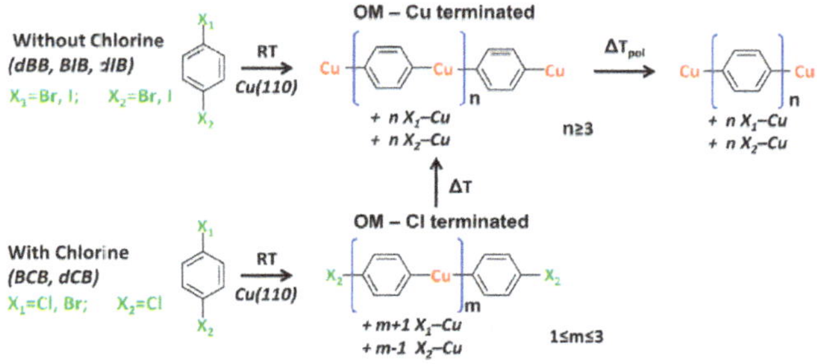

Fig. 1 Reaction schemes for dihalobenzene polymerizations. dBB, BIB and dIB on Cu(110) follow a two-step reaction forming an OM intermediate at RT, and polymers at higher temperatures. Chlorine-containing precursors (BCB and dCB) first form short OM chains due to incomplete dehalogenation of Cl at RT. Increasing the temperature results in a complete dehalogenation, forming longer OM chains. The polymerization step of the reaction is the same for all the five precursors. *m* and *n* represent the OM chain size.

at RT, in which the partial dehalogenation of the chlorine-containing molecules results in short chlorine-terminated OM chains, which change to Cu-terminated chains after dechlorination is completed at 150 °C. By contrast, the bromine- and iodine-containing molecules produce longer Cu-terminated chains at RT (Fig. 1).

The chemical states of the different carbon species involved in the dehalogenation and polymerization processes can be identified by measurement of the C 1s photoemission spectra, as reported in Fig. 2. At RT, the C 1s spectra are dominated by a main peak corresponding to the four closely equivalent sp^2 carbon atoms (C–C) at a binding energy (BE) of approximately 284.0 eV. The dehalogenated carbon atoms bound to Cu surface adatoms (C–Cu), give their contribution at a lower BE (283.3 eV). In the case of chlorinated molecules, the whole C 1s spectrum is shifted (0.3–0.4 eV) towards a higher BE, in which an additional minority component is detected at 285.5 eV, which we attribute to the intact C–Cl component. Similar to the bromine-containing molecules, the C 1s spectrum of the BCB precursor contains a shoulder at a lower BE, which is characteristic of Br dehalogenation. In the case of the doubly chlorinated molecule (dCB), the similar C–Cu component at the lower BE is clearly detected, thus suggesting a partial Cl dehalogenation. This weak component is more resolved at higher temperature (150 °C) when Cl dehalogenation is completed.

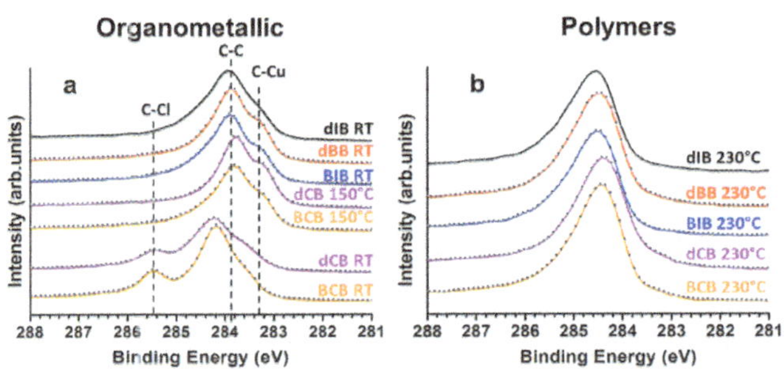

Fig. 2 C 1s spectra for saturated coverage of each precursor on Cu(110) for the organometallic (a) and polymer (b) phases.

This observation is consistent with the Cl 2p and Br 3d spectra of BCB reported in Fig. 3. The doublet Br 3d peak at 68.4 eV BE is in agreement with the dehalogenated C–Br at RT and 150 °C, whereas the (mostly) undissociated C–Cl is confirmed by only one doublet at ~200.0 eV (Cl $2p_{3/2}$) at RT. The dCB, on the other hand, has undergone nearly 50% of C–Cl dehalogenation at RT. as indicated by the additional doublet at a much lower BE of 198.0 eV, associated with Cl atoms directly bound to the Cu surface.

The evolution from the OM to polymeric phase for the five studied precursors was examined by acquiring fast-XPS spectra of the C 1s core level as a function of temperature. The fast-XPS maps presented in Fig. 4 show the differences for every precursor, indicating a halogen-dependent effect on the activation energy (starting temperature) and the kinetics (interval) of the corresponding reaction. All of

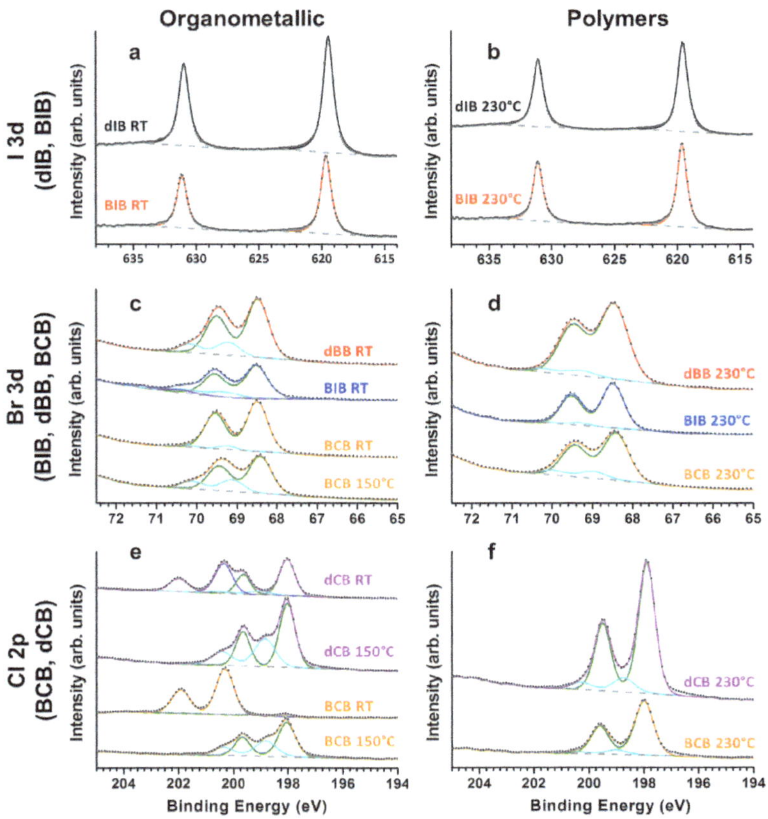

Fig. 3 XPS spectra (I 3d, Br 3d, Cl 2p) of organometallic (OM) (left panels, a, c, e) and polymer (right panels, b, d, f) chains for the saturated coverage of five precursors on Cu(110) at different temperatures. The component in green is attributed to the dissociated halogens, also observed for the spectra of the polymer chains. The component in light blue, observed for the long OM chains (dBB and BIB at RT, and BCB and dCB at 150 °C), is significantly reduced after polymerization. This component likely originates from the different adsorption sites of the halogens when long chains are present. For the Br 3d peak corresponding to BIB at RT, a weak component (in purple) is observed at 70.7 eV, attributed to a small fraction of intact C–Br bonds.

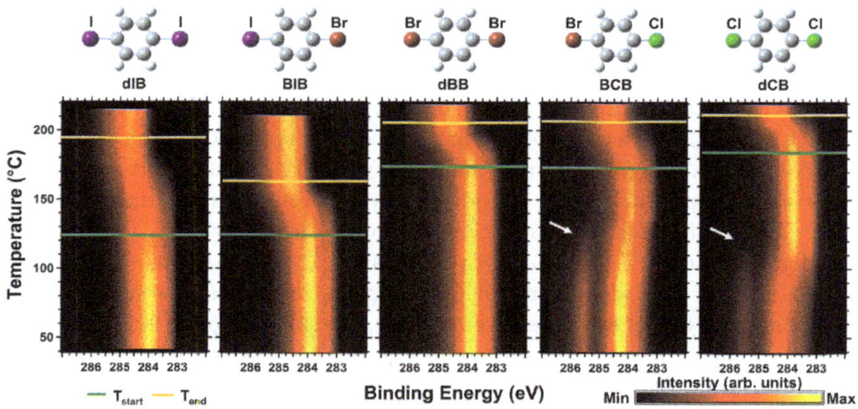

Fig. 4 Fast-XPS measurements of C 1s during annealing (0.2 °C s^{-1} heating rate) of a saturated monolayer of each precursor on Cu(110) dosed at RT. The temperatures at which 10% and 90% of the polymerization reaction is completed are marked with T_{start} (green) and T_{end} (orange) lines, respectively. White arrows indicate the change in the C 1s spectra due to C–Cl dissociation.

the fast-XPS maps exhibit a shift in the C 1s spectra towards a higher BE above 140 °C. This is a signature of polymerization, and is correlated with the progressive vanishing of the C–Cu component (at 283.3 eV) and growth of the C–C component at a higher BE,[28,37] shown in Fig. 2. The correlation of polymerization and the halogen type is seen in the starting temperature (T_{start}) of the conversion of OM to polymer, at 125 ± 2 °C for the dIB and BIB, at 175 ± 2 °C for dBB and BCB, and at 185 ± 2 °C for dCB, showing that T_{start} increases from the iodine- to the bromine- to the chlorine-containing molecules (green lines, Fig. 4). While for every molecule the reaction is completed within an approximately 40 °C range, the reaction for dIB proceeds across an 80 °C range (Table S1†). The final BE position of the main C 1s peak is found at 284.5 eV with small variations of ±0.1 eV between the various precursors, which remains within the experimental resolution.

The maps for chlorine-containing molecules present an earlier transition (than the OM-to-polymer conversion temperature range) at about 120 °C, witnessed by the C 1s shift towards lower BE, as highlighted by the white arrows in Fig. 4, which we can ascribe to the complete C–Cl dissociation. Full C–Br and C–I bond dehalogenation takes place below RT and, therefore, does not appear in the fast-XPS maps of the non-chlorinated precursors. The fast-XPS maps not only identify the critical temperature of the on-surface chemical reactions, but also highlight that the C–C coupling step of the reaction is halogen-dependent. For the iodine- and bromine-containing molecules, the scheme is similar to previous reports,[28,29,37,52] where a fully dehalogenated phase of OM chains forms at RT, and remains stable up to the polymerization temperature. For chlorine-containing molecules, the RT phase, which is stable up to 120 °C, is rather associated with the incomplete dehalogenation of dCB and BCB precursors. As a result, most of the phenyl groups would be Cl-terminated and form short OM chains on the surface (2 to 4 phenyls long). The XPS spectra for Cl 2p presented in Fig. 3e shows that approximately 50% of the Cl–C bonds are dissociated for dCB, but only a small fraction for BCB.

C–Cl bond dissociation is completed at temperatures above 120 °C (Fig. 4), and longer chains are formed (>20 phenylenes long, as shown in Fig. 5). For dBB, BIB and dIB, the C–Br and C–I bonds are dissociated upon adsorption at RT on Cu(110), as can be seen from the Br 3d (Fig. 3c) and I 3d (Fig. 3a) spectra, in which their $3d_{5/2}$ components at 68.3 eV and 619.5 eV, respectively, are attributed to Cu–Br[28,29] and Cu–I bonds.[18] A single chemical state was observed for I 3d at every temperature during the reaction. For Br 3d and Cl 2p, the dominant states are found at 68.3 eV ($3d_{5/2}$) and 198.0 eV ($2p_{5/2}$), respectively, and additional components at higher BEs were observed at 69.1 eV (Br $3d_{5/2}$) and 198.8 eV (Cl $2p_{5/2}$) (Fig. 3c and e). In the case of dBB and BIB, these components were present at RT, while for the chlorine-containing molecules, they became apparent only after annealing to 150 °C, when dehalogenation is complete (Fig. 3e). The other chemical states (0.7 eV higher BE shift) in the Br 3d and Cl 2p spectra can be tentatively attributed to different adsorption sites of the halogen within the OM structure. The BIB spectra at RT exhibit a weak peak at 70.7 eV (Br $3d_{5/2}$), attributable to a minor fraction of Br–C bonds still being intact at RT.

Cleavage of the C–Br and C–I bonds occurs at lower temperatures, below the temperature range investigated in this study. For example, it is known that the dehalogenation of iodobenzene on Cu(110) takes place at approximately −100 °C.[23] Although the dissociation temperature of Br for bromobenzene or dBB

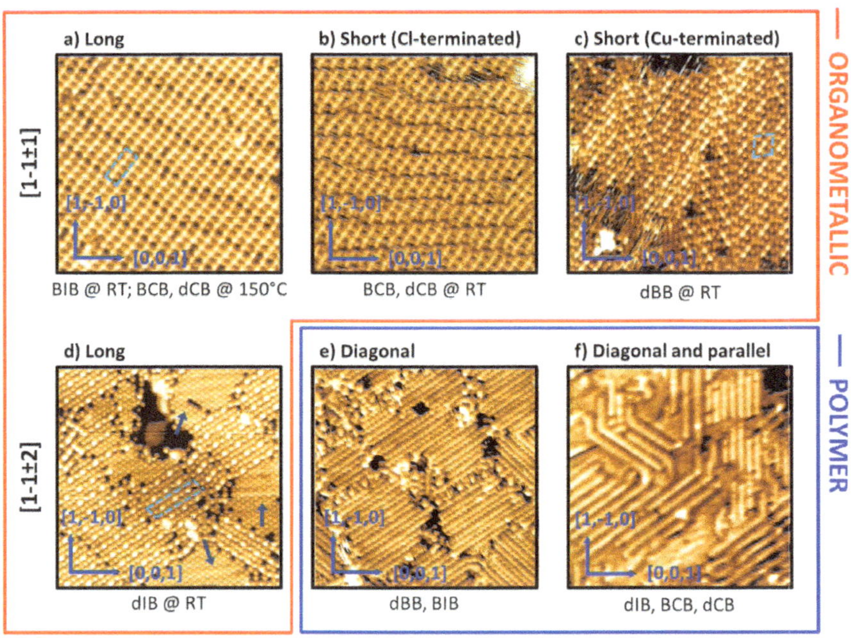

Fig. 5 STM images (20 × 20 nm^2) obtained for the five precursors on Cu(110). The OM phases (a–d) are shown in the red box; the blue arrows in (d) indicate the *c*(2 × 2) iodine superstructure; light blue dashed lines represent the molecular unit cells. BCB and dCB show two different OM phases at RT (b) and after annealing at 150 °C (a). The polymer phases (e and f), formed when annealing above 200 °C are presented in the blue box.

on Cu(110) is not available in the literature, the aryl halides (Ar–X) have shown a reactivity trend of Ar–I > Ar–Br > Ar–Cl. The dissociation of Br is therefore expected to happen at a higher temperature than iodine.[54,55] This trend agrees with DFT calculated dissociation energies of iodobenzene and bromobenzene,[20] and with the experimental values of the dissociation enthalpy of a Ph–halogen bond: 67 ± 2 kcal mol^{-1} for Ph–I, 84 ± 1 kcal mol^{-1} for Ph–Br and 97 ± 1 kcal mol^{-1} for Ph–Cl.[56]

A comparison of fast-XPS maps and XPS spectra between dIB and dBB is presented in Fig. S1.† dBB is taken as the representative of the other bromine- and chlorine-containing molecules (BIB, BCB and dCB) because their fast-XPS maps show similar behavior for the polymerization process, both in the temperature range of the polymerization reaction (Table S1,† temperature range) and in the OM chains, all along the same directions (Fig. 5). Fig. S1a and c† show the fast-XPS maps, where the C 1s line profiles at various temperatures through the ramp are shown every 20 °C in panels (b, d) (identical profiles have been omitted). While for dIB the shift is continuous and gradual toward higher BE values, it is sharp and sudden for dBB. STM images (Section 2) also show a clear difference between dBB and dIB structures on the surface, likely due to the fact that for dIB the OM chains are oriented along the same direction as the polymer, whereas for all other molecules this is not the case, as discussed in Section 2. Therefore, dIB coupling can happen for two monomers in an OM chain without affecting the other adjacent ones, *i.e.* the reaction can take place gradually, while for the other precursors a rotation is required for the phenyl–phenyl coupling, affecting the neighboring chains, which, we speculate, may require an additional barrier to be overcome for the monomers to couple.

(2) Structural characterization by STM

Deposition of each precursor on Cu(110) at RT results in distinctive self-assembled OM structures, oriented along different lattice directions, which produce poly(*para*-phenylene) (PPP) polymers upon annealing. The OM (red frame, Fig. 5a–d) and polymer (blue frame, Fig. 5e and f) structures obtained for the five precursors at different temperatures are shown in Fig. 5. The assignment of these phases is based on the distances between the protrusions in the STM data for the OM and polymer chains (6.3 Å and 4.4 Å, respectively),[28] together with the C 1s analysis described above.

Organometallic chains. All of the OM phases (Fig. 5a–d), appearing in two domains related by mirror symmetry with respect to the [1−10] direction, are made of chains of phenylene groups linked through copper atoms, and stacked into a 2D structure. The OM chains are oriented along either the [1−1±1] directions (Fig. 5a–c, dCB, BCB, dBB, BIB) or the [1−1±2] directions (Fig. 5d, dIB), and are interdigitated by rows of protrusions, identified as halogens.[28,57] For iodine-containing molecules, a portion of the surface is covered by the $c(2 \times 2)$ iodine superstructure (Fig. 5d). The directions of the molecular domains, together with the iodine superstructure, are shown in Fig. S2.† Higher resolution STM images of every phase for each precursor are presented in Fig. 6, superimposed with the proposed molecular structure. OM chains along the [1−1±2] directions are observed only for dIB, and are composed of ordered domains of long chains (limited only by the size of the terraces, defects and other domains), with

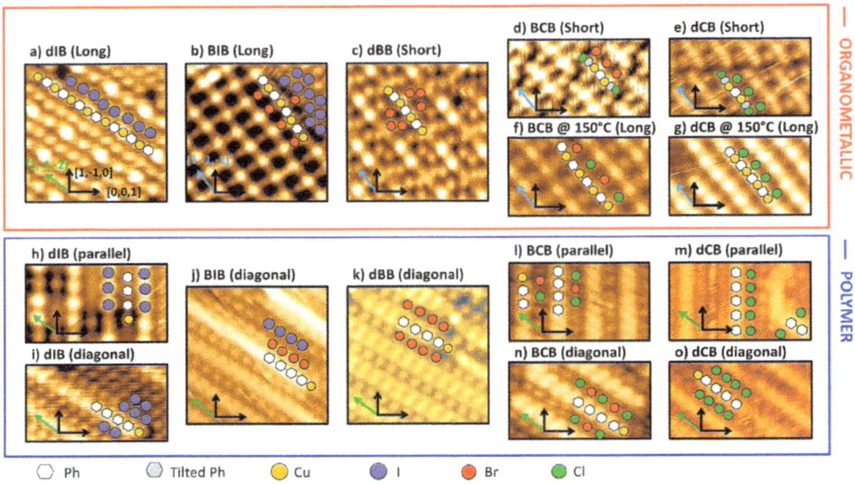

Fig. 6 STM images (square 5 × 5 nm^2, rectangular 2.5 × 5.0 nm^2) of the OM (a–e at RT and f and g at 150 °C) and polymeric phases (h–o at 230 °C) observed for each precursor, superimposed with molecular structures and copper lattice directions. OM chains for dIB (a) are orientated along the [1−1−2] direction, the same direction as the diagonal polymeric chains, while all the other OM chains are along the [1−1−1] direction.

a periodic strain-relieving kink every seven phenyls (Fig. S3†), and a surface reconstruction corresponding to the epitaxy matrices (1, −4|11, 10) and (1, 4|−11, 10). For the structures that produce [1−1±1]-oriented OMs, we observe two different types of chains: long (for BIB) and short (for dBB, BCB and dCB) at RT. The short chain structures are made of a smaller number of phenyls, between two to four for BCB and dCB (Fig. 5b) and exclusively three for dBB ('chevron' phase, Fig. 5c).[57] For dBB, the short chains are always Cu-terminated (Fig. 6c), while for BCB and dCB, they are Cl-terminated (Fig. 6d and e), as inferred from the XPS spectra in Fig. 3 showing complete debromination for dBB, as opposed to dechlorination for BCB and dCB. The presence of Cl at the chain-end forces the molecular aggregate to bend the terminal phenyls upwards, resulting in an apparent projected length in the STM profiles obtained along the [1−1−1] direction of the Cu–Ph–Cl group equal to 4.9 Å, which is shorter than the 6.4 Å known length of a Cu–Ph–Cu, as shown in Fig. S4a and b.† The reduced dichroism in the NEXAFS measurements support the hypothesis of partial lifting of some benzene rings (see Section 3). The RT structure for dBB is characterized by a longer-range order and described by epitaxy matrices (1, −4|6, 0) and (1, −1|4, 1).[37] An epitaxy matrix cannot be written for BCB and dCB at RT, due to a lack of long-range order.

The short OM phase of the chlorine-containing molecules presents a high mobility even at RT, as observed by consecutive STM images (one every minute), presented in ESI Video S1,† in which the orientation and domain size changes.

The dBB chevron phase is stable up to the polymerization temperature (175 °C), while the chain length of BCB and dCB increases when the dehalogenation process is completed at 150 °C, and long OM chains made of a higher number of phenyls (>20) are observed on the surface (Fig. S4d†). As for BIB at RT, these chains comprise two domains along the [1−1±1] directions and have a periodic strain-relieving kink every four phenyls (Fig. S3†) and are described by the epitaxy matrices (2, −2|4, 9) and (2, 2|−4, 9).

The difference between the OM chain direction for dIB and that of the other molecules (BIB, dBB, BCB and dCB) can be inferred from the halogens' preferred adsorption site on Cu(110); iodine atoms occupy hollow sites,[18] while Br and Cl atoms are located at short-bridge sites (Fig. S3†).[28,29,58] STM line profiles of the halogen atoms between the OM chains show that for dIB, iodine atoms adsorb along the [1−1±2] directions, with a 4.4 Å periodicity and are located in hollow positions; while for the other molecules, Br and Cl atoms follow the [1−1±1] directions, with a 6.3 Å periodicity and are located at short-bridge positions. We can therefore suppose that in the BIB case, bromine atoms decorate the OM chains, while iodine atoms are segregated into the $c(2 \times 2)$ domains. At the boundaries between the $c(2 \times 2)$ domains and the OM chains, both Br and I atoms are located along the chains, with two atoms in hollow positions and one atom at a short-bridge position (Fig. S5†).

Polymer chains. The polymer phase (Fig. 5e and f), is made of linear chains of π-conjugated phenyls decorated with halogen atoms. dBB and BIB exclusively produce polymers along the direction [1−1±2] (the 'diagonal' direction of the Cu(110) surface unit cell), while BCB, dCB and dIB show polymers along both the diagonal [1−1±2] and 'parallel' (to the Cu(110) close-packed row, [1−10]) directions. The phenyl–phenyl distance is 4.4 Å for all of the polymers, a distance which is commensurate with the diagonal direction and incommensurate for the parallel direction.[57] Commensurability with the surface permits diagonal polymers to grow with no strain penalty, and results in the formation of large islands of polymers with a mean length greater than 20 phenyl units. However, the electronic properties are strictly related to the polymer length,[14,57] and whether a particular halogen or an intermediate phase favors longer chains is therefore important for selecting the right building block.

(3) NEXAFS and molecular orientation

C K-edge NEXAFS spectra obtained at RT as a function of the photon's polarization direction for all precursors (Fig. 7) present two π^* transitions, π_1^* and π_2^* at ~284.5 and ~288.5 eV, respectively, in agreement with the cases of dBB on Cu(110)[28] and halobenzenes on Cu(111).[59] These spectra show that the phenyl rings are mostly flat on the surface, supporting the orientational hypotheses derived from the STM images. The C K-edge NEXAFS spectra at RT for chlorine-containing molecules show higher tilt of the phenyl ring, according to the change in the intensity of the π^* states as a function of the polarization vector direction of incoming photons with respect to the surface normal, particularly for π_2^* resonance (black line in Fig. 7), where a peak is visible for $\theta = 0°$.[60] This feature is not present for the non-chlorinated molecules at RT (Fig. 7a), and is consistent with a fraction of the molecules being tilted away from the surface. We additionally note the presence of a shoulder on the high-energy side of the π_1^* resonance at $\theta = 90°$ (Fig. 7b), a feature not observed for the other precursors. After annealing to 150 °C, the intensity of the π_2^* resonance at $\theta = 0°$ for dCB and BCB is strongly suppressed, and the spectra in all polarizations become qualitatively indistinguishable from the RT spectra observed for the other molecules. This is

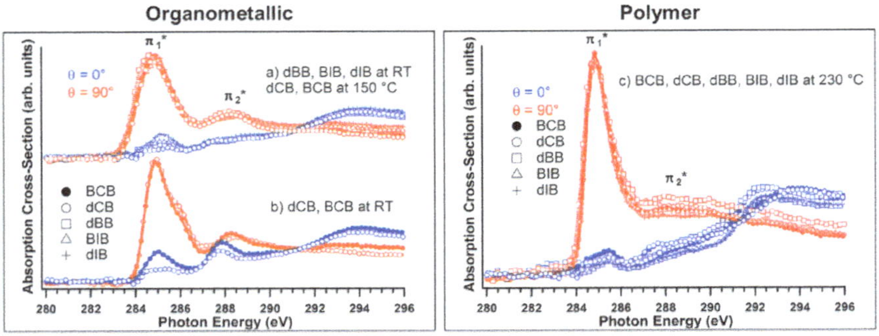

Fig. 7 Polarization-dependent C K-edge NEXAFS spectra for the five studied precursors; (a) dBB, BIB and dIB at RT and BCB and dCB at 150 °C; (b) dCB and BCB at RT and (c) all precursors at 230 °C. For each sample we report $\theta = 90°$ (red, TM geometry, close to p-polarization) and $\theta = 0°$ (blue, TE geometry, s-polarization) with the incident radiation falling in the plane containing the normal sample and the [1−10] lattice direction. Single spectra for each precursor are presented in Fig. S6 and S7.†

consistent with the STM data, where OM structures along the same directions are observed for all molecules but dIB, and the C 1s XPS data, where the spectra are identical for dCB and BCB at 150 °C and dIB, BIB and dBB at RT (Fig. 2). Annealing all of the precursors to 230 °C completely suppresses the π_2^* resonance intensity, indicating the planarity of the aromatic rings, expected for the PPP polymers (Fig. 7 and S7†).[28,29,61]

(4) Role of the halogen in the reaction

The halogen might affect the reaction mechanism and the barrier energies in a variety of ways. Here, we present two hypotheses based on (i) the effect of differing diffusion barriers for the halogens, and (ii) the possible presence of the halogen atoms on-top of the copper atoms in the OM structure. The hypotheses are not mutually exclusive and the observed behaviour could contain contributions from both effects.

To explore the first hypothesis, NEB calculations for the diffusion of the halogens along different directions of the Cu(110) surface were performed ([001], [1−10], [1−11] and [1−12]). These simulations consider the diffusion of the halogen alone and use a hollow position as the starting position for I atoms and a short-bridge position as the starting position for Br and Cl atoms, known to be the respective preferred adsorption site for the halogens.[58,62–64] The overall trend of the diffusion energy as I < Br < Cl (Fig. 8) is consistent with the starting temperature of the OM-to-polymer conversion reaction found in the fast-XPS maps (Fig. 4), where the polymerization of dCB starts at the highest temperature, and at the lower temperature for iodine-containing molecules. Unlike chlorine and bromine, the diffusion barrier for iodine along the [1−12] direction is higher than the other two directions. This could be explained by the diffusion of iodine along the [1−12] direction, which needs to pass from a stable hollow adsorption site to an on-top location on the Cu surface. Therefore, having the lower diffusion energy barriers along the other two directions, the iodine atoms can make a two-step movement, along [1−10] and [001], instead of diffusing

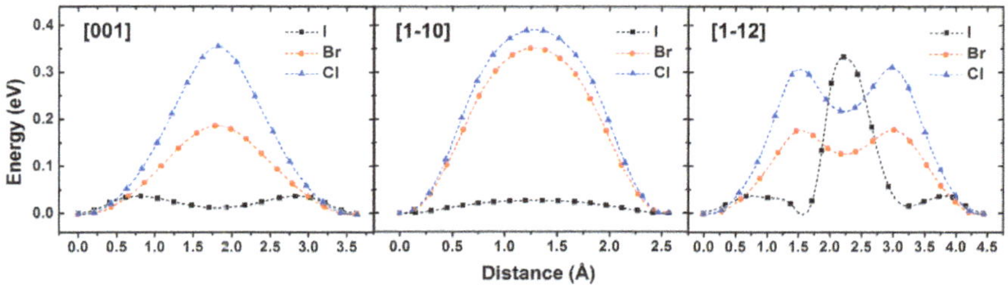

	E_d[001] (eV)	E_d[1-10] (eV)	E_d[1-12] (eV)
Cl	0.36	0.39	0.31
Br	0.19	0.35	0.18
I	0.04	0.03	0.33

Fig. 8 (Top panels) NEB plots of the minimum energy path for the diffusion of iodine, bromine and chlorine along one unit cell in the [1–10], [001] and [1–12] directions of Cu(110). For Br and Cl, the starting equilibrium position is short-bridge, while for I it is hollow. (Bottom panel) Diffusion energies (E_d) for I, Br and Cl along the studied directions.

directly through the [1−12]. The molecular structures for local maxima and minima of the paths are presented in the ESI (Fig. S8–11†), together with the calculations for the diffusion along the [1−11] direction. We can further speculate that a lower diffusion barrier for the halogens also facilitates monomer diffusion, thereby reducing the transition temperature. Conversely, a higher diffusion barrier for the halogens requires an overall higher temperature for the polymerization to occur, as observed experimentally.

The second hypothesis is connected with the halogen being integrated into portions of the OM chains, as proposed by Di Giovannantonio *et al.*[28,29] In this work Br atoms were proposed to occupy two different positions of the unit cell of OM chains for dBB on Cu(110), four per unit cell on the short-bridge lattice site between adjacent OM chains, and four adsorbed on top of the Cu atoms forming the OM (*i.e.* between the phenyls). This hypothesis is based on simulated STM images, which show a better qualitative match with the experimental STM images when top Br atoms are included. To complete the polymerization reaction, the copper atoms in the OM structure need to be expelled from the chains to permit covalent binding between adjacent phenyls. If a halogen atom is present on top of these Cu atoms, the energies involved in the polymerization reaction must depend on the type of halogen, and will result in a different temperature for the reaction depending on the type of precursor.

Conclusions and perspectives

The role of the halogen in the Ullmann coupling polymerization of five 1,4-dihalobenzene precursors containing Cl, Br, and I on Cu(110) was studied using combined fast-XPS, NEXAFS and STM analysis. On one hand, the nature of the halogen atom, hence the strength of its bond to carbon atoms, affects the temperature of the dehalogenation step. On the other hand, the type of halogen drives the geometry of the OM structures obtained at RT as well as the orientation of the polymers. In fact, (i) the length of the OM chains depends on the carbon–halogen bond dissociation energy, and (ii) the OM structures for precursors containing Br and Cl (dBB, dCB, BCB, BIB) are aligned along the [1−1±1] directions, as opposed to alignment along the [1−1±2] directions for precursors containing only iodine (dIB). The OM structures follow different reaction kinetics, exhibiting a gradual transition for dIB while the others undergo a sudden transition into PPP polymers. A key finding of this study is that the temperature range of the polymerization is affected by the type of halogen, and is qualitatively correlated to the halogen's diffusion energy. Our results show that the halogen is not merely a reaction by-product, but rather an important parameter governing the on-surface Ullmann polymerization reaction.

Statement of author contributions

J. L. D, D. F. P., F. R. and G. C. conceived the project. G. G., M. D. G. and S. T. performed the STM experiments, and with the supervision of J. L. D. and M. E. Y. F. R. participated in the analysis of the STM images. G. G., M. D. G., J. L. D. and G. C. performed NEXAFS and fast-XPS experiments at the Elettra synchrotron. A. V. and L. F. assisted with synchrotron measurements at the Aloisa beamline. M. E. performed the theoretical calculations. G. G. wrote the manuscript, together with M. E. and G. C. All of the authors commented critically on the manuscript.

Conflict of interest

The authors declare no competing financial interest.

Acknowledgements

This work is partially supported by the Italy-France International Program of Scientific Cooperation (PICS CNR-CNRS). G. C. and F. R. acknowledge support from CNR through Short-term Mobility (STM) Fellowships. The authors acknowledge beamtime access and support from the Elettra light source in Italy. F. R. and D. F. P. are supported by NSERC through individual Discovery Grants as well as the Fonds québécois de la recherche sur la nature et les technologies (FQNRT) through a Team Grant, and the Ministère du Développement Économique, de l'Innovation et de l'Exportation (MDEIE) through a Collaboration Grant. F. R. is also grateful to the Canada Research Chair program for funding and partial salary support. G. G. is grateful to G. Vasseur and M. El Garah for their participation in collecting the STM images. Computations were performed on the tcs supercomputer at the SciNet HPC Consortium, funded by the Canada Foundation for Innovation under Compute Canada, the Government of Ontario, Ontario Research Fund – Research Excellence, and the University of Toronto. Some of the simulated structures in the SI are presented using VESTA software.

Notes and references

1 K. S. Novoselov, A. K. Geim, S. V. Morozov, D. Jiang, Y. Zhang, S. V. Dubonos, I. V. Grigorieva and A. A. Firsov, *Science*, 2004, **306**, 666–669.
2 C. Lee, X. Wei, J. W. Kysar and J. Hone, *Science*, 2008, **321**, 385–388.

3 A. H. Castro Neto, F. Guinea, N. M. R. Peres, K. S. Novoselov and A. K. Geim, *Rev. Mod. Phys.*, 2009, **81**, 109–162.

4 F. Bonaccorso, Z. Sun, T. Hasan and A. C. Ferrari, *Nat. Photonics*, 2010, **4**, 611–622.

5 M. Ebrahimi and F. Rosei, *Nature*, 2017, **542**, 423–424.

6 W. Liu, X. Luo, Y. Bao, Y. P. Liu, G.-H. Ning, I. Abdelwahab, L. Li, C. T. Nai, Z. G. Hu, D. Zhao, B. Liu, S. Y. Quek and K. P. Loh, *Nat. Chem.*, 2017, **9**, 563–570.

7 F. J. Uribe-Romo and W. R. Dichtel, *Nat. Chem.*, 2012, **4**, 244–245.

8 S. E. Thompson and S. Parthasarathy, *Mater. Today*, 2006, **9**, 20–25.

9 M. El Garah, J. M. MacLeod and F. Rosei, *Surf. Sci.*, 2013, **613**, 6–14.

10 D. F. Perepichka and F. Rosei, *Science*, 2009, **323**, 216–217.

11 J. Sakamoto, J. van Heijst, O. Lukin and A. D. Schlüter, *Angew. Chem., Int. Ed.*, 2009, **48**, 1030–1069.

12 N. R. Champness, *Nat. Chem.*, 2012, **4**, 149–150.

13 J. W. Colson and W. R. Dichtel, *Nat. Chem.*, 2013, **5**, 453–465.

14 R. Gutzler and D. F. Perepichka, *J. Am. Chem. Soc.*, 2013, **135**, 16585–16594.

15 J. Liu, B.-W. Li, Y.-Z. Tan, A. Giannakopoulos, C. Sanchez-Sanchez, D. Beljonne, P. Ruffieux, R. Fasel, X. Feng and K. Müllen, *J. Am. Chem. Soc.*, 2015, **137**, 6097–6103.

16 R. Gutzler, *Phys. Chem. Chem. Phys.*, 2016, **18**, 29092–29100.

17 M. Di Giovannantonio, T. Kosmala, B. Bonanni, G. Serrano, N. Zema, S. Turchini, D. Catone, K. Wandelt, D. Pasini, G. Contini and C. Goletti, *J. Phys. Chem. C*, 2015, **119**, 19228–19235.

18 J. Lipton-Duffin, O. Ivasenko, D. Perepichka and F. Rosei, *Small*, 2009, **5**, 592–597.

19 L. Grill, M. Dyer, L. Lafferentz, M. Persson, M. V. Peters and S. Hecht, *Nat. Nanotechnol.*, 2007, **2**, 687–691.

20 J. Björk, F. Hanke and S. Stafström, *J. Am. Chem. Soc.*, 2013, **135**, 5768–5775.

21 L. Lafferentz, V. Eberhardt, C. Dri, C. Africh, G. Comelli, F. Esch, S. Hecht and L. Grill, *Nat. Chem.*, 2012, **4**, 215–220.

22 A. Basagni, L. Colazzo, F. Sedona, M. DiMarino, T. Carofiglio, E. Lubian, D. Forrer, A. Vittadini, M. Casarin, A. Verdini, A. Cossaro, L. Floreano and M. Sambi, *Chem.–Eur. J.*, 2014, **20**, 14296–14304.

23 M. Xi and B. E. Bent, *J. Am. Chem. Soc.*, 1993, **115**, 7426–7433.

24 S.-W. Hla, L. Bartels, G. Meyer and K.-H. Rieder, *Phys. Rev. Lett.*, 2000, **85**, 2777–2780.

25 S. Schlögl, W. M. Heckl and M. Lackinger, *Surf. Sci.*, 2012, **606**, 999–1004.

26 R. Gutzler, H. Walch, G. Eder, S. Kloft, W. M. Heckl and M. Lackinger, *Chem. Commun.*, 2009, 4456–4458, DOI: 10.1039/b906836h.

27 M. Bieri, M. T. Nguyen, O. Groning, J. M. Cai, M. Treier, K. Ait-Mansour, P. Ruffieux, C. A. Pignedoli, D. Passerone, M. Kastler, K. Mullen and R. Fasel, *J. Am. Chem.Soc.*, 2010, **132**, 16669–16676.

28 M. Di Giovannantonio, M. El Garah, J. Lipton-Duffin, V. Meunier, L. Cardenas, Y. Fagot Revurat, A. Cossaro, A. Verdini, D. F. Perepichka, F. Rosei and G. Contini, *ACS Nano*, 2013, **7**, 8190–8198.

29 M. Di Giovannantonio, M. El Garah, J. Lipton-Duffin, V. Meunier, L. Cardenas, Y. Fagot-Revurat, A. Cossaro, A. Verdini, D. F. Perepichka, F. Rosei and G. Contini, *ACS Nano*, 2014, **8**, 1969–1971.

30 J. Eichhorn, T. Strunskus, A. Rastgoo-Lahrood, D. Samanta, M. Schmittel and M. Lackinger, *Chem. Commun.*, 2014, **50**, 7680–7682.

31 L. Cardenas, R. Gutzler, J. Lipton-Duffin, C. Fu, J. L. Brusso, L. E. Dinca, M. Vondráček, Y. Fagot-Revurat, D. Malterre and F. Rosei, *Chem. Sci.*, 2013, **4**, 3263–3268.

32 L. Leung, T. Lim, Z. Ning and J. C. Polanyi, *J. Am. Chem. Soc.*, 2012, **134**, 9320–9326.

33 A. Eisenstein, L. Leung, T. Lim, Z. Ning and J. C. Polanyi, *Faraday Discuss.*, 2012, **157**, 337–353.

34 M. Chen, J. Xiao, H.-P. Steinrück, S. Wang, W. Wang, N. Lin, W. Hieringer and J. M. Gottfried, *J. Phys. Chem. C*, 2014, **118**, 6820–6830.

35 M. Xi and B. E. Bent, *Surf. Sci.*, 1992, **278**, 19–32.

36 J. M. Meyers and A. J. Gellman, *Surf. Sci.*, 1995, **337**, 40–50.

37 M. Di Giovannantonio, M. Tomellini, J. Lipton-Duffin, G. Galeotti, M. Ebrahimi, A. Cossaro, A. Verdini, N. Kharche, V. Meunier, G. Vasseur, Y. Fagot-Revurat, D. F. Perepichka, F. Rosei and G. Contini, *J. Am. Chem. Soc.*, 2016, **138**, 16696–16702.

38 J. Björk, *J. Phys.: Condens. Matter*, 2016, **28**, 083002.

39 I. Horcas, R. Fernandez, J. Gomez-Rodriguez, J. Colchero, J. Gómez-Herrero and A. Baro, *Rev. Sci. Instrum.*, 2007, **78**, 013705.

40 N. Fairley, *CasaXPS Manual 2.3.15: Introduction to XPS and AES*, Casa Software, 2009.

41 G. Contini and S. Turchini, *Comput. Phys. Commun.*, 1996, **94**, 49–52.

42 L. Floreano, A. Cossaro, R. Gotter, A. Verdini, G. Bavdek, F. Evangelista, A. Ruocco, A. Morgante and D. Cvetko, *J. Phys. Chem. C*, 2008, **112**, 10794–10802.

43 G. Kresse and J. Hafner, *Phys. Rev. B: Condens. Matter Mater. Phys.*, 1993, **47**, 558–561.

44 G. Kresse and J. Furthmüller, *Phys. Rev. B: Condens. Matter Mater. Phys.*, 1996, **54**, 11169–11186.

45 L. Chris, G. Daniel, G. Leslie, P. Richard, B. Neil, C. Michael, H. Teresa, D. Jillian, Y. Ching-Hsing, C. Joseph, L. J. Dursi, C. Jason, N. Scott, P. Jaime, K. Neil and Z. Ramses Van, *J. Phys.: Conf. Ser.*, 2010, **256**, 012026.

46 J. P. Perdew, M. Ernzerhof and K. Burke, *J. Chem. Phys.*, 1996, **105**, 9982–9985.

47 P. E. Blöchl, *Phys. Rev. B: Condens. Matter Mater. Phys.*, 1994, **50**, 17953–17979.

48 G. Kresse and D. Joubert, *Phys. Rev. B: Condens. Matter Mater. Phys.*, 1999, **59**, 1758–1775.

49 S. Grimme, J. Antony, S. Ehrlich and H. Krieg, *J. Chem. Phys.*, 2010, **132**, 154104.

50 G. Henkelman, B. P. Uberuaga and H. Jónsson, *J. Chem. Phys.*, 2000, **113**, 9901–9904.

51 G. Henkelman and H. Jónsson, *J. Chem. Phys.*, 2000, **113**, 9978–9985.

52 W. Wang, X. Shi, S. Wang, M. A. Van Hove and N. Lin, *J. Am. Chem. Soc.*, 2011, **133**, 13264–13267.

53 K. J. Shi, D. W. Yuan, C. X. Wang, C. H. Shu, D. Y. Li, Z. L. Shi, X. Y. Wu and P. N. Liu, *Org. Lett.*, 2016, **18**, 1282–1285.

54 T. D. Sheppard, *Org. Biomol. Chem.*, 2009, **7**, 1043–1052.

55 J. L. Lin and B. E. Bent, *J. Phys. Chem.*, 1992, **96**, 8529–8538.

56 S. J. Blanksby and G. B. Ellison, *Acc. Chem. Res.*, 2003, **36**, 255–263.

57 G. Vasseur, Y. Fagot-Revurat, M. Sicot, B. Kierren, L. Moreau, D. Malterre, L. Cardenas, G. Galeotti, J. Lipton-Duffin, F. Rosei, M. Di Giovannantonio, G. Contini, P. Le Fèvre, F. Bertran, L. Liang, V. Meunier and D. F. Perepichka, *Nat. Commun.*, 2016, **7**, 10235.
58 B. V. Andryushechkin, K. N. Eltsov and V. M. Shevlyuga, *Surf. Sci.*, 2005, **584**, 278–286.
59 M. X. Yang, M. Xi, H. Yuan, B. E. Bent, P. Stevens and J. M. White, *Surf. Sci.*, 1995, **341**, 9–18.
60 J. Stöhr, *NEXAFS spectroscopy*, Springer-Verlag, Berlin, Heidelberg, 1992.
61 J. Novak, M. Oehzelt, S. Berkebile, M. Koini, T. Ules, G. Koller, T. Haber, R. Resel and M. G. Ramsey, *Phys. Chem. Chem. Phys.*, 2011, **13**, 14675–14684.
62 J. L. Stickney, C. B. Ehlers and B. W. Gregory, *Langmuir*, 1988, **4**, 1368–1373.
63 L.-J. Wan and K. Itaya, *J. Electroanal. Chem.*, 1999, **473**, 10–18.
64 B. V. Andryushechkin, in *Surf. Interface Sci.*, ed. K. Wandelt, Wiley-VCH Verlag GmbH & Co. KGaA, Weinheim, Germany, 2016, vol. 5, ch. 35, pp. 207–254.

Faraday Discussions

PAPER

The dynamics of benzene on Cu(111): a combined helium spin echo and dispersion-corrected DFT study into the diffusion of physisorbed aromatics on metal surfaces

M. Sacchi, *[a] P. Singh,[b] D. M. Chisnall,[c] D. J. Ward, [c] A. P. Jardine,[c] W. Allison,[c] J. Ellis[c] and H. Hedgeland[d]

Received 8th March 2017, Accepted 27th April 2017
DOI: 10.1039/c7fd00095b

We use helium spin-echo spectroscopy (HeSE) to investigate the dynamics of the diffusion of benzene adsorbed on Cu(111). The results of these measurements show that benzene moves on the surface through an activated jump-diffusion process between the adsorption sites on a Bravais lattice. Density Functional Theory (DFT) calculations with van der Waals (vdW) corrections help us understand that the molecule diffuses by jumping through non-degenerate hollow sites. The results of the calculations shed light on the nature of the binding interaction between this prototypical aromatic molecule and the metallic surface. The highly accurate HeSE experimental data provide a quantitatively stringent benchmark for the vdW correction schemes applied to the DFT calculations and we compare the performances of several dispersion interaction schemes.

1. Introduction

The self-assembly of aromatic molecules and polymers on the surface of semi-conductors and metals is a key component in the design and production of organo-electronics and photovoltaic devices, as well as being central in corrosion protection and coating technologies. The self-assembly process is characterised by three main factors: the surface-molecule binding, the intermolecular interactions between adsorbed precursors, and the dynamic behaviour (*i.e.*, surface diffusion, rotational entropy and conformational mobility) of the adsorbate. Among these factors, the surface dynamics of the adsorbed precursor are by far

[a]*Department of Chemistry, University of Surrey, Guildford, GU2 7XH, UK. E-mail: m.sacchi@surrey.ac.uk*
[b]*The Perse School, Cambridge, CB2 8QF, UK*
[c]*Cavendish Laboratory, University of Cambridge, JJ Thomson Avenue, Cambridge, CB3 0HE, UK*
[d]*School of Physical Sciences, The Open University, Walton Hall, Milton Keynes, MK7 6AA, UK*

the most difficult to study because the complexity of the molecular motion, once it has been activated, and the timescale then involved (around 10^{-8} to 10^{-12} s) mean that commonly employed microscopy techniques are unsuited to the task.[1] In this study we investigate the adsorption and diffusion of benzene on Cu(111). Benzene is the most fundamental building block in Polycyclic Aromatic Hydrocarbons (PAHs) and forms a part of several of the π-conjugated molecules and polymers employed in organic electronics, such as PTCDA, PTCDI, rubrene, polyphenylene and derivatives.[2–5] The ability to characterise and predict the self-assembly properties of this large class of molecules intrinsically depends on our understanding of the interactions and dynamic behaviour of this iconic six-membered aromatic ring when confined on the surface of single crystals. In previous works we have explored the dynamics of the cyclopentadienyl anion (Cp), pyrrole and thiophene adsorbed on the same Cu(111) surface.[6–9] The combination of state of the art helium spin-echo spectroscopy (HeSE) and density functional theory calculations has provided insight into the role of charge transfer and ionic binding in aromatic adsorption,[8,9] and the contribution of rotational,[6] vibrational[7] and other external degrees of freedom[10] to molecular transport on surfaces.

Notwithstanding the central role of benzene adsorption in surface and interface chemistry, the number of low-coverage, atomically resolved experimental studies of the adsorption of this aromatic molecule on copper and other coinage metal surfaces is relatively limited.[11–20] Early STM and DFT studies by Komeda *et al.*[20] and by Lorente *et al.*[19] show that benzene adsorbs in a flat orientation on the Cu(100) surface, with the hollow sites being the preferred adsorption sites.[19] On Cu(111) the molecule adsorbs up to a single monolayer coverage with the aromatic ring parallel to the surface[17,18] and is able to diffuse freely and form stable islands near the steps at a temperature of 77 K.[17,21] The long series of previous computational studies of benzene adsorption on Cu(111), employing both standard DFT generalised gradient approximation (GGA) exchange–correlation (XC) functionals and dispersion-corrected XC functionals, are motivated by the general lack of agreement on some fundamental results, such as the preferred adsorption site, the adsorption energies and the molecule-surface distance at the equilibrium.[12] The previous experimental and theoretical studies generally focused on the properties of the C_6H_6–Cu(111) system in equilibrium at a given surface temperature and coverage. The dynamical aspects of the self-assembly process, in particular the surface mobility, have not been quantitatively explored. The main reason for the lack of information on the barriers for surface diffusion, as well as the friction coefficients, resides in the difficulty in determining the minimum energy adsorption site and, particularly, in measuring with accuracy the diffusion rates of the molecule. For instance, in the case of thiophene on Cu(111),[8] three different degrees of freedom – molecular translation, rotation and vertical motion – were individually characterized by DFT and HeSE, giving accurate, quantitative insight into the atomic-scale motion of a heteroaromatic on a surface. In this work we present new theoretical and experimental insights into the adsorption and diffusion of benzene on Cu(111), a paradigmatic surface science system for benchmarking weak dispersion interactions in 2D.

2. Methods

2.1 Theoretical method

The potential energy surface of benzene on Cu(111) has been mapped by performing a series of first-principles DFT calculations for adsorption on high-symmetry sites: top, bridge and hollow (fcc and hcp). Since the experiments were performed in a low-coverage regime, only the in-plane rotational degree of freedom (the rotation around the central C_6 molecular axis) was considered and the geometry of the benzene was optimised at the two different initial angular orientations (see Fig. 1), 0 and 30 degrees (with respect to the $<1\bar{1}0>$ direction). In the adsorption of aromatics on noble and coinage metal surfaces, the molecule-surface binding is dominated by the long-range dispersion forces, *i.e.*, van der Waals (vdW) forces. It is well-known that DFT, in its classic local density approximation (LDA) or generalised gradient approximation (GGA), is local (or semi-local) in nature; therefore a major effort has been undertaken in the last decade or so to overcome this restriction, by applying correction schemes or *ad hoc* modifications to popular functionals such as PBE or B88. A review of the general strengths and weaknesses of the various approaches is beyond the scope of this paper, but one can draw the case that, as for "non-dispersion corrected" DFT, a classic Ladder[22] or more modern Stairway[23] of chemical accuracy *versus* computational cost exists. The right method for a specific system needs to be selected by balancing the desired level of accuracy and overall cost of the calculations. For the present study we have employed first and second step methods,[23] following the classification of Klimeš and Michaelides, and in order of increasing accuracy these are: DFT-D2 (G06) by Grimme and co-authors,[24] Ortmann–

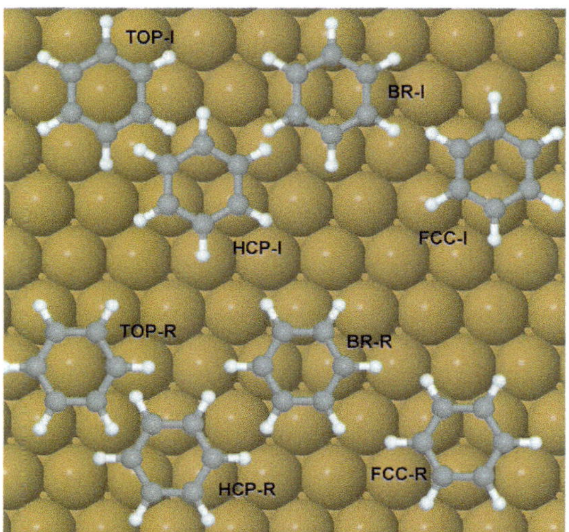

Fig. 1 Schematic representation of benzene adsorbed on high symmetry adsorption sites in the two angular orientations (0 and 30 degrees with respect to the $<1\bar{1}0>$ direction) considered in this work. In our notation, I = 0 degrees rotation and R = 30 degrees rotation.

Bechstedt–Schmidt (OBS) in ref. 25, Tkatchenko–Scheffler (TS),[26] and the Self-Consistent Tkatchenko–Scheffler (TSSCS) correction[27] schemes. The exchange–correlation functional of choice is the Perdew–Burke–Ernzerhof (PBE)[28] functional with the G06, TS, TSSCS corrections and the Perdew–Wang'91 (PW91)[29] functional for the OBS correction. The PW91 XC-functional is very similar in accuracy to the PBE functional. It has been employed for testing the OBS scheme because, in the original formulation and benchmarking, the functional used was PW91, therefore the C_6 coefficients available are more compatible with PW91 than with PBE. Given the relatively large dimensions of the $(2\sqrt{3} \times 2\sqrt{3})$ surface unit cell needed to simulate the experimental conditions, we note the extremely good performance of the TS and related schemes in combination with the GGA exchange–correlation functional, both in terms of the cost and chemical accuracy.

The remaining details of the calculations and convergence criteria have been previously reported and discussed in several recent reports,[6–9] but we summarise here the most important computational parameters. The surface has been modelled as a seven-layer slab, with a vacuum layer of ~20 Å vertically separating the periodically repeated supercells. The cut-off energy of the plane wave basis set was fixed at 11 Hartrees, while the Brillouin zone was sampled with a $4 \times 4 \times 1$ k-point Monkhorst–Pack[30] grid. The bottom three layers were kept fixed in the bulk crystal positions while the top four atomic layers were allowed to relax during structural optimisation.

2.2 Experimental method

Helium spin-echo spectroscopy uses a beam of neutral, nuclear spin-polarised helium-3 atoms to measure the time-dependent correlation of surface adsorbates. Magnetic fields are used to divide the helium beam into spin-dependent wavepackets, which arrive at the surface with a known temporal separation. The interference of the scattered wavepackets results in a measurable polarisation of the beam which varies with the separation time of the wavepackets, reflecting the change that occurred in the correlation of the surface adsorbate molecules over that time. The measured normalised beam polarisation, $P(t)/P(t = 0)$, is proportional to the intermediate scattering function (ISF), $I(\Delta K,t)$. As this is a scattering technique, the measurements are in the reciprocal space and the ISF is the Fourier transform of the spatial correlation function describing the surface, $G(R,t)$. The ISF hence tells us how quickly the adsorbate molecules "dephase" from their initial positions on picosecond timescales, and the reciprocal length scales are given by $2\pi/\Delta K$. By considering the variation of the rate of dephasing with temperature and scattering momentum transfer, ΔK, we gain a description of the adsorbate motion and can quantify details such as the energy landscape and dissipation rate. More detailed descriptions of the technique are given elsewhere.[1,31]

The experiments were performed on a single-crystal Cu(111) sample, mounted in a scattering chamber with a base pressure of 1×10^{-10} mbar and cleaned by repeated cycles of Ar^+ sputtering (800 eV, 10 μA, 20 min at 300 K) and annealing (800 K, 30 s). The surface quality was monitored using the sample's reflectivity to the helium beam (>20%) and the shape of the helium scattering specular peak. Benzene (>99.9% purity, Aldrich), purified by several freeze–evacuate–thaw cycles, was deposited on the surface by backfilling the chamber and monitoring the drop in the specularly reflected helium beam during adsorption. Using the same method as in our recent study of benzene on Cu(001),[10] we estimate absolute coverages of 0.03 and 0.1 monolayers (ML) for the low and high coverage measurements, respectively.

3. Experimental results

The slow decay of the measured polarisation that dominates the ISFs, illustrated in Fig. 2, is indicative of surface diffusion. We can quantify the slow dephasing rate, α, by fitting an exponential decay of the form $ae^{-\alpha t} + c$. However, such a fit of the single exponential decay is only appropriate to quantify motion on a Bravais lattice and, in the case of a non-Bravais lattice, multiple exponential terms are to be expected.[32] On the Cu(111) surface benzene can sit on a Bravais lattice of adsorption sites in the case where the top site is the energetic minimum, or on a non-Bravais lattice if bridge site adsorption is preferred. The hollow sites form a non-Bravais lattice in the case that the hcp and fcc hollow sites are degenerate in energy, but revert to a Bravais lattice when the hcp and fcc are not degenerate and the adsorbate preferentially adsorbs to one of the two types. By considering whether the measured polarisation shows evidence of multiple decays, we can attribute the likely adsorption sites. Referring to the analytical models detailed by Tuddenham *et al.*,[32] for bridge or degenerate hollow site adsorption, we would expect the second exponential term to be most apparent at a momentum transfer of $Q/(2\pi/a) \sim 0.6$ in the $[11\bar{2}]$ direction, corresponding to $\Delta K \sim 1.5$ Å^{-1} on a Cu(111) surface, with the magnitude of this term increasing strongly in the

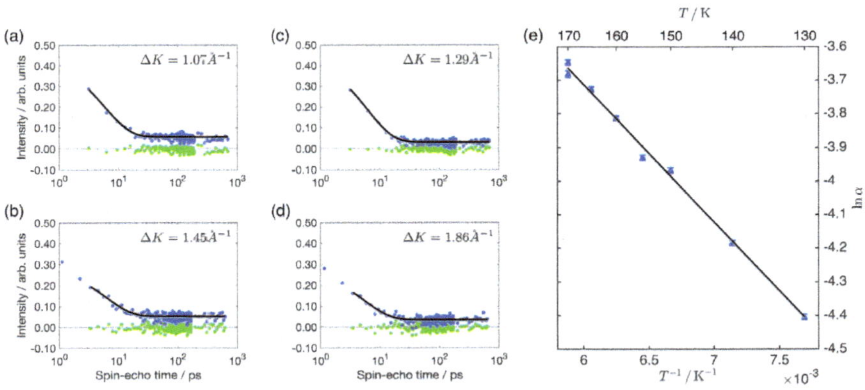

Fig. 2 Polarisation measurements as a function of the spin-echo time, *t*, for 0.03 ML (panels a, c) and 0.1 ML (panels b, d) of benzene/Cu(111). The different panels show the measurements at the indicated momentum transfers and a temperature of 170 K. The experimental data are plotted as blue points, a single exponential of the form $f(t) = a\exp(-\alpha t) + c$ from 3 ps is shown as a black line and the residual after subtracting the exponential from the data as green points. The data at times smaller than 3 ps is excluded as it is influenced by the initial fast decay described in the text. Panel (e) shows the temperature dependence of the dephasing rate for a coverage of 0.03 ML at $\Delta K = 0.3$ Å^{-1} as an Arrhenius plot. The solid line illustrates the fit to an Arrhenius form, resulting in an effective activation energy of 35 ± 1 meV. All the data were measured along the [112] direction.

range above 1 $Å^{-1}$. In Fig. 2c, we show the ISF for 0.1 ML coverage at 1.45 $Å^{-1}$, along with a single exponential fit and residuals. It can be seen that there is no evidence of the second decay that would be needed to support a model of bridge or degenerate hollow site adsorption. The presence of a single exponential decay at similar values of momentum transfer is illustrated more widely in panels a–d of Fig. 2, for coverages of both 0.03 ML and 0.1 ML, confirming that there is no evidence of adsorption on a non-Bravais lattice. The experimental data hence suggests adsorption on the top sites or non-degenerate hollow sites.

In Fig. 2e, we present the temperature dependency of the dephasing rate as an Arrhenius plot, at a coverage of 0.03 ML and momentum transfer of 0.3 $Å^{-1}$ in the $[11\bar{2}]$ direction. From the Arrhenius plot, we find an effective activation energy of 35 ± 1 meV, a value approximately one third of that measured for benzene adsorbed on the (001) facet of copper, but very similar to the 41 ± 1 meV measured for five-membered aromatic cyclopentadienyl on the same Cu(111) surface.[9]

In addition to the slow decay that is caused by the diffusive motion of the adsorbate, and which we have analysed above, a typical polarisation measurement also shows an initial rapid drop within the first one or two picoseconds. The much smaller and more rapid decay here is associated with the intracellular motion of the adsorbate molecule when it is localised within a particular adsorption hollow.[32] In the polarisation curves we present in Fig. 2, it may be noticed that the data at 0.1 ML has two data points taken at smaller times than those shown in the 0.03 ML curves. These two data points are excluded from the subsequent exponential fit illustrated by the black lines, in order to ensure that there is no influence from the initial drop. Once these points are excluded, the slow decay can be considered independently from the fast decay as it occurs on such a different time scale.

4. Computational results

As we discussed in the previous section, the experimental data suggest adsorption on top or non-degenerate hollow sites, with a barrier to diffusion of around 35 meV. We now discuss how our computational results point to a global energy minimum for physisorbed benzene on hollow sites, and a weaker local minimum at top sites, which are therefore not likely to be involved in the diffusion process. In order to put the calculations into context, it is useful to summarise here the three main classes of vdW (or dispersion force) correction schemes most commonly employed in surface chemistry. Historically, long-range dispersion-correction schemes have considered differently formulated pairwise atomic potentials that are added to the Kohn–Sham potential energy term. The computational cost of the DFT calculation is therefore essentially unchanged, since the corrective term can be evaluated analytically, and the calculation follows simple approximations derived from the classical electrostatic description of the instantaneous dipole–dipole interactions in London's formula. The vdW energy term is therefore added at the end of each self-consistent energy minimisation cycle, inside a structural optimisation or transition state search. One can generally classify the semiempirical vdW-corrections according to the transferability and accuracy of the approximations involved in the calculations of the so-called C_6 coefficients that tune the intensity of the attractive interactions between a given pair of atoms. The simplest approximations (the DFT-D and DFT-D2 schemes developed by Grimme and co-authors[24,34]) assume that the C_6 coefficients are constant during the calculations and independent of the local chemical environment in which the atomic pairs are situated at a given time. More accurate methods (DFT-D3 or TS, for instance) introduce C_6 coefficients that are dependent on the atomic coordination and local environment (through the atomic polarisabilities and vdW volumes) and therefore offer greater accuracy and predictive power than the older corrections, especially when the corrections are applied to molecular and surface systems in which the same atomic element has several hybridisation states or a varying oxidation state. Finally, the most computationally challenging approach is through the development of vdW-enabled functionals (vdW-DF). These can be described as modified XC functionals that tend to model closely the problematic trend in the asymptotic long-range electron–electron interaction (or lack thereof) in classic LDA and GGA XC-functionals. To do so, they add extra terms in the density derived, for instance, by perturbation theory, Random Phase Approximation (RPA) or local frequency responses (effective plasma frequency).[35] vdW-DF functionals and DFT-D3 or TS schemes offer a similar level of accuracy for a given benchmark set, with a slight advantage of vdW-DF for some specific combination of XC functional and subsystem. For surface chemistry, the major advantage of the DFT-D3 and TS method in particular is that by adding a negligible amount of additional computing cost to the original DFT calculation, they obtain in most cases exceedingly accurate results, specifically for the reaction and diffusion barriers, although the absolute adsorption energies tend to be overestimated (by about ~0.2–0.4 eV for small molecules).[36]

In our previous work we benchmarked the performance of the original TS scheme, as implemented in CASTEP, on several aromatic systems: pyrrole/Cu(111),[7] thiophene/Cu(111)[6] and benzene/Cu(001).[6] For pyrrole and thiophene the surface translational/rotational barriers are in very good agreement with the experimental results obtained by HeSE spectroscopy. Taking into account that supramolecular self-assembled systems on metal surfaces have to be modelled with supercells containing hundreds of atoms, the computational-cost advantage of the TS-scheme, when compared to the vdW-DFT functionals, is significant. In this work we have employed the TS and the TSSCS method, which adds self-consistent screening to the original correction scheme, and is thought to be capable of generally improving the performance of the method in adsorbate systems.

The adsorption energy of benzene on the high-symmetry sites on a Cu(111)-$(2\sqrt{3} \times 2\sqrt{3})$ surface cell is reported in Fig. 3 and Table 1 and compared with selected DFT results from the recent studies of Carter and Rohl.[37] First of all, it is clear that both the TS and the TSSCS methods agree in indicating the hollow site as the minimum energy adsorption site for benzene. The diffusion barrier is slightly lower for the TSSCS calculations (18 meV instead of 21 meV) and both the TS and TSSCS fall below the experimental results (35 meV), although the difference might be considered to be within reasonable chemical accuracy. The oldest and most primitive correction scheme (G06, ref Grimme *et al.*[24]) also assigns the position of the global minimum to the HCP-R site, but shows a more pronounced difference in the angular dependence of the adsorption energy on TOP sites (0.111 eV). In general, one would not expect Zero Point Energy (ZPE) corrections to be

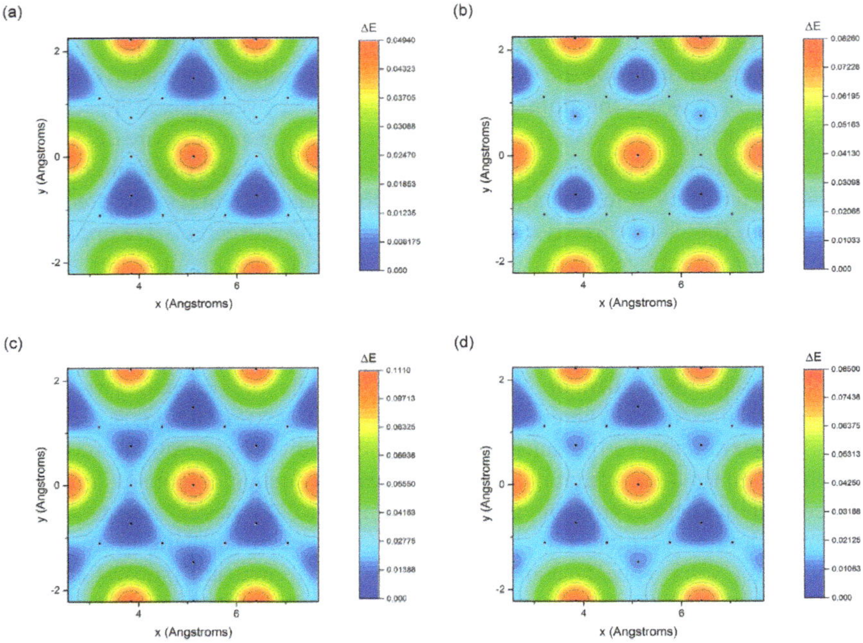

Fig. 3 Potential energy surfaces derived from the DFT calculations with (a) G06,[24] (b) OBS,[25] (c) TS[26] and (d) TSSCS[33] vdW corrections.

important in the case of a physisorbed aromatic molecule in a completely flat configuration. For this study we calculated the ZPE contributions to the barrier height using DFT with TS vdW corrections. The effect of the ZPE contribution is essentially to shift the minimum energy adsorption site from the HCP to FCC sites, marginally lowering the barrier height between the hollows and BR sites

Table 1 Energy corrugation, in eV, of the different dispersion energy corrections employed in this study for benzene adsorbed on the high symmetry sites of Cu(111). BR is the bridge site; FCC and HCP the fcc and hcp hollow sites, respectively; TOP is the top site. *I* and *R* indicate the rotational configurations, which differ by a 30 degree rotation around the *z*-axis (Fig. 1). G06 is the vdW correction scheme proposed by Grimme,[24] OBS is the correction scheme of Ortmann, Bechstedt and Schmidt,[25] TS and TSSCS are the original Tkatchenko–Scheffler correction scheme and the modified TS scheme with self-consistent screening[26,33]

Adsorption site	TS	TSSCS	G06	OBS
BR-I	0.023	0.018	0.018	0.026
BR-R	0.021	0.018	0.011	0.045
FCC-I	0.033	0.028	0.023	*0.014*
FCC-R	*0.010*	*0.012*	*0.012*	0.032
HCP-I	0.021	0.014	0.011	**0.000**
HCP-R	**0.000**	**0.000**	**0.000**	0.023
TOP-I	0.125	0.165	0.160	0.082
TOP-R	0.111	0.085	0.049	0.157

from 21 meV to 17 meV. The HCP and TOP sites are now 86 meV and 180 meV higher in energy than the FCC site. In this paper we will not discuss further the effects of ZPE corrections and the relative accuracy of each pairwise correction scheme for evaluating vibrational frequencies since this will be the subject of future works.

Table 2 reports the optimised structure of benzene on the hollow, bridge and top sites. The distance between the molecule and surface at the proposed adsorption site (2.958 Å) is in very good agreement with the reported height of 0.29 nm derived from the work function change measurements.[38] The accurate position of the molecule on the surface, and therefore the geometrical corrugation of the PES along the x, y plane,[39] is the single most important factor in determining the quality of a molecule/surface potential. Our calculations confirm, as previously observed by Carrasco *et al.*[40] for benzene on transition metal surfaces, that the interaction between benzene and the terraces of the single crystal metal surface is reasonably accurately captured by the original TS scheme and by its related methods. We note that, without vdW corrections, the potential energy curve will become completely unrealistic (see $E_{\text{ads-GGA}}$ results in Table 2); all the calculated adsorption energies are positive, therefore no binding between benzene and the surface is predicted. The TSSCS corrections, by including the self-consistent screening, also result in good agreement with the experimentally derived adsorption height, while the internal structure of the adsorbate (average C–C bond length, and C–H angle) is essentially identical to the TS values. The moderate downward tilting of the essentially unpolarised C–H bonds does not contribute overall to the stabilisation of the system, by reducing the surface-molecule dipole system, as this is dominated by the binding of the aromatic ring. As described by Witte *et al.*,[41] the so-called cushion effect is clearly visible in Fig. 4, which shows the electron density-difference plot for benzene adsorbed on the HCP-R site, with a region of electron depletion just above the surface atoms underneath the adsorbate. This quantum chemical effect governs the electronic density of states redistribution between the organic and metal surfaces and it is caused by the Pauli repulsion between the electron density in the layer immediately above the topmost (111) plane and the approaching electron-rich π-system of the aromatic ring. The displaced electron density cloud forms a three-dimensional "cushion" that holds the molecule at approximately 3 Å,

Table 2 Adsorption energy with ($E_{\text{ads-vdW}}$) and without ($E_{\text{ads-GGA}}$) vdW correction, adsorbate height and average hydrogen–carbon bond angles obtained with PBE + TS corrections for benzene adsorbed on the high symmetry sites of Cu(111)

Surface Site	$E_{\text{ads-vdW}}$ (eV)	$E_{\text{ads-GGA}}$ (eV)	Height (Å)	Charge (e)	C–H angle (degrees)
BR-I	−1.027	0.023	2.959	−0.59	−0.663
BR-R	−1.029	0.038	2.972	−0.56	−0.183
FCC-I	−1.018	0.009	2.955	−0.59	−0.777
FCC-R	−1.040	0.031	2.967	−0.60	−0.211
HCP-I	−1.030	0.016	2.960	−0.59	−0.380
HCP-R	−1.050	0.003	2.958	−0.58	−0.313
TOP-I	−0.926	0.098	2.971	−0.48	−1.444
TOP-R	−0.940	0.093	3.174	−0.60	−1.437

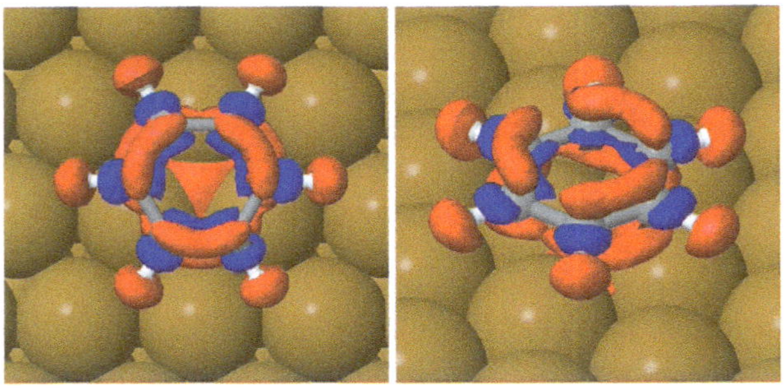

Fig. 4 Top view (left) and side view (right) of the charge density difference plots showing charge transfer upon adsorption of benzene on the Cu(111) hcp site in the HCP-R rotational configuration. The red contours indicate an electron density increase by 0.005 electrons per Å^3; the blue contours indicate an electron density decrease by 0.005 electrons per Å^3.

causing a total shift of the work-function in the opposite direction to the predictions of the simplistic electrostatic models. A comparison between our results and those reported by Witte *et al.*[41] shows that the cluster model used in early calculations clearly overestimates the total charge redistribution due to the localised nature of the cluster compared with the periodically repeated surface model employed in the present study. Nevertheless, the understanding of the binding mechanism of benzene on Cu(111) captured by Witte and co-authors within the cushion effect model is essentially correct, especially in predicting the sign of the dipole moment change and consequently the lowering of the work function of the metal surface upon the adsorption of benzene. Witte *et al.* calculated a work function change of −1.08 eV, in excellent quantitative agreement with our calculations (Table 3) at low coverage (1/12 ML), where we found that benzene causes a shift of the work function of 0.99 eV, within 6% of the experimental measurements.[41] Comparing these results with the work function change induced by the adsorption of Cp on the same surface, as calculated by Sacchi *et al.*,[8] we notice some similarities, but also a striking difference. We find that the change in the work function does not significantly change when the benzene coverage is increased above 0.11 ML, which is similar to Cp adsorption, where $\Delta\Phi$ changed by only 5% when the coverage was increased from 0.11 to 0.14 ML. Again, we observe some degree of depolarisation in the benzene overlayer when the molecules are compressed beyond a certain critical distance, although it

Table 3 Calculated surface dipole moment, $\Delta\mu$ (D) and the work function change $\Delta\Phi$ (eV) for benzene adsorbed on Cu(111) at increasing coverage (ML)

Coverage (ML)	$\Delta\mu$ (D)	$\Delta\Phi$ (eV)
0.08	1.8	−0.99
0.11	2.3	−1.70
0.14	1.7	−1.62

is not clear how far the molecules could be compressed before a critical change in the adsorption angle may occur.[18]

The position of the lowest energy structure and the measured barrier for jump diffusion are entirely consistent with the previous measurements on Cp/Cu(111). In principle, one would not expect such a similarity in the behaviour of these two molecules, given the different symmetry groups (D_{6h} and D_{5h}). It is observed that the substrate reduces the symmetry of both molecules and the extent of charge backtransfer in Cp more than compensates for the larger dispersion interaction between the six-membered ring and surface. The slightly higher barrier for the jump diffusion of cyclopentadienyl on the same surface can be partly rationalised by an increase in the ionic binding, combined with the reduced symmetry compatibility between the pentagonal aromatics and hexagonal substrate. The local adsorption site (hollow hcp) has a reduced three-fold symmetry with respect to the six-fold rotational symmetry of the clean surface and the reduced overlap between the d_z orbitals of the three Cu atoms with the π-system of benzene is clearly visible in the density difference plot of Fig. 4. The molecular orbital mismatch between the substrate and molecule (Fig. 4 and 5) reduces the height of the rotational barrier compared to what we observed for benzene on Cu(001), where the experimental and theoretical activation energy are a factor of 3 and 35 higher than for the Cu(111) surface, respectively. The much greater energetic corrugation of the Cu(100) surface is therefore captured by the vdW-corrected DFT simulations, but with a clear DFT overestimation of the barrier height for the jump diffusion (hollow to hollow site) for Cu(100). It is also worth discussing the charge transfer between the substrate and adsorbate in different locations on the

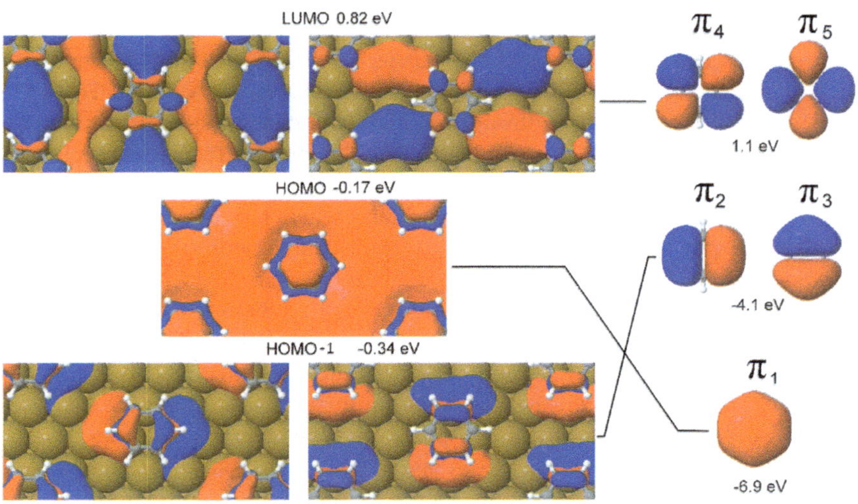

Fig. 5 Left, the LUMO (0.82 eV), HOMO (−0.17 eV) and HOMO−1 (−0.34 eV) Kohn–Sham orbitals (KS-MOs) at the gamma point for benzene adsorbed on a hollow hcp site. Right, from top to bottom, the symmetry related pi orbitals for isolated benzene. The KS-MOs of the surface are joined by solid lines with the symmetry equivalent pi orbitals of benzene. The π_3, π_2 and π_1 orbitals are much lower in energy than the correspondent surface mixed-character KS-MOs and are represented beside the latter for the purpose of easy visualisation.

PES. For smaller molecules, the extent of donation and backdonation determines the intensity of the binding energy. However, for benzene on Cu(111), the charge transfer from the surface to the molecule does not seem to be correlated to either the distance between the molecule and the surface, or the adsorption energy. Still, it is worth noting that the highest energy local minimum (TOP-I) does show the lowest degree of electron backdonation (−0.4 e) from the substrate.

It is evident from Fig. 5, which shows the Kohn–Sham orbitals (KS-MOs) at the gamma point for benzene adsorbed on a hollow hcp site, that the extremely low chemical interaction between the Cu(111) surface and the adsorbate does not translate into a negligible mixing between the surface and molecular orbitals. In fact, if we concentrate our attention on the frontier orbitals of the system (a selection of these, the HOMO, HOMO−1 and LUMO are represented in Fig. 5), the high degree of mixing between the surface state and molecular π states is immediately visible in the Kohn–Sham orbital representations of these states. For instance, in the HOMO−1 the lower lobes of the π_2 and π_3 orbitals extend to the copper atoms surrounding the hollow site, while the character of the π_4 and π_5 orbitals in the LUMO is completely mixed with the surface states connecting adjacent molecules. Furthermore, the calculations show that the π_1 and (π_2, π_3) orbitals are blue shifted in energy by 6.7 eV and 3.8 eV respectively to become part of the HOMO and HOMO−1 bands.

Comparing the results of the calculations with those reported by Carter *et al.*, we can observe a substantial agreement between our data and the results of optB88-vdW, revPBE (vdW-DF) and rPW86 (vdW-DF2). In particular, the molecule-surface height is very similar to that which Carter *et al.* reported for the vdW-DF (optB88-vdW) calculations (2.93 Å) and the barrier height is also entirely consistent with the results of the vdW-DF calculations (in the 0–30 meV range).[37] The binding energy calculated by the TS and TSSCS methods overestimates the experimental adsorption energy (by about 0.4 eV) in the same way as less accurate correction schemes OBS and G06. The barrier for diffusing from HCP over the BR site is about 20 meV, with FCC lying around 10 meV higher than HCP. The energetics of the diffusion pathway do not seem to change significantly with the choice of the vdW corrections with TS, TSSCS, G06 and OBS schemes resulting in barriers in the 11–26 meV range.

All four schemes also result in energetic differences between the HCP and FCC hollows of only 10–14 meV. The experimental data supports adsorption to the non-degenerate hollow sites as the ISFs did not show the additional decay that would be expected in the degenerate case. However, there is a further subtlety. The measurements were carried out at 170 K, corresponding to a thermal energy of 15 meV. If the energetic differences were indeed as small as 10–14 meV, we would expect to have to consider the contribution from jumps from the FCC sites, which would then be reasonably occupied. The ratio of jump rates from the HCP to FCC and FCC to HCP sites is given by $\lambda = \exp(\Delta E/k_BT)$ where ΔE is the energy difference of the two sites. For the calculated energies, λ is in the range of 2–3. Tuddenham *et al.* illustrate that for $\lambda = 2$ (equivalent to a 10 meV difference), a second exponential term would still be expected, but would become most apparent at slightly higher momentum transfers (around 1.8 $Å^{-1}$), an effect that is not seen in Fig. 2d. Due to the exponential nature of λ, the strength of the second exponential term rapidly falls away in magnitude as ΔE increases, which suggests that the calculations are probably underestimating the difference between the two sites to a similar degree to the diffusion barrier, where the calculations are 10–25 meV lower than the experiment. A FCC–HCP difference in energy of just 25 meV would be indiscernible within the present experimental data from the greater difference.

The small energetic difference between the I and R rotational states (11–23 meV on the energetically preferred site) would lead to the occurrence of both rotational states at the temperatures of the experiment. However, if we consider the energies presented in Table 2, although these states will be present frequently, they do not affect the rate-limiting barrier, which is from HCP-R to BR-R for TS and TSSCS, HCP-R to FCC-R for G06 and HCP-I to BR-I for OBS. The population of the rotated states would be expected to affect the pre-exponential factor for diffusion, rather than the Arrhenius activation energy that was considered in this work. A typical molecule's trajectory across the surface would be expected to involve rotation as well as translation, but without the degree of coupling between the two that might be described in terms of quasi-static steering[10] where the molecule must rotate to pass over the rate-limiting barrier.[42]

5. Conclusions

Benzene adsorption on Cu(111) has been established as a benchmark system for exploring the 2D dynamics of self-assembled aromatics and conjugated molecules. In this work we have shown that benzene adsorbs on the hollow sites of the Cu(111) surface and undergoes jump-diffusion with an extremely low barrier height (35 ± 1 meV, according to the HeSE results, compared with 12–26 meV for the DFT calculations with different vdW correction schemes). The potential energy landscape for this system is well characterised by the combination of a GGA XC-functional (PBE) and vdW corrections with an accuracy that is comparable with that of modern DFT-DF functionals. The HeSE results clearly exclude that benzene might adsorb, as previously suggested, on bridge sites and the DFT calculations help to understand that atop sites, although compatible with the jump-diffusion mechanism, are not energetically the most favourable. Rotation around the molecular central C_6 axis is also an activated process that shows a barrier height (12–23 meV) that is essentially identical to that of the jump-diffusion. By analysing the electron density redistribution upon adsorption and the change in the ordering and energetics of the molecular orbitals interacting with the surface bands, we rationalise the observed small, but finite, barriers for diffusion in terms of the symmetry mismatch between the D_{6h} symmetry of the MOs and the three-fold symmetry of the charge backdonated to the molecule.

Acknowledgements

MS would like to thank the Royal Society for his University Research Fellowship. HH acknowledges the support of the Leverhulme Trust. Additionally, we thank Dr Barbara Lechner for her helpful discussions during the analysis of the experimental data. This work used the ARCHER UK National Supercomputing Service *via* our membership of the UK's HEC Materials Chemistry Consortium, which is funded by EPSRC (EP/L000202).

References

1 A. P. Jardine, H. Hedgeland, G. Alexandrowicz, W. Allison and J. Ellis, *Prog. Surf. Sci.*, 2009, **84**, 323–379.
2 A. Zykov, S. Bommel, C. Wolf, L. Pithan, C. Weber, P. Beyer, G. Santoro, J. P. Rabe and S. Kowarik, *J. Chem. Phys.*, 2017, **146**, 052803.
3 V. G. Ruiz, W. Liu and A. Tkatchenko, *Phys. Rev. B*, 2016, **93**, 035118.
4 Y. He, F. Bussolotti, Q. Xin, J. Yang, S. Kera, N. Ueno and S. Duhm, *J. Phys. Chem. C*, 2016, **120**, 14568–14574.
5 A. Rastgoo-Lahrood, J. Björk, M. Lischka, J. Eichhorn, S. Kloft, M. Fritton, T. Strunskus, D. Samanta, M. Schmittel, W. M. Heckl and M. Lackinger, *Angew. Chem., Int. Ed.*, 2016, **55**, 7650–7654.
6 B. A. J. Lechner, M. Sacchi, A. P. Jardine, H. Hedgeland, W. Allison, J. Ellis, S. J. Jenkins, P. C. Dastoor and B. J. Hinch, *J. Phys. Chem. Lett.*, 2013, **4**, 1953–1958.
7 B. A. J. Lechner, H. Hedgeland, J. Ellis, W. Allison, M. Sacchi, S. J. Jenkins and B. J. Hinch, *Angew. Chem., Int. Ed.*, 2013, **52**, 5085–5088.
8 M. Sacchi, S. J. Jenkins, H. Hedgeland, A. P. Jardine and B. J. Hinch, *J. Phys. Chem. C*, 2011, **115**, 16134–16141.
9 H. Hedgeland, B. A. J. Lechner, F. E. Tuddenham, A. P. Jardine, W. Allison, J. Ellis, M. Sacchi, S. J. Jenkins and B. J. Hinch, *Phys. Rev. Lett.*, 2011, **106**, 186101.
10 H. Hedgeland, M. Sacchi, P. Singh, A. J. McIntosh, A. P. Jardine, G. Alexandrowicz, D. J. Ward, S. J. Jenkins, W. Allison and J. Ellis, *J. Phys. Chem. Lett.*, 2016, **7**, 4819–4824.
11 S. J. Jenkins, *Proc. R. Soc. A*, 2009, **465**, 2949–2976.
12 K. Tonigold and A. Gross, *J. Chem. Phys.*, 2010, **132**, 224701–224710.
13 J. Eichhorn, W. M. Heckl and M. Lackinger, *Chem. Commun.*, 2013, **49**, 2900–2902.
14 S. Simpson and E. Zurek, *J. Phys. Chem. C*, 2012, **116**, 12636–12643.
15 S. Katano, S. Ushioda and Y. Uehara, *J. Phys. Chem. Lett.*, 2010, **1**, 2763–2768.
16 J. R. Lomas and G. Pacchioni, *Surf. Sci.*, 1996, **365**, 297–309.
17 S. J. Stranick, M. M. Kamna and P. S. Weiss, *Surf. Sci.*, 1995, **338**, 41–59.
18 M. Xi, M. X. Yang, S. K. Jo, B. E. Bent and P. Stevens, *J. Chem. Phys.*, 1994, **101**, 9122–9131.
19 N. Lorente, M. F. G. Hedouin, R. E. Palmer and M. Persson, *Phys. Rev. B*, 2003, **68**, 155401.
20 T. Komeda, Y. Kim, Y. Fujita, Y. Sainoo and M. Kawai, *J. Chem. Phys.*, 2004, **120**, 5347–5352.
21 H. Lesnard, N. Lorente and M. L. Bocquet, *J. Phys.: Condens. Matter*, 2008, **20**, 224012.
22 S. F. Sousa, P. A. Fernandes and M. J. Ramos, *J. Phys. Chem. A*, 2007, **111**, 10439–10452.
23 J. Klimes and A. Michaelides, *J. Chem. Phys.*, 2012, **137**, 120901–120912.
24 S. Grimme, *J. Comput. Chem.*, 2006, **27**, 1787–1799.
25 F. Ortmann, F. Bechstedt and W. G. Schmidt, *Phys. Rev. B*, 2006, **73**, 205101.
26 A. Tkatchenko and M. Scheffler, *Phys. Rev. Lett.*, 2009, **102**, 073005.

27 A. Tkatchenko, R. A. DiStasio Jr, R. Car and M. Scheffler, *Phys. Rev. Lett.*, 2012, **108**, 236402.

28 J. P. Perdew, K. Burke and M. Ernzerhof, *Phys. Rev. Lett.*, 1996, **77**, 3865–3868.

29 J. P. Perdew and Y. Wang, *Phys. Rev. B*, 1992, **45**, 13244.

30 H. J. Monkhorst and J. D. Pack, *Phys. Rev. B*, 1976, **13**, 5188.

31 F. Mezei, *Z. Phys. A: Hadrons Nucl.*, 1972, **255**, 146–160.

32 F. E. Tuddenham, H. Hedgeland, A. P. Jardine, B. A. J. Lechner, B. J. Hinch and W. Allison, *Surf. Sci.*, 2010, **604**, 1459–1475.

33 V. G. Ruiz, W. Liu, E. Zojer, M. Scheffler and A. Tkatchenko, *Phys. Rev. Lett.*, 2012, **108**, 146103.

34 S. Grimme, *J. Comput. Chem.*, 2004, **25**, 1463–1473.

35 K. Berland, V. R. Cooper, K. Lee, E. Schroder, T. Thonhauser, P. Hyldgaard and B. I. Lundqvist, *Rep. Prog. Phys.*, 2015, **78**, 066501.

36 M. Sacchi and S. Jenkins, *Phys. Chem. Chem. Phys.*, 2014, **16**, 6101–6107.

37 D. J. Carter and A. L. Rohl, *J. Comput. Chem.*, 2014, **35**, 2263–2271.

38 K. Toyoda, Y. Nakano, I. Hamada, K. Lee, S. Yanagisawa and Y. Morikawa, *Surf. Sci.*, 2009, **603**, 2912–2922.

39 G. R. Darling and S. Holloway, *Rep. Prog. Phys.*, 1995, **58**, 1595–1672.

40 J. Carrasco, W. Liu, A. Michaelides and A. Tkatchenko, *J. Chem. Phys.*, 2014, **140**, 084704.

41 G. Witte, S. Lukas, P. S. Bagus and C. Wöll, *Appl. Phys. Lett.*, 2005, **87**, 263502–263503.

42 M. Sacchi, D. J. Wales and S. J. Jenkins, *J. Phys. Chem. C*, 2011, **115**, 21832–21842.

Faraday Discussions

PAPER

Complex molecular surfaces and interfaces: concluding remarks

David B. Amabilino *[ab] and Steven L. Tait *[c]

Received 12th September 2017, Accepted 12th September 2017
DOI: 10.1039/c7fd90075a

This paper is derived from our concluding remarks presentation and the ensuing conversations at the *Faraday Discussions* meeting on *Complex Molecular Surfaces and Interfaces*, Sheffield, UK, 24th–26th July 2017. This meeting was comprised of sessions on understanding the interaction of molecules with surfaces and their subsequent organisation, reactivity or properties from both experimental and theoretical perspectives. This paper attempts to put these presentations in the wider context and focuses on topics that were debated during the meeting and where we feel that opportunities lie for the future development of this interdisciplinary research area.

Where have we been?

The adsorption of molecules onto surfaces has been understood in thermodynamic terms for a century; different types of molecular film growth modes depend on intermolecular interactions and give rise to the classic isotherm models.[1] Truly, we stand on the shoulders of giants. More recently, and thanks to the analytical tools at our disposal, the bonding between adsorbate and surface have been understood, apart from the trivial case of covalent grafting onto surfaces that is perhaps intuitive. Now, the non-covalent[2,3] and covalent bonds[4,5] that are formed between molecules at various kinds of interfaces can be used to prepare self-organized networks on surfaces, making use of chemistry (supramolecular and molecular) copied from assemblies in the solid and solution states.[6] Complex molecular surfaces and interfaces can be interpreted in a number of ways, and it is clear from the papers and discussion in this volume that even the more minimalist combination of adsorbent and adsorbate comprises a great depth of subtlety and detail.[7]

Discussions reflect where we have been and where we are going: it has been shown how complex networks are formed, and can transform, and deeper

[a]*School of Chemistry, The University of Nottingham, NG7 2RD Nottingham, UK. E-mail: david.amabilino@nottingham.ac.uk*
[b]*GSK Carbon Neutral Laboratories for Sustainable Chemistry, The University of Nottingham, Jubilee Campus, Triumph Road, Nottingham, NG7 2TU, UK*
[c]*Department of Chemistry, Indiana University, Bloomington, Indiana, USA. E-mail: tait@indiana.edu*

understanding of these systems is now greater than ever before. These new levels of understanding come through in the papers and discussions contained in this magnificent volume and lead to new and more complex questions, particularly as we look to apply these principles to functional materials.

In this article we have highlighted six topics where opportunities lie for the development of this area. The huge advances made in surface characterisation have allowed the path behind us to be made, and it seems highly likely that further developments coupled with theoretical advances will help guide us forward.[8] One of the key challenges in this field, that we identify later, is to use the pieces of information from each technique, including theory, to come up with specific and general rules for assembly. And it is on the topic of these general rules where we begin our remarks.

Toward a rulebook for molecular adsorption, diffusion, and assembly

We have already indicated that the overall adsorption isotherm types and a collection of selected interactions between surface and adsorbents have been quantified and established, mainly for relatively simple systems,[9–11] because the interpretation of more complex systems is beyond the state of the art. Various specific adsorbate–adsorbent pairs have been studied, under varying conditions (such as temperature and environment – meaning in vacuum or solvent, where for certain we could be more adventurous as noted in the Discussion) such that a sense of rules and guiding principles for molecular adsorption, diffusion, and assembly of more complicated situations is being developed. An all-encompassing "rulebook" for guiding a user as to how a given molecule will adsorb, diffuse and assemble on a surface would allow the design of any number of complex systems that might assemble through non-covalent or covalent bonds, or eventually hierarchically using a combination of both.

In combination, the already established but evolving guides for hydrogen bonding directionality and strength (from solution,[12] solid state[13] and theory[14,15]) and general criteria for interactions between (ionic) metal atoms and ligands (hard/soft Lewis acid/base, Pearson's rules) have guided a relatively high level of understanding of the self-assembly of supramolecular networks on surfaces, although fine details are often not fully explicable. In general, the strength and directionality of chemical bonds follows the schematic graph shown in Fig. 1. This representation includes the metal–adsorbate atom interactions specifically, as they are quite unique to surfaces that are often used as substrates for mono-layers and films, whether "flat" or in nanoparticles, as discussed in the literature for the case of gold.[16,17] Even as the complexity of the molecular adsorbate increases, the chemistry, charge transfer, and interaction of the surface plays a significant role in determining the chemical form and organization of the adsorbate (DOI: 10.1039/c7fd00116a).[18]

Often, a qualitative model of relative interaction importance is all that can be obtained, but this allows for predicting trends. Empirical guidelines can be useful. Ideally, one would want to improve feedback between experiment and theory to eventually achieve quantitatively accurate and predictive models. We need to move from system specific knowledge toward a predictive modelling and

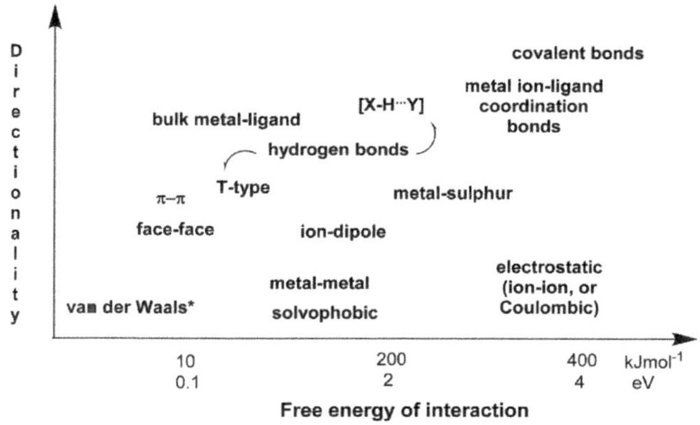

Fig. 1 A schematic guide to the strength and directionality of bonds and interactions that are important in the structure of molecular surfaces and interfaces.

design that allows the preparation of (eventually functional) multicomponent systems. Although bonding at interfaces has been largely understood for some time, current research is still revealing important aspects of the interplay of these interactions that leads to ordering and structure of molecular systems at interfaces, through advanced experimental methods (DOI: 10.1039/c7fd00089h) and through the coupling of experiment and theory (DOI: 10.1039/c7fd00095b). The hierarchy of interactions that leads to the complexity of these systems is implicit. The predictive aspect of hierarchical system assembly in homogeneous solution is itself beyond the current state of the art, and perhaps studies of hierarchical growth at interfaces can provide insight, in a more general sense, for this challenge.

Up to now, we have seen that many examples of networks on surfaces are inspired by known chemistries (using non-covalent and covalent bonds) previously seen in solution or the solid state, where single crystal structures are solved. We asked before this meeting "Can two-dimensional structure be correlated with three dimensional order?". In this regard, the rules of topology and connectivity of chemical linkers and nodes to form networks has, to a degree, reached a point where two dimensional frameworks can be designed at will, for coordination networks,[19] purely hydrogen bonded systems,[20] and even those where van der Waals interactions and interdigitation of alkyl chains are responsible for the patterns.[21] It has been shown that tritopic ligands (such as 1,3,5-tripyridylbenzene) form regular honeycomb or triangular lattices on the (111) surfaces of gold and copper.[22,23] The emergence of a demiregular lattice with C_3 tritopic ligands arises in the present discussion because of the presence of both two-fold and three-fold coordination modes of the metal atom that links the ligands (DOI: 10.1039/c7fd00088j). The flexible nature of the linking in this particular case shows the relative flexibility that monodentate ligands can provide to metal atoms, a feature that is very difficult to predict.

Complexity also arises when the adsorption geometry deviates from the plane of the interface, such as the case of bowl-shaped hydrocarbons, where the concave surface is usually oriented away from the surface.[24,25] Extending the π surface of the molecules makes the molecules stand on their rim no matter what the coverage on copper (111), a somewhat counter-intuitive situation caused by dominant electrostatic interactions between adsorbate and surface (DOI: 10.1039/c7fd00109f). This orientation of the molecules allows weaker interactions between aromatic surfaces as well as between edges and faces to orient the molecules as extended packing grows. Transitions in growth orientation are observed as coverage increases so that bilayer films exhibit structures very different from the monolayer.[26] There is clearly much to explore in the formation of layers of molecules with persistent shapes at interfaces that compete with adsorbate–surface interactions.

The situation concerning the interpretation of adsorption geometry is particularly complicated for the case of molecules containing several polar moieties, such as amino acids and peptides.[27] Indeed, the surface assembly of biomolecules in general is a complicated yet promising area of research.[28] While certain interactions are apparently retained, such as the interaction between the carboxylic acid groups and copper (where the acid deprotonates to form the carboxylate more often than not[29]), the variation of molecule and metal surface leads to a complex structural space. It has been shown that the footprints these molecules have on surfaces can be influenced by the adsorbing group and the stereogenic centres that define the "hands" of the adsorbates, in terms of chirality.[30] Moving to dipeptides (comprising two amino acid moieties, Fig. 2) the binding to metal surfaces is different depending on the nature of the adsorbent, as demonstrated elegantly for the glycine–proline dipeptide.[31] This molecule presents very real challenges in the interpretation of its adsorption to the (110) surfaces of gold and silver (DOI: 10.1039/c7fd00116a). The reasons for this situation are the competing interactions of carbonyl, amido and amino groups for the surface and the possibility that the molecule can exist in different forms (neutral or zwitterion) that depend on the amount of adsorbate on the surface. This system, and surely others of the multitude possible from combinations of amino acids, offer a huge chemical space to explore and discover rules for adsorption.

The binding of molecules to metal surfaces can be so strong that restructuring takes place at the interface. This effect is seen dramatically for a heterotripeptide insulin growth factor (IGF, gly–pro–glu) on the gold (111) surface.[32] The reordering of the metal atoms does not only arise at kinks and step-edges, but

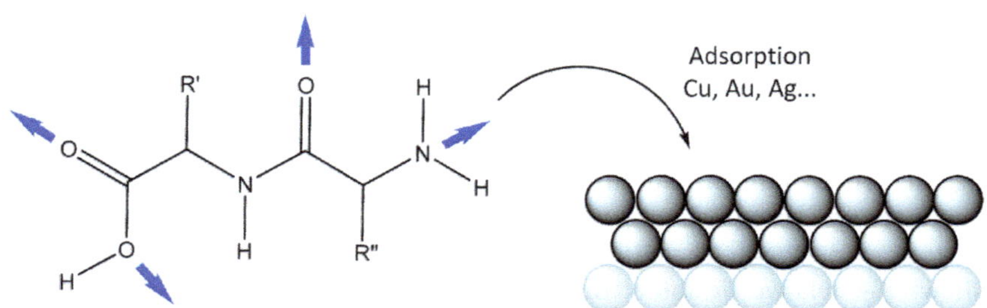

Fig. 2 A generic dipeptide which, when adsorbed onto a variety of metals (represented by the grey spheres), has the potential to generate a combinatorial space where rules concerning adsorption could be explored. The blue arrows indicate the Lewis basic sites on the dipeptide skeleton, where *R′* and *R″* could be the same or different, and the stereochemistry likewise.

also in apparently defect-free terraces. "Fingers" of covered metal, ten nanometres wide grow across the surface. It appeared that the origin of these features was a result of delicate interactions of the bound surface atoms with a scanning tunnelling microscope (STM) tip. The observation shows how a strong adsorbate weakens the interaction between the bound metal in the top layer and the neighbouring adsorbent atoms. The high adsorption energy of the tripeptide results from its principally zwitterionic form on the surface (shown by Reflection Absorption Infrared Spectroscopy (RAIRS) and X-ray photoelectron spectroscopy, XPS). On the basis of this result, could we devise systems where surface atoms are reconstructed by "intelligent" adsorbate molecules to give functional shapes? Certainly for this IGF tripeptide, a similar phenomenon involving adsorbate "walking" also restructures the surface of Au(110).[33]

Electrostatic interactions are potentially the strongest non-covalent force between molecules,[34] if in general poorly directional, and it has been shown in solution that they can complement weaker forces.[35] In this *Faraday Discussion*, it has been shown that regular striped patterns can be formed by molecules on a surface, and it is believed that electrostatic interactions are responsible for this effect (Kühnle, DOI: 10.1039/C7FD00089H).

Apart from this mainly thermodynamic perspective, kinetics can play an important role in the formation of surface-based systems, and can even can be dominant.[36] Metallooctaethylporphyrins adsorbed onto gold(111) display sluggish exchange with their solution in 1-phenyloctane.[37] Mixtures of the cobalt(II) and nickel(II) complexes of these macrocycles adsorb giving a monolayer (within seconds at room temperature) that is a perfect solid solution, that is a result of the almost identical structures of the layers and the strong adsorbate–adsorbent interaction. Heating the system to 135 °C and imaging at room temperature after cooling revealed that the nickel porphyrins are replaced by the cobalt(II) homologue. Remarkably, certain thiolates on gold exchange more slowly than the porphyrins.[38] This example shows the importance of kinetics, driven by very strong non-covalent interactions, in the formation of layers where metastable states exist, and others have reported analogous phenomena in a case that is completely different chemically.[39] In some cases, this can lead to very different structure and properties, as for star shaped molecules that give open or dense monolayer patterns.[40]

It is perhaps useful to consider ideas in confinement to isolate molecules and reveal key characteristics of their adsorbed state for the rulebook. Surface analysis often confines species to a 2D plane, but we have also seen examples of further confinement to 1D motion or to small nanowell regions (referring to the in plane dimensions).[41,42] For example, network systems (of covalent or non-covalent nature) with "yoctowells" (referring to the well volume[43]) offer the possibility of both studying the fundamental science of interactions between surface and guest in a defined space, but also to study the effects of interactions between the host networks and the guest in the monolayer (this could be considered a kind of second sphere coordination[44]). The use of variable temperature studies on the mobility of molecules confined to the wells and subsequent Arrhenius analysis are bound to provide quantitatively significant values of the interaction energy between molecules and surfaces. An indication as to how this might be done is seen in the motion of a "walker" molecule on Cu(110), whose motion was restricted by pre-formed molecular "fences" constructed from oligomeric porphyrin molecules.[45] While the mobile molecules diffuse freely along the copper rows (the [1$\bar{1}$0] axis of the surface) when they are unimpeded in this system, moving freely between barriers when the latter are separated by the length of more than a few molecules. However, when the "walkers" are "corralled" by two fences, they only have the perpendicular direction available, and move much more slowly than in the [1$\bar{1}$0]. While this difference in mobility has not been quantified to date, it is clear that confinement of this type combined with quantitative analysis of the thermal barriers associated with the process will allow unique insight into the interactions taking place and experimental determination of the energies of interaction between surface and adsorbate.

Directional diffusion is often observed due to adsorbent structure, but sometimes also as a result of the adsorbate structure.[46,47] The most common surfaces on which directional movement of adsorbed molecules is seen are ridged, such as the (110) surfaces of metals. The directional diffusion of molecules on a surface can lead to specific chemical reactions taking place, as shown for porphyrin derivatives that are activated by C–H bond scission on a crystal of copper.[48] Theory was absolutely central to the explanation of the observed covalent bond-forming reaction.

Atoms removed from the surface of adsorbents whose outermost elements – such as metals – are relatively labile and can give rise to surface element-mediated structure formation. How can we know when these elements will sit between them? The flippant answer, of course, is when it is thermodynamically favourable for the system. From what we have seen so far, it is clear that this incorporation of mobile metal atoms is relatively easy, but no numbers have been put on this process so far. Furthermore, for low symmetry metal surfaces, the reconstruction of the adsorbent can occur (as we see in Fig. 1, the interactions between metal atoms themselves and between metal atoms and adsorbent molecules is of similar order). For example, the (110) surface of a gold crystal will restructure when doped with low coverage sulphur[49] or metal nanoislands.[50] Great progress can be made in the understanding of these interface phenomena through thorough investigation, giving an idea not only of adsorption, but interface structural energy minimisation after the event.

Polymorphism

An important principle that came up throughout the Discussion was the broad range of strength and directionality in the various interactions at play in molecular self-assemblies (Fig. 1). This variety inherently leads to the possibility that a system may have multiple states that are very near each other in energy and also similar in population. Thus, a natural feature of self-assembly in this scenario is polymorphism. In some experiments, we attempt to drive the system toward one uniform thermodynamically favoured state through solvent annealing or thermal annealing, but we also saw examples in the Discussion where multiple structures co-existed, such as multiple Moiré patterns from a system as simple as trimesic acid on graphite.[51] Multiple domain orientation angles were observed, indicating that there was not a strong energetic distinction between the domain orientations (DOI: 10.1039/c7fd00113d). Another example of polymorphs was reported in terephthalic acid and its derivatives on graphite (DOI: 10.1039/c7fd00112f). It was argued in the Discussion that we are likely, as a community, to encounter more examples of polymorphism as we push toward more complex systems (see below), especially toward biomimetic systems.

Developing an understanding of polymorphism and the interactions at play in the formation of structures may provide an opportunity to program systems to select a desirable architecture. One could envision the possibility to artificially bias a naturally polymorphic system at a critical point to steer it toward a desired morphological state, *i.e.*, tailor the system into one low energy state by design. This could be achieved by the addition of an additional assembly component (dynamic library model[52]), by thermodynamic control,[53] or by the application of an external bias or field.[54,55] This is a key motivation for understanding self-assembly and exploring the application of that understanding for control and design. It also highlights the need to understand crystallization and nucleation (see below).

We should also consider that the polymorphism may not be inherently undesirable, even though it is often treated this way. Polymorphic systems offer the opportunity to achieve multiple functions in a single system, which may be a route to achieve tandem catalysts or bifunctional or switching functions. This was elaborated on in the opening lecture of our Discussion (DOI: 10.1039/C7FD90072D). This may require a new paradigm for self-assembled systems in which we embrace the opportunities that a polymorphic system may offer rather than instinctively trying to anneal it toward a uniform structure. An important discussion point was that the community would benefit from learning more about the variety of structures observed in the course of a study, rather than only being presented with the "best" structure in the published work. This would help us all to better understand the characteristics of self-assembly and to identify and characterize polymorphic states that offer opportunities for multiple functionalities in a system. Developing polyfunctionality from polymorphic systems will test the capacity of our analytical tools as well as our ability to build greater complexity into the systems we study.

Nucleation – observing and exploring

Nucleation is a kinetically determined phenomenon that cannot be observed directly in the present day.[56] The two stage nucleation of solid phases, whereby a glassy-type condensed phase[57] is formed and then transforms to the well-defined solid phase that subsequently grows, is favoured over the classical nucleation theory that is currently not thought to be active in many systems.[58] Control over this step in the crystallisation process can give rise to phases with different structural characteristics, polymorphs, which were covered in the previous section. The structure, and ultimately the function, of a molecular system therefore depends on these nascent stages of formation of the ordered phase. It is present in every self-assembled system we study, but is frequently elusive experimentally. The reason for this situation is that, in general, a relatively poorly ordered state transforms into a periodic ordered one in a fraction of a second.

Research into nucleation of molecules with little possibility of rearrangement, such as fullerenes,[59] using STM have shown the formation of layers of molecules in a stepwise manner. Although these studies often see growth phenomena, the nucleation event is not actually seen, but inferred afterwards. The same is true for work on chiral compounds where the growth of segregated chiral layers is impeded by the attachment of opposite handed molecules at the edge of small domains.[60] Thus, apparent nuclei are observed, although they are frozen in a matrix after the event. In order to gain further understanding, it will be necessary to probe with temperature control and high time resolution, focusing on the early stages of aggregation where molecules transform from a glassy aggregate to a viable nucleus with a periodic array of molecules. Studies in ultra-high vacuum offer the possibility to control the kinetic rates by wide temperature control, but under ambient conditions one would need to design an assembly system with just the right kinetic features to directly observe nucleation events. Theoretical studies, particularly molecular dynamics simulations, may be able to provide insight until experimental advances are achieved, but there is an inherent difficulty in such studies due to self-assembly phases typically varying in energy by only tenths of an electron-volt. Thus, there is a critical need for highly accurate force field development.

Multicomponent networks formed by alkoxylated dehydrobenzo[12]annulene (DBA) derivatives – which generate hexagonal nanowell frameworks thanks to the interaction between their alkyl chains – and guest coronene, or coronene with isophthalic acid are achiral on Au(111), and on HOPG are only chiral.[61] This difference is believed to result from lower diffusional barriers for guests on graphite, that allow them to act as nucleation sites for the formation of achiral pores in the gold case where they are adsorbed more robustly. After nucleation of achiral pores, assembly of a regular superlattice proceeds. Studies of chiral information transfer from chiral solvent molecules in a DBA assembly provided an opportunity to identify and evaluate nucleation of chiral domains at interfaces (DOI: 10.1039/c7fd00103g). The induction of chirality into monolayers of achiral DBAs by the solvent was specific, with the chain length of the chiral secondary alcohol determining the handedness from the nucleation point, although this was only observed living on in the assembly after the event. Molecular modelling provided a possible mechanism for this process, although the observation (by experiment or theory) of stochastic nucleation events remains a very real challenge awaiting suitors.

Polymer crystal nucleation and growth[62] is perhaps ideally suited to study at interfaces, because the emergence of functioning crystal nuclei (from the glassy condensed core) must be slower than for small molecules that diffuse and reorient on a much shorter timescale. In the Discussion, it was demonstrated by molecular dynamics simulations that surfaces can favour the crystallization of a model polymer (DOI: 10.1039/c7fd00071e). It was shown that, unlike molecular systems in which nucleation can be favoured by rough surfaces, in the present polymer case the "flattening" of a crystal plane onto the smooth interface region actually promoted growth, while theory-introduced roughness impeded the process. The possibility of using flow to study and favour the nucleation of polymers[63,64] is also a tempting possibility at an interface, as already shown for thin films of molecules where molecular alignment over the multilayer was achieved.[65] The control of nucleation to determine structure and control orientation is sure to find further discoveries in the coming years. At this stage there is inherent competition between growth modes and domains, which provides an opportunity to design and program bias into the system to steer the system toward a desired architecture and function.

Going from simple well defined systems to complex ill-defined systems

Although great achievements have been demonstrated using well-defined systems and carefully controlled experimental conditions, we heard many examples of research groups moving toward more complex systems. We saw complexity in many different directions, including length scale, multiple components, hierarchy, environment, and stimuli. Some groups are focusing on systems with greater reversibility of bonding, such as work to improve domain size and order in covalent organic frameworks.[66] Large oligopeptides can be grown into films on gold using a layer by layer approach.[67] The complexity of the electrostatically driven assembly of two molecules in films allowed the system to be steered by the pH between two growth modes (DOI: 10.1039/c7fd00100b). The kinetically trapped copolymers behave as a viscoelastic material, which can be dissolved by treatment with acid or base, leading perhaps to stimuli responsive coatings with optical or sensor properties.

Natural systems are poorly defined on a macroscopic level, they are soft, malleable, and can take on a number of forms, but within this often amorphous form of living matter lies exquisite organisation and order. A particular length scale of interest is in the range of 10 nm to 1 μm; modern technologies struggle to synthesize and manufacture materials with strict control in this size range, yet this is the same size range where nature is the most interesting[68] – due to molecular self-assembly! Can we take on this kind of "fuzzy logic" in the design of interfaces? To truly understand these systems we will need new approaches to analyse them. Part of this will include facing the challenge of moving from simple, well-defined systems to complex, ill-defined systems. There are many experimental dimensions to explore; consider reversibility, temperature, time, size, stimuli, history, bonding, topography, length scales, multiple components, environment, hierarchy, *etc.* As pointed out in our Discussion, we are moving into deep experimental water!

Studies of complex systems not only moves the field into research questions that are close to applications, but can also reveal new fundamental insights that would not be accessible in simplified systems. For example, the Discussion included a study of epidermal cells at a liquid interface that was able to reveal the importance of nanoscale mechanical properties and how these impact the proliferation of the cells (DOI: 10.1039/c7fd00091j). Studies of peptide binding at metal oxide nanoparticle surfaces builds understanding of adsorptive interactions at the heart of even more complex protein/inorganic interfaces (DOI: 10.1039/c7fd00105c). Increasing complexity by moving toward multi-component systems allows for the study of intermolecular interactions and dynamics that will be essential at complex molecular interfaces, and the hydrogen bonded framework formed by perylenetetracarboxylic diimide and melamine allows the growth of defined chemisorbed layers[69] as well as to the nesting radial molecules that demonstrate dynamic characteristics in the wells (DOI: 10.1039/c7fd00115k).

Just as defects exist in assembled molecules, so too in molecules that are made on surfaces under different conditions. For example, the dehydrogenation of alkanes or aromatics leads to aligned structures on certain anisotropic surfaces,[70,71] and absolute precise control and understanding of the scope of these reactions is under rapid development. The formation of polymers can be guided when using the halide elimination (often referred to as Ullmann-type) to form covalently bonded systems, and the choice of halogen is very influential (DOI: 10.1039/c7fd00099e). The formation of graphene nanoribbons from perylene tetracboxylic dianhydride *via* decarboxylation is a case in point (DOI: 10.1039/c7fd00129k). The formation of ribbons proceeds *via* a cuprate, and these organometallic polymers were treated with gases in order to try and release the ribbons. The results show the challenges faced in this approach. In general, the disorder that can arise in certain cases at least, could be cleaned up by post-assembly annealing,[72] a method that might be interesting, in a general sense, for complex surfaces.

Rationalising multiple analytical inputs with theory

Stunning experimental results, particularly from scanning tunnelling microscopy, were presented throughout our Discussion, demonstrating the achievements that are being made by pressing analytical tools to new applications and better results. While specialization in one technique adds tremendous value to understanding a molecular system and is often essential to achieve new milestones in resolution and development of new analytical methods, we also recognized the value of bringing together multiple experimental techniques or experiment and theory. This was demonstrated in examples of multiple analytical methods by one group, but quite often these involve collaboration between different research teams, often including teams from different traditional training background fields.

There were many examples of experiment and theory collaboration or of the combination of different experimental analysis methods, such as structural and chemical analysis or local and global techniques, in our Discussion. For example, a detailed DFT study of metal–organic complexes that had been experimentally studied by STM and XPS was presented (DOI: 10.1039/c7fd00097a).[73] This provided detailed descriptions of atomic positions and charge transfer that would be impossible experimentally, but that provide important ideas about the design of other systems. In addition, the usefulness of Normal Incidence X-ray Standing Wave (NIXSW) was clearly revealed in conversations, an example being butylthiolate on gold shows that an adatom of the metal is involved in the structure.[74] The NIXSW measurements are quite unique in the information they provide. DART-MS was shown to provide information complementary to STM, even in ambient environments.[75] DART can analyse the composition of a surface layer (DOI: 10.1039/c7fd00068e) as demonstrated using fluorinated tags on the adsorbate layer. NMR relaxation methods were also presented as a tool for surface analysis to probe the reorientation dynamics near a surface as a probe of the adsorbed surface layer (DOI: 10.1039/c7fd00098g). It was evident throughout our Discussion that rationalizing multiple analytical inputs with theory leads to much deeper insight and better characterization of systems. In this case, the whole can be greater than the sum of the parts.

A key value in bringing together multiple analysis approaches to a complex problem such as molecular assemblies is that the different approaches will often reveal apparent contradictions. As researchers revisit the system to resolve and reconcile these contradictions, they can often develop a better understanding of the system and produce deeper insight that improves the overall quality of the research results. There are significant challenges in interdisciplinary collaboration, particularly when these span traditional disciplinary boundaries. True collaboration with intellectual engagement from multiple perspectives (not just routine analysis as a service) will enrich the research.

Structure and function

The properties of molecular systems at interfaces are generally addressed from two perspectives, either the single molecule or a continuous monolayer located between two bulk phases, including interphases. The function of the system depends on the composition and constitution of the molecules as well as on their supramolecular arrangement in both cases, and the most subtle changes in adsorption configuration can lead to very significant modulation in behaviour.

This feature has been seen in the conductivity exhibited by molecules chemisorbed on gold, where the putative molecular wires are quasi-perpendicular to the plane of the interface.[76] The flexibility in parts of these molecules can lead to slightly different conformations through, for example, the presence of fractional charge.[77] In the present discussion it was demonstrated through theory that the tilt angle due to packing can charge transport in molecules that have conformational flexibility (DOI: 10.1039/c7fd00101k). Transport properties of porphyrin assemblies were demonstrated to depend critically on the structure (DOI: 10.1039/c7fd00118e).[78] The results are important, because the slightest lack of control in the orientation and conformation of molecules at interfaces was proven to have very significant effects on the passage of electrons through them.

The complexity of the early stages of self-assembled monolayer growth can be made use of to effectively isolate secondary adsorbates in the holes left in this metastable primary adsorbate layer.[79] The approach facilitates the insulation of secondary adsorbates from like neighbours, permitting the study of their properties without interference from close contact effects that might modulate electronic conductivity. This idea is analogous to that proposed above for the study of the diffusion of molecules at surfaces, except here the studies can be performed in a liquid. The preparation method yields a completely different monolayer structure to the monolayer self-assembly using two thiols at the same time from the same solution. In this latter case, disordered monolayers are usually formed with a random distribution of both components. Molecular wires have had their conductance measured using this approach, combined with the attachment of gold nanoparticles to their outward-facing "end" to ensure effective contact.[80] The bottom-up assembly of this adsorbate-controlled system employing multiple supramolecular concepts is an elegant example of how complexity at interfaces can be harnessed.

Beyond the measurement of the electronic transport properties of molecules at interfaces, one of the questions raised at the start of this Discussion was: how can we make use of single molecule properties at an interface? Perhaps one of the most tried and tested is probing of the electronic transport properties of molecules at interfaces, with interest ranging from the preparation of rectifiers[81] to molecular switches.[82,83]

The mobility of ions to and from interfaces is of primordial importance in biological systems and man-made devices. Anions have been shown to bind, and be removed from cavities of bespoke supramolecular hosts held in monolayers.[84,]

[85] A model has been developed to explain the movement of the anions as changes in an STM tip (used to image the monolayers) that is able to cause motion in the layers (DOI: 10.1039/c7fd00104e). It is evident that the motion of ions in junctions such as these raises many fundamentally important questions, and the complexity of this and related systems awaits further exploration in coming years.

The localisation of charges within layers in compounds that are useful for their molecular electronic properties can be achieved at metal surfaces, as shown for the electron acceptor tetracyanoquinodimethane (TCNQ).[86] In the Discussion, precise measurements on the layer formed by the potassium salt of the reduced acceptor revealed that the anion presents a twist, similar to the case of the compound in the solid state (DOI: 10.1039/c7fd00093f). This conclusion was reached on the basis of NIXSW measurements, which proved essential in determining the precise nature of the adsorbed layer. This level of detail is surely not possible using microscopy techniques alone, and shows the care that is required in interpreting electronic states at interfaces.

From a practical viewpoint, one of the most important areas of study in complex interfaces is in the understanding and prevention of corrosion, or in general surface modification under certain environmental conditions. This could be done by the adsorption of surfactants to the affected surface,[87] or by the use of inorganic additives, such as sodium thiosulphate, that are used to prevent steel corrosion (DOI: 10.1039/c7fd00093f). The morphology of the films of the salt on steel, where a hierarchical assembly of the salt determines performance as an inhibitor of corrosion, was explored by electron microscopy, and the Discussion revealed that the transfer of ideas from the complex molecular materials area to this field is an interesting avenue to explore.

Perhaps one of the most promising areas for specific complex function at interfaces is in the interaction of synthetic materials with biological ones.[88] The control of interfaces in the diagnosis and treatment of maladies is increasingly studied and exploited. The variety of functions exposed by supramolecular materials[89] are available to study with the deepest precision using surface science techniques, and there is surely a window of opportunity to explore some of these fascinatingly complex systems at interfaces where their behaviour might deviate from that of the bulk.

Outlook and opportunities

Self-assembling systems were once regarded as highly complicated, as they are, but perhaps more complicated still are the cases where precise self-assembly of highly periodic and uniform structures does not occur. The localized information gathered from scanning probe microscopy can often give insight into poorly organized systems and allow for their redesign. In our Discussion, it was suggested that we need to pay more attention to the poorly ordered structures in our experiments and discuss those as they may teach us as much about intermolecular interactions and spontaneous organization as more visually-appealing ordered networks.

In general, adsorption to flat and ordered surfaces simplifies the task of understanding interactions between adsorbent and surface, and in general adsorption to oxide surfaces[90] has received less attention because of their poorer long range order and the difficulty of molecular resolution imaging, compared

Fig. 3 A photograph from the city centre in Sheffield, showing a regular flat façade we would like to have on a surface on the left, and an imperfect covering on the right. ©David Amabilino.

with metals. However, the practical importance of adsorption to metal oxides in areas ranging from catalysis to electronic devices and sensors[91] means that these complex interfaces are sure to receive increasing attention in years to come. In general, one could summarise the situation as shown in the photograph in Fig. 3! In a way, and difficult as it is, the scientific community in the area will gain enormously by seeing what is important rather than seeing what is easy to see. The most active sites on a surface regarding catalysis, or the properties of a molecule near a defect, might be most important and yet presently they are tremendously hard to observe. In order to advance and make our science useful, we must learn how to see and understand the complicated diversity in practical surfaces, or generate in a controlled way the most active or interesting regions of a surface in those systems. Evidently, interdisciplinary vision and understanding, from experiment to theory, are essential in order to address these extremely challenging goals.

Many of the studies required to populate a rule book for adsorption and diffusion are likely to require meticulous experiments and monotonous measurement of structural parameters at variable temperature. As far as the authors know, the automation of measurement and interpretation of data is not implemented as yet in the area of scanning probe microscopy, but these tools would allow more reliable data collection because of the possibility of huge data generation thereby reducing errors associated with the stochastic nature of many of the systems under study.

The mathematician Alan Turing expressed a sentiment very true of the area we have discussed: "We can only see a short distance ahead, but we can see plenty there that needs to be done". The outstanding interactions and participation in this meeting revealed the richness of the field before us.

Acknowledgements

We would like to thank deeply all the contributors to this Faraday Discussion. SLT acknowledges research support from the National Science Foundation (DMR 1533988). DBA thanks GSK, EPSRC and the School of Chemistry in Nottingham for supporting his present post.

References

1 I. Langmuir, *Chem. Rev.*, 1933, **13**, 147.
2 A. G. Slater, P. H. Beton and N. R. Champness, *Chem. Sci.*, 2011, **2**, 1440.
3 D. B. Amabilino, *Supramolecular Chemistry at Surfaces*, RSC, 2016.
4 P. A. Held, H. Fuchs and A. Studer, *Chem.–Eur. J.*, 2017, **23**, 5874.
5 R. Lindner and A. Kühnle, *ChemPhysChem*, 2015, **16**, 1582.
6 O. Ivasenko and D. F. Perepichka, *Chem. Soc. Rev.*, 2011, **40**, 191.
7 S. M. Barlow and R. Raval, *Surf. Sci. Rep.*, 2003, **50**, 201.
8 M. Linares, A. Minoia, P. Brocorens, D. Beljonne and R. Lazzaroni, *Chem. Soc. Rev.*, 2009, **38**, 806.
9 L. Kampschulte, T. L. Werblowsky, R. S. K. Kishore, M. Schmittel, W. M. Heckl and M. Lackinger, *J. Am. Chem. Soc.*, 2008, **130**, 8502.
10 W. Song, N. Martsinovich, W. M. Heckl and M. Lackinger, *Phys. Chem. Chem. Phys.*, 2014, **16**, 13239.
11 U. Mazur and K. W. Hipps, *Chem. Commun.*, 2015, **51**, 4737.
12 C. L. Perrin and J. B. Hielson, *Annu. Rev. Phys. Chem.*, 1997, **48**, 511.
13 T. Steiner, *Angew. Chem., Int. Ed.*, 2002, **41**, 48.
14 K. Brandhorst and J. Grunenberg, *Chem. Soc. Rev.*, 2008, **37**, 1558.
15 S. K. Pandey, D. Manogaran, S. Manogaran and H. F. Schaefer, *J. Phys. Chem. A*, 2017, **121**, 6090.
16 E. Pensa, E. Cortés, G. Corthey, P. Carro, C. Vericat, M. H. Fonticelli, G. Benítez, A. A. Rubert and R. C. Salvarezza, *Acc. Chem. Res.*, 2012, **45**, 1183.
17 E. Villarreal, G. G. Li, Q. Zhang, X. Fu and H. Wang, *Nano Lett.*, 2017, **17**, 4443.
18 T.-C. Tseng, C. Urban, *et al.*, *Nat. Chem.*, 2010, **2**, 374.
19 J. V. Barth, *Annu. Rev. Phys. Chem.*, 2007, **58**, 375.
20 J. A. Theobald, N. S. Oxtoby, M. A. Phillips, N. R. Champness and P. H. Beton, *Nature*, 2003, **424**, 1029.
21 K. Tahara, H. Yamaga, E. Ghijsens, K. Inukai, J. Adisoejoso, M. O. Blunt, S. De Feyter and Y. Tobe, *Nat. Chem.*, 2011, **3**, 714.
22 Z. Shi, J. Liu, T. Lin, F. Xia, P. N. Liu and N. Lin, *J. Am. Chem. Soc.*, 2011, **133**, 6150.
23 J. Liu, T. Lin, Z. Shi, F. Xia, L. Dong, P. N. Liu and N. Lin, *J. Am. Chem. Soc.*, 2011, **133**, 18760.
24 L. Zoppi, T. Bauert, J. S. Siegel, K. K. Baldridge and K. H. Ernst, *Phys. Chem. Chem. Phys.*, 2012, **14**, 13365.
25 L. Merz, M. Parschau, L. Zoppi, K. K. Baldridge, J. S. Siegel and K. H. Ernst, *Angew. Chem., Int. Ed.*, 2009, **48**, 1966.
26 D. Skomski and S. L. Tait, *J. Phys. Chem. C*, 2013, **117**, 2959.
27 A. Vallee, V. Humblot and C. M. Pradier, *Acc. Chem. Res.*, 2010, **43**, 1297.
28 A. González-Campo and D. B. Amabilino, *Top. Curr. Chem.*, 2013, **333**, 109.
29 V. Humblot, C. Méthivier, R. Raval and C. M. Pradier, *Surf. Sci.*, 2007, **601**, 4189.
30 A. G. Mark, M. Forster and R. Raval, *ChemPhysChem*, 2011, **12**, 1474.
31 C. Méthivier, H. Cruguel, D. Costa, C. M. Pradier and V. Humblot, *Langmuir*, 2016, **120**, 13759.
32 V. Humblot, A. Vallée, A. Naitabdi, F. Tielens and C.-M. Pradier, *J. Am. Chem. Soc.*, 2012, **134**, 6579.

33 V. Humblot, A. Tejeda, J. Landoulsi, A. Vallée, A. Naitabdi, A. Taleb and C.-M. Pradier, *Surf. Sci.*, 2014, **628**, 21.

34 Y. Lvov, G. Decher and H. Mohwald, *Langmuir*, 1993, **9**, 481.

35 J. A. Subirana and M. Soler-Lopez, *Annu. Rev. Biophys. Biomol. Struct.*, 2003, **32**, 27.

36 U. Mazur and K. W. Hipps, *Chem. Commun.*, 2015, **51**, 4737.

37 A. Bhattaral, U. Mazur and K. W. Hipps, *J. Am. Chem. Soc.*, 2014, **136**, 2142.

38 D. S. Karpovich and G. J. Blanchard, *Langmuir*, 1994, **10**, 3315.

39 R. Gutzler, L. Cardenas and F. Rosei, *Chem. Sci.*, 2011, **2**, 2290.

40 A. Bellec, C. Arrigoni, G. Schull, L. Douillard, C. Fiorini-Debuisschert, F. Mathevet, D. Kreher, A.-J. Attias and F. Charra, *J. Chem. Phys.*, 2011, **134**, 124702.

41 M. Wahl, M. Stöhr, H. Spillmann, T. A. Jung and L. H. Gade, *Chem. Commun.*, 2007, 1349.

42 D. Kühne, F. Klappenberger, W. Krenner, S. Klyatskaya, M. Ruben and J. V. Barth, *Proc. Natl. Acad. Sci. U. S. A.*, 2010, **107**, 21332.

43 T. Wang, S. Bhosale, S. Bhosale, G. Li and J. H. Fuhrhop, *Acc. Chem. Res.*, 2006, **39**, 498.

44 F. M. Raymo and J. F. Stoddart, *Chem. Ber.*, 1996, **129**, 981.

45 S. Haq, B. Wit, H. Sang, A. Floris, Y. Wang, J. Wang, L. Pérez-García, L. Kantorovitch, D. B. Amabilino and R. Raval, *Angew. Chem., Int. Ed.*, 2015, **54**, 7101.

46 S. C. Jensen, A. Shank, R. J. Madix and C. M. Friend, *ACS Nano*, 2012, **6**, 2925.

47 X. Bouju, F. Cherioux, S. Coget, G. Rapenne and F. Palmino, *Nanoscale*, 2013, **5**, 7005.

48 A. Floris, S. Haq, M. In't Veld, D. B. Amabilino, R. Raval and L. Kantorovich, *J. Am. Chem. Soc.*, 2016, **138**, 5837.

49 H. Walen, D.-J. Liu, J. Oh, H. J. Yang, Y. Kim and P. A. Thiel, *J. Phys. Chem. C*, 2015, **119**, 21000.

50 C. D. Tempas, D. Skomski and S. L. Tait, *Surf. Sci.*, 2016, **654**, 33.

51 M. Lackinger, S. Griessl, W. M. Heckl, M. Hietschold and G. W. Flynn, *Langmuir*, 2005, **21**, 4984.

52 J. M. Lehn, *Chem.–Eur. J.*, 1999, **5**, 2455.

53 Y. Yang and M. B. Zimmt, *J. Phys. Chem. B*, 2015, **119**, 7740.

54 B. E. Hirsch, K. P. McDonald, A. H. Flood and S. L. Tait, *J. Chem. Phys.*, 2015, **142**, 101914.

55 J. Wang, A. Kulago, W. R. Browne and B. L. Feringa, *J. Am. Chem. Soc.*, 2010, **132**, 4191.

56 S. Karthika, T. K. Radhakrishnan and P. Kalaichelvi, *Cryst. Growth Des.*, 2016, **16**, 6663.

57 N. Karpukhina, R. G. Hill and R. V. Law, *Chem. Soc. Rev.*, 2014, **43**, 2174.

58 D. B. Amabilino and X. Obradors, *Chem. Soc. Rev.*, 2014, **43**, 2009.

59 E. I. Altman and R. J. Colton, *Surf. Sci.*, 1992, **279**, 49.

60 A. Robin, P. Iavicoli, K. Wurst, M. S. Dyer, S. Haq, D. B. Amabilino and R. Raval, *Cryst. Growth Des.*, 2010, **10**, 4516.

61 T. Balandina, K. Tahara, N. Sändig, M. O. Blunt, J. Adisoejoso, S. Lei, F. Zerbetto, Y. Tobe and S. De Feyter, *ACS Nano*, 2012, **6**, 8381.

62 G. Reiter, *Chem. Soc. Rev.*, 2014, **43**, 2055.

63 A. Jabbarzadeh and R. I. Tanner, *Macromolecules*, 2010, **43**, 8136.

64 G. Lamberti, *Chem. Soc. Rev.*, 2014, **43**, 2240.
65 S.-L. Lee, Z. Yuan, L. Chen, K. S. Mali, K. Müllen and S. De Feyter, *J. Am. Chem. Soc.*, 2014, **136**, 4117.
66 Y. Hu, N. Goodeal, Y. Chen, A. M. Ganose, R. G. Palgrave, H. Bronstein and M. O. Blunt, *Chem. Commun.*, 2016, **52**, 9941.
67 H. Frisch, E.-C. Fritz, F. Stricker, L. Schmüser, D. Spitzer, T. Weidner, B. J. Ravoo and P. Besenius, *Angew. Chem., Int. Ed.*, 2016, **55**, 7242.
68 Scale of Things Chart, U. S. Department of Energy, https://science.energy.gov/bes/community-resources/scale-of-things-chart/, accessed 16 Sept. 2017.
69 R. Madueno, M. T. Räisänen, C. Silien and M. Buck, *Nature*, 2008, **454**, 618.
70 S. Haq, F. Hanke, M. S. Dyer, M. Persson, P. Iavicoli, D. B. Amabilino and R. Raval, *J. Am. Chem. Soc.*, 2011, **133**, 12031.
71 D. Zhong, J.-H. Franke, S. K. Podiyanachari, T. Blömker, H. Zhang, G. Kehr, G. Erker, H. Fuchs and L. Chi, *Science*, 2011, **334**, 213.
72 H. Wang, C. Zhang, J. Zhu, F. Fontein, Y. Wang, Y. Wang, H. Fuchs, W. Wang and L. Chi, *J. Mater. Chem. C*, 2017, **5**, 6672.
73 D. Skomski, C. D. Tempas, K. A. Smith and S. L. Tait, *J. Am. Chem. Soc.*, 2014, **136**, 9862.
74 A. Chaudhuri, D. C. Jackson, T. J. Lerotholi, R. G. Jones, T.-L. Lee, B. Detlefs and D. P. Woodruff, *Phys. Chem. Chem. Phys.*, 2010, **12**, 3229.
75 R. Sen, J. Escorihuela, M. M. J. Smulders and H. Zuilhof, *Langmuir*, 2016, **32**, 3412.
76 I. Baldea, Z. Xie and C. D. Frisbie, *Nanoscale*, 2015, **7**, 10465.
77 I. Baldea, *RSC Adv.*, 2016, **6**, 111903.
78 A. E. Schuckman, B. W. Ewers, L. H. Yu, J. P. C. Tomé, L. M. Pérez, C. M. Drain, J. G. Kushmerick and J. D. Batteas, *J. Phys. Chem. C*, 2015, **119**, 13569.
79 M. T. Cygan, T. D. Dunbar, J. J. Arnold, L. A. Bumm, N. F. Shedlock, T. P. Burgin, L. Jones, D. L. Allara, J. M. Tour and P. S. Weiss, *J. Am. Chem. Soc.*, 1998, **120**, 2721.
80 X. D. Cui, A. Primak, X. Zarate, J. Tomfohr, O. F. Sankey, A. L. Moore, T. A. Moore, D. Gust, G. Harris and S. M. Lindsay, *Science*, 2001, **294**, 571.
81 R. M. Metzger, *Chem. Rev.*, 2015, **115**, 5056.
82 V. Balzani, A. Credi and M. Venturi, *ChemPhysChem*, 2008, **9**, 202.
83 E. Coronado, P. Gavina and S. Tatay, *Chem. Soc. Rev.*, 2009, **38**, 1674.
84 B. E. Hirsch, S. Lee, B. Qiao, C. H. Chen, K. P. McDonald, S. L. Tait and A. H. Flood, *Chem. Commun.*, 2014, **50**, 9827.
85 S. Lee, B. E. Hirsch, Y. Liu, J. R. Dobscha, D. W. Burke, S. L. Tait and A. H. Flood, *Chem.–Eur. J.*, 2016, **22**, 560.
86 N. Abdurakhmanova, A. Floris, T. C. Tseng, A. Comisso, S. Stepanow, A. De Vita and K. Kern, *Nat. Commun.*, 2012, **3**, 940.
87 V. Pandarinathan, K. Lepkova, S. I. Bailey, T. Becker and R. Gubner, *Ind. Eng. Chem. Res.*, 2014, **53**, 5858.
88 *Nanostructured thin films and surfaces*, ed. C. Kumar, Wiley-VCH, 2010.
89 D. B. Amabilino, D. K. Smith and J. W. Steed, *Chem. Soc. Rev.*, 2017, **46**, 2404.
90 S. P. Pujari, L. Scheres, A. T. M. Marcelis and H. Zuilhof, *Angew. Chem., Int. Ed.*, 2014, **53**, 6322.
91 U. Diebold, *Surf. Sci. Rep.*, 2003, **48**, 53.

Faraday Discussions

DISCUSSIONS

Probing properties of molecule-based interface systems: general discussion and Discussion of the Concluding Remarks

David Amabilino, Ioan Bâldea, James Batteas, Peter Beton, Nerea Bilbao, Giovanni Costantini, Jonathan Davidson, Steven De Feyter, Yuri Diaz Fernandez, Karl-Heinz Ernst, Brandon Hirsch, Ahmad Jabbarzadeh, Robert Jones, Angelika Kühnle, Markus Lackinger, Zhi Li, Nian Lin, Trolle R. Linderoth, Natalia Martsinovich, Martin Nalbach, Claire-Marie Pradier, Talat Rahman, Rasmita Raval, Neil Robinson, Federico Rosei, Marco Sacchi, Mario Samperi, Ana Sanz Matias, Alex Saywell, Sebastian Schwaminger and Steven L. Tait

DOI: 10.1039/C7FD90077E

Ioan Bâldea opened a general discussion of the paper by Angelika Kühnle: Congratulations on the nice work. You mentioned the case of elevated temperatures, did you study more systematically the impact of temperature (T)? And how T-variations possibly affect the distance histogram?

Angelika Kühnle replied: Thanks for the question. We are currently investigating the impact of temperature on the distance distributions. In principle, there are two effects: at low temperatures, the molecules might not be mobile enough and the stripes will not adopt their optimum position. Thus, the distribution will have a larger standard deviation. For elevated temperatures, "excited states" can also be accessed, *i.e.*, a small modulation in the energy landscape due to repulsion might not be detectable above a certain temperature.

Ioan Bâldea said: Maybe in the experimental histograms (*e.g.* the blue bars in Fig. 3a–c, DOI: 10.1039/C7FD00089H) you can disentangle the physical effect(s) from the geometrical effects (grey bars in those figures) and focus on the (T-dependence of the) former. Anyway, presenting quantitative results on the effect of temperature on distance histograms would be interesting also for a theorist aiming to compare with your data.

Angelika Kühnle replied: The grey bars represent the stripe distance distribution that would be detected for randomly placed, non-interacting stripes. We show the comparison with our results to demonstrate that a repulsive force must exist between the stripes. We use the experimental data (blue bars) to get an estimate for the dipole moment strength that is needed to arrive at the experimentally obtained distance distribution. Regarding the temperature-dependent experiments, you are absolutely right. We are currently working on this point.

Markus Lackinger asked: Would you please explain how the geometric distribution is derived in more detail?

Angelika Kühnle answered: The geometric distribution is the distribution to be applied when you have two possible outcomes (in our case "row" *versus* "no row") and the probability of finding a row (and of not finding a row) is the same for all lattice sites. You can find a detailed derivation in the Supporting Information of the publication by Julia Neff.[1]

1 J. L. Neff, H. Söngen, R. Bechstein, P. Maass and A. Kühnle, Long-Range Order Induced by Intrinsic Repulsion on an Insulating Substrate, *J. Phys. Chem. C*, 2015, **119**, 24927.

Nian Lin queried: Have you done coverage-dependent analysis of chain–chain separation distribution to see how the separation changes?

Angelika Kühnle replied: Yes, we have studied this.[1] The stripe-to-stripe distance distribution is a direct function of the coverage. The more molecules you put on the surface, the closer the stripes get. What you also see in this analysis (Fig. 3 in the above-mentioned publication) is that the standard deviation of the distribution gets broader for decreasing coverage. However, the standard deviation is always significantly smaller than the standard deviation of the corresponding distribution of randomly placed, non-interacting stripes (Fig. 4a in the above-mentioned publication), which is by the way an alternative way of identifying the presence of a repulsive interaction in our system.

1 J. L. Neff, H. Söngen, R. Bechstein, P. Maass and A. Kühnle, Long-Range Order Induced by Intrinsic Repulsion on an Insulating Substrate, *J. Phys. Chem. C*, 2015, **119**, 24927.

Nian Lin continued: You have this dipole–dipole repulsion; how does this affect the length of the chains? Would the repulsion shorten the chains?

Angelika Kühnle replied: This is a very interesting point. Yes, both the width and the length of the molecular aggregates are steered by the balance between the short-range attraction and long-range repulsion in the respective directions. Therefore, as mentioned in my presentation, analyzing the island width and length distributions should allow us to elucidate the details of the (often anisotropic) molecule interaction potential. We currently follow this idea with the structures formed from 3-EBA molecules.

Steven De Feyter commented: You have looked at quite a large number of molecules, and you selected four of them to discuss. Do you have any idea about their orientation with respect to the substrate? What is the effect of different substituents? How does their orientation affect the repulsive inter-row interactions?

Angelika Kühnle responded: For some of the molecules tested, we have by now a rather detailed understanding of the binding configuration to the surface thanks to the efforts of our theory partners.[1–3] However, for the 3-HBA presented here, the present DFT calculations remain inconclusive so far. For the point I want to make here, however, I think it is important to note that the specific adsorption of the molecule towards the surface is not so decisive. Any binding towards the surface, no matter how simple or complex, will be associated with the formation of an electrical dipole perpendicular to the surface. This is the only point that matters.

Regarding the effect of different substituents, we aim to gain insights by systematically changing both number and nature of functional groups on a benzene core. We have, for example, studied all six isomers of dihydroxybenzoic acid.[4] The result shows that although the molecules are seemingly similar, the self-assembled structure can vary substantially for many different reasons. For example, the acidity of the molecule is an important parameter as a low pK_a value is often associated with a structural transition, which we associate to a change in binding from dominating molecule–molecule to a balanced molecule–molecule and molecule–surface interaction.[1]

1 C. Paris, A. Floris, S. Aeschlimann, M. Kittelmann, F. Kling, R. Bechstein, L. Kantorovich and A. Kühnle, Increasing the Templating Effect on a Bulk Insulator Surface: From a Kinetically Trapped to a Thermodynamically More Stable Structure, *J. Phys. Chem. C*, 2016, **120**, 17546.
2 C. Guo, Y. Wang, M. Kittelmann, L. Kantorovitch, A. Kühnle and A. Floris, Mechanisms of covalent assembly on a bulk insulating surface, *J. Phys. Chem. C*, 2017, **121**, 10053.
3 A. Richter, V. Haapasilta, C. Venturini, R. Bechstein, A. Gourdon, A. S. Foster and A. Kühnle, Diacetylene polymerization on a bulk insulator surface, *Phys. Chem. Chem. Phys.*, 2017, **19**, 15172.
4 J. L. Neff, M. Kittelmann, R. Bechstein and A. Kühnle, Decisive Influence of Substitution Positions in Molecular Self-Assembly, *Phys. Chem. Chem. Phys.*, 2014, **16**, 15437.

Steven De Feyter asked: Have you tried deposition of the molecules at the water/calcite interface? If so, what is the effect? Did you observe any adsorption?

Angelika Kühnle answered: Thanks for this very relevant question. Yes, we investigated many of the molecules that formed beautiful structures on calcite (10.4) kept in UHV also at the calcite–water interface. Actually, it was Martin Nalbach sitting here in the audience who tried a lot of them. However, for most of these small benzoic acid derivatives, Martin showed that none of them could form a nice self-assembled structure at the calcite–water interface. This finding again underlines the impact of the solvent on molecular self-assembly. The poster presented by Martin showing the assembly of benzopurpurine (BPP) at the calcite–water interface illustrates the complexity of the situation in liquids. Ongoing molecular dynamics simulations carried out by Paolo Raiteri suggest that BPP, which forms a self-assembled structure, does not anchor to the calcite surface directly. So, you cannot understand the self-assembly process at the solid–liquid interface without considering the impact of the solvent and, in particular, the interfacial solvation structure. This makes the situation rather complex and hampers a direct transfer of knowledge gain in UHV to the situation in liquids. We clearly lack understanding and predictive power when it comes to solid–liquid interfaces.

Giovanni Costantini queried: Have you studied the effect of molecular coverage on the width of the molecular chains?

Do the molecular chains thicken before they are "pushed together" by increasing coverage? I believe this could give you information on the interplay between repulsive and attractive interactions regulating the self-assembly of these systems.

Angelika Kühnle answered: Yes, we have studied the coverage dependence. You find that the molecular rows move closer as a function of coverage as the average stripe-to-stripe distance is a direct function of the coverage. The molecular double rows of 3-HBA do not get wider because the structure is defined by the internal structure. We do see, however, that the rows are pushed together.

In this context, the result from 3-EBA is interesting, because this molecule shows islands with (1×1) internal structure. As mentioned in my talk, we want to use this molecule to do exactly what you propose, namely get insights into the balance of short-range attraction and long-range repulsion by analyzing the island length and width distribution.

1 J. L. Neff, H. Söngen, R. Bechstein, P. Maass and A. Kühnle, Long-Range Order Induced by Intrinsic Repulsion on an Insulating Substrate, *J. Phys. Chem. C*, 2015, **119**, 24927.

Trolle R. Linderoth commented: Your data demonstrate that there is repulsion between the 1D molecular chains on the calcite surface. But could you be more specific about what is being repelled? In Fig. 2 (DOI: 10.1039/C7FD00089H), you show a series of AFM images which demonstrate that the chains do not move on the surface but rather there is a continuous decay/growth process with new nucleation events occurring where presumably the repulsive interactions involve the individual molecular species during this process and create, for instance, an enhanced nucleation possibility at some optimum distance from another chain. Could you clarify if this is the way we should think about the underlying dynamics?

Angelika Kühnle responded: We cannot follow the nucleation event in our AFM images, however, I think it is important to note that we did an experiment where we deposited additional molecules to a sample where stripes were already present (with a stripe-to-stripe distance distribution associated with the initial coverage). In such a case, the stripes have to re-assemble to adjust to the increased coverage after the second deposition. Indeed, after the second deposition, the stripes have adopted their stripe-to-stripe distance distribution from the increased coverage. This experiment indicates that the rows can move. We do not, however, think that the rows move as a whole. From the AFM movie, our understanding is that the rows form by attachment and detachment of diffusing monomers (or perhaps clusters consisting of a certain critical monomer number). Thus, when considering the existence of a row at a given position, the electrostatic potential built up by this row will dictate the nucleation of a new row in the neighborhood. We used this picture for our simplified Monte Carlo simulations explained in the Supporting Information of the publication by Julia Neff.[1]

1 J. L. Neff, H. Söngen, R. Bechstein, P. Maass and A. Kühnle, Long-Range Order Induced by Intrinsic Repulsion on an Insulating Substrate, *J. Phys. Chem. C*, 2015, **119**, 24927.

Rasmita Raval asked: Question 1, what is understood about the impact that molecules have on calcite surfaces? In biomineralisation, calcite surfaces can be restructured substantially by adsorbing molecules.

Question 2, do Monte Carlo simulations give an indication of the strength of the interactions that could yield the self-assembly stripes? Does one have to assume strong electrostatic repulsion? Presumably, weaker repulsive interactions would yield the same behaviour?

Angelika Kühnle answered: Regarding your first question, many of these benzoic acid derivatives are known to restructure the surface when added to the solid–liquid interface. In UHV, however, we have not observed massive restructuring of the surface upon adsorption of these molecules.

Regarding your second question, correct, the electrostatic interaction needed can be explained by dipole–dipole repulsion of dipoles perpendicular to the surface that are reasonable for adsorption-induced dipoles. Our MC simulations yield a value for $\sqrt{n^*p}$ of 30 Debye, which can be achieved, *e.g.*, with molecules forming a dipole moment of 5 Debye and a critical stripe length of 7 nm.

David Amabilino said: Beautiful work, thanks for the presentation. Regarding the last point, you said you find it hard to think of something other than electrostatics, but in one way that's surprising – there's a whole range of length scales between your strips, but on the other hand, charges are very localised on your surface. This reminds me of a case on much larger molecules,[1] they saw stripes, direction of growth of an aggregate, and they saw a very strict dependence on concentration. There, their explanation of that was a dewetting phenomenon from the surface, which in a way could come from nucleation and directional growth. Could that phenomenon (bearing in mind the relatively hydrophobic thing on a quite polar surface you've got here) explain your results?

1 R. van Hameren, P. Schön, A. M. van Buul, J. Hoogboom, S. V. Lazarenko, J. W. Gerritsen, H. Engelkamp, P. C. M. Christianen, H. A. Heus, J. C. Maan, T. Rasing, S. Speller, A. E. Rowan, J. A. A. W. Elemans and R. J. M. Nolte, *Science*, 2006, **314**, 1433–1436.

Angelika Kühnle replied: Thanks for this suggestion. I want to stress that we take care to exclude correlated growth effects. The explanation I presented is based on thermodynamics, not kinetic effects. We verify this for example by deposition of further molecules on a sample were there are already stripes present. When you add more molecules, the stripe distance distribution is changed to the new coverage, demonstrating sufficient mobility for the stripes to adapt to the new situation. Therefore, the observed stripe-to-stripe distance distribution is most likely not an effect that is caused by kinetic effects during growth.

David Amabilino continued: The formula you present explains the phenomenon, I agree, perhaps you might put a number of factors in there; given you're using a surface with very localised charges, can you justify that something on top of that very charge localised surface (that is insulating) can mediate an electrostatic interaction over such a long range?

Angelika Kühnle replied: I do not want to claim that the ordering we observe must originate from electrostatic repulsion of dipoles. The point I want to make is, however, that electrostatic repulsion can be a possible explanation. In the beginning, I thought that electrostatic repulsion would not be sufficient to explain repulsion over a 16 nm distance. But a simple back-of-the-envelope estimate[1] shows that the dipole moment needed to explain the ordering seen for 3-HBA can be explained by dipole moments with reasonable dipole moment strength. So, this explanation appears most appealing to me.

At present, I cannot think of another equally obvious origin. But they might be there.

1 J. L. Neff, H. Söngen, R. Bechstein, P. Maass and A. Kühnle, Long-Range Order Induced by Intrinsic Repulsion on an Insulating Substrate, *J. Phys. Chem. C*, 2015, **119**, 24927.

Robert Jones opened a general discussion of the paper by Karl-Heinz Ernst: The terphenylcorannulene (TPC) shown in Fig. 1b and c (DOI: 10.1039/C7FD00109F), has three outer edge benzo rings, as well as six C_6 rings at the bottom of the bowl. For isolated adsorbed molecules, where there are no interactions between TCP molecules, STM showed that the cup shaped molecule adsorbs with the middle of the outer edge benzo ring parallel to the surface. *i.e.* the cup adsorbs on the middle of its edge and stands up on the surface. You say that a constraint of the central terphenyl C_6 ring is required in an AMBER force field calculation to achieve the experimentally observed tilting of the molecule. Could you comment a little more on why the bonding of the isolated molecule occurs *via* this middle outer edge benzo group, rather than *via* one of the inner C_6 rings?

Karl-Heinz Ernst responded: I can only speculate. The middle ring of the terphenyl group being parallel is strongly suggested by the STM contrast. The AMBER calculations were performed to see if the π–π interaction, as suggested by the geometric alignment in the dimer, is indeed also the result of the calculation. But with AMBER we got a little bit "less" tilted configuration, not compatible with the observation. Hence the constraint (Surface–molecule interaction is often not well described in force field calculations, but in few cases works quite well).

Now why the middle ring? All I can say is that it might be due to minimalization of the Pauli repulsion. There must be optimum binding of the aromatic molecule and the electrons of a metal pushing away from the adsorbate spot. Certainly a matter of polarizability.

Marco Sacchi commented: Wonderful results. In order to understand the relative importance of vdW interactions and dipole–dipole interactions, what is the intermolecular distance in the calculations? I agree that buckybowls are fairly big molecules, therefore I suppose that the unit cell has been chosen to be as large as possible. Also, is there strong charge transfer from the surface to the molecule?

Karl-Heinz Ernst responded: See Fig. 1. It was a 16 by 20 angstrom cell.

No, there is no charge transfer (judged by the density of states calculation), only rearrangement of the charge due to Pauli repulsion.

Brandon Hirsch asked: How do these molecules begin to stack into the 3D structure?

Karl-Heinz Ernst answered: Some of these bowls (corannulene and pentamethylcorannulene) were found to show surface-assisted bowl-in-bowl stacking. This has not been observed as 3D crystals.[1]

1 T. Bauert, K. K. Baldridge, J. S. Siegel and K.-H. Ernst, Surface-assisted bowl-in-bowl stacking of nonplanar aromatic hydrocarbons, *Chem. Commun.*, 2011, **47**, 7995–7997.

Giovanni Costantini said: You were saying that if you increase the coverage, at some point the molecules are forced together. Is it possible that this is happening together with a rearrangement of the molecule? Maybe a situation where the molecules are slightly less (or more) parallel to the surface, thereby effectively reducing the Pauli repulsion induced dipole?

Karl-Heinz Ernst responded: Actually, the π–π interaction is already stronger than the dipole–dipole interaction, because we see them joining together at low coverage. Within the possibilities of STM, we did not see a reorientation. From calculations with helicenes, we also see, that π–π interaction is stronger than dipole–dipole repulsion. Interface dipoles induced by Pauli repulsion depend strongly on polarizability. For metals this is usually large. But the same applies to depolarization, when two molecules come closer. So I do not see a strong dipole–dipole repulsion mechanism here.

Giovanni Costantini continued: Have you done DFT calculations for a monolayer? I think it would be extremely interesting to see whether the closeness of neighbouring molecules, and thus the increased dipole–dipole repulsion, induces

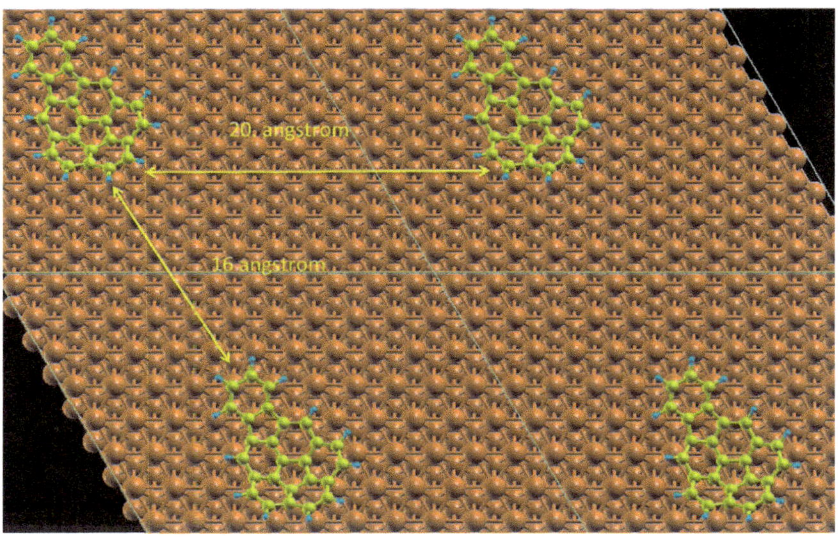

Fig. 1 Representation of the 16 by 20 Å unit cell of indenocorannulene used in the DFT-D calculations.

a (potentially slightly) different adsorption configuration that reduces the Pauli repulsion and thus also the magnitude of the dipole.

Karl-Heinz Ernst answered: No we haven't. I believe that is a very expensive calculation, if feasible at all.

Because we see already for small clusters and dimers a very close arrangement in experiments, dipole–dipole repulsion (of this type of Pauli repulsion induced interface dipole) does not seem to come into play. Calculations for pentahelicene show a similar effect for monomers (4.5 Debye). Pentahelicene dimers just show a push-back effect for the entire dimer (7.5 Debye), as if the two entities would fuse together.

Peter Beton commented: I was also going to ask about whether some of the effects we heard in the previous talk were relevant to this ordering in view of the induced dipole. You mentioned in the introduction that these materials are potentially relevant for organic transistors or photovoltaics. Have they actually been used in that form? Is the material available in quantities that might allow some kind of systematic study? Also, what might be the advantages? Replacing PCBM?

Karl-Heinz Ernst replied: The better solubility should make them more interesting than PCBM. However, the matching of electronic levels of acceptor and donor in photovoltaics has not been successful so far. We tested corannulene *versus* pentacene in an organic transistor set-up. Pentacene was much better. Corannulene might become available in sufficient quantities, because an upscaled synthesis scheme has been presented recently.

Robert Jones remarked: Fig. 4(a) (DOI: 10.1039/C7FD00109F) shows a beautiful rectangular array of over 100 TPC molecules surrounded by a large area of disordered molecules. It is striking that within the image, there appears to be no other highly ordered arrays of TPC molecules but with smaller numbers of molecules. One might have expected to see a distribution of sizes for the ordered structures. Could you comment on why there is such a large difference between random structure and one large, very highly ordered structure?

Karl-Heinz Ernst replied: This can be explained by a low rate of nucleation. The probability of having a nucleus that allows further growth was, at that coverage, low.

In the area shown, only one nucleus formed at a certain temperature. (Note that is was a cooling experiment, so the temperature was falling permanently). With temperature falling further, the 2D gas then solidified into a disordered "glass". Due to limited mobility, at that temperature, no further nucleation and growth was possible. We have seen this often for other systems at low or intermediate coverages.

Rasmita Raval opened a general discussion of the paper by Neil Robinson: Thank you for your nice talk. You talk about T_1 timescales. Did you consider T_2 timescales? Are they not important in this system?

Neil Robinson responded: T_1 and T_2 are sensitive to different frequencies of molecular motion. In particular, T_1 is sensitive to motions occurring at the Larmor frequency ($\omega_0 \propto$ magnetic field strength) and above, $T_1^{-1} \propto J(\omega_0) + 4J(2\omega_0)$, while T_2 is sensitive to slower motions, such as those occurring at zero frequency, $T_2^{-1} \propto 3J(0) + 5J(\omega_0) + 2J(2\omega_0)$. In the present work we have chosen to observe T_1 relaxation at high magnetic field in order to access fast molecular dynamics, which may be associated with molecular tumbling motions. We have not probed T_2 relaxation in the current experiments. T_2 is an interesting parameter in terms of probing the dynamics of molecules at catalyst surfaces; for example, the sensitivity of T_2 relaxation to slow molecular motion gives measurements access to the translational motions of molecules across the pore surface. In particular, T_2 measurements may be combined with T_1 relaxation analysis – either through separate experiments on the same system, or through the use of two-dimensional relaxation correlation experiments – in order to generate a T_1/T_2 ratio. It has recently been shown that such a metric provides a quantitative measure of surface affinity.[1] However, the measurement of T_2 relaxation times associated with adsorbed species is far from trivial at high magnetic field. A particular problem is the existence of internal magnetic field gradients, which are generated by differences in magnetic susceptibility across the solid–liquid interface. Diffusion through these internal gradients leads to enhanced rates of T_2 relaxation.[2] As a result, it is typical that only an effective T_2 may be measured for liquid-saturated porous media investigated at high field. In turn, T_1/T_2 ratios are typically established at low or intermediate magnetic field strengths, rather than the high field used in this particular work. Although techniques to correct for the presence of these internal gradients exist, such methods are highly non-trivial.[3] It would be very interesting to perform low field T_1/T_2 measurements on the systems I have described in order to assess the differences in surface affinity upon hydroxyl passivation.

1 C. D'Agostino, J. Mitchell, M. D. Mantle and L. F. Gladden, Interpretation of NMR Relaxation as a Tool for Characterising the Adsorption Strength of Liquids inside Porous Materials, *Chem.–Eur. J.*, 2014, **20**, 13009–13015.
2 J. Mitchell, T. C. Chandrasekera, M. L. Johns, L. F. Gladden and E. J. Fordham, magnetic resonance relaxation and diffusion in the presence of internal gradients: The effect of magnetic field strength, *Phys. Rev. E*, 2010, **81**, 026101.
3 J. Mitchell and T. C. Chandrasekera, Understanding generalized inversions of nuclear magnetic resonance transverse relaxation time in porous media, *J. Chem. Phys.*, 2014, **141**, 224201.

David Amabilino commented: Thanks for the talk. You've shown nicely how this works for porous materials. A big problem, I guess you've picked up over the last couple of days, is that a lot of us are working with monolayers. Looking at those spectroscopically, *e.g.* with XPS, is very challenging. I know people who have looked at monolayers with, *e.g.* EPR. How could your techniques help study adsorption in the kind of system that's more a flat surface than a porous one?

Neil Robinson replied: Although the general theory of surface sensitive relaxation in saturated porous media relies on the fast exchange of molecules between bulk and adsorbed states, there is nothing I can see, theoretically, to prevent the investigation of monolayer coverage over flat surfaces. In this case, the observed relaxation would move from the biphasic fast exchange description of Brownstein and Tarr (as described in our paper, DOI: 10.1039/C7FD00098G), to the observation of only surface relaxation. A similar condition would be found in liquid-saturated microporous media, where very small pore diameters mean only surface-bound species are present. That being said, NMR is an inherently insensitive spectroscopic technique. The limiting factor in performing NMR measurements is often gaining adequate signal (or more correctly, signal-to-noise ratio, SNR) during the experimental time available. SNR is a complex function of NMR-active nuclear spin density within the sample, magnetic field strength, gyromagnetic ratio of the spins under observation, and the number of repeat scans used throughout a given experiment. The high porosity of fluid-saturated meso- and microporous media is necessary to provide adequate SNR in systems where we have an interest in the relaxation of adsorbed species, and the high surface-to-volume ratios associated with such structures, combined with the enhanced rates of relaxation at the pore surface, ensures the surface sensitivity of such measurements. The investigation of monolayer and sub-monolayer coverages on flat surfaces may therefore prove difficult, simply due to the lack of nuclear spins present across the surface of interest. The polarization achieved by a given magnetic field strength may, however, be dramatically enhanced beyond thermodynamically accessible states by so-called hyperpolarization techniques, leading to greater SNR.[1] The proton (^{1}H) hyperpolarization of hydrocarbon species would typically involve synthesis of the molecule of interest through a heterogeneous hydrogenation reaction, involving the reaction of a gaseous feed with parahydrogen using an immobilized paramagnetic catalyst.[2] Furthermore, the SNR of ^{13}C NMR experiments – the relaxation of which has been shown to be useful in determining adsorbate orientation[3] – may be enhanced by transferring polarization to the carbon nucleus from neighbouring protons.[4,5]

One condition I would like to stress is that the surfaces investigated should not be ferromagnetic or contain significant quantities of paramagnetic ions, as interactions with paramagnetic species will have dramatic effects on the relaxation characteristics of molecules under investigation.[6,7]

1 D. A. Barskiy *et al.*, NMR Hyperpolarization Techniques of Gases, *Chem.–Eur. J.*, 2017, **23**, 725–751.
2 L.-S. Bouchard *et al.*, Para-Hydrogen-Enhanced Hyperpolarized Gas-Phase Magnetic Resonance Imaging, *Angew. Chem., Int. Ed.*, 2007, **46**, 4064–4068.
3 P. A. Vecino *et al.*, Determining adsorbate configuration on alumina surfaces with ^{13}C nuclear magnetic resonance relaxation time analysis, *Phys. Chem. Chem. Phys.*, 2015, **17**, 20830–20839.
4 G. A. Morris and R. Freeman, Enhancement of nuclear magnetic resonance signals by polarization transfer, *J. Am. Chem. Soc.*, 1979, **101**, 760–762.
5 D. M. Doddrell, D. T. Pegg and M. R. Bendall, Distortionless enhancement of NMR signals by polarization transfer, *J. Magn. Reson.*, 1982, **48**, 323–327.
6 P. J. McDonald, J.-P. Korb, J. Mitchell and L. Monteilhet, Surface relaxation and chemical exchange in hydrating cement pastes: A two-dimensional NMR relaxation study, *Phys. Rev. E*, 2005, **72**, 011409.
7 C. D'Agostino, P. Bräuer, P. Charoen-Rajapark, M. D. Crouch and L. F. Gladden, Effect of paramagnetic species on T_1, T_2 and T_1/T_2 NMR relaxation times of liquids in porous $CuSO_4/Al_2O_3$, *RSC Adv.*, 2017, **7**, 36163–36167.

David Amabilino remarked: In EPR, you cut the square surface into lots of pieces and put one on top of the other and then put the stack into the spectrometer. I wonder if that could be used with a kind of NMR spectroscopy, perhaps using T_2, but for sure it is a challenge, but could be useful.

Neil Robinson answered: Stacking surfaces of interest and trapping molecules between them sounds like a porous structure is being generated. If this is the case, then relaxation experiments may well be applied in order to probe the dynamics of adsorbate molecules in such a system. The issue of low SNR may again arise due to the small number of spins present within each molecular monolayer; however, this limitation may be circumvented to some extent by using many layers stacked upon one another, leading to a reasonable number of spins being present within the sample as a whole. I would be more tempted to use T_1 rather than T_2, however, as the T_2 relaxation associated with such a system may well be too rapid to measure.

Sebastian Schwaminger said: You now work with a completely saturated system, but usually do mass transport effects also affect the tumbling of molecules in your measurements? Concerning your system, do pore sizes have a stronger effect on the molecular movement than the surface groups?

Neil Robinson replied: Interestingly, it has been observed in previous work (using unpassivated oxides) that changes to T_1 relaxation and the molecular self-diffusion coefficient are not independent upon imbibition of a liquid throughout a mesoporous network.[1] While relaxation in the bulk liquid does indeed contain an intermolecular contribution associated with molecular self-diffusion, this contribution is not necessarily present within saturated mesoporous solids. Here we typically observed only relaxation associated with the adsorbed surface layer, diffusion through which will be slow.

To a large extent, it is not the pore size which affects the effective self-diffusion coefficient of liquids through a porous system. Rather, such dynamics are far more susceptible to the interconnectivity of the porous network – which may be somewhat dependent on the pore sizes present – defined by a dimensionless metric known as the tortuosity. Non-negligible adsorption interactions occurring between functional molecules and hydroxylated pore surfaces have been shown to significantly alter the observed self-diffusivities of such molecules, relative to the tortuosity of the porous structure,[1] and we are currently performing pulsed-field-gradient NMR diffusion experiments on the materials discussed in order to quantify the influence of passivating alkyl groups on such interactions.

1 C. D'Agostino, J. Mitchell, L. F. Gladden and M. D. Mantle, Hydrogen Bonding Network Disruption in Mesoporous Catalyst Supports Probed by PFG-NMR Diffusometry and NMR Relaxometry, *J. Phys. Chem. C*, 2012, **116**, 8975–8982.

Robert Jones commented: The DRIFTS spectra in Fig. 1 (DOI: 10.1039/C7FD00098G) clearly show that the OH stretch reduces in intensity when the surfaces have been passivated. However, as the scans are relative changes in absorbance, it is not clear whether this is a partial or a complete removal of the peak. Could you comment on whether the OH signals were completely extinguished, implying complete passivation of the surface, or whether some OH signal remained after passivation, implying that there was still some concentration of surface OH within the pores.

Neil Robinson replied: Making decisive conclusions concerning whether complete passivation was achieved is unfortunately not a trivial task using DRIFTS, even from examination of the individual spectra before and after passivation, due to the complex nature of the baseline. I include, as an example, the raw absorption data gathered from our anatase titania samples in the attached Fig. 2. Indeed, quantitative analysis using DRIFTS is notoriously difficult, requiring reproducibly-milled and packed particles below 1 μm in diameter, high sample dilution, and conversion of the acquired absorbance data using the Kulbeka–Mulk equation. In the present work, samples were ground by hand using a pestle and mortar, and without detailed analysis of the resulting particle size distribution. It is therefore ill-advised to attempt a quantitative analysis of our spectra, which we present only as a qualitative confirmation of successful hydroxyl passivation. It is worth noting that the currently established theory of surface relaxation predicts a subtle dependence on surface hydroxyl coverage. In particular, the influence of surface hydroxyl density on our observed relaxation is likely to be magnetic field dependent. At high magnetic field strength, as used in the present work, T_1 relaxation is deemed to be dominated by fast motions such as molecular tumbling. The prefactors in eqn (1) from our paper (DOI: 10.1039/C7FD00098G) are therefore related to the intramolecular dipole–dipole coupling of protons in the liquid phase (although it should be noted that at no point have we used this equation analytically). At low magnetic field strength, however, relaxation is sensitive to slower motions. In the case of liquid-saturated mesoporous solids, the two-dimensional translation of adsorbed molecules across hydroxylated pore surfaces becomes a more significant source of relaxation. A general spectral density expression for relaxation rates under such conditions has been developed by Korb and co-workers (see, for example, ref. 1) and applied to catalyst surfaces by Mitchell *et al.*[2] Here, the spectral density prefactor is directly dependent on the number of relaxation sinks (adsorption

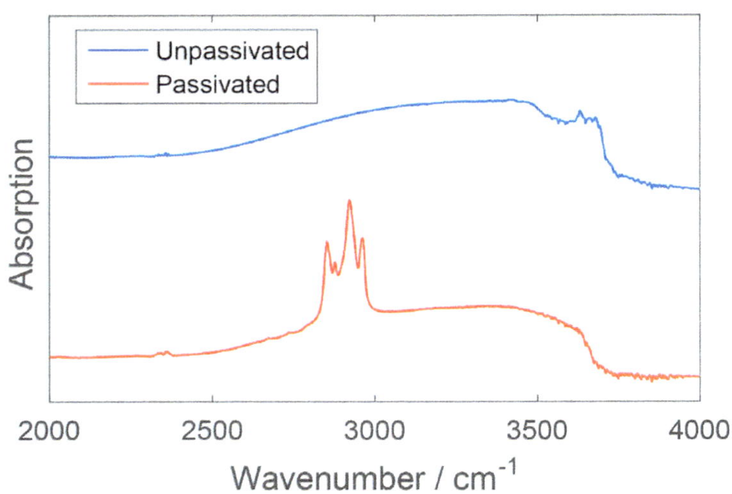

Fig. 2 Diffuse reflectance infrared spectra of anatase titania pre- and post-passivation.

sites) at the surface, which will be related to the number of surface hydroxyl groups (although not necessarily directly proportional, due to the possibility of polydentate adsorption sites involving rotationally mobile surface hydroxyls). This being said, it was not the purpose of the present work to explore relaxation as a continuous function of hydroxyl coverage. Such work would best be performed on a single oxide rather than across a range of different materials. Although beyond the scope of the current work, a low field investigation of the influence of surface hydroxyl density on methanol relaxation would be enlightening.

1 S. Godefroy, J.-P. Korb, M. Fleury and R. G. Bryant, *Phys. Rev. E*, 2001, **64**, 021605.
2 J. Mitchell, L. M. Broche, T. C. Chandrasekera, D. J. Lurie and L. F. Gladden, *J. Phys. Chem. C*, 2013, **117**, 17699–17706.

Karl-Heinz Ernst opened a general discussion of the topics discussed during the session: Coming back to calcite. I call myself a stereochemist and I wonder how exactly modification of surface morphology is achieved by modifiers, like tartaric acid. Is it just the footprint of the molecules or rather the free energy situation at step edges? And wouldn't it be very interesting to study this kind of molecules?

Angelika Kühnle responded: Many people study the impact of organic modifiers (amino acids, peptides, polymers and others) on the surface morphology of calcite. You can find a wide range of results depending on the specific molecule used, *e.g.*, polar step edges that get stabilized[1,2] or surface reconstructions with etch pits that greatly differ from the thermodynamically most stable rhombohedral shape.

So far, however, we largely lack predictive power. We have not studied tartaric acid yet. However, my guess would be that we do not see a change in the calcite–water interface when adding tartaric acid. But – as I said – we need to test as we do not have predictive power so far.

1 M. Nalbach, S. Klassen, R. Bechstein and A. Kühnle, Molecular Self-Assembly Versus Surface Restructuring During Calcite Dissolution, *Langmuir*, 2016, **32**, 9975.
2 R. Momper, M. Nalbach, K. Lichtenstein, R. Bechstein and A. Kühnle, Stabilization of polar step edges on calcite (10.4) by the adsorption of Congo Red, *Langmuir*, 2015, **31**, 7283.

Angelika Kühnle commented: We have studied a large range of organic molecules at the calcite–water interface. A major finding is that results obtained from UHV investigations cannot be directly transferred to the solid–liquid interface. The presence of the water molecule severely alters the situation. We have to consider effects like molecule deprotonation depending on the solution pH, complexation with calcium ions that inevitably exist in solution when immersing a calcite crystal. Also, the details of the hydration structure at the surface is decisive. Again, the example of benzopurpurine shown on Martin Nalbach's poster nicely illustrates this: at the calcite–water interface, the molecules do not even anchor to the calcite directly but appear to "float" on top of the second hydration layer as indicated by ongoing molecular dynamics simulations. Moreover, at the calcite–water interface, we often observe a massive restructuring of the surface by the presence of additives (*e.g.* in ref. 1). Many of these effects are poorly understood and we are currently not in the status to predict what to expect when adding a given molecule to the calcite–water interface.

1 Nalbach, S. Klassen, R. Bechstein and A. Kühnle, Molecular Self-Assembly Versus Surface Restructuring During Calcite Dissolution, *Langmuir*, 2016, **32**, 9975.

Karl-Heinz Ernst noted: Tartaric acid is a strong calcite modifier, like other dicarboxylic acids such as maleic acid, aspartic acid or gamma carboxyl glutamic acid, as has been shown in the work of Stephen Mann *et al.*[1]

1 S. Mann, J. M. Didymus, N. P. Sanderson, B. R. Heywood and E. J. A. Samper, *J. Chem. Soc., Faraday Trans.*, 1990, **86**, 1873–1880.

Karl-Heinz Ernst remarked: We and others have shown that the metal-UHV is a very nice model system for macroscopic rearrangements of the system. Amino acids on Cu, for example, induce upon annealing, strong stereochemical effect by restructuring and stabilizing only certain facets. Similar effects have been observed in the biomineralization of calcite and calcium oxalate crystallizing in the environment of tartaric acid with certain peptides.

Angelika Kühnle said: When discussing the impact of organic molecules on the calcite morphology, we might also want to differentiate between work where an existing flat (10.4) surface is reconstructed by the presence of organic molecules and studies in which crystals grow from a supersaturated calcium carbonate solution in the presence of organic additives. Also for the latter, a large body of literature exists.

Karl-Heinz Ernst contributed: But it is hard to control purity and avoid crystallization effects due to impurities.

Zhi Li opened a general discussion of the paper by Federico Rosei: I have a question about the substrate effect. You mentioned that the organometallic chains follow certain directions, therefore, have you evaluated or what do you expect the impact of the substrate, with different orientation index, to be on the polymerization efficiency, based on the polymer length?

Federico Rosei responded: While systematic studies on the effect of the substrate are not available, various molecules have been studied on different substrate orientations, such as 1,4-diiodobenzene (dIB), which we studied on Cu(111), Cu(100) and Cu(110).[1] In all cases, polyparaphenylene chains were obtained, however, as also reported in the present work (DOI: 10.1039/C7FD00099E) and by Vasseur *et al.*,[2] the orientation and length of the PPP chains on the surface can strongly affect their properties, such as band gap. The surface is therefore to be considered as a parameter controlling the reaction, not only due to its catalytic effect but also in terms of the directions of the surface atoms in driving the molecular assembly. In this perspective, vicinal surfaces could allow the chains to grow in only one direction,[3] while different geometries (square or hexagonal) would have more symmetrical outcomes. However, we must stress two points: first, the symmetry of the substrate is usually taken as equal to the symmetry of the formed polymer, in order to make the surface a good template for the polymerization reaction; second, we showed that the diagonal polymers are exactly commensurate with the surface (both having single units distant 0.44 nm along the [1–12] direction). Being commensurate with the surface reduces the mismatch, which is favourable to obtain longer-range defect-free structures.

1 J. Lipton-Duffin, O. Ivasenko, D. Perepichka and F. Rosei, *Small*, 2009, **5**, 592–597.
2 G. Vasseur, Y. Fagot-Revurat, M. Sicot, B. Kierren, L. Moreau, D. Malterre, L. Cardenas, G. Galeotti, J. Lipton-Duffin, F. Rosei, M. Di Giovannantonio, G. Contini, P. Le Fèvre, F. Bertran, L. Liang, V. Meunier and D. F. Perepichka, *Nat. Commun.*, 2016, **7**, 10235.
3 A. Saywell, J. Schwarz, S. Hecht and L. Grill, *Angew. Chem., Int. Ed.*, 2012, **51**, 5096–5100.

Zhi Li continued: If you don't use Cu or Ag substrates, but use a different one (*e.g.* h-BN, graphite) for the polymerization reaction. Do you expect that the reaction can still proceed as well as on Cu or Ag substrates by adding external Cu or Ag.

Federico Rosei responded: Nudge-Elastic Band calculation of the Ullmann reaction from single molecules to organometallic dimers or networks suggests that the reaction can proceed both *via* incorporating the surface adatoms or with the molecules "extracting"atoms from the substrate.[1,2] Adding Cu or Ag atoms to a less reactive surface (such as HOPG) could therefore lead to dehalogenation, as it was demonstrated by Zhou *et al.*,[3] who reported the formation of various organometallic species on Au(111) by depositing extrinsic Ni and Ag atoms. The reaction, however, would be controlled and limited by the number of available catalytic adatoms, which would also remain on the surface as unwanted side products of the reaction after polymers are formed. In addition, depending on the interaction between the surface and the extrinsic metal atoms, the polymerization reaction could also be hindered. The size, arrangement and order of these added catalytic centers/clusters to the surface could also play a key factor in limiting the order of the formed polymer. Another possible way would be to avoid the use of metallic catalysts in the reaction, for example by directly obtaining the polymerization on bulk insulators, *via* thermal- or UV-induced dehalogenation. An example has been shown by Kittelmann *et al.*,[4] who reported the formation of polymers using an Ullmann-like reaction on $CaCO_3$. This path could, however, introduce different challenges, for example a lower adsorption energy or sticking coefficients of the molecular layer on the insulating surface compared to metallic substrates, or inversely mediating new reaction paths which could lead to the dissociation of the molecules before they can undergo the desirable reaction.

1 J. Björk, F. Hanke and S. Stafström, *J. Am. Chem. Soc.*, 2013, **135**, 5768–5775.
2 D. Barton, H.-Y. Gao, P. A. Held, A. Studer, H. Fuchs, N. L. Doltsinis and J. Neugebauer, *Chem.–Eur. J*, 2017, **23**, 6190–6197.
3 X. Zhou, F. Bebensee, Q. Shen, R. Bebensee, F. Cheng, Y. He, H. Su, W. Chen, G. Q. Xu, F. Besenbacher, T. R. Linderoth and K. Wu, *Mater. Chem. Front.*, 2017, **1**, 119–127.
4 M. Kittelmann, M. Nimmrich, R. Lindner, A. Gourdon and A. Kühnle, *ACS Nano*, 2013, **7**, 5614–5620.

Markus Lackinger asked: Could you explain the shifts of C 1s that you observe with Fast-XPS? I would be particularly interested in the origin of the overall shift that is observed upon conversion from the organometallic to the covalent state.

Federico Rosei answered: The overall shift is due to the conversion of C–Cu to C–C bonds. Shown by Di Giovannantonio *et al.*,[1,2] the C 1s core level spectrum of 1,4-dibromobenzene (dBB) on Cu(110) at RT can be deconvoluted into three different components, one at lower binding energy (BE), assigned to carbon atoms bound to copper (283.2 eV), and the peaks at the higher BE (283.8 eV and 284.2 eV) attributed to the remaining carbon atoms in the phenyl group. These peaks originate from the non-equivalent positions of phenyl carbons with respect to the copper substrate. After polymerization, the peak at 283.2 eV disappears, and two new components at 284.4 eV and 285.1 eV are found. These components are at higher BE compared to their organometallic counterparts, as expected and calculated by Björk *et al.*[3] Each spectrum (a single line of the Fast-XPS map) presents the combination of these five peaks depending on the progress of the reaction. The observed shift of the Fast-XPS is therefore due to the transition from the organometallic to polymer. The shape of the curve is related to the kinetic of the reaction, which for dBB on Cu(110) follows a nucleation-and-growth mechanism.[4] While the same overall shift is found for all the studied molecules, it can be seen that the shape of this transition differs depending on the halogen (in particular for iodine-containing molecules), which therefore affects the reaction kinetics.

1 M. Di Giovannantonio, M. El Garah, J. Lipton-Duffin, V. Meunier, L. Cardenas, Y. Fagot Revurat, A. Cossaro, A. Verdini, D. F. Perepichka, F. Rosei and G. Contini, *ACS Nano*, 2013, 7, 8190–8198.
2 M. Di Giovannantonio, M. El Garah, J. Lipton-Duffin, V. Meunier, L. Cardenas, Y. Fagot-Revurat, A. Cossaro, A. Verdini, D. F. Perepichka, F. Rosei and G. Contini, *ACS Nano*, 2014, **8**, 1969–1971.
3 J. Björk, F. Hanke and S. Stafström, *J. Am. Chem. Soc.*, 2013, **135**, 5768–5775.
4 M. Di Giovannantonio, M. Tomellini, J. Lipton-Duffin, G. Galeotti, M. Ebrahimi, A. Cossaro, A. Verdini, N. Kharche, V. Meunier, G. Vasseur, Y. Fagot-Revurat, D. F. Perepichka, F. Rosei and G. Contini, *J. Am. Chem. Soc.*, 2016, **138**, 16696–16702.

Jonathan Davidson commented: A recent area of research interest has been using halogen bonding to create supramolecular assemblies. An example of a molecule used would be 1,4-diiodotetrafluorobenzene. Ullmann coupling presents a possible unwanted side reaction in studies of halogen bonding assemblies on metal surfaces. Have you performed or are you aware of any work studying similar fluorinated halobenzenes on coinage metals?

Federico Rosei responded: The use of halogen bonding as a driving force to grow and stabilize supramolecular networks is a vivid field of research, which allows the formation of both 2D and 3D self-assembled crystals.[1] The halo terminal groups could, however, be subjected to various reactions, such as nucleophilic substitutions, elimination-addition and Grignard reactions. The dehalogenation which drives the Ullmann reaction makes it undesirable for forming supramolecular assemblies. The dehalogenation step of the Ullmann reaction is dependent on both the substrate/metallic species and halogen, while only some of these combinations yield room temperature dehalogenation. In the present work, we observe that Cl–C bonds are mostly intact at RT on Cu, while I–C and Br–C are readily dissociated. Regarding the case of fluorinated halobenzenes on transition metals, Blake *et al.*[2] studied the interaction of 1-fluoro-4-bromobenzene (FBB) with Cu(111), and found out that C–Br dissociates at RT, while the C–F bonds remain intact. NEXAFS analysis also showed that the fluorophenyls were tilted from the surface, confirming that the reactivity of fluorinated halobenzenes on copper is lower than halobenzenes with other halogens. A conscious choice of the involved halogens and metal substrates could therefore prevent such unfavourable side reactions.

1 L. C. Gilday, S. W. Robinson, T. A. Barendt, M. J. Langton, B. R. Mullaney and P. D. Beer, *Chem. Rev.*, 2015, **115**, 7118–7195.
2 M. M. Blake, S. U. Nanayakkara, S. A. Claridge, L. C. Fernández-Torres, E. C. H. Sykes and P. S. Weiss, *J. Phys. Chem. A*, 2009, **113**, 13167–13172.

Alex Saywell queried: With regards to the "Fast XPS" data that you present; what is the temporal resolution that you are able to obtain? In your opinion, what is the limiting factor for time-resolved XPS and do you see scope for developing this as a way to access information on reaction kinetics for various on-surface systems?

Federico Rosei responded: The reported Fast-XPS spectra were achieved at the ALOISA beamline on the Elettra synchrotron, in the snapshot mode with a 2D delay-line detector. The data were collected with a rate of one spectrum per one second, with a photon flux of 10^{12} ph s^{-1} 0.02% BW^{-1}. In this system, the limiting factor to achieve a statistically good signal is given by the small aperture angle (2° FWHM) of the electron analyzer. Increasing the size and the luminosity of the analyzer would easily allow the acquisition rate of 10 Hz. Faster rates, however, require a flux increase, with the drawback of radiation damage of the molecules, which becomes a problem at 10^{13} ph s^{-1}. From the technical point of view, the 2D delay-line detector is currently limited by the deadtime of 10 msec for emptying its buffer memory, whereas the electronics themselves (reading and conversion) would allow rates in the kHz range. Fast-XPS could therefore be applied to study on-surface reactions of different molecules, for example NH_3 dissociation on Si[1] and hydrogenation of graphene on Ir(111).[2] The physical limits of this technique allows us to study the reactions which take place in the order of 10–100 msec, therefore strongly reducing the range of applications. Faster reaction kinetics would require a different experimental approach, *e.g.* pump-and-probe techniques.

1 M. Satta, R. Flammini, A. Goldoni, A. Baraldi, S. Lizzit and R. Larciprete, *Phys. Rev. Lett.*, 2012, **109**, 036102.
2 R. Balog, M. Andersen, B. Jørgensen, Z. Sljivancanin, B. Hammer, A. Baraldi, R. Larciprete, P. Hofmann, L. Hornekær and S. Lizzit, *ACS Nano*, 2013, **7**, 3823–3832.

Markus Lackinger said: Relatively swift acquisition of XP spectra relies on the high radiation intensities available at synchrotron facilities. How can you exclude effects of radiation damage in such a experimental setting? Do you have to take special precautions such as, *e.g.*, constantly moving the sample during spectra acquisition?

Federico Rosei replied: At the ALOISA beamline a photon flux of 10^{12} ph s^{-1} 0.02% BW^{-1} was used to perform the experiments, while keeping the sample at a grazing angle of 4°, which further spreads the beam over an illuminated area of 3×0.1 mm^2. Such flux density is not extremely damaging for our molecules and we could therefore analyze our samples without the need to constantly move the sample during acquisition. For every molecule, however, damage tests were performed, by repeating and comparing the same C 1s scan after long expositions, without any indication of beam damaging in the data. In particular, the beam could be problematic when studying molecular multilayers or intact halobenzenes, as it is shown in Di Giovannantonio *et al.*,[1,2] where beam exposition caused a partial dehalogenation of a multilayer of dBB on Cu(110).

1 M. Di Giovannantonio, M. El Garah, J. Lipton-Duffin, V. Meunier, L. Cardenas, Y. Fagot-Revurat, A. Cossaro, A. Verdini, D. F. Perepichka, F. Rosei and G. Contini, *ACS Nano*, 2013, **7**, 8190–8198.
2 M. Di Giovannantonio, M. El Garah, J. Lipton-Duffin, V. Meunier, L. Cardenas, Y. Fagot-Revurat, A. Cossaro, A. Verdini, D. F. Perepichka, F. Rosei and G. Contini, *ACS Nano*, 2014, **8**, 1969–1971.

Nian Lin asked: Comparing the NEXAFS data of the organometallic state with the polymerized chains, do you conclude that the phenyl rings in the organometallic state are more tilted with respect to the substrate? Why?

Federico Rosei answered: The NEXAFS spectrum of a flat benzene should present only one π^* transition.[1] Nevertheless, our NEXAFS data shows that the organometallic systems present two clear π^* transitions, labeled as π_1^* and π_2^* in our work (DOI: 10.1039/C7FD00099E). Geometrical distortion of the ring could produce a symmetry reduction, which, as a consequence, would give rise to the π_2^* components.[2] The two resonance components for chemisorbed phenyl rings (or derivatives), due to geometrical distortion[3] or to the formation of new electronic states,[4] have been previously reported in the literature.

1 M. X. Yang, M. Xi, H. Yuan, B. E. Bent, P. Stevens and J. M. White, *Surf. Sci.*, 1995, **341**, 9–18.
2 L. G. M. Pettersson, H. Ågren, Y. Luo and L. Triguero, *Surf. Sci.*, 1998, **408**, 1–20.
3 J. Stöhr, *NEXAFS spectroscopy*, Springer-Verlag Berlin Heidelberg, 1992.
4 M. Weinelt, N. Wassdahl, T. Wiell, O. Karis, J. Hasselström, P. Bennich, A. Nilsson, J. Stöhr and M. Samant, *Phys. Rev. B*, 1998, **58**, 7351–7360.

Markus Lackinger commented: In your NEXAFS spectra, the resonance at higher photon energies was assigned to a C 1s to π_2^* transition. Yet, we observe similar features on Ag(111) that however appear to be related to the organometallic state. How confident are you that this assignment is correct and do you think acquisition of additional spectra at various incidence angles would be suitable to clarify this?

Federico Rosei replied: The observed π_2^* resonance is connected to a non-planarity of the phenyl rings, and is observed more prominently for 1-bromo-4-chlorobenzene (BCB) and 1,4-dichlorobenzene (dCB) at RT, for which we know that the organometallic chains contain out of plane Ph–Cl segments (from STM data). This resonance is strongly reduced after annealing at 150 °C, with the curves being virtually indistinguishable from the RT spectra of the other precursors. After annealing at 230 °C polymers are formed, and the observed π_2^* resonance is completely suppressed, suggesting the planarity of the PPP polymers and the absence of C–metal bond. Therefore, these features seem to be related to organometallic states, and their reduction upon polymerization for all the five studied systems strengthen this point. Acquiring additional spectra at different incidence angles could not provide additional important information.

Steven De Feyter remarked: The fact you observe a halogen dependent behavior, is this good or bad in view of controlling the synthesis?

Federico Rosei responded: Identification of the parameters which control a reaction is to be seen as a positive finding. The dependence of dehalogenation and polymerization temperatures on the halogen presents in the reactants could be used to lead distinct pathways with various products. For example, as shown by Lafferentz *et al.*[1] a hierarchical growth can be obtained by controlling the key factors in the reaction. This is the scope of reaction kinetics and its benefits in fundamental and applied sciences, as if "playing with the knob" to control the outcome. However, while opening possibilities, having more parameters in the ground adds more complexity to the study, to control and *a priori* predict, which might be challenging and possibly unfavorable.

1 L. Lafferentz, V. Eberhardt, C. Dri, C. Africh, G. Comelli, F. Esch, S. Hecht and L. Grill, *Nat. Chem.*, 2012, **4**, 215–220.

Robert Jones noted: One hypothesis put forward for the influence of the halogens on the reaction, is that the halogen adsorbs on top of the Cu atom that links two phenyl groups together. Can you outline what the experimental evidence is for this? In particular, as each phenyl group produces one halogen atom when the –Cu– linkage is formed, only half the available halogen atoms can be located on the linking Cu, the other half must be elsewhere. Does this accord with the STM images?

Federico Rosei replied: The hypothesis of halogen atoms adsorbed on top of the organometallic copper bridges was proposed by Di Giovannantonio *et al.*[1] for the adsorption of dBB on Cu(110), based on qualitative matching between the experimental STM data and the DFT simulated images. Furthermore, the on-top halogens accounted for the 2-to-1 stoichiometric ratio of Br to Ph in the unit cell (four Br on-top and four Br adjacent to the four phenyls in the OM chains). Low temperature STM images, however, show that Br and Cl atoms can also be found outside the molecular networks, on areas of the surface free of molecules, which weaken the idea of keeping the on-top halogens to maintain the stoichiometry. A possible experimental hint of the presence of the on-top halogens could be the additional components at RT which are present in the Br 3d and Cl 2p spectra, and are strongly reduced after polymerization. Core-level shift calculations are currently ongoing to further explain this point.

Yuri Diaz Fernandez opened a general discussion of the paper by Marco Sacchi: It's quite interesting that you have found consistent results for the adsorption configurations for 3 of the corrections and not for 1? Can you comment in the wider context of the selection of the modelling conditions?

Marco Sacchi answered: Yes, we were also quite surprised by the relative consistency of the DFT results across the various vdW correction schemes employed. The TS and TSSCS methods are essentially very similar, with exception of the screening term, therefore we assume that, to a first approximation, the screening corrections are not essentials for capturing the corrugation of the PES of benzene on Cu(111).

Ana Sanz Matias asked: According to the computational results, the differences in adsorption energy between the adsorption sites are very small. For instance, in Table 2 (DOI: 10.1039/C7FD00095B) the differences between the bridge and the HCP and FCC hollow sites are well below 0.05 eV. To which extent can you draw conclusions (*i.e.* regarding the adsorption site or the diffusion mechanisms) from these values?

Marco Sacchi responded: It is very difficult to determine *a priori* the "error bar" for DFT calculations, given the fact that the performance and accuracy of a DFT calculation generally varies according to the specific surface system investigated. In our experience (see for instance: ref. 1–4), the precision of DFT calculations for aromatics adsorbed on metal surfaces is surprisingly good for most systems, therefore we believe that DFT is able to quantify the small energy corrugation of benzene/Cu(111).

1 M. Sacchi, S. J. Jenkins, H. Hedgeland, A. P. Jardine and B. J. Hinch, *J. Phys. Chem. C*, 2011, **115**, 16134–16141.
2 B. A. J. Lechner, H. Hedgeland, J. Ellis, W. Allison, M. Sacchi, S. J. Jenkins and B. J. Hinch, *Angew. Chem., Int. Ed.*, 2013, **52**, 5085–5088.
3 B. A. J. Lechner, M. Sacchi, A. P. Jardine, H. Hedgeland, W. Allison, J. Ellis, S. J. Jenkins, P. C. Dastoor and B. J. Hinch, *J. Phys. Chem. Lett.*, 2013, **4**, 1953–1958.
4 H. Hedgeland, M. Sacchi, P. Singh, A. J. McIntosh, A. P. Jardine, G. Alexandrowicz, D. J. Ward, S. J. Jenkins, W. Allison and J. Ellis, *J. Phys. Chem. Lett.*, 2016, 7, 4819–4824.

Ana Sanz Matias commented: You have calculated benzene diffusion barriers and adsorption energies, and future work on vibrational frequencies is also mentioned in the manuscript. Have you considered combining this information into a kinetic Monte Carlo model to investigate the dynamics of the diffusion process?

Marco Sacchi answered: Yes, we have indeed considered this line of research for other surface systems, such as diffusion of glycine on Cu(110),[1] but not for benzene on Cu(111), since the dynamics of this system are relatively simple and other approaches (such as molecular dynamics) could be employed.

1 M. Sacchi, D. J. Wales and S. J. Jenkins, *Phys. Chem. Chem. Phys.*, 2017, **19**, 16600–16605.

Rasmita Raval queried: Does the direction in which the beams are aligned with regard to the surface matter and, if so, can information on diffusion along particular axes of the surface be captured?

Marco Sacchi answered: Yes, HeSE measures only along one crystallographic direction, corresponding to the azimuthal scattering angle, at a time.[1] By rotating the crystal you can measure in other directions, and the ratio of the diffusion rates in the different directions will give an idea of the ratio of the variation in the corrugation heights. In this case, we only measured in the $\langle 11\bar{2} \rangle$ direction. The measured Arrhenius barrier, however, isn't expected to vary between the azimuths (as motion in the $\langle 1\bar{1}0 \rangle$ directions also project on the $\langle 11\bar{2} \rangle$ for instance).

1 G. Alexandrowicz, A. P. Jardine, P. Fouquet, S. Dworski, W. Allison and J. Ellis, *Phys. Rev. Lett.*, 2004, **93**, 156103.

Natalia Martsinovich remarked: You have calculated diffusion barriers in quite good agreement with experiment, but found less good agreement in adsorption energies. In the well-known Grimme's dispersion correction, you can tune the strength of the dispersion interaction by changing the value of the s_6 parameter. In your opinion, what is the way forward in achieving accuracy in calculations of weak interactions? Is it useful to make this semi-empirical correction even more empirical by tuning the coefficients, *e.g.* by fitting to known experimental data, or is it preferable to develop new functionals with dispersion interactions embedded in them?

Marco Sacchi responded: From a philosophical point of view, I think we should strive to keep the level of empiricism in the vdW corrections as minimal as possible, therefore I would not recommend to tune the C_6 coefficient to obtain a desired adsorption energy, but to use more advanced vdW treatments, such as MBD or vdW+DF.

Neil Robinson said: I have a general question about the use and application of dispersion corrections in DFT calculations containing multiple adsorption environments. How transferable is the use of a specific dispersion correction across surfaces with different electronic structures? For example, in an example calculation containing a supported metal nanoparticle embedded within an oxide surface, are there different considerations which must be made for the influence of the dispersion correction on the interaction of adsorbates with the two different surfaces present within the system?

Marco Sacchi replied: The intensity of the vdW corrections should depend only on the specific elements that compose the surfaces and, generally, on the local chemical environment or oxidation state of each atom present in the system. I think that the presence of multiple adsorption environments, for instance in metal oxides with embedded metal nanoparticles, should not be an obstacle for most modern dispersion correction schemes or long-range corrected XC-functionals. Clearly, accounting for many body corrections and screening contributions would also be important for these systems.

Robert Jones commented: The density functional theory (DFT) calculations with van der Waals corrections seem to give the work function change of the surface after adsorption rather accurately. Can these be applied in a straightforward way to other molecular adsorbate systems on metals to provide similar work function change predictions? Or is a certain level of parameterisation needed for each system to achieve this level of accuracy?

Marco Sacchi responded: Yes, as far as I know, the work function and dipole moment calculations can be applied successfully to other molecular adsorbate systems without parameterisation.

Giovanni Costantini queried: Could you comment on what the problems are with molecules which are less planar, therefore typically lying less close to the surface and being less hybridised with the states of the substrate? In particular, on the importance of the self-interaction problem in DFT and on the ways to overcome it.

Marco Sacchi answered: I do not have specific experience with self-interaction problems in large organic molecules adsorbed on surfaces. Historically, vdW corrections have been tested for aromatic dimers as well as for graphite layer interactions, therefore I do not think that the size of the adsorbate or its orientation would necessarily cause huge problems. Clearly, charge transfer and multipole effects could become more important for vertically oriented molecules at high coverage, therefore one would need to accurately test DFT against quantitative experimental measurements in order to insure that the precision of the vdW corrections is maintained.

Discussion of the Concluding Remarks

Steven L. Tait opened the discussion of the Concluding Remarks: A question to all: In the closing remarks, David Amabilino and I have discussed five themes to summarize our meeting: developing a rulebook for molecular adsorption, diffusion, and assembly; polymorphism; nucleation – observing and exploring; going from simple, well-defined systems to complex, ill-defined systems; and rationalizing multiple analytical inputs with theory. As we look toward the future development of this research field, which challenges and opportunities are exciting and engaging to you? We would particularly like to have the students and postdocs contribute to this discussion to identify the key themes that seem exciting to them.

Neil Robinson responded: I would like to see a greater emphasis placed on the investigation of applied systems, as well as potential applications of fundamental topics. Much of the work discussed here has been focused on very ideal adsorbate systems. As has been alluded to in the closing remarks, I feel that we should not, as a community, be put off by the idea of confronting the messy, non-idealised systems found in the many industrial applications of complex surfaces; indeed, quite the opposite. As someone who is currently writing a PhD thesis in this area, I certainly feel the need to have a strong argument for the possible applications of the surfaces and adsorbate structures I have investigated. A concentration on the applied elements of surface science is also favourable in terms of attracting funding, which is something our community has struggled with in recent years.

David Amabilino asked: Could you give an example of an applied system that is in need of analytical probing and solutions from the knowledge we have of complex molecular systems at surfaces and interfaces?

Neil Robinson replied: Polar isomers exhibiting identical physical properties can have very different responses to functionalised surfaces exhibiting hydrophilic or hydrophobic properties. Thus, the separation of complex mixtures using porous membrane materials with tailored surfaces is an area that could show real promise in terms of industrial applications.

Brandon Hirsch addressed David Amabilino and Steven L. Tait: The future of surface science will likely expose valuable insights into kinetically trapped states and concepts on non-equilibrium systems. This work, and the individual molecular detail that our field provides, is one of the remaining frontiers of science. However, I believe that in the future the field will also begin to recognise the extent to which our measurements are perturbing the final state of the system. In particular at liquid/solid interfaces, where dynamics are incredibly important, these concepts are likely to become much more prevalent than the few cases already seen in literature and those that were highlighted at the conference by Prof. Flood and Prof. De Feyter.

David Amabilino answered: The variation of temperature in assembly of molecules on surfaces is likely to reveal a great deal about "trapped" states. Ostwald ripening has been observed in monolayer systems that have not reached a place near their thermodynamic minimum. That is true for physisorbed systems. But these cases can be the "easy" ones. For chemisorbed systems, the kinetic trapping determined by covalent bond formation means that trapped states abound, and cannot mature, and in these cases the conditions of deposition are very important. They are somewhat easier to study in the sense that they are less prone to measurement perturbation.

The question actually raises an interesting point about chemisorption and physisorption, that comes back to the question of the rulebook for interactions. There is clearly a grey area where deposition under certain conditions leads to stable structures that are characteristic of chemisorbed states, but that can evolve under temperature variation, more characteristic of a physisorbed state. Finding the sweet spot for this kind of annealing is a challenge and great opportunity to prepare complex structures.

Steven L. Tait said: A question addressed to all, but especially the students and postdocs: we have discussed the idea of developing a rulebook for molecular assembly at surfaces at several points during the meeting. How can we do a better job at distilling observations about specific systems into generally useful principles about molecular assembly that could aid others in the predictive design of new systems? When you study the research literature, what are key pieces of information that you would expect should be available, but have not been reported or compiled?

Nerea Bilbao commented: I find it important to keep working and focusing our efforts on completing and developing a rulebook on the effects of surfaces on molecules and molecular assemblies. A better understanding on the forces that govern surface reactions, or surface-supported self-assembly, will help us control and design, in a better way, the chemical systems we are after.

Sebastian Schwaminger noted: It's sometimes challenging to find the right analytics for every system. Therefore, it is necessary to develop new techniques to get new insights and new points of view towards a system. Coming back to what

K.-H. Ernst said, how nobody has imagined the development of an ambient pressure XPS in the past and right now it is unimaginable to think about techniques like ambient pressure PEEM, but who knows what is going to happen in the future? Furthermore, it is very challenging to find someone who can actually perform the analytics needed to describe a certain system which might be completely different to what he is doing. Thus the knowledge transfer between different disciplines/analysis techniques is crucial to the understanding of many phenomena.

Martin Nalbach remarked: For studying molecules at the solid–liquid interface, in my case particularly the calcite (10.4)–water interface, *in situ* AFM does not tell you the exact arrangement of the molecules at step edges or on terraces. Further experimental techniques, as well as support from theory, would be very helpful to achieve more insights into the adsorption behavior on the molecules.

The solvent itself is a molecule and can strongly interact with the substrate surface. Therefore, it is very important to also consider the surface–solvent interaction as well as the arrangement and order of the solvent molecules at the interface when studying the influence of molecules on the substrate surface.

Jonathan Davidson said: My supervisor, Stuart Clarke, made a comment about this in the 2nd Discussion session (DOI: 10.1039/C7FD90074K), but I would like to reiterate that diffraction techniques are often underutilised in surface science. STM and related scanning probe techniques are inherently invasive. For example, in DFT simulations using periodic boundary conditions, such as in the paper presented by Marco Sacchi (DOI: 10.1039/C7FD00095B), a "vacuum gap" of 15–20 angstroms between the bottom of one layer and the top of another is generally considered necessary to avoid interference between adjacent layers. However, as Brandon Hirsch's paper (DOI: 10.1039/C7FD00104E) pointed out, the tip–sample distance in STM experiments can generally be considered as being on the order of 10 angstroms, and can be even less. Thus, even in the absence of a significant bias voltage, the tip will be perturbing the surface relative to simulations. Diffraction and related techniques provide a non-invasive method to evaluate surface structure and can help validate results from scanning probe techniques, bridging the gap between these techniques and simulations.

David Amabilino responded: That is true, diffraction techniques can be useful, mainly for seeing periodic structures I would say (although that is often the case for STM too), and especially LEED has proven vital as shown by people in this discussion and others (for example ref. 1–3). Helium diffraction has also been used for the study of SAMs.[4]

1 J. Stettner and A. Winkler, Characterization of Alkanethiol Self-Assembled Monolayers on Gold by Thermal Desorption Spectroscopy, *Langmuir*, 2010, **26**, 9659–9665.

2 T. Yokoyama, M. Kawasaki, T. Asari, S. Ohno, M. Tanaka and Y. Yoshimoto, Adsorption and self-assembled structures of sexithiophene on the Si(111)-3—$\sqrt{}\times$3—$\sqrt{3}\times$3-Ag surface, *J. Chem. Phys.*, 2015, **142**, 204701.

3 M. Sanchez-Arenillas and E. Mateo-Marti, Pyrite surface environment drives molecular adsorption: cystine on pyrite(100) investigated by X-ray photoemission spectroscopy and low energy electron diffraction, *Phys. Chem. Chem. Phys.*, 2016, **18**, 27219–27225.

4 E. Albayrak and M. F. Danisman, Helium Diffraction Study of Low Coverage Phases of Mercaptoundecanol and Octadecanethiol Self-Assembled Monolayers on Au(111)

Prepared by Supersonic Molecular Beam Deposition, *J. Phys. Chem. C*, 2013, **117**, 9801–9811.

Yuri Diaz Fernandez commented: One of the recurrent themes over the conference was the importance of disorder and how it could drive the formation of ordered phases. We are currently lacking complementary analytical tools to characterize the dynamics of disordered phases and phase boundaries in 2-dimensions. Translating into 2D probing techniques already established for 3-dimensional material chemistry systems (*e.g.* total scattering pair distribution function analysis) may bring a step change in our ability to understand complex molecular processes at surfaces.

Karl-Heinz Ernst noted: New methods are not only important to understand the properties or influence of disordered phases. Over the last 100 years, after XRD was revolutionarily introduced and told us there are actually atoms, we thought we had a single crystal. Electron microscopies then revealed recently that we don't have a single crystal, although XRD suggested this, but rather a poly-crystalline sample, but all crystallites are more or less aligned equally. See the aragonite example in ref. 1.

1 X. Li and Z. Huang, *Phys. Rev. Lett.*, 2009, **102**, 075502.

Karl-Heinz Ernst continued: And defects especially play a role at surfaces. After preparation of surfaces with much lower step density, for example, the dissociative sticking probability of molecules like water on Pt(111) became orders of magnitude smaller.

Rasmita Raval remarked: What is quite clear is that the field of complex molecules at surfaces will require a broad and deep discipline base in order to flourish and reach its fullest potential. For example, the need for the rigour of surface science and a multi-technique approach is vital to capture the multifaceted aspects of these systems. The need for the methodology, insights and concepts of organic and supramolecular chemistry is also clear. In addition, we need to bring in the functionality focus that the materials scientist and engineer have, alongside the complexity that is inherent in the biological system. Theory will be instrumental in capturing the structure, kinetics and thermodynamics of these systems. Finally, we will need to embrace the big data and information space that science is currently generating. I feel that there is a real need to broaden the minds of all researchers in the field, from professor to student, so that we possess sufficient breadth and depth to address the challenges in the field. This would allow us to grasp the advances in other fields and lever that knowledge into ours, and will avoid reinventing the wheel that keeps occurring across science.

Angelika Kühnle added: Along this line, I think we still have to learn to better combine the worlds of physicists and chemists. There is still a rather different view on many things and a different way of describing the situation. For example, I have the impression that when considering disorder and entropic effects in molecular self-assembly physicists can learn a lot from chemists and *vice versa*.

Steven L. Tait asked: Much of the work that we have discussed in this meeting is truly inter-disciplinary. This creates a real challenge in the design of our training and education programs as the student researchers (and senior researchers) need to have a depth of knowledge in more than one traditional field of study. What have you found to be effective methods in your training programs to prepare researchers to work successfully in inter-disciplinary research fields?

Trolle R. Linderoth noted: Regarding the previous comment concerning better integration between chemistry and physics. At the Interdisciplinary Nanoscience Center at Aarhus University (iNANO), since 2002 we have been running interdisciplinary educational bachelor and master programmes in nanoscience, encompassing both physics, chemistry and molecular biology.[1] I just wish to comment that the experience shows that this concept really works and the students are able to pursue very interesting projects taking advantage of this interdisciplinary breadth.

1 T. R. Linderoth, T. Fredsted and F. Besenbacher, *Mat. Today*, 2008, **11**, 6.

Marco Sacchi said: In my personal experience, I've collaborated with physicists for more than 10 years and I enjoyed working on interdisciplinary projects. I think it is fundamentally a question of keeping a flexible mental attitude and being able to enjoy the communication challenges given by the use of different scientific language and terminologies in different communities. Also, I would like to add another great challenge we face in the surface science community. From a theoretical point of view we have a theoretical "software" challenge that is the rationalisation and quantification of entropy, disorder, complexity, solvent effects. But we also have a physical, "hardware", challenge, that is the limited amount of computer power available, especially on large national HPC facilities. In the last decade, USA, Europe and China have consistently increased the investment in HPC resources. The UK needs to urgently increase its investment in computational power dedicated to materials and surface science in order to remain competitive in these areas.

David Amabilino asked: Many examples of interfaces we've seen in this meeting comprise molecules co-parallel with the adsorbent on which they are deposited, but for many potential applications in different areas, like catalysis, electronic devices, sensors and so on, the molecules are perpendicular to the plane, and it is really tough to analyse that, although theory can model orientation and order. Are there any ideas about how you could analyse, experimentally, these areas of interfaces?

James Batteas commented: We are now investing heavily in high precision, *in situ* techniques at TAMU, in particular *in situ* TEM methods that will allow us to look at complex nanoparticle systems (such as catalysts), where it has been shown that the interactions of solvents with nanoparticles can be followed. We have also added a number of *in situ* nanomechanical testing tools. In general, we feel that without moving more of our techniques to *in situ* based methods, we can't really address the issues we're after at the molecular level.

Mario Samperi remarked: I'd like to spend few words on two different topics. Firstly, I strongly believe that we should always keep in mind how extremely important the role of the solvent is on governing both the molecule–molecule and molecule–substrate interactions. In this context, solvent mixtures can represent an incredibly wide range of opportunities for achieving properties and features that one solvent is not able to show in itself. As human beings behave and interact with each other because of the environment in which they are in, similarly the behavior of molecules is strongly imposed by the medium composition. Secondly, I was considering how we approach supramolecular chemistry differently when we move from solution to surface. One of the greatest challenges for future research is to erect a solid bridge between chemistry on the surface and in solution. The scientists awarded the Nobel Prize in Chemistry in 2016 have shown us how to make catenanes, rotaxanes, how to utilize molecular recognition and molecular self-assembly processes for the design and synthesis of devices at the nanoscopic scale. Now we need to put them on surfaces, and make them work together to create supramolecular machines.

Steven De Feyter noted: One of the important challenges is understanding the role liquids or solvents play in supramolecular self-assembly at the liquid–solid interface. The choice of the liquid is, in many cases, motivated by technical considerations. For instance, only a small set of solvents are considered for STM imaging at the liquid–solid interface, and few systematic studies are carried out. More in depth studies are needed, involving other techniques, to shine light on the different roles the liquid environment plays.

Natalia Martsinovich commented: We also need to expand our toolkit of theoretical methods. At the moment, most studies of self-assembled systems in the literature use DFT. However, the use of DFT is limited by the size of the systems. Force fields are used much less in this field, but they are able to treat much larger length scales and time scales. Notably, the DFT community and the force field/molecular dynamics community think in different ways, and ask and answer different kinds of questions: while DFT calculations are largely focused on obtaining structures and energies, molecular dynamics users thinks in terms of order parameters and time correlation functions, and even diffusion coefficients[1] and entropy[2] can be calculated. Therefore, force fields and molecular dynamics calculations can be very useful, especially if we want to describe complex systems, such as solid/liquid interfaces.

1 J. Wang and T. Hou, *J. Comput. Chem.*, 2011, **32**, 3505–3519.
2 J. Gaberle, D. Z. Gao, M. B. Watkins and A. L. Shluger, *J. Phys. Chem. C*, 2016, **120**, 3913–3921.

Talat Rahman added: We should also not forget that DFT itself needs to be reformed to include effects such as electron correlations which can have a large impact on the characteristics of metal coordination centers, such as V or Fe, or on oxide surfaces and nanostructures. Such developments are in the literature (for example: ref. 1). Additionally, as has been discussed at this workshop, both kinetic and dynamical (vibrational) effects need to be taken into consideration together with insights from electronic structure calculations from DFT.

1 S. R. Acharya *et al.*, *Computation*, 2016, **4**, 34.

Claire-Marie Pradier addressed Steven L. Tait: May we come back to the first challenge you listed in the concluding remarks, what governs molecular adsorption, diffusion and assembly. You showed a very interesting diagram with different types of interaction between molecules on the surface. We should consider interactions between molecules and the substrate too. There may be competition or synergy when considering interactions between molecules, and between molecule and substrate. Techniques to measure these interactions, and their strength, apart from AFM, would be useful.

Steven L. Tait concluded: David Amabilino and I thank you all for your active participation in this concluding discussion. We especially appreciate the participation of so many students and postdoctoral researchers. There are many exciting challenges that lie before us in our study of molecular systems at surfaces and interfaces. We look forward to meeting again in the future to continue our discussion of the ongoing progress that we make as a community in developing new discoveries, new insights, and new applications of complex molecular surfaces and interfaces.

Poster titles

Physisorbed halogen-bonded monolayers studied by STM and PXRD, **J. Davidson, A. Brewer, C. Truscott, M. Sacchi and S. M. Clarke**, *University of Cambridge, UK*

The morphology of C_{60} and C_{60}-PTCDI heterostructures on hexagonal boron nitride, **J. Kerfoot, V. Korolkov, S. Svatek, T. Taniguchi, K. Watanabe and P. Beton**, *University of Nottingham, UK*

Supramolecular heterostructures – extending molecular self-assembly into 3D, **V. Korolkov, M. Baldoni, N. Besley, E. Besley, K. Watanabe, T. Taniguchi and P. Beton**, *University of Nottingham, UK*

The adsorption structure of TCNQ on Ag(111): a quantitative determination, **P. Blowey, L. Rochford, D. Duncan, D. Warr, T.-L. Lee, G. Costantini and P. Woodruff**, *University of Warwick, UK*

2D self-assembly of DBA in confined spaces: competition between the assembled structures in nanocorrals, **Y. Hu, A. M. Bragança, B. E. Hirsch, L. Verstraete, K. Tahara, Y. Tobe and S. De Feyter**, *KU Leuven, Belgium*

Preserving graphitic surfaces from grafting of aryl radicals by ultra thin organic layer protection, **Z. Li, H. Van Gorp, P. Walke, T. H. Phan, Y. Fujita, J. Greenwood, O. Ivasenko, K. Tahara, Y. Tobe, H. Uji-i, S. Mertens and S. De Feyter**, *KU Leuven, Belgium*

Molecular grafting as a tool for chiral surface modification, **J. Seibel, L. Verstraete, Y. Kubo, K. Tahara, Y. Tobe and S. De Feyter**, *KU Leuven, Belgium*

Interplay of chain-chain versus Coulombic interactions in the structure of [OMIM][Tf_2N] adsorbed on Au(110), **R. Fouston, S. Gangopadhyay, C. Chiutu, P. Moriarty and R. Jones**, *University of Nottingham, UK*

Enhancement of the chemical stability of single layer covalent organic frameworks (sCOFs) through a surface confined chemical reaction, **N. Goodeal, H. Bronstein and M. Blunt**, *University College London, UK*

Chemisorption by electrochemical reduction of diazonium cations: fundamentals and applications, **B. Hirsch, L. Verstraete, A. M. Bragança, H. Van Gorp, A. Brown, P. Walke, K. Tahara, Y. Tobe and S. De Feyter**, *KU Leuven, Belgium*

Study of the self-assembly of novel diketopyrrolopyrrole H-bonding chromophores by scanning tunneling microscopy and scanning electron microscopy, **J. Humphreys, D. B. Amabilino, P. H. Beton and V. V. Korolkov**, *University of Nottingham, UK*

Nanoscale control over the mixing behavior of surface-confined bi-component supramolecular networks using an oriented external electric field, **G. Velpula, J. Teyssandier, K. S. Mali and S. De Feyter**, *KU Leuven, Belgium*

Supramolecular macrocyclic systems based on Watson-Crick base-pairing at the liquid-solid interface, **N. Bilbao, I. Destoop, S. De Feyter and D. González-Rodríguez**, *KU Leuven, Belgium*

Surface interaction on gold nanoparticles evaluated by terahertz spectroscopy, **H. Kawazumi and S. Yamada**, *Kindai University, Kyushu, Japan*

Systematic modification of surface chemistry to probe the self-assembly architecture of thin gel films, **M. Angelerou, M. Marlow and M. Zelzer**, *University of Nottingham, UK*

Ullmann coupling reactions on Ag(111) and Ag(110); influence of substrate on the formation of covalently coupled products and intermediate metal-organic structures, **C. J. Judd, S. L. Haddow, N. R. Champness and A. Saywell**, *University of Nottingham, UK*

Organisation and ordering of 1D porphyrin polymers synthesised by on-surface Glaser coupling, **A. Saywell, A. S. Browning, P. Rahe, H. L. Anderson and P. H. Beton**, *University of Nottingham, UK*

The growth of porphyrin-based two dimensional covalent organic frameworks (2D-COFs) via the Schiff base reaction, **C. Nowicka-Dylag, N. Goodeal, M. Blunt, H. Bronstein and T. Clarke**, *University College London, UK*

"Dark contours": chiral modifiers as STM markers, **A. Sanz Matias, J. N. Harvey, S. De Feyter and O. Ivasenko**, *KU Leuven, Belgium*

Where is the most hydrophobic spot? Benzopurpurine self-assembly at the calcite-water interface, **M. Nalbach, P. Raiteri, S. Klassen, S. Schäfer, J. D. Gale, R. Bechstein and A. Kühnle**, *Johannes Gutenberg University Mainz, Germany*

A supramolecular xerogel on a surface, **M. Samperi, L. Pérez-García and D. Amabilino**, *University of Nottingham, UK*

Modular approach for photoswitchable antimicrobial surfaces, **Y. A. Diaz Fernandez, I. Sorzabal-Bellido, M. Moritz, A. Susarrey-Arce, P. Pallavicini, A. Taglietti and R. Raval**, *University of Liverpool, UK*

Harwell XPS: EPSRC's National XPS Research Facility, **P. Davies**, *Cardiff University*

The Faraday Discussions Poster Prize for the best poster was jointly awarded to Miss Yi Hu of *KU Leuven, Belgium*, for her poster on "2D self-assembly of DBA in confined spaces: competition between the assembled structures in nanocorrals" and Mr Niall Goodeal of *University College London, UK*, for his poster on "Enhancement of the chemical stability of single layer covalent organic frameworks (sCOFs) through a surface confined chemical reaction".

List of participants

Professor David Amabilino, *University of Nottingham, United Kingdom*
Ms Maria Angelerou, *University of Nottingham, United Kingdom*
Dr Ioan Bâldea, *Heidelberg University, Germany*
Mrs Olga Barykina, *Eink, USA*
Professor James D. Batteas, *Texas A. & M. University, USA*
Professor Pol Besenius, *Johannes Gutenberg-Universität Mainz, Germany*
Professor Peter Beton, *University of Nottingham, United Kingdom*
Dr Nerea Bilbao, *KU Leuven, Belgium*
Miss Yvonne Billard, *York University, United Kingdom*
Mr Philip Blowey, *University of Warwick, United Kingdom*
Dr Matthew Blunt, *University College London (UCL), United Kingdom*
Professor Manfred Buck, *University of St Andrews, United Kingdom*
Professor Neil Champness, *University of Nottingham, United Kingdom*
Professor Lifeng Chi, *Soochow University, P. R. China*
Dr Stuart Clarke, *University of Cambridge, United Kingdom*
Professor Giovanni Costantini, *University of Warwick, United Kingdom*
Dr Marie Cote, *Royal Society of Chemistry, United Kingdom*
Mr Jonathan Davidson, *University of Cambridge, United Kingdom*
Professor Philip Davies, *Cardiff University, United Kingdom*
Professor Steven De Feyter, *KU Leuven, Belgium*
Dr Yuri A. Diaz Fernandez, *University of Liverpool, United Kingdom*
Mr Deepak Dwivedi, *Curtin University, Australia*
Professor Karl-Heinz Ernst, *Empa, Switzerland*
Professor Amar Flood, *Indiana University, USA*
Dr Julien Gautrot, *Queen Mary University of London, United Kingdom*
Mr Niall Goodeal, *University College London, United Kingdom*
Dr Brandon Hirsch, *KU Leuven, Belgium*
Miss Yi Hu, *KU Leuven, Belgium*
Dr Vincent Humblot, *Laboratoire de Réactivité de surface, UMR CNRS 7197, France*
Mr Joshua Humphreys, *University of Nottingham, United Kingdom*
Professor Ahmad Jabbarzadeh, *University of Sydney, Australia*
Professor Robert Jones, *University of Nottingham, United Kingdom*
Mr Chris Judd, *University of Nottingham, United Kingdom*
Professor Hirofumi Kawazumi, *Kindai University, Kyushu, Japan*
Mr James Kerfoot, *University of Nottingham, United Kingdom*
Dr Vladimir Korolkov, *University of Nottingham, United Kingdom*
Professor Angelika Kühnle, *Johannes Gutenberg University Mainz, Germany*
Professor Markus Lackinger, *Technische Universität München / Deutsches Museum, Germany*
Mr Zhi Li, *KU Leuven, Belgium*
Professor Nian Lin, *Hong Kong University of Science and Technology, Hong Kong*
Professor Trolle Linderoth, *Aarhus University, Denmark*
Professor Natalia Martsinovich, *University of Sheffield, United Kingdom*

Professor Andrew Mount, *University of Edinburgh, United Kingdom*
Ms Lynn Murphy, *Royal Society of Chemistry, United Kingdom*
Mr Martin Nalbach, *Johannes Gutenberg University Mainz, Germany*
Ms Caroline Nowicka-Dylag, *University College London, United Kingdom*
Dr Claire-Marie Pradier, *Université Pierre et Marie Curie, CNRS, France*
Professor Talat Rahman, *University of Central Florida, USA*
Professor Rasmita Raval, *University of Liverpool, United Kingdom*
Mr Neil Robinson, *University of Cambridge, United Kingdom*
Professor Federico Rosei, *INRS, Canada*
Dr Marco Sacchi, *University of Surrey, United Kingdom*
Dr Mario Samperi, *University of Nottingham, United Kingdom*
Ms Ana Sanz Matias, *KU Leuven, Belgium*
Dr Alex Saywell, *University of Nottingham, United Kingdom*
Mr Sebastian Schwaminger, *Technical University of Munich, Germany*
Dr Johannes Seibel, *KU Leuven, Belgium*
Dr Steven Tait, *Indiana University, USA*
Dr Joan Teyssandier, *KU Leuven, Belgium*
Mrs Sarah Thirkell, *Royal Society of Chemistry, United Kingdom*
Professor Phil Woodruff, *University of Warwick, United Kingdom*
Dr Ruth Zadik, *Royal Society of Chemistry, United Kingdom*
Professor Han Zuilhof, *Wageningen University, Netherlands*

